Abstract Chiral Polytopes

Abstract polytopes are partially ordered sets that satisfy some key aspects of the face lattices of convex polytopes. They are chiral if they have maximal symmetry by combinatorial rotations, but none by combinatorial reflections. Aimed at graduate students and researchers in combinatorics, group theory or Euclidean geometry, this text gives a self-contained introduction to abstract polytopes and specialises in chiral abstract polytopes.

The first three chapters are introductory and mostly contain basic concepts and results. The fourth chapter talks about ways to obtain chiral abstract polytopes from other abstract polytopes, while the fifth discusses families of chiral polytopes grouped by common properties such as their rank, their small size or their geometric origin. Finally, the last chapter relates chiral polytopes with geometric objects in Euclidean spaces. This material is complemented by a number of examples, exercises and figures, and a list of 75 open problems to inspire further research.

DANIEL PELLICER is Investigador Titular B at the National Autonomous University of Mexico. He was awarded the Marsden postdoctoral fellowship by the Fields Institute (2011), and a Young Affiliate Fellowship by TWAS (2013–2018).

NEW MATHEMATICAL MONOGRAPHS

All the titles listed below can be obtained from good booksellers or from Cambridge University Press. For a complete series listing visit www.cambridge.org/mathematics.

11. B. Bekka, P. de la Harpe and A. Valette *Kazhdan's Property (T)*
12. J. Neisendorfer *Algebraic Methods in Unstable Homotopy Theory*
13. M. Grandis *Directed Algebraic Topology*
14. G. Michler *Theory of Finite Simple Groups II*
15. R. Schertz *Complex Multiplication*
16. S. Bloch *Lectures on Algebraic Cycles (2nd Edition)*
17. B. Conrad, O. Gabber and G. Prasad *Pseudo-reductive Groups*
18. T. Downarowicz *Entropy in Dynamical Systems*
19. C. Simpson *Homotopy Theory of Higher Categories*
20. E. Fricain and J. Mashreghi *The Theory of H(b) Spaces I*
21. E. Fricain and J. Mashreghi *The Theory of H(b) Spaces II*
22. J. Goubault-Larrecq *Non-Hausdorff Topology and Domain Theory*
23. J. Śniatycki *Differential Geometry of Singular Spaces and Reduction of Symmetry*
24. E. Riehl *Categorical Homotopy Theory*
25. B. A. Munson and I. Volić *Cubical Homotopy Theory*
26. B. Conrad, O. Gabber and G. Prasad *Pseudo-reductive Groups (2nd Edition)*
27. J. Heinonen, P. Koskela, N. Shanmugalingam and J. T. Tyson *Sobolev Spaces on Metric Measure Spaces*
28. Y.-G. Oh *Symplectic Topology and Floer Homology I*
29. Y.-G. Oh *Symplectic Topology and Floer Homology II*
30. A. Bobrowski *Convergence of One-Parameter Operator Semigroups*
31. K. Costello and O. Gwilliam *Factorization Algebras in Quantum Field Theory I*
32. J.-H. Evertse and K. Győry *Discriminant Equations in Diophantine Number Theory*
33. G. Friedman *Singular Intersection Homology*
34. S. Schwede *Global Homotopy Theory*
35. M. Dickmann, N. Schwartz and M. Tressl *Spectral Spaces*
36. A. Baernstein II *Symmetrization in Analysis*
37. A. Defant, D. García, M. Maestre and P. Sevilla-Peris *Dirichlet Series and Holomorphic Functions in High Dimensions*
38. N. Th. Varopoulos *Potential Theory and Geometry on Lie Groups*
39. D. Arnal and B. Currey *Representations of Solvable Lie Groups*
40. M. A. Hill, M. J. Hopkins and D. C. Ravenel *Equivariant Stable Homotopy Theory and the Kervaire Invariant Problem*
41. K. Costello and O. Gwilliam *Factorization Algebras in Quantum Field Theory II*
42. S. Kumar *Conformal Blocks, Generalized Theta Functions and the Verlinde Formula*
43. P. F. X. Müller *Hardy Martingales*
44. T. Kaletha and G. Prasad *Bruhat–Tits Theory*
45. J. Schwermer *Reduction Theory and Arithmetic Groups*
46. J. Kotus and M. Urbański *Meromorphic Dynamics I*
47. J. Kotus and M. Urbański *Meromorphic Dynamics II*
48. S. Garibaldi, H. P. Petersson and M. L. Racine *Albert Algebras over Commutative Rings*
49. D. Pellicer *Abstract Chiral Polytopes*
50. C. Xu *K-Stability of Fano Varieties*

Abstract Chiral Polytopes

DANIEL PELLICER
National Autonomous University of Mexico

CAMBRIDGE
UNIVERSITY PRESS

CAMBRIDGE
UNIVERSITY PRESS

Shaftesbury Road, Cambridge CB2 8EA, United Kingdom

One Liberty Plaza, 20th Floor, New York, NY 10006, USA

477 Williamstown Road, Port Melbourne, VIC 3207, Australia

314–321, 3rd Floor, Plot 3, Splendor Forum, Jasola District Centre,
New Delhi – 110025, India

103 Penang Road, #05–06/07, Visioncrest Commercial, Singapore 238467

Cambridge University Press is part of Cambridge University Press & Assessment,
a department of the University of Cambridge.

We share the University's mission to contribute to society through the pursuit of education, learning and research at the highest international levels of excellence.

www.cambridge.org
Information on this title: www.cambridge.org/9781108493246

DOI: 10.1017/9781108695046

When citing this work, please include a reference to the DOI 10.1017/9781108695046

First published 2025

A catalogue record for this publication is available from the British Library

Library of Congress Cataloging-in-Publication Data
Names: Pellicer, Daniel, author.
Title: Abstract chiral polytopes / Daniel Pellicer, Universidad Nacional Autónoma de México.
Description: Cambridge, United Kingdom ; New York, NY : Cambridge University Press, 2025. | Series: New mathematical monographs | Includes bibliographical references and index.
Identifiers: LCCN 2024045987 (print) | LCCN 2024045988 (ebook) | ISBN 9781108493246 (hardback) | ISBN 9781108695046 (epub)
Subjects: LCSH: Polytopes.
Classification: LCC QA691 .P45 2025 (print) | LCC QA691 (ebook) | DDC 516/.158–dc23/eng/20241209
LC record available at https://lccn.loc.gov/2024045987
LC ebook record available at https://lccn.loc.gov/2024045988

ISBN 978-1-108-49324-6 Hardback

Contents

Preface

The present book was inspired by the harmony and elegance exhibited by highly symmetric polytopes like the Platonic and Archimedean solids. This work is meant to be self-contained, although it can be understood as a natural sequel of the monograph *Abstract Regular Polytopes* [188]. Part of the material presented here has a close relationship with that of the recently published *Geometric Regular Polytopes* [179]. These three books, published in the first quarter of this century, reflect the continuous interest on these topics in the past decades.

The readers of this book will find in its first chapters the basic definitions and results on abstract polytopes and chirality. The exposition of those chapters is intended to be suitable for graduate students; it is complemented with examples and exercises that may help to develop intuition. These features make this work suitable as a text book for a graduate course on chiral polytopes.

The later chapters rely on previously published work. Some of the ideas and proofs are omitted, and instead the reader is pointed to the relevant publications. While adding more and more material, it became clear how difficult it would be to mention every result on chiral polytopes stated to this date. At some point it became necessary to prioritise the results that will represent better the ideas developed to this day. We hope that the readers will regard all references as valuable and worth exploring, even if they are cited here only once, and this is done in general terms. We also face the fact that papers including material on chiral polytopes are published every year. It is expected that by the time of publication of this book new results will have become available, but these could not be included in these chapters.

Graduate students may find this book particularly useful when looking for directions of research. We have included 75 problems that were open at the time of conclusion of this work; on top of their appearance in their respective sections, they are also listed in Appendix C. We encourage whoever attempts to solve any of them to first make a thorough search to determine whether that specific problem is still open, and to learn about recent partial results that may prove useful in reaching the final answer.

In several parts we rely on other branches of mathematics such as algebraic topology, Euclidean or hyperbolic geometry, and group theory. We do not expect all readers to be experts on these, and we made some effort to provide explanations that avoid deep concepts and results from those areas. Nevertheless, we want to share our opinion that mathematics is more pleasant and fulfilling when we understand it from various viewpoints. Abstract polytopes are objects that allow interpretations from various angles, and we encourage the reader to make an effort to absorb at least a general idea when reading about aspects of chiral polytopes that are described using notions and terminology that are not to their taste.

This book is a product of many factors of different natures. The author wishes to thank the Departamento de Matemáticas y Estadística, Universidad de la Frontera in Temuco, Chile, for its hospitality during the sabbatical year in which this work was finished. Several colleagues encouraged the efforts to make this book a reality. The author is particularly grateful to Gabe Cunningham, who volunteered to look into a large part of the early drafts, and made a number of suggestions that translated into significant improvements to the original content.

1

Introduction

In everyday language, an object is said to be chiral whenever it lacks mirror symmetry. Human hands and shells of snails are examples of objects regarded as chiral. Formally, it is understood that a subset of an orientable geometric space is chiral if no orientation-reversing isometry of the space preserves it. This definition is more restrictive than just discarding symmetries acting locally as mirror reflections.

The notion of 'polytope' arises as a generalisation of the polygon and polyhedron. In the classical theory, polygons are two-dimensional objects bounded by edges, whereas polyhedra are three-dimensional objects bounded by polygons. One can continue by considering four-dimensional objects bounded by polyhedra (these are the 4-polytopes). The intuitive idea is that an n-polytope is an n-dimensional object bounded by $(n-1)$-polytopes. If we know some $(n-1)$-polytope $\mathcal{P}$ then we can easily understand some n-polytopes like the prism or the pyramid over $\mathcal{P}$.

The present book studies combinatorial objects known as *abstract polytopes*, with particular attention to those possessing a high degree of symmetry. Abstract polytopes are an abstraction of the initial geometric notion of polytopes. They consist of elements (called 'faces') of different ranks, and an incidence relation satisfying some of the main properties of the faces of convex polytopes. In particular, the face-lattice of every convex polytope constitutes an abstract polytope. Naturally, abstract polytopes of ranks 2 and 3 are called *abstract polygons* and *abstract polyhedra*, respectively. In this context, the term *chiral* denotes not just all orientable abstract polytopes lacking orientation-reversing automorphisms, but we also require the polytopes to be symmetric by all possible *abstract rotations*. The term 'chiral' for such highly symmetric abstract polytopes was first used by Schulte and Weiss in [247]; it had a fast acceptance by the community, which was reflected by the number of subsequent related papers. It is important to keep in mind, however, that in other areas of mathematics and in other sciences chirality does not imply any particular symmetry.

The study of chiral abstract polytopes is a natural follow-up to the theory of the most symmetric class of abstract polytopes, called *regular polytopes*, which are extensively treated in the monograph [188]. It has become clear since the definitions and first results that it is impossible to talk in detail about chiral polytopes without talking about regular polytopes as well.

Abstract regular polytopes admit all possible symmetry by *abstract reflections*. In the same way as their geometric counterparts, composition of two abstract reflections fixing a point yields an abstract rotation; this ensures that regular polytopes are also symmetric by all abstract rotations. Moreover, precisely half of the symmetries of a regular orientable polytope are products of abstract rotations. In this sense, the degree of symmetry of regular polytopes doubles the degree of symmetry of chiral polytopes. The only constraint for an abstract chiral polytope to become regular is precisely the lack of orientation-reversing symmetry.

Regular polytopes in their geometric version have been studied since antiquity, including convex regular polygons and the five Platonic solids. This is best illustrated by Euclid's *Elements* [85, Book XIII], although some features of regular polytopes were known even before Euclid's time. Since then, several mathematicians developed ideas leading to the study of the combinatorial structure of regular polytopes, leaving aside convexity and other geometric properties. The first big step in this direction was taken by Johannes Kepler (1571–1630), who allowed polyhedra to have star-shaped faces [152]. He observed that the symmetry of the small stellated dodecahedron and the great stellated dodecahedron have much in common with that of the Platonic solids, despite the fact that Kepler's polyhedra are not convex. This idea was taken further by Louis Poinsot (1777–1859), who considered faces meeting in a star-shape manner around vertices [232]. Ludwig Schläfli (1814–1895) and Edmund Hess (1843–1903) explored polytopes outside the plane and space, and classified the higher-dimensional analogues of regular convex and star polyhedra [126, 237].

In the twentieth century, J. F. Petrie and H. S. M. Coxeter admitted the possibility of polyhedra having infinitely many vertices, edges and faces [61]. Branko Grünbaum considered non-planar faces as building blocks, also allowing faces to meet around a vertex in a non-planar arrangement [104]. From this viewpoint, the faces of polyhedra are no longer associated to particular surfaces embedded in 3-space; instead, they become the embedding of combinatorial structures consisting of vertices and edges. The final list of 48 regular polyhedra satisfying the definition dictated by Grünbaum was given by Andreas W. M. Dress [83, 84]. In recent years Peter McMullen has made a lot of progress in the description of regular polytopes in higher-dimensional Euclidean spaces, extending the key ideas of Grünbaum's concept of polyhedron [172–174, 176, 177]. The final jump to a purely combinatorial definition was done in [76], where L. Danzer and E. Schulte established the axioms of abstract polytopes. The monograph [188] comprises the necessary theory to understand the most symmetric among these objects, that is, abstract regular polytopes. Many others have made contributions to the construction of the current theory of abstract regular polytopes. A more detailed account of historical remarks can be found in [188, Section 1A].

The history of chiral abstract polytopes is considerably younger. There are classical polyhedra with a large group of rotations allowing no orientation-reversing symmetries. The snub cube and the snub dodecahedron have as full symmetry groups the rotation group of the cube and the rotation group of the dodecahedron, respectively (see Figure 1.1). However, these polyhedra fail our definition of chirality, since no non-trivial rotation around the axis through a vertex and the centre of the polyhedron acts like a symmetry. Also, in both cases there are triangles whose centres do not belong to axes of 3-fold rotations preserving the polyhedron. Informally, we could say that the snub cube and the snub dodecahedron are not chiral because it is not possible to rotate around all vertices and centres of faces.

Under certain sets of conditions, imposing all possible symmetry by rotations to polytopes forces symmetry by reflections. The objects listed below are witnesses of this phenomenon.

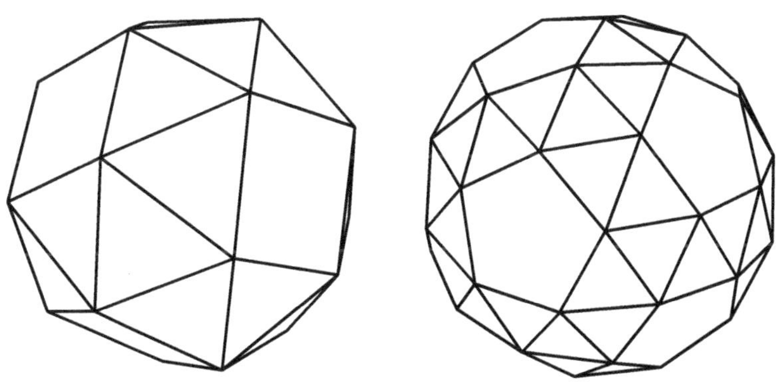

(a) (b)

Figure 1.1 (a) Snub cube and (b) snub dodecahedron.

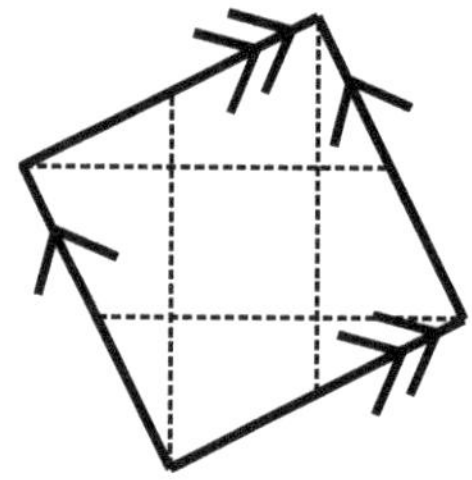

Figure 1.2 The toroidal polyhedron $\{4, 4\}_{(2,1)}$.

- Convex polytopes [168].
- Tessellations of Euclidean or hyperbolic spaces [48].
- Finite polyhedra in the Euclidean space $\mathbb{E}^3$ [245].
- Planar polyhedra [245].
- Apeirohedra (infinite polyhedra) in $\mathbb{E}^3$ with planar faces [246].
- Tessellations of the d-dimensional torus for $d \geq 3$ [122].

Thus, the existence of chiral polytopes in geometric spaces is restricted to several constraints. As a result, chiral polytopes did not appear in the classical theory of polytopes. For instance, the discovery and full classification of chiral polyhedra in $\mathbb{E}^3$ took place only in the current century [245, 246].

Figure 1.2 shows the simplest chiral polytope; it is the polyhedron called $\{4,4\}_{(2,1)}$, and it is represented as a map on the torus. It contains 5 vertices, 10 edges and 5 faces. Later in this book it will become clear that it allows no reflection, but it is symmetric under all 4-fold rotations around vertices and centres of faces.

The formal study of abstract chiral polytopes started in the first half of the twentieth century with the works of Brahana and Coxeter on chiral maps on the torus (see [13, 49]). Most attention was given to polytopes of rank 3, in part because of the difficulty of working combinatorially with these objects in higher ranks, and in part because rank-3 chiral abstract polytopes are a particular case of chiral maps on surfaces, which are interesting objects on their own. In Section 2.3 we present a detailed summary of relevant discoveries on chiral maps and polytopes.

It is a striking contrast that numerous mathematicians since the early Greeks knew about regular polytopes, whereas the study of chiral polytopes was initiated only in the past century. When trying to construct examples of chiral polytopes we usually find that among all polytopes that are invariant under all possible combinatorial rotations, those with a simple description tend to be also invariant under reflections, and so they are regular instead of chiral. As a result, few chiral polytopes were known before the computer era, in comparison to the wealth of regular polytopes that had been studied by that time. Even with the help of computers, only finitely many chiral polytopes of ranks 6 and higher have been described; while on the other hand, cubes and tessellations by cubes are well-understood examples of regular polytopes that exist in arbitrarily large ranks.

Despite the initial lack of examples of chiral polytopes, they are the most-studied abstract polytopes among those that are not regular. The fact that they are invariant under all possible abstract rotations makes them similar in many aspects to their regular counterparts, and provides tools to study them. This is a good reason to consider chiral polytopes as the natural class to investigate after the class of the regular ones.

Thus, the study of abstract chiral polytopes constitutes an area of recent development that has a wide interaction with the topic of maps on surfaces. These objects can be defined in multiple equivalent ways, and motivate pleasant interactions of distinct branches of mathematics, like combinatorics of partially ordered sets, finitely presented groups, algebraic topology of simplicial complexes, as well as Euclidean, projective and hyperbolic geometry. Mathematicians looking for fields of research may be encouraged by several interesting open questions about chiral polytopes, listed in Appendix C (some of them taken from [220, 250]). In Chapters 2 and 3 the reader will find the basic theory of chiral polytopes, whereas the main directions of research are suggested in Chapters 4, 5 and 6. Throughout, the concepts and results are illustrated by plenty of examples.

2

Abstract Regular and Chiral Polytopes

In this chapter we aim to provide an introduction to abstract polytopes and their symmetries, suitable for graduate students and researchers starting in this field. Here the reader will find many examples and exercises accompanying the explanations. Much of the material in this chapter can also be found in [188, Chapter 2] in a more condensed manner. Nevertheless, we encourage those interested in acquiring a deep understanding of the nature of abstract polytopes to compare the definitions, results and proofs presented in Sections 2.1 and 2.2 with those provided in [188].

The chapter consists of four sections. The first of them is dedicated to the definitions and main results concerning abstract polytopes and their symmetries. Then, in Section 2.2 we start the study of polytopes with a high degree of symmetry by establishing the basics of regular and two-orbit polytopes. Chiral polytopes, the main topic of this book, are introduced in Section 2.3. Finally, in Section 2.4 we describe many regular and chiral polyhedra that can be visualised as maps on the torus; they are natural examples to consider when developing the theory of highly symmetric polytopes in subsequent chapters of this book.

2.1 Abstract Polytopes

We intend to provide a combinatorial definition of a polytope not relying on geometry or topology. For example, there is a bijection between the vertices of the cube and the vertices of a prism over a rhombus that maps adjacent vertices to adjacent vertices, and maps the four vertices of each face of the cube to the four vertices of some face of the prism. Hence, from a combinatorial point of view, a cube and a prism over a rhombus are isomorphic objects. Similarly,

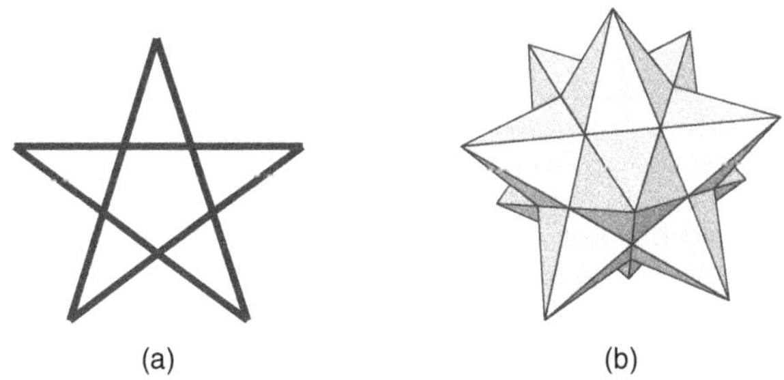

Figure 2.1 (a) The pentagram and (b) the small stellated dodecahedron.

a pentagon is combinatorially equivalent to a *pentagram* (star pentagon as in Figure 2.1(a)).

Our main interest is in the symmetry of combinatorial structures, and therefore our definition captures the main combinatorial features of regular convex polygons, the Platonic solids, and other well-known highly symmetric bodies. In doing so we must deviate from convexity to keep in mind objects like the small stellated dodecahedron (see Figure 2.1(b)), a non-convex body that can be appreciated as a decoration of Saint Mark's Basilica in Venice, and that was later regarded by Kepler as a regular polyhedron (see [152]).

With this in mind, a polytope will consist of sets of vertices, edges, etc., equipped with some partial order replacing the idea of inclusion in the geometric setting. Before stating the definition of an abstract polytope, we shall introduce some preliminary concepts on partially ordered sets (or *posets* for short).

2.1.1 Partial Orders

Two elements F and G of a poset $(X, \leq)$ are said to be *incident* whenever $F \leq G$ or $G \leq F$. The *incidence graph* of $(X, \leq)$ is the graph with vertex set X where two vertices are adjacent if and only if they are incident. A *chain* is a totally ordered subset of X. We say that F is a *least* (resp. *greatest*) element if $F \leq G$ for all $G \in X$ (resp. if $F \geq G$ for all $G \in X$).

Remark 2.1.1 A poset $(X, \leq)$ has at most one least element, and at most one greatest element.

Example 2.1.2 If $(X, \leq)$ is a totally ordered set then every pair of elements are incident, every subset constitutes a chain, and its incidence graph is complete.

We say that G is a *cover element* of F whenever $F < G$ but there is no $H \in X$ such that $F < H < G$. A function $r\colon X \to \mathbb{Z}$ is a *rank function* whenever

(1) $F < G$ implies that $r(F) < r(G)$, and
(2) if G is a cover element of F then $r(G) = r(F) + 1$.

If X has a rank function r, we call $r(F)$ the *rank* of F. Posets with rank functions are also called *graded posets*.

Exercise 2.1.3 Determine which of the following posets are graded:

- $\mathbb{Z}$ with the order given by divisibility,
- the power set of the set $\{1,2,3,4,5\}$, with the order given by inclusion,
- the power set of $\mathbb{Z}$, with the order given by inclusion,
- $\mathbb{Q}$ with the usual total order $\leq$.

The following lemma is a direct consequence of [8, Lemma 1].

Lemma 2.1.4 *A poset X with least and greatest elements has a rank function if and only if all maximal chains have the same number of elements.*

In this work the rank of the least elements will be -1, and the *rank of the poset* $(X, \leq)$ is defined as the rank of the greatest element.

The *Hasse diagram* of a graded poset $(X, \leq)$ is a graph with vertex set X where two vertices are adjacent if and only if one of them is a cover element of the other. By convention, all elements of the same rank are represented on a horizontal line in the Hasse diagram, and, for every i, the elements with rank i are placed in a line below the elements with rank $i + 1$.

As an example, we take X to be the poset consisting of a greatest element M of rank 1, a least element m of rank -1, and two non-incident elements y and z of rank 0. Then the cover elements of m are x and y; each of x and y has M as its unique cover element, and M has no cover elements. The Hasse diagram is the one shown in Figure 2.2.

Given two elements F and G of $(X, \leq)$ with $F \leq G$, the *section* $[F,G]$ is defined as $\{H \mid F \leq H \leq G\}$. If $(X, \leq)$ has a rank function r and a least element of rank -1, then $[F,G]$ has a rank function induced by that of $(X, \leq)$, and the rank of $[F,G]$ is $r(G) - r(F) - 1$.

A poset with rank function satisfies the *diamond condition* whenever every section $[F,G]$ of rank 1 consists of precisely four elements: F, G, H_1 and H_2 with $F < H_i < G$ for $i \in \{1,2\}$. The name of this property comes from the

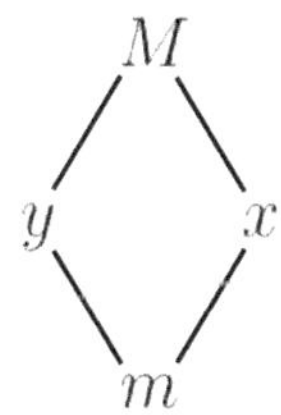

Figure 2.2 Hasse diagram of a poset with four elements.

fact that the Hasse diagram of each rank-1 section resembles a diamond. This diagram is precisely that in Figure 2.2 with $F = m$, $G = M$, $H_1 = x$, $H_2 = y$.

We say that a poset $(X, \leq)$ is *connected* whenever it has a greatest element M and a least element m, and either it is a poset of rank $n \leq 1$ or for every $F, G \in X \setminus \{M, m\}$ there exist $H_1, \ldots, H_k \in X \setminus \{M, m\}$ such that F is incident to H_1, G is incident to H_k and H_i is incident to H_{i+1} for every $i \in \{1, \ldots, k-1\}$. In other words, the incidence graph of $X \setminus \{M, m\}$ is connected. A poset is said to be *strongly connected* if all its sections are connected.

Exercise 2.1.5 Find a graded poset having least and greatest elements that is not connected.

Exercise 2.1.6 Find a graded poset having least and greatest elements that is connected, but not strongly connected.

2.1.2 Abstract Polytopes

We now have the necessary elements to define an abstract polytope.

Definition 2.1.7 An *abstract polytope* of rank n (or n-polytope) is a poset $(\mathcal{P}, \leq)$ satisfying the following conditions:

(a) it has a greatest element F_n and a least element F_{-1},
(b) every maximal chain has exactly $n + 2$ elements including F_n and F_{-1},
(c) it satisfies the diamond condition,
(d) it is strongly connected.

In the remainder of this work we shall refer to abstract polytopes simply by 'polytopes'. We shall precede 'polytopes' with descriptive adjectives such as 'convex' or 'geometric' when refering to specific previously defined classes of mathematical objects.

Definition 2.1.7 may not point directly to familiar objects like a triangle or a cube. In part this is because it does not carry any geometric meaning. Nevertheless, this definition takes into account many combinatorial aspects of the usual geometric polytopes.

Example 2.1.8 A triangle has three vertices and three edges. Every vertex is contained in two edges and every edge contains two vertices. If we label the vertices and edges as in Figure 2.3(a), then we can represent the order relation given by the inclusion in the Hasse diagram in Figure 2.3(b). The minimum corresponds to $\varnothing$ and the maximum to the triangle itself.

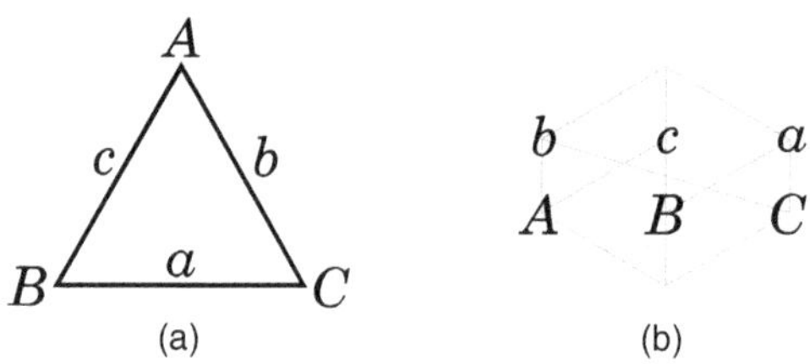

Figure 2.3 (a) Labels of the faces of a triangle, (b) triangle as an abstract polytope.

Example 2.1.8 illustrates the way to represent geometric polytopes as abstract polytopes. We will take all i-faces of the geometric polytope together with $\varnothing$ and the polytope itself as elements of a poset, ordered by inclusion. The fact that the face-lattice of every convex polytope is an abstract polytope will be proven in Section 2.1.3.

From the combinatorial point of view, every two triangles are equivalent; they have three vertices and three edges, and their incidence relation can be described by the Hasse diagram in Figure 2.3(b). This motivates the following definition of isomorphism.

We say that two polytopes $\mathcal{P}$ and $\mathcal{Q}$ are *isomorphic* (denoted by $\mathcal{P} \cong \mathcal{Q}$) if there exists a bijection $\beta \colon \mathcal{P} \to \mathcal{Q}$ such that β and β^{-1} preserve the incidence; in that case we say that β is an *isomorphism* from $\mathcal{P}$ to $\mathcal{Q}$. In other words, two polytopes are isomorphic if their posets are the same up to renaming the elements. When restricting ourselves to combinatorial aspects, we may regard isomorphic polytopes as equal.

The elements of any abstract polytope are called *faces* and can be organised by rank (see Lemma 2.1.4). We shall denote the rank of a face F by $r(F)$. Faces of rank i are also called *i-faces*. We say that F_{-1} and F_n are *improper faces*, and all other faces are *proper*. In correspondence to the geometric origin of abstract polytopes, we shall refer to the 0-faces as *vertices*, to the 1-faces as *edges*, to the $(n-2)$-faces as *ridges* and to the $(n-1)$-faces as *facets*.

In ranks 0, 1 and 2, all but two abstract polytopes correspond to convex polytopes. A poset with only one element of rank -1 and one of rank 0 trivially satisfies the four axioms of Definition 2.1.7, and hence it is the only 0-polytope; it corresponds to the face-lattice of a single point. The diamond condition implies that there is only one polytope of rank 1. It has one (-1)-face, two 0-faces and one 1-face, and it corresponds to the face-lattice of a line segment. Polytopes of rank 2 are also called *polygons*; the diamond condition implies that every polygon has at least two vertices. For every integer $m \geq 2$, there is a unique polygon with m vertices and m edges; the details are left as an exercise.

Exercise 2.1.9 Show that for every integer $m \geq 3$, there is a unique polygon with m vertices and m edges, and that it is isomorphic to the face-lattice of a convex m-gon.

Exercise 2.1.10 Show that there is a unique polygon with two vertices and a unique polygon with infinitely many vertices.

There are many abstract *polyhedra* (polytopes of rank 3) that are not the face-lattice of any convex polyhedron. In particular, any abstract polyhedron with multiple edges cannot be the face-lattice of a convex polyhedron since in Euclidean space there is only one line segment between any pair of points. The polygon with two vertices mentioned in Exercise 2.1.10 is an example of an abstract polytope with double edges. Example 2.1.11 shows polyhedra with an arbitrarily large number of edges between a pair of vertices. We shall give a few more examples of polyhedra with multiple edges in Section 2.4.

Example 2.1.11 Let $\mathcal{P}$ be an *n-gon* (that is, a polygon with n vertices). We construct a poset $\mathcal{Q}$ of rank 3 whose 1-faces are the vertices of $\mathcal{P}$ and whose 2-faces are the edges of $\mathcal{P}$ with the incidences induced from $\mathcal{P}$. To complete the poset $\mathcal{Q}$ we add a unique maximum and a unique minimum, together with two 0-faces, each of which is incident to all 1-faces.

It is easy to verify that $\mathcal{Q}$ satisfies the first three items of Definition 2.1.7. The fourth item is a consequence of the facts that $\mathcal{Q}$ itself is connected, both sections between a 0-face and the 3-face of $\mathcal{Q}$ are isomorphic to $\mathcal{P}$ (but shifted by 1 in rank) and all facets are polygons with two vertices. Therefore, $\mathcal{Q}$ is a polyhedron.

Finally, $\mathcal{Q}$ has n-tuple edges since the two vertices are incident to the n edges.

This polyhedron can be visualised as a beach ball with n slices, and it is also known as *hosohedron* (see [55, Section 1.7]).

In general, for every $n \geq 3$, the class of abstract polytopes isomorphic to the face-lattices of convex n-polytopes is not a representative sample of the entire class of abstract polytopes. One indicator of this is that there are infinitely many finite chiral polyhedra (see Section 2.3 for definitions and Section 2.4 for an infinite family); however, we shall see in Chapter 6 that none of them is the face-lattice of a convex polyhedron.

The third and fourth items of Definition 2.1.7 are hereditary; that is, if a poset satisfies this condition, then it is also satisfied by each of its sections. This fact, together with the notion of rank in a poset, implies the following remark.

Remark 2.1.12 Every section of a polytope is a polytope.

In the theory of polytopes it is customary to denote the section $[F, G]$ by G/F. In many instances it is convenient to consider a given face F of rank j as the j-polytope F/F_{-1}. In those cases we shall abuse notation and refer to both mathematical objects simply by F.

The section F_n/F is called the *coface* at F. In the case where F is a vertex of a polytope $\mathcal{P}$ then F_n/F is called the *vertex-figure* at F. The vertices of the vertex-figure at F are the edges containing F, and in general the $(k-1)$-faces of the vertex-figure are the k-faces of the polytope incident to F. If $\mathcal{P}$ has no multiple edges, in some instances it is convenient to reinterpret the vertex-figure so that its vertices are the neighbours of F in the underlying graph of $\mathcal{P}$, since the neighbours are in a one-to-one correspondence with the edges incident to F.

The diamond condition and strong connectivity are restrictive enough to discard the possibility of a polytope with uncountably many faces. They also make it possible to give a simpler definition of isomorphism of polytopes.

Exercise 2.1.13 Show that the number of faces of every polytope is countable.

Exercise 2.1.14 Show that a polytope $\mathcal{P}$ is isomorphic to a polytope $\mathcal{Q}$ if and only if there exists a bijection β from the face set of $\mathcal{P}$ to that of $\mathcal{Q}$ that preserves incidence. In other words, show that the fact that β^{-1} preserves incidence follows from the axioms of abstract polytopes together with the fact that β preserves incidence.

It is time to introduce some concrete examples. The following three n-polytopes are defined as convex regular polytopes in Euclidean n-space $\mathbb{E}^n$

(for more details see [55, Section 7]). In all cases, every face is completely determined by its vertices; this makes it possible to represent the faces of the corresponding abstract polytope as subsets of vertices, ordered by inclusion.

Example 2.1.15 The regular *n-simplex* is the convex hull of $n+1$ pairwise equidistant points in $\mathbb{E}^n$. Every subset of $k+1$ vertices determines a k-face.

Example 2.1.16 The *n-cube* is the convex hull of the points in $\mathbb{E}^n$ whose n coordinates are in $\{1, -1\}$. For every $I \subset \{1, \dots, n\}$ with size k and every vertex $\bar{y} = (y_1, \dots, y_n)$, the set of 2^k vertices

$$\{\bar{x} = (x_1, \dots, x_n) \in \{1, -1\}^n : x_i = y_i \quad \text{if } i \notin I\}$$

constitutes a k-face of the n-cube.

Example 2.1.17 The *cross polytope* of rank n is the convex hull of the points $\pm e_i$, where the ith coordinate of e_i is 1 and all other coordinates are 0. For $k < n$, the k-faces are given by the sets of $k+1$ vertices that do not contain $\{e_i, -e_i\}$ for any i.

Remark 2.1.18 The cross polytope of rank 2 is a square, and the one of rank 3 is an octahedron.

It will become clear in Section 2.1.3 that the structures defined in Examples 2.1.15, 2.1.16 and 2.1.17 are abstract polytopes. The reader is encouraged to try to prove directly that they satisfy the axioms in Definition 2.1.7.

Exercise 2.1.19 Verify the following statements.

1. The k-faces of the n-simplex are k-simplices.
2. The vertex-figures of the n-simplex are $(n-1)$-simplices.
3. The k-faces of the n-cube are k-cubes.
4. The vertex-figures of the n-cube are $(n-1)$-simplices.
5. For each $k < n$, the k-faces of the n-dimensional cross polytope are k-simplices.
6. The vertex-figures of the n-dimensional cross polytope are $(n-1)$-dimensional cross polytopes.

Exercise 2.1.20 It is well known that the complete graph K_6 is 5 edge-colourable; Figure 2.4 shows the edges of each colour separately. Assign colour 1 to an arbitrary perfect matching, and assign colours 2, 3, 4 and 5 to the remaining edges incident to any given vertex.

(a) Show that the colours given so far can be extended to a proper 5 edge colouring in precisely two ways.
(b) For any proper 5 edge colouring of K_6, consider the five octahedra with 1-skeleton determined by the complements of the matchings. Show that the poset consisting of the 6 vertices, 15 edges, 20 triangles and 5 octahedra is a 4-polytope.

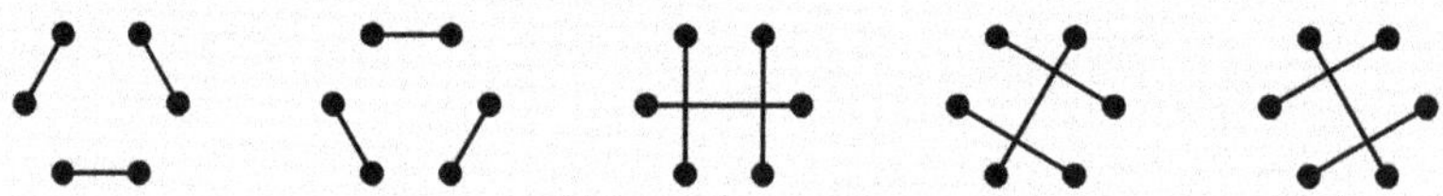

Figure 2.4 A proper 5 edge-colouring of K_6.

Exercise 2.1.21 Consider the proper 5 edge-colouring c of the complete graph K_6 indicated in Figure 2.4. For each colour $i \in \{1,2,3,4,5\}$, consider another colouring c_i in the same set as follows. The edges of colour i are the same in both colourings. For $j \neq i$, we define the edges of colour j as explained next. With respect to the colouring c, there are precisely two triangles whose edges have colours different from i and j; these two triangles are disjoint, and the remaining three edges with colours not in $\{i,j\}$ form a perfect matching M_j. The edges of colour j in the colouring c_i are precisely the edges of M_j.

(a) Show that every colouring c_i induces a polytope isomorphic to that obtained in Exercise 2.1.20 (b).
(b) The 15 perfect matchings of K_6 occur among the chromatic classes of the c_i. Therefore the colourings c and c_i induce 15 octahedra. Show that every triangle of K_6 belongs to six of these octahedra. These octahedra can be grouped in two sets of three octahedra in such a way that if two such octahedra are determined by the same colouring (c or c_i for some i), then they belong to the same set.
(c) From (b) we know that each of the 20 triangles of K_6 may be interpreted as a pair of triangles on the same vertex set in such a way that each new triangle belongs to precisely three octahedra, induced by three of the colourings. This gives a total of 40 triangles. Show that

the six 4-polytopes constructed from the colourings c_i and c, together with the 6 vertices, 15 edges, 40 triangles and 15 octahedra constitute a 5-polytope.

(d) Show that if in (c) we consider only the 20 triangles of K_6, then the resulting structure fails to be a polytope.

Often it is convenient to work with sets of faces, not necessarily of the same ranks. Sections of polytopes are examples of such sets.

The *k-skeleton* of a polytope is the poset consisting of all faces with rank at most k. The 0-skeleton is the vertex set, and the 1-skeleton is the underlying graph.

Another example of sets of elements of distinct ranks are the maximal chains, also called *flags*. They already appeared in Definition 2.1.7(b), and will play a main role in upcoming sections. For simplicity, when working with specific polytopes we may omit the minimum and maximum when listing the elements of the flags. We shall denote the set of flags of a polytope $\mathcal{P}$ by $\mathcal{F}l(\mathcal{P})$.

The flags of some polytopes $\mathcal{P}$ can be visualised as follows. Whenever $\mathcal{P}$ is represented as a tessellation $\mathcal{T}$ of some geometric space $\mathbb{S}$, the flags of $\mathcal{P}$ are in a one-to-one correspondence with the maximal simplices of the *barycentric subdivision* of $\mathcal{T}$; that is, every flag can be identified with the simplex with corners at the centroids of all its proper faces. With this identification, adjacent flags correspond to neighbouring simplices.

Convex n-polytopes are examples of such a visualisation; they can be represented as tessellations of an n-sphere $\mathbb{S}^n$, by projecting all proper faces from an interior point to a circumscribed n-sphere. When convenient, the flags can be projected back to the boundary of the convex polytope. In this way, the flags of a convex polygon can be pictured as line segments from an edge midpoint to an incident vertex, and those of a convex polyhedron as triangles with corners in the centre of a face F_2, the midpoint of an edge F_1 incident to F_2 and a vertex of F_1. Figure 2.5(a) shows the barycentric subdivision of a hexagonal face of a polyhedron, and Figure 2.5(b) shows that of the square face of a polyhedron.

Let $\Phi = \{F_{-1}, \ldots, F_n\}$ be a flag of an n-polytope, where F_i is the face of rank i. The diamond condition has as a consequence that for every $i \in \{0, \ldots, n-1\}$, the i-face F_i can be replaced by the unique face $G_i \neq F_i$ that satisfies that $F_{i-1} < G_i < F_{i+1}$, and form a new flag Φ^i. The flags Φ and Φ^i are said to be *adjacent*, and when we want to emphasise the rank we say that Φ^i is *i-adjacent* to Φ.

We shall extend recursively the notation of adjacency of flags so that for every i, if w is a word in $\{0, \ldots, n-1\}$, then the flag $\Phi^{w,i}$ is $(\Phi^w)^i$. In

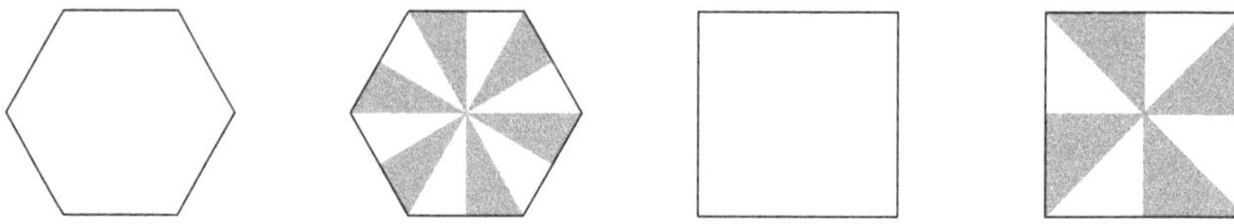

Figure 2.5 Representation of the flags of (a) a hexagon and of (b) a square.

particular, $\Phi^{i,j}$ is the j-adjacent flag of Φ^i, and hence it is the j-adjacent flag of the i-adjacent flag of Φ. When convenient we shall also denote the juxtaposition of two identical words w by w^2, and recursively for $k > 2$ we write w^k for $w^{k-1}w$.

Remark 2.1.22 The flags of a p-gon can be coloured black and white so that every black flag is 0-adjacent and 1-adjacent to white flags, and vice versa (see Figure 2.5).

If Φ is a black flag then $\Phi^{0,1}, \Phi^{(0,1)^2}, \dots, \Phi^{(0,1)^k}$ are the black flags in the k edges following the edge containing Φ according to some local orientation. The situation is similar if Φ is a white flag, but the orientation in which we consider the edges must be reversed. It follows that for every flag Φ we have that $\Phi^{(0,1)^k} = \Phi$ if and only if k divides p.

The adjacency of flags satisfies the following properties.

Proposition 2.1.23 *Let Φ be a flag of an n-polytope $\mathcal{P}$. Then*

$$\Phi^{i,i} = \Phi \qquad \text{for each } i \in \{0, \dots, n-1\}, \tag{2.1}$$

$$\Phi^{i,j} = \Phi^{j,i} \qquad \text{whenever } |i - j| > 1. \tag{2.2}$$

Proof Equation (2.1) follows directly from the diamond condition (see Figure 2.6(a)).

Let $i, j \in \{0, \dots, n-1\}$ such that $|i - j| > 1$. For $\ell \in \{0, \dots, n-1\}$ let F_ℓ be the ℓ-face of Φ, and for $k \in \{i, j\}$ let G_k be the k-face of Φ^k. Then $\Phi^{i,j}$ and $\Phi^{j,i}$ have the same faces as Φ except for the i-faces and the j-faces. By definition of i-adjacency, the i-face of Φ^i is G_i; furthermore, since the $(j-1)$- and $(j+1)$-faces of Φ^i are the $(j-1)$- and $(j+1)$-faces of Φ then the j-face of $\Phi^{i,j}$ is G_j (see Figure 2.6(b)). It can be shown in the same way that the i-face of $\Phi^{j,i}$ is G_i and the j-face of $\Phi^{j,i}$ is G_j. This proves that (2.2) holds. ■

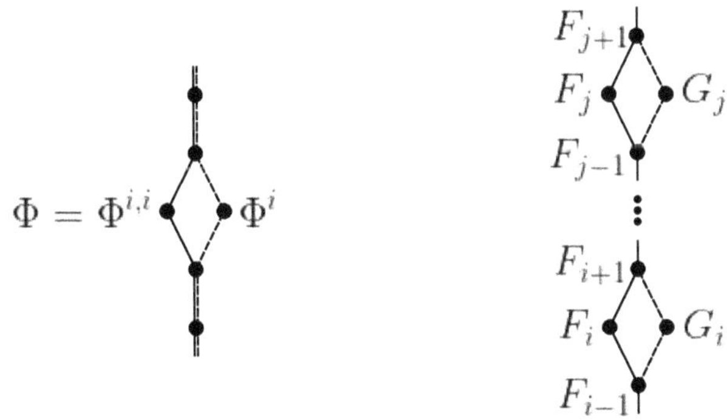

(a) (b)

Figure 2.6 (a) The i-adjacent of the i-adjacent of a flag, (b) the i-adjacent of the j-adjacent of a flag.

Item (d) of Definition 2.1.7 constitutes a connectivity property of abstract polytopes. We shall explore other connectivity properties next.

Let $(\mathcal{P}, \leq)$ be a poset satisfying items (a), (b) and (c) of Definition 2.1.7. We say that $(\mathcal{P}, \leq)$ is *flag-connected* if for every two flags Φ and Ψ there exists a sequence of flags

$$\Phi = \Phi_0, \Phi_1, \ldots, \Phi_k = \Psi, \tag{2.3}$$

where Φ_{i-1} and Φ_i are adjacent for $i \in \{1, \ldots, k\}$. Furthermore, $(\mathcal{P}, \leq)$ is *strongly flag-connected* if in addition in the sequence in (2.3) we require that $\Phi \cap \Psi \subseteq \Phi_i$ for every $i \in \{1, \ldots, k-1\}$.

A strongly flag-connected poset is clearly flag-connected. The converse is not true. For example, if we identify two opposite vertices u and w of a cube $\mathcal{C}$ without identifying edges or 2-faces then we obtain a poset $\mathcal{Q}$ with 7 vertices, 12 edges and 6 squares. It still has as many flags as the cube, since there is an obvious bijective mapping η from the set of flags of $\mathcal{C}$ to that of $\mathcal{Q}$. We can easily verify that $\mathcal{Q}$ satisfies items (a), (b) and (c) of Definition 2.1.7. Furthermore, $\mathcal{Q}$ is flag-connected since this property is inherited from $\mathcal{C}$. However, if Φ is a flag of $\mathcal{C}$ containing u and Ψ is a flag of $\mathcal{C}$ containing w then every sequence of adjacent flags from Φ to Ψ must include flags whose vertices are neither u nor w. Hence, there cannot be a sequence of adjacent flags from $\eta(\Phi)$ to $\eta(\Psi)$ where every flag contains the identified vertex $\{u, w\}$ of $\mathcal{Q}$. We conclude that $\mathcal{Q}$ is not strongly flag-connected.

Proposition 2.1.24 *Every polytope is strongly flag-connected.*

Proof We proceed by induction over the rank n of $\mathcal{P}$. The result is clearly true for $n < 2$. Assume that $\mathcal{P}$ is an n-polytope with $n \geq 2$, and let Φ and

Ψ be flags of $\mathcal{P}$. In what follows we shall show the existence of a sequence $S = \{\Phi_i\}_{i=0,\dots,k}$ of flags such that $\Phi = \Phi_0$, $\Psi = \Phi_k$, Φ_i is adjacent to Φ_{i+1} and $\Phi \cap \Psi \subseteq \Phi_i$ for every $i \in \{0, \dots, k-1\}$.

We first consider the case where $\Phi \cap \Psi$ contains a proper face F. Let Φ' and Ψ' be the flags of F/F_{-1} contained in Φ and Ψ, respectively; and let Φ'' and Ψ'' be the flags of F_n/F contained in Φ and Ψ, respectively. Then, by the inductive hypothesis, there exists a sequence S_1' of flags in F/F_{-1} from Φ' to Ψ', where every two consecutive flags are adjacent and $\Phi' \cap \Psi'$ is contained in every flag. There also exists a sequence S_2'' of flags in F_n/F from Φ'' to Ψ'', where every two consecutive flags are adjacent and $\Phi'' \cap \Psi''$ is contained in every flag. Now we construct sequences S_1 and S_2 of flags of $\mathcal{P}$ by extending the flags in S_1' with the j-faces of Φ for $j \geq r(F)$, and the flags in S_2' with the j-faces of Ψ for $j \leq r(F)$. The desired sequence S is the concatenation of S_1 and S_2.

If $\Phi \cap \Psi = \{F_{-1}, F_n\}$, let G_Φ and G_Ψ be proper faces in Φ and Ψ, respectively. Since $\mathcal{P}$ is connected, there exist $H_1, \dots, H_m$ proper faces such that G_Φ is incident to H_1, G_Ψ is incident to H_m, and H_i is incident to H_{i+1} for every i. Let Φ_0 be a flag containing G_Φ and H_1, for $i \in \{1, \dots, m-1\}$ let Φ_i be a flag containing H_i and H_{i+1}, and let Φ_m be a flag containing H_m and F_Ψ. By the previous case, we know that for $i \in \{1, \dots, m\}$ there is a sequence S_i of flags from Φ_{i-1} to Φ_i where any two consecutive flags are adjacent. The concatenation $S_1 \cdots S_m$ is the desired sequence S. ■

The notion of adjacency of flags motivates the following definition.

The *flag graph* of a polytope $\mathcal{P}$ is the graph whose vertex set is the set $\mathcal{F}l(\mathcal{P})$ of flags of $\mathcal{P}$, and two of them are adjacent if and only if the corresponding flags are adjacent. It is also customary to attach the label i to every edge of the flag graph whose endpoints correspond to i-adjacent flags. Figure 2.7 shows the flag graph of a square. The square and its flags are indicated in grey.

Exercise 2.1.25 Draw the flag graph of the tetrahedron.

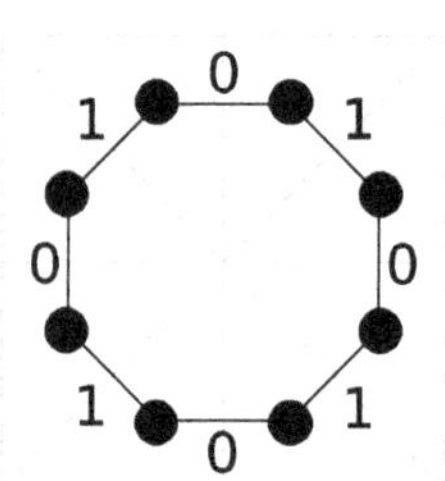

Figure 2.7 The flag graph of a square.

We shall see in Proposition 2.1.46 that the partial order of $\mathcal{P}$ can be completely recovered from its flag graph.

Proposition 2.1.24 implies that every polytope is flag-connected, which leads us to the following remark.

Remark 2.1.26 The flag graph of every polytope is connected.

Isomorphisms of polytopes preserve rank, and since they preserve incidence they map flags into flags. Furthermore, the diamond condition ensures that adjacent flags are mapped to adjacent flags. The following proposition states an equivalence of the definition of isomorphism in terms of flags.

Proposition 2.1.27 *Let $\mathcal{P}$ and $\mathcal{Q}$ be two n-polytopes. Then $\mathcal{P} \cong \mathcal{Q}$ if and only if there exists a bijection β from $\mathcal{F}l(\mathcal{P})$ to $\mathcal{F}l(\mathcal{Q})$ mapping i-adjacent flags to i-adjacent flags for every $i \in \{0, \ldots, n-1\}$.*

Proof If η is an isomorphism between $\mathcal{P}$ and $\mathcal{Q}$ then the diamond condition guarantees that η maps i-adjacent flags of $\mathcal{P}$ to i-adjacent flags of $\mathcal{Q}$ for all i.

From now on we assume that there is a bijection β from $\mathcal{F}l(\mathcal{P})$ to $\mathcal{F}l(\mathcal{Q})$ that preserves i-adjacency for every i. Consequently, if Ψ is a flag of $\mathcal{Q}$ and $\Psi = \beta(\Phi)$ then $\beta(\Phi^i) = \Psi^i$, or equivalently, $\Phi^i = \beta^{-1}(\Psi^i)$. It follows that β^{-1} maps adjacent flags of $\mathcal{Q}$ to adjacent flags of $\mathcal{P}$.

We next define a function $\hat{\beta}$ from the set of faces of $\mathcal{P}$ to the set of faces of $\mathcal{Q}$. For a j-face F of $\mathcal{P}$ we define $\hat{\beta}(F)$ as a face F' of $\mathcal{Q}$ with the properties that $r(F') = j$, and there exists a flag Φ of $\mathcal{P}$ containing F such that $F' \in \beta(\Phi)$. We claim that there is a unique j-face F' of $\mathcal{Q}$ satisfying those properties. Indeed, the strong flag-connectivity of $\mathcal{P}$ implies that, for every pair of flags Φ and Ψ containing F, there exists a sequence S of flags starting in Φ and ending in Ψ where any two consecutive flags are adjacent, and all of them contain F; in particular, two consecutive flags are i-adjacent for some $i \neq j$. Since β preserves i-adjacencies, the images under β of any two consecutive flags in S are i-adjacent for some $i \neq j$. This implies that the j-faces of $\beta(\Phi)$ and $\beta(\Psi)$ are the same, proving our claim.

The function $\hat{\beta}$ just defined is surjective, since every face of $\mathcal{Q}$ is contained in a flag and β is bijective. Furthermore, if two j-faces F and G of $\mathcal{P}$ have the same image F' under β then there exist flags Φ and Ψ of $\mathcal{P}$ with $F \in \Phi$ and $G \in \Psi$ such that $F' \in \beta(\Phi) \cap \beta(\Psi)$. By the strong flag-connectivity of $\mathcal{Q}$ there must exist a sequence of flags $\beta(\Phi) = \Psi_0, \Psi_1, \ldots, \Psi_m = \beta(\Psi)$ where any two consecutive flags are i-adjacent for some $i \neq j$. By letting $\Phi_k = \beta^{-1}(\Psi_k)$ we can see that $\Phi_0 = \Phi, \Phi_1, \ldots, \Phi_m = \Psi$ is a sequence of flags of $\mathcal{P}$ where any two are i-adjacent for some $i \neq j$. It follows that $F = G$ and $\hat{\beta}$ is a bijection.

Finally, if $F \leq G$ are faces of $\mathcal{P}$ then they are contained in a flag Φ. It follows that $\hat{\beta}(F)$ and $\hat{\beta}(G)$ are contained in $\beta(\Phi)$. Therefore $\hat{\beta}$ preserves incidence. ■

In the previous proof we observed that if $\beta \colon \mathcal{F}l(\mathcal{P}) \to \mathcal{F}l(\mathcal{Q})$ maps i-adjacent flags to i-adjacent flags, then β^{-1} also posseses this property. The function from the faces of $\mathcal{Q}$ to the faces of $\mathcal{P}$ associated to β^{-1} is precisely $(\hat{\beta})^{-1}$.

Isomorphisms are a special kind of a more general class of mappings that preserve the incidence of polytopes.

Definition 2.1.28 We say that an n-polytope $\mathcal{P}$ *covers* an n-polytope $\mathcal{Q}$ whenever there is a rank and incidence preserving function $\eta \colon \mathcal{P} \to \mathcal{Q}$ that also preserves adjacency of flags. In this case we also say that η is a *covering*, that $\mathcal{P}$ is a *cover* of $\mathcal{Q}$, and that $\mathcal{Q}$ is a *quotient* of $\mathcal{P}$.

Coverings of polytopes are called *rap-maps* (rank-and-adjacency preserving maps) in [188, Section 2D]. There are other natural definitions of quotients of polytopes, although this one is the most broadly used. For other definitions, as well as examples of their differences, see [191].

Example 2.1.29 Let $\mathcal{P}$ be an n-gon. Assume that the vertices of $\mathcal{P}$ in cyclic order are $0, \ldots, n-1$.

Let $m \geq 2$ be a proper divisor of n and let $\mathcal{Q}$ be an m-gon, with vertices labelled $0, \ldots, m-1$ in cyclic order. Then the function η that maps vertex i to j, where j is the class of i modulo m, induces a covering from $\mathcal{P}$ to $\mathcal{Q}$. Under this covering, the edge of $\mathcal{P}$ between vertices u and w is mapped to the edge of $\mathcal{Q}$ between $u\eta$ and $w\eta$.

This covering can be understood as an (n/m)-fold wrapping of $\mathcal{P}$ along its rotation centre. It is illustrated for $n = 8$ and $m = 4$ in Figure 2.8, where the two grey vertices of the octagon are mapped to the grey vertex of the square.

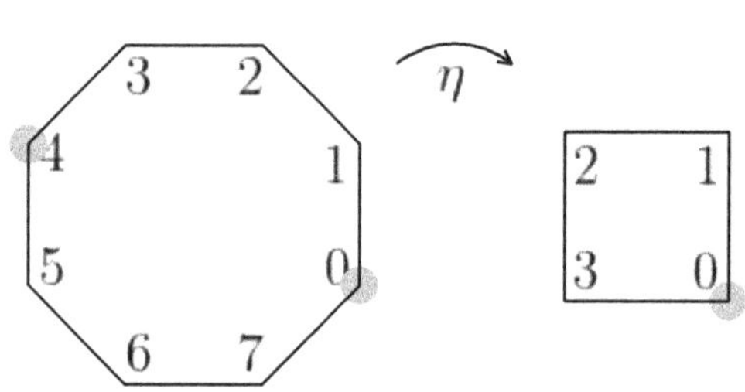

Figure 2.8 An octagon covers a square.

Example 2.1.30 Let $m \geq 3$ be a proper divisor of n. Then the prism over an n-gon covers the prism over an m-gon as an (n/m)-fold wrapping of $\mathcal{P}$ along its n-fold rotation axis. The covering restricted to the top and bottom of the n-prism is precisely that of Example 2.1.29, while n/m quadrilaterals of the n-prism are mapped to a single quadrilateral of the m-prism.

Coverings of polytopes have a concept of degree similar to that of coverings in algebraic topology.

Lemma 2.1.31 *Let η be a covering from $\mathcal{P}$ to $\mathcal{Q}$ and let $\Phi, \Psi \in \mathcal{F}l(\mathcal{Q})$. Then the cardinalities of the preimages of Φ and Ψ under η are the same.*

Proof We shall denote the preimage of a flag Υ by $\eta^{-1}(\Upsilon)$. Let $i_1, \ldots, i_k \in \{0, \ldots, n-1\}$ such that $\Psi = \Phi^{i_1,\ldots,i_k}$.

For each $\hat{\Phi} \in \eta^{-1}(\Phi)$ we verify that $\eta(\hat{\Phi}^{i_1,\ldots,i_k}) = (\eta(\hat{\Phi}))^{i_1,\ldots,i_k} = \Phi^{i_1,\ldots,i_k} = \Psi$. This allows us to define the function $h\colon \eta^{-1}(\Phi) \to \eta^{-1}(\Psi)$ that maps $\hat{\Phi}$ to $\hat{\Phi}^{i_1,\ldots,i_k}$ for our particular choice of $i_1, \ldots, i_k$. The uniqueness of the i-adjacent flags for each i forces h to be injective. Furthermore, the function $h'\colon \eta^{-1}(\Psi) \to \eta^{-1}(\Phi)$ that maps $\hat{\Psi}$ to $\hat{\Psi}^{i_k,\ldots,i_1}$ is the inverse of h, and the cardinalities of the two preimages are the same. ■

Definition 2.1.32 The *degree* of a covering η from $\mathcal{P}$ to $\mathcal{Q}$ is the cardinality of the preimage of any flag of $\mathcal{Q}$ under η.

Example 2.1.33 The degree of the covering from the octagon to the square in Example 2.1.29 is 2, since the preimages of the flag consisting of the vertex 0 and the edge between 0 and 1 consist of the two flags below. The vertex must be congruent to 0 modulo 4, and the other endpoint of the edge must be congruent to 1 modulo 4. It follows that the two flags are the flag with vertex 0 and the edge between 0 and 1, and the flag with vertex 4 and the edge between 4 and 5.

Reasoning in a similar fashion it is straightforward to see that the degrees of the coverings in Examples 2.1.29 and 2.1.30 are in both cases n/m.

The name ‘covering’ suggests that this kind of mappings are surjective. This is indeed the case.

Proposition 2.1.34 *Every covering of n-polytopes is surjective.*

Proof Let η be a convering from a polytope $\mathcal{P}$ to a polytope $\mathcal{Q}$. Let Φ be a flag of $\mathcal{P}$, let G be a face of $\mathcal{Q}$ and Ψ be a flag of $\mathcal{Q}$ containing G. We want to show that there exists a face F of $\mathcal{P}$ such that $\eta(F) = G$.

Since $\mathcal{Q}$ is flag-connected, there exists a sequence $S := \{\Psi_0 = \eta(\Phi), \Psi_1, \dots, \Psi_m = \Psi\}$ of flags of $\mathcal{Q}$, where for $i > 0$ there exists j_i such that $\Psi_i = \Psi_{i-1}^{j_i}$. In other words, $\Psi = (\eta(\Phi))^{j_1,\dots,j_m}$. It follows that $\eta(\Phi^{j_1,\dots,j_m}) = (\eta(\Phi))^{j_1,\dots,j_m} = \Psi$, because η is a covering. We can conclude that the desired face F is the face of $\Phi^{j_1,\dots,j_m}$ of the same rank as G. ■

The next proposition shows that coverings of polytopes can also be defined in terms of flags. Its proof uses the same technique as that of Proposition 2.1.27 and we omit it.

Proposition 2.1.35 *Let $\mathcal{P}$ and $\mathcal{Q}$ be two n-polytopes. Then there is a covering from $\mathcal{P}$ to $\mathcal{Q}$ if and only if there exists a surjection β from $\mathcal{F}l(\mathcal{P})$ to $\mathcal{F}l(\mathcal{Q})$ mapping i-adjacent flags to i-adjacent flags for each $i \in \{0, \dots, n-1\}$.*

Every n-polytope $\mathcal{P}$ can be naturally associated to a simplicial complex (and hence to the corresponding topological space) by considering the proper faces of $\mathcal{P}$ as vertices, and every chain of size $k+1$ as a k-simplex. This simplicial complex is called the *order complex* of $\mathcal{P}$. This is precisely the underlying topological space explained in [258, Section 2] of the 'combinatorial map' given by the flag graph.

Example 2.1.36 The order complex of a square has eight vertices (0-simplices): the four vertices, and the four edges of the square. It also has eight line segments (1-simplices), corresponding to the eight flags (pairs of incident vertex and edge). Two 1-simplices intersect in a 0-simplex if and only if the corresponding flags share a vertex or an edge. Therefore, the order complex corresponds to the subdivision of the circle $\mathbb{S}^1$ into eight segments.

Similarly, the order complex of an m-gon is the subdivision of $\mathbb{S}^1$ into $2m$ segments.

In fact, the order complex of every convex n-polytope $\mathcal{P}$ is a simplicial subdivision of the $(n-1)$-sphere, since the flags of $\mathcal{P}$ correspond to the simplices in the barycentric subdivision of the tessellation induced by $\mathcal{P}$ when projected to a circumscribed $(n-1)$-sphere.

Motivated by the nature of the topological spaces, polytopes whose order complex is a subdivision of a sphere (resp. projective space, torus) are called *spherical polytopes* (resp. *projective polytopes*, *toroidal polytopes*).

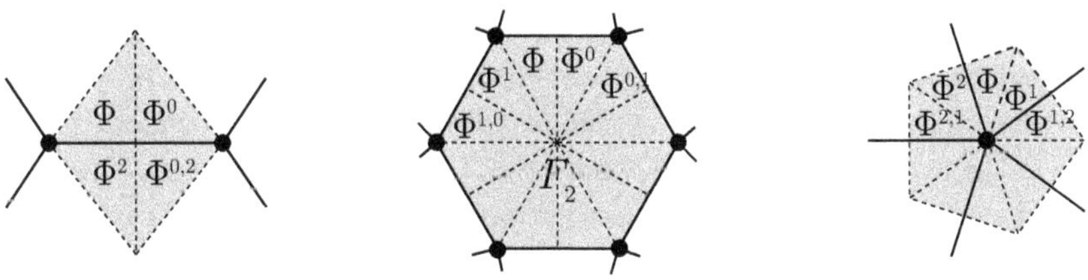

Figure 2.9 The flags containing (a) an edge, (b) a face and (c) a vertex.

If $\mathcal{P}$ is a polyhedron then each maximal simplex of its order complex is a triangle whose corners correspond to the vertex, edge and 2-face of some flag Φ. Each side ℓ of that triangle belongs to a unique other triangle, corresponding to the adjacent flag that differs from Φ precisely on the element not contained in ℓ. The following features are a consequence of the strong flag-connectivity of $\mathcal{P}$. The corner of the triangle representing the edge of Φ belongs precisely to the four triangles representing Φ, Φ^0, $\Phi^{0,2}$ and Φ^2 (see Figure 2.9(a)). The triangles representing the flags of a finite 2-face F_2 form a polygon having the corner representing F_2 as an interior point (see Figure 2.9(b)). Finally, every corner of a triangle representing a vertex with finite degree k is surrounded by $2k$ triangles that form a k-gon having that corner as an interior point (see Figure 2.9(c)). Hence, the order complex of $\mathcal{P}$ is a manifold, as long as $\mathcal{P}$ has finite faces and finite vertex-figures. We have proven the following proposition.

Proposition 2.1.37 *The order complex of every polyhedron with finite faces and vertex-figures is a connected surface without boundary. Furthermore, if the polyhedron is finite then the surface is compact.*

A *map* on a surface S is a 2-cell embedding of a connected graph (possibly with multiple edges and loops) into S. That is, the connected components of the surface after removing the image of the graph are homeomorphic to disks (see for example [7, Chapter 5] and [149, 257, 269]). Two maps $\mathcal{M}_1$ and $\mathcal{M}_2$ on surfaces $\mathcal{S}_1$ and $\mathcal{S}_2$ are *isomorphic* if and only if there is an homeomorphism from $\mathcal{S}_1$ to $\mathcal{S}_2$ such that when restricted to the underlying graph of $\mathcal{M}_1$ it becomes an isomorphism to the underlying graph of $\mathcal{M}_2$ (see for example [251] for an equivalent definition).

Proposition 2.1.37 associates a map $\mathcal{M}_\mathcal{P}$ on some surface to every polyhedron $\mathcal{P}$ with finite faces and vertex-figures in the following way. The underlying surface $\mathcal{S}_\mathcal{P}$ of $\mathcal{M}_\mathcal{P}$ is the topological space defined by the order complex $\mathcal{C}_\mathcal{P}$ of $\mathcal{P}$: the vertices of the underlying graph of $\mathcal{M}_\mathcal{P}$ are the points of $\mathcal{S}_{\mathcal{P}}$ that are corners of triangles of $\mathcal{C}_{\mathcal{P}}$ that correspond to vertices of $\mathcal{P}$; and the edges of the underlying graph of $\mathcal{M}_{\mathcal{P}}$ are arcs between pairs of vertices (of the underlying graph) that are unions of the two sides of triangles T_1 and T_2 of $\mathcal{C}_{\mathcal{P}}$ that are opposite to the corners of T_1 and T_2 representing a 2-face of $\mathcal{P}$. However, not every map on a surface represents a polyhedron; for example, maps with loops have edges containing only one vertex, and hence they do not posses the diamond condition.

Exercise 2.1.38 Let $\mathcal{P}$ and $\mathcal{Q}$ be polyhedra. Show that $\mathcal{P} \cong \mathcal{Q}$ if and only if the maps $\mathcal{M}_{\mathcal{P}}$ and $\mathcal{M}_{\mathcal{Q}}$ defined above are isomorphic.

In higher rank, not every polytope can be represented by a manifold. As we shall see soon, there are 4-polytopes whose vertex-figures can be represented as maps on the torus. In their order complexes, the neighbourhoods of the corners representing vertices of those polytopes are not spheres, but tori.

The order complex is a motivation to define orientability on abstract polytopes. A polytope is said to be *orientable* if its flag graph is bipartite. It is not within the scope of this section to justify that the notion of orientability just defined is equivalent to requiring that the order complex is orientable as a topological space (for definitions of orientability see [123]). To develop intuition, we suggest to the reader the following exercise.

Exercise 2.1.39 Show that a polyhedron is orientable if and only if its order complex (as a surface) does not contain a Möbius strip.

Given a polytope $\mathcal{P}$, its *dual* $\mathcal{P}^{\delta}$ consists of the set of faces of $\mathcal{P}$ with the reverse order. Clearly, $(\mathcal{P}^{\delta})^{\delta} = \mathcal{P}$. If $\mathcal{P} \cong \mathcal{P}^{\delta}$ then we say that $\mathcal{P}$ is *self-dual*.

Example 2.1.40 The triangle is a self-dual polygon. This can be easily seen from Figure 2.3(b), by drawing the Hasse diagram upside down. In fact, every polygon is self-dual.

The tetrahedron is a self-dual polyhedron; the dual of the cube is isomorphic to the octahedron. The details in a more general setting are left as an exercise.

Exercise 2.1.41 Use the definitions in Examples 2.1.15, 2.1.16 and 2.1.17 to show that the n-simplex is self-dual, and that the dual of the n-cube is isomorphic to the cross polytope of rank n.

If $\mathcal{P}$ is a polytope then all items of Definition 2.1.7 are satisfied by $\mathcal{P}^\delta$, and so the dual of every polytope is a polytope. The roles of F_n and F_{-1} in $\mathcal{P}^\delta$ are the minimum and maximum, respectively; and if F is an i-face of $\mathcal{P}$ then it is an $(n-1-i)$-face of $\mathcal{P}^\delta$. In particular, vertices of $\mathcal{P}$ correspond to facets of $\mathcal{P}^\delta$ and vice versa. The set $\mathcal{F}l(\mathcal{P}^\delta)$ of flags of $\mathcal{P}^\delta$ is the set $\mathcal{F}l(\mathcal{P})$, in the understanding that the flags are unordered sets. Two such flags are i-adjacent in $\mathcal{P}$ if and only if they are $(n-1-i)$-adjacent in $\mathcal{P}^\delta$.

Duality preserves many properties of polytopes like the ones in the following remark, whose proof is straightforward and we omit.

Remark 2.1.42 Let $\mathcal{P}$ be a polytope.

- If $\mathcal{P}$ covers $\mathcal{Q}$ then $\mathcal{P}^\delta$ covers $\mathcal{Q}^\delta$.
- The order complex of $\mathcal{P}^\delta$ is isomorphic to the order complex of $\mathcal{P}$ (as simplicial complexes).
- The flag graph of $\mathcal{P}^\delta$ can be obtained from the flag graph of $\mathcal{P}$ by replacing label i with label $n-1-i$ for every i.
- $\mathcal{P}$ is orientable if and only if $\mathcal{P}^\delta$ is orientable.

Often we are interested on studying abstract polytopes by determining a particular flag as base of operations. This will certainly be the idea when studying the automorphism groups of regular and chiral polytopes. In the cases where the polytope is not regular, the choice of flag is particularly relevant. A polytope $\mathcal{P}$ is said to be *rooted* when a base flag Φ is specified, and we shall denote it by $\mathcal{P}_{(\Phi)}$.

We say that two rooted polytopes are *isomorphic* when there is an isomorphism between them that maps the base flag of one to that of the other. Clearly, if two rooted polytopes $\mathcal{P}_{(\Phi)}$ and $\mathcal{Q}_{(\Psi)}$ are isomorphic then $\mathcal{P} \cong \mathcal{Q}$. The converse is not true. If $\mathcal{P}$ is the prism over a triangle, Φ is a flag containing a triangle and Ψ is a flag containing a rectangle, then no automorphism of $\mathcal{P}$ will map Φ to Ψ and so $\mathcal{P}_{(\Phi)}$ is not isomorphic to $\mathcal{P}_{(\Psi)}$.

2.1.3 Equivalent Definitions

Definition 2.1.7 was motivated by the face-lattice of convex polytopes. In this section we present some equivalent definitions that prove useful in certain contexts. First we relate strong connectivity with strong flag-connectivity.

Theorem 2.1.43 *Let $\mathcal{P}$ be a poset satisfying items (a), (b) and (c) of Definition 2.1.7. Then $\mathcal{P}$ is a polytope if and only if it is strongly flag-connected.*

Proof Proposition 2.1.24 establishes strong flag-connectivity for polytopes. It only remains to show that if $\mathcal{P}$ is strongly flag-connected then it is strongly connected.

We first show that $\mathcal{P}$ is connected. Assume that $\mathcal{P}$ has rank 2 or higher and let F and G be distinct proper faces of $\mathcal{P}$. Let Φ and Ψ be flags containing F and G, respectively. Then there is a sequence

$$\Phi = \Phi_0, \Phi_1, \ldots, \Phi_k = \Psi, \tag{2.4}$$

where Φ_{i-1} and Φ_i are adjacent for $i \in \{1, \ldots, k\}$. If $k = 0$ then F is incident to G and the smallest path between them in the incidence graph has length 1. We proceed by induction. Let $k \geq 1$ and assume that if H is a face of $\mathcal{P}$ belonging to flag Ψ_{k-1} and there is a sequence $\Psi_0, \ldots, \Psi_{k-1}$ of flags where Ψ_{i-1} is adjacent to Ψ_i, then there is a path in the incidence graph from F to H. Then we choose any face H in the flag Φ_{k-1} in (2.4) different from the j-face, where $j \in \{0, \ldots, n-1\}$ is such that $\Phi_{k-1} = \Phi_k^j$. By the induction hypothesis, there is a path in the incidence graph from F to H, and this is the desired path when $G = H$. If $G \neq H$ then $G \in \Phi_k$ and it is incident to H; in that case we can concatenate a path from F to H with the edge between H and G to obtain the desired path in the incidence graph.

Now, if $F \leq G$ then the section G/F is strongly flag-connected, since any two flags of G/F can be extended to flags of $\mathcal{P}$ that differ between them only in faces greater than F and smaller than G. By the arguments in the previous paragraph each section must be connected, implying that $\mathcal{P}$ is strongly connected. ∎

In some instances a recursive definition of an abstract polytope is particularly helpful. The most natural one is to interpret an n-polytope as an collection of $(n-1)$-polytopes carefully assembled.

Proposition 2.1.44 *Let $\mathcal{P}$ be a poset with maximum F_n, minimum F_{-1} and rank function on the set $\{-1, 0, \ldots, n\}$ for $n \geq 3$. Then $\mathcal{P}$ is a polytope if and only if all sections of rank $n-1$ are $(n-1)$-polytopes and $\mathcal{P}$ is flag-connected.*

Proof The forward implication follows from Remark 2.1.12 and Proposition 2.1.24. In the reminder of the proof we assume that all sections of $\mathcal{P}$ of rank $n-1$ are $(n-1)$-polytopes and that $\mathcal{P}$ is flag-connected; our goal is to show that $\mathcal{P}$ satisfies the third and fourth items of Definition 2.1.7.

Every diamond (1-section) of $\mathcal{P}$ belongs to a section of rank $n-1$ and so the diamond condition holds.

Since $\mathcal{P}$ is flag-connected, between any two flags Φ, Ψ there is a sequence of flags from one to the other where any two consecutive flags are adjacent.

Furthermore, if Φ and Ψ have a proper face F in common then we can restrict them first to F/F_{-1} and then to F_n/F. In each of these two sections we can find a sequence of flags between the restrictions $\tilde{\Phi}$ and $\tilde{\Psi}$ of Φ and Ψ where any two consecutive flags are adjacent, and $\tilde{\Phi} \cap \tilde{\Psi}$ is contained in every flag of the sequence. (Note here that each of F/F_{-1} and F_n/F belongs to a section of rank $n-1$, and by hypothesis and Proposition 2.1.24 they are strongly flag-connected.) We can extend each of the flags of the sequences to suitable flags of $\mathcal{P}$ to obtain the desired sequence, as we did in the proof of Proposition 2.1.24. This proves that $\mathcal{P}$ is strongly flag-connected. ■

Proposition 2.1.44 says that when assembling together a family of $(n-1)$-polytopes to construct an n-polytope we need to ensure that the vertex-figures are also polytopes. We will use this alternative definition to show that the face-lattices of convex polytopes are abstract polytopes (see [107] for definitions).

Theorem 2.1.45 *The face-lattice of every convex polytope, including the empty set and the convex polytope itself as improper faces, is an abstract polytope.*

Proof We will proceed by induction. The result holds for convex polytopes of dimensions 0, 1, 2.

Let $\mathcal{Q}$ be a convex n-polytope for $n \geq 3$. The facets and vertex-figures of $\mathcal{Q}$ are defined to be convex $(n-1)$-polytopes, and by our induction hypothesis, their face-lattices are abstract polytopes. According to Proposition 2.1.44 it only remains to show that convex-polytopes are flag-connected.

Let Φ and Ψ be flags of $\mathcal{Q}$. We can identify them with simplices whose corners are at the centroids of each of their proper faces. We may project $\mathcal{Q}$ to a circumscribed $(n-1)$-sphere $\mathbb{S}^{n-1}$, and take the barycentric subdivision $\mathcal{T}$ of the tessellation of $\mathbb{S}^{n-1}$ induced by this projection. Every $(n-2)$-face of $\mathcal{T}$ belongs to two facets of $\mathcal{T}$ that correspond to two adjacent flags of $\mathcal{Q}$. We choose a point x_Φ and a point x_Ψ in the interiors of the simplices of $\mathcal{T}$ corresponding to Φ and Ψ, respectively. These two points can be chosen in such a way that the geodesic g between them does not intersect j-faces of $\mathcal{T}$ for $j < n-2$. Therefore the desired path in the flag graph is given by the sequence of facets of $\mathcal{T}$ intersecting g when the latter is traversed from Φ to Ψ. ■

The concept of abstract polytope was largely motivated by geometric structures like convex polytopes or the polyhedra described in [104], where the faces play the main role. When studying some aspects of abstract polytopes the flags are the relevant substructures, and it is convenient to have a definition of polytopes in terms of their flags instead of their faces. One of the most natural ways to do this is through the flag graph.

Proposition 2.1.46 *Every polytope can be recovered from its flag graph.*

Proof Let $\mathcal{P}$ be an n-polytope, and for each $i \in \{0, \ldots, n-1\}$, let Δ_i be the subgraph of the flag graph of $\mathcal{P}$ obtained by removing all edges with label i. A connected component X of Δ_i consists of some flag Φ together with all flags that can be reached from Φ by a path whose edges have labels in $\{0, \ldots, n-1\} \setminus \{i\}$. It follows that if F_i is the i-face of Φ then it is also the i-face of all flags in X. This allows us to define a function f that maps each such component X to such a face F_i. The strong flag-connectivity tells us that if F_i is the i-face of a flag Ψ then there is a sequence of flags between Φ and Ψ where consecutive flags are adjacent, and the i-face of every flag in the sequence is F_i. This forces Ψ to be in X, and this in turn implies that f is injective. Clearly f is surjective, and therefore there is a one-to-one correspondence between the set of i-faces of $\mathcal{P}$ and the components of Δ_i.

So far we have established that we can recover the set of faces of $\mathcal{P}$ and their ranks from the flag graph of $\mathcal{P}$. In what follows we shall recover the incidence. An i-face F_i is incident to a j-face F_j if and only if there is a flag containing both of them. In the flag graph this is viewed as a non-empty intersection of the connected components of Δ_i and Δ_j corresponding to F_i and F_j, respectively. Therefore, the incidence of the faces of $\mathcal{P}$ is obtained from the non-empty intersection of the connected components of the graphs Δ_i. ■

The previous proposition gives relevance to n-regular graphs with a proper edge-colouring with n colours. Following [273] we may define a maniplex in many ways, including the following.

Definition 2.1.47 An *n-maniplex* $\mathcal{M}$ is a connected n-regular graph with a proper edge colouring with set of colours $\{0, \ldots, n-1\}$ such that for every $i, j \in \{0, \ldots, n-1\}$ satisfying that $|i-j| \geq 2$ every connected component of the graph induced by the edges labelled i and j is a square.

In order to avoid confusion, we shall call *flags* the vertices of a maniplex $\mathcal{M}$, and reserve the term 'vertex' for the 0-faces defined below.

An n-maniplex can be defined in an alternative way as the flag set $\mathcal{M}_0$ of $\mathcal{M}$ together with an n-tuple of fixed-point-free involutions in the symmetric group on $\mathcal{M}_0$. The edges of label i are replaced by the fixed-point-free involution r_i that maps one flag to another if and only if these two flags were adjacent by an edge labelled i. Then, the fact that the connected components of the graph induced by edges with labels i and j are squares if $|i-j| \geq 2$, can be reinterpreted as $r_i r_j = r_j r_i$. We shall study the group $\langle r_0, \ldots, r_{n-1} \rangle$ in Section 3.2.

In [273] the rank appears shifted by one, so that n-maniplexes are $(n+1)$-regular graphs. In subsequent works the rank was adapted to fit with the notation of polytopes, and we shall continue this trend. Indeed, with this convention it follows from (2.2) that the flag graph of every n-polytope is an n-maniplex.

Maniplexes are very general structures devised so that every map on a surface can be seen as a 3-maniplex, and vice versa. In some works it has been convenient to include certain degeneracies of the maps, like having faces with only one edge, or lying on surfaces with boundaries. These degeneracies impact the definition of the maniplex by admitting multiple edges and semi-edges; such graphs are called *premaniplexes* in [133]. Since the main topic of this text is abstract polytopes, we shall not need such generality, and in the definition of the maniplex in terms of a permutation group of the flags we shall require that if $i \neq j$ then the involutions r_i and r_j have no common transposition.

In the remainider of this section we compare the concepts of n-polytope and n-maniplex. The main source for this discussion is [98].

In the spirit of Proposition 2.1.46, for $i \in \{0, \ldots, n-1\}$ we define the *i-faces* of $\mathcal{M}$ as the connected components of the subgraph induced by all edges with labels $j \neq i$. The following lemma provides a useful combinatorial property of each face of a maniplex.

Lemma 2.1.48 *Let F_j be a j-face of an n-maniplex $\mathcal{M}$, and let Φ and Ψ be two flags of $\mathcal{M}$ contained in F_j. Then there is a flag Υ of $\mathcal{M}$ contained in F_j with the property that there is a path from Φ to Υ whose edges have colours in $\{j+1, \ldots, n-1\}$, and a path from Υ to Ψ whose edges have colours in $\{0, \ldots, j-1\}$ (see Figure 2.10).*

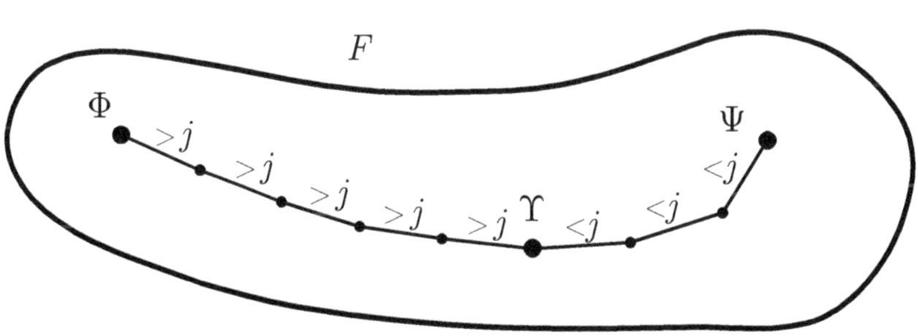

Figure 2.10 The path from Φ to Ψ through the flag Υ.

Proof Since $\Phi, \Psi \in F_j$, there exist paths from Φ to Ψ, none of whose edges are labelled j. Given such a path ($\Phi = \Phi_0, \Phi_1, \ldots, \Phi_m = \Psi$), we let ℓ_r be the label of the edge between Φ_{r-1} and Φ_r, for $r \in \{1, \ldots, m\}$. If $\ell_1 < j$ and $\ell_2 > j$ then we can substitute Φ_1 by the flag Φ_1' that forms the square with edges labelled ℓ_1 and ℓ_2 that contains the flags Φ_0, Φ_1 and Φ_2. This effectively switches ℓ_1 and ℓ_2 in the list of labels of the paths from Φ to Ψ. More generally, if $\ell_r < j$ and $\ell_{r+1} > j$ for some r then we can replace Φ_r by the flag Φ_r' that forms the square with edges labelled ℓ_r and ℓ_{r+1} that contains the flags Φ_{r-1}, Φ_r and Φ_{r+1}; this switches ℓ_r and ℓ_{r+1} in the lists of labels of the paths between Φ and Ψ. A simple inductive procedure shows that a path from Φ to Ψ with minimal possible number of edges with labels in $\{0, \ldots, j-1\}$ appearing before the last edge with label in $\{j+1, \ldots, n-1\}$ must have all its edges with labels greater than j appearing before all edges with labels less than j. In this situation Υ is the only flag of the path incident to an edge with label in $\{0, \ldots, j-1\}$ and an edge with label in $\{j+1, \ldots, n-1\}$. ∎

From our understanding of the flag graphs of polytopes, it is natural to define for an i-face F_i and a j-face F_j of a maniplex that $F_i \leq F_j$ if and only if $i \leq j$ and $F_i \cap F_j \neq \varnothing$. We show next that this relation $\leq$ defines a partial order.

Lemma 2.1.49 *The relation $\leq$ just defined is a partial order.*

Proof It is straightforward to show that the relation is reflexive and antisymmetric; we only need to show that it is also transitive.

Let F_i, F_j and F_k be an i-face, a j-face and a k-face, respectively, such that $F_i \leq F_j$ and $F_j \leq F_k$. Let Φ be a flag in $F_i \cap F_j$ and Ψ a flag in $F_j \cap F_k$.

Since $\Phi, \Psi \in F_j$, Lemma 2.1.48 implies the existence of a flag $\Upsilon \in F_j$, a path from Φ to Υ whose edges have labels in $\{j+1, \ldots, n-1\}$, and a path from Υ to Ψ whose edges have labels in $\{0, \ldots, j-1\}$.

The flag Υ belongs to F_i, since it is connected to Φ by a path consisting of edges with labels greater than j and $j > i$. Similarly, it must belong to F_k. It follows that $F_i \cap F_k \neq \varnothing$, and hence $\leq$ is transitive. ∎

By adjoining a unique minimum and a unique maximum to the set of faces of a maniplex $\mathcal{M}$, we obtain a poset, denoted $\mathcal{P}(\mathcal{M})$, that satisfies Items (a) and (b) in Definition 2.1.7. From Proposition 2.1.46 we know that if $\mathcal{M}$ is the flag graph of a polytope $\mathcal{P}$, then $\mathcal{P} \cong \mathcal{P}(\mathcal{M})$. We say that $\mathcal{M}$ is *polytopal* whenever it is the flag graph of a polytope.

Not every maniplex is polytopal, as shown next.

Example 2.1.50 The graph in Figure 2.11 represents a maniplex $\mathcal{M}_0$, where all flags belong to the unique 0-face and to the unique 2-face. It has two 1-faces, each of them incident to the 0-face and to the 2-face. It follows that $\mathcal{M}_0$ cannot be the flag graph of a polyhedron since every 1-face belongs to only one 2-face.

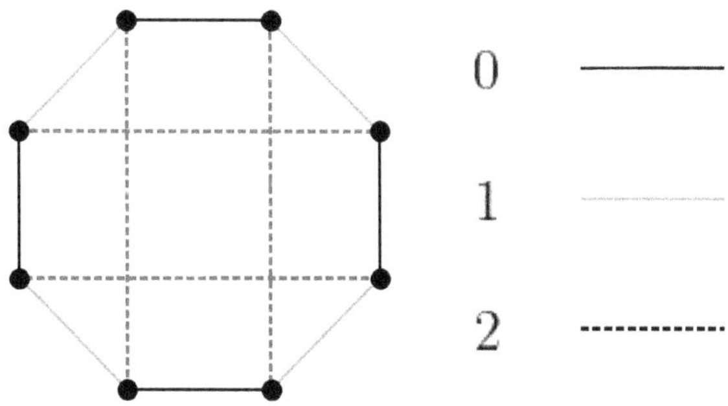

Figure 2.11 A maniplex that is not the flag graph of a polytope.

We can examine further the maniplex $\mathcal{M}_0$ in Example 2.1.50. Its poset $\mathcal{P}(\mathcal{M}_0)$ contains only two flags, each completely determined by the 1-face it contains. Hence, the eight flags of $\mathcal{M}_0$ (here we mean the vertices of the labelled graph) do not represent faithfully the two flags of $\mathcal{P}(\mathcal{M}_0)$ (as a poset). Following [98, Section 3], we say that the poset $\mathcal{P}(\mathcal{M})$ of a maniplex $\mathcal{M}$ is *faithful* if there is a bijection between the flags of $\mathcal{P}(\mathcal{M})$ (as a poset) and the flags of $\mathcal{M}$ (vertices of the graph). It follows from the strong flag-connectivity that the poset of the flag graph of every polytope is faithful.

The maniplex $\mathcal{M}_0$ in Figure 2.11 is an example of a maniplex whose poset is not faithful because each maximal chain of $\mathcal{P}(\mathcal{M}_0)$ corresponds to more than one flag of $\mathcal{M}_0$. As we shall see, every flag of a maniplex $\mathcal{M}$ corresponds to at least one maximal chain of $\mathcal{P}(\mathcal{M})$, and hence the only way in which the poset of a maniplex can fail to be faithful is if two or more flags of $\mathcal{M}$ correspond to one maximal chain of $\mathcal{P}(\mathcal{M})$. The idea of the next proposition is to show that each maximal chain of $\mathcal{P}(\mathcal{M})$ corresponds to one or more flags of $\mathcal{M}$, but it is stated in a more general form.

Proposition 2.1.51 *Let $k \in \{1, \dots, n\}$, let $\mathcal{M}$ be an n-maniplex, and for $j \in \{1, \dots, k\}$ let F_j be an i_j-face of $\mathcal{M}$. If for each j and j' the faces F_j and $F_{j'}$ are incident then $\bigcap_{j \in \{1,\dots,k\}} F_j$ is non-empty.*

Proof If for each j and j' the faces F_j and $F_{j'}$ are incident then the set $\{F_j\colon j \in \{1,\ldots,k\}\}$ constitutes a chain of $\mathcal{P}(\mathcal{M})$. The statement of this proposition establishes that the intersection of every chain of $\mathcal{P}(\mathcal{M})$ contains at least one flag of $\mathcal{M}$.

We may assume that $i_1 < i_2 < \cdots < i_k$. We proceed by induction over k.

There is nothing to prove for $k = 1$, while the case $k = 2$ is the definition of incidence of faces of maniplexes. The case $k = 3$ was shown in the proof of Lemma 2.1.49.

Assume that $k \geq 4$. By inductive hypothesis, we know that there is a flag $\Phi \in \bigcap_{j\in\{1,\ldots,k-1\}} F_j$ and a flag $\Psi \in \bigcap_{j\in\{2,\ldots,k\}} F_j$. Since $\Phi \in F_1 \cap F_{k-1}$ and $\Psi \in F_{k-1} \cap F_k$, Lemma 2.1.48 implies that there exist a flag Υ in F_{k-1}, a path from Φ to Υ whose edges have labels in $\{i_{k-1}+1,\ldots,n-1\}$ and a path from Υ to Ψ whose edges have labels in $\{0,\ldots,i_{k-1}-1\}$. Then Υ must belong to $\bigcap_{j\in\{1,\ldots,k-1\}} F_j$ (same as Φ) and to F_k (same as Ψ). Therefore $\Upsilon \in \bigcap_{j\in\{1,\ldots,k\}} F_j$. ■

A couple of necessary and sufficient conditions for a maniplex to be polytopal can be found in [98]. We summarise them in the remainder of this section.

We say that an n-maniplex $\mathcal{M}$ satisfies the *component intersection property* whenever the intersection of every family of mutually incident faces of $\mathcal{M}$ is connected. Moreover, it satisfies the *path intersection property* if for every $0 \leq i < j \leq n-1$, whenever there is a path from some flag Φ to some flag Ψ whose edges have labels greater than i and a path from Φ to Ψ whose edges have labels less than j, there is a path from Φ to Ψ whose edges have labels in $\{i+1,\ldots,j-1\}$.

The following result is a corollary of Proposition 2.1.51.

Corollary 2.1.52 *If a maniplex $\mathcal{M}$ satisfies the component intersection property then $\mathcal{P}(\mathcal{M})$ is faithful.*

Proof If X is a maximal chain of $\mathcal{P}(\mathcal{M})$ then by Proposition 2.1.51 we know that there is at least one flag of $\mathcal{M}$ contained in every element of X; this in turn implies that X must contain an element of every rank. If there were two or more flags in the intersection of X, the component intersection property would force them to belong to a connected component; but since X has an element of every rank, the resulting graph has no edges. We can conclude that X corresponds to only one flag of $\mathcal{M}$. ■

Exercise 2.1.53 Show that the maniplex in Figure 2.11 satisfies neither the component intersection property, nor the path intersection property.

The path intersection property is hereditary in the following sense.

Lemma 2.1.54 *Let $\mathcal{M}$ be an n-maniplex satisfying the path intersection property, let $\{F_i\}_{i\in\{0,\dots,n-1\}}$ be a family of mutually incident faces of $\mathcal{M}$, where F_i has rank i, and let $0 \leq k \leq n-1$. Then $\bigcap_{i\in\{k,\dots,n-1\}} F_i$ and $\bigcap_{i\in\{0,\dots,k\}} F_i$ are maniplexes (after shifting the edge labels if necessary) satisfying the path intersection property. Furthermore, if $j > k$ then $F_j \cap (\bigcap_{i\in\{0,\dots,k\}} F_i)$ is a face of $\bigcap_{i\in\{0,\dots,k\}} F_i$.*

Proof To show that $\bigcap_{i\in\{k,\dots,n-1\}} F_i$ is a maniplex we only need to prove that it is connected, since the other properties are inherited from those of $\mathcal{M}$. In that case, the path intersection property for the intersections is also inherited from that of $\mathcal{M}$, since every path in $\bigcap_{i\in\{k,\dots,n-1\}} F_i$ is a path in $\mathcal{M}$. The arguments for $\bigcap_{i\in\{0,\dots,k\}} F_i$ are dual to those for $\bigcap_{i\in\{k,\dots,n-1\}} F_i$, which allow us to consider only the latter.

In order to prove connectivity of $\bigcap_{i\in\{k,\dots,n-1\}} F_i$ we proceed by induction over $m := n-k-1$. If $m = 0$ then $k = n-1$ and $\bigcap_{i\in\{k,\dots,n-1\}} F_i$ is an $(n-1)$-face of $\mathcal{M}$, which is connected by definition.

Assume that $\bigcap_{i\in\{k+1,\dots,n-1\}} F_i$ is connected for some $0 < k < n-1$, and therefore that $\bigcap_{i\in\{k+1,\dots,n-1\}} F_i$ is a maniplex satisfying the path intersection property. The case $m = 0$ applies now to show that $\bigcap_{i\in\{k,\dots,n-1\}} F_i$ is connected, which finishes the first part of the statement.

To prove the second part we only need to show that $F_j \cap (\bigcap_{i\in\{0,\dots,k\}} F_i)$ is connected. Let Φ and Ψ be two flags of F_j that are also in $\bigcap_{i\in\{0,\dots,k\}} F_i$. From Lemma 2.1.48 we know that for some flag Υ of F_j there is a path π_1 from Φ to Υ whose edges have labels in $\{j+1,\dots,n-1\}$, and a path π_2 from Υ to Ψ whose edges have labels in $\{0,\dots,j-1\}$. On the other hand, there is a path π_3 from Φ to Ψ whose edges have labels in $\{k+1,\dots,n-1\}$. Now the path intersection property with π_2 and the concatenation of π_1^{-1} and π_3 implies the existence of a path π_4 from Υ to Ψ whose edges have labels in $\{k+1,\dots,j-1\}$. The concatenation of π_1 and π_4 is a path from Φ to Ψ in $F_j \cap (\bigcap_{i\in\{0,\dots,k\}} F_i)$, and therefore this last graph is connected. ∎

In the next two lemmas we exhibit some implications of the component intersection property.

Lemma 2.1.55 *If a maniplex $\mathcal{M}$ satisfies the component intersection property then $\mathcal{P}(\mathcal{M})$ satisfies the diamond condition.*

Proof Let $i \in \{1,\dots,n-1\}$, let F_{i-1} and F_{i+1} be incident $(i-1)$- and $(i+1)$-faces of $\mathcal{M}$ and let $G = F_{i-1} \cap F_{i+1}$. The component intersection property implies that G is connected.

We claim that given any two flags of G there is a path between them that has at most one edge labelled i. Indeed, a simple induction allows us to 'remove' pairs of edges labelled i from any path as follows. If a path with length k between flags Φ and Ψ starts and ends with edges labelled i and has no other edge labelled i, then we may denote by ℓ_j the label of the jth edge, and then the unique path with length $k-2$ starting at Φ where the jth edge has label ℓ_{j+1} ends at Ψ because $|i-\ell_{j+1}| \geq 2$, and this path has no edge labelled i.

The previous claim implies that the subgraph obtained from G by removing all edges labelled i has precisely two connected components: a component C_0 consisting of all flags that can be reached from an arbitrary flag Φ_0 by a path without edges labelled i, and the component C_1 consisting of all flags that can be reached from Φ_0 by a path that uses precisely one edge labelled i. The component intersection property implies that these two components extend to two distinct i-faces F_i and F_i' of $\mathcal{M}$. Indeed, if C_0 and C_1 were in the same i-face F_i then they would be two distinct connected components of $G \cap F_i = F_{i-1} \cap F_i \cap F_{i+1}$, contradicting the component intersection property. We conclude that F_i and F_i' are distinct, and they are the only i-faces that are simultaneously incident to F_{i-1} and F_{i+1}.

The proof of the cases $i = 0$ and $i = n-1$ is the same, starting just with the face F_1 or F_{n-1} instead of $F_{i-1} \cap F_{i+1}$. ∎

Lemma 2.1.56 *If $\mathcal{M}$ satisfies the component intersection property then $\mathcal{P}(\mathcal{M})$ is strongly flag-connected.*

Proof We know from Corollary 2.1.52 that the poset $\mathcal{P}(\mathcal{M})$ is faithful. Then, every flag $\hat{\Phi}$ of $\mathcal{P}(\mathcal{M})$ is represented by a unique flag Φ of $\mathcal{M}$.

Any two flags $\hat{\Phi}_0$ and $\hat{\Psi}_0$ of $\mathcal{P}(\mathcal{M})$ that contain some set $\{\hat{F}_i\}_{i \in I}$ of faces, correspond to two flags Φ_0 and Ψ_0 of $\mathcal{M}$ that are contained in the set $\{F_i\}_{i \in I}$ of faces, where F_i is the i-face with vertex set given by the flags Φ such that $\hat{\Phi}$ contains $\hat{F}_i$. Let a_i be the rank of F_i. The component intersection property tells us that $\cap_{i \in I} F_i$ is connected and so there is a path from Φ_0 to Ψ_0

$$(\Phi_0, \Phi_1, \ldots, \Phi_k = \Psi_0)$$

none of whose edges have labels in $\{a_i\}_{i \in I}$. This path corresponds to a sequence

$$\hat{\Phi}_0, \hat{\Phi}_1, \ldots, \hat{\Phi}_k = \hat{\Psi}_0$$

of flags of $\mathcal{P}(\mathcal{M})$ such that $\hat{\Phi}_{j-1}$ is ℓ-adjacent to $\hat{\Phi}_j$ if and only if Φ_{j-1} is ℓ-adjacent to Φ_j. It follows that any two consecutive flags in this sequence are adjacent, and that every flag contains the set $\{\hat{F}_i\}_{i \in I}$. Therefore $\mathcal{P}(\mathcal{M})$ is strongly flag-connected. ∎

We are ready to state the main result on polytopality of maniplexes (see [98, Theorem 5.3]).

Theorem 2.1.57 *The following are equivalent for a maniplex $\mathcal{M}$:*

1. *$\mathcal{M}$ is polytopal,*
2. *$\mathcal{M}$ satisfies the component intersection property,*
3. *$\mathcal{M}$ satisfies the path intersection property.*

Proof Assume first that $\mathcal{M}$ is polytopal; we shall show that it satisfies the path intersection property. Let $i < j$, let Φ and Ψ be two flags such that there exist a path from Φ to Ψ with labels in $\{i+1, \dots, n-1\}$ and a path from Φ to Ψ with labels in $\{0, \dots, j-1\}$, and let $\hat{\Phi}$ and $\hat{\Psi}$ be the corresponding flags in $\mathcal{P}(\mathcal{M})$. Then $\hat{\Phi}$ and $\hat{\Psi}$ contain the same k-faces for every $k \in \{0, \dots, i\} \cup \{j, \dots, n-1\}$. Owing to the strong flag-connectivity, there exists a sequence of flags from $\hat{\Phi}$ to $\hat{\Psi}$ where any two consecutive are adjacent, and such that all flags contain $\hat{\Phi} \cap \hat{\Psi}$. It follows that two consecutive flags in the sequence are ℓ-adjacent for some $\ell \in \{i+1, \dots, j-1\}$, and hence it induces a path in $\mathcal{M}$ from Φ to Ψ whose edges have labels in $\{i+1, \dots, j-1\}$. We can now conclude that $\mathcal{M}$ satisfies the path intersection property.

Next assume that $\mathcal{M}$ satisfies the path intersection property, let $k \geq 1$ and for $i \in \{1, \dots, k\}$ let F_i be a j_i-face of $\mathcal{M}$. In what follows we show that if the faces F_i are mutually incident then $\bigcap_{i \in \{1,\dots,k\}} F_i$ is connected and therefore $\mathcal{M}$ satisfies the component intersection property. We proceed by induction over k. The base case is when $k = 1$; in this situation the intersection consists of a j_1-face, which is connected by definition.

Let $k \geq 2$; without loss of generality we assume that if $i < \ell$ then $j_i < j_\ell$. By Proposition 2.1.51, $\bigcap_{i \in \{1,\dots,k\}} F_i \neq \varnothing$; if this intersection has only one flag then it is connected, hence we may assume that it contains at least two flags Φ and Ψ. Our goal is to construct a path from Φ to Ψ in $\bigcap_{i \in \{1,\dots,k\}} F_i$, and we shall do so by introducing two intermediate flags Υ_1 and Υ_2 as follows.

We use Lemma 2.1.48 for F_1 and F_2 to define flags $\Upsilon_1 \in F_1$ and $\Upsilon_2 \in F_2$ such that there are paths

- π_1 from Φ to Υ_1 whose edges have colours in $\{j_1 + 1, \dots, n-1\}$,
- π_2 from Υ_1 to Ψ whose edges have colours in $\{0, \dots, j_1 - 1\}$,
- π_3 from Φ to Υ_2 whose edges have colours in $\{j_2 + 1, \dots, n-1\}$,
- π_4 from Υ_2 to Ψ whose edges have colours in $\{0, \dots, j_2 - 1\}$,

as shown in Figure 2.12.

By concatenating π_1^{-1} with π_3 we obtain a path from Υ_1 to Υ_2 that uses only edges with labels in $\{j_1 + 1, \dots, n-1\}$. (Here π_1^{-1} stands for the path that traverses π_1 backwards.) Similarly, the concatenation of π_2 with π_4^{-1} gives

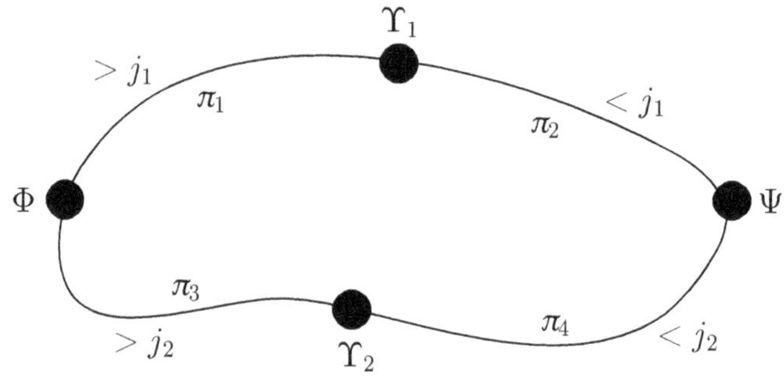

Figure 2.12 The paths π_1, π_2, π_3 and π_4.

us a path from Υ_1 to Υ_2 whose edges have labels in $\{0, \dots, j_2 - 1\}$. By the path intersection property we conclude that there exists a path $\hat{\pi}$ from Υ_1 to Υ_2 whose edges have labels in $\{j_1 + 1, \dots, j_2 - 1\}$.

So far we have paths $\hat{\pi}^{-1}$ from Υ_2 to Υ_1 and π_2 from Υ_1 to Ψ that do not use edges with colours in $\{j_1, \dots, j_k\}$. In order to construct a path from Φ to Ψ whose edges have no colours in $\{j_1, \dots, j_k\}$ it only remains to construct a path with that property from Φ to Υ_2.

Let $\mathcal{M}'$ be the intersection of all j-faces containing Φ for $j < j_2$. Then Φ and Υ_2 belong to $\mathcal{M}'$, since they are connected by the path π_3. Now we use Lemma 2.1.54 to see that $\mathcal{M}'$ is a maniplex satisfying the path intersection property, and by induction hypothesis, the restriction of $\cap_{j \geq 2} F_j$ to $\mathcal{M}'$ is connected. On the other hand, Φ and Ψ were chosen so that they belong to $\cap_{j \geq 2} F_j$. Since Υ_2 is connected to Ψ by the concatenation of $\hat{\pi}^{-1}$ and π_2, the flag Υ_2 also belongs to $\cap_{j \geq 2} F_j$, and we can conclude that Φ and Υ_2 are in $\cap_{j \geq 2} F_j$ restricted to $\mathcal{M}'$. This means that there exists a path $\tilde{\pi}$ from Φ to Υ_2 whose edges have labels greater than j_2 and distinct to j_i for every $i \geq 3$. The concatenation of $\tilde{\pi}$ with $\hat{\pi}^{-1}$ and π_2 is the desired path in $\cap_{i \in \{1, \dots, k\}} F_i$, and we can conclude that the component intersection property is satisfied in $\mathcal{M}$.

It remains to show that if $\mathcal{M}$ satisfies the component intersection property then it is polytopal. It follows from Proposition 2.1.49 and Corollary 2.1.52 that $\mathcal{P}(\mathcal{M})$ is faithful and is a poset satisfying the first two items of Definition 2.1.7. The diamond condition and strong flag-connectivity were shown in Lemma 2.1.55 and Lemma 2.1.56, respectively. ■

When convenient, we may consider a polytope $\mathcal{P}$ as a maniplex, in the understanding that the maniplex is actually the flag graph of $\mathcal{P}$.

Theorem 2.1.57 tells us that if a map $\mathcal{M}$ on a surface is polytopal then, when modelling $\mathcal{M}$ as a maniplex, it must satisfy the component intersection property. In particular, if $i < j$ and an i-face F_i is incident to a j-face F_j then their intersection must be connected. However, the connected component of a flag Φ in $F_i \cap F_j$ consists only of Φ, Φ^k and the edge labelled k between them, where $k \in \{0, 1, 2\} \setminus \{i, j\}$. If $i = 0$ and $j = 1$ this implies that the edge F_j must have two distinct vertices; similarly, if $i = 1$ and $j = 2$ this tells us that the edge F_i belongs to two distinct 2-faces. Finally, if $i = 0$ and $j = 2$ then the interpretation from the viewpoint of a map on a surface is that if the 2-face F_2 is an m-gon then it must have m distinct vertices. The converse is also true for maps with more than four flags; the details are left as an exercise.

Exercise 2.1.58 Show that a map $\mathcal{M}$ with more than four flags is polytopal whenever it satisfies the following two properties.

1. Each of its 2-faces has at least two vertices.
2. Each 2-face has twice as many simplices in its barycentric subdivision as vertices incident to it.

Exercise 2.1.59 Find a 3-maniplex $\mathcal{M}$ such that $\mathcal{P}(\mathcal{M})$ is faithful, but $\mathcal{M}$ is not polytopal.

As an implication of Theorem 2.1.57, the concept of maniplex is more general than the concept of polytope. Hence, every result on maniplexes immediately translates to a result on polytopes. Some maniplexes that are not polytopal, like the one in Example 2.1.50, fit nicely in the general theory (see Section 2.4). They justify the existence of results on maniplexes that otherwise could be stated just for polytopes.

On the other hand, the additional restriction of strong connectivity forbids every section to intersect itself in non-trivial ways, and in some settings avoids degenerate (perhaps uninteresting) cases. Moreover, as we shall see in Chapter 3, highly symmetric polytopes admit elegant correspondences with certain classes of groups. Hence, we encourage the reader to understand these two concepts and use the one more appropriate in each case.

2.1.4 Symmetry of Polytopes

We often talk about symmetry of polytopes in two distinct settings. *Local symmetry* refers to isomorphic small substructures out of which the polytope (or a large portion of it) can be reconstructed. On the other hand, *global symmetry* involves automorphisms. Global symmetry will be addressed later in this section; first we shall focus on local symmetry.

One of the most common ways to define the Platonic solids is as convex polyhedra satisfying that

- their 2-faces are all congruent regular polygons,
- their vertices all belong to the same number of edges.

The two properties above constitute examples of local symmetry. They may be generalised to higher dimensions in order to impose local symmetry to convex n-polytopes by requiring regular isomorphic facets and isomorphic vertex-figures. These two requirements are not very strong in the abstract setting (see Section 2.4), and some other idea will be used to define regularity. Nevertheless, when restricting ourselves to convex polytopes, the generalisation of those two properties suffices to characterise all regular ones. This motivates the study of local symmetry and the extent of its impact on global symmetry. We next define an index of local symmetry of abstract polytopes, weaker than the two properties mentioned above.

Assume that $\mathcal{P}$ is an n-polytope such that for each $k \in \{1, \ldots, n-1\}$, all sections between a pair of incident $(k-2)$-face F_{k-2} and $(k+1)$-face F_{k+1} is a p_k-gon for some fixed $p_k \in \{2, \ldots, \infty\}$, regardless of the choice of F_{k-2} and F_{k+1}. Then we say that $\mathcal{P}$ is *equivelar* and that it has *Schläfli type* $\{p_1, \ldots, p_{n-1}\}$. For simplicity, we refer to the Schläfli type just by 'type'.

Example 2.1.60 An n-gon has type $\{n\}$. The types of the tetrahedron, cube and octahedron are $\{3, 3\}$, $\{4, 3\}$ and $\{3, 4\}$, respectively.

The number p_i in the type of an equivelar polytope $\mathcal{P}$ describes an aspect of the local structure of $\mathcal{P}$. Any $(i+1)$-face F_{i+1} of $\mathcal{P}$ can be constructed by suitably gluing p_i i-faces at each $(i-2)$-face, in such a way that every $(i-1)$-face belongs to precisely two i-faces. This is the way in which we build a cube out of cardboard; we join squares, two at every edge, in such a way that there are three squares at every vertex.

Exercise 2.1.61 Let $\mathcal{P}$ be an n-polytope and let $\mathcal{R}$ be the poset obtained from $\mathcal{P}$ by removing the maximal face and adding three faces: two faces of rank n, each of which is incident to every $(n-1)$-face of $\mathcal{P}$, and one maximal face of rank $n+1$ incident to both faces of rank n.

(a) Show that $\mathcal{R}$ is a polytope.
(b) Show that if $\mathcal{P}$ has type $\{p_1, \ldots, p_{n-1}\}$ then $\mathcal{R}$ has type $\{p_1, \ldots, p_{n-1}, 2\}$.

This polytope is often called the *trivial extension* of $\mathcal{P}$.

Polygons are examples of polytopes completely determined by their types, since a polytope with type $\{p\}$ must have rank 2 and p vertices. In general there is more than one equivelar polyhedra with a given type.

Exercise 2.1.62 Prove that the tetrahedron is the only polyhedron with type $\{3, 3\}$.

Exercise 2.1.63 Prove that there are exactly two polyhedra with type $\{4, 3\}$.

If $\mathcal{P}$ is a polytope with type $\{p_1, \ldots, p_{n-1}\}$ and F_i is an i-face of $\mathcal{P}$ then the section F_i/F_{-1} is equivelar with type $\{p_1, \ldots, p_{i-1}\}$. More generally, if F_j is a j-face and $j \leq i - 3$ then the section F_i/F_j is equivelar with type $\{p_{j+2}, \ldots, p_{i-1}\}$. This, together with Exercises 2.1.19 and 2.1.63, shows that the type of the simplex is $\{3, \ldots, 3\}$, while that of the n-cube is $\{4, 3, \ldots, 3\}$.

Every section with rank 2 of $\mathcal{P}$ is dual to a section with rank 2 of the dual polytope $\mathcal{P}^\delta$. Furthermore, if F_{i+1} and F_{i-2} are an $(i + 1)$-face and an $(i - 2)$-face, respectively, then the section F_{i-2}/F_{i+1} of $\mathcal{P}^\delta$ is determined by the $(n - i + 1)$-face F_{i-2} and the $(n - i - 2)$-face F_{i+1}. Hence, if $\mathcal{P}$ has type $\{p_1, \ldots, p_{n-1}\}$ then the type of $\mathcal{P}^\delta$ is $\{p_{n-1}, \ldots, p_1\}$. In particular, the type of the cross polytopes is $\{3, \ldots, 3, 4\}$.

Now let us consider two i-faces F_i and G_i, and two j-faces F_j and G_j of an equivelar polytope, such that $F_j \leq F_i$ and $G_j \leq G_i$. If $j = i - 3$ then the sections F_i/F_j and G_i/G_j are both isomorphic to the polygon $\{p_{i-1}\}$; however, if $i - j \geq 4$ then even though F_i/F_j and G_i/G_j have the same type, they need not be isomorphic.

Exercise 2.1.64 Find an equivelar 4-polytope that has non-isomorphic facets.

Elsewhere, polytopes containing sections of rank 2 isomorphic to 2-gons are considered degenerate. If an entry of the type of a polytope is 2 then it must have many such sections. The following result illustrates a peculiarity of such polytopes.

Lemma 2.1.65 *Let $\mathcal{P}$ be an equivelar n-polytope with type $\{p_1, \ldots, p_{n-1}\}$. If $p_i = 2$ then for every $j \leq i - 1$ and $k \geq i$, every j-face is incident with every k-face.*

Proof We proceed by induction over n. The base case is when $n = 2$, where $\mathcal{P}$ is a 2-gon, and every pair of faces of distinct ranks are incident.

Assume that $n \geq 3$ and that the result is true for $(n-1)$-polytopes. If $i < n-1$ then, by the induction hypothesis, in each facet of $\mathcal{P}$ every j-face is incident with every k-face for every $j \leq i-1$ and $k \geq i$. Let F_{n-1} and F'_{n-1} be facets of $\mathcal{P}$ that are incident to a common ridge F_{n-2}. Then, if $j \leq i-1$, all j-faces of F_{n-1} are incident with every $(n-2)$-face of F_{n-1}, since $n-2 \geq i$. In particular, all j-faces of F_{n-1} are incident to F_{n-2}. Since F_{n-2} is a facet of F'_{n-1}, we also have that all j-faces of F_{n-1} are j-faces of F'_{n-1}. We can conclude that if $j \leq i$ then every j-face of F_{n-1} is contained in every facet that shares a ridge with F_{n-1}.

Now, if F_{n-1} and F''_{n-1} are arbitrary facets of $\mathcal{P}$ then the strong connectivity of $\mathcal{P}$ can be used to find a sequence $F_{n-1} = G_0, H_0, G_1, H_1, \ldots, G_{m-1}, H_{m-1}, G_m = F''_{n-1}$ where for every i we have that G_i is a facet, H_i is a ridge and $H_i \leq G_i$, $H_i \leq G_{i+1}$. An inductive procedure on the length m of the sequence generalises the discussion in the previous paragraph so that if $j \leq i-1$ then every j-face of F_{n-1} is a j-face of F''_{n-1}. This allows us to conclude that every j-face is in every facet. Therefore, if $j \leq i-1$, $k \geq i$ and F_k is a k-face contained in some facet F_{n-1} then every j-face F_j of $\mathcal{P}$ is contained in F_{n-1}, and the induction hypothesis applied to F_{n-1} implies that $F_j \leq F_k$.

If $i = n-1$ then the type of $\mathcal{P}^\delta$ is $\{2, p_{n-2}, \ldots, p_1\}$ and we can apply the case $i < n-1$ to conclude that every vertex of $\mathcal{P}^\delta$ is contained in every face of rank 1 or higher. The dual statement is that in $\mathcal{P}$ every facet contains every face with rank $n-2$ or lower. ∎

It is easy to understand why the conclusion of Lemma 2.1.65 can be seen as a degeneracy, since it is satisfied by no convex polytope of rank 2 or higher.

We next introduce another form of local symmetry, this time in terms of the lengths of some special paths.

Definition 2.1.66 A *Petrie path* π of an n-polytope $\mathcal{P}$ with $n \geq 3$ is a closed or infinite path with edges $e_1, \ldots, e_m$ or $\ldots, e_{-1}, e_0, e_1, \ldots$ on the 1-skeleton of $\mathcal{P}$ satisfying that

- for each $k \in \{2, \ldots, n-1\}$ and every i there exists a k-face $F_{k,i}$ containing $e_i, \ldots, e_{i+k-1}$ (subindices taken modulo m in the finite case);
- for every i, $F_{2,i}$ and $F_{2,i+1}$ are the only 2-faces contained in $F_{3,i}$ (or in $\mathcal{P}$ if $n = 3$) that contain e_{i+1};
- for every i, $F_{n-1,i}$ and $F_{n-1,i+1}$ are the two facets containing $F_{n-2,i+1}$;
- for $k \in \{3, \ldots, n-2\}$, $F_{k-1,i+1} \leq F_{k,i} \leq F_{k+1,i}$, and $F_{k,i+1}$ is the unique k-face other than $F_{k,i}$ incident to $F_{k-1,i+1}$ and to $F_{k+1,i}$.

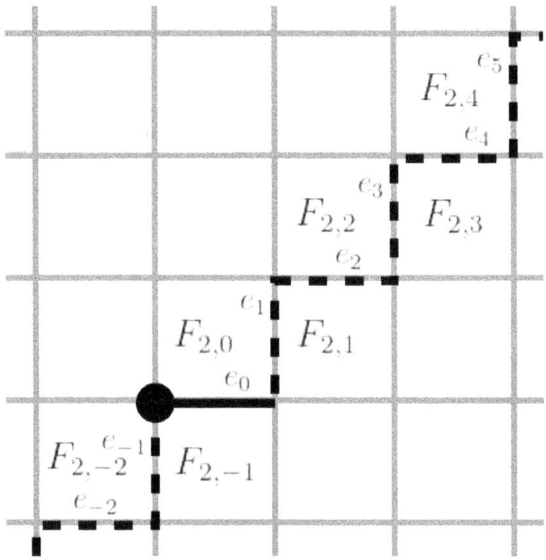

Figure 2.13 A Petrie path of the square tiling.

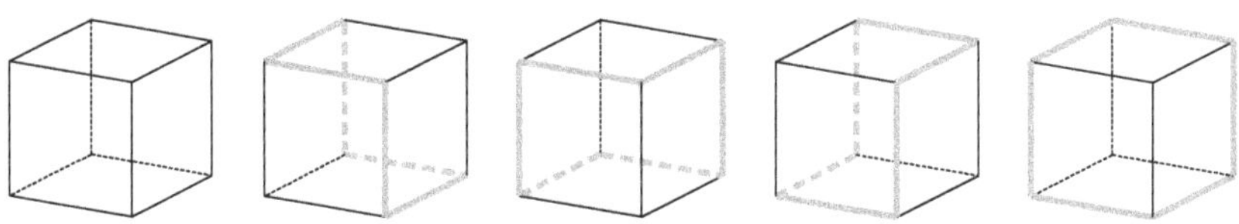

Figure 2.14 The four Petrie paths of the cube.

The previous definition suggests that Petrie paths are not easy to understand. We recommend developing intuition on them in a recursive manner.

Informally, a Petrie path of a polyhedron is a path where we traverse along some edge, then traverse the next edge in the 2-face at the left, then the next edge in the 2-face that at that point is located at the right, and so on, always alternating local orientations. When picturing the polyhedron as a map on a surface, we can think of the Petrie path as a zigzag. Figure 2.13 illustrates a Petrie path of the square tiling indicating the notation of Definition 2.1.66.

The cube has four Petrie paths, illustrated in grey in Figure 2.14. Each of them can be obtained by removing a pair of antipodal vertices together with their incident edges.

In higher ranks, a Petrie path takes $n-1$ edges of a Petrie path of a facet F_{n-1}, where the last $n-2$ edges $e_2, \ldots, e_{n-1}$ belong to a ridge F_{n-2}. Then it traverses an edge e_n such that $e_2, \ldots, e_n$ is a subpath of a Petrie path of the unique facet other than F_{n-1} incident to F_{n-2}. Under some circumstances the faces $F_{k,i}$ in Definition 2.1.66 can be determined directly from the sequence of edges.

Exercise 2.1.67 Show that if a polyhedron has no section of rank 2 isomorphic to a 2-gon then its Petrie paths can be defined as closed or infinite paths where every two, but not three, consecutive edges belong to the same 2-face.

Exercise 2.1.68 Exhibit a polytope $\mathcal{P}$, a Petrie path π of $\mathcal{P}$ and a sequence of consecutive k edges of π for some $k \geq 2$ such that the sequence belongs to more than one k-face.

The last three items of Definition 2.1.66 suggest an inductive recipe to reconstruct the set of faces $F_{k,i}$ from the faces $F_{k,1}$. The next remark follows.

Remark 2.1.69 With the notation of Definition 2.1.66, the edge e_1, the vertex between e_m (e_0 if the path is infinite) and e_1 and the faces $F_{k,1}$ for $k \in \{2, \ldots, n-1\}$ completely determine the Petrie path.

Petrie paths may fail to be polygons, since they can repeat vertices. When they are polygons they are also referred to as *Petrie polygons*.

Exercise 2.1.70 Determine the natural numbers $n \geq 3$ such that the n-prism has Petrie polygons, and the natural numbers $m \geq 3$ such that the m-bipyramid has Petrie polygons.

Exercise 2.1.71 Find the length of the Petrie paths of the n-simplex, n-cube and cross polytope of rank n. Determine if they are polygons. Do all have the same length?

We may require a polytope to have as local symmetry that all its Petrie paths have the same length. Alternatively, we may say that polytopes whose Petrie paths have distinct lengths fail to possess the highest degree of symmetry. In Section 2.2 we shall see that such polytopes cannot satisfy the definition of regular polytopes.

Exercise 2.1.72 Find an equivelar polyhedron that has Petrie paths of two or more distinct sizes.

So far in this section we have explored properties of polytopes that rely on conditions satisfied by several of its substructures, like sections or Petrie paths. We now turn our attention to global symmetry, where the key notion is that of automorphisms, defined next.

Definition 2.1.73 An *automorphism* of a polytope is a bijection of its elements that preserves the order.

Example 2.1.74 The triangle in Figure 2.3(a) is invariant under a geometric reflection fixing vertex A and edge a, while swapping vertices B and C, and edges b and c. This induces the permutation $(B,C)(b,c)$ of its proper elements, which can be easily verified to preserve incidence in the partial order in Figure 2.3(b).

Similarly, the reflection fixing vertex B and edge b while swapping A and C induces an automorphism represented by the permutation $(A,C)(a,c)$. The two non-trivial automorphisms exhibited so far generate the symmetric group S_3 (acting separately on the sets $\{A,B,C\}$ and $\{a,b,c\}$). Since every edge is determined by its two vertices, any permutation of the vertices determines a unique automorphism. Therefore the triangle has precisely six automorphisms, corresponding to the six permutations of its vertices.

The set of automorphisms of a polytope $\mathcal{P}$ forms a group (called the *automorphism group*) with composition as its operation. This group is denoted by $\Gamma(\mathcal{P})$, and we follow the convention that it induces a right action on the faces, sections and flags of $\mathcal{P}$.

Every section of $\mathcal{P}$ must be isomorphic to its image under an automorphism of $\mathcal{P}$, but not every pair of isomorphic sections determined by faces of the same ranks are in the same orbit under $\Gamma(\mathcal{P})$.

Exercise 2.1.75 Find a polyhedron $\mathcal{P}$ having two 2-faces F, G with the same number of vertices, but such that $F\gamma \neq G$ for every $\gamma \in \Gamma(\mathcal{P})$.

Exercise 2.1.76 Find a polyhedron with trivial automorphism group.

As a direct consequence of Definition 2.1.73, the image of any flag under any automorphism is another flag. Furthermore, if F_i and F_i' are the i-faces of the flags Φ and Φ^i, respectively, and F_j is the j-face of Φ for some $j \neq i$, then $F_j\gamma$ is a face of $\Phi\gamma$ and of $\Phi^i\gamma$. In particular, when considering the cases

$j = i - 1$ and $j = i + 1$, we see that $F_{i-1}\gamma \leq F_i\gamma, F_i'\gamma \leq F_{i+1}\gamma$, and the diamond condition implies that $F_i\gamma$ and $F_i'\gamma$ are the only proper faces in the section $F_{i+1}\gamma/F_{i-1}\gamma$. This proves the following lemma.

Lemma 2.1.77 *Let $\mathcal{P}$ be an n-polytope, let Φ be a flag of $\mathcal{P}$, and let $\gamma \in \Gamma(\mathcal{P})$. Then, $\Phi^i\gamma = (\Phi\gamma)^i$ for each $i \in \{0, \ldots, n-1\}$.*

The previous lemma generalises to the following stronger result.

Proposition 2.1.78 *Let $\mathcal{P}$ be an n-polytope, let $i_1, \ldots, i_k \in \{0, \ldots, n-1\}$, let Φ be a flag of $\mathcal{P}$, and let $\gamma \in \Gamma(\mathcal{P})$. Then, $\Phi^{i_1,\ldots,i_k}\gamma = (\Phi\gamma)^{i_1,\ldots,i_k}$.*

Proof We proceed by induction over k. The base case is when $k = 1$ and it is precisely Lemma 2.1.77.

Assume that the proposition is true for some $k \geq 1$ and let $i_1, \ldots, i_{k+1} \in \{0, \ldots, n-1\}$. Then

$$\begin{aligned}\Phi^{i_1,\ldots,i_{k+1}}\gamma &= (\Phi^{i_1,\ldots,i_k})^{i_{k+1}}\gamma = (\Phi^{i_1,\ldots,i_k}\gamma)^{i_{k+1}} \\ &= ((\Phi\gamma)^{i_1,\ldots,i_k})^{i_{k+1}} = (\Phi\gamma)^{i_1,\ldots,i_{k+1}}.\end{aligned}$$

■

In many instances it is convenient to interpret automorphisms in terms of flags instead of doing it in terms of faces. The next lemma follows from Proposition 2.1.27.

Lemma 2.1.79 *Every bijection γ of the flags of an n-polytope $\mathcal{P}$ satisfying that $\Phi^i\gamma = (\Phi\gamma)^i$ for every $i \in \{0, \ldots, n-1\}$ and for every flag Φ, induces an automorphism of $\mathcal{P}$ acting on $\mathcal{F}l(\mathcal{P})$ like γ.*

Lemma 2.1.79 can be restated in terms of the flag graph as follows. Every automorphism of a polytope $\mathcal{P}$ corresponds to a colour-preserving automorphism of the flag graph of $\mathcal{P}$, and every such automorphism of the flag graph corresponds to an automorphism of $\mathcal{P}$.

Recall that an action of a group on a set is said to be *free* or *semiregular* if id is the only group element that has fixed points. Proposition 2.1.78 has as a useful consequence that an automorphism can be defined by specifying the image of one single flag, and therefore the automorphism in Lemma 2.1.79 is unique. This is equivalent to saying that $\Gamma(\mathcal{P})$ acts freely on $\mathcal{F}l(\mathcal{P})$.

Proposition 2.1.80 *Let $\mathcal{P}$ be a polytope and Φ be a flag of $\mathcal{P}$.*

1. *If $\Phi\gamma = \Phi$ for some $\gamma \in \Gamma(\mathcal{P})$ then $\gamma = \mathrm{id}$.*
2. *For each flag Ψ there is at most one automorphism mapping Φ to Ψ.*

Proof Let $\Phi, \Psi \in \mathcal{F}l(\mathcal{P})$ and assume that $\gamma \in \Gamma(\mathcal{P})$ is such that $\Phi\gamma = \Phi$. From the flag connectivity we know that there exist $i_1, \ldots, i_m \in \{0, \ldots, n-1\}$ such that $\Psi = \Phi^{i_1,\ldots,i_m}$ (here n is the rank of $\mathcal{P}$). It follows that

$$\Psi\gamma = \Phi^{i_1,\ldots,i_m}\gamma = (\Phi\gamma)^{i_1,\ldots,i_m} = \Phi^{i_1,\ldots,i_m} = \Psi.$$

We can conclude that γ acts trivially on $\mathcal{F}l(\mathcal{P})$, and this is possible only if $\gamma = \mathrm{id}$. This proves the first item.

To prove the second item let $\alpha, \beta \in \Gamma(\mathcal{P})$ be such that $\Phi\alpha = \Phi\beta$. Then $\Phi\alpha\beta^{-1} = \Phi$, and the first item implies that $\alpha = \beta$. ■

We have as a consequence of Proposition 2.1.80 that all flag orbits under $\Gamma(\mathcal{P})$ of a polytope $\mathcal{P}$ have size $|\Gamma(\mathcal{P})|$. However, $\mathcal{P}$ may have orbits of i-faces of distinct sizes.

Exercise 2.1.81 Find a polyhedron with two vertex orbits of distinct sizes under $\Gamma(\mathcal{P})$.

An n-polytope $\mathcal{P}$ is *i-face transitive* if for every two i-faces F and G there exists $\gamma \in \Gamma(\mathcal{P})$ such that $G = F\gamma$. We say that $\mathcal{P}$ is *fully transitive* if it is i-face transitive for every $i \in \{0, \ldots, n-1\}$. The n-simplex, n-cube and cross polytope of rank n are examples of fully transitive polytopes.

Exercise 2.1.82 Find a polyhedron that is vertex-transitive but not face-transitive, one that is vertex-transitive but not edge transitive, and one that is edge-transitive but not vertex-transitive.

Exercise 2.1.83 Let $\mathcal{P}$ be the 4-polytope defined in Exercise 2.1.20.

(a) Show that $\mathcal{P}$ has type $\{3, 4, 4\}$.
(b) Show that it is is vertex-transitive.
(c) Show that it is also facet-transitive.
(d) Show that every permutation of the facets of the polytope induces an automorphism.
(e) Show that $\Gamma(\mathcal{P})$ acts faithfully on its facets. Conclude that $\Gamma(\mathcal{P})$ is isomorphic to S_5.

The automorphism group of the dual $\mathcal{P}^\delta$ of a polytope $\mathcal{P}$ is completely determined by that of $\mathcal{P}$.

Remark 2.1.84 Since automorphisms preserve incidence, as bijections of the faces the groups $\Gamma(\mathcal{P})$ and $\Gamma(\mathcal{P}^\delta)$ are equal.

A *duality* of a self-dual polytope $\mathcal{P}$ is a bijection that reverses the partial order. This, together with Definition 2.1.73 implies directly the following remark.

Remark 2.1.85 1. The composition of two dualities is an automorphism.
2. The composition of a duality and an automorphism (in any order) is a duality.

As a consequence of Remark 2.1.85, the set of all automorphisms and dualities of a polytope $\mathcal{P}$ forms a group with the composition. This group is called the *extended group* and is denoted by $\overline{\Gamma}(\mathcal{P})$. Remark 2.1.85 also implies that the index of $\Gamma(\mathcal{P})$ in $\overline{\Gamma}(\mathcal{P})$ is at most 2, and it is 2 if and only if $\mathcal{P}$ is self-dual.

Example 2.1.86 The triangle in Figure 2.3 has precisely six dualities represented by the permutations

$$\begin{array}{ll} (a,A)(b,B)(c,C), & (a,A)(b,C)(c,B), \\ (a,B,c,A,b,C), & (a,B)(b,A)(c,C), \\ (a,C,b,A,c,B), & (a,C)(b,B)(c,A). \end{array}$$

It is straightforward to verify that each of these permutations induces a duality. The list is compete, since the triangle, being self-dual, must have the same number of automorphisms and dualities (see Example 2.1.74).

Dualities also act on flags of an n-polytope in a natural way. If δ is a duality that maps the flag Φ into a flag Ψ then the i-face of Ψ is the image of the $(n-i-1)$-face of Φ under δ. In other words,

$$(\Phi^i)\delta = (\Phi\delta)^{n-i-1}. \tag{2.5}$$

A *polarity* is a duality of order 2. Not every self-dual polytope possesses a polarity. The first known self-dual polyhedron with no polarities was exhibited in [143], as a map on a surface.

Exercise 2.1.87 For every $n \geq 2$ find a self-dual polyhedron containing a duality of order $2n$.

2.2 Regular and Two-Orbit Polytopes

Most of the theory of abstract polytopes has been devoted to those with a large automorphism group with respect to the number of flags. In this section we study the polytopes with the property that their number of automorphisms is at least half the number of flags.

2.2.1 Regular Polytopes

As a consequence of the second item of Proposition 2.1.80, the order of the automorphism group of a polytope $\mathcal{P}$ is at most its number of flags.

Definition 2.2.1 We say that $\mathcal{P}$ is *regular* if $|\Gamma(\mathcal{P})| = |\mathcal{F}l(\mathcal{P})|$.

The word 'regular' is one of the most overused words in the mathematical world. In this case it is consistent with the regular action of $\Gamma(\mathcal{P})$ on the set of flags, but the reader should not assume that regularity in other contexts implies regularity in the context of polytopes. (For example, a regular graph is one where all vertices have the same degree, regardless of the automorphism group of the graph. However, there are non-regular polytopes, like the cuboctahedron in Figure 2.15(a), whose 1-skeleton is regular as a graph.)

Intuitively, Definition 2.2.1 says that a regular polytope looks the same regardless of the side from where we are looking. This definition is more restrictive than other natural generalisations of the properties of the Platonic solids, like the one that requires only isomorphic regular facets and isomorphic regular vertex-figures. The latter definition is satisfied by some classes of two-orbit polytopes, defined in the next section, and also by many less symmetric polytopes. Nevertheless, Definition 2.2.1 is enough to construct a rich theory, a large part of which can be found in [188].

The triangle is an example of a regular polygon. It has six flags and, as pointed out in Example 2.1.74, it has also six automorphisms. In fact, every polygon is regular as an abstract polytope.

Exercise 2.2.2 Show that every polygon is regular.

Example 2.2.3 The cross polytope $\mathcal{O}_n$ of rank n, defined in Example 2.1.17, is regular. We claim that its number of flags is $2^n \cdot n!$. This is true for $n = 2$ and $n = 3$, since the square and the octahedron have 8 and 48 flags, respectively. For $n \geq 4$, the number of flags of $\mathcal{O}_n$ can be computed recursively, since it has $2n$ vertices and its vertex-figures are $(n-1)$-dimensional cross polytopes (see Exercise 2.1.19).

Every isometry of $\mathbb{E}^n$ preserving the vertex set of $\mathcal{O}_n$ induces an automorphism of $\mathcal{O}_n$. Furthermore, this automorphism is trivial if and only if the isometry acts trivially on the vertex set of $\mathcal{O}_n$ (it has four vertices that are not coplanar). With this in mind, we next define two subgroups of $\Gamma(\mathcal{O}_n)$.

Every permutation σ of the coordinate axes induces an isometry γ_σ that preserves $\mathcal{O}_n$ by preserving the positive and negative directions of the axes when being permuted. Then

$$\Gamma_1 := \{\gamma_\sigma : \sigma \text{ is a permutation of the coordinate axes}\}$$

is isomorphic to a subgroup of $\Gamma(\mathcal{O}_n)$. On the other hand, given $I \subseteq \{1,\ldots,n\}$, the isometry γ_I that preserves every coordinate axis, while swapping positive and negative ends of the ith axis if and only if $i \in I$, induces an automorphism of $\mathcal{O}_n$. Then

$$\Gamma_2 := \{\gamma_I : I \subseteq \{1,\ldots,n\}\}$$

is isomorphic to a subgroup of $\Gamma(\mathcal{O}_n)$.

The groups Γ_1 and Γ_2 have trivial intersection. The former has $n!$ elements while the latter has 2^n. It follows that $|\Gamma(\mathcal{O}_n)| \geq 2^n \cdot n!$. Since $\Gamma(\mathcal{O}_n)$ acts semiregularly on the flags, we can conclude that $\mathcal{O}_n$ has as many flags as automorphisms, and hence it is regular.

Exercise 2.2.4 Let $\mathcal{P}$ be a regular n-polytope and let $\mathcal{R}$ be the $(n+1)$-polytope mentioned in Exercise 2.1.61 containing precisely two facets isomorphic to $\mathcal{P}$. Show that $\mathcal{R}$ is regular.

Regular polytopes satisfy many nice properties mentioned next. Given a regular polytope $\mathcal{P}$ with i-faces F_i and G_i, and j-faces F_j and G_j such that $F_i \leq F_j$ and $G_i \leq G_j$, we can complete the chains $\{F_i, F_j\}$ and $\{G_i, G_j\}$ to flags Φ_F and Φ_G. Flag-transitivity implies that there exists $\gamma \in \Gamma(\mathcal{P})$ such that $\Phi_G = \Phi_F\gamma$; it follows that $G_i = F_i\gamma$ and $G_j = F_j\gamma$. The following remark lists several direct consequences of this fact.

Remark 2.2.5 Let $\mathcal{P}$ be a regular polytope. Then

1. $\mathcal{P}$ is fully transitive,
2. for every i and j, every two sections determined by an i-face and an incident j-face are isomorphic,
3. $\mathcal{P}$ is equivelar.

The unique 0-polytope is trivially regular, since it only has one flag. The unique 1-polytope is also regular since there is an automorphism that swaps its two flags. Exercise 2.2.2 states that every polygon is regular. It follows that non-regular polytopes have rank at least 3. The previous remark makes it easy to give examples of non-regular polytopes. For example the snub cube, whose 2-faces are 6 squares and 32 triangles (Figure 1.1(a)), is not 2-face transitive and hence cannot be regular.

Remark 2.2.5 lists some consequences of regularity on local symmetry of polytopes. The next proposition and corollary extend that list.

Proposition 2.2.6 *Let $n \geq 3$, let $\mathcal{P}$ be a regular n-polytope, and let π_1 and π_2 be two Petrie paths of $\mathcal{P}$. Then there exists $\gamma \in \Gamma(\mathcal{P})$ such that $\pi_2 = \pi_1\gamma$.*

Proof Let $e_{1,1}, \ldots, e_{n-1,1}$ and $e_{1,2}, \ldots, e_{n-1,2}$ be two subpaths of π_1 and π_2 starting at vertices $F_{0,1}$ and $F_{0,2}$, respectively. For $i \in \{1, 2\}$ we contruct a flag Φ_i as follows. The vertex of Φ_i is $F_{0,i}$, the edge of Φ_i is $e_{1,i}$, and for $k \geq 2$, the k-face of Φ_i is the k-face containing $e_{1,i}, \ldots, e_{k,i}$ mentioned in Definition 2.1.66.

Since $\mathcal{P}$ is regular, there exists $\gamma \in \Gamma(\mathcal{P})$ such that $\Phi_2 = \Phi_1\gamma$. Remark 2.1.69 now implies that $\pi_2 = \pi_1\gamma$. ■

Corollary 2.2.7 *The lengths of all Petrie paths of a regular polytope are the same.*

Regularity is a hereditary property, as shown next.

Proposition 2.2.8 *Let $\mathcal{P}$ be an n-polytope, let $0 \leq i < j \leq n-1$ and let F_i be an i-face and F_j be a j-face such that $F_i \leq F_j$. Then*

1. *F_j/F_i is regular,*
2. *there exists $\Delta \leq \Gamma(\mathcal{P})$ with $\Delta \cong \Gamma(F_j/F_i)$ such that Δ and $\Gamma(F_j/F_i)$ have isomorphic actions on F_j/F_i.*

Proof Let $\hat{\Upsilon}$ be a flag of $\mathcal{P}$ containing F_i and F_j. For each flag Φ of F_j/F_i we construct a flag $\hat{\Phi}$ of $\mathcal{P}$ by adjoining all faces of $\hat{\Upsilon}$ of ranks less than i and greater than j.

Let Φ_0 and Ψ_0 be two flags of F_j/F_i. Regularity of $\mathcal{P}$ implies the existence of $\gamma \in \Gamma(\mathcal{P})$ such that $\hat{\Psi}_0 = \hat{\Phi}_0\gamma$. Such a γ fixes F_i and F_j; thus, it preserves F_j/F_i. The restriction of γ to F_j/F_i is an automorphism mapping Φ_0 to Ψ_0. This proves the first item.

To prove the second item let $\{\mathcal{F}_\Upsilon\}$ be the set of flags of $\mathcal{P}$ whose k-faces are those of Υ for $k \leq i$ and $k \geq j$. The group consisting of all automorphisms mapping Υ to any flag in $\{\mathcal{F}_\Upsilon\}$ acts on F_j/F_i like $\Gamma(F_j/F_i)$. ■

In order to prove that a polytope is regular it is not necessary to verify the existence of an automorphism mapping a given flag to each other flag. The following theorem constitutes the standard way to prove regularity of polytopes.

Theorem 2.2.9 *An n-polytope $\mathcal{P}$ is regular if and only if for some flag Φ there exist automorphisms $\rho_0, \ldots, \rho_{n-1}$ (that depend on Φ) such that $\Phi\rho_i = \Phi^i$.*

Proof The flag-transitivity of $\Gamma(\mathcal{P})$ implies the existence of the automorphisms ρ_i, regardless of the choice of Φ.

Conversely, if for some flag Φ there are automorphisms ρ_i for $i \in \{0, \ldots, n-1\}$, and Ψ is another flag then, by strong flag-connectivity, there exist $i_1, \ldots, i_m \in \{0, \ldots, n-1\}$ such that $\Psi = \Phi^{i_1,\ldots,i_m}$. Then

$$\Psi = \Phi^{i_1,\ldots,i_m} = (\Phi^{i_1})^{i_2,\ldots,i_m} = (\Phi\rho_{i_1})^{i_2,\ldots,i_m} = \Phi^{i_2,\ldots,i_m}\rho_{i_1}. \tag{2.6}$$

An inductive argument then shows that $\Psi = \Phi\rho_{i_m} \cdots \rho_{i_1}$. Therefore $\Gamma(\mathcal{P})$ acts transitively on the flags and $\mathcal{P}$ is regular. ■

The flag Φ in Theorem 2.2.9 is called the *base flag*, and the automorphisms ρ_i are called the *standard reflections* with respect to Φ. The generators ρ_i depend on the choice of base flag, as can be easily seen in the case of a polygon. However, in most cases the base flag is either irrelevant or specified beforehand, and we shall not incude Φ as a parameter in the notation for the standard reflections.

The standard reflections satisfy the following properties.

Proposition 2.2.10 *If $\mathcal{P}$ is a regular polytope with type $\{p_1, \ldots, p_{n-1}\}$ and $\rho_0, \ldots, \rho_{n-1}$ are its standard reflections with respect to some base flag, then*

1. *$\Gamma(\mathcal{P}) = \langle \rho_0, \ldots, \rho_{n-1} \rangle$,*
2. *if $|i-j| > 2$ then $\rho_i\rho_j = \rho_j\rho_i$,*
3. *for $i \in \{1, \ldots, n-1\}$, the order of $\rho_{i-1}\rho_i$ is p_i,*
4. *if Ψ is the subchain of the base flag consisting of the faces with ranks $i \in I$ for some $I \subseteq \{0, \ldots, n-1\}$ then the stabiliser of Ψ in $\Gamma(\mathcal{P})$ is $\langle \rho_j : j \notin I \rangle$.*

Proof Let Φ be the base flag of $\mathcal{P}$.

Assume that $\gamma \in \Gamma(\mathcal{P})$ maps Φ to some flag Ψ. Let $i_1, \ldots, i_m \in \{0, \ldots, n-1\}$ such that $\Phi^{i_1,\ldots,i_m} = \Psi$. The first item follows from (2.6) as in the proof of Theorem 2.2.9; in this situation $\gamma = \rho_{i_m} \cdots \rho_{i_1}$.

If Φ is the base flag and $|i-j| > 2$, (2.2) implies that

$$\Phi\rho_i\rho_j = \Phi^i\rho_j = (\Phi\rho_j)^i = \Phi^{j,i} = \Phi^{i,j} = (\Phi\rho_i)^j = \Phi^j\rho_i = \Phi\rho_j\rho_i.$$

It follows from Proposition 2.1.80 (2) that $\rho_i\rho_j = \rho_j\rho_i$.

To prove the third item we note that $\Phi\rho_{i-1}\rho_i = \Phi^{i-i}\rho_i = (\Phi\rho_i)^{i-1} = \Phi^{i,i-1}$, which implies that $\Phi(\rho_{i-1}\rho_i)^k = \Phi^{(i,i-1)^k}$. Since the section between the $(i-2)$-face and the $(i+1)$-face of Φ is a p_i-gon, Remark 2.1.22 implies that the order of $\rho_{i-1}\rho_i$ is p_i.

Finally, if Ψ is the subchain of the base flag consisting of the faces with ranks $j \notin I$ then $\Psi\rho_i = \Psi$ for every $i \in I$, which tells us that $\langle \rho_i : i \in I\rangle \subseteq \mathrm{Stab}_{\Gamma(\mathcal{P})}(\Psi)$. Now, if $\gamma \in \mathrm{Stab}_{\Gamma(\mathcal{P})}(\Psi)$ then the strong flag-connectivity of $\mathcal{P}$ implies that $\Phi\gamma = \Phi^{i_1,\ldots,i_m}$ for some $i_1, \ldots, i_m \in I$. Proceeding as in (2.6) we can see that $\Phi\gamma = \Phi\rho_{i_m}\cdots\rho_{i_1}$, and by Proposition 2.1.80 we can conclude that $\gamma \in \langle \rho_i : i \in I\rangle$. Therefore, $\mathrm{Stab}_{\Gamma(\mathcal{P})}(\Psi) \subseteq \langle \rho_i : i \in I\rangle$, and the fourth item holds. ■

Exercise 2.2.11 Use Theorem 2.2.9 to show that the n-simplex, the n-cube and the cross polytope are regular n-polytopes.

The face-lattice of every convex regular polytope is an abstract regular polytope. Also all regular tessellations of Euclidean n-space can be seen as abstract regular polytopes. The main reference for these facts is [55].

Regular polytopes that are orientable are also called *orientably regular*. Elsewhere the term *directly regular* has also been used to denote this class of polytopes. The reader should be aware that the term *orientably regular* applied for maps on surfaces was earlier used to denote maps on orientable surfaces where the group of orientation preserving automorphisms acts transitively on the arcs (pairs of incident vertex and edge; see for example [49, 59, 94]). This includes a larger class of maps than the one consisting of those transitive on their flags (here 'flag' should be understood as a maximal simplex in the baricentric subdivision of the map). For consistency with the literature on abstract polytopes, we reserve the term 'orientably regular' exclusively for those polytopes that are orientable and regular.

A regular polytope that is not orientable is called *non-orientably regular*. Polygons and the Platonic solids are examples of orientably regular polytopes. Many examples of non-orientably regular polyhedra can be found in [41, 116].

Exercise 2.2.12 Let $\mathcal{Q}$ be the quotient of the (three-dimensional) cube by identifying every pair of antipodal vertices, every pair of antipodal edges, and every pair of antipodal 2-faces.

1. Show that the cube covers $\mathcal{Q}$.
2. Show that $\mathcal{Q}$ is a polyhedron.
3. Show that $\mathcal{Q}$ is non-orientably regular.

This polyhedron is called the *hemi-cube*.

By definition, it is possible to colour the flags of an orientably regular polytope with two colours in such a way that adjacent flags have distinct colours. This property is inherited to the sections (see also Proposition 2.2.8).

Remark 2.2.13 Every section of an orientably regular polytope is orientably regular.

The automorphisms of an orientably regular polytope $\mathcal{P}$ preserving the bipartition of the flags form a subgroup of index 2 of $\Gamma(\mathcal{P})$ called the *even subgroup*, and we denote it by $\Gamma^+(\mathcal{P})$. When there is no possibility of confusion with the rotations of a geometric space we shall refer to even subgroups also as *rotation groups*. As will become clear in Proposition 2.2.14, if $\mathcal{P}$ is non-orientably regular it is convenient to denote $\Gamma(\mathcal{P})$ also by $\Gamma^+(\mathcal{P})$.

The even subgroup of $\mathcal{P}$ has a standard generating set, defined in a similar way as the standard reflections for $\Gamma(\mathcal{P})$.

Proposition 2.2.14 *Let $\mathcal{P}$ be a regular n-polytope.*

1. *For any given flag Φ there exist $\sigma_1, \dots, \sigma_{n-1} \in \Gamma^+(\mathcal{P})$ (that depend on Φ) such that $\Phi\sigma_i = \Phi^{i,i-1}$.*
2. *If $\Psi = \Phi^{j,i}$ then $\Psi = \Phi\sigma_{i+1}\cdots\sigma_j$ if $i < j$ and $\Psi = \Phi(\sigma_{i+1}\cdots\sigma_j)^{-1}$ if $j < i$.*
3. *$\Gamma^+(\mathcal{P}) = \langle\sigma_1, \dots, \sigma_{n-1}\rangle$.*

Proof The existence of the automorphisms σ_i in the first item is guaranteed by the regularity of $\mathcal{P}$. Moreover, if $\mathcal{P}$ is orientably regular then the generators σ_i must belong to $\Gamma^+(\mathcal{P})$ since they map Φ to a flag in its same part of the bipartition.

We prove the second item by induction over $\ell := |j - i|$; the base case is $\ell = 1$, when $\Psi = \Phi\sigma_j$ if $j = i + 1$ and $\Psi = \Phi\sigma_i^{-1}$ if $i = j + 1$, since

$$\Phi^{i-1,i} = \Phi^{i-1,i}\sigma_i\sigma_i^{-1} = (\Phi\sigma_i)^{i-1,i}\sigma_i^{-1} = (\Phi^{i,i-1})^{i-1,i}\sigma_i^{-1} = \Phi\sigma_i^{-1}.$$

If $\ell \geq 2$ and $j > i$ then, by the inductive hypothesis,

$$\Psi = \Phi^{j,j-1,j-1,i} = (\Phi\sigma_j)^{j-1,i} = \Phi^{j-1,i}\sigma_j = \Phi\sigma_{i+1}\cdots\sigma_j.$$

The case $i > j$ is analogous.

It only remains to show the third item. Assume first that $\mathcal{P}$ is orientably regular and let $\gamma \in \Gamma^+(\mathcal{P})$. Then Φ and $\Phi\gamma$ are in the same part of the bipartition of flags, and by flag-connectivity, $\Phi\gamma$ can be written as $\Phi^{i_1,\dots,i_{2m}}$ for some $i_1, \dots, i_{2m} \in \{0, \dots, n-1\}$. In order to show that $\gamma \in \langle\sigma_1, \dots, \sigma_{n-1}\rangle$ we proceed by induction over m; the base case $m = 1$ is given by the second item. If $m > 1$ then we use the inductive hypothesis to see that

$$\Phi\gamma = (\Phi^{i_1,\ldots,i_{2(m-1)}})^{i_{2m-1},i_{2m}} = (\Phi\gamma')^{i_{2m-1},i_{2m}} = \Phi^{i_{2m-1},i_{2m}}\gamma'$$

for some $\gamma' \in \Gamma^+(\mathcal{P})$. The case $m = 1$ can now be used to conclude that $\gamma \in \langle \sigma_1, \ldots, \sigma_{n-1} \rangle$.

On the other hand, if $\mathcal{P}$ is non-orientably regular then every flag can be written as $\Phi^{i_1,\ldots,i_{2m}}$ for some $i_1, \ldots, i_{2m} \in \{0, \ldots, n-1\}$, and the procedure above shows that $\Gamma(\mathcal{P}) = \langle \sigma_1, \ldots, \sigma_{n-1} \rangle$. ∎

When $\mathcal{P}$ is a p-gon, σ_1 acts on the vertex and edge sets as a 1-step rotation of the regular p-gon with respect to its centre. In general, σ_i acts in the p_i-gonal section between the $(i-2)$-face and the $(i+1)$-face of the base flag as a rotation of the p_i-gon. This motivates us to call the automorphisms σ_i *standard rotations*, and to refer to $\Gamma^+(\mathcal{P})$ also as the *rotation subgroup*.

Exercise 2.2.15 Embed the n-simplex in $\mathbb{E}^{n+1}$ as the convex hull of the $(n+1)$ points with n entries equal to 0 and the remaining one equal to 1. Choose as base flag the one containing as i-face the i-simplex with vertices having 1 in the coordinates $\{1, \ldots, i+1\}$, for $i \in \{0, \ldots, n-1\}$.

1. Show that the n-simplex is regular, explicitly computing the standard reflections ρ_i as isometries of $\mathbb{E}^{n+1}$.
2. Show that the abstract rotation σ_i of the n-simplex is the isometry that permutes the coordinates $i-1$, i and $i+1$ while fixing the others, mapping $(x_1, \ldots, x_{n+1})$ to $(x_1, \ldots, x_{i-2}, x_{i+1}, x_{i-1}, x_i, x_{i+2}, \ldots, x_{n+1})$.

The standard rotations satisfy the following properties.

Proposition 2.2.16 *Let $\mathcal{P}$ be a regular n-polytope with type $\{p_1, \ldots, p_{n-1}\}$ and let $\sigma_1, \ldots, \sigma_{n-1}$ be the standard rotations. Then*

1. $\sigma_i = \rho_{i-1}\rho_i$,
2. $(\sigma_i \cdots \sigma_j)^2 = \mathrm{id}$ *for every* $1 \le i < j \le n-1$,
3. $\sigma_i\sigma_j = \sigma_j\sigma_i$ *if* $|i-j| \ge 3$,
4. *the order of* σ_i *is* p_i.

Proof Let Φ_0 be the base flag of $\mathcal{P}$. We know that $\Phi_0\sigma_i = \Phi_0^{i,i-1} = (\Phi_0\rho_i)^{i-1} = \Phi_0^{i-1}\rho_i = \Phi_0\rho_{i-1}\rho_i$, which proves the first item.

The second item follows from (2.2) and the second item of Proposition 2.2.14.

If $|i-j| \ge 3$ then $|(j-1)-i| \ge 2$ and

$$\Phi_0\sigma_i\sigma_j = \Phi_0^{j,j-1,i,i-1} = \Phi_0^{i,i-1,j,j-1} = \Phi_0\sigma_j\sigma_i,$$

and so the third item holds.

Finally, from the first item of this proposition and the third item of Proposition 2.2.10 we see that the order of $\sigma_i = \rho_{i-1}\rho_i$ is p_i. ■

Exercise 2.2.17 Show that $\Gamma^+(\mathcal{P})$ consists of the orientation-preserving symmetries of $\mathcal{P}$ if $\mathcal{P}$ is the n-simplex, the n-cube or the n-dimensional cross polytope.

2.2.2 Two-Orbit Polytopes

When trying to understand highly symmetric non-regular polytopes, one of the natural choices is to fix the number of flag orbits under the automorphism group. The simplest of these choices is when there are two such orbits. Maniplexes with three flag orbits under their automorphism groups were studied in [69].

Definition 2.2.18 We say that $\mathcal{P}$ is a *two-orbit polytope* if $\Gamma(\mathcal{P})$ induces two orbits on the flags of $\mathcal{P}$.

Some two-orbit polytopes were known centuries ago, like the cuboctahedron, the rhombic dodecahedron and the Archimedean tiling 3.6.3.6 of the plane with two hexagons and two triangles surrounding every vertex (see Figure 2.15(c)). In one or another context, some two-orbit maps were studied in the twentieth century. This includes many articles on chiral maps (see Section 2.3) and some others on edge-transitive maps (see for example [101, 108, 261]). The systematic study of two-orbit polytopes started with [139] and it has contributed many elements to understand polytopes that are neither regular nor chiral. The main source for this section is [129].

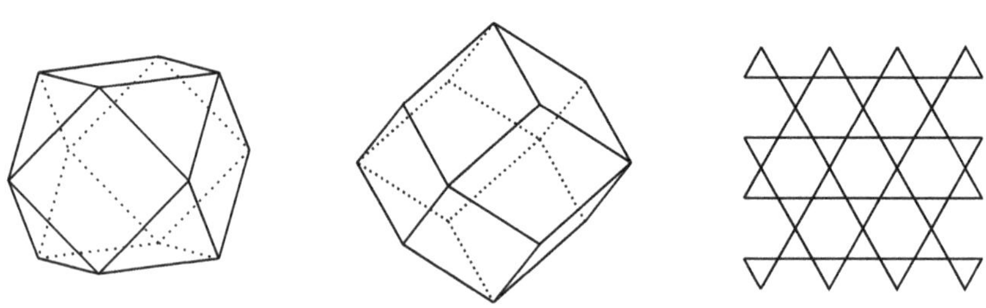

(a) (b) (c)

Figure 2.15 (a) Cuboctahedron, (b) rhombic dodecahedron and (c) tessellation 3.6.3.6.

The flag-connectivity of a two-orbit polytope $\mathcal{P}$ implies that for some i there exists a flag in one orbit that is i-adjacent to a flag in the other orbit. One nice feature of two-orbit polytopes is that the determination of such an i does not depend on the choice of flag or the choice of flag orbit.

Proposition 2.2.19 *Let $\mathcal{P}$ be a two-orbit n-polytope and let $i \in \{0, \ldots, n-1\}$. If there exist two i-adjacent flags in distinct orbits under $\Gamma(\mathcal{P})$ then Ψ and Ψ^i are in distinct orbits for every flag Ψ.*

Proof Let Φ be a flag such that Φ and Φ^i are in distinct orbits and let Ψ be an arbitrary flag of $\mathcal{P}$. Without loss of generality we assume that Φ and Ψ are in the same flag orbit, and hence there exists $\gamma \in \Gamma(\mathcal{P})$ such that $\Psi = \Phi\gamma$. Then $\Psi^i = (\Phi\gamma)^i = \Phi^i\gamma$ belongs to the same flag orbit as Φ^i, and so Ψ and Ψ^i are in distinct flag orbits. ∎

Exercise 2.2.20 Give an example of a polyhedron with three flag orbits where the conclusion of Proposition 2.2.19 does not hold.

Thanks to Proposition 2.2.19 we may classify two-orbit polytopes according to the indices i for which i-adjacent flags are in the same flag orbit. In order to do this it suffices to consider the flags adjacent to only one fixed flag.

Definition 2.2.21 Let $I \subset \{0, \ldots, n-1\}$, let $\mathcal{P}$ be a two-orbit n-polytope, and let Φ be any flag of $\mathcal{P}$. We say that $\mathcal{P}$ is in class 2_I whenever Φ and Φ^i belong to the same orbit if and only if $i \in I$.

For example, the cuboctahedron and the tessellation 3.6.3.6 in Figure 2.15(a) and (c) belong to class $2_{\{0,1\}}$. To see this it suffices to note that each flag orbit consists of all flags whose face has a certain fixed number of edges; in particular, two flags containing the same face belong to the same flag orbit. This implies that the 0- and 1-adjacent flags are in the same orbit, but 2-adjacent flags belong to distinct orbits, since every edge belongs to faces with distinct numbers of edges.

Similarly, the rhombic dodecahedron in Figure 2.15(b) is in class $2_{\{1,2\}}$. It has vertices of degrees 3 and 4; one flag orbit consists of those flags whose vertices have degree 3 while the other of those flags whose vertices have degree 4.

When there is no possibility of confusion, in the subindex we may list the elements of the set I instead of writing I. In this way, the cuboctaedron is in class 2_{01} and the rhombic dodecahedron is in class 2_{12}. Also, the class $2_{\emptyset}$ is frequently referred to just as class 2.

Example 2.2.22 Figure 2.16 shows two maps of type $\{4,4\}$ on the torus (opposite sides of the rectangle are identified in an orientable manner), whose vertices, edges and 2-faces form polyhedra (see Exercise 2.2.23). In Section 2.4 we will develop techniques to better understand their automorphism groups. Here we will use permutations of the edge sets to describe automorphisms. Clearly the image of a vertex F_0 (resp. 2-face F_2) under an automorphism will be the unique vertex contained in (resp. 2-face containing) the images of the edges containing F_0 (resp. contained in F_2).

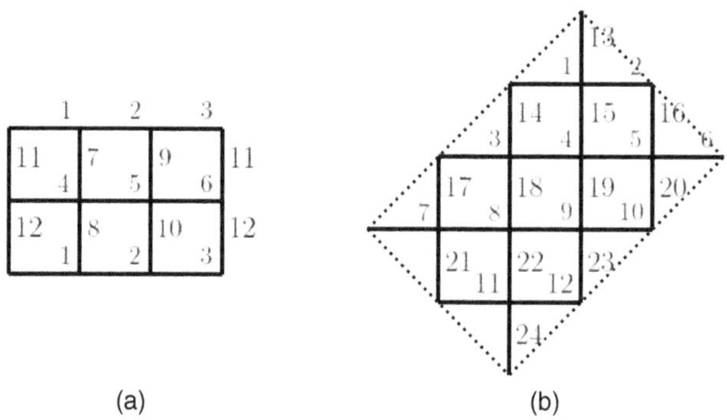

Figure 2.16 (a) A two-orbit polyhedron with 6 faces, (b) a two-orbit polyhedron with 12 faces.

The map on the torus in Figure 2.16(a) consists of a 2×3 grid. It has 48 flags and 12 edges, 6 of which are in a horizontal direction (labelled from 1 to 6) while the other 6 are vertical. It admits the automorphisms

$$(1,2,3)(4,5,6)(7,9,11)(8,10,12),$$
$$(1,4)(2,5)(3,6)(7,8)(9,10)(11,12),$$
$$(1,3)(4,6)(7,9)(8,10),$$
$$(7,8)(9,10)(11,12),$$

given as permutations of edges. The group generated by these permutations has order 24 and hence the polyhedron has at most two flag orbits. However, there is no automorphism swapping edges 1 and 12 while fixing the common vertex and facet of these two edges, since otherwise it would be forced to:

- swap edges 4 and 8, since they are the remaining edges of the only square containing edges 1 and 12;
- swap edges 5 and 7, since they are the remaining edges that contain the vertex F_0 at the intersection of edges 4 and 8;

- swap edges 1 and 10, and edges 2 and 11, since edges 1 and 11 complete the only square containing edges 4 and 7 while edges 2 and 10 complete the only square containing edges 5 and 8.

This causes a contradiction, since it was supposed to swap 1 and 12. Hence, there is no automorphism mapping the flag containing edge 1, the vertex contained in edges 1 and 12 and the 2-face containing edges 1, 4, 8 and 12 to its 1-adjacent flag. Therefore, the map in Figure 2.16(a) represents a two-orbit polyhedron.

Informally speaking, the edges of the polyhedron in Figure 2.16(a) are arranged in (vertical) straight lines containing two edges, and in (horizontal) straight lines containing three edges. An edge in a straight line of the first kind cannot be mapped to an edge in a straight line of the second kind, and vice versa. By examining the flag orbits we see that two flags belong to the same orbit if and only if either both contain horizontal edges or both contain vertical edges. We can conclude that this map is in class 2_{02}, since 0- and 2-adjacencies preserve the edges, while 1-adjacency changes from a horizontal edge to a vertical edge and vice versa.

The map in Figure 2.16(b) contains 12 vertices, 24 edges and 96 flags. It is invariant under the automorphisms

$$(1,2,11,12,3,4,5,6,7,8,9,10)(13,21,22,23,14,15,16,24,17,18,19,20),$$
$$(1,20)(2,16)(3,23)(4,19)(5,15)(6,13)(7,24)(8,22)(9,18)(10,14)(11,21)(12,17),$$
$$(1,17)(2,21)(3,14)(4,18)(5,22)(6,24)(7,13)(8,15)(9,19)(10,23)(11,16)(12,20),$$

as permutations of edges. The group generated by these permutations has order 48 and hence this polyhedron has at most two flag orbits. It is possible to use an argument like the one for the polyhedron (a) to verify that there is no automorphism mapping the flag containing the vertex contained in edges 8 and 9, edge 9 and the face containing edges 4, 9, 18 and 19 to its 2-adjacent flag. Here the intuition is that two flags are in the same flag orbit if and only if the segment between the vertex and the centre of 2-face of one of them is parallel to the corresponding segment of the other flag. This characterisation of the flag orbits makes it easy to determine that the polyhedron in Figure 2.16(b) belongs to class 2_1.

The reader is encouraged to make a picture indicating all flags in one of the flag orbits to better visualise this example.

Exercise 2.2.23 Show that the partial orders of the faces of the two maps in Figure 2.16 are polyhedra.

There is no two-orbit n-polytope in class $2_{\{0,\ldots,n-1\}}$, since flag-connectivity and Proposition 2.2.19 would force it to be regular, not two-orbit. There are no polygons in classes 2_0, 2_1 or 2, since every polygon is regular. In principle, for every $n \geq 3$ two-orbit n-polytopes in class 2_I could exist for every proper subset I of $\{0, \ldots, n-1\}$, and hence there are at most $2^n - 1$ classes of two-orbit n-polytopes. This bound is achieved for $n = 3$, and was recently proven to be achieved for every n in the case of maniplexes in [229]. Further progress in this direction was achieved in [190, Theorem 5.3.8].

Problem 1 Determine whether there exist two-orbit n-polytopes in class 2_I for every $n \geq 4$ and for every proper subset I of $\{0, \ldots, n-1\}$.

Transitivity on the i-faces of a two-orbit polytope is completely determined by the class of the polytope.

Proposition 2.2.24 *Let $I \subsetneq \{0, \ldots, n-1\}$, let $\mathcal{P}$ be a two-orbit n-polytope in class 2_I, and let $j \in \{0, \ldots, n-1\}$. Then $\mathcal{P}$ is j-face transitive if and only if $I \neq \{0, \ldots, n-1\} \setminus \{j\}$.*

Proof Assume first that $\mathcal{P}$ is j-face transitive. Some j-faces F_1 and F_2 must contain flags in the first and second flag orbits, respectively. Transitivity in the j-faces implies that F_1 also has flags in the second flag orbit and that F_2 also has flags in the first one. Now, if there are two flags in distinct orbits containing a j-face, strong flag-connectivity implies that there must be $i \neq j$ such that $i \notin I$ and two flags Φ and Φ^i are in distinct orbits. In this situation we can conclude that $I \neq \{0, \ldots, n-1\} \setminus \{j\}$.

Conversely, if $I \neq \{0, \ldots, n-1\} \setminus \{j\}$ then Φ and Φ^i are in distinct orbits for some $i \in \{0, \ldots, n-1\} \setminus \{j\}$, and therefore these two flags contain the same j-face. It follows from Proposition 2.2.19 that every j-face is contained in flags in each orbit. In this case $\mathcal{P}$ is j-face transitive. ■

Corollary 2.2.25 *A two-orbit n-polytope is fully transitive if and only if it is in class 2_I with $|I| < n-1$.*

Corollary 2.2.26 *A two-orbit n-polytope in class $2_{\{0,\ldots,n-1\}\setminus\{i\}}$ is j-face transitive if and only if $j \neq i$.*

The reader can compare these last results with the polyhedra in Figures 2.15 and 2.16.

Not every class of two-orbit polytopes admits self-dual elements. In fact, (2.5) implies that if $\mathcal{P}$ is a self-dual n-polytope in class 2_I then $i \in I$ if and only if $n - i - 1 \in I$.

Example 2.2.27 The n-polytopes in class $2_{\{0,\dots,n-2\}}$ have i-adjacent flags in the same flag orbit if and only if $i \leq n-2$. It follows from the strong flag-connectivity that all flags containing the same facet belong to the same flag orbit. Moreover, if two flags are $(n-1)$-adjacent then they belong to the two distinct orbits, implying that facets that share a ridge are contained in flags in distinct orbits. From this we can conclude that there exists a bipartition of the facets where two facets that share a ridge are in distinct parts. The flag orbits 1 and 2 consist of all flags containing a facet in parts 1 and 2, respectively.

From this characterisation it is easy to see that there cannot be a self-dual n-polytope in class $2_{\{0,\dots,n-2\}}$. Such a polytope would not be facet-transitive, but it would be vertex-transitive, since every vertex is contained in flags Φ and Φ^{n-1} in distinct flag orbits. The dual would then be facet-transitive but not vertex-transitive.

The bipartition of the facets in Example 2.2.27 is obvious in the cuboctahedron and the tiling 3.6.3.6, since the two kinds of facets are non-isomorphic.

Two-orbit polyhedra in class 2_{01} whose facets are all isomorphic do exist. One way to construct them is through the well-known medial operation on regular polyhedra of type $\{p, p\}$ that are not self-dual (see [214, Table 2]). Specifically, we may consider the medial of the polyhedron $\{6,6\} * 72b$ in [116] (see also the map R4.8 in [41]), which is not self-dual.

On the other hand, little is known about two-orbit polytopes in class $2_{\{0,\dots,n-2\}}$ for $n \geq 4$.

Problem 2 For every $n \geq 4$ determine whether there exists a two-orbit polytope in class $2_{\{0,\dots,n-2\}}$ whose facets are all isomorphic.

The next problem is related to Problem 1.

Problem 3 Determine if there are self-dual n-polytopes in class 2_I for every set I satisfying that $i \in I$ if and only if $n-i-1 \in I$.

Given a self-dual two-orbit polytope $\mathcal{P}$, a duality δ maps a flag Φ of $\mathcal{P}$ into the flag $\Phi\delta$. Since the flags of $\mathcal{P}$ and $\mathcal{P}^\delta$ are the same when considered merely as sets of faces, we may ask ourselves whether Φ and $\Phi\delta$ belong to the same flag orbit under $\Gamma(\mathcal{P})$. In an affirmative case we say that $\mathcal{P}$ is *properly self-dual*, and it is *improperly self-dual* whenever no flag Φ and duality δ satisfy that Φ and $\Phi\delta$ are in the same flag orbit. As we shall see next, the definition of proper self-duality is independent from the choices of flag Φ and duality δ.

Lemma 2.2.28 *A two-orbit n-polytope $\mathcal{P}$ is properly self-dual if and only if for every flag Φ and every duality δ there exists $\gamma_{\delta,\Phi} \in \Gamma(\mathcal{P})$ such that $\Phi\delta = \Phi\gamma_{\delta,\Phi}$.*

Proof Assume that $\mathcal{P}$ is properly self-dual. Then there exist a flag Φ_0, a duality δ_0 and $\gamma_0 \in \Gamma(\mathcal{P})$ such that $\Phi_0\delta_0 = \Phi_0\gamma_0$. Let Φ be an arbitrary flag and δ an arbitrary duality.

If Φ and Φ_0 are in the same flag orbit then there exists $\beta \in \Gamma(\mathcal{P})$ such that $\Phi = \Phi_0\beta$. Since $\Gamma(\mathcal{P})$ is an index two subgroup of $\overline{\Gamma}(\mathcal{P})$, we can write

$$\Phi\delta = \Phi_0\beta\delta = \Phi_0\delta_0\beta' = \Phi_0\gamma_0\beta' = \Phi(\beta')^{-1}\gamma_0\beta'$$

for some $\beta' \in \Gamma(\mathcal{P})$, and $(\beta')^{-1}\gamma_0\beta'$ can be taken as $\gamma_{\delta,\Phi}$.

We have proven that δ preserves the flag orbit containing Φ_0. It follows that it also preserves the other flag orbit, and the forward implication follows.

The converse follows directly from the definition of proper self-duality. ■

Example 2.2.29 Figure 2.17 shows in solid black lines the maps on the torus in Figure 2.16, and in grey lines a representation of the dual maps, where every vertex of the original map is the centre of a 2-face of the dual map, and vice versa. The formal verification that in each case the black and grey maps are dual and isomorphic to each other is left as an exercise. The triangle in each map represents a flag for both the primal and the dual maps.

In the map (a), the edge in the flag of the primal map is horizontal, whereas the edge of the flag of the dual map is vertical. According to Example 2.2.22, these two flags cannot be in the same orbit, and hence the corresponding polyhedron is improperly self-dual.

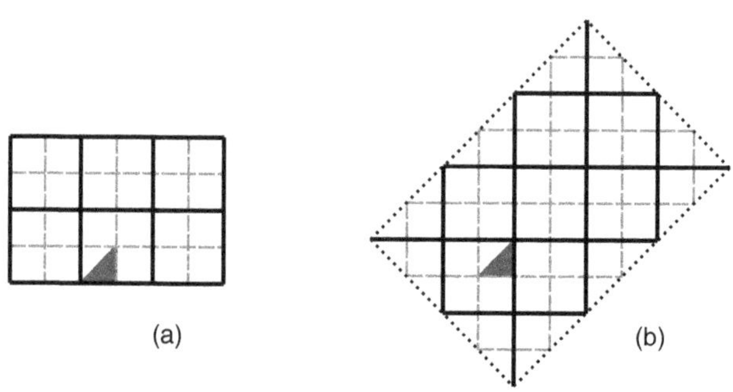

Figure 2.17 (a) An improperly self-dual two-orbit polyhedron, (b) a properly self-dual two-orbit polyhedron.

The segment between the vertex and the centre of the square in the flag of map (b) is parallel to the long vector determining the fundamental region, regardless of the flag being considered in the primal or the dual map. According to Example 2.2.22, the flag is in the same orbit when interpreted as a flag in the primal or dual map. It follows that the corresponding polyhedron is properly self-dual.

Exercise 2.2.30 For each of the two maps $\mathcal{P}$ in Figure 2.16 do the following.

1. Show that the identity on the edge set of $\mathcal{P}$ induces a duality, by exhibiting explicitly the bijection between the set of vertices of the primal and the set of 2-faces of the dual.
2. Show that $\mathcal{P}$ is isomorphic to its dual by explicitly exhibiting an isomorphism between them.
3. Conclude that $\mathcal{P}$ is isomorphic to the overlapped map with grey edges in Figure 2.17.

Lemma 2.2.31 *A two-orbit polytope is properly self-dual if and only if for every flag Φ there exists a duality δ_Φ such that $\Phi\delta_\Phi = \Phi$.*

Proof Let $\mathcal{P}$ be a properly self-dual polytope, let Φ be a flag of $\mathcal{P}$ and let δ be a duality. Then by Lemma 2.2.28 there exists an automorphism γ such that $\Phi\delta\gamma = \Phi$, and Remark 2.1.85 (2) tells us that $\delta_\Phi := \delta\gamma$ is a duality. The converse follows directly from the definition of proper self-duality. ■

Given the characterisation of properly self-dual two-orbit polytopes in Lemma 2.2.31, elsewhere a *properly self-dual* polytope has been defined as a self-dual polytope (with an arbitrary number of flag orbits under its automorphism group) whose extended group contains a duality that fixes a flag.

We conclude this section pointing out how little we know about higher-rank two-orbit polytopes. Up to now there are very few constructions of two-orbit polytopes of rank $n \geq 4$. One of them is through a universal construction of two-orbit n-polytopes in class $2_{\{0,\dots,n-2\}}$ whose facets are regular of two different kinds, and whose ridges belong to one facet of each kind (see [196]). In almost all cases the resulting polytopes are infinite.

The only well-understood family of finite two-orbit polytopes containing one element in each rank $n \geq 3$ was published in [221] (see also [180, Section 7], where related families of two-orbit polytopes with one member of each even rank are also mentioned). The representative of rank n consists of a two-orbit polytope in class $2_{\{0,\dots,n-2\}}$ whose facets are:

- the $(n-1)$-dimensional cross polytopes obtained by intersecting the n-dimensional cross polytope with each coordinate hyperplane, and
- those simplicial facets of the n-dimensional cross polytope with an even number of vertices having a coordinate -1.

Exercise 2.2.32 Show that the structure just defined is indeed a two-orbit n-polytope in class $2_{\{0,\ldots,n-2\}}$.

Problems 2 and 3 are just a sample of the aspects of two-orbit polytopes that have not been explored in depth. Problem 1, which was settled for maniplexes only in 2019, deserves special mention. Even though we have more than one way to construct two-orbit n-polytopes in class $2_{\{0,\ldots,n-2\}}$, we still are unable to give a full description of any two-orbit n-polytope in class 2 for large values of n. Polytopes in class 2 are precisely the chiral polytopes and we shall study them next.

2.3 Chiral Polytopes

A *chiral polytope* is a two-orbit polytope in class 2. This means that it has two flag orbits under its automorphism group, with adjacent flags always in distinct orbits.

The word 'chiral' is usually applied to three-dimensional objects whose mirror image is not superimposable with the original one by a continuous rigid motion, like some specific molecules, or the shells of snails. This adjective was introduced for polytopes in [247] and may be understood as an abbreviation of 'the most symmetric among all polytopes exhibiting no mirror symmetry'. A mirror symmetry in the order complex of an n-polytope is understood to fix point-wise an $(n-1)$-dimensional subspace (the *mirror*), and it would map those flags that are placed next to the mirror to the adjacent flags across the mirror. Such a polytope would not be chiral as defined above.

Regular and chiral polytopes have several aspects in common, and it is convenient to have a term denoting polytopes in either of these two classes. In many works on maps the term 'orientably regular' was used to refer to maps which are either orientably regular or chiral in our sense. To avoid confusion, we follow [81] and say that a polytope is *rotary* whenever it is either regular or chiral. Then, a polytope is *rotary orientable* if it is either orientably regular or chiral. Rotary orientable polytopes are characterised as those whose flag graphs are bipartite and have automorphism groups that act transitively on each part. In particular, rotary-orientable maps are maps on orientable surfaces that have the largest possible number of orientation preserving automorphisms, that is, half as many as their flags.

Chiral polytopes have attracted more attention that other classes of two-orbit polytopes, and there may be more than one reason for this. On one hand, class $2_{\emptyset}$ could be regarded as slightly more symmetric than any other class of two-orbit polytopes, since the empty set is the only proper subset of $\{0, \ldots, n-1\}$ invariant under every permutation of that set. When colouring one flag orbit in white and the other one in black, chiral polytopes would have adjacent flags of distinct colours, giving the impression of a stronger local symmetry than two-orbit polytopes in other classes, where some adjacent flags would share colour while others would not.

The main reason why chiral polytopes have been studied more widely than other classes of two-orbit polytopes has to do with orientability. Clearly, the flag graph of every chiral polytope is bipartite, which directly implies the following remark.

Remark 2.3.1 Every chiral polytope is orientable.

In particular, every chiral map lies on an orientable surface, and its group of orientation-preserving automorphisms acts transitively on each flag orbit. Many authors have been interested exclusively on orientation-preserving map automorphisms, in which case the highest degree of symmetry is achieved by both orientably regular and chiral maps (see for example [13, 49, 146]). In this situation the term 'regular' is consistent with the regular action of the orientation-preserving automorphism group on the arcs of the underlying graph. Analogues for other classes of two-orbit polytopes to the relationship between orientability and chirality can be found in [153].

In any case, before the computer era chiral maps and polytopes were harder to find than their regular counterparts. There are neither chiral convex polytopes nor chiral tessellations of a Euclidean space (see [168]). Moreover, many of the natural constructions for rotary maps or polytopes turn out to yield exclusively regular objects. In order to stress the difficulties in constructing chiral polytopes, before providing basic results for them we include a brief history of the search for such objects.

2.3.1 Brief History of Chiral Polytopes

A group-theoretical construction of a chiral map with type $\{p-1, p-1\}$ is given in [124] for each prime $p \geq 5$ congruent to 1 modulo 4. It has p vertices (the elements of $\mathbb{Z}_p$) and p faces. Given a generator g of the multiplicative group $\mathbb{Z}_p^*$, the faces of the map are the sets

$$(0) = \{g^0, g^1, \dots, g^{p-2}\}, \tag{2.7}$$

$$(g^i) = \{g^0 + g^i, g^1 + g^i, \dots, g^{p-2} + g^i\} \quad i \in \{0, \dots, p-2\}, \tag{2.8}$$

where the vertices are listed in the cyclic order in which they appear around the face according to a fixed orientation. However, the emphasis of that work is not on the automorphism group of these maps, and there were no subsequent publications investigating similar symmetry properties in other maps.

In the first half of the twentieth century, Brahana and Coxeter studied rotary maps on the two-dimensional torus (see [13, 14, 49]). Their full classification can also be found in [59, Section 3.2], together with the standard criterion to determine if a given rotary map on the torus is regular or chiral. These maps will be described in Section 2.4.

Looking for chiral maps on surfaces with negative Euler characteristic by hand is not an easy task. It was several years after the classification of rotary maps on the torus that new chiral maps were found and described. J. R. Edmonds is credited with the first of these discoveries, a map with type $\{7,7\}$ on an orientable surface with genus 7 (see [53, Section 21.3]). Shortly after, Sherk devised a construction of chiral maps with type $\{6,6\}$ from maps on the torus with type $\{3,6\}$ (see [253]). With this construction it became clear that there are chiral maps on compact surfaces with arbitrarily large genus. Nevertheless, not every compact orientable surface of positive genus carries chiral maps. This was established when Garbe showed that there are no chiral maps on surfaces of genera in the set $\{2,3,4,5,6\}$ (see [92]).

In the second half of the twentieth century many chiral maps on surfaces with genera greater than 6 were found. It is known that only finitely many of them lie on a given surface; nevertheless there are infinitely many of them with a given type $\{p,q\}$ whenever $\frac{1}{p} + \frac{1}{q} < \frac{1}{2}$ (this will be explained in Section 5.1; see also [42, 259]).

In present days we can access lists generated by computer with thousends of chiral maps (see [41, 118, 233]); the group-theoretical general tools to construct these maps will be described in Chapter 3. Such lists are useful to propose hypotheses, later to be refuted or to become theorems. One of these theorems states that there are no chiral maps on (orientable) surfaces with genus g where $g-1$ is prime and $g-2$ is not divisible by 3, 5 or 8 (see [43]). From these results we can conclude that infinitely many orientable surfaces carry chiral maps, and infinitely many orientable surfaces do not carry chiral maps. Nevertheless, we still seem to be far away from determining the complete list of surfaces of either of these kinds (see [220, Problem 19]).

Problem 4 Determine all positive integers g such that there exists a chiral map on a surface of genus g.

Most chiral maps are polytopal, but not all of them are. Consequently, Problem 4 can be posed for polyhedra instead of maps. It would also be interesting to characterise all chiral maps that are not polytopal (see [220, Problem 18]).

Problem 5 Characterise all chiral maps on surfaces that are not polytopal.

Much less is known about chiral polytopes of ranks higher than 3.

In [262] two 3-manifolds are constructed by identifying opposite sides of a dodecahedron in two distinct ways. We can think that the manifolds admit a tiling with only one (dodecahedral) tile, and by doing this the result is two chiral 4-maniplexes. They cannot be polytopal since they have only one 3-face. However, the symmetry aspect of the manifolds was not mentioned in that work.

Coxeter was the pioneer in the study of chiral 4-polytopes. He considered some specific quotients of the regular tessellations of hyperbolic three-dimensional space $\mathbb{H}^3$ (see [54]). Given a regular tessellation of $\mathbb{H}^3$, he studied the largest rotary quotient that would have some Petrie paths of a given size. As it turned out, sometimes the resulting structure would have Petrie paths of two distinct sizes, in which case it was forced to be chiral. Such objects were called *twisted honeycombs*. These quotients can also be understood as tilings of hyperbolic manifolds; in particular, facets and vertex-figures are isomorphic to convex regular polyhedra. Not all of these objects were polytopal, but some of them were. Coxeter's idea was continued in [60, 209, 264] to construct more twisted honeycombs. Twisted honeycombs will be addressed in Section 5.3.2.

The formal definition of chiral polytope was established in 1991 by Schulte and Weiss in [247], and a few years later other chiral 4-polytopes were constructed by reductions of distinct groups of matrices (see [194, 199, 201, 210, 248]). In 1995 a universal construction of a chiral $(n+1)$-polytope with facets isomorphic to any given chiral n-polytope $\mathcal{P}$ with regular facets was given in [249]; the resulting polytope is infinite and any other chiral polytope with facets isomorphic to $\mathcal{P}$ must be a quotient of the universal one. As a consequence, it was possible to construct infinite chiral 5-polytopes and 6-polytopes.

The first known finite chiral 5-polytopes were found in [38] with the help of a computer software. In that work it was also established which are the smallest chiral polyhedra (those containing fewest flags) in each of the three following categories: not self-dual, properly self-dual and improperly self-dual. Further small examples of chiral 4- and 5-polytopes in those three categories were also mentioned.

It was in 2010 that it was finally established that chiral n-polytopes exist for every $n \geq 3$. A recursive construction for chiral n-polytopes with regular faces was given in [218], where the recursive step is composed of two substeps. First, given a chiral $(n-1)$-polytope $\mathcal{P}_0$ satisfying certain properties, a regular $(n-1)$-polytope $\mathcal{P}_1$ that covers $\mathcal{P}_0$ is constructed; the second step consists of finding a chiral n-polytope $\mathcal{Q}_0$ whose facets are isomorphic to $\mathcal{P}_1$ and that satisfies the same properties required for $\mathcal{P}_0$, so that the recursive step can be applied once again.

There are several drawbacks to the construction in [218], some of them related to how little can be said of the resulting n-polytope $\mathcal{Q}_0$. For instance, its automorphism group is not described, and it is unclear whether the vertex-figures are regular or chiral. Even the number of flags of $\mathcal{Q}_0$ is unknown, although it is much larger than the number of flags of the chiral $(n-1)$-polytope $\mathcal{P}_0$. In fact, the main drawback is that the number of flags of the smallest n-polytope constructed in this way, starting from a chiral polyhedron, seems to grow as a function of n that cannot be bounded above by any finite composition of exponentials. This makes the construction impractical for computing concrete examples.

Other constructions of chiral n-polytopes require a smaller chiral n-polytope as a starting point, this includes those in [44, 78]. Some computational constructions of chiral n-polytopes, like the ones outlined in [38, 82, 275, 276] are severely constrained for ranks $n > 5$ due to the size of the groups to consider.

Up to now, we know the smallest chiral n-polytopes only for $n \leq 6$. Their numbers of flags are 40, 240, 1 440 and 18 432 for ranks 3, 4, 5 and 6, respectively (see [38, 45]). It transpires that the ratio between the numbers of flags of the smallest chiral 6-polytopes and their counterparts in rank 5 is about twice as large as the analogous ratio involving ranks 4 and 3, or ranks 5 and 4. The evidence so far suggests that the number of flags of the smallest chiral n-polytope divided by the number of flags of the smallest chiral $(n-1)$-polytope is a rapidly growing function of n.

Some lower bounds for the numbers of flags of chiral n-polytopes are given in [66]. If $n \geq 8$ then every chiral polytope has at least $48(n-2)(n-2)!$ flags, and it is unlikely that this bound is tight. By contrast, for $n \geq 9$ the smallest regular non-degenerate n-polytope (in the sense that its type has no entry 2) has only $2 \cdot 4^{n-1}$ flags (see [32]). The difficulties encountered so far to find small examples of chiral polytopes with ranks 7 and higher suggest the following question (see [220, Problem 35]).

Problem 6 Determine the number of flags of the smallest chiral n-polytope for every n.

In summary, the study of chiral polytopes is still young and leads to interesting questions of several kinds. There are many questions, like Problem 6, for which we have answers for regular polytopes but not for chiral ones. In particular, we are still in the search for families of chiral polytopes with at least one polytope of each rank $n \geq 3$, where we can say many specific details about the representatives in each rank (like their types, numbers of flags and automorphism groups).

In the author's opinion, the understanding of chiral polytopes is relevant as an example of a much deeper concept. In topics of mathematics where symmetry is relevant, it is natural to try to get a handle of those objects possessing most symmetry among all objects being studied. However, it is not always possible to require a certain degree of symmetry but not more than that. A remarkable example of this is the result by McMullen stating that every equivelar convex polytope is combinatorially regular (see [168]), and so we cannot have chiral or any other kind of equivelar two-orbit polytopes realised in a convex way. The search for chiral polytopes includes as one of its main difficulties guaranteeing that the resulting object is not regular, even though we are also asking the automorphism group to contain all possible combinatorial rotations. In other words, we are simultaneously requiring a large degree of symmetry and forbidding any extra symmetry. Can we say general conditions under which there exist objects testifying this phenomenon (like abstract chiral polytopes) and conditions under which there are no such objects (like convex chiral polytopes)?

2.3.2 Basic Results

In what follows we recall the basic definitions and results on chiral polytopes. The main source for this section is [247].

Definition 2.3.2 A polytope is *chiral* whenever its automorphism group induces two orbits on flags, in such a way that every two adjacent flags belong to distinct orbits.

Chiral polytopes are precisely those two-orbit polytopes in class 2, as defined in the previous section.

Example 2.3.3 Our first example of a chiral polyhedron is the one represented in Figure 2.18(a) as a map on the torus. To this end, opposite sides of the dotted square must be identified orientably. It has 5 vertices (labelled 1 to 5), 10 edges and 5 squares. This polyhedron is called $\{4,4\}_{(2,1)}$, a name that will be clarified in the next section.

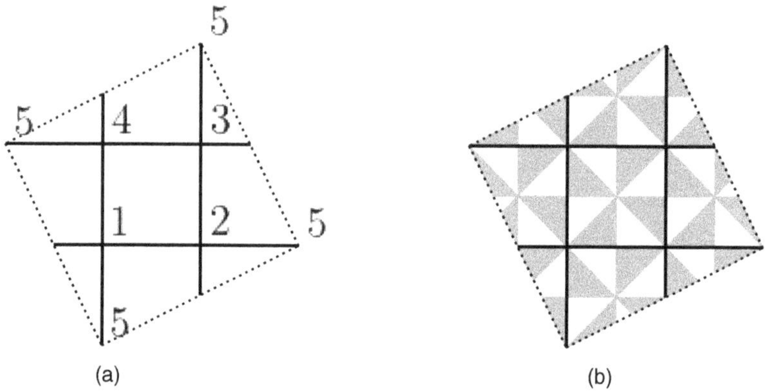

Figure 2.18 The polyhedron $\{4,4\}_{(2,1)}$ with (a) labels on its vertices and with (b) its flag orbits.

To see that $\{4,4\}_{(2,1)}$ is chiral we first note that every edge is determined by its two endpoints (no multiple edges), and every square is determined by its four vertices. Hence, the permutation of the vertices given by any automorphism $\gamma \in \Gamma(\{4,4\}_{(2,1)})$ completely determines γ.

The permutation $(1,2,3,4)$ induces an automorphism σ of $\{4,4\}_{(2,1)}$ that is visualised as a 4-fold counterclockwise rotation about the centre of the square $(1,2,3,4)$ in Figure 2.18(a). The grey and white flags in the central square in Figure 2.18(b) are two orbits of flags under the group generated by this automorphism.

The permutation $(1,2,5,4,3)$ also induces an automorphism γ; this can be verified directly by confirming that the images of two endpoints of an edge are the endpoints of an edge and that squares are also mapped to squares. This automorphism acts transitively on the squares, implying that the group $\langle\sigma,\gamma\rangle$ has at least 20 elements. In fact, this group induces two orbits on the flags of $\{4,4\}_{(2,1)}$, corresponding precisely to the sets of white flags and of grey flags in Figure 2.18(b). This proves that $\{4,4\}_{(2,1)}$ is rotary.

It remains to be shown that $\{4,4\}_{(2,1)}$ is not regular. Let Φ be the grey flag containing vertex 1 and the edge between 1 and 2, and let $\Psi = \Phi^2$. Suppose to the contrary that there is an automorphism ρ mapping Φ to Ψ. Since $\Phi\rho^2 = \Phi^2\rho = (\Phi\rho)^2 = (\Phi^2)^2 = \Phi$, ρ must be an involution. On the other hand, ρ maps the face $(1,2,3,4)$ to the face $(1,2,4,5)$ by fixing vertices 1 and 2, while sending 3 to 4 and 4 to 5. It follows that $(3)\rho^2 = (4)\rho = 5$, contradicting that ρ is involutory.

Although chiral polytopes are not regular, they satisfy several symmetry properties, some of which are listed in the next proposition (compare with Remark 2.2.5).

Proposition 2.3.4 *If $\mathcal{P}$ is a chiral polytope, then*

1. *$\mathcal{P}$ is fully transitive,*
2. *every two sections determined by an i-face and an incident j-face are isomorphic,*
3. *$\mathcal{P}$ is equivelar.*

Proof The first item is a consequence of Corollary 2.2.25. Let F_i and F_j be an i-face and a j-face, respectively, such that $F_i < F_j$. If $j - i \leq 2$ then the section F_j/F_i is isomorphic to the unique polytope of rank $j - i - 1$. If $j - i \geq 3$ and Φ is a flag of $\mathcal{P}$ containing F_i and F_j, then Φ and Φ^{i+1} are in distinct flag orbits. It follows that if F_i' and F_j' are an i-face and an incident j-face, and Φ' is a flag containing F_i' and F_j' then there exists $\gamma \in \Gamma(\mathcal{P})$ such that $\Phi'\gamma \in \{\Phi, \Phi^{i+1}\}$. The automorphism γ determines an isomorphism between the sections F_j/F_i and F_j'/F_i', proving the second item. The third item follows from the second one by considering the sections of rank 2. ■

Chiral n-polytopes have many other transitivity properties. In particular, their automorphism groups act transitively on the set of chains containing faces with ranks in any proper subset of $\{0, \ldots, n - 1\}$. The proof is left as an exercise.

Exercise 2.3.5 Show that if $\mathcal{P}$ is a chiral n-polytope and $j \in \{0, \ldots, n-1\}$ then $\Gamma(\mathcal{P})$ acts transitively on the chains consisting of faces with ranks in $\{-1, \ldots, n\} \setminus \{j\}$.

Next we determine that the sections of chiral polytopes are highly symmetric themselves.

Proposition 2.3.6 *Every section of a chiral polytope is rotary.*

Proof Let $\mathcal{P}$ be a chiral polytope, let F_i and F_j be respectively an i-face and a j-face, and assume that $F_i < F_j$.

We colour white the flags of $\mathcal{P}$ in one flag orbit and we colour black the flags on the other orbit, and say that a flag of F_j/F_i is white (resp. black) if it is contained in a white (resp. black) flag of $\mathcal{P}$. All adjacent flags to a white flag of F_j/F_i must be black and vice versa, since that is the case in $\mathcal{P}$. Since every automorphism of $\mathcal{P}$ that fixes F_i and F_j induces an automorphism of F_j/F_i, all white (or black) flags of F_j/F_i are in the same flag orbit, and hence F_j/F_i is rotary.

It may happen that a flag of F_j/F_i is both white and black. In that case, it is in the same flag orbit of any other flag of F_j/F_i, and so the section is regular. The section F_j/F_i can only be chiral if no flag is simultaneously black and white. ■

The previous proposition together with the second item of Proposition 2.3.4 imply the following corollary.

Corollary 2.3.7 *The facets (or vertex-figures) of any chiral n-polytope are isomorphic rotary $(n-1)$-polytopes.*

If F_{n-1} is a facet of a chiral polytope $\mathcal{P}$, then its dual F_{n-1}^{δ} is isomorphic to a vertex-figure of $\mathcal{P}^{\delta}$. Therefore $\mathcal{P}$ has regular facets if and only if $\mathcal{P}^{\delta}$ has regular vertex-figures, and vice versa.

A chiral polytope may not have any chiral proper section (see Example 2.3.3); however, every chiral polytope has regular sections, as shown next.

Proposition 2.3.8 *The $(n-2)$-faces and the cofaces at an edge of a chiral n-polytope are regular.*

Proof Let $\mathcal{P}$ be a chiral n-polytope. We know from Proposition 2.3.6 that the $(n-2)$-faces of $\mathcal{P}$ are rotary. We only need to show that if F_{n-2} is an $(n-2)$-face and Φ is a flag of F_{n-2} then Φ and Φ^0 belong to the same flag orbit. The result for the cofaces at an edge then will follow by duality.

Let $\hat{\Phi}$ be a flag of $\mathcal{P}$ containing Φ and denote by F_i the i-face of $\hat{\Phi}$. Then $\hat{\Phi}^{0,n-1}$ and $\hat{\Phi}$ are in the same flag orbit under $\Gamma(\mathcal{P})$. Restricting $\hat{\Phi}^{0,n-1}$ and $\hat{\Phi}$ to the section F_{n-2}/F_{-1}, we conclude that Φ and Φ^0 belong to the same flag orbit of F_{n-2}. ■

Since all polygons are regular, all chiral polyhedra have regular facets and regular vertex-figures. Chiral polytopes with chiral facets or chiral vertex-figures must have rank 4 or higher.

In [66] it is pointed out that in ranks $n = 4$ and $n = 5$, the number of flags of the smallest chiral n-polytopes with regular facets and vertex-figures is larger than that of the smallest chiral n-polytopes with chiral facets and regular vertex-figures, while this in turn is larger than or equal to the number of flags of the smallest chiral n-polytopes with chiral facets and chiral vertex-figures. This suggests the following problem (see [66, Problems 4 and 5]).

Problem 7 Let $f_{rr}(n)$, $f_{cr}(n)$ and $f_{cc}(n)$ denote the number of flags of the smallest chiral n-polytopes with regular facets and regular vertex-figures, chiral facets and regular vertex-figures, and chiral facets and chiral vertex-figures, respectively.

1. Determine whether $f_{rr}(n) \geq f_{cr}(n) \geq f_{cc}(n)$ for every $n \geq 3$. Are the inequalities strict for every $n \geq 5$?
2. Determine the functions $f_{rr}(n)$, $f_{cr}(n)$ and $f_{cc}(n)$.

Determining whether a polytope is rotary or not does not require finding automorphisms mapping a given flag to all other flags in its flag orbit. The following theorem is the analogue for rotary polytopes of Theorem 2.2.9.

Theorem 2.3.9 *An n-polytope $\mathcal{P}$ is rotary if and only if for any flag Φ there exist automorphisms $\sigma_1, \ldots, \sigma_{n-1}$ such that $\Phi\sigma_i = \Phi^{i,i-1}$. Moreover, $\mathcal{P}$ is regular if and only if $\Gamma(\mathcal{P})$ also contains any of the standard reflections.*

Proof If $\mathcal{P}$ is chiral then the flags Φ and $\Phi^{i,i-1}$ are in the same part of the partition given by the flag orbits, guaranteeing the existence of the automorphisms σ_i. Clearly, such automorphisms exist also if $\mathcal{P}$ is regular.

Conversely, assume that $\Gamma(\mathcal{P})$ contains the automorphisms $\sigma_1, \ldots, \sigma_{n-1}$. Then, in the same way as in the proof Proposition 2.2.14 we can show that for any $i_1, \ldots, i_{2m} \in \{0, \ldots, n-1\}$ there is an automorphism $\gamma \in \langle \sigma_1, \ldots, \sigma_{n-1} \rangle$ such that $\Phi\gamma = \Phi^{i_1,\ldots,i_{2m}}$. It follows that $\mathcal{P}$ has at most two flag orbits under $\langle \sigma_1, \ldots, \sigma_{n-1} \rangle$. In case $\mathcal{P}$ is not regular then one of the orbits consists of flags Φ^w where w is a word of even length in the alphabet $\{0, \ldots, n-1\}$, and the other orbit of those where w is a word of odd length. We conclude that $\mathcal{P}$ is rotary.

If $\mathcal{P}$ is chiral then adjacent flags belong to distinct orbits, and so $\Gamma(\mathcal{P})$ cannot contain any of the standard reflections. This concludes the proof. ■

It becomes clear now that the automorphism groups of chiral polytopes have certain similarity with the even subgroups of the automorphism groups of regular polytopes. Since the automorphisms σ_i are defined in the same way for chiral polytopes as they were for the regular ones, we may call them *standard rotations* in either case. When convenient, if $\mathcal{P}$ is chiral we shall denote $\Gamma(\mathcal{P})$ by $\Gamma^+(\mathcal{P})$, and refer to it also as the *rotation subgroup*.

The following result is a consequence of Proposition 2.2.14 and Theorem 2.3.9.

Corollary 2.3.10 *If $\mathcal{P}$ is a rotary polytope then $\Gamma^+(\mathcal{P}) = \langle \sigma_1, \ldots, \sigma_{n-1} \rangle$.*

Example 2.3.11 In Example 2.3.3 we showed that the polyhedron $\{4,4\}_{(2,1)}$ is rotary by exhibiting two automorphisms that generate a group that acts transitively on the set of grey flags. Theorem 2.3.9 suggests that this could have been done just by determining the existence of σ_1 and σ_2.

We choose as base flag Φ_0 the one containing vertex 1, the edge between 1 and 2, and the central face (in Figure 2.18(b) this is a grey flag). Then $\Phi_0^{1,0}$ is the flag containing vertex 4, the edge between 1 and 4, and the central face. The permutation $(1,4,3,2)$ induces an automorphism mapping Φ_0 to $\Phi_0^{1,0}$. Similarly, $\Phi_0^{2,1}$ is the flag containing vertex 1, the edge between 1 and 5, and the square with vertex set $\{1,5,4,2\}$. The permutation $(2,5,3,4)$ induces an automorphism mapping Φ_0 to $\Phi_0^{2,1}$. We conclude that σ_1 and σ_2 are defined by the permutations $(1,4,3,2)$ and $(2,5,3,4)$, respectively. This is an alternative way to conclude that $\{4,4\}_{(2,1)}$ is rotary.

We emphasise that the standard rotations $\sigma_1,\ldots,\sigma_{n-1}$ depend on the choice of base flag, and in general are not unique for a given rotary polytope.

Exercise 2.3.12 Determine the permutations defining σ_1 and σ_2 for $\{4,4\}_{(2,1)}$ with vertices labelled as in Example 2.3.3, if the base flag is chosen to be the one containing vertex 2, the edge between 1 and 2 and the central square in Figure 2.18.

Exercise 2.3.13 Show that the structures defined by Heffter, constructed from the polygons in (2.7) and (2.8), are indeed chiral maps. Are they polyhedral? What happens if the same construction is applied to primes p congruent to 3 modulo 4?

Some further similarities between the automorphism groups of chiral polytopes and the even subgroups of regular polytopes are made evident in the next proposition (compare with Proposition 2.2.16).

Proposition 2.3.14 *Let $\mathcal{P}$ be a chiral polytope with type $\{p_1,\ldots,p_{n-1}\}$ and let $\sigma_1,\ldots,\sigma_{n-1}$ be the standard rotations. Then*

1. $(\sigma_i\cdots\sigma_j)^2 = \mathrm{id}$ *for every* $1 \le i < j \le n-1$,
2. $\sigma_i\sigma_j = \sigma_j\sigma_i$ *if* $|i-j| \ge 3$,
3. *the order of* σ_i *is* p_i.

Proof The first two items are proven in the same way as in Proposition 2.2.16. If Φ_0 is the base flag then

$$\Phi_0\sigma_i^k = \Phi_0^{i,i-1}\sigma_i^{k-1} = (\Phi_0\sigma_i^{k-1})^{i,i-1} = \cdots = \Phi_0^{(i,i-1)^k}.$$

Now the third item follows from Remark 2.1.22. ■

Despite all the similarities shown so far between the automorphism groups of chiral polytopes and the even subgroups of regular polytopes, there are some concrete results that apply to only one of these two classes of polytopes. These differences are exemplified by the following remark and lemma (compare the lemma with Exercise 2.1.61).

Remark 2.3.15 Item (4) in Proposition 2.2.10 says that the stabiliser of a subset Ψ of the base flag of a regular polytope is the group generated by the standard reflections that stabilise Ψ. The direct analogue does not hold for the even subgroup of an orientable rotary polytope and the generators σ_i.

Indeed, if $\mathcal{P}$ is an orientable rotary polyhedron then the stabiliser in $\Gamma^+(\mathcal{P})$ of the base edge F_1 is the group generated by the automorphism that maps the base flag Φ_0 to $\Phi_0^{0,2}$, so this stabiliser is not trivial. On the other hand, none of the standard rotations fix F_1.

Lemma 2.3.16 *Let $\mathcal{P}$ be a chiral polytope with type $\{p_1, \dots, p_{n-1}\}$. Then $p_i \geq 3$ for every $i \in \{1, \dots, n-1\}$.*

Proof Assume to the contrary that $p_i = 2$ for some $i \in \{1, \dots, n-1\}$. Then Lemma 2.1.65 implies that for every $j \leq i-1$ and every $k \geq i$, every j-face is incident to every k-face.

Suppose first that $i < n-1$. By Proposition 2.3.8, every i-face is a regular polytope of rank at least 1. Let F_i be an i-face and let $\rho \in \Gamma(F_i)$ such that $\Phi_0\rho = \Phi_0^0$ for some flag Φ_0 of F_i. Then we can define a bijection γ of $\mathcal{P}$ as $F\gamma = F\rho$ if F has rank less than i, and $F\gamma = F$ if the rank of F is at least i. This bijection is indeed an automorphism because every i-face contains every j-face for $j < i$. If Ψ is a flag of $\mathcal{P}$ containing Φ_0 then $\Psi\gamma = \Psi^0$, contradicting that $\mathcal{P}$ is chiral.

If $i = n-1$ then we replace $\mathcal{P}$ by $\mathcal{P}^\delta$ in the argument above to conclude that $\mathcal{P}^\delta$ (and hence, also $\mathcal{P}$) is not chiral. ■

Even though Remark 2.3.15 establishes that there is no analogue for rotary polytopes (and their even subgroups) to the fourth item of Proposition 2.2.10, there is a weaker version of that item that still holds.

Proposition 2.3.17 *Let $\mathcal{P}$ be an orientable rotary polytope with standard rotations $\sigma_1, \dots, \sigma_{n-1}$ with respect to the base flag Φ. For some $j \in \{0, \dots, n-1\}$ let $\Phi_{<j}$ and $\Phi_{>j}$ denote the subsets of Φ consisting of faces with ranks less* *than j and greater than j, respectively. Then the stabilisers of $\Phi_{<j}$ and of $\Phi_{>j}$ in $\Gamma^+(\mathcal{P})$ are $\langle\sigma_{j+1},\dots,\sigma_{n-1}\rangle$ and $\langle\sigma_1,\dots,\sigma_j\rangle$, respectively.*

Proof Let F_{j-1} be the $(j-1)$-face of Φ. Then the set of flags of $\mathcal{P}$ containing $\Phi_{<j}$ consists precisely of the flags of the section $\mathcal{P}/F_{j-1}$, each of them augmented by the faces of Φ with ranks less than $j-1$. We know from Remark 2.2.13 and Proposition 2.3.6 that the section $\mathcal{P}/F_{j-1}$ is orientable rotary. Furthermore, every $\gamma \in \Gamma^+(\mathcal{P})$ that fixes $\Phi_{<j}$ becomes an automorphism $\hat{\gamma} \in \Gamma^+(\mathcal{P}/F_{j-1})$ when restricted to this section. On the other hand, every $\hat{\gamma} \in \Gamma^+(\mathcal{P}/F_{j-1})$ can be extended to some $\gamma \in \Gamma^+(\mathcal{P})$ by establishing that the image of Φ under γ is $\Phi_{>j-2}\hat{\gamma}$ augmented by the faces of Φ with ranks less than $j-1$. Proposition 2.1.80 (2) implies that the restriction of γ to $\mathcal{P}/F_{j-1}$ equals $\hat{\gamma}$. Hence, the stabiliser of $\Phi_{<j}$ in $\Gamma^+(\mathcal{P})$ is generated by the standard rotations of $\Gamma^+(\mathcal{P}/F_{j-1})$ extended so that they act trivially on $\Phi_{<j}$, and therefore this stabiliser is $\langle\sigma_{j+1},\dots,\sigma_{n-1}\rangle$.

An analogous argument shows that the stabiliser of $\Phi_{>j}$ in $\Gamma^+(\mathcal{P})$ is $\langle\sigma_1,\dots,\sigma_j\rangle$. ■

Corollary 2.3.18 *Let $n \geq 3$, let $\mathcal{P}$ be an orientable rotary n-polytope, and let $\Gamma^+(\mathcal{P}) = \langle\sigma_1,\dots,\sigma_{n-1}\rangle$. Then the stabilisers in $\Gamma^+(\mathcal{P})$ of the base facet and of the base vertex are $\langle\sigma_1,\dots,\sigma_{n-2}\rangle$ and $\langle\sigma_2,\dots,\sigma_{n-1}\rangle$, respectively.*

We shall discuss in detail the stabiliser in $\Gamma^+(\mathcal{P})$ of a subset of the base flag in Section 3.1.

Standard rotations are not the only automorphisms that have been used as generators of the rotation subgroup of a rotary orientable polytope. Under some circumstances the following generating set seems more convenient.

Let $\mathcal{P}$ be an orientable rotary polytope and let Φ be its base flag. For $k \in \{1,\dots,n-1\}$ we define $\tau_k \in \Gamma^+(\mathcal{P})$ as the automorphism such that

$$\Phi\tau_k = \Phi^{0,k}. \tag{2.9}$$

In particular, $\tau_1 = \sigma_1^{-1}$, while (2.2) implies that for every $k \geq 2$ the automorphism τ_k is an involution.

Proposition 2.3.19 *Let $\mathcal{P}$ be an orientable rotary polytope. Then $\Gamma^+(\mathcal{P}) = \langle\tau_k : k \in \{1,\dots,n-1\}\rangle$.*

Proof We already know that $\sigma_1 = \tau_1^{-1}$. On the other hand, for $i \geq 2$,

$$\Phi\tau_{i-1}\tau_i = \Phi^{0,i-1}\tau_i = (\Phi\tau_i)^{0,i-1} = \Phi^{i,0,0,i-1} = \Phi^{i,i-1},$$

which implies that $\sigma_i = \tau_{i-1}\tau_i$. Therefore $\Gamma^+(\mathcal{P}) = \langle\sigma_1,\dots,\sigma_{n-1}\rangle \subseteq \langle\tau_k : k \in \{1,\dots,n-1\}\rangle$. The proposition follows now from the fact that $\tau_i \in \Gamma^+(\mathcal{P})$ for each i. ■

Example 2.3.20 The polyhedron $\{4,4\}_{(2,1)}$, with the labelling of its vertices as in Figure 2.18, can be shown to be chiral by exhibiting the generators

$$\tau_1 = (1,2,3,4),$$
$$\tau_2 = (1,2)(3,5),$$

as permutations of the vertices, with respect to the base flag consisting of vertex 1, the edge between 1 and 2, and the 2-face containing 1, 2, 3 and 4. Since the base flag here is the base flag chosen in Example 2.3.11, τ_1 is precisely σ_1^{-1}, while τ_2 is $\sigma_1\sigma_2$.

The proof of Proposition 2.3.19 provides us with a simple expression of σ_i in terms of the generators τ_j, which allows us to transform any expression in terms of the standard rotations to an expression in terms of the τ_j. The next lemma can be used to perform such a transformation.

Lemma 2.3.21 *Let $\mathcal{P}$ be an orientable rotary n-polytope with standard rotations $\sigma_1, \ldots, \sigma_{n-1}$, and let τ_k be as in (2.9). Then $\tau_k = \sigma_1 \cdots \sigma_k$ for every $k \in \{2, \ldots, n-1\}$.*

Proof If Φ is the base flag of $\mathcal{P}$ then it was shown in the proof of Proposition 2.2.16 that $\Phi\sigma_1 \cdots \sigma_k = \Phi^{0,k}$, and the lemma follows from Proposition 2.1.80. ∎

The generating set $\{\tau_1, \ldots, \tau_{n-1}\}$ just defined was used in [37]. Its main advantage over the standard rotations is that most of the generators τ_i are involutions, which makes them easier to manipulate in many instances. They also satisfy some relations, like the ones in the next exercise.

Exercise 2.3.22 Let $\tau_1, \ldots, \tau_{n-1}$ be defined as in (2.9). Show that

1. $\tau_i \tau_1 \tau_i = \tau_1^{-1}$ for every $i \geq 3$,
2. $\tau_i \tau_j = \tau_j \tau_i$ if $1 < i < j - 1$.

Exercise 2.3.23 Express the stabilisers under the action of $\Gamma^+(\mathcal{P})$ of the base facet and of the base vertex of an orientable rotary polytope $\mathcal{P}$, in terms of the generators τ_i.

In the definition of τ_k in (2.9) the superindex 0 plays a special role. This could be generalised so that the superindex i is privileged instead, by defining a generating set $\{\tau_{j,i}\}_{j\in\{0,\dots,n-1\}\setminus\{i\}}$ such that if Φ is the base flag then $\Phi\tau_{j,i} = \Phi^{i,j}$. There are ways to express each of the generators $\tau_{j,i}$ in terms of the standard rotations, and vice versa, similar to those in the proof of Proposition 2.3.19 and in Lemma 2.3.21. Furthermore, hybrid generating sets that include some standard rotations and some generators $\tau_{j,i}$ have already been used (see [230, Section 6]). For instance, generating sets including all but one standard rotation together with the extra generator $\tau_{n-3,n-1}$ were used in [70, 218].

Informally, chiral polytopes are those polytopes that have all possible symmetry by rotations (generated by the standard rotations), but none by reflections. Objects in Euclidean 3-space that have only symmetry by orientation-preserving isometries cannot be superimposed onto their image under a reflection in a mirror. The snub cube in Figure 1.1 and the helices discussed in Section 6.1 are examples of this kind of object; our hands and shells of snails are examples of bodies with no non-trivial symmetry. For each of these examples it is possible to establish a convention of which one is left-handed and which one is right-handed. In the reflection of the mirror, right-handed objects look like their left-handed version and vice versa. Since we cannot perform physically a reflection of an object in real life, the left-handed and right-handed representatives are considered distinct, and may have very different functional properties.

Isomorphisms of polytopes are defined independently from the orientability properties of the polytopes, and hence we regard two isomorphic polytopes as (combinatorially) equal, even if one of them is thought as left-handed and the other one as right-handed. To illustrate this we include two representations of the polyhedron $\{4,4\}_{(2,1)}$ in Figure 2.19. One representation is the mirror image of the other, making evident the isomorphism between them. One way to define the isomorphism is by the mapping from the vertices of one representation to the other given by the trivial permutation of $\{1,2,3,4,5\}$.

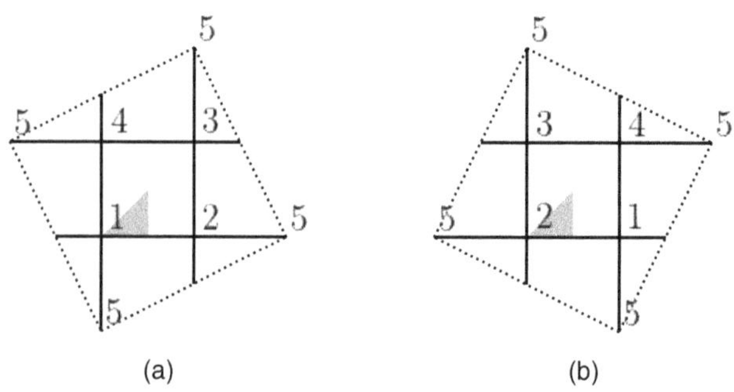

Figure 2.19 (a) Left-handed and (b) right-handed versions of the same polyhedron.

We can distinguish left- and right-handed versions of the same chiral polytope $\mathcal{P}$ by choosing a base flag Φ, and hence thinking of the rooted polytope $\mathcal{P}_{(\Phi)}$. If we choose the grey flags in Figure 2.19 as base flags then the two representations of $\{4,4\}_{(2,1)}$ are non-isomorphic as rooted polytopes. Indeed, if we intend to give an isomorphism ϕ from the map in (a) to the map in (b) mapping the grey flag of the one in (a) to the grey flag of the one in (b), then ϕ maps vertex 1 to vertex 2, the edge between 1 and 2 to the edge between 1 and 2, and the face containing vertices 1, 2, 3 and 4 to the face containing vertices 1, 2, 3 and 4. From here it follows that $3\gamma = 4$. On the other hand, the edge between 1 and 5 must be mapped to the only edge at vertex 2 (the vertex of the base flag) in the 2-face containing the base edge (between vertices 1 and 2) that is not the 2-face in the base flag. In the map in Figure 2.19(b) this is the edge between vertices 2 and 4, implying that $5\gamma = 4$. This contradicts that $3\gamma = 4$.

Formally speaking, the left and right-handed versions of a chiral polytope $\mathcal{P}$ correspond to choices of base flags in one or another flag orbit. If two flags Φ and Ψ belong to distinct orbits then the rooted polytopes $\mathcal{P}_{(\Phi)}$ and $\mathcal{P}_{(\Psi)}$ are non-isomorphic, and the reflection on a mirror can be understood as looking the polytope from an adjacent flag (the 0-adjacent flag in the case of Figure 2.19). From this new flag the rooted polytope may have some distinct properties; this will become evident in Section 3.1.

The left- and right-handed versions of a chiral polytope were called *enantiomorphic forms* in [248], but there the understanding is that the enantiomorphic form of a left-handed chiral polytope is the right-handed chiral polytope, and vice versa. We prefer to understand the form of the left hand to be the left one, and say that the *chirality form* of a rooted chiral polytope $\mathcal{P}_{(\Phi)}$ is precisely the one determined by the choice of Φ as base flag. Then we shall refer to $\mathcal{P}_{(\Phi^0)}$ as the *enantiomorphic image*, or simply the *enantiomorph*, of $\mathcal{P}_{(\Phi)}$.

According to the discussion in Section 2.2, the dual of a chiral polytope is again a chiral polytope, and so the class of chiral polytopes may admit self-dual elements. This is the case, since the polyhedron $\{4,4\}_{(2,1)}$ is self-dual. Indeed, when labeling the vertices and 2-faces of $\{4,4\}_{(2,1)}$ as in Figure 2.20, the bijection

$$1 \mapsto a, \quad 2 \mapsto b, \quad 3 \mapsto d, \quad 4 \mapsto c, \quad 5 \mapsto e,$$

determines a duality in which faces a, b, c, d and e are mapped to vertices 3, 1, 5, 4 and 2, respectively.

In fact, $\{4,4\}_{(2,1)}$ is improperly self-dual. With the duality given above, the flag consisting of vertex 1, the edge F_1 between vertices 1 and 2, and the 2-face

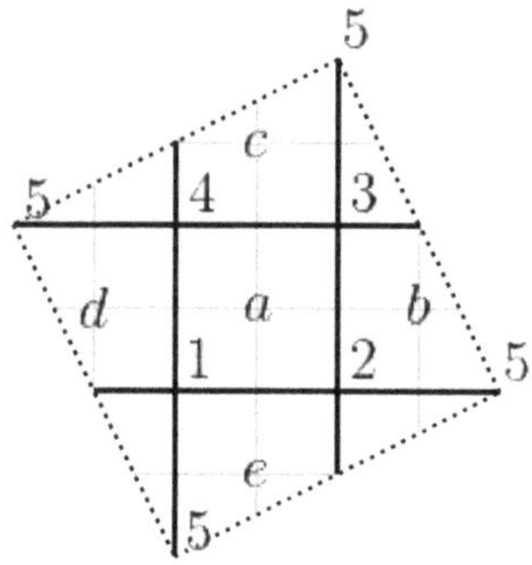

Figure 2.20 The polyhedron $\{4,4\}_{(2,1)}$ and its dual.

a is mapped to the 2-face a, the edge between 2 and 3, and vertex 3. As can be seen in Figure 2.18, those two flags are in distinct orbits.

Exercise 2.3.24 Determine if the chiral maps defined by Heffter, constructed from the polygons in (2.7) and (2.8) are self-dual. If they are, determine if they are properly or improperly self-dual.

Chiral properly self-dual polytopes always posses a polarity, as shown next.

Proposition 2.3.25 *Let $\mathcal{P}$ be a chiral polytope with base flag Φ.*

(a) If $\mathcal{P}$ is properly self-dual then $\overline{\Gamma}(\mathcal{P})$ contains a polarity.
(b) If $\mathcal{P}$ is improperly self-dual then $\overline{\Gamma}(\mathcal{P})$ contains a duality δ mapping the base flag Φ to Φ^0. Moreover, the order of δ in $\overline{\Gamma}(\mathcal{P})$ is 4.

Proof By Lemma 2.2.31, if $\mathcal{P}$ is properly self-dual then for any given flag Φ there exists a duality δ such that $\Phi\delta = \Phi$. This implies that δ^2 is an automorphism that fixes Φ, and therefore δ is a polarity.

If $\mathcal{P}$ is improperly self-dual then for any flag Φ and any duality η we have that Φ and $\Phi\eta$ belong to distinct orbits. It follows that Φ^0 and $\Phi\eta$ are in the same orbit, and hence there exists $\gamma \in \Gamma(\mathcal{P})$ such that $\Phi\eta\gamma = \Phi^0$. Then the duality $\delta := \eta\gamma$ satisfies that $\Phi\delta = \Phi^0$, as desired. Now, $\Phi\delta^2 = \Phi^0\delta = (\Phi\delta)^{n-1} = \Phi^{0,n-1}$. Since $\mathcal{P}$ is chiral, its rank is at least 3. This, together with the fact that δ^2 is an automorphism, implies that

$$\Phi\delta^4 = \Phi(\delta^2)^2 = \Phi^{0,n-1}\delta^2 = (\Phi\delta^2)^{0,n-1} = \Phi^{0,n-1,0,n-1} = \Phi.$$

We conclude that the order of δ is 4. ■

Not every improperly self-dual chiral polytope possesses a polarity (see [138, p. 132]), but many of them do.

Proposition 2.3.26 *If $\mathcal{P}$ is an improperly self-dual chiral n-polytope for some odd n then it possesses a polarity.*

Proof By replacing 0 by $\frac{n-1}{2}$ in the proof of the second item of Proposition 2.3.25 we can show that for a given flag Φ, $\overline{\Gamma}(\mathcal{P})$ contains a duality δ mapping a given flag Φ to $\Phi^{\frac{n-1}{2}}$. Then

$$\Phi(\delta^2) = \left(\Phi^{\frac{n-1}{2}}\right)\delta = (\Phi\delta)^{\frac{n-1}{2}} = \Phi^{\frac{n-1}{2},\frac{n-1}{2}} = \Phi,$$

and therefore δ is a polarity. ■

Other conditions for improperly self-dual chiral n-polytopes to possess a polarity are mentioned in [138]. However, our lack of examples of self-dual n-polytopes for $n \geq 6$ limits the motivation on a further study of improperly self-dual chiral n-polytopes with no polarities.

Problem 8 Exhibit a properly self-dual and an improperly self-dual chiral n-polytope for every $n \geq 6$.

Problem 9 Determine all values of n for which there exist self-dual chiral n-polytopes that possess no polarity.

2.4 Toroidal Rotary Polytopes

Before continuing with other structural results we want to introduce more examples of chiral polytopes. With this in mind, in this section we describe the families of rotary $(n+1)$-polytopes whose order complex is the n-torus. For a full understanding of the ideas presented here some knowledge on geometry is required. The main references for this section are [59, 185, 234].

2.4.1 Toroidal Polytopes as Quotients of Tessellations

The n-torus $\mathbb{T}_n$ is the orbit space of $\mathbb{E}^n$ under the action of a group generated by n translations by linearly independent vectors. Such groups are called *lattice groups*.

If $\bar{v}_1, \ldots, \bar{v}_n$ are the translation vectors of the generators of a lattice group Λ then the most common way to represent the n-torus $\mathbb{E}/\Lambda$ is as the parallelepiped $\Pi = \Pi(\bar{v}_1, \ldots, \bar{v}_n)$ determined by the vectors $\bar{v}_1, \ldots, \bar{v}_n$. In this parallelepiped, each pair of opposite facets consists of translates by the generator of Λ that is not a direction vector of these facets, and $\mathbb{T}_n$ as a topological space arises by identifying every point in a facet of Π with its translate in the opposite facet. Points in the boundary of Π that belong to more than one facet are identified multiple times.

The representation of $\mathbb{T}_n$ as a parallelepiped is easy to visualise when $n = 2$, where Π is a parallelogram with opposite edges identified orientably; that is, a point in the edge with direction vector $\bar{v}_1$ is identified with its image under the translation by $\pm\bar{v}_2$. The four corners of Π become identified as one point of $\mathbb{T}_n$. For more details on compact manifolds as quotients of Euclidean, spherical or hyperbolic spaces see [234].

We call *toroidal polytopes* those whose order complex is $\mathbb{T}_n$ for some n. If a polytope $\mathcal{P}$ has order complex $\mathbb{T}_n$ then its rank must be $n + 1$, and the union of the simplices having as a corner a given facet or a given vertex must be homeomorphic to Euclidean n-dimensional balls. It follows that the order complexes of the facets and vertex-figures of $\mathcal{P}$ must be homeomorphic to $(n - 1)$-spheres, and therefore the facets of $\mathcal{P}$ can be seen as simply connected tiles of the torus. In this way $\mathcal{P}$ can be understood as a tessellation of $\mathbb{T}_n$, which abusing notation we may also call $\mathcal{P}$.

Every tessellation $\mathcal{P}$ of $\mathbb{T}_n = \mathbb{E}^n/\Lambda$ is the quotient of a tessellation $\mathcal{T}_{\mathcal{P}}$ of $\mathbb{E}^n$ invariant under the lattice group Λ. The tessellation $\mathcal{T}_{\mathcal{P}}$ can be constructed by representing $\mathcal{P}$ as a tiling of the parallelepiped Π determined by the translation vectors $\bar{v}_1, \dots, \bar{v}_n$ of a generating set of Λ (perhaps with partial i-faces in the boundary), and then piling up copies of the tessellation of Π in the n directions given by $\bar{v}_1, \dots, \bar{v}_n$. Figure 2.21 illustrates the tessellation $\mathcal{T}_{\mathcal{P}}$ of $\mathbb{E}^2$ by squares, constructed by piling up copies of the parallelogram representing $\{4,4\}_{(2,1)}$ in the two directions determining this fundamental region.

The tessellation $\mathcal{T}_{\mathcal{P}}$ has Λ as a group of translations, and is a *normal* tessellation in the sense of [109], that is, there exist $r, R > 0$ such that every facet of $\mathcal{T}_{\mathcal{P}}$ contains a ball of radius r and is contained in a ball of radius R.

In order to study rotary toroidal polytopes we want to consider only the tessellations of $\mathbb{T}_n$ that are (combinatorially) rotary. In particular, they must have rotary facets, rotary vertex-figures, rank $n + 1 \geq 3$, and at most two flag orbits. Such objects satisfy the hypothesis of the following theorem.

Theorem 2.4.1 *Let $\mathcal{P}$ be an equivelar toroidal $(n + 1)$-polytope. Then $\mathcal{P}$ is isomorphic to the quotient of a regular tessellation of $\mathbb{E}^n$ by a translation group.*

It is outside the scope of this section to prove Theorem 2.4.1. For a full understanding of the topological, geometric and combinatorial aspects of this topic we suggest [55, 77, 80, 122, 151, 163, 165, 168]. Henceforth we may

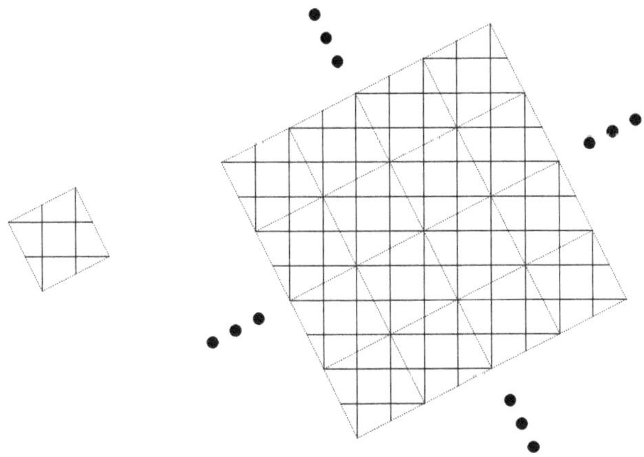

Figure 2.21 The polyhedron $\{4,4\}_{(2,1)}$ induces the tessellation by squares of the plane.

assume that $\mathcal{T}_\mathcal{P}$ is a tessellation of $\mathbb{E}^n$ by tiles isometric to convex regular polytopes with regular convex vertex-figures.

According to Theorem 2.4.1 and the results in [55], if $n = 3$ or $n \geq 5$ then the only possible type for toroidal $(n+1)$-polytopes is $\{4, 3^{n-2}, 4\}$. (Here, and subsequently, we use the convention that if an entry p of the type is repeated k consecutive times we substitute the string $p, \ldots, p$ with p^k.) There are the additional possibilities of types $\{3,6\}$ and $\{6,3\}$ for $n = 2$, and of types $\{3,4,3,3\}$ and $\{3,3,4,3\}$ for $n = 4$.

Each automorphism γ of $\mathcal{P}$ can be understood as an isometry of $\mathbb{T}_n$ that maps i-faces of $\mathcal{P}$ to i-faces of $\mathcal{P}$. Then γ sends any given ball $\mathcal{B}$ of $\mathbb{T}_n$ to some other ball $\mathcal{B}\gamma$. We shall abuse notation and use the same notation for the ball $\mathcal{B}$ in $\mathbb{T}_n$ and the region of $\mathbb{E}^n$ contained in the parallelepiped Π corresponding to the points of $\mathcal{B}$.

If $\mathcal{B}$ and $\Pi = \Pi(\bar{v}_1, \ldots, \bar{v}_n)$ are chosen so that $\mathcal{B}$ and $\mathcal{B}\gamma$ are contained in the interior of Π then there is a unique isometry $f_{\gamma,\mathcal{B}}$ of $\mathbb{E}^n$ that maps $\mathcal{B}$ to $\mathcal{B}\gamma$ sending every $x \in \mathcal{B}$ to $x\gamma$. The isometry $f_{\gamma,\mathcal{B}}$ must map the facets of $\mathcal{T}_\mathcal{P}$ intersecting $\mathcal{B}$ to other facets, and since there is only one way to reconstruct $\mathcal{T}_\mathcal{P}$ from one given facet, $f_{\gamma,\mathcal{B}}$ must be a symmetry of $\mathcal{T}_\mathcal{P}$.

In $\mathbb{T}_n$, γ sends geodesics of finite length to geodesics of the same length. If a geodesic ℓ is closed and has finite length d then it must be the projection to Π_n of a line segment l with length d. Furthermore, the endpoints of that segment must differ by some $\bar{u} \in \Lambda$ of norm d. On the other hand, $\ell\gamma$ is also a closed geodesic of Π_n of length d, implying that the endpoints of $\ell f_{\gamma,\mathcal{B}}$ also differ by a vector of Λ of norm d; in fact, they must differ by $\bar{u} f_{\gamma,\mathcal{B}}$.

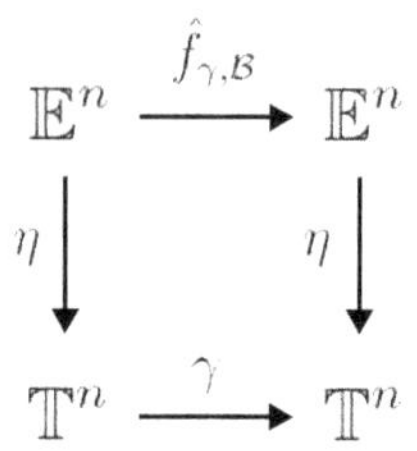

Figure 2.22 Every automorphism of $\mathcal{T}_{\mathcal{P}}/\Lambda$ induces an isometry of $\mathbb{E}^n$ that respects the projection η.

In particular, if $\bar{v}_1, \ldots, \bar{v}_n$ are the generators of Λ then each $\bar{v}_i f_{\gamma,\mathcal{B}}$ is an integer combination of $\bar{v}_1, \ldots, \bar{v}_n$. It follows that the parallelepiped $\Pi f_{\gamma,\mathcal{B}}$ is itself a fundamental region for Λ. As a consequence, if $\eta\colon \mathbb{E}^n \to \mathbb{T}_n$ is the quotient by the lattice group Λ then the diagram in Figure 2.22 commutes. Furthermore, if $\mathcal{B}'$ is another ball then $f_{\gamma,\mathcal{B}'} = f_{\gamma,\mathcal{B}t}$ for some translation $t \in \Lambda$.

Exercise 2.4.2 Let $\gamma \in \Gamma(\{4,4\}_{(2,1)})$ be given by the permutation (2, 3) (4, 5) of the vertices, when labelled as in Figure 2.18(a). Choose a ball $\mathcal{B}$ of $\mathbb{T}_2$ and find the isometry $f_{\gamma,\mathcal{B}}$ of $\mathbb{E}^2$.

So far we have assigned to any $\gamma \in \Gamma(\mathcal{P})$ an isometry $f_{\gamma,\mathcal{B}}$ of $\mathbb{E}^n$ that preserves $\mathcal{T}_{\mathcal{P}}$ and maps fundamental regions of Λ to fundamental regions of Λ. On the other hand, if an isometry f of $\mathbb{E}^n$ maps fundamental regions of Λ to fundamental regions of Λ then the assignment $\gamma_f\colon \mathbb{T}_n \to \mathbb{T}_n$ that for $x \in \mathbb{E}^n$ sends $x\Lambda$ to $xf\Lambda$ is well defined. If in addition f preserves $\mathcal{T}_{\mathcal{P}}$ then γ_f is an automorphism of $\mathcal{P}$. Furthermore, two isometries f and $\tilde{f}$ of $\mathbb{E}^n$ induce the same automorphism $\alpha_f = \alpha_{\tilde{f}}$ if and only if for each $x \in \mathbb{E}^n$, $x\tilde{f} = xft$ for some translation $t \in \Lambda$ independent of the choice of x; continuity guarantees that $\tilde{f} = ft$. The next theorem follows.

Theorem 2.4.3 *Let Λ be a lattice group of $\mathbb{E}^n$ preserving the tessellation $\mathcal{T}$ and let $\mathcal{P} = \mathcal{T}/\Lambda$. Let $K(\mathcal{T},\Lambda)$ denote the group of isometries of $\mathbb{E}^n$ that preserve $\mathcal{T}$ and map fundamental regions of Λ to fundamental regions of Λ. Then $\Gamma(\mathcal{P}) \cong K(\mathcal{T},\Lambda)/\Lambda$.*

Example 2.4.4 For convenience, the regular tessellation by squares is denoted by its type $\{4,4\}$ (a reasoning behind this convention is given in Chapter 3), and we assume that its vertices are located in the points of $\mathbb{E}^2$ with integer coordinates. We shall denote the translation by vector $\bar{u}$ by $t_{\bar{u}}$.

Let $\mathcal{P}$ be the polyhedron $\{4,4\}_{(2,1)}$. As illustrated in Figure 2.21, $\mathcal{P}$ can be understood as a quotient of the regular tessellation by squares of $\mathbb{E}^2$ by the lattice group $\langle t_{(2,1)}, t_{(-1,2)} \rangle$.

Every translation by a vector with integer coordinates is in the group $K(\{4,4\},\Lambda)$ defined in Theorem 2.4.3. If we denote by H the group of all translations by vectors with integer coordinates then $H/\Lambda \subseteq \Gamma(\mathcal{P})$. The group H/Λ is cyclic with five elements, and is generated by $t_{(1,0)}\Lambda$. It acts regularly on the vertex set of $\mathcal{P}$.

In order to determine the remaining automorphisms of $\mathcal{P}$ we only need to find a vertex stabiliser. Owing to the transitivity of H on the vertex set of $\{4,4\}$, it suffices to determine the stabiliser in $\Gamma(\mathcal{P})$ of $o\Lambda$, where o is the origin of $\mathbb{E}^2$. To this end we recall that the group of isometries of $\mathbb{E}^2$ that fixes o and preserves $\{4,4\}$ is the dihedral group D generated by a 4-fold rotation about o and the reflection about the x-axis. The four rotations in D map fundamental regions of Λ to fundamental regions of Λ; for example, the rotation f given by $(x,y)f = (-y,x)$ maps the generating set $\{(2,1),(-1,2)\}$ to the generating set $\{(-1,2),(-2,-1)\}$ (see Figure 2.23(a)).

On the other hand, none of the reflections in D map fundamental regions of Λ to fundamental regions of Λ. The images of $(2,1)$ under these reflections are $(2,-1)$, $(1,2)$, $(-2,1)$ and $(-1,-2)$, and a standard procedure shows that none of them are the translation vector of an element in Λ. If f denotes one of these reflections then γ_f would assign more than one image to the point $o\Lambda \in \mathbb{T}_2$, and so it does not induce an automorphism of $\mathcal{P}$. The image of one fundamental region of Λ under one of the reflections is illustrated in Figure 2.23(b).

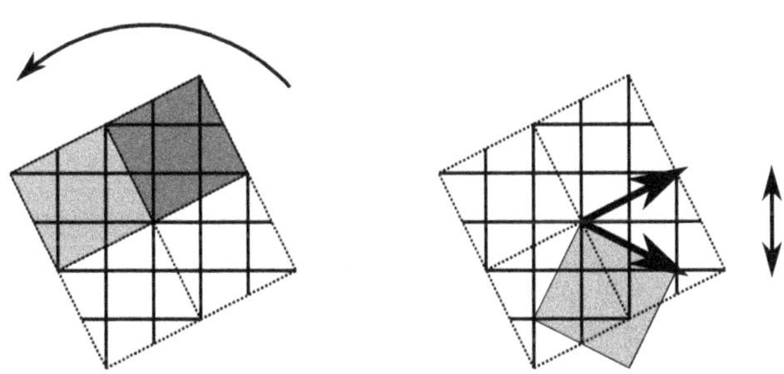

(a) (b)

Figure 2.23 (a) A 4-fold rotation maps fundamental regions to fundamental regions; (b) a horizontal reflection does not map fundamental regions to fundamental regions.

This shows that $\Gamma(\mathcal{P})$ contains the 20 elements that arise as compositions of an element in $C\Lambda/\Lambda$ and an element in H/Λ, where C is the cyclic index 2 subgroup of D. By colouring the flags of $\{4,4\}$ in one orbit under the action of the group $K(\{4,4\},\Lambda)$, we can see that this group consists precisely of all orientation preserving symmetries of $\{4,4\}$.

Exercise 2.4.5 Find the lattice group Λ and the group $K(\{4,4\},\Lambda)$ defined in Theorem 2.4.3 for the two-orbit polyhedra in Example 2.2.22.

The automorphism group of every equivelar toroidal polyhedron $\mathcal{P} = \mathcal{T}_\mathcal{P}/\Lambda$ is rather large. Every translation and every half-turn of $\mathbb{E}^2$ map fundamental regions of Λ to fundamental regions of Λ. As a consequence, all translations and half-turns in the symmetry group of $\mathcal{T}_\mathcal{P}$ belong to the group $K(\mathcal{T}_\mathcal{P},\Lambda)$. Similarly, if $\mathcal{P}$ is an $(n+1)$-polytope in $\mathbb{E}^n$ for some $n \geq 3$ then every translation and every central inversion about a vertex or about a centre of facet belong to $K(\mathcal{T}_\mathcal{P},\Lambda)$. We state next a couple of consequences of this fact.

Proposition 2.4.6 *Every equivelar toroidal polyhedron is vertex-transitive.*

Proof The translation groups of the regular tessellations with types $\{3,6\}$ and $\{4,4\}$ act vertex-transitively. If $\mathcal{P}$ is an equivelar toroidal polytope of one of these types, each such translation t induces an automorphism γ_t of $\mathcal{P}$, and the group

$$\langle \gamma_t : t \text{ is a translation} \rangle$$

acts vertex-transitively on $\mathcal{P}$.

The translation group of the regular tessellation with type $\{6,3\}$ induces two orbits on vertices, which are interchanged by every half-turn (about the midpoint of an edge or about the centre of a hexagon). Here the group

$$\langle \gamma_t : t \text{ is a translation or a half-turn} \rangle$$

acts vertex-transitively on every equivelar toroidal polytope of type $\{6,3\}$. ■

Lemma 2.4.7 *With the notation above, if the translation group of $\mathcal{T}_\mathcal{P}$ acts vertex-transitively and the vertex stabilisers of $\mathcal{T}_\mathcal{P}$ in $K(\mathcal{T}_\mathcal{P},\Lambda)$ have index k in the vertex stabilisers in $\Gamma(\mathcal{T}_\mathcal{P})$ then $\mathcal{P}$ is a k-orbit polytope.*

Proof Since every flag of $\mathcal{P}$ is a flag orbit under the action of Λ, the number of flags of $\mathcal{P}$ is the index $[\Gamma(\mathcal{T}_\mathcal{P})\colon \Lambda]$ of Λ in $\Gamma(\mathcal{T}_\mathcal{P})$. On the other hand, $\Gamma(\mathcal{P}) = K(\mathcal{T}_\mathcal{P},\Lambda)/\Lambda$. It follows that the number of flag orbits of $\mathcal{P}$ is the index

$$[\Gamma(\mathcal{T}_{\mathcal{P}})/\Lambda : K(\mathcal{T}_{\mathcal{P}},\Lambda)/\Lambda] = [\Gamma(\mathcal{T}_{\mathcal{P}}) : K(\mathcal{T}_{\mathcal{P}},\Lambda)].$$

Vertex-transitivity of the groups $\Gamma(\mathcal{T}_{\mathcal{P}})$ and $K(\mathcal{T}_{\mathcal{P}},\Lambda)$ guarantees that the number of flag orbits of $\mathcal{P}$ is the desired quotient of the sizes of the vertex stabilisers. ∎

Some techniques similar to those discussed in this section have been used in [30, 136] to work on toroidal polytopes that are not rotary.

2.4.2 Chiral Toroidal Polyhedra

There are three regular tessellations of $\mathbb{E}^2$. They are the tessellation $\{4,4\}$ by squares, the tessellation $\{3,6\}$ by equilateral triangles and the tessellation $\{6,3\}$ by regular hexagons. These tessellations are universal in a sense that will be explained in Chapter 3, and hence we denote these tessellations by their Schläfli types. Theorem 2.4.1 implies that every rotary toroidal polyhedron must have type $\{4,4\}$, $\{3,6\}$ or $\{6,3\}$, and must be isomorphic to a quotient of one of these tessellations by a translation group.

The tessellations $\{3,6\}$ and $\{6,3\}$ of $\mathbb{E}^2$ are dual to each other in the sense that one can be obtained from the other by placing a vertex in the centre of each 2-face and joining two of them by a line segment whenever the corresponding two 2-faces share an edge. If $\mathcal{P}$ is a toroidal polyhedron with type $\{6,3\}$ we can represent its combinatorial dual by the above geometric procedure to obtain a tessellation $\mathcal{P}^\delta$ with type $\{3,6\}$ of $\mathbb{T}^2$. In this situation, if $\mathcal{P} = \{6,3\}/\Lambda$ for some lattice group Λ then $\mathcal{P}^\delta = \{3,6\}/\Lambda$, since the underlying torus is geometrically the same for both polyhedra. It follows that every rotary toroidal polyhedron with type $\{6,3\}$ is the dual of a rotary toroidal polyhedron with type $\{3,6\}$. In what follows we shall restrict ourselves to the study of rotary toroidal polyhedra of types $\{4,4\}$ and $\{3,6\}$. The translation groups of the tessellations $\{4,4\}$ and $\{3,6\}$ are vertex-transitive, implying that the automorphism group of every toroidal polyhedra of these types acts vertex-transitively. According to Lemma 2.4.7, to determine the symmetry type of each toroidal polyhedron of one of these types it suffices to study the vertex stabilisers in their automorphism groups.

We first consider the rotary toroidal polyhedra with type $\{4,4\}$. To this end we assume that the vertices of $\{4,4\}$ are the points in $\mathbb{Z}^2$. The stabiliser in $\Gamma(\{4,4\})$ of the origin o is a dihedral group $\mathcal{D}_4$ with eight elements. Four of the elements of $\mathcal{D}_4$ are line reflections, each of which maps some flag Φ of $\{4,4\}$ to some adjacent flag Φ^i. Hence, if any of these reflections is in $K(\{4,4\},\Lambda)$ then $\mathcal{P}$ has adjacent flags $\Phi\Lambda$ and $\Phi^i\Lambda$ in the same orbit. Therefore, if $\mathcal{P}$ is chiral then the stabiliser of o in $K(\{4,4\},\Lambda)$ must be contained in the rotation

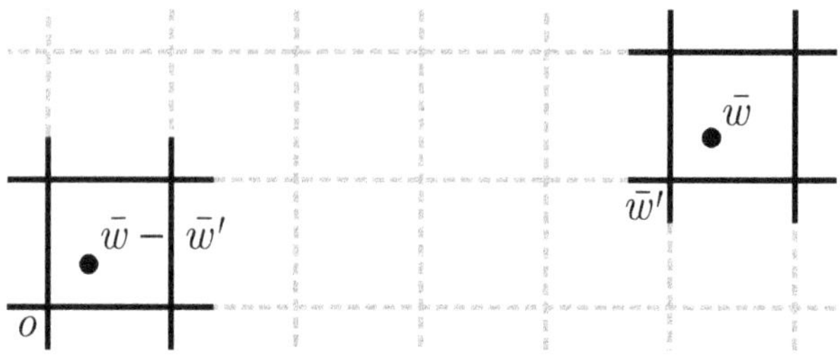

Figure 2.24 The vectors $\bar{w}$ and $\bar{w} - \bar{w}'$.

subgroup $\mathcal{D}_4^+$ of $\mathcal{D}_4$, and by Lemma 2.4.7, it must be precisely $\mathcal{D}_4^+$. If $\mathcal{P}$ is regular then the stabiliser of o in $K(\{4,4\},\Lambda)$ must be $\mathcal{D}_4$. In either of these two cases, Λ must be invariant under every element in $\mathcal{D}_4^+$.

We let $\mathcal{P} = \{4,4\}/\Lambda$ be a rotary toroidal polyhedron determined by the lattice group Λ, denote by $t_{\bar{u}}$ the translation by any vector $\bar{u}$, and let $\bar{v} = (a,b)$ be a non-zero translation vector of smallest norm among those of the translations in Λ. The images under $\mathcal{D}_4^+$ of $\bar{v}$ are $\bar{v}$, $(-b,a)$, $(-a,-b)$ and $(b,-a)$. This forces $\langle t_{\bar{v}}, t_{(-b,a)}\rangle \subset \Lambda$. The orbit of o under $\langle t_{\bar{v}}, t_{(-b,a)}\rangle$ is the vertex set of a tessellation by squares having some of its edges parallel to $\bar{v}$ and some perpendicular to $\bar{v}$.

We claim that $\Lambda = \langle t_{\bar{v}}, t_{(-b,a)}\rangle$. Assume to the contrary that there exists $t_{\bar{w}} \in \Lambda \setminus \langle t_{\bar{v}}, t_{(-b,a)}\rangle$ and let $\bar{w}' \in o\langle t_{\bar{v}}, t_{(-b,a)}\rangle$ such that $d(\bar{w},\bar{w}') \leq d(\bar{w},\bar{u})$ for every $\bar{u} \in o\langle t_{\bar{v}}, t_{(-b,a)}\rangle$. Then $t_{\bar{w}} t_{\bar{w}'}^{-1} \in \Lambda$ is the translation by the vector $\bar{w} - \bar{w}'$, which is closer to o than to any other translation vector of Λ (see Figure 2.24). Since the points $o\langle t_{\bar{v}}, t_{(-b,a)}\rangle$ are the vertices of a tessellation by squares then the distance $d(o, \bar{w} - \bar{w}')$ must be at most $\frac{1}{\sqrt{2}} d(o,\bar{v})$, contradicting the choice of $\bar{v}$.

As we just saw, every rotary toroidal polyhedron with type $\{4,4\}$ is the quotient of the tessellation $\{4,4\}$ by a translation group of the form $\Lambda_{(a,b)} = \langle t_{(a,b)}, t_{(-b,a)}\rangle$ for some non-zero vector (a,b). It is customary to denote $\{4,4\}/\Lambda_{(a,b)}$ by $\{4,4\}_{(a,b)}$, since for rotary polyhedra $\{4,4\}/\Lambda$ one vector suffices to determine the lattice group Λ.

In the other direction, for any non-zero vector $\bar{v} = (a,b)$ we may consider the lattice group $\Lambda = \langle t_{\bar{v}}, t_{(-b,a)}\rangle$. We denote by Φ_0 the flag of $\{4,4\}$ with vertex at o, edge beween o and $(0,1)$ and 2-face the square determined by o, $(1,0)$, $(0,1)$ and $(1,1)$. The isometries S_1 and S_2 mapping (x,y) respectively to $(1-y,x)$ and $(-y,x)$ send Φ_0 to $\Phi_0^{1,0}$ and $\Phi_0^{2,1}$. They are both rotations by an angle of $\pi/2$, and hence they map square fundamental regions of Λ to other square fundamental regions. Then S_1 and S_2 induce automorphisms of

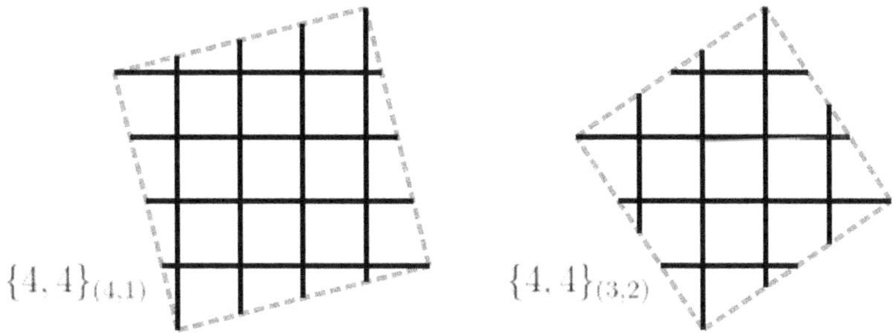

Figure 2.25 The polyhedra $\{4,4\}_{(4,1)}$ and $\{4,4\}_{(3,2)}$.

$\{4,4\}/\Lambda$ mapping $\Phi_0\Lambda$ to $(\Phi_0\Lambda)^{1,0}$ and $(\Phi_0\Lambda)^{2,1}$, respectively, implying that $\{4,4\}/\Lambda$ is rotary. Moreover, $\{4,4\}/\Lambda$ is a polyhedron as long as its faces are squares and its vertex-figures are squares. For this condition to be satisfied, it suffices to discard the possibilities that $\bar{v}$ is a vector determining an edge or a diagonal of a square of $\{4,4\}$, or in other words, to require that $|\bar{v}| \geq 2$. Figure 2.25 shows the rotary polyhedra $\{4,4\}_{(4,1)}$ and $\{4,4\}_{(3,2)}$.

The quotient $\{4,4\}/\Lambda$ is regular if and only if it is rotary and the reflection R_x about the x-axis is in $K(\{4,4\},\Lambda)$. Assume that $\Lambda = \langle t_{(a,b)}, t_{(-b,a)}\rangle$; then $R_x \in K(\{4,4\},\Lambda)$ if and only if $(a,-b)$ is an integer combination of (a,b) and $(-b,a)$, or equivalently, if and only if $(a,-b)$ is a vertex of the tessellation by squares whose vertices are the integer combinations of (a,b) and $(-b,a)$. In this tessellation, the four neighbours of o are (a,b), $(-b,a)$, $(-a,-b)$ and $(b,-a)$, and these are the only points with norm a^2+b^2. It follows that $\{4,4\}/\Lambda$ is regular if and only if $(a,-b)$ is one of these four points, which occurs if and only if either $a = 0$, $b = 0$ or $a = \pm b$.

We summarise the previous discussion in the next theorem.

Theorem 2.4.8 *Let Λ be a lattice group preserving the regular tessellation $\{4,4\}$ by squares of $\mathbb{E}^2$. The quotient $\{4,4\}/\Lambda$ is rotary if and only if $\Lambda = \langle t_{(a,b)}, t_{(-b,a)}\rangle$ for some non-zero vector (a,b), and it is regular if and only if $ab(b-a) = 0$. Furthermore, it is a polyhedron if and only if $|(a,b)| \geq 2$.*

If the group Λ collapses two or more vertices of a square into the same vertex of $\{4,4\}/\Lambda$ the resulting structure can still be understood as a map on the torus, where the closure of a 2-face is no longer a disk. This identification occurs precisely when Λ is the lattice group induced either by the translation $t_{(1,0)}$ or the translation $t_{(1,1)}$. The corresponding maps are regular and are illustrated in Figure 2.26.

There are two natural ways to give coordinates to the tessellation $\{3,6\}$. One of them is by using two perpendicular axes as for the tessellation $\{4,4\}$, so that

Figure 2.26 The maps $\{4,4\}_{(1,0)}$ and $\{4,4\}_{(1,1)}$ are not polyhedral.

the vertices of one of the triangles are $(0,0)$, $(1,0)$ and $\left(\frac{1}{2}, \frac{\sqrt{3}}{2}\right)$. In this case many vertices have non-integer coordinates, and perhaps that is the reason why the other coordinate system, described next, is used in most publications.

When dealing with the tessellations $\{3,6\}$ and $\{6,3\}$ we shall use the ordered basis $\beta := \left\{(1,0), \left(\frac{1}{2}, \frac{\sqrt{3}}{2}\right)\right\}$, so that the coordinate axes describe an angle of $\pi/3$. We may assume that the coordinates of the vertices of $\{3,6\}$ are the points of $\mathbb{E}^2$ with integer coordinates with respect to the basis β.

The stabiliser in $\Gamma(\{3,6\})$ of o is a dihedral group $\mathcal{D}_6$ with 12 elements, generated by a 6-fold rotation about o and the reflection about the x-axis. Since we are using the basis β, the matrices of these isometries are respectively

$$\begin{pmatrix} 0 & -1 \\ 1 & 1 \end{pmatrix} \quad \text{and} \quad \begin{pmatrix} 1 & 1 \\ 0 & -1 \end{pmatrix},$$

even though they are not orthogonal matrices.

For the same reason given when studying toroidal polyhedra with type $\{4,4\}$, the map $\{3,6\}/\Lambda$ is rotary if and only if the stabiliser of o in the group $K(\{3,6\},\Lambda)$ (as defined in Theorem 2.4.3) contains the orientation-preserving subgroup $\mathcal{D}_6^+$ of $\mathcal{D}_6$, and it is regular if and only if the stabiliser of o in $K(\{3,6\},\Lambda)$ is $\mathcal{D}_6$.

If the stabiliser of o in $K(\{3,6\},\Lambda)$ contains $\mathcal{D}_6^+$ and $\bar{v} := (a,b)$ is a non-zero translation vector of an element of Λ with smallest norm, then the images under $\mathcal{D}_6^+$ of $\bar{v}$ are $\bar{v}$, $(-b,a+b)$, $(-a-b,a)$, $-\bar{v}$, $-(-b,a+b)$ and $-(-a-b,a)$. This forces $\langle t_{(a,b)}, t_{(-b,a+b)}\rangle \subseteq \Lambda$. In fact, $\langle t_{(a,b)}, t_{(-b,a+b)}\rangle = \Lambda$, and the argument for this is very similar to that for the case of type $\{4,4\}$. In the present case the set $o\langle t_{(a,b)}, t_{(-b,a+b)}\rangle$ is the vertex set of a tessellation by equilateral triangles with edges parallel to (a,b), $(-b,a+b)$ and $(-a-b,a)$ (see Figure 2.27(a) for the case when $(a,b) = (2,1)$). If there were a vector $\bar{w}$ such that $t_{\bar{w}} \in \Lambda \setminus \langle t_{(a,b)}, t_{(-b,a+b)}\rangle$, we may assume that $\bar{w}$ has smallest norm possible among such vectors. We can take as the fundamental region of the group Λ the union of two triangles that share an edge with direction vector $(-a-b,a)$ (see Figure 2.27(b) for two fundamental regions in the case when

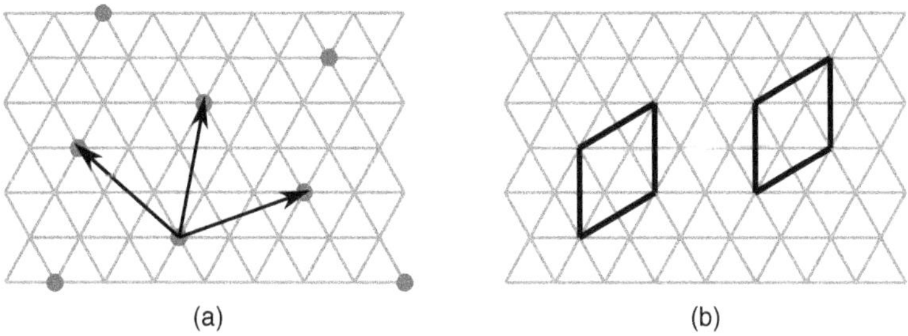

Figure 2.27 (a) The vectors $(2,1)$, $(-1,3)$ and $(-3,2)$, and (b) two fundamental regions of the translation group generated by $(1,1)$ and $(-1,2)$.

$(a,b) = (1,0)$). The distance of any point of such a fundamental region to the closest corner of the region is at most $\frac{\sqrt{3}}{2}|\bar{v}|$. By taking the composition of $t_{\bar{w}}$ with some translation in $\langle t_{(a,b)}, t_{(-b,a+b)}\rangle$ we may find $t_{\bar{u}} \in \Lambda$ (the analogue of $t_{\bar{w}-\bar{w}'}$ in the case of type $\{4,4\}$) such that $|\bar{u}| \leq \frac{\sqrt{3}}{2}|\bar{v}|$, contradicting the choice of $\bar{v}$. Therefore, every rotary map with type $\{3,6\}$ is $\{3,6\}/\langle t_{(a,b)}, t_{(-b,a+b)}\rangle$ for some non-zero vector (a,b).

We show next that the converse is also true, that is, if (a,b) is a non-zero vector then the map $\{3,6\}/\langle t_{(a,b)}, t_{(-b,a+b)}\rangle$ is rotary. We define Φ_0 to be the flag containing o, the edge between o and $(0,1)$, and the triangle containing the base edge and the vertex $(1,0)$. The automorphisms S_1 and S_2 of $\{3,6\}$ mapping Φ_0 to $\Phi_0^{1,0}$ and $\Phi_0^{2,1}$ send an arbitrary point (x,y) to $(1-x-y,x)$ and $(-y,x+y)$, respectively. These are a 3-fold and a 6-fold rotation, respectively. To verify that they send fundamental regions of $\langle t_{(a,b)}, t_{(-b,a+b)}\rangle$ to fundamental regions of $\langle t_{(a,b)}, t_{(-b,a+b)}\rangle$ it suffices to note that the latter group has hexagonal fundamental regions as those in Figure 2.28(a). The image of one such hexagon H under S_1 or S_2 is a translate of H. Hence, for $i \in \{1,2\}$ the automorphism σ_i of $\{3,6\}/\langle t_{(a,b)}, t_{(-b,a+b)}\rangle$ mapping $\Phi\Lambda$ to $\Phi S_1 \Lambda = (\Phi\Lambda)^{i,i-1}$ is the ith standard rotation, concluding the proof that this map is rotary. Consistently with the notation for rotary maps with type $\{4,4\}$, the map $\{3,6\}/\langle t_{(a,b)}, t_{(-b,a+b)}\rangle$ is denoted by $\{3,6\}_{(a,b)}$.

In Figure 2.28(b) and (c) we illustrate rhombic fundamental regions for the maps $\{3,6\}_{(3,1)}$ and $\{3,6\}_{(2,2)}$, respectively. Hexagonal fundamental regions for the same maps are shown in Figure 2.29.

We summarise the results on rotary maps of type $\{3,6\}$ in the next theorem.

Theorem 2.4.9 *Let Λ be a lattice group preserving the regular tessellation $\{3,6\}$ by triangles of $\mathbb{E}^2$. The quotient $\{3,6\}/\Lambda$ is rotary if and only if $\Lambda = \langle t_{(a,b)}, t_{(-b,a+b)}\rangle$ for some non-zero vector (a,b), and it is a regular if and*

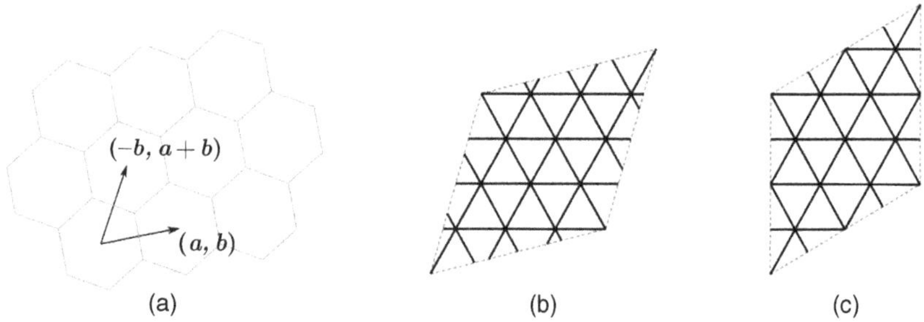

Figure 2.28 (a) Hexagonal fundamental regions of the group $\langle t_{(a,b)}, t_{(-b,a+b)}\rangle$ of translations of $\{3,6\}$; (b) rhombic fundamental region for the map $\{3,6\}_{(3,1)}$ and (c) rhombic fundamental region for the map $\{3,6\}_{(2,2)}$.

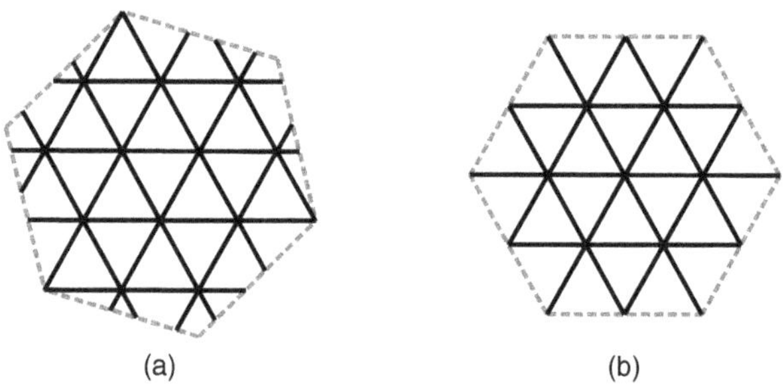

Figure 2.29 Hexagonal fundamental regions for the maps $\{3,6\}_{(3,1)}$ (a) and $\{3,6\}_{(2,2)}$ (b).

only if $ab(b-a) = 0$. *Furthermore, it is a polyhedron if and only if* $(a,b) \notin \{\pm(1,0), \pm(0,1), \pm(-1,1)\}$.

As mentioned above, the maps on the torus with type $\{6,3\}$ are duals of those of type $\{3,6\}$. We denote $\{3,6\}^{\delta}_{(a,b)}$ by $\{6,3\}_{(a,b)}$. The map $\{6,3\}_{(a,b)}$ is rotary (resp. regular, polytopal) if and only if $\{3,6\}_{(a,b)}$ is rotary (resp. regular, polytopal). The polyhedra $\{6,3\}_{(3,1)}$ and $\{6,3\}_{(2,2)}$ are shown in Figure 2.30 with rhombic and hexagonal fundamental regions.

The toroidal polyhedra $\{4,4\}_{(2,0)}$, $\{3,6\}_{(1,1)}$ and $\{3,6\}_{(2,0)}$ have double, triple and double edges, respectively (see Figure 2.31). This is a consequence of the fact that there are translations of the corresponding lattice groups Λ that send a neighbour of a given vertex of the Euclidean tessellation to another of its neighbours. These three lattice groups are the only ones satisfying this property, and hence $\{4,4\}_{(2,0)}$, $\{3,6\}_{(1,1)}$ and $\{3,6\}_{(2,2)}$ are the only toroidal rotary polyhedra with multiple edges, up to isomorphism.

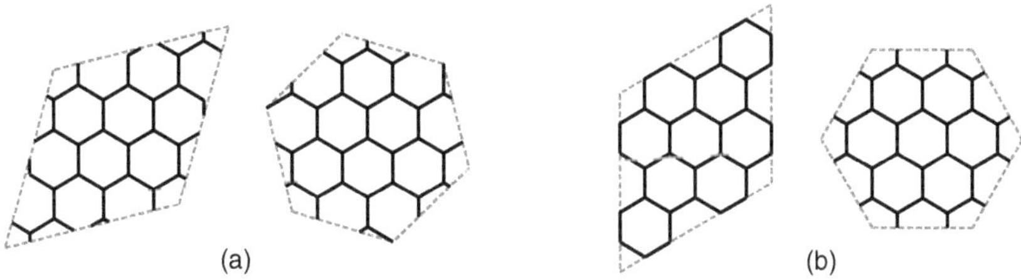

Figure 2.30 Toroidal rotary polyhedra $\{6,3\}_{(3,1)}$ (a) and $\{6,3\}_{(2,2)}$ (b).

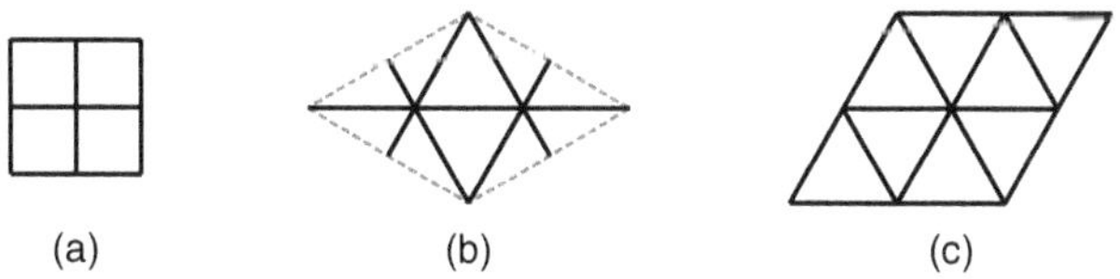

Figure 2.31 The toroidal poyhedra $\{4,4\}_{(2,0)}$ (a), $\{3,6\}_{(1,1)}$ (b) and $\{3,6\}_{(2,0)}$ (c) have multiple edges.

Two rotary toroidal polyhedra $\{p,q\}_{(a,b)}$ and $\{p,q\}_{(a',b')}$ are isomorphic if and only if the lattice groups $\Lambda_{(a,b)}$ and $\Lambda_{(a',b')}$ generated by (a,b) and by (a',b') satisfy that $g^{-1}\Lambda_{(a,b)}g = \Lambda_{(a',b')}$, for some isometry g preserving the tessellation $\{p,q\}$. The mapping ϕ_g from the faces of $\{p,q\}_{(a,b)}$ to the faces of $\{p,q\}_{(a',b')}$ defined by $\phi_g(F\Lambda_{(a,b)}) = Fg\Lambda_{(a',b')}$ determines an isomorphism. If $\{p,q\} = \{4,4\}$ then the two polyhedra are isomorphic if and only if $(a',b') \in \{\pm(a, \pm b), \pm(b, \pm a)\}$; if $\{p,q\} \in \{\{3,6\},\{6,3\}\}$ then the two polyhedra are isomorphic if and only if $(a',b') \in \{\pm(a,b), \pm(b,a), \pm(-b,a+b), \pm(a+b,-b), \pm(a+b,-a), \pm(-a,a+b)\}$.

If $ab(a-b) \neq 0$ and we fix a base flag Φ_0 of the regular tessellation $\{p,q\}$ of $\mathbb{E}^2$ then we determine the chirality form of the map $\{p,q\}_{(a,b)}$ by understanding it as the rooted polytope $(\{p,q\}_{(a,b)})_{(\Phi_0\Lambda_{(a,b)})}$. For example, if we choose as base flag of $\{4,4\}$ the one containing o, the edge between o and $(1,0)$ and the 2-face centred at $\left(\frac{1}{2},\frac{1}{2}\right)$, then the polyhedra $\{4,4\}_{(2,1)}$ and $\{4,4\}_{(1,2)}$ become rooted at translates of the flags illustrated in Figure 2.19. Even though these two polyhedra are isomorphic, as rooted polyhedra they are non-isomorphic.

As a convention, when speaking about the rotary polyhedron $\{p,q\}_{(a,b)}$ we assume that $a,b \geq 0$. If it is chiral, then its enantiomorph is $\{p,q\}_{(b,a)}$, in the understanding the base flags are images of one given flag of $\{p,q\}$ under the quotients by $\Lambda_{(a,b)}$ and $\Lambda_{(b,a)}$.

The study of rotary toroidal polyhedra has a long history; for further information on this topic see [13, 14, 19, 59].

Exercise 2.4.10 Show that the vertex-figures of the polytope in Exercises 2.1.20 and 2.1.83 are isomorphic to $\{4,4\}_{(2,1)}$. Conclude that the polytope has type $\{3,4,4\}$.

Exercise 2.4.11 Show that the polytope in Exercises 2.1.20 and 2.1.83 is chiral.

Exercise 2.4.12 Show that the polytope in Exercise 2.1.21 is chiral of type $\{3,4,4,3\}$.

Exercise 2.4.13 Show that the chiral polytope in Exercise 2.1.21 is improperly self-dual.

Exercise 2.4.14 Show that the automorphism group of the polytope in Exercise 2.1.21 is isomorphic to the symmetric group on its facets.

2.4.3 Higher Ranks

Equivelar toroidal $(n+1)$-polytopes with type $\{4,3^{n-2},4\}$ are quotients by lattice groups of the Euclidean tessellation $\{4,3^{n-2},4\}$ by n-dimensional cubes of $\mathbb{E}^n$.

The number of flag orbits of toroidal polyhedra with type $\{4,4\}$ under their automorphism groups may be one, two or four. This is a consequence of Lemma 2.4.7, since the vertex stabilisers of $\{4,4\}$ in the group $K(\{4,4\},\Lambda)$ contain at least one non-trivial element, while the vertex stabilisers of $\{4,4\}$ in $\Gamma(\{4,4\})$ contain precisely eight elements. As the dimension n increases, toroidal $(n+1)$-polytopes may have larger numbers of flag orbits.

Exercise 2.4.15 Find a family $\{\mathcal{Q}_m\}_{m\geq 3}$ of equivelar toroidal polytopes with type $\{4,3^{m-2},4\}$, where $\mathcal{Q}_m$ has rank m and at least $m+1$ flag orbits.

Here our interest is on those toroidal polytopes $\{4,3^{n-2},4\}/\Lambda$ that are rotary, which severely restricts the lattice groups Λ.

Example 2.4.16 Let $s \in \mathbb{Z}^+$ and $\Lambda_{(s,0,\dots,0)}$ denote the lattice group consisting of vectors in $s\mathbb{Z}^n$. We may choose as a fundamental region for $\Lambda_{(s,0,\dots,0)}$ a cube with edge length s and edges parallel to the coordinate axes.

We claim that the quotient $\{4,3^{n-2},4\}/\Lambda_{(s,0,\dots,0)}$ is regular. To convince ourselves of this fact it suffices to note that the automorphisms of $\{4,3^{n-2},4\}$ correspond precisely to the isometries of $\mathbb{E}^n$ that preserve $\mathbb{Z}^n$. Every such isometry preserves the set of directions of the coordinate axes, and hence it maps the chosen fundamental region of $\Lambda_{(s,0,\dots,0)}$ to some fundmental region of $\Lambda_{(s,0,\dots,0)}$. It follows that if f is the automorphism of $\{4,3^{n-2},4\}$ that maps a flag Φ to a flag Ψ, then f induces an automorphism of $\{4,3^{n-2},4\}/\Lambda_{(s,0,\dots,0)}$ that maps $\Phi\Lambda_{(s,0,\dots,0)}$ to $\Psi\Lambda_{(s,0,\dots,0)}$.

The facets are genuine n-cubes and the vertex-figures are cross polytopes whenever $s \geq 2$. Proposition 2.1.44 implies that this is a sufficient and necessary condition for $\{4,3^{n-2},4\}/\Lambda_{(s,0,\dots,0)}$ to be polytopal. If $s = 1$ then every edge is incident to only one vertex; therefore the resulting structure fails the diamond condition and it is not a polytope.

In order to enumerate rotary toroidal $(n+1)$-polytopes with type $\{4,3^{n-2},4\}$ we first relate the even subgroups $\Gamma^+(\{4,3^{n-2},4\})$ and $\Gamma^+(\{4,3^{n-2},4\}/\Lambda)$.

Lemma 2.4.17 *Let $\sigma_1,\dots,\sigma_n$ be the standard rotations of $\{4,3^{n-2},4\}$ with respect to the base flag Φ_0, and let Λ be a lattice group such that $\mathcal{P} := \{4,3^{n-2},4\}/\Lambda$ is a rotary toroidal polytope. Then the functions mapping a face $F\Lambda$ of $\mathcal{P}$ to $F\sigma_i\Lambda$ are the standard rotations of $\mathcal{P}$ with respect to the base flag $\Phi_0\Lambda$.*

Proof Let F_i be the i-face of Φ_0, let $\sigma'_1,\dots,\sigma'_n$ be the standard rotations of $\mathcal{P}$, let Π be a fundamental region of Λ and let $\mathcal{B}_i$ be a ball containing $F_1\sigma_2^{-1}$ such that $\mathcal{B}_i$ and $\mathcal{B}_i\sigma_i$ are contained in the interior of Π.

For every $i \in \{1,\dots,n\}$ there exists an isometry $f_{\sigma'_i,\mathcal{B}_i}$ of $\mathbb{E}^n$ that maps $\mathcal{B}_i \subseteq \Pi$ to $\mathcal{B}_i\sigma_i \subseteq \Pi$, preserves $\{4,3^{n-2},4\}$, maps fundamental regions of Λ to fundamental regions of Λ, and makes the diagram in Figure 2.22 commute. This means that if F is a face of $\{4,3^{n-2},4\}$ then $(Ff_{\sigma'_i,\mathcal{B}_i})\Lambda = (F\Lambda)\sigma'_i$. It only remains to show that $f_{\sigma'_i,\mathcal{B}_i} = \sigma_i$.

Let $F'_1 = F_1\sigma_1$, that is, the edge in F_2 that contains F_0 other than F_1; and let $F'_0 = F_0\sigma_1$, that is, the endpoint of F'_1 other than F_0. Our choice of $\mathcal{B}_1$ implies that the isometry $f_{\sigma'_1,\mathcal{B}_1}$ maps F_0 to F'_0. From the commutativity of the diagram in Figure 2.22 we get that $F_1 f_{\sigma'_1,\mathcal{B}_1} = F'_1$, since this is the only edge in

$F'_1\Lambda$ that contains F'_0. The commutativity of the same diagram and the strong flag-connectivity of $\mathcal{P}$ imply that $F_i f_{\sigma'_1,\mathcal{B}_1} = F_i$ for $i \geq 2$, since F_i is the only representative of $F'_i\Lambda$ that contains F_0. Hence, $\Phi_0 f_{\sigma'_1,\mathcal{B}_1} = \Phi_0^{1,0} = \Phi_0\sigma_1$, and therefore $f_{\sigma'_1,\mathcal{B}_1} = \sigma_1$.

The last argument shows that if $i \geq 2$ and $j \notin \{i, i-1\}$ then $F_j f_{\sigma'_i,\mathcal{B}_i} = F_j$. Moreover, if $F'_i = F_i\sigma_i$ and $F'_{i-1} = F_{i-1}\sigma_i$ then we use that F'_i and F'_{i-1} are the only i- and $(i-1)$-faces that contain F_0 in $F'_i\Lambda$ and $F'_{i-1}\Lambda$, respectively, to show that $\Phi_0 f_{\sigma'_i,\mathcal{B}_i} = \Phi_0^{i,i-1} = \Phi_0\sigma_i$. Therefore $f_{\sigma'_i,\mathcal{B}_i} = \sigma_i$ for every i, as desired. ∎

The vertex-figures of $\{4, 3^{n-2}, 4\}$ are n-dimensional cross polytopes. We denote by $\mathcal{O}_n$ the vertex-figure at o, which can be represented by the convex hull of the neigbours of o in $\{4, 3^{n-2}, 4\}$.

According to Exercise 2.2.17, $\Gamma^+(\mathcal{O}_n)$ consists precisely of the orientation-preserving automorphisms of $\mathcal{O}_n$. Furthermore, from Example 2.2.3 we can extract that every automorphism of $\mathcal{O}_n$ is a composition of a permutation of the coordinates with the change of sign of some of the coordinates.

The isometry that changes the sign of one coordinate x_i is the reflection with respect to the hyperplane orthogonal to the x_i-axis. Hence, the isometry that changes signs to all coordinates x_i for every i in some set I preserves orientation if and only if $|I|$ is even. On the other hand, the isometry that permutes the ith and jth coordinates is the reflection with respect to the hyperplane $x_i = x_j$. Hence, the isometry that permutes the coordinates according to some $\sigma \in S_n$ preserves orientation if and only if σ is an even permutation.

The previous discussion and Lemma 2.4.17 imply the following lemma.

Lemma 2.4.18 *Let $\mathcal{P} = \{4, 3^{n-2}, 4\}/\Lambda$ be a rotary toroidal polytope. If f is either*

- *a composition of an isometry changing the sign of an even number of coordinates, with an isometry inducing an even permutation on the coordinates, or*
- *a composition of an isometry changing the sign of an odd number of coordinates, with an isometry inducing an odd permutation on the coordinates,*

then f sends fundamental regions of Λ to fundamental regions of Λ. Furthermore, if $\mathcal{P}$ is regular then the reflection with respect to the hyperplane $x_i = 0$ sends fundamental regions of Λ to fundamental regions of Λ.

Corollary 2.4.19 *Let* $\mathcal{P} = \{4, 3^{n-2}, 4\}/\Lambda$ *be a rotary toroidal polytope, let* $\mathbf{\Lambda}$ *denote the set of translation vectors of elements in* Λ *and let* $\bar{x} \in \mathbf{\Lambda}$*. Then* $\bar{x}f \in \mathbf{\Lambda}$ *for every* f *that performs an even permutation of the entries, and for every* f *that changes the sign of an even number of entries.*

In view of Lemma 2.4.18 and Corollary 2.4.19 we make a brief parenthesis to introduce some relevant lattice groups of $\mathbb{E}^n$ and some of their properties.

Following [188, Section 6D], we denote by $(s^k, 0^{n-k})$ the vector in $\mathbb{E}^n$ with the first k entries equal to s and the remaining entries equal to 0, and by $\mathbf{\Lambda}_{(s^k,0^{n-k})}$ the set of integer combinations of vectors with k entries in $\{s, -s\}$ and $n-k$ entries equal to 0. Then $\mathbf{\Lambda}_{(s,0^{n-1})}$ consists of all vectors whose entries are multiples of s. Finally, we denote by $\Lambda_{(s^k,0^{n-k})}$ the lattice group with translation vectors in $\mathbf{\Lambda}_{(s^k,0^{n-k})}$. The next lemma states some relationships among the lattice groups $\Lambda_{(s^k,0^{n-k})}$.

Lemma 2.4.20 *With the previous notation,*

1. $\Lambda_{(2s,0^{n-1})} \subseteq \Lambda_{(s^k,0^{n-k})}$ *for* $k \in \{1, 2, n\}$,
2. $\Lambda_{(2s,0^{n-1})}$ *has index* 2^n *in* $\Lambda_{(s,0^{n-1})}$,
3. $\Lambda_{(s^2,0^{n-2})}$ *has index* 2 *in* $\Lambda_{(s,0^{n-1})}$,
4. $\Lambda_{(s^n)}$ *has index* 2^{n-1} *in* $\Lambda_{(s,0^{n-1})}$.

Proof To prove the first item we shall look at vectors $\bar{x} \in \mathbf{\Lambda}_{(s^k,0^{n-k})}$ with k entries equal to s and the remaining $n-k$ entries equal to 0. For any given i we may choose $\bar{x}$ so that its ith entry is s. By definition, the vector $\bar{x}'$ obtained from $\bar{x}$ by changing the sign of the ith coordinate is also in $\mathbf{\Lambda}_{(s^k,0^{n-k})}$. It follows that $\bar{x} - \bar{x}' \in \mathbf{\Lambda}_{(s^k,0^{n-k})}$, and this is the vector with ith entry equal to $2s$ and the remaining $n-1$ entries equal to 0. This shows that $\mathbf{\Lambda}_{(s^k,0^{n-k})}$ contains all generators of $\mathbf{\Lambda}_{(2s,0^{n-1})}$.

The sets of points $\mathbf{\Lambda}_{(2s,0^{n-1})} + (\epsilon_1, \ldots, \epsilon_n)$ where $(\epsilon_1, \ldots, \epsilon_n) \in \{0, s\}^n$ form a partition of $\mathbf{\Lambda}_{(s,0^{n-1})}$. The translations by the vectors on the parts of this partition are precisely the cosets of $\Lambda_{(2s,0^{n-1})}$ in $\Lambda_{(s,0^{n-1})}$. Since there are precisely 2^n parts in the partition, the index of $\Lambda_{(2s,0^{n-1})}$ in $\Lambda_{(s,0^{n-1})}$ is 2^n, and the second item holds.

Since all entries of the translation vectors of elements in $\Lambda_{(s^2,0^{n-2})}$ and $\Lambda_{(s^n)}$ are multiples of s, both groups are contained in $\Lambda_{(s,0^{n-1})}$. To determine the indices of these two subgroups we shall give an alternative characterisation of $\mathbf{\Lambda}_{(s^2,0^{n-2})}$ and $\mathbf{\Lambda}_{(s^n)}$.

The sum of the coordinates of every generator of $\mathbf{\Lambda}_{(s^2,0^{n-2})}$ is a multiple of $2s$. Conversely, every $\bar{x} \in s\mathbb{Z}^n$ whose sum of entries is a multiple of $2s$ is in $\mathbf{\Lambda}_{(s^2,0^{n-2})}$ (the details are left as an exercise). Hence, $\mathbf{\Lambda}_{(s^2,0^{n-2})}$ consists precisely of the points in $\mathbb{E}^n$ whose coordinates are all multiples of s and whose sum of coordinates is a multiple of $2s$. It follows that $\mathbf{\Lambda}_{(s^2,0^{n-2})}$ is the union of the 2^{n-1} parts $\mathbf{\Lambda}_{(2s,0^{n-1})} + (\epsilon_1, \ldots, \epsilon_n)$ where $\epsilon_i = s$ for an even number of indices i. This, together with the second item, proves the third item.

The fourth item can be shown in a way analogous to the third item. In this case, $\mathbf{\Lambda}_{(s^n)}$ consists of the points satisfying that either all their entries are odd multiples of s, or all their entries are even multiples of s (one inclusion is left as an exercise). It follows that $\mathbf{\Lambda}_{(s^n)}$ is the union of the parts $\mathbf{\Lambda}_{(2s,0^{n-1})} + (0, \ldots, 0)$ and $\mathbf{\Lambda}_{(2s,0^{n-1})} + (s, \ldots, s)$, and so the index of $\Lambda_{(s^n)}$ in $\Lambda_{(s,0^{n-1})}$ is 2^{n-1}. ∎

Exercise 2.4.21 Show that if $\bar{x} = (x_1, \ldots, x_n) \in s\mathbb{Z}^n$ satisfies that $\sum x_i$ is a multiple of $2s$ then $\bar{x} \in \mathbf{\Lambda}_{(s^2,0^{n-2})}$.

Exercise 2.4.22 Show that if $\bar{x} = (x_1, \ldots, x_n) \in s\mathbb{Z}^n$ satisfies that either x_i is an odd multiple of s for every i, or x_i is an even multiple of s for every i, then $\bar{x} \in \mathbf{\Lambda}_{(s^n)}$.

The quotients $\{4, 3^{n-2}, 4\}/\Lambda_{(s^k,0^{n-k})}$ are costumarily denoted by $\{4, 3^{n-2}, 4\}_{(s^k,0^{n-k})}$, and they are natural candidates when searching for rotary polytopes.

Proposition 2.4.23 *The polytope* $\{4, 3^{n-2}, 4\}_{(s^k,0^{n-k})}$ *is regular. Furthermore, it is polytopal if and only if* $s \geq 2$.

Proof We showed in Example 2.4.16 that $\{4, 3^{n-2}, 4\}/\Lambda_{(s,0^{n-1})}$ is regular.

According to Lemma 2.4.7, to show regularity of $\{4, 3^{n-2}, 4\}/\Lambda_{(s^k,0^{n-k})}$ we only need to verify that the stabiliser of o in $\Gamma(\{4, 3^{n-2}, 4\})$ maps fundamental regions of $\Lambda_{(s^k,0^{n-k})}$ to fundamental regions of $\Lambda_{(s^k,0^{n-k})}$. This is clear, since all automorphisms of the vertex-figure at o are compositions of permutations of the entries and sign changes of entries, all of which preserve the generating set of $\mathbf{\Lambda}_{(s^k,0^{n-k})}$. Therefore, $\mathbf{\Lambda}_{(s^k,0^{n-k})}$ is regular.

If $s = 1$ then the translation by the vector $(s^k, 0^{n-k})$ maps some vertex F_0 of a cube C of $\{4, 3^{n-2}, 4\}$ to another vertex F_0' of C. If $k = 1$ then every edge of $\{4, 3^{n-2}, 4\}/\Lambda_{(s^k,0^{n-k})}$ has only one vertex and the diamond condition fails. If $k > 1$ then every edge has two distinct vertices. In this situation there is a path in the flag graph of $\{4, 3^{n-2}, 4\}$ from a flag Φ in C containing F_0 to a flag Φ' in C containing F_0' completely contained in C; this path induces a path in the flag graph of $\{4, 3^{n-2}, 4\}/\Lambda_{(s^k,0^{n-k})}$ from $\Phi\Lambda_{(s^k,0^{n-k})}$ to $\Phi'\Lambda_{(s^k,0^{n-k})}$, none of whose edges has label n. There is also a path from $\Phi\Lambda_{(s^k,0^{n-k})}$ to $\Phi'\Lambda_{(s^k,0^{n-k})}$ that never changes vertex (and so none of its edges has label 0). However, there is no path between these two flags in the flag graph of $\{4, 3^{n-2}, 4\}/\Lambda_{(s^k,0^{n-k})}$ with edges with labels only in $\{1, \dots, n-1\}$, since otherwise it would induce a path from Φ to Φ' with no edges labelled 0 or n. It follows that $\{4, 3^{n-2}, 4\}/\Lambda_{(s^k, 0^{n-k})}$ does not satisfy the path intersection property and Theorem 2.1.57 implies that it is not polytopal.

On the other hand, if $s = 2$ then all facets are n-cubes and all vertex-figures are cross polytopes, and it follows from Proposition 2.1.44 that $\{4, 3^{n-2}, 4\}/\Lambda_{(s^k, 0^{n-k})}$ is polytopal. ∎

Example 2.4.24 The sets $\mathbf{\Lambda}_{(1,0,0)}$, $\mathbf{\Lambda}_{(1,1,0)}$ and $\mathbf{\Lambda}_{(1,1,1)}$ consist of the points of the well-known cubic lattice, face-centred cubic lattice and body-centred cubic lattice, respectively. The *cubic lattice* is the set of vertices of the tessellation by cubes, usually identified with $\mathbb{Z}^3$. The *face-centred cubic lattice* consists of all vertices and all centres of squares of a tessellation of cubes. Finally, the *body-centred cubic lattice* consists of all vertices and all centres of cubes of the tessellation of cubes. If the cubes have edge length 2 then we can assign a point with integer coordinates to every vertex of the face-centred and body-centred cubic lattices.

We already met in Example 2.4.16 the regular polytopes $\{4,3,4\}_{(s,0,0)}$, which can be visualised as $s \times s \times s$ grids in the 3-torus.

We may consider parallelepipeds as fundamental regions for the lattice groups $\Lambda_{(s,s,0)}$ and $\Lambda_{(s,s,s)}$ in order to visualise the regular polytopes $\{4,3,4\}_{(s,s,0)}$ and $\{4,3,4\}_{(s,s,s)}$, respectively. However, such fundamental regions do not exhibit much symmetry. Alternatively, we may take the Voronoi cells of $\mathbf{\Lambda}_{(s,s,0)}$ and $\mathbf{\Lambda}_{(s,s,s)}$ as fundamental regions for $\Lambda_{(s,s,0)}$ and $\Lambda_{(s,s,s)}$. These are rhombic dodecahedra and truncated octahedra, respectively. These choices make it easier to visualise the fact that the stabiliser of o in $\Gamma(\{4,3,4\})$ sends fundamental regions of Λ to fundamental regions of Λ.

The rotary toroidal 4-polytopes $\{4,3,4\}_{(2,0,0)}$, $\{4,3,4\}_{(2,2,0)}$ and $\{4,3,4\}_{(2,2,2)}$ are illustrated with these fundamental regions in Figure 2.32.

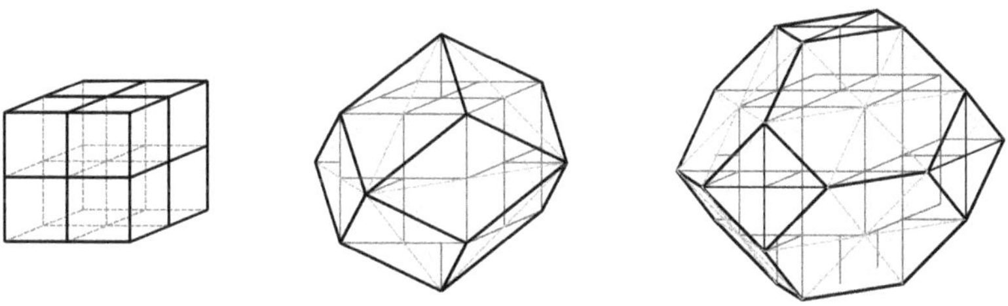

(a) (b) (c)

Figure 2.32 The toroidal 4-polytopes $\{4,3,4\}_{(2,0,0)}$ (a), $\{4,3,4\}_{(2,2,0)}$ (b) and $\{4,3,4\}_{(2,2,2)}$ (c).

We are ready to determine all rotary toroidal polytopes with type $\{4,3^{n-2},4\}$.

Theorem 2.4.25 *For $n \geq 3$, the toroidal rotary $(n+1)$-polytopes with type $\{4,3^{n-2},4\}$ are precisely the quotients $\{4,3^{n-2},4\}_{(s^k,0^{n-k})}$ for $k \in \{1,2,n\}$ and integers $s \geq 2$.*

Proof Let $\mathcal{P} = \{4,3^{n-2},4\}/\Lambda$ be a rotary toroidal polytope and let $\mathbf{\Lambda}$ be the set of translation vectors of elements in Λ. For convenience we denote by $\bar{e}_i$ the ith canonical vector of $\mathbb{E}^n$.

Let $\bar{x} = (x_1, \ldots, x_n) \in \mathbf{\Lambda}$ be such that it contains a positive entry x_i minimal among all positive entries of all vectors in $\mathbf{\Lambda}$. By Corollary 2.4.19, the vector obtained from $\bar{x}$ by performing any even permutation of its entries is also in $\mathbf{\Lambda}$, and so we may assume that $s := x_1$ is the smallest positive entry of all positive entries of vectors in $\mathbf{\Lambda}$.

Again, by Corollary 2.4.19, the vector

$$\bar{x}' = \begin{cases} (s, -x_2, -x_3, -x_4 \ldots, -x_n) & \text{if } n \text{ is odd,} \\ (s, x_2, -x_3, -x_4, \ldots, -x_n) & \text{if } n \text{ is even} \end{cases}$$

is also in $\mathbf{\Lambda}$. By adding $\bar{x}$ and $\bar{x}'$ we obtain that $2s\bar{e}_1 \in \mathbf{\Lambda}$ if n is odd, and that $2s\bar{e}_1 + 2x_2\bar{e}_2 \in \mathbf{\Lambda}$ if n is even. In what follows we shall show that $2s\bar{e}_1 \in \mathbf{\Lambda}$ also in the case when n is even.

Let $\bar{y} = 2s\bar{e}_1 + 2x_2\bar{e}_2 \in \mathbf{\Lambda}$. By changing signs of the second and third coordinates, and adding the resulting vector to $\bar{y}$ we obtain that $4s\bar{e}_1 \in \mathbf{\Lambda}$. Since $\mathbf{\Lambda}$ is invariant under even permutations of its coordinates, $4s\bar{e}_i \in \mathbf{\Lambda}$ for every i. We can add suitable multiples of these vectors to $\bar{x}$ to obtain a vector $\bar{x}'' = (s, x_2'', \ldots, x_n'')$, where $-2s < x_i'' \leq 2s$. The choice of s forces $x_i'' \in \{-s, 0, s, 2s\}$ for every i. (Recall that $\bar{z} \in \mathbf{\Lambda}$ if and only if $-\bar{z} \in \mathbf{\Lambda}$.)

If $x_i'' = 0$ for some i then we may change signs in the first and ith coordinates of $\bar{x}''$ and substract the resulting vector from $\bar{x}$ to conclude that $2s\bar{e}_1 \in \mathbf{\Lambda}$. If $x_i'' = 2s$ we proceed in the same way to imply that $2s\bar{e}_1 + 4s\bar{e}_i \in \mathbf{\Lambda}$, and since $4s\bar{e}_i \in \mathbf{\Lambda}$, we conclude that $2s\bar{e}_1 \in \mathbf{\Lambda}$.

If $x_i'' \in \{-s, s\}$ for every $i \in \{2, \ldots, n\}$ then we may assume that $x_j = s$ for some j, since otherwise we replace $\bar{x}$ with $-\bar{x}$ to obtain a vector with all but one entries equal to s. Now we transpose the first and jth entries and then change the sign of the first entry (see Lemma 2.4.18); the resulting vector $\bar{y}'$ must be in $\mathbf{\Lambda}$, and hence $\bar{x} - \bar{y}' = 2s\bar{e}_1 \in \mathbf{\Lambda}$.

We now use that $(2s, 0, \ldots, 0) \in \mathbf{\Lambda}$ regardless of the parity of n. By permuting coordinates we get that $\mathbf{\Lambda}$ contains $2s\bar{e}_i$ for every i. By adding suitable multiples of these vectors to $\bar{x}$ we obtain a vector $\bar{x}''' = (s, x_2''', \ldots, x_n''')$, where

$0 \le x_i''' < 2s$. The choice of s forces $x_i''' \in \{0, s\}$ for every i. Among all vectors in $\mathbf{\Lambda}$ whose first entry is s and all whose entries are in $\{0, s\}$ we choose $\bar{y} = (y_1, \dots, y_n)$ with a minimal number k of entries equal to s.

If $k = 1$ then $\mathbf{\Lambda} = \mathbf{\Lambda}_{(s,0^{n-1})}$. If $k = 2$ then Lemma 2.4.20 forces $\mathbf{\Lambda} = \mathbf{\Lambda}_{(s^2,0^{n-2})}$, since $\mathbf{\Lambda} \subsetneq \mathbf{\Lambda}_{(s,0^{n-1})}$. If $k = n$ then the fact that $2s\bar{e}_i \in \mathbf{\Lambda}$ for every i forces $\mathbf{\Lambda} = \mathbf{\Lambda}_{(s^n)}$. Proposition 2.4.23 confirms that these values of k yield rotary polytopes for $s \ge 2$, and Lemma 2.4.20 confirms that the resulting polytopes are distinct for distinct values of k.

It remains to show that k cannot take any value in $\{3, \dots, n-1\}$. Assume to the contrary that $k \in \{3, \dots, n-1\}$, and so the vectors $\bar{y}_1 = s\bar{e}_1 + \cdots + s\bar{e}_k$ and $\bar{y}_2 = -s\bar{e}_2 - \cdots - s\bar{e}_{k+1}$ are in $\mathbf{\Lambda}$ (see Lemma 2.4.18). Then $\bar{y}_1 + \bar{y}_2 = \bar{e}_1 - \bar{e}_{k+1} \in \mathbf{\Lambda}$, contradicting the assumption that $k > 2$. ■

Theorem 2.4.25, together with Example 2.4.16 and Proposition 2.4.23, implies the non-existence of chiral toroidal polytopes with cubical facets for ranks higher than 2.

Corollary 2.4.26 *There are no chiral toroidal $(n+1)$-polytopes of type $\{4, 3^{n-2}, 4\}$ for $n \ge 3$.*

The classification of toroidal chiral polyhedra in the previous section and Corollary 2.4.26 tell us that dimension $n = 2$ is exceptional in being the only one where the cubic tessellation admits chiral quotients. As a reason for this we point out that the restrictions of isometries f in Lemma 2.4.18 for $n \ge 3$ are restrictive enough to force the lattice group Λ to be $\Lambda_{(s^k,0^{n-k})}$ for some k. However, if $n = 2$ there are no non-trivial even permutations of the entries, and by performing an even number of sign changes we can only obtain id and the central inversion about o. In this situation, the only non-trivial constraint for Λ is that its set of translation vectors must be invariant under the two isometries that transpose the two coordinates while changing the sign of only one of them. These isometries are a 4-fold rotation about o and its inverse, both of which preserve the sets of points $\mathbf{\Lambda}$ associated to the lattice groups Λ described in the previous section.

Besides the regular tessellations of $\mathbb{E}^2$ and the regular tessellations by cubes of $\mathbb{E}^n$ there are only two other regular tessellations of Euclidean spaces, namely the tessellations $\{3,4,3,3\}$ and $\{3,3,4,3\}$ of $\mathbb{E}^4$, with 24-cells and cross polytopes as facets, respectively (see [55, Chapter 7] for details). Theorem 2.4.3 implies that the only rotary toroidal polytopes that we have not considered so far are those with types $\{3,4,3,3\}$ and $\{3,3,4,3\}$. We shall not assume that the reader is familiar with these tessellations, and will present the main result without proof. Detailed discussions on the topic can be found in [185] and [188, Section 6H].

Theorem 2.4.27 *The only chiral toroidal polytopes are the polyhedra* $\{4,4\}_{(a,b)}$, $\{3,6\}_{(a,b)}$ *and* $\{6,3\}_{(a,b)}$ *with* $ab(a-b) \neq 0$.

In Section 5.3 we shall mention the possibility of other rotary quotients of regular Euclidean tessellations.

3

Groups Related to Chiral Polytopes

The definition of chiral polytopes directly involves the action of the automorphism group on the flags. As we shall see, this allows us to obtain results about chiral polytopes through group-theoretical techniques. To date, the most common group-theoretical tool to study chiral polytopes has been a one-to-one correspondence between the class of chiral polytopes and a certain class of groups with a prescribed generating set. The actions of two other groups on the sets of flags of chiral polytopes have been studied more recently. They are the connection groups and the chirality groups. These three groups are related by several results, which provide many group-theoretical resources to analyse properties of chiral polytopes.

Group-theoretical techniques to work with rotary polytopes have proven very fruitful. In particular, to this date the only known chiral n-polytopes for $n \geq 6$ have been obtained in this way. Nevertheless, those readers interested in rotary polytopes are encouraged always to keep in mind that these objects have a combinatorial nature with a natural topological space associated to them. The author strongly believes that the interaction of the algebraic, combinatoric, geometric and topologic aspects of polytopes will give us a much deeper understanding of these objects, in comparison with the one that we would get by restricting ourselves to just one or two flavours of techniques.

This chapter is organised in three sections. The first one is devoted to the structure of the automorphism groups of rotary polytopes, and their correspondence with a class of groups with prescribed generators. The connection and chirality groups of chiral polytopes are discussed in the second and third sections, respectively.

3.1 Automorphism Group

In the previous chapter we mentioned that an n-polytope can be understood as a collection of $(n-1)$-polytopes (the facets), suitably glued so that two of them meet on each of their facets, in such a way that the vertex-figure at every vertex is an $(n-1)$-polytope and the entire structure is connected (see Proposition 2.1.44).

In order to construct examples of n-polytopes, the previous idea is not easy to put into practice, even for $n = 3$. As an exercise, the reader may try to find the smallest equivelar polyhedron with type $\{4,5\}$ by imagining certain supply of squares, and identifying two of them at every edge in such a way that five of them meet at every vertex producing vertex-figures isomorphic to pentagons. This approach becomes even harder if we require the polyhedron to have specific symmetry properties, like being rotary, regular or chiral.

Constructing polytopes by describing the poset explicitly is also impractical. With this in mind we need other techniques to describe polytopes.

In this section we shall use group theory to encode rotary polytopes in an economical fashion. The power of this technique can be seen in the fact that up to date all finite chiral n-polytopes with $n \geq 6$ have been found with this group-theoretical approach. Moreover, only one 5-polytope has a purely combinatorial description, devised only after its discovery in an algebraic way.

Besides the group-theoretical approach, there are other ways to construct n-polytopes for large values of n. One of them is through operations, and will be discussed in Chapter 4. Polytopes of rank n can also be obtained as quotients of larger n-polytopes; this idea was used in Section 2.4. Nevertheless, it has the disadvantage that it requires sufficient knowledge of large n-polytopes, which is not always easy to obtain. Polytopes of large ranks have also been described through their realisations on geometric spaces, like the Euclidean d-space. However, none of the three approaches just mentioned has been helpful to find chiral polytopes of ranks higher than 5. This highlights the relevance of the correspondence between chiral polytopes and groups with prescribed generators described in this section.

3.1.1 Coset Representation of Polytopes

We intend to represent the faces of a given polytope $\mathcal{P}$ with substructures of $\Gamma(\mathcal{P})$. One of the most natural ways to do this is as follows.

Definition 3.1.1 The *coset representation* of a rooted polytope $\mathcal{P}_{(\Phi)}$ is a function cr from the set of proper faces of $\mathcal{P}$ to the power set $2^{\Gamma(\mathcal{P})}$ defined by

$$cr(F) := \{\gamma \in \Gamma(\mathcal{P}) \colon F \in \Phi\gamma\}.$$

Example 3.1.2 In Figure 3.1 we interpret the grey flag as the base flag Φ_0 of the triangle $\mathcal{P}$. The flag Φ is labelled with $\gamma \in \Gamma(\mathcal{P})$ whenever $\Phi = \Phi_0\gamma$. Then

$$\begin{aligned} &cr(A) = \{\mathrm{id}, \rho_1\}, && cr(B) = \{\rho_0\rho_1, \rho_0\rho_1\rho_0\}, && cr(C) = \{\rho_0, \rho_1\rho_0\}, \\ &cr(AC) = \{\mathrm{id}, \rho_0\}, && cr(AB) = \{\rho_1, \rho_0\rho_1\}, && cr(BC) = \{\rho_1\rho_0, \rho_0\rho_1\rho_0\}, \end{aligned}$$

where the edges are denoted by their endpoints.

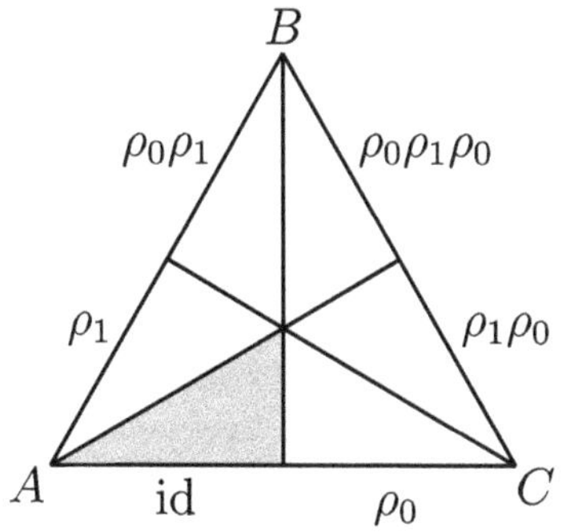

Figure 3.1 Coset representation of the triangle.

The assignment of faces in Definition 3.1.1 may not be one-to-one. For example, if $\Gamma(\mathcal{P})$ is a two-orbit n-polytope in class $2_{\{0,\ldots,n-1\}\setminus\{i\}}$ then there may be many i-faces that are assigned the empty set. In some cases the coset representation admits simple modifications that yield to one-to-one correspondences between the faces of $\mathcal{P}$ and certain subsets of $\Gamma(\mathcal{P})$. This is the case of two-orbit polyhedra that are not fully transitive (see [129, Subsection 5.2]). However, if $\Gamma(\mathcal{P})$ is trivial then there is little hope of finding suitable modifications to the coset representation that will enable it as a useful tool to study $\mathcal{P}$. Since rotary polytopes are fully transitive, we shall have no need for any major modification of the coset representation.

The following proposition justifies the name of the coset representation.

Proposition 3.1.3 *Let $\mathcal{P}_{(\Phi)}$ be a fully transitive rooted polytope, let F_i be the i-face in Φ and let G_i be any i-face of $\mathcal{P}$. Then the coset representation of*

$\mathcal{P}_{(\Phi)}$ assigns to F_i the stabiliser $\Gamma_i := \mathrm{Stab}_{\Gamma(\mathcal{P})}(F_i)$ of F_i, and it assigns to G_i the coset $\Gamma_i \gamma$ where $\gamma \in \Gamma(\mathcal{P})$ is any automorphism such that $G_i = F_i \gamma$.

Proof We know that $cr(F_i) = \{\phi \in \Gamma(\mathcal{P}) \colon F_i \in \Phi\phi\}$. It follows that every automorphism in $cr(F_i)$ fixes F_i, and every automorphism that fixes F_i is in $cr(F_i)$. We conclude that $cr(F_i) = \Gamma_i$.

If $G_i = F_i\gamma$ then the coset $\Gamma_i\gamma$ consists of automorphisms all of which map F_i to G_i. Conversely, if $G_i = F_i\gamma'$ then $F_i = G_i\gamma^{-1} = F_i\gamma'\gamma^{-1}$. It follows that $\gamma' = (\gamma'\gamma^{-1})\gamma \in \Gamma_i\gamma$, which tells us that $\Gamma_i\gamma$ is precisely the set of automorphisms that map F_i to G_i. By definition, $cr(G_i)$ is the set of automorphisms that map F_i to G_i, which concludes the proof. ■

Corollary 3.1.4 *If $\mathcal{P}$ is a fully transitive n-polytope then for every $i \in \{0, \ldots, n-1\}$ the restriction of the coset representation to the set of i-faces is a bijection to the right cosets of the stabiliser Γ_i of the i-face in the base flag.*

Corollary 3.1.4 is illustrated in Example 3.1.2, where $cr(A)$ is precisely the stabiliser of vertex A, while $cr(AC)$ is the stabiliser of the edge AC. Then $cr(B)$, $cr(C)$, $cr(AB)$ and $cr(BC)$ are respectively $cr(A)\rho_0\rho_1$, $cr(A)\rho_0$, $cr(AC)\rho_1$ and $cr(AC)\rho_1\rho_0$.

The main idea behind the coset representation is to codify a given polytope $\mathcal{P}$ in a simple way that is suitable for algebraic manipulations. With this in mind, when convenient we may adapt the representation to take into account only some proper subgroup of $\Gamma(\mathcal{P})$.

Remark 3.1.5 If $\Delta \subseteq \Gamma(\mathcal{P})$ is a fully transitive group of automorphisms of $\mathcal{P}$ then the coset representation cr_Δ of $\mathcal{P}$ with respect to the subgroup Δ satisfies the conclusions of Proposition 3.1.3.

Exercise 3.1.6 Determine the coset representation of the triangle with the notation in Figure 3.1, with respect to its rotation subgroup.

Proposition 3.1.3 tells us that we may understand the set of i-faces of $\mathcal{P}$ as the set of right cosets of Γ_i. In order to reconstruct $\mathcal{P}$ from its coset representation we only need to recover the incidence of faces. The next result describes how to do this for a large class of highly symmetric polytopes.

Theorem 3.1.7 *Let $\mathcal{P}_{(\Phi)}$ be a rooted polytope and let $\Delta \le \Gamma(\mathcal{P})$. Assume that for every $0 \le i < j \le n-1$ the group Δ acts transitively on the set of chains $\{F_i, F_j\}$ where F_i has rank i, F_j has rank j, and $F_i < F_j$. Then*

1. *for every i the coset representation cr_Δ induces a bijection between the set of i-faces and the right cosets of the stabiliser of the i-face of Φ,*

2. *if* G_i *is an* i*-face and* G_j *is a* j*-face then* $G_i \le G_j$ *if and only if* $i \le j$ *and* $cr_\Delta(G_i) \cap cr_\Delta(G_j) \ne \varnothing$.

Proof The first item is a consequence of Corollary 3.1.4, since under the hypothesis of this theorem Δ must act fully transitively on $\mathcal{P}$.

Let B_i and B_j be the i- and j-faces of Φ, respectively. If $G_i \le G_j$ then the hypothesis on Δ implies that there exists $\gamma \in \Delta$ such that $B_i\gamma = G_i$ and $B_j\gamma = G_j$. It follows that $\gamma \in cr_\Delta(G_i) \cap cr_\Delta(G_j)$.

Conversely, if F and G are faces such that $cr_\Delta(F) \cap cr_\Delta(G) \ne \varnothing$ then there exists $\gamma \in \Gamma(\mathcal{P})$ such that $\Phi\gamma$ contains F and G. In this case F and G must be incident. ■

For convenience, we call 2*-chain transitive* those fully transitive polytopes whose automorphism groups satisfy the extra hypothesis of Theorem 3.1.7. Clearly the automorphism groups of 2-chain transitive polytopes do not act transitively on all chains with 2-elements, but they do if we prescribe the ranks of the elements in the chains to be the same. The previous definition can be generalised to increase the size of the chains where the group acts transitively, and so we say that an n-polytope $\mathcal{P}$ is k*-chain transitive* if for every $i_1, \dots, i_k \in \{0, \dots, n-1\}$ and every two chains Ψ, Υ consisting of an i-face for each $i \in \{i_1, \dots, i_k\}$, there exists $\gamma \in \Gamma(\mathcal{P})$ such that $\Psi\gamma = \Upsilon$.

Exercise 3.1.8 For every $n \ge 3$ and every $I \subset \{0, \dots, n-1\}$, determine the largest k such that all two-orbit polytopes in class 2_I are k-chain transitive.

There are some fully transitive polytopes that are not 2-chain transitive. This is the case of the polyhedron in Figure 2.16(b). This polyhedron is in class 2_1, and according to Corollary 2.2.25 it is fully transitive. We choose as base flag Φ the grey flag indicated in Figure 3.2; we denote by F_0 and F_2 the base vertex and 2-face of Φ, respectively, and by F_2' the 2-face of Φ^2. Then $F_0 \le F_2'$. On the other hand, the images of Φ containing F_2' are the four flags indicted in Figure 3.2 as light grey triangles, none of which contains F_0. We can conclude that there is no automorphism mapping Φ to a flag containing F_0 and F_2' simultaneously, and therefore $cr(F_0) \cap cr(F_2') = \varnothing$.

The previous difficulty regarding the identification of the incidence of faces with non-trivial intersection of cosets was solved in [129, Subsection 5.1] for the polyhedron in Figure 3.2, as well as for all polyhedra in classes 2_0, 2_1 and 2_2. This was achieved by suitably modifying the coset representation, so that the incidence of faces does correspond to a non-empty intersection. To this day, little work has been done in connection with the coset representation of

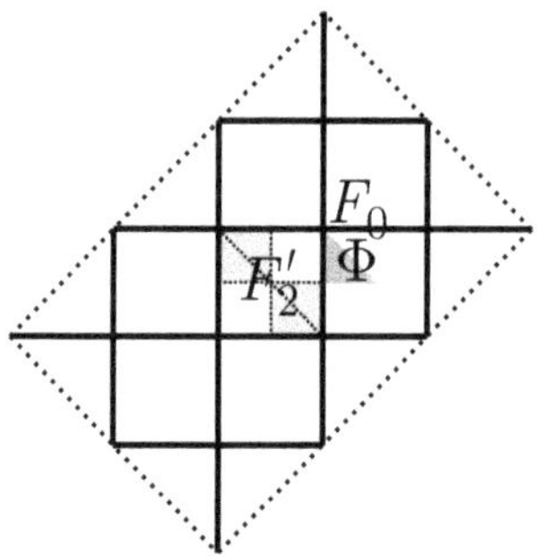

Figure 3.2 Incidence in the coset representation of this map cannot be defined through non-empty intersection.

polytopes with more than two flag orbits, and so it is unclear to what extent it is possible to alter the coset representation of arbitrary fully transitive polytopes that are not 2-chain transitive, so that the modified representation constitutes a useful tool to study these polytopes. Recently, an alternative approach to group-theoretical constructions of arbitrary polytopes appeared in [133], where the stabilisers of faces do not represent a central ingredient.

Problem 10 Develop the theory of the coset representation of fully transitive polytopes that are not 2-chain transitive.

Next we are interested in the action of $\Gamma(\mathcal{P})$ on the coset representation of $\mathcal{P}$, inherited from the action on $\mathcal{P}$.

Proposition 3.1.9 *Let $\mathcal{P}_{(\Phi)}$ be a rooted fully transitive polytope. Then the action of $\Gamma(\mathcal{P})$ on the faces of $\mathcal{P}$ is translated to the action by right multiplication on its coset representation; that is,*

$$cr(F\gamma) = cr(F)\gamma$$

for every face F and every $\gamma \in \Gamma(\mathcal{P})$.

Proof Let F_i be the i-face of the base flag Φ. We know that $cr(F_i) = \Gamma_i$, and that if $G_i = F_i\gamma$ for some $\gamma \in \Gamma(\mathcal{P})$ then $cr(G_i) = \Gamma_i\gamma$. If $\alpha \in \Gamma(\mathcal{P})$ and $H_i = G_i\alpha$ then $H_i = F_i\gamma\alpha$ and we compute $cr(H_i)$ as

$$cr(G_i\alpha) = cr(H_i) = \Gamma_i\gamma\alpha = cr(G_i)\alpha.$$

■

The previous discussion tells us that the coset representation is a powerful tool in the following sense. If we want to work with a 2-chain transitive n-polytope $\mathcal{P}$, we do not need to determine the Hasse diagram of $\mathcal{P}$, nor any other representation of the poset. Instead, it suffices to have in mind the pair

$$cr(\mathcal{P}) := (\Gamma(\mathcal{P}), (\Gamma_0, \ldots, \Gamma_{n-1})), \tag{3.1}$$

where for each i, the subgroup Γ_i is the stabiliser in $\Gamma(\mathcal{P})$ of the i-face of the base flag. Theorem 3.1.7 guarantees that $\mathcal{P}$ can be recovered from $cr(\mathcal{P})$. In many instances it is easy to work with a pair like that in (3.1), making the coset representation of $\mathcal{P}$ more suitable for manipulation than the polytope $\mathcal{P}$ itself. That is certainly the case if $\mathcal{P}$ is infinite.

One of the main features of the coset representation of 2-chain transitive polytopes relies on the following standard result from group actions.

Lemma 3.1.10 *Let Γ be a group acting on a set X, and for every $C \subseteq X$ denote the pointwise stabiliser $\{\gamma \in \Gamma \colon c\gamma = c \text{ for every } c \in C\}$ of C by Fix_C. Then $Fix_A \cap Fix_B = Fix(A \cup B)$ for any $A, B \subseteq X$.*

Note that pointwise stabilisers of chains under the automorphism group of a graded poset are precisely the stabilisers of the chains, since automorphisms preserve ranks. Thus, in our context Lemma 3.1.10 will be used simply with stabilisers.

The following proposition is a general version of the *intersection property*, which is key in much of the theory of rotary polytopes developed so far, and is a direct consequence of Lemma 3.1.10.

Proposition 3.1.11 *Let $\mathcal{P}_{(\Phi)}$ be a rooted 2-chain transitive polytope. Let Ψ and Υ be two chains contained in Φ. Then*

$$\Gamma_\Psi \cap \Gamma_\Upsilon = \Gamma_{\Psi \cup \Upsilon},$$

where Γ_X denotes the stabilisers of the set X.

Remark 3.1.12 Let $\mathcal{P}_{(\Phi)}$ be a rooted 2-chain transitive n-polytope, and let Ψ be the subchain of Φ containing the elements whose ranks are in some index set $I \subseteq \{0, \ldots, n-1\}$. Then the stabiliser of Ψ is expressed in terms of the stabilisers Γ_i of the faces of Φ as $\Gamma_\Psi = \cap_{i \in I} \Gamma_i$.

Next we will focus on how to construct polytopes from groups. The arguments presented below work well for constructing 4-chain transitive polytopes, although some ideas apply for other 2-chain transitive polytopes. The following proposition justifies our choice of the class of polytopes to consider.

Proposition 3.1.13 *Let $\mathcal{P}_{(\Phi)}$ be a 4-chain transitive rooted polytope, and let F_i and F_j be the i- and j-faces of Φ for some $i < j$. Denote by Γ_k the stabiliser of the k-face of Φ in $\Gamma(\mathcal{P})$, and denote by 'cr' the mapping in the coset representation of $\mathcal{P}$. Then*

1. *$\Gamma_i \cap \Gamma_j \cap cr(G) \neq \varnothing$ for every face G of F_j/F_i,*
2. *the section F_j/F_i is isomorphic to the poset obtained by intersecting with $\Gamma_i \cap \Gamma_j$ each element in the coset representation of $\mathcal{P}$ restricted to F_j/F_i, with incidence determined by non-empty intersection.*

Proof Given $G \in F_j/F_i$ of rank ℓ there exists $\gamma \in \Gamma(\mathcal{P})$ that maps the chain $\{F_i, F_\ell, F_j\}$ to the chain $\{F_i, G, F_j\}$. In particular, $\gamma \in cr(G)$, and clearly $\gamma \in \Gamma_i \cap \Gamma_j$. This proves the first item.

Let $\mathcal{Q}$ be the poset obtained by intersecting with $\Gamma_i \cap \Gamma_j$ each element in the coset representation of $\mathcal{P}$ restricted to F_j/F_i. For $i < k < j$ the elements of $\mathcal{Q}$ of rank $k - i - 1$ are the intersections with $\Gamma_i \cap \Gamma_j$ of the cosets $\Gamma_k\gamma$ for $\gamma \in \Gamma$. Since these cosets are disjoint and their intersections with $\Gamma_i \cap \Gamma_j$ are non-empty, there is a natural bijection between the set of k-faces in the section F_j/F_i and the intersections with $\Gamma_i \cap \Gamma_j$ of the cosets of Γ_k.

If $(cr(G) \cap \Gamma_i \cap \Gamma_j) \cap (cr(H) \cap \Gamma_i \cap \Gamma_j) \neq \varnothing$ for faces G and H of F_j/F_i then $cr(G) \cap cr(H) \neq \varnothing$, implying that G and H are incident. If instead $(cr(G) \cap \Gamma_i \cap \Gamma_j) \cap (cr(H) \cap \Gamma_i \cap \Gamma_j) = \varnothing$, we claim that G and H cannot be incident in $\mathcal{P}$. Assume on the contrary that $F_i \leq G \leq H \leq F_j$, where G and H are m- and ℓ-faces, respectively. Since $\mathcal{P}$ is 4-chain transitive, there exists $\gamma \in \Gamma(\mathcal{P})$ such that $F_i\gamma = F_i$, $F_j\gamma = F_j$, $G = F_m\gamma$ and $H = F_\ell\gamma$. It follows that $\gamma \in \Gamma_i \cap \Gamma_j$, that $cr(G) = \Gamma_m\gamma$ and that $cr(H) = \Gamma_\ell\gamma$, and so $\gamma \in (cr(G) \cap \Gamma_i \cap \Gamma_j) \cap (cr(H) \cap \Gamma_i \cap \Gamma_j)$, contradicting our hypothesis. This shows the equivalence of the incidences in $\mathcal{Q}$ and F_j/F_i, which proves the second item. ■

In order to use candidate coset representations to construct 4-chain transitive n-polytopes, we need candidate groups Γ and $\Gamma_i \subseteq \Gamma$ for $i \in \{0, \ldots, n-1\}$, with the hope that the relation given by

$$\Gamma_i\alpha \leq \Gamma_j\beta \quad \text{if and only if} \quad i \leq j \quad \text{and} \quad \Gamma_i\alpha \cap \Gamma_j\beta \neq \varnothing, \tag{3.2}$$

constitutes a partial order. Then we need to verify that the so-constructed incidence structure, after the addition of a least element and a greatest element, is indeed a 4-chain transitive polytope $\mathcal{P}(\Gamma)$. In that case Γ may equal $\Gamma(\mathcal{P}(\Gamma))$, or may be a proper subgroup. The next proposition provides sufficient conditions for $(\Gamma, \{\Gamma_0, \ldots, \Gamma_{n-1}\})$ to be a connected poset with respect to the relation (3.2).

Proposition 3.1.14 *Let $n \geq 2$, let Γ be a group, for $i \in \{0, \dots, n-1\}$ let $\Gamma_i \leq \Gamma$, let $\Gamma_{-1} = \Gamma_n = \Gamma$, and let $\leq$ be defined as in* (3.2). *If*

1. $\Gamma = \langle \Gamma_0, \dots, \Gamma_{n-1} \rangle$,
2. *every time that $\Gamma_i \alpha \cap \Gamma_j \beta \neq \varnothing$ and $\Gamma_j \beta \cap \Gamma_k \delta \neq \varnothing$ for some $-1 \leq i < j < k \leq n$ and some $\alpha, \beta, \delta \in \Gamma$, there exists $\gamma \in \Gamma$ such that $\alpha\gamma \in \Gamma_i$, $\beta\gamma \in \Gamma_j$ and $\delta\gamma \in \Gamma_k$,*

then $((\Gamma, \{\Gamma_0, \dots, \Gamma_{n-1}\}, \leq))$ is a connected poset.

Proof The relation $\leq$ is clearly reflexive and antisymmetric. Let $i < j < k$ and let $\alpha, \beta, \delta \in \Gamma$ such that $\Gamma_i \alpha \cap \Gamma_j \beta \neq \varnothing$ and $\Gamma_j \beta \cap \Gamma_k \delta \neq \varnothing$; that is, $\Gamma_i \alpha \leq \Gamma_j \beta$ and $\Gamma_j \beta \leq \Gamma_k \delta$. The second item tells us that there exists $\gamma \in \Gamma$ such that $\Gamma_i \alpha = \Gamma_i \gamma^{-1}$, $\Gamma_j \beta = \Gamma_j \gamma^{-1}$ and $\Gamma_k \delta = \Gamma_k \gamma^{-1}$. It follows that $\gamma^{-1} \in \Gamma_i \alpha \cap \Gamma_k \delta$, and hence $\Gamma_i \alpha \leq \Gamma_k \delta$, proving transitivity. Therefore $\leq$ is an order relation.

Now we prove that $((\Gamma, \{\Gamma_0, \dots, \Gamma_{n-1}\}, \leq))$ is connected. Let F be a proper element, then $F = \Gamma_{k_0} \alpha$ for some $k_0 \in \{0, \dots, n-1\}$ and some $\alpha \in \Gamma$. It suffices to show that F is connected to Γ_j in the incidence graph of the poset, for some $j \in \{0, \dots, n-1\}$. The first item implies that $\alpha = \gamma_1 \cdots \gamma_m$ where $\gamma_\ell \in \Gamma_{k_\ell}$ for some $k_1, \dots, k_m \in \{0, \dots, n-1\}$. The existence of a path from $\Gamma_{k_0} \alpha$ to Γ_j will be shown by induction over m. The base case is when $m = 1$, where $\alpha \in \Gamma_{k_1}$; this tells us that $\Gamma_{k_0} \alpha$ is incident to Γ_{k_1}, since both cosets contain α, and hence the desired path in the incidence graph can be chosen to have one edge. For $m \geq 2$ we assume that there is a path in the incidence graph from $\Gamma_{k_1} \gamma_2 \cdots \gamma_m$ to Γ_j. This path can be extended with the edge between the nodes representing $\Gamma_{k_0} \alpha$ and $\Gamma_{k_1} \alpha = \Gamma_{k_1} \gamma_2 \cdots \gamma_m$, since these two cosets contain α. We can conclude that the poset is connected. ■

We are now ready for our main result on coset representations of 4-chain transitive polytopes.

Theorem 3.1.15 *Let $n \geq 4$, let Γ be a group, for $i \in \{0, \dots, n-1\}$ let $\Gamma_i \leq \Gamma$, and let $\Gamma_{-1} = \Gamma_n = \Gamma$. Then $(\Gamma, \{\Gamma_0, \dots, \Gamma_{n-1}\})$ is the coset representation of a 4-chain transitive n-polytope $\mathcal{P}(\Gamma)$ if and only if*

1. *every time that $\Gamma_i \alpha \cap \Gamma_j \beta \neq \varnothing$, $\Gamma_j \beta \cap \Gamma_k \delta \neq \varnothing$ and $\Gamma_k \delta \cap \Gamma_\ell \zeta \neq \varnothing$ for some $i < j < k < \ell$ and some $\alpha, \beta, \delta, \zeta \in \Gamma$, there exists $\gamma \in \Gamma$ such that $\alpha\gamma \in \Gamma_i$, $\beta\gamma \in \Gamma_j$, $\delta\gamma \in \Gamma_k$ and $\zeta\gamma \in \Gamma_\ell$;*
2. *for every $i \in \{0, \dots, n-1\}$ there exist precisely two cosets $\Gamma_i \gamma$ such that $\Gamma_i \gamma \cap \Gamma_{i-1} \neq \varnothing$ and $\Gamma_i \gamma \cap \Gamma_{i+1} \neq \varnothing$ simultaneously;*
3. $$\Gamma_i \cap \Gamma_j = \big\langle \Gamma_i \cap \Gamma_j \cap \Gamma_k \colon k \in \{i+1, \dots, j-1\} \big\rangle$$

   for every $i, j \in \{-1, 0, \dots, n\}$.

In this situation, $\bigcap_{i\in\{0,\dots,n-1\}}\Gamma_i = \{\mathrm{id}\}$ *if and only if* Γ *acts as a subgroup of* $\Gamma(\mathcal{P}(\Gamma))$ *through right multiplication.*

Proof If $(\Gamma, \{\Gamma_0, \dots, \Gamma_{n-1}\})$ is the coset representation of a 4-chain transitive polytope $\mathcal{P}(\Gamma)$ then the four cosets mentioned in the first item correspond to faces F_i, F_j, F_k and F_ℓ with $F_i \leq F_j \leq F_k \leq F_\ell$. Since $\mathcal{P}(\Gamma)$ is 4-chain transitive, there must be an automorphism γ mapping them to the four faces of the base flag having the same ranks as F_i, F_j, F_k and F_ℓ. Equivalently, the four cosets can be multiplied on the right by γ to obtain subgroups Γ_t, implying that the groups Γ and Γ_t satisfy the first item.

The second item is a direct translation to the context of the coset representation of the diamond condition of $\mathcal{P}(\Gamma)$ for the $(i-1)$- and $(i+1)$-faces of the base flag, and hence it must be satisfied by the groups Γ and Γ_t.

Finally, the third item is a consequence of connectivity of all sections of $\mathcal{P}(\Gamma)$. Indeed, if for some $i < j$ the group $\Lambda := \langle \Gamma_i \cap \Gamma_j \cap \Gamma_k : k \in \{i+1, \dots, j-1\}\rangle$ were a proper subgroup of $\Gamma_i \cap \Gamma_j$, and $\alpha \in (\Gamma_i \cap \Gamma_j) \setminus \Lambda$ then $(\Gamma_\ell \cap \Gamma_i \cap \Gamma_j)\alpha \subseteq (\Gamma_i \cap \Gamma_j) \setminus \Lambda$ for every $\ell \in \{i+1, \dots, j-1\}$. On the other hand, if $\beta \in \Lambda$ then $(\Gamma_m \cap \Gamma_i \cap \Gamma_j)\beta \subseteq \Lambda$ for every $m \in \{i+1, \dots, j-1\}$. For each such choice of α, β, ℓ and m we must have that $(\Gamma_\ell \cap \Gamma_i \cap \Gamma_j)\alpha \cap (\Gamma_m \cap \Gamma_i \cap \Gamma_j)\beta = \emptyset$, and therefore the set of faces

$$\{\Gamma_k\gamma : k \in \{i+1, \dots, j-1\}, \gamma \in \Lambda\}$$

would be disconnected from the remaining faces of the section determined by the faces corresponding to Γ_i and Γ_j. We can conclude that the groups Γ and Γ_t satisfy the third item.

In what follows assume that $(\Gamma, \{\Gamma_0, \dots, \Gamma_{n-1}\})$ satisfies the three items. We shall show that the relation $\leq$ defined in (3.2) is an order relation, and that the set of cosets of all subgroups Γ_i together with a least element and a greatest element constitute a 4-chain transitive polytope $\mathcal{P}(\Gamma)$.

To show that the relation $\leq$ induces a connected poset we only need to verify that $(\Gamma, \{\Gamma_0, \dots, \Gamma_{n-1}\})$ satisfies the two items of Proposition 3.1.14. The first item of that proposition follows immediately from the third one of this theorem with $i = -1$ and $j = n$, whereas the second item of the proposition is a direct consequence of the first item of this theorem. Here we observe that right multiplication by elements in Γ preserves the partial order.

Let $\mathcal{P}(\Gamma)$ denote the set of cosets of the subgroups Γ_i, together with a least element and a greatest element. To see that the partial order $(\mathcal{P}(\Gamma), \leq)$ is an abstract polytope we only need to show that it satisfies the diamond condition and that it is strongly connected.

If $i \in \{0, \ldots, n-1\}$ and $\Gamma_{i-1}\alpha \leq \Gamma_{i+1}\beta$ for some $\alpha, \beta \in \Gamma$ then we use the fist item and there must exist $\delta \in \Gamma$ such that $\Gamma_{i-1}\alpha = \Gamma_{i-1}\delta$ and $\Gamma_{i+1}\beta = \Gamma_{i+1}\delta$, implying that the section $\Gamma_{i+1}\beta/\Gamma_{i-1}\alpha$ can be rewritten as $\Gamma_{i+1}\delta/\Gamma_{i-1}\delta$. The second item tells us that there are precisely two cosets $\Gamma_{i_1}\gamma$ and $\Gamma_{i_2}\gamma$ in the section $\Gamma_{i+1}/\Gamma_{i-1}$, and since right multiplication by δ preserves the partial order, there must be precisely two cosets $\Gamma_{i_1}\gamma\delta$ and $\Gamma_{i_2}\gamma\delta$ in the section $\Gamma_{i+1}\beta/\Gamma_{i-1}\alpha$. Therefore $(\mathcal{P}(\Gamma), \leq)$ satisfies the diamond condition.

To show that $(\mathcal{P}(\Gamma), \leq)$ is strongly connected we need to show that every section $\Gamma_j\beta/\Gamma_i\alpha$ is connected. As before, the first item, together with the fact that right multiplication by elements of Γ preserves the incidence, implies that it suffices to verify that every section Γ_j/Γ_i (determined by elements of the base flag) is connected. The first item implies that the section Γ_j/Γ_i is isomorphic to the poset whose elements are the cosets of $\Gamma_i \cap \Gamma_j \cap \Gamma_k$ in $\Gamma_i \cap \Gamma_j$ for $k \in \{i, \ldots, j\}$ with the order defined in (3.2). Then the first and third items suffice to show that the section Γ_j/Γ_i satisfies both items of Proposition 3.1.14. We conclude that every section of rank at least 2 is connected, and this finishes the proof of the polytopality of $(\mathcal{P}(\Gamma), \leq)$. The fact that $\mathcal{P}(\Gamma)$ is 4-chain transitive follows directly from the first item.

We already know that Γ acts as a group of automorphisms by right multiplication; on the other hand, if $\bigcap_{i\in\{0,\ldots,n-1\}}\Gamma_i = \{\mathrm{id}\}$ then the group homomorphism from Γ to $\Gamma(\mathcal{P}(\Gamma))$ determined by this action is injective. We can conclude that in that case Γ is isomorphic to a subgroup of $\Gamma(\mathcal{P}(\Gamma))$. Conversely, if right multiplication of elements in Γ determines the action of a subgroup of $\Gamma(\mathcal{P}(\Gamma))$ isomorphic to Γ then Proposition 2.1.80 implies that the only element fixing the base flag is id, which forces $\bigcap_{i\in\{0,\ldots,n-1\}}\Gamma_i = \{\mathrm{id}\}$. ∎

Results similar to Theorem 3.1.15 have been used mainly in the following way. In order to work with all polytopes in a given class $\mathcal{C}_{\mathcal{P}}$, there has been established a one-to-one correspondence between class $\mathcal{C}_{\mathcal{P}}$ and a certain class $\mathcal{C}_{\Gamma}$ of groups, in such a way that the automorphism groups of polytopes in $\mathcal{C}_{\mathcal{P}}$ are groups in $\mathcal{C}_{\Gamma}$ with the subgroups Γ_i prescribed in one way or another. As a consequence of these correspondences many theorems have been proven with group-theoretical techniques, and many constructions of polytopes have been achieved through the coset representation.

The correspondence where $\mathcal{C}_{\mathcal{P}}$ is the class of regular polytopes can be found in [188, Section 2E], while that where $\mathcal{C}_{\mathcal{P}}$ is the class of chiral polytopes was given in [247]. Both correspondences were done without the requirement of 4-chain transitivity, and will appear later in this section. Similar correspondences when $\mathcal{C}$ is the class of two-orbit polyhedra in class 2_I were given in [129]. The coset representation of polytopes described here can be understood as a particular case of the broader approach for the description of more general combinatorial structures in [256].

3.1.2 Coxeter Groups and Regular Polytopes

Two classes of groups play a major role when studying the automorphism groups of regular polytopes: the string Coxeter groups and the string C-groups, defined next.

A *Coxeter group* is a group generated by involutions with defining relations given exclusively by the orders of the products of every two generators. It is allowed for there to be an infinite order of the product of a pair (or more pairs) of generators. The *rank* of a Coxeter group is the size of its generating set. Coxeter groups can be efficiently encoded in a *diagram*, which consists of a complete graph with labelled edges, where every generator r_i is represented by a node, and two of them are adjacent by an edge labelled k if the order of their product is k. By convention, the edge between nodes representing r_i and r_j is omitted if the order of $r_i r_j$ is 2, and its label is omitted if the order of $r_i r_j$ is 3. A Coxeter group is *irreducible* if its diagram (with the established conventions) is connected. Every reducible Coxeter group Γ is isomorphic to the direct product of the Coxeter groups whose diagrams are the connected components of the diagram of Γ. We shall not need theory about general Coxeter groups; for details on this topic see [28, 55, 127, 140].

A *string Coxeter group* is a Coxeter group with a string diagram (that is, the underlying graph is a path). It has presentation

$$\langle r_0, \ldots, r_{n-1} \colon (r_i r_j)^{p_{i,j}} = \mathrm{id} \rangle, \tag{3.3}$$

where $p_{i,i} = 1$, $p_{i,i-1} = p_{i-1,i} = p_i$ and $p_{i,j} = 2$ if $|i - j| \geq 2$. Its diagram is the one in Figure 3.3, and this group is often denoted by $[p_1, \ldots, p_{n-1}]$.

The cyclic group with two elements satisfies the definition of a string Coxeter group; however, for many purposes it is required that the rank of a Coxeter group is at least 2, and this shall be assumed henceforth. The simplest examples of string Coxeter groups are the dihedral groups, which are

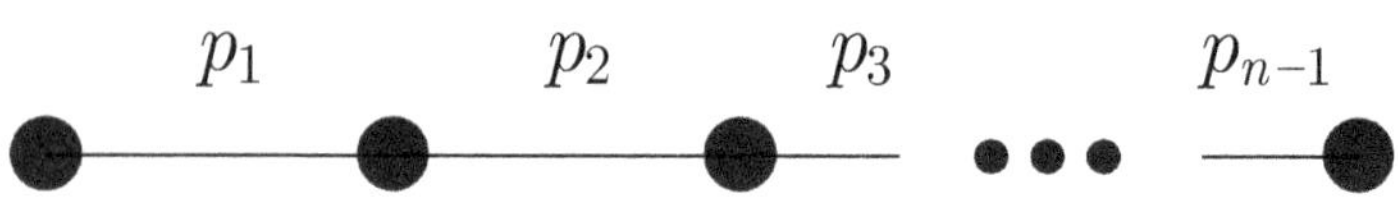

Figure 3.3 Coxeter diagram of the Coxeter group $[p_1, \ldots, p_{n-1}]$.

frequently presented as $\langle x, y \colon x^2 = y^2 = (xy)^{p_1} = \mathrm{id}\rangle$. Their diagrams are those in Figure 3.3, with only two nodes and an edge labelled p_1.

Coxeter realised that every discrete group of spherical or Euclidean isometries generated by hyperplane reflections is a Coxeter group (see [48, Theorem 8]). Furthermore, the symmetry group of the convex regular polytope with type $\{p_1, \dots, p_{n-1}\}$ is the string Coxeter group $[p_1, \dots, p_{n-1}]$, and the symmetry group of the tessellation of Euclidean n-dimensional space with type $\{p_1, \dots, p_n\}$ is the string Coxeter group $[p_1, \dots, p_n]$ (see [55, Chapter XI]). In both cases, the generators of the symmetry group are taken to be hyperplane reflections whose mirrors bound a flag, where the latter is realised geometrically as the convex hull of the centroids of all proper faces, together with the centroid of the polytope itself in the convex case.

It is remarkable that every finite irreducible string Coxeter group is isomorphic to the symmetry group of a convex regular polytope. It follows that the only finite irreducible string Coxeter groups are the infinite families

- $[3, \dots, 3]$ (symmetry group of the n-simplex);
- $[3, \dots, 3, 4]$ (symmetry group of the n-dimensional cross polytope);
- $[p]$, for $p \geq 5$ (symmetry group of the regular convex p-gon;

together with the sporadic string Coxeter groups $[3, 5]$, $[3, 4, 3]$ and $[3, 3, 5]$, which are the symmetry groups of the icosahedron, 24-cell and 600-cell, respectively. Dual polytopes have the same Coxeter group as their symmetry groups, but with the subindices reversed; for instance, the symmetry groups of the cube and the octahedron are often written as $[4, 3]$ and $[3, 4]$, respectively, although as groups of isometries they are the same.

The string Coxeter groups corresponding to the regular tessellations of Euclidean spaces are $[\infty]$, $[3, 6]$, $[3, 3, 4, 3]$ together with the infinite family $[4, 3, 3, \dots, 3, 4]$. The group in the last family having $d - 2 \geq 0$ entries equal to 3 corresponds to the symmetry group of the tessellations of cubes in $\mathbb{E}^d$. The sporadic three groups are isomorphic to the symmetry groups of the tessellation of $\mathbb{E}^1$ by equal segments, the face-to-face tessellations of $\mathbb{E}^2$ by equilateral triangles, and the tessellation of $\mathbb{E}^4$ by cross polytopes, respectively. The groups in the infinite family and the three sporadic groups are known as the string Coxeter groups of *Euclidean type*.

All Coxeter groups, and in particular those with string diagram, satisfy the following *intersection property* (see [140, Chapter 5]).

Lemma 3.1.16 *If* $r_0, \dots, r_{n-1}$ *are the standard generators of a Coxeter group and* $I, J \subseteq \{0, \dots, n-1\}$ *then*

$$\langle r_i \colon i \in I\rangle \cap \langle r_i \colon i \in J\rangle = \langle r_i \colon i \in I \cap J\rangle. \tag{3.4}$$

It is not true that the automorphism group of every abstract regular polytope is a string Coxeter group. For instance, there are infinitely many finite regular toroidal polyhedra (see Section 2.4), but as mentioned above, there exist only finitely many finite string Coxeter groups of rank 3. Nevertheless, the automorphism group of every regular polytope is strongly related to a string Coxeter group.

Proposition 3.1.17 *The automorphism group of every regular n-polytope $\mathcal{P}$ is a quotient of a string Coxeter group. Furthermore, the generators ρ_i of $\Gamma(\mathcal{P})$ satisfy the intersection property* (3.4) *for every $I, J \subseteq \{0, \ldots, n-1\}$.*

Proof The first three items of Proposition 2.2.10 imply that the automorphism group of every regular polytope of type $\{p_1, \ldots, p_{n-1}\}$ is a quotient of the string Coxeter group $[p_1, \ldots, p_{n-1}]$, since the former satisfies all the defining relations of the latter.

The fourth item of Proposition 2.2.10 describes the stabiliser of a subchain Ψ of the base flag Φ as $\Gamma_\Psi = \langle \rho_j : j \notin I' \rangle$, where I' is the set of ranks of the faces in Ψ. If Υ is the subchain of Φ consisting of the faces of ranks in an index set J' then $\Gamma_{\Psi \cup \Upsilon} = \langle \rho_j : j \notin I' \cup J' \rangle$, since $\Psi \cup \Upsilon$ consists of the subchain of Φ whose faces have ranks in $I' \cup J'$. The intersection property (3.4) now follows from Proposition 3.1.11 by taking I' and J' as the complements in $\{0, \ldots, n-1\}$ of I and J, respectively. ∎

A group generated by involutions satisfying all the defining relations of a string Coxeter group (perhaps with some extra defining relations) as well as the intersection property (3.4) is called a *string C-group*. The 'C' stands for Coxeter, but there are many string C-groups that are not string Coxeter groups. Proposition 3.1.17 can be rewritten as follows.

Remark 3.1.18 The automorphism group of every regular polytope is a string C-group.

Not every group is a string C-group, since not every group is generated by involutions (for instance, groups of odd order are not string C-groups). On the other hand, if a group Γ is a string C-group, it may be a string C-group with respect to more than one generating set. Since our interest is to construct n-polytopes $\mathcal{P}(\Gamma)$ from string C-groups Γ through the construction in Section 3.1.1, we shall consider string C-groups not simply as groups, but as pairs consisting of a group and a generating set of involutions $\rho_0, \ldots, \rho_{n-1}$. From this set we may proceed with the construction by defining Γ_i as $\langle \rho_j : j \neq i \rangle$. The *rank* of a string C-group is the cardinality of its generating set.

Example 3.1.19 The symmetric group S_5 is a string C-group with respect to more than one generating set. One of them is the set of transpositions

$$\{(1,2),(2,3),(3,4),(4,5)\},$$

and the poset constructed from

$$\Gamma_0 := \langle (2,3),(3,4),(4,5)\rangle$$
$$\Gamma_1 := \langle (1,2),(3,4),(4,5)\rangle$$
$$\Gamma_2 := \langle (1,2),(2,3),(4,5)\rangle$$
$$\Gamma_3 := \langle (1,2),(2,3),(3,4)\rangle$$

is isomorphic to the face-lattice of the 4-simplex.

The group S_5 is also a string C-group with respect to the generating set

$$\{\rho_0 := (4,5), \rho_1 := (1,2)(3,4), \rho_2 := (2,3)\},$$

and the resulting poset is the coset representation of the polyhedron $\{6,4\} * 120$ in [116] (this is the dual of map N7.1 in [41]).

The next theorem states that string C-groups are precisely the automorphism groups of abstract regular polytopes (see [188, Theorem 2E11]).

Theorem 3.1.20 *Let Γ be a string C-group with respect to the generating set $\{\rho_0, \ldots, \rho_{n-1}\}$, and let $\mathcal{P}(\Gamma)$ be the poset constructed as in Section 3.1.1. Then $\mathcal{P}(\Gamma)$ is a regular n-polytope and $\Gamma(\mathcal{P}(\Gamma)) = \Gamma$.*

Here we do not present a proof of Theorem 3.1.20. One proof can be found in [188, Section 2E]. Alternatively, it can be proved for string C-groups with rank at least 4 using Theorem 3.1.15. The details, and the case of string C-groups of rank at most 3 are left as exercises. Later in this section we shall present a full proof for the analogous result for rotary polytopes.

Exercise 3.1.21 Show that if $\mathcal{P}$ is a regular n-polytope, $I \subseteq \{0, \ldots, n-1\}$, $\Gamma_i = \langle \rho_j : j \neq i\rangle$ and $\Gamma(\mathcal{P}) = \langle \rho_0, \ldots, \rho_{i-1}\rangle$ then

$$\langle \rho_i : i \notin I\rangle = \cap_{i\in I}\Gamma_i.$$

Exercise 3.1.22 Let Γ be a string C-group with respect to the generating set $\{\rho_0, \ldots, \rho_{n-1}\}$, and let $\Gamma_i = \langle \rho_j : j \neq i \rangle$.

1. Show that if $\alpha \in \langle \rho_0, \ldots, \rho_{i-1} \rangle$ and $\beta \in \langle \rho_{i+1}, \ldots, \rho_{n-1} \rangle$ then $\alpha\beta = \beta\alpha$.
2. Show that if $\gamma \in \Gamma_i$ then $\gamma = \gamma_1\gamma_2$ for some $\gamma_1 \in \langle \rho_0, \ldots, \rho_{i-1} \rangle$ and $\gamma_2 \in \langle \rho_{i+1}, \ldots, \rho_{n-1} \rangle$.
3. Show that if $i < j < k$ and $\Gamma_j \cap \Gamma_k \gamma \neq \emptyset$ for some $\gamma \in \Gamma$ then $\Gamma_i \cap \Gamma_k \gamma \neq \emptyset$.

Exercise 3.1.23 Use Exercise 3.1.22 to show that if Γ is a string C-group with rank at least 4 then it satisfies the three items in the statement of Theorem 3.1.15.

Exercise 3.1.24 Show that for $n \leq 3$ every string C-group Γ of rank n is the automorphism group of a regular n-polytope $\mathcal{P}(\Gamma)$ where the generators of Γ act on $\mathcal{P}(\Gamma)$ as the standard reflections.

As a direct consequence of Theorem 3.1.20, the string Coxeter group $[p_1, \ldots, p_{n-1}]$ is the automorphism group of a regular n-polytope, usually denoted by its Schläfli type $\{p_1, \ldots, p_{n-1}\}$. The regular polytope $\{p_1, \ldots, p_{n-1}\}$ is said to be *universal of type* $\{p_1, \ldots, p_{n-1}\}$ because it satisfies the following universal property (see Corollary 3.2.42 for a more general version).

Proposition 3.1.25 *If $\mathcal{P}$ is a regular polytope with type $\{p_1, \ldots, p_{n-1}\}$ then it is a quotient of the universal polytope $\{p_1, \ldots, p_{n-1}\}$.*

Proof Let $[p_1, \ldots, p_{n-1}] = \langle \rho_0, \ldots, \rho_{n-1} \rangle$ and $\Gamma(\mathcal{P}) = \langle \rho'_0, \ldots, \rho'_{n-1} \rangle$. Then the generators $\rho'_0, \ldots, \rho'_{n-1}$ satisfy all defining relations of the generators of $[p_1, \ldots, p_{n-1}]$ (compare (3.3) with Proposition 2.2.10). It follows that there is an epimorphism η from $[p_1, \ldots, p_{n-1}]$ to $\Gamma(\mathcal{P})$ mapping ρ_i to ρ'_i for every i. If N is the kernel of η then $\Gamma(\mathcal{P}) \cong [p_1, \ldots, p_{n-1}]/N$ and we may identify each ρ'_i with the coset $N\rho_i$, and the element $\gamma' = \rho'_{i_1} \cdots \rho'_{i_k} \in \Gamma(\mathcal{P})$ with the coset $N\rho_{i_1} \cdots \rho_{i_k}$.

For $i \in \{0, \ldots, n-1\}$ let $\Gamma_i = \langle \rho_j : j \neq i \rangle \subseteq [p_1, \ldots, p_{n-1}]$ and $\Gamma'_i = \langle \rho'_j : j \neq i \rangle \subseteq \Gamma(\mathcal{P})$. Recall from the automorphism theorems that $\Gamma'_i = (\Gamma_i N)/N \cong \Gamma_i/(N \cap \Gamma_i)$. To show that $\mathcal{P}$ is a quotient of $\{p_1, \ldots, p_{n-1}\}$ we define ζ as the natural mapping from the coset representation of $\{p_1, \ldots, p_{n-1}\}$

to that of $\mathcal{P}$, that is, ζ sends $\Gamma_i\gamma$ to $\Gamma'_i\gamma'$. This is well defined, since if $\Gamma_i\alpha = \Gamma_i\beta$ then $\alpha\beta^{-1} \in \Gamma_i$, and therefore $\alpha'(\beta')^{-1} \in \Gamma'_i$, implying that $\Gamma'_i\alpha' = \Gamma'_i\beta'$.

As a mapping of posets ζ preserves rank. It also preserves incidence, since any two incident faces $\Gamma_i\alpha$ and $\Gamma_j\beta$ can be expressed as $\Gamma_i\gamma$ and $\Gamma_j\gamma$ for some $\gamma \in \Gamma_i\alpha \cap \Gamma_j\beta$, and then

$$\zeta(\Gamma_i\alpha) \cap \zeta(\Gamma_j\beta) = \zeta(\Gamma_i\gamma) \cap \zeta(\Gamma_j\gamma) = \Gamma'_i\gamma' \cap \Gamma'_j\gamma'$$

contains γ'.

To show that ζ preserves adjacencies of flags we use that both $\{p_1, \ldots, p_{n-1}\}$ and $\mathcal{P}$ are regular polytopes. Let $\Phi_0 = \{\Gamma_0, \ldots, \Gamma_{n-1}\}$ be the base flag of (the coset representation of) $\{p_1, \ldots, p_{n-1}\}$, let Φ be any flag of $\{p_1, \ldots, p_{n-1}\}$ and let $\gamma \in [p_1, \ldots, p_{n-1}]$ such that $\Phi = \Phi_0\gamma$. Then $\Phi'_0 := \zeta(\Phi_0) = \{\Gamma'_0, \ldots, \Gamma'_{n-1}\}$ is the base flag of $\Gamma(\mathcal{P})$. For every $i \in \{0, \ldots, n-1\}$,

$$\zeta(\Phi^i) = \zeta(\Phi_0^i\gamma) = \zeta(\Phi_0\rho_i\gamma) = \Phi'_0\rho'_i\gamma' = (\Phi'_0)^i\gamma' = (\Phi'_0\gamma')^i = \zeta(\Phi)^i.$$

Since every polytope is a poset isomorphic to its coset representation, ζ induces a covering from $\{p_1, \ldots, p_{n-1}\}$ to $\mathcal{P}$. ■

The covering ζ in the proof of Proposition 3.1.25 belongs to a special class of coverings defined next.

Definition 3.1.26 Let $\mathcal{P}$ and $\mathcal{Q}$ be polytopes with $\mathcal{P}$ covering $\mathcal{Q}$. We say that $\mathcal{Q}$ is a *quotient* of $\mathcal{P}$ *by automorphisms* whenever there exists $N \leq \Gamma(\mathcal{P})$ such that $\mathcal{Q}$ is the orbit space of $\mathcal{P}$ under N. The partial order is given by $FN \leq GN$ if and only if there exist $F' \in FN$ and $G' \in GN$ such that $F' \leq G'$. In such cases we write $\mathcal{Q} = \mathcal{P}/N$.

The requirement of $\mathcal{P}$ covering $\mathcal{Q}$ in Definition 3.1.26 is no accident, as explained next.

Remark 3.1.27 It may be the case that a polytope $\mathcal{Q}$ is obtained as the orbit space of $\mathcal{P}$ under the action of some $N \leq \Gamma(\mathcal{P})$, and still the corresponding mapping is not a quotient as defined in Definition 2.1.28. For an example of this, see [191, Example 2.13].

Remark 3.1.28 It is not true that for every $N \leq \Gamma(\mathcal{P})$ the structure $\mathcal{P}/N$ is a polytope. Identifying orbits of flags into a single flag may destroy the diamond condition or the strong flag-connectivity. For instance, if $\mathcal{P}$ is fully transitive then $\mathcal{P}/\Gamma(\mathcal{P})$ consists of only one face of each rank, failing the diamond condition.

Several examples of quotients by group actions were given in Section 2.4 from the string Coxeter groups [3,6], [4,4] and [6,3] (see [191, Example 2.15] for an example of a covering of polyhedra not induced by such an action by automorphisms). From the proof of Proposition 3.1.25 we conclude the following.

Proposition 3.1.29 *If $\mathcal{P}$ is a regular polytope with type $\{p_1, \ldots, p_{n-1}\}$ then $\mathcal{P}$ is the quotient of the universal polytope $\{p_1, \ldots, p_{n-1}\}$ by the group action of some $N \leq [p_1, \ldots, p_{n-1}]$.*

The nature of the group $N \leq [p_1, \ldots, p_{n-1}]$ impacts the kind of symmetry of the quotient $\{p_1, \ldots, p_{n-1}\}/N$.

Proposition 3.1.30 *Let $N \leq [p_1, \ldots, p_{n-1}]$ and assume that $\mathcal{P}$ is the quotient of $\{p_1, \ldots, p_{n-1}\}$ determined by the group action of N. Then $\mathcal{P}$ is regular if and only if $N \triangleleft [p_1, \ldots, p_{n-1}]$.*

Proof In the proof of Proposition 3.1.25 it was shown that if $\mathcal{P}$ is regular then N is the kernel of the group epimorphism mapping the standard generators of $[p_1, \ldots, p_{n-1}]$ to those of $\Gamma(\mathcal{P})$. We can conclude that in that case $N \triangleleft [p_1, \ldots, p_{n-1}]$.

Now assume that $N \triangleleft [p_1, \ldots, p_{n-1}]$ and let F be an i-face of $\{p_1, \ldots, p_{n-1}\}$, so that FN is an i-face of $\mathcal{P}$. Then, for every $\alpha \in [p_1, \ldots, p_{n-1}]$ the coset $N\alpha$ acts like an automorphism of $\mathcal{P}$, since $FN(N\alpha) = (F\alpha)N$ is an i-face of $\mathcal{P}$.

Let Φ_0 and Φ'_0 be the base flags of $\{p_1, \ldots, p_{n-1}\}$ and $\mathcal{P}$, respectively, and denote by ζ the covering from $\{p_1, \ldots, p_{n-1}\}$ to $\mathcal{P}$. By definition, ζ must preserve adjacencies, and therefore

$$\Phi'_0 \rho_i N = \zeta(\Phi_0 \rho_i) = \zeta(\Phi_0^i) = (\Phi'_0)^i. \tag{3.5}$$

Consequently, the automorphisms $\rho_i N$ serve as standard reflections for $\mathcal{P}$ and it is a regular polytope. ■

Remark 3.1.31 Given a group $N \leq [p_1, \ldots, p_{n-1}]$ it is possible that the orbit space $\{p_1, \ldots, p_{n-1}\}/N$ has the structure of a polytope, without being a quotient as defined in Section 2.1 (see [191, Example 2.13]). In Proposition 3.1.30 we are assuming that the mapping from $\{p_1, \ldots, p_{n-1}\}$ to $\{p_1, \ldots, p_{n-1}\}/N$ is adjacency preserving for the validity of (3.5), which for arbitrary N may not hold.

The even subgroup of the universal polytope $\{p_1, \ldots, p_{n-1}\}$ is the subgroup of $[p_1, \ldots, p_{n-1}]$ of elements that are expressed as products of even numbers of standard generators. It is an index two subgroup (see [140, Proposition 5.1]), and it is customarily denoted by $[p_1, \ldots, p_{n-1}]^+$. As we shall see, it plays a major role in the coset representations of rotary polytopes.

The correspondence between the classes of regular polytopes and string C-groups (with specified generators) makes it possible to analyse combinatorial aspects of regular polytopes through group-theoretical properties of string C-groups. The following result illustrates this fact with orientability of regular polytopes (see Corollary 3.2.61 for a group-theoretical determination of the lengths of Petrie paths of regular polytopes). For this result we recall that the *normal closure* of a subset $X \leq \Gamma$ is the smallest normal subgroup of Γ that contains X, and it is generated by all conjugates of X.

Lemma 3.1.32 *Let $\mathcal{P}$ be a regular n-polytope of type $\{p_1, \ldots, p_{n-1}\}$ with $\Gamma(\mathcal{P})$ given by the quotient of the Coxeter group $[p_1, \ldots, p_{n-1}]$ by the normal closure of the set $\{w_1, \ldots, w_m\}$ of words in $\{\rho_0, \ldots, \rho_{n-1}\}$. Then $\mathcal{P}$ is orientable if and only if $w_j \in [p_1, \ldots, p_{n-1}]^+$ for all $j \in \{1, \ldots, m\}$.*

Proof We abuse notation and assume that $\mathcal{P} = \{p_1, \ldots, p_{n-1}\}/N$, where N is the normal closure of $\{w_1, \ldots, w_m\}$. We denote by ζ the covering from $\{p_1, \ldots, p_{n-1}\}$ to $\mathcal{P}$.

Assume first that $\mathcal{P}$ is orientable, let $j \in \{1, \ldots, m\}$ and suppose that $w_j = \rho_{i_1} \cdots \rho_{i_k}$. If Φ_0 is the base flag of $\{p_1, \ldots, p_{n-1}\}$ then

$$\Phi_0 w_j = \Phi_0 \rho_{i_1} \cdots \rho_{i_k} = \Phi_0^{i_k, \ldots, i_1}.$$

Since $\mathcal{P}$ is a quotient of $\{p_1, \ldots, p_{n-1}\}$,

$$\zeta(\Phi_0) = \zeta(\Phi_0 w_j) = \zeta(\Phi_0^{i_k, \ldots, i_1}) = \zeta(\Phi_0)^{i_k, \ldots, i_1}.$$

Orientability of $\mathcal{P}$ implies that k is even, and therefore $w_j \in [p_1, \ldots, p_{n-1}]^+$.

Conversely, if $w_j \in [p_1, \ldots, p_{n-1}]^+$ for all $j \in \{1, \ldots, m\}$ then N is contained in $[p_1, \ldots, p_{n-1}]^+$. It follows that every $\nu \in N$ preserves the colours of the bicolouring of the flags of $\{p_1, \ldots, p_{n-1}\}$. This makes it possible to define a bicolouring of flags of $\mathcal{P}$ by assigning a flag $\zeta(\Phi)$ the colour of any of its preimages $\Phi\nu$ ($\nu \in N$). Since quotients preserve adjacencies, $\mathcal{P}$ is orientable. ■

String C-groups are a particular case of a more general class of groups. A *string group generated by involutions* (or *sggi*, for short) is a group generated by distinct involutions $\rho_0, \ldots, \rho_{n-1}$ such that if $|i - j| \geq 2$ then $(\rho_i \rho_j)^2 = \mathrm{id}$. Note that the class of sggi of rank n consists precisely of the quotients of string Coxeter groups of rank n where the kernels do not contain generators or products of two distinct generators. It follows that string C-groups are sggi that satisfy the intersection property (3.4).

An sggi Γ is *orientable* whenever each defining relation of Γ has an even number of factors when written as a word in the generating set $\{\rho_0, \ldots, \rho_{n-1}\}$. By definition all string Coxeter groups are orientable sggi.

The poset $\mathcal{P}(\Gamma)$ described in Section 3.1.1 can be constructed for sggi Γ that are not string C-groups. In those cases $\mathcal{P}(\Gamma)$ will not be a polytope, but may still have an interesting structure. For convenience we keep the notation $\mathcal{P}(\Gamma)$ for the general case when Γ is an sggi.

Exercise 3.1.33 Let $\Gamma = \langle x, y :: x^2 = y^2 = (xy)^4 = \mathrm{id}\rangle$ be the dihedral group with eight elements.

1. Show that Γ does not have a generating set that makes it a string C-group of rank 3, but that it is an sggi with respect to the generating set $\{x, y, yxy\}$.
2. Show that if $\{x, y, yxy\}$ is chosen as the generating set then $\mathcal{P}(\Gamma)$ is isomorphic to the face-lattice of the toroidal map $\{4, 4\}_{(1,0)}$.

It is relatively easy to construct sggi, either as arbitrary quotients of string Coxeter groups, or by carefully choosing involutions of a given group to form a generating set $\{\rho_0, \ldots, \rho_{n-1}\}$ such that $(\rho_i\rho_j)^2 = \mathrm{id}$ whenever $|i - j| \geq 2$. In comparison, it is generally hard to determine whether a given sggi satisfies the intersection property (3.4). In principle, the verification of the intersection property would require a number of comparisons of groups that grows exponentially with the rank of the sggi in question. The following two results provide easier procedures to determine when an sggi satisfies the intersection property (see [188, Proposition 2E16]).

Proposition 3.1.34 *Let* $\Gamma = \langle \rho_0, \ldots, \rho_{n-1}\rangle$ *be an sggi and assume that* $\Gamma_{n-1} = \langle \rho_0, \ldots, \rho_{n-2}\rangle$ *is a string C-group. If* $\Gamma_{n-1} \cap \langle \rho_j, \ldots, \rho_{n-1}\rangle = \langle \rho_j, \ldots, \rho_{n-2}\rangle$ *for every* $j \in \{1, \ldots, n-1\}$ *then* Γ *is a string C-group.*

Proof For any $I \subseteq \{0, \ldots, n-1\}$ we denote $\langle \rho_j : j \notin I\rangle$ by $\Gamma_{\overline{I}}$. Since Γ is an sggi, the group Γ_i can be expressed as the direct product

$$\Gamma_{\{0,\ldots,i-1\}} \times \Gamma_{\{i+1,\ldots,n-1\}} \tag{3.6}$$

(see Exercise 3.1.22).

Clearly $\Gamma_{\overline{I}} \subseteq \bigcap_{i\in I}\Gamma_i$; we shall prove that $\bigcap_{i\in I}\Gamma_i \subseteq \Gamma_{\overline{I}}$. Assume that $m \geq 2$ and $I = \{i_1, \ldots, i_m\}$ where $i_j < i_k$ if $j < k$. If $\gamma \in \bigcap_{i\in I}\Gamma_i$ then $\gamma \in \Gamma_{i_m}$ and therefore it can be expressed as $\gamma = \alpha\beta$ for some $\alpha \in \langle \rho_0, \ldots, \rho_{i_m-1}\rangle$ and $\beta \in \langle \rho_{i_m+1}, \ldots, \rho_{n-1}\rangle$. In particular, $\beta \in \Gamma_{\overline{I}}$ and it remains to show that $\alpha \in \Gamma_{\overline{I}}$. We know that $\alpha \in \langle \rho_0, \ldots, \rho_{i_m-1}\rangle \subseteq \Gamma_{n-1}$ and we also know that

$\beta, \gamma \in \Gamma_i$ for every $i \in I$. It follows that $\alpha \in \Gamma_i$ for every $i \in I$. For each $i \in I$ we express $\alpha = \delta_i \zeta_i$ with $\delta_i \in \langle \rho_0, \ldots, \rho_{i-1} \rangle$ and $\zeta_i \in \langle \rho_{i+1}, \ldots, \rho_{n-1} \rangle$. Since $\delta_i \in \Gamma_{n-1}$, we must have that $\zeta_i \in \Gamma_{n-1}$ as well. The hypothesis of this proposition implies that $\zeta_i \in \langle \rho_{i+1}, \ldots, \rho_{n-2} \rangle$ and hence $\alpha \in \Gamma_{\overline{\{i,n-1\}}}$. But Γ_{n-1} is a string C-group that contains $\Gamma_{\overline{\{i,n-1\}}}$ for every $i \in I$, and therefore

$$\alpha \in \cap_{i \in I} \Gamma_{\overline{\{i,n-1\}}} = \Gamma_{\overline{I \cup \{n-1\}}} \subseteq \Gamma_{\overline{I}}$$

(see Exercise 3.1.21). We conclude that $\Gamma_{\overline{I}} = \cap_{i \in I} \Gamma_i$.

Now, if $I, J \subseteq \{0, \ldots, n-1\}$ then

$$\Gamma_I \cap \Gamma_J = (\cap_{i \notin I} \Gamma_i) \cap (\cap_{i \notin J} \Gamma_i) = \cap_{i \notin I \cap J} \Gamma_i = \Gamma_{I \cap J}.$$

■

Proposition 3.1.35 *Let $\Gamma = \langle \rho_0, \ldots, \rho_{n-1} \rangle$ be an sggi and assume that $\Gamma_{n-1} = \langle \rho_0, \ldots, \rho_{n-2} \rangle$ and $\Gamma_0 = \langle \rho_1, \ldots, \rho_{n-1} \rangle$ are string C-groups. If $\Gamma_{n-1} \cap \Gamma_0 = \langle \rho_1, \ldots, \rho_{n-2} \rangle$ then Γ is a string C-group.*

Proof We shall show that $\Gamma_{n-1} \cap \langle \rho_j, \ldots, \rho_{n-1} \rangle = \langle \rho_j, \ldots, \rho_{n-2} \rangle$ for every $j \in \{1, \ldots, n-1\}$, and then this proposition follows from Proposition 3.1.34. Given $j \in \{1, \ldots, n-1\}$ we can express $\Gamma_{n-1} \cap \langle \rho_j, \ldots, \rho_{n-1} \rangle$ as

$$\begin{aligned} \Gamma_{n-1} \cap \langle \rho_j, \ldots, \rho_{n-1} \rangle &= \Gamma_{n-1} \cap \Gamma_0 \cap \langle \rho_j, \ldots, \rho_{n-1} \rangle \\ &= \langle \rho_1, \ldots, \rho_{n-2} \rangle \cap \langle \rho_j, \ldots, \rho_{n-1} \rangle. \end{aligned}$$

Since Γ_0 is a string C-group containing $\langle \rho_1, \ldots, \rho_{n-2} \rangle$ and $\langle \rho_j, \ldots, \rho_{n-1} \rangle$, we have that $\langle \rho_1, \ldots, \rho_{n-2} \rangle \cap \langle \rho_j, \ldots, \rho_{n-1} \rangle = \langle \rho_j, \ldots, \rho_{n-2} \rangle$ as desired. ■

Proposition 3.1.35 could also be proven through the poset $\mathcal{P}(\Gamma)$ with the help of Proposition 2.1.44.

Exercise 3.1.36 Let Γ, Γ' be the subgroups of the symmetric group S_7 with generating sets

$$\{(1,2), (2,3)(4,5), (3,4)(5,6), (4,5)(6,7)\},$$

$$\{(1,2), (2,3)(4,5)(6,7), (3,4)(5,6), (4,7)(5,6)\},$$

respectively. Show that with respect to the given generating sets Γ and Γ' are sggi, that Γ is a string C-group, but Γ' is not.

We conclude this section with another useful criterion to determine that some sggi are string C-groups, in terms of coverings (see [188, Lemma 2E17]).

Lemma 3.1.37 *Let* $\Gamma = \langle \rho_0, \ldots, \rho_{n-1} \rangle$ *be an sggi and let* $\Delta = \langle \rho'_0, \ldots, \rho'_{n-1} \rangle$ *be a string C-group. If there exists an epimorphism from* Γ *to* Δ *mapping* ρ_i *to* ρ'_i *for every* i*, with the property that restricted to* $\Gamma_{n-1} = \langle \rho_0, \ldots, \rho_{n-2} \rangle$ *or to* $\Gamma_0 = \langle \rho_1, \ldots, \rho_{n-1} \rangle$ *is an isomorphism, then* Γ *is a string C-group.*

Proof Let ζ be the mapping from Γ to Δ. We assume that $\zeta_{|\Gamma_{n-1}}$ is an isomorphism. A dual argument can be used for the case when $\zeta_{|\Gamma_0}$ is an isomorphism.

According to Proposition 3.1.34, we only need to verify that

$$\Gamma_{n-1} \cap \langle \rho_j, \ldots, \rho_{n-1} \rangle = \langle \rho_j, \ldots, \rho_{n-2} \rangle$$

for every $j \in \{1, \ldots, n-1\}$. Let $j \in \{1, \ldots, n-1\}$, let $\gamma \in \Gamma_{n-1} \cap \langle \rho_j, \ldots, \rho_{n-1} \rangle$, and let $\gamma' = \zeta(\gamma)$. Then $\gamma' \in \langle \rho'_0, \ldots, \rho'_{n-2} \rangle \cap \langle \rho'_j, \ldots, \rho'_{n-1} \rangle$. Since Δ is a string C-group, γ' must be in $\langle \rho'_j, \ldots, \rho'_{n-2} \rangle$, and since ζ is one-to-one on Γ_{n-1} then $\gamma \in \langle \rho_j, \ldots, \rho_{n-2} \rangle$. This shows that

$$\Gamma_{n-1} \cap \langle \rho_j, \ldots, \rho_{n-1} \rangle \subseteq \langle \rho_j, \ldots, \rho_{n-2} \rangle,$$

and the other inclusion always holds. ■

3.1.3 Rotary Polytopes

We turn our attention to orientable rotary polytopes $\mathcal{P}$ and their coset representations with respect to their rotation subgroups $\Gamma^+(\mathcal{P})$.

We start by recalling from Theorem 2.3.9 as well as from Propositions 2.2.14, 2.2.16 and 2.3.14 that $\Gamma^+(\mathcal{P})$ is generated by the standard rotations $\sigma_1, \ldots, \sigma_{n-1}$ and that they satisfy

$$(\sigma_i \sigma_{i+1} \cdots \sigma_j)^2 = \text{id} \tag{3.7}$$

for all $i < j$.

In order to obtain the coset representation of $\mathcal{P}$ with respect to $\Gamma^+(\mathcal{P})$ we need a description of the stabiliser Γ_i of the i-face of the base flag Φ_0. The strong flag-connectivity implies that Γ_i consists of all automorphisms $\gamma \in \Gamma^+(\mathcal{P})$ such that $\Phi_0\gamma = \Phi_0^{j_1, \ldots, j_{2m}}$ for some $j_1, \ldots, j_{2m} \in \{0, \ldots, n-1\} \setminus \{i\}$ and some non-negative integer m. In particular, Γ_i contains all standard rotations σ_j such that $j-1, j \neq i$. More generally, Γ_i contains all automorphisms $\tau_{j,k}$ defined by

$$\Phi_0 \tau_{j,k} = \Phi_0^{k,j} \tag{3.8}$$

for $j \neq k \in \{0, \ldots, n-1\} \setminus \{i\}$, of which the automorphisms τ_k in (2.9) are the particular case $\tau_{k,0}$. Here we deviate from the notation in [247], where $\Phi_0 \tau_{j,k}$

was defined to be $\Phi_0^{k,j-1}$ for $j \leq k$; the notation here is more flexible since any $j,k \in \{0,\dots,n-1\}$ may appear as first or second subindex, regardless of how j and k compare.

We observe that $\tau_{j,j} = \mathrm{id}$, that $\tau_{j,k}^{-1} = \tau_{k,j}$, and that for every $j < k$ the automorphism $\tau_{j,k}$ can be expressed as $\sigma_{j+1}\cdots\sigma_k$ (see the second item of Proposition 2.2.14). It follows that if $|k-j| \geq 2$ then $\tau_{j,k}$ is an involution and it satisfies that $\tau_{k,j} = \tau_{j,k}$. Furthermore, σ_j can be written also as $\tau_{j-1,j}$, implying that we have translations of expressions in terms of the generators σ_i into expressions in terms of the elements $\tau_{i,j}$, and vice versa. These simple observations allow us to find natural generating sets in terms of the elements $\tau_{i,j}$ for some subgroups of $\Gamma^+(\mathcal{P})$.

Remark 3.1.38 Let $\mathcal{P}$ be a rotary n-polytope and let $1 \leq i < j \leq n-1$. Then

$$\langle \sigma_i, \dots, \sigma_j \rangle = \langle \tau_{k,\ell} \colon i-1 \leq k \leq \ell \leq j \rangle.$$

A useful generating set for Γ_i will be shown in Corollary 3.1.40 as a consequence of the next more general lemma.

Lemma 3.1.39 *Let $\mathcal{P}$ be a rotary n-polytope with base flag Φ_0 and standard rotations $\sigma_1, \dots, \sigma_{n-1}$, let $\tau_{j,k}$ be defined as in* (3.8)*, and let Ψ be the subset of Φ_0 consisting of the faces with ranks in some set $I \subseteq \{0,\dots,n-1\}$. Then the stabiliser Γ_I in $\Gamma^+(\mathcal{P})$ of Ψ is*

$$\langle \tau_{j,k} \colon j,k \in \{0,\dots,n-1\} \setminus I \rangle.$$

Proof Let Λ_I denote $\langle \tau_{j,k} \colon j,k \in \{0,\dots,n-1\} \setminus I\rangle$. It is clear that every $\tau_{j,k}$ with $j,k \notin I$ belongs to Γ_I, and so we only need to show that $\Gamma_I \subseteq \Lambda_I$.

Let $\gamma \in \Gamma_I$. The strong flag-connectivity implies that there exist $j_1,\dots,j_m \in \{0,\dots,n-1\} \setminus I$ such that $\Phi_0\gamma = \Phi_0^{j_1,\dots,j_m}$. Since $\gamma \in \Gamma^+(\mathcal{P})$, m must be even.

Now,

$$\Phi_0\gamma = \Phi_0^{j_1,\dots,j_m} = (\Phi_0\tau_{j_2,j_1})^{j_3,\dots,j_m} = \Phi_0^{j_3,\dots,j_m}\tau_{j_2,j_1},$$

and continuing in an inductive manner we see that $\Phi_0\gamma = \Phi_0\tau_{j_m,j_{m-1}}\cdots\tau_{j_2,j_1}$. We can conclude that $\gamma \in \Lambda_I$, and the lemma holds. ■

Corollary 3.1.40 *Let $\mathcal{P}$ be a rotary n-polytope with base flag Φ_0 and standard rotations $\sigma_1,\dots,\sigma_{n-1}$. Then the stabiliser in $\Gamma^+(\mathcal{P})$ of the i-face in Φ_0 is*

$$\Gamma_i = \langle \tau_{j,k} \colon j,k \in \{0,\dots,n-1\} \setminus \{i\} \rangle. \tag{3.9}$$

Corollary 3.1.41 *Let $\mathcal{P}$ be a rotary n-polytope, for each i let Γ_i be the stabiliser in $\Gamma^+(\mathcal{P})$ of the base i-face, and for $I \subseteq \{0, \ldots, n-1\}$ let $\Gamma_{\overline{I}}$ be $\langle \tau_{j,k} \colon j, k \notin I\rangle$. Then*

$$\Gamma_{\overline{I}} = \cap_{i \in I} \Gamma_i.$$

Proof This follows from Proposition 3.1.11, since Γ_i is the stabiliser of the base i-face (see Corollary 3.1.40), and $\Gamma_{\overline{I}}$ is the stabiliser of the chain consisting of the i-faces with $i \in I$ (see Lemma 3.1.39). ■

The generating set in (3.9) is redundant when $n \geq 4$. Corollary 2.3.18 makes this clear for Γ_0 and Γ_{n-1}. We next present simpler generating sets for the stabiliser Γ_i.

Lemma 3.1.42 *Let $n \geq 3$, let $i \in \{1, \ldots, n-2\}$, let $\mathcal{P}$ be a rotary n-polytope and let Γ_i denote the stabiliser in $\Gamma^+(\mathcal{P})$ of the base i-face. Then*

$$\Gamma_i = \langle \sigma_1, \ldots, \sigma_{i-1}, \tau_{i-1,i+1}, \sigma_{i+2}, \ldots, \sigma_{n-1}\rangle,$$

where $\tau_{k,j}$ is defined as in (3.8).

Furthermore, if $n \geq 4$ and $k \in \{1, \ldots, n-2\} \setminus \{i\}$ then

$$\Gamma_i = \langle \tau_{k,j} \colon j \in \{0, \ldots, n-1\} \setminus \{i, k\}\rangle.$$

Proof We let Φ be the base flag and denote

$$\Lambda := \langle \sigma_1, \ldots, \sigma_{i-1}, \tau_{i-1,i+1}, \sigma_{i+2}, \ldots, \sigma_{n-1}\rangle,$$
$$\Delta := \langle \tau_{k,j} \colon j \in \{0, \ldots, n-1\} \setminus \{i, k\}\rangle.$$

Corollary 3.1.40 directly implies that $\Lambda, \Delta \subseteq \Gamma_i$.

For any $j, j' \in \{0, \ldots, n-1\} \setminus \{i\}$ satisfying that $j' < j$, we have that

$$\begin{aligned}
\Phi\sigma_j \cdots \sigma_{j'} &= \Phi^{j,j-1}\sigma_{j+1} \cdots \sigma_{j'} = (\Phi\sigma_{j+1} \cdots \sigma_{j'})^{j,j-1} \\
&= (\Phi^{j+1,j}\sigma_{j+2} \cdots \sigma_{j'})^{j,j-1} = \cdots \\
&= \Phi^{j',j'-1,j'-1,j'-2,\ldots,j+1,j,j,j-1} \\
&= \Phi^{j',j-1}.
\end{aligned}$$

It follows that if $j' < i < j$ then

$$\tau_{j',j} = \sigma_{j'+1} \cdots \sigma_{i-1}\tau_{i-1,i+1}\sigma_{i+2} \cdots \sigma_j.$$

This shows that all generators $\tau_{j,j'}$ of Γ_i are in Λ, and we can conclude that $\Gamma_i \subseteq \Lambda$.

To finish the proof it suffices to show that $\tau_{i-1,i+1} \in \Delta$, and that for every $j \in \{0, \ldots, n-1\} \setminus \{i, i+1\}$, we also have that $\sigma_j \in \Delta$. This is easily verified as follows:

$$\Phi\tau_{i-1,i+1} = \Phi^{i+1,i-1} = \Phi^{i+1,k,k,i-1} = \Phi\tau_{k,i-1}^{-1}\tau_{k,i+1},$$
$$\Phi\sigma_j = \Phi^{j,j-1} = \Phi^{j,k,k,j-1} = \Phi\tau_{k,j-1}^{-1}\tau_{k,j},$$

where Φ denotes the base flag. ∎

As a consequence of Proposition 3.1.11 and Lemma 3.1.39, the rotation subgroup of every orientable rotary polytope must satisfy the following *intersection property*.

Proposition 3.1.43 *Let $\mathcal{P}$ be an orientable rotary polytope and let $\tau_{i,j}$ be as defined in* (3.8). *Then*

$$\langle \tau_{i,j} : i, j \in I \rangle \cap \langle \tau_{i,j} : i, j \in J \rangle = \langle \tau_{i,j} : i, j \in I \cap J \rangle. \tag{3.10}$$

The intersection property (3.10) cannot be replaced by its analogue

$$\langle \sigma_i : i \in I \rangle \cap \langle \sigma_i : i \in J \rangle = \langle \sigma_i : i \in I \cap J \rangle \tag{3.11}$$

in terms of the standard rotations to that of regular polytopes in (3.4). This is illustrated by the next example.

Example 3.1.44 The rotation group of the 4-simplex is isomorphic to the alternating group A_5 with generators $\sigma_1 = (1,3,2)$, $\sigma_2 = (2,4,3)$ and $\sigma_3 = (3,5,4)$ (compare with Exercise 2.2.15). Conjugating σ_1 with σ_3 and σ_3^{-1} we get that $(1,3,2), (1,4,2), (1,5,2) \in \langle \sigma_1, \sigma_3 \rangle$. It is well known that these three 3-cycles generate A_5; consequently,

$$\langle \sigma_1, \sigma_3 \rangle \cap \langle \sigma_2 \rangle = \langle \sigma_2 \rangle.$$

On the other hand, we know that the 4-simplex is a rotary polytope, implying that (3.11) does not constitute a valid intersection condition on the groups generated by the standard rotations.

Next we investigate the structure of groups $\Gamma = \langle \sigma_1, \ldots, \sigma_{n-1} \rangle$ satisfying (3.7) and the intersection property (3.10). For convenience we shall refer by *string even groups* to groups $\langle \sigma_1, \ldots, \sigma_{n-1} \rangle$ where $\sigma_i \neq \text{id}$ for every i, and such that they satisfy (3.7). The even subgroups of string C-groups, and of sggi in general, are string even groups. For every string even group we define

$$\tau_{i,i} := \text{id}, \qquad \tau_{i,j} := \sigma_{i+1} \cdots \sigma_j, \qquad \tau_{j,i} := \tau_{i,j}^{-1} \tag{3.12}$$

for $0 \leq i < j \leq n-1$, so that if Γ is the rotation group of a rotary polytope then the definitions of $\tau_{i,j}$ in (3.8) and (3.12) are equivalent. In particular, $\sigma_i = \tau_{i-1,i}$. The following lemma states some elementary properties of the elements σ_i and $\tau_{i,j}$, and its proof is left as an exercise.

Lemma 3.1.45 *Let $\Gamma = \langle \sigma_1, \ldots, \sigma_{n-1} \rangle$ be a string even group and for $i, j \in \{0, \ldots, n-1\}$ let $\tau_{i,j}$ be defined as in* (3.12). *Then*

(a) if $|i - j| \geq 3$ then $\sigma_i \sigma_j = \sigma_j \sigma_i$;
(b) if $|i - j| \geq 2$ then $\tau_{i,j}$ is an involution;
(c) $\tau_{i,j}\tau_{j,k} = \tau_{i,k}$ for every $i, j, k \in \{0, \ldots, n-1\}$;
(d) if $k, \ell \notin \{i-1, i+1, j-1, j+1\}$ then $\tau_{i,j}\tau_{k,\ell} = \tau_{k,\ell}\tau_{i,j}$.

Exercise 3.1.46 Prove Lemma 3.1.45.

In what follows we discuss the subgroups of Γ that are natural candidates to be chosen as the subgroups Γ_i to construct a combinatorial structure from Γ. For $i \in \{0, \ldots, n-1\}$ we define Γ_i as $\langle \tau_{j,k} \colon j, k \in \{0, \ldots, n-1\} \setminus \{i\} \rangle$, so that in the case that Γ is the rotation group of a rotary polytope then this definition coincides with the description in Corollary 3.1.40. Among other useful properties, the next lemma provides another familiar generating set for Γ_i.

Lemma 3.1.47 *Let $\Gamma = \langle \sigma_1, \ldots, \sigma_{n-1} \rangle$ be a string even group, let $\tau_{j,k}$ be defined as in* (3.12)*, let $\Gamma_i := \langle \tau_{j,k} \colon j, k \in \{0, \ldots, n-1\} \setminus \{i\} \rangle$, and denote $\Gamma_{<i} := \langle \sigma_1, \ldots, \sigma_{i-1} \rangle$ and $\Gamma_{>i+1} := \langle \sigma_{i+2}, \ldots, \sigma_{n-1} \rangle$. Then*

1. $\Gamma_{<i}\Gamma_{>i+1} = \Gamma_{>i+1}\Gamma_{<i}$.
2. $\Gamma_{<i}\tau_{i-1,i+1} = \tau_{i-1,i+1}\Gamma_{<i}$ and $\tau_{i-1,i+1}\Gamma_{>i+1} = \Gamma_{>i+1}\tau_{i-1,i+1}$.
3. $\Gamma_i = \langle \sigma_1, \ldots, \sigma_{i-1}, \tau_{i-1,i+1}, \sigma_{i+2}, \ldots, \sigma_{n-1} \rangle = \Gamma_{<i}\langle \tau_{i-1,i+1} \rangle \Gamma_{>i+1}$.

Proof The first item is a direct consequence of the first item of Lemma 3.1.45.

If $j \leq i - 3$ then from the first item of Lemma 3.1.45, we know that

$$\sigma_j \tau_{i-1,i+1} = \sigma_j \sigma_i \sigma_{i+1} = \sigma_i \sigma_{i+1} \sigma_j = \tau_{i-1,i+1} \sigma_j. \tag{3.13}$$

As an application of (3.7) we obtain

$$\begin{aligned} \sigma_{i-1}\tau_{i-1,i+1} &= \sigma_{i-1}\sigma_i\sigma_{i+1} = \sigma_{i+1}^{-1}\sigma_i^{-1}\sigma_{i-1}^{-1} = \sigma_i\sigma_{i+1}\sigma_{i-1}^{-1} \\ &= \tau_{i-1,i+1}\sigma_{i-1}^{-1}. \end{aligned} \tag{3.14}$$

Combining the first item of Lemma 3.1.45 with (3.7) it follows that

$$\begin{aligned} \sigma_{i-2}\tau_{i-1,i+1} &= \sigma_{i-2}\sigma_i\sigma_{i+1} = \sigma_{i-2}\sigma_{i+1}^{-1}\sigma_i^{-1} = \sigma_{i+1}^{-1}\sigma_{i-2}\sigma_i^{-1} \\ &= \sigma_{i+1}^{-1}\sigma_{i-2}\sigma_{i-1}\sigma_{i-1}^{-1}\sigma_i^{-1} = \sigma_{i+1}^{-1}\sigma_{i-2}\sigma_{i-1}\sigma_i\sigma_{i-1} \\ &= \sigma_{i+1}^{-1}\sigma_i^{-1}\sigma_{i-1}^{-1}\sigma_{i-2}^{-1}\sigma_{i-1} = \sigma_i\sigma_{i+1}\sigma_{i-1}^{-1}\sigma_{i-2}^{-1}\sigma_{i-1} \\ &= \tau_{i-1,i+1}\sigma_{i-1}^{-1}\sigma_{i-2}^{-1}\sigma_{i-1}. \end{aligned} \tag{3.15}$$

From (3.13), (3.14) and (3.15) we conclude that $\Gamma_{<i}\tau_{i-1,i+1} = \tau_{i-1,i+1}\Gamma_{<i}$. The equality $\tau_{i-1,i+1}\Gamma_{>i+1} = \Gamma_{>i+1}\tau_{i-1,i+1}$ can be proven in an analogous way.

The inclusions

$$\Gamma_{<i}\langle\tau_{i-1,i+1}\rangle\Gamma_{>i+1} \subseteq \langle\sigma_1,\ldots,\sigma_{i-1},\tau_{i-1,i+1},\sigma_{i+2},\ldots,\sigma_{n-1}\rangle \subseteq \Gamma_i$$

are immediate. In order to prove the third item it remains to show that $\Gamma_i \subseteq \Gamma_{<i}\langle\tau_{i-1,i+1}\rangle\Gamma_{>i+1}$. Let $j,k \in \{0,\ldots,n-1\} \setminus \{i\}$ with $j < k$. If $k < i$ then $\tau_{j,k} = \sigma_{j+1}\cdots\sigma_k \in \Gamma_{<i}$. Similarly, if $j > i$ then $\tau_{j,k} = \sigma_{j+1}\cdots\sigma_k \in \Gamma_{>i+1}$. Finally, if $j < i < k$ then

$$\tau_{j,k} = \sigma_{j+1}\cdots\sigma_{i-1}\tau_{i-1,i+1}\sigma_{i+2}\cdots\sigma_k \in \Gamma_{<i}\langle\tau_{i-1,i+1}\rangle\Gamma_{>i+1}.$$

It follows that all generators of Γ_i are in $\Gamma_{<i}\langle\tau_{i-1,i+1}\rangle\Gamma_{>i+1}$, concluding the proof. ■

As we have seen, string even groups possess some key properties of the rotation groups of rotary polytopes. Other such properties are exhibited in the next results.

Lemma 3.1.48 *Let $i \in \{1,\ldots,n-1\}$ and let $\Gamma = \langle\sigma_1,\ldots,\sigma_{n-1}\rangle$ be a string even group. If $k \in \{i-1,i\}$ and either $\ell < i-2$ or $\ell > i+1$ then $\langle\sigma_i\rangle\tau_{k,\ell} = \tau_{k,\ell}\langle\sigma_i\rangle$.*

Proof If $\ell < i-2$ and $k = i-1$ then

$$\sigma_i^{-1}\tau_{k,\ell} = \sigma_i^{-1}\sigma_{i-1}^{-1}\cdots\sigma_{\ell+1}^{-1} = \sigma_{\ell+1}\cdots\sigma_{i-1}\sigma_i = \tau_{k,\ell}\sigma_i,$$

and in the case that $\ell < i-2$ and $k = i$, we have that

$$\sigma_i\tau_{k,\ell} = \sigma_i\sigma_i^{-1}\cdots\sigma_{\ell+1}^{-1} = \sigma_{i-1}^{-1}\cdots\sigma_{\ell+1}^{-1} = \sigma_{\ell+1}\cdots\sigma_{i-1} = \tau_{k,\ell}\sigma_i^{-1}.$$

In both cases the statement follows. The cases when $\ell > i+1$ are analogous. ■

Lemma 3.1.49 *Let $\Gamma = \langle\sigma_1,\ldots,\sigma_{n-1}\rangle$ be a string even group. Let $i,j,k \in \{0,\ldots,n-1\}$ such that $i < j < k$, and let $\Gamma_0,\ldots,\Gamma_{n-1}$ be as in (3.9). If $\Gamma_j \cap \Gamma_k\gamma \neq \varnothing$ for some $\gamma \in \Gamma$ then $\Gamma_i \cap \Gamma_k\gamma \neq \varnothing$. Furthermore, if $\Gamma_j \cap \Gamma_k\gamma \neq \varnothing$ and $n \geq 4$ then $\Gamma_i \cap \Gamma_j \cap \Gamma_k\gamma \neq \varnothing$.*

Proof If $\Gamma_j \cap \Gamma_k\gamma \neq \varnothing$ then $\Gamma_k\gamma = \Gamma_k\gamma'$ for some $\gamma' \in \Gamma_j$. We may use the third item of Lemma 3.1.47 to rewrite γ' as $\alpha\tau_{j-1,j+1}^c\beta$ for some $c \in \{0,1\}$, $\alpha \in \Gamma_{<j}$ and $\beta \in \Gamma_{>j+1}$. We define $\gamma'' := \alpha^{-1}\gamma = \tau_{j-1,j+1}^c\beta$.

In the case that $i < j-1$ or $c = 0$ we can use Corollary 3.1.40 to imply that $\gamma'' = \tau_{j-1,j+1}^c\beta \in \Gamma_i$ and that $\gamma'' = \alpha^{-1}\gamma' \in \Gamma_k\alpha^{-1}\gamma' = \Gamma_k\gamma$.

On the other hand, if $i+1 = j$ and $c = 1$ then

$$\sigma_j^{-1}\gamma'' = \sigma_j^{-1}\tau_{j-1,j+1}\beta = \sigma_{j+1}\beta = \sigma_{i+2}\beta \in \Gamma_i,$$

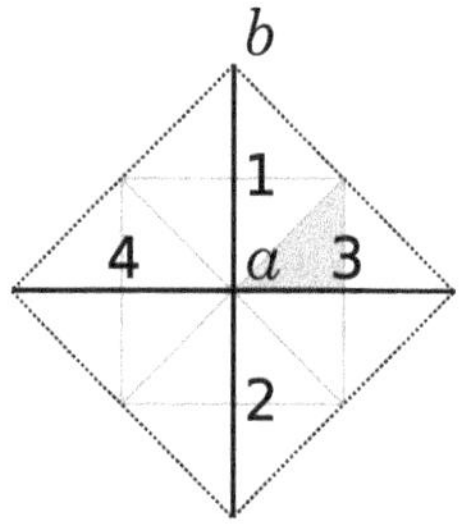

Figure 3.4 The rotation group of the map $\{4,4\}_{(1,1)}$ does not satisfy the intersection condition.

and $\sigma_j^{-1}\gamma'' = \sigma_j^{-1}\alpha^{-1}\gamma' \in \Gamma_k\sigma_j^{-1}\alpha^{-1}\gamma' = \Gamma_k\gamma$. In both cases we can conclude that $\Gamma_i \cap \Gamma_k\gamma \neq \varnothing$.

If $n \geq 4$ and $i < j-1$ then $\gamma'' \in \Gamma_i \cap \Gamma_j \cap \Gamma_k\gamma$. In the case that $n \geq 4$, $i = j-1$ and $c = 1$ we conclude from Lemma 3.1.45 (c) that $\tau_{\ell,i}\gamma'' = \tau_{\ell,j+1}\beta$ for any $\ell \in \{0,\ldots,n-1\} \setminus \{i,j,k\}$, and so $\tau_{\ell,i}\gamma'' \in \Gamma_i \cap \Gamma_j \cap \Gamma_k\gamma$. ■

For a group $\Gamma = \langle\sigma_1,\ldots,\sigma_{n-1}\rangle$ to be the rotation group of a rotary n-polytope it is not enough to be a string even group. For example, the rotation group of the map $\{4,4\}_{(1,1)}$ in Figure 3.4 is faithfully represented as a permutation group on the chains consisting of a vertex and an edge. With respect to the grey flag, the generators σ_1 and σ_2 correspond to the permutations $(a,b)(3,1,4,2)$ and $(3,2,4,1)$, respectively. Then $\sigma_1\sigma_2$ corresponds to the permutation (a,b), implying that the generators of the rotation group of $\{4,4\}_{(1,1)}$ satisfy (3.7), although it is not a polyhedron. This last statement can be verified alternatively from the fact that it does not satisfy the intersection condition (3.10). From Lemma 3.1.42 we can write $\Gamma_0 = \langle\sigma_2\rangle$ and $\Gamma_2 = \langle\sigma_1\rangle$, but

$$\Gamma_0 \cap \Gamma_2 \neq \langle\tau_{i,j} : i,j \in \{1\}\rangle = \{\mathrm{id}\}$$

since $\sigma_1^2 = \sigma_2^2$ is represented by the permutation $(1,2)(3,4)$.

It is now evident that we need to study string even groups that satisfy the intersection condition (3.10). We start with the following lemma (compare with the last conclusion of Lemma 3.1.42).

Lemma 3.1.50 *Let $n \geq 4$, let $\Gamma = \langle\sigma_1,\ldots,\sigma_{n-1}\rangle$ be a string even group that satisfies the intersection condition* (3.10), *and for each i let Γ_i be defined as in* (3.9). *Then for any three distinct elements i,j,k of $\{0,\ldots,n-1\}$,*

$$\Gamma_i \cap \Gamma_j = \langle\tau_{k,\ell} : \ell \in \{0,\ldots,n-1\} \setminus \{i,j,k\}\rangle,$$

where $\tau_{k,\ell}$ is defined as in (3.8).

Proof Let Δ denote $\langle \tau_{k,\ell} \colon \ell \in \{0,\dots,n-1\} \setminus \{i,j,k\}\rangle$. We only need to show that $\Gamma_i \cap \Gamma_j \subseteq \Delta$, since every time that $k,\ell \notin \{i,j\}$ we have that $\tau_{k,\ell} \in \Gamma_i \cap \Gamma_j$.

Let $\gamma \in \Gamma_i \cap \Gamma_j$, and without loss of generality assume that $i < j$. From the third item of Lemma 3.1.47 we have that $\gamma = \alpha \tau_{i-1,i+1}^c \beta$ for some $c \in \{0,1\}$, $\alpha \in \Gamma_{<i}$ and $\beta \in \Gamma_{>i+1}$. The intersection condition implies that

$$\begin{aligned}\tau_{i-1,i+1}^c \beta &\in \Gamma_j \cap \langle \tau_{i-1,i+1}, \tau_{r,s} \colon r,s \geq i+1\rangle \\ &\subseteq \langle \tau_{r,s} \colon r,s \in \{i-1\} \cup (\{i+1,\dots,n-1\} \setminus \{j\})\rangle.\end{aligned}$$

To conclude the proof it suffices to note that if $r,s \notin \{i,j\}$ then $\tau_{r,s} = \tau_{k,r}^{-1} \tau_{k,s} \in \Delta$; indeed, α and $\tau_{i-1,i+1}^c \beta$ are generated by such elements. ■

We are ready to prove that if Γ is a string even group that satisfies the intersection condition (3.10) then $(\Gamma, \{\Gamma_0, \dots, \Gamma_{n-1}\})$ is the coset representation of a polytope.

Proposition 3.1.51 *Let $n \geq 3$, let $\Gamma = \langle \sigma_1, \dots, \sigma_{n-1}\rangle$ be a string even group that satisfies the intersection condition* (3.10), *and let Γ_i be as in* (3.9). *Then $(\Gamma, \{\Gamma_0, \dots, \Gamma_{n-1}\})$ is the coset representation of an n-polytope.*

Proof We first prove that $(\mathcal{P}(\Gamma), \leq)$ is a poset, where $\leq$ is defined as in (3.2). If $\Gamma_i\alpha \cap \Gamma_j\beta \neq \varnothing$ and $\Gamma_j\beta \cap \Gamma_k\delta \neq \varnothing$ we let $\zeta \in \Gamma_i\alpha \cap \Gamma_j\beta$. Then $\Gamma_j\beta \cap \Gamma_k\delta = \Gamma_j\zeta \cap \Gamma_k\delta \neq \varnothing$, implying that $\Gamma_j \cap \Gamma_k\delta\zeta^{-1} \neq \varnothing$. From Lemma 3.1.49 we know that $\Gamma_i \cap \Gamma_k\delta\zeta^{-1} \neq \varnothing$, which implies that $\Gamma_i\alpha \cap \Gamma_k\delta = \Gamma_i\zeta \cap \Gamma_k\delta \neq \varnothing$. This shows transitivity of $\leq$; clearly it is also reflexible and anti-symmetric.

Now we know that $\mathcal{P}(\Gamma)$ is a poset, and it is clear that right multiplication by elements in Γ preserves the incidence. To prove that $\mathcal{P}(\Gamma)$ is a polytope we only need to show that it satisfies the diamond condition and strong connectivity.

In order to show that $\mathcal{P}(\Gamma)$ satisfies the diamond condition we take two incident cosets $\Gamma_{i-1}\alpha$ and $\Gamma_{i+1}\beta$. Then there exists $\delta \in \Gamma_{i-1}\alpha \cap \Gamma_{i+1}\beta$. Therefore, there exist precisely two right cosets of Γ_i incident to both $\Gamma_{i-1}\alpha$ and $\Gamma_{i+1}\beta$ if and only if there exist two right cosets of Γ_i incident to both $\Gamma_{i-1} = \Gamma_{i-1}\alpha\delta^{-1}$ and $\Gamma_{i+1} = \Gamma_{i+1}\beta\delta^{-1}$. Hence it suffices to show the diamond condition for the subgroups Γ_{i-1} and Γ_{i+1}.

The intersection condition (3.10) tells us that

$$\Gamma_{i-1} \cap \Gamma_{i+1} = \langle \tau_{j,k} \colon j,k \in \{0,\dots,n-1\} \setminus \{i-1,i+1\}\rangle.$$

On the other hand, Lemmas 3.1.45 and 3.1.47 can be used to show that

$$\langle \tau_{j,k} \colon j,k \in \{0,\dots,n-1\} \setminus \{i-1,i+1\}\rangle = \Gamma_{<i-1}\langle \tau_{i-2,i}\rangle\langle \tau_{i,i+2}\rangle\Gamma_{>i+2},$$

where $\tau_{i-2,i} = \mathrm{id}$ if $i \leq 1$ and $\tau_{i,i+2} = \mathrm{id}$ if $i \geq n-2$. Then

$$\begin{aligned}\Gamma_{<i-1}\langle\tau_{i-2,i}\rangle\langle\tau_{i,i+2}\rangle\Gamma_{>i+2} &= \Gamma_{<i-1}\langle\tau_{i-2,i}\rangle\langle\tau_{i-2,i+2}\rangle\Gamma_{>i+2}\\ &= \Gamma_{<i-1}\langle\tau_{i-2,i+2}\rangle\Gamma_{>i+2}\langle\tau_{i-2,i}\rangle \subseteq \Gamma_i\langle\tau_{i-2,i}\rangle,\end{aligned}$$

and hence the only two cosets of Γ_i that are incident simultaneously with Γ_{i-1} and Γ_{i+1} are Γ_i and $\Gamma_i\tau_{i-2,i}$, if $i \geq 2$. It can be proven similarly that if $i \leq 1$ (and hence $i < n-2$) then the only two such cosets of Γ_i are Γ_i and $\Gamma_i\tau_{i,i+2}$. Since $\tau_{i-2,i}, \tau_{i,i+2} \notin \Gamma_i$, the two cosets are distinct.

To show strong connectivity we let G/F be a section of $\mathcal{P}(\Gamma)$ of rank at least 2. It follows from the definition of incidence that there exists $\gamma \in \Gamma$ such that $F = \Gamma_i\gamma$ and $G = \Gamma_j\gamma$, for some $i, j \in \{1, \ldots, n-1\}$ with $i \leq j-3$. Then the sections G/F and Γ_j/Γ_i are isomorphic. Hence, it suffices to show that every section of type Γ_j/Γ_i is connected, whenever $j - i \geq 3$.

A coset $\Gamma_k\gamma$ is in the section Γ_j/Γ_i if and only if it intersects both Γ_i and Γ_j. If $n \geq 4$ and $j-i \geq 3$ we let $\gamma' \in \Gamma_i \cap \Gamma_k\gamma$, so that $\Gamma_i \leq \Gamma_k \leq \Gamma_j(\gamma')^{-1}$. In this situation Lemma 3.1.49 guarantees that $\Gamma_i \cap \Gamma_k \cap \Gamma_j(\gamma')^{-1} \neq \varnothing$, and therefore $\Gamma_i \cap \Gamma_k\gamma \cap \Gamma_j = \Gamma_i \cap \Gamma_k\gamma' \cap \Gamma_j \neq \varnothing$. It follows that the section Γ_j/Γ_i is precisely the poset

$$(\Gamma_i \cap \Gamma_j, \{\Gamma_i \cap \Gamma_k \cap \Gamma_j : i < k < j\}).$$

This poset satisfies the hypotheses of Proposition 3.1.14, so we can conclude that it is connected.

For $n = 3$, connectivity can be proved exactly as in Proposition 3.1.14 (note that the second item of that proposition is not used to show connectivity).

It only remains to show that for arbitrary n the sections of rank 2 are connected.

If $n = 3$ then the sections of rank 2 are Γ_2/Γ_{-1} and Γ_3/Γ_0. The cosets of $\Gamma_0 = \langle\sigma_2\rangle$ incident to Γ_2 must contain σ_1^k for some k. Furthermore, if $\sigma_1^{k_1} \neq \sigma_1^{k_2}$ then the intersection condition forces $\Gamma_0\sigma_1^{k_1} \neq \Gamma_0\sigma_1^{k_2}$. Similarly, the cosets of Γ_1 incident to Γ_2 are precisely those of the form $\Gamma_1\sigma_1^k$. So far we have in the section Γ_2/Γ_{-1} as many cosets of Γ_0 and as many cosets of Γ_1 as the order of σ_1. Now $\Gamma_1\sigma_1^k = \{\sigma_1^k, \tau_{0,2}\sigma_1^k\}$, but $\tau_{0,2}\sigma_1^k = \sigma_2^{-1}\sigma_1^{-1}\sigma_1^k = \sigma_2^{-1}\sigma_1^{k-1}$, and so this coset is incident to $\Gamma_0\sigma_1^\ell$ if and only if $\ell \in \{k-1, k\}$. This gives precisely the incidence relation of an abstract polygon and hence it is connected. The connectivity of the section Γ_3/Γ_0 can be verified analogously.

Finally, if $n \geq 4$ and $\Gamma_{i-2} \leq \Gamma_j\gamma \leq \Gamma_{i+1}$, as mentioned above, we have that there exists $\gamma' \in \Gamma_{i-2} \cap \Gamma_j\gamma \cap \Gamma_{i+1}$. We choose k as the element in $\{i-1, i\} \setminus \{j\}$. Lemma 3.1.50 implies that $\gamma' \in \langle\tau_{k,\ell} : \ell \notin \{i-2, i+1\}\rangle$. Almost all generators of the latter group are in Γ_j, with the only exception of $\tau_{k,j} \in \{\sigma_i, \sigma_i^{-1}\}$. Furthermore, if $\ell < i-2$ or $\ell > i+1$ then $\langle\sigma_i\rangle\tau_{k,\ell} = \tau_{k,\ell}\langle\sigma_i\rangle$

(see Lemma 3.1.48). It follows that $\gamma' \in \Gamma_j \sigma_i^m$ for some integer m. To conclude the proof we proceed as in the case $n = 3$ to conclude that the section $\Gamma_{i+1}/\Gamma_{i-2}$ is an abstract polygon, and hence, it is connected. ■

As we could see in the proof of the previous proposition, sometimes it is necessary to treat the cases where the rank is low separately. If we restrict ourselves to ranks 5 and higher, the prevous proposition can be proved using Theorem 3.1.15.

Exercise 3.1.52 Show that if $n \geq 5$ then $(\Gamma, \{\Gamma_0, \ldots, \Gamma_{n-1}\})$ is the coset representation of a 4-chain transitive polytope, using Theorem 3.1.15.

In Proposition 3.1.51 we required n to be at least 3. If $n = 2$ the situation is a little different. The group Γ is cyclic, generated by σ_1. Additionally, $\Gamma_0 = \Gamma_1 = \{id\}$, whereas for $n \geq 3$ we have that $\Gamma_i \neq \Gamma_j$ if $i \neq j$. Since Γ acts as the rotation group of a polygon, there is still a bijection between Γ and each of the flag orbits. There is also a bijection between the cosets of Γ_i and the set of i-faces, for $i \in \{0, 1\}$, but incidence is no longer given by a non-empty intersection, since the coset representing each edge contains only one element and so it would be incident to only one vertex. This is an example of the necessity of 2-chain transitivity in Theorem 3.1.7.

The next theorem is the analogue to Theorem 3.1.20 for rotary polytopes and their rotation subgroups.

Theorem 3.1.53 *Let $n \geq 3$, let $\Gamma = \langle \sigma_1, \ldots, \sigma_{n-1} \rangle$ be a string even group satisfying the intersection condition* (3.10), *and let Γ_i be as in* (3.9). *Then $(\Gamma, \{\Gamma_0, \ldots, \Gamma_{n-1}\})$ is the coset representation of a unique orientable rotary n-polytope $\mathcal{P}(\Gamma)$ with $\Gamma^+(\mathcal{P}(\Gamma)) = \Gamma$, and standard rotations σ_i acting by right multiplication. Furthermore, $\mathcal{P}(\Gamma)$ is regular if and only if for some $i \in \{0, \ldots, n-1\}$ there is an involutory group automorphism η of Γ such that*

$$\sigma_j \eta = \begin{cases} \sigma_{i-1}\sigma_i^2 & \text{if } j = i-1, \\ \sigma_j^{-1} & \text{if } j \in \{i, i+1\}, \\ \sigma_{i+1}^2 \sigma_{i+2} & \text{if } j = i+2, \\ \sigma_j & \text{otherwise.} \end{cases}$$

(Here we understand that whenever a subindex does not belong to $\{1, \ldots, n-1\}$, the row may be disregarded.)

Proof We already know from Proposition 3.1.51 that $(\Gamma, \{\Gamma_0, \ldots, \Gamma_{n-1}\})$ is the coset representation of an n-polytope $\mathcal{P}(\Gamma)$ in which the elements of Γ act as automorphisms by right multiplication. In the proof of Proposition 3.1.51

we established 2-transitivity of $\mathcal{P}(\Gamma)$; uniqueness of $\mathcal{P}(\Gamma)$ is a consequence of Theorem 3.1.7. It remains to be shown that $\mathcal{P}(\Gamma)$ is rotary with the σ_i acting like the standard rotations, that it is orientable, and the regularity criterion.

Let $\Phi_0 = \{\Gamma_0, \dots, \Gamma_{n-1}\}$ be the base flag of $\mathcal{P}(\Gamma)$. To show that $\mathcal{P}(\Gamma)$ is a rotary polytope with the σ_i serving as standard rotations we first observe that $\sigma_i \in \Gamma_j$ for every $j \notin \{i-1, i\}$, implying that if $i \in \{1, \dots, n-1\}$ then the automorphism σ_i of $\mathcal{P}(\Gamma)$ fixes all faces of Φ, except the $(i-1)$-face and the i-face. In the proof of Proposition 3.1.51 we saw that the i-face incident to Γ_{i-1} and Γ_{i+1} other than Γ_i is either $\Gamma_i \tau_{i-2,i}$ or $\Gamma_i \tau_{i,i+2}$. Since

$$\Gamma_i \tau_{i-2,i} = \Gamma_i \sigma_{i-1}^{-1} \tau_{i-2,i} = \Gamma_i \sigma_{i-1}^{-1} \sigma_{i-1} \sigma_i = \Gamma_i \sigma_i,$$
$$\Gamma_i \tau_{i,i+2} = \Gamma_i \tau_{i-1,i+2} \tau_{i+2,i} = \Gamma_i \sigma_i,$$

and $\Gamma_{i-1} \neq \Gamma_{i-1}\sigma_i$, we can conclude that the i-face of $\Phi_0\sigma_i$ is that of Φ_0^i. Since $\Gamma_{i-1}\sigma_i \neq \Gamma_{i-1}$, the diamond condition forces $\Phi_0\sigma_i = \Phi_0^{i,i-1}$. This shows that σ_i acts like the ith standard rotation.

Orientability of $\mathcal{P}(\Gamma)$ is a consequence of the intersection property. Indeed, if $\mathcal{P}(\Gamma)$ were non-orientable then it would be regular, and there would exist an automorphism γ mapping the base flag Φ_0 to $\Phi_0^0 = \{\Gamma_0\tau_{0,2}, \Gamma_1, \dots, \Gamma_{n-1}\}$. Then γ would belong to $\bigcap_{i\in\{1,\dots,n-1\}} \Gamma_i$, but the intersection property implies that this group is $\langle \tau_{j,k} \colon j,k \in \{0\} \rangle$, which is trivial. This would force $\gamma = \mathrm{id}$, but then $\Phi_0\gamma \neq \Phi_0^0$, a contradiction.

If $\mathcal{P}(\Gamma)$ is regular then conjugation by ρ_i acts like the desired group automorphism η. It remains to show that if Γ has such a group automorphism η then $\Gamma(\mathcal{P}(\Gamma))$ contains an element ρ acting like the standard reflection ρ_i. To do this we shall use Proposition 2.1.27 and define ρ as the bijection of flags

$$\Phi_0\alpha\rho := \Phi_0^i \eta(\alpha),$$
$$\Phi_0^i\alpha\rho := \Phi_0 \eta(\alpha)$$

for every $\alpha \in \Gamma$. Since η is a involution, the bijection ρ is also involutory.

If $(\Phi_0\alpha)^j = \Phi_0^i\beta$ for some $\alpha, \beta \in \Gamma$ then we use that $\Phi_0^i = \Phi_0^{i,j,j} = (\Phi_0\tau_{j,i})^j$ to conclude that $\alpha = \tau_{j,i}\beta$, and therefore $\beta = \tau_{i,j}\alpha$. Hence,

$$(\Phi_0\alpha)^j\rho = \Phi_0^i\tau_{i,j}\alpha\rho = \Phi_0\eta(\tau_{i,j}\alpha) = \Phi_0\eta(\tau_{i,j})\eta(\alpha),$$
$$(\Phi_0\alpha\rho)^j = (\Phi_0^i\eta(\alpha))^j = \Phi_0^{i,j}\eta(\alpha) = \Phi_0\tau_{j,i}\eta(\alpha).$$

In view of the previous equations and Proposition 2.1.27, we will prove that ρ is an automorphism of $\mathcal{P}(\Gamma)$ if we can show that for every $j \in \{0, \dots, n-1\}$ we have that $\eta(\tau_{i,j}) = \tau_{j,i}$. This is immediate if $j \in \{i, i+1\}$. If $j < i-1$ then

$$\eta(\tau_{i,j}) = \eta(\tau_{j,i}) = \eta(\sigma_{j+1} \cdots \sigma_i) = \sigma_{j+1} \cdots \sigma_{i-2}(\sigma_{i-1}\sigma_i^2)(\sigma_i^{-1})$$
$$= \sigma_{j+1} \cdots \sigma_i = \tau_{i,j}.$$

The proof in the case when $j > i$ is analogous.

The definition of the bijection ρ implies that $\Phi_0\rho = \Phi_0^i$, and so the induced automorphism acts like ρ_i. This forces $\mathcal{P}(\Gamma)$ to be regular. ∎

The condition for regularity in Theorem 3.1.53 becomes simpler for some values of i. The next corollaries state the case when $i = 0$ as well as the case when $n = 3$ and $i = 1$.

Corollary 3.1.54 *With the notation of Theorem 3.1.53, $\mathcal{P}(\Gamma)$ is regular if and only if there exists a group automorphism η such that*

$$\sigma_j\eta = \begin{cases} \sigma_1^{-1} & \text{if } j = 1, \\ \sigma_1^2\sigma_2 & \text{if } j = 2, \\ \sigma_j & \text{otherwise.} \end{cases}$$

Corollary 3.1.55 *A rotary polyhedron $\mathcal{P}$ with $\Gamma^+(\mathcal{P}) = \langle\sigma_1,\sigma_2\rangle$ is regular if and only if Γ has an involutory group automorphism mapping σ_i to σ_i^{-1} for $i \in \{1,2\}$.*

Remark 3.1.56 The polytope constructed in Theorem 3.1.53 from the coset representation $(\Gamma, \{\Gamma_0, \ldots, \Gamma_{n-1}\})$ is orientable, and it is the unique orientable rotary polytope $\mathcal{Q}$ with that coset representation with respect to $\Gamma^+(\mathcal{P})$. If $\mathcal{Q}$ is regular, it is possible that a non-orientably regular polytope $\mathcal{P}$ also has $(\Gamma, \{\Gamma_0, \ldots, \Gamma_{n-1}\})$ as its coset representation. The relationship between $\mathcal{P}$ and $\mathcal{Q}$ in such cases will be addressed in Section 4.4.

The coset representation of orientable, rotary polytopes in terms of their rotation groups has the following hereditary property.

Proposition 3.1.57 *Let Γ and $\mathcal{P}(\Gamma)$ be as in Theorem 3.1.53, and let $i, j \in \{-1, \ldots, n\}$ such that $j - i \geq 4$. Then the section Γ_j/Γ_i is isomorphic to $\mathcal{P}(\langle\sigma_{i+2}, \ldots, \sigma_{j-1}\rangle)$.*

Proof Propositions 2.2.8 and 2.3.6 tell us that Γ_j/Γ_i is an orientable rotary polytope. Furthermore, if $i + 2 \leq k \leq j - 1$ then σ_k is one of its standard rotations. Hence, $\Gamma^+(\Gamma_j/\Gamma_i) = \langle\sigma_{i+2}, \ldots, \sigma_{j-1}\rangle$ and Theorem 3.1.53 can be used to conclude that $\Gamma_j/\Gamma_i \cong \mathcal{P}(\langle\sigma_{i+2}, \ldots, \sigma_{j-1}\rangle)$. ∎

As we shall see in Section 3.1.4, Theorem 3.1.53 is a powerful tool to construct rotary polytopes. In order to use it, we first need a suitable group Γ. When n is large, the intersection property as stated in (3.10) is too hard to compute, since the number of intersections to compute grows exponentially with respect to n. In what follows we provide tools to prove that some groups satisfy the intersection property.

The intersection property (3.10) is simplified considerably when $\mathcal{P}$ is a polyhedron.

Proposition 3.1.58 *Let $\Gamma = \langle \sigma_1, \sigma_2 \rangle$ be a string even group. If*

$$\langle \sigma_1 \rangle \cap \langle \sigma_2 \rangle = \{\text{id}\}, \tag{3.16}$$

then Γ satisfies the intersection condition (3.10).

Proof If I has at most one element then Γ_I is trivial, while if $I = \{0, 1, 2\}$ then $\Gamma_i = \Gamma$. In both cases the intersection condition holds. It remains to show that (3.16) implies the intersection property in the cases when I and J have two elements each.

Equation (3.16) is precisely the case when $I = \{0, 1\}$ and $J = \{1, 2\}$, since $\langle \tau_{i,j} : i, j \in I \rangle = \langle \sigma_1 \rangle$ and $\langle \tau_{i,j} : i, j \in J \rangle = \langle \sigma_2 \rangle$. If (3.16) holds, $I = \{0, 2\}$ and $J \in \{\{0, 1\}, \{1, 2\}\}$, then $\langle \tau_{i,j} : i, j \in I \rangle = \{\text{id}, \tau_{0,2}\}$. But $\tau_{0,2} = \sigma_1 \sigma_2$ cannot belong to $\langle \sigma_1 \rangle$ or to $\langle \sigma_2 \rangle$ since that would force $\sigma_2 \in \langle \sigma_1 \rangle$ or $\sigma_1 \in \langle \sigma_2 \rangle$, respectively. It follows that

$$\langle \tau_{i,j} : i, j \in I \rangle \cap \langle \tau_{i,j} : i, j \in J \rangle = \{\text{id}\} = \langle \tau_{i,j} : i, j \in I \cap J \rangle.$$

This shows that the intersection property holds for all subsets I and J of $\{0, 1, 2\}$. ■

For ranks $n > 3$, we shall state in Proposition 3.1.60 analogues of Propositions 3.1.34 and 3.1.35 to determine that some groups satisfy the intersection condition. We shall need the following lemma.

Lemma 3.1.59 *Let $\Gamma = \langle \sigma_1, \dots, \sigma_{n-1} \rangle$ be a string even group, let Γ_i be as in* (3.9), *assume that Γ_{n-1} satisfies the intersection condition* (3.10) *and also assume that $\Gamma_{n-1} \cap \langle \sigma_j, \dots, \sigma_{n-1} \rangle = \langle \sigma_j, \dots, \sigma_{n-2} \rangle$ for every $j \in \{2, \dots, n-1\}$. Then for any $0 \leq j \leq k \leq n-1$, every element in $\cap_{i=j}^{k} \Gamma_i$ can be expressed as $\alpha \tau_{j-1,k+1}^{c} \beta$ for some $c \in \{0, 1\}$, $\alpha \in \langle \sigma_1, \dots, \sigma_{j-1} \rangle$ and $\beta \in \langle \sigma_{k+2}, \dots, \sigma_{n-1} \rangle$.*

Proof We proceed by induction over $k - j$, the base case being the third item of Lemma 3.1.47 when $k - j = 0$. Assume that the result is true for some j and k, also that $k < n - 1$, and let $\gamma \in \cap_{i=j}^{k+1} \Gamma_i$.

If $k + 1 = n - 1$ then from the inductive hypothesis we know that $\gamma = \alpha \tau_{j-1,n-1}^{c}$ for some $c \in \{0, 1\}$ and $\alpha \in \langle \sigma_1, \dots, \sigma_{j-1} \rangle$. Since $\alpha, \gamma \in \Gamma_{n-1}$ we must also have that $\tau_{j-1,n-1}^{c} \in \Gamma_{n-1}$. Since $\tau_{j-1,n-1}^{c} \in \langle \sigma_j, \dots, \sigma_{n-1} \rangle$, the hypothesis of this lemma implies that $\tau_{j-1,n-1}^{c} \in \langle \sigma_j, \dots, \sigma_{n-2} \rangle$.

If $c = 1$ then $\tau_{j-1,n-1} = \sigma_j \cdots \sigma_{n-1} \in \langle \sigma_j, \ldots, \sigma_{n-2} \rangle$, which forces $\sigma_{n-1} \in \langle \sigma_j, \ldots, \sigma_{n-2} \rangle$, contradicting our hypothesis. Hence $c = 0$ and $\gamma = \alpha \in \langle \sigma_1, \ldots, \sigma_{j-1} \rangle$; the lemma is satisfied.

When $k+1 < n-1$ we use the inductive hypothesis to write

$$\gamma = \alpha \tau_{j-1,k+1}^{c} \beta$$

for some $c \in \{0,1\}$, $\alpha \in \langle \sigma_1, \ldots, \sigma_{j-1} \rangle$ and $\beta \in \langle \sigma_{k+2}, \ldots, \sigma_{n-1} \rangle$. We also use the case $k - j = 0$ to write

$$\gamma = \delta \tau_{k,k+2}^{d} \zeta$$

for some $d \in \{0,1\}$, $\delta \in \langle \sigma_1, \ldots, \sigma_k \rangle$ and $\zeta \in \langle \sigma_{k+3}, \ldots, \sigma_{n-1} \rangle$. It follows that

$$\begin{aligned}
\delta^{-1} \alpha \tau_{j-1,k+1}^{c} = \tau_{k,k+2}^{d} \zeta \beta^{-1} &\in \langle \sigma_1, \ldots, \sigma_{k+1} \rangle \cap \langle \sigma_{k+1}, \ldots, \sigma_{n-1} \rangle \\
&= (\langle \sigma_1, \ldots, \sigma_{k+1} \rangle \cap \Gamma_{n-1}) \cap \langle \sigma_{k+1}, \ldots, \sigma_{n-1} \rangle \\
&= \langle \sigma_1, \ldots, \sigma_{k+1} \rangle \cap (\Gamma_{n-1} \cap \langle \sigma_{k+1}, \ldots, \sigma_{n-1} \rangle) \\
&= \langle \sigma_1, \ldots, \sigma_{k+1} \rangle \cap \langle \sigma_{k+1}, \ldots, \sigma_{n-2} \rangle = \langle \sigma_{k+1} \rangle.
\end{aligned}$$

If $c = 0$ then $\delta^{-1}\alpha \in \langle \sigma_1, \ldots, \sigma_k \rangle \cap \langle \sigma_{k+1} \rangle$ and since Γ_{n-1} satisfies the intersection condition we must have that $\delta^{-1}\alpha = \mathrm{id}$, so $\delta = \alpha$. With an argument similar to those used above we can conclude that

$$\tau_{k,k+2}^{d} = \zeta \beta^{-1} \in \langle \sigma_{k+1}, \sigma_{k+2} \rangle \cap \langle \sigma_{k+2}, \ldots, \sigma_{n-1} \rangle = \langle \sigma_{k+2} \rangle,$$

and also that $d = 0$. Therefore $\gamma = \alpha\zeta$ satisfying the conclusion of the lemma.

Finally, if $c = 1$ then we note that $\delta^{-1}\alpha\tau_{j-1,k+1} = \delta^{-1}\alpha\tau_{j-1,k}\sigma_{k+1}$, implying that $\delta^{-1}\alpha\tau_{j-1,k} \in \langle \sigma_1, \ldots, \sigma_k \rangle \cap \langle \sigma_{k+1} \rangle = \{\mathrm{id}\}$. It follows that $\delta = \alpha\tau_{j-1,k}$ and $\delta^{-1}\alpha\tau_{j-1,k+1} = \tau_{k,k+2}^{d}\zeta\beta^{-1} = \sigma_{k+1}$. But now

$$\tau_{k,k+2}^{d} \sigma_{k+1} = \zeta \beta^{-1} \in \langle \sigma_{k+1}, \sigma_{k+2} \rangle \cap \langle \sigma_{k+2}, \ldots, \sigma_{n-1} \rangle,$$

and the latter group equals $\langle \sigma_{k+2} \rangle$. In follows that $d = 1$ and that $\gamma = \alpha\tau_{j-1,k}\tau_{k,k+2}\zeta = \alpha\tau_{j-1,k+2}\zeta$, which is the desired expression. ■

Proposition 3.1.60 *Let $n \geq 4$, let $\Gamma = \langle \sigma_1, \ldots, \sigma_{n-1} \rangle$ be a string even group and assume that $\Gamma_{n-1} = \langle \sigma_1, \ldots, \sigma_{n-2} \rangle$ satisfies the intersection condition* (3.10).

(a) If $\Gamma_{n-1} \cap \langle \sigma_j, \ldots, \sigma_{n-1} \rangle = \langle \sigma_j, \ldots, \sigma_{n-2} \rangle$ for every $j \in \{2, \ldots, n-1\}$ then Γ also satisfies the intersection condition (3.10).

(b) If $\Gamma_0 = \langle \sigma_2, \ldots, \sigma_{n-1} \rangle$ satisfies the intersection condition and $\Gamma_{n-1} \cap \Gamma_0 = \langle \sigma_2, \ldots, \sigma_{n-2} \rangle$ then Γ also satisfies the intersection condition (3.10).

Proof To prove (a) we adapt the proof of Proposition 3.1.34, for which we denote by $\Gamma_{\overline{I}}$ the group $\langle \tau_{i,j} : i, j \notin I \rangle$. For an alternative proof see Exercise 3.1.61.

Let $I = \{i_1, \ldots, i_m\}$ with $i_j < i_k$ if $j < k$. First we establish that $\bigcap_{i \in I} \Gamma_i = \Gamma_{\overline{I}}$, for which we only need to show that $\bigcap_{i \in I} \Gamma_i \subseteq \Gamma_{\overline{I}}$. Let ℓ be the maximum $k < i_m$ such that $k \notin I$, and let $\gamma \in \bigcap_{i \in I} \Gamma_i$. Lemma 3.1.59 tells us that $\gamma = \alpha \tau^c_{\ell, i_m+1} \beta$ for some $c \in \{0, 1\}$, $\alpha \in \langle \sigma_1, \ldots, \sigma_\ell \rangle$ and $\beta \in \langle \sigma_{i_m+2}, \ldots, \sigma_{n-1} \rangle$. We know that $\tau^c_{\ell, i_m+1}, \beta \in \Gamma_i$ for every $i \in I$, so it must be true that $\alpha \in \Gamma_i$ as well.

From the third item of Lemma 3.1.47 we know that for each $j \in I$ with $j < \ell$ we can write $\alpha = \delta \tau^d_{j-1, j+1} \zeta$ with $d \in \{0, 1\}$, $\delta \in \langle \sigma_1, \ldots, \sigma_{j-1} \rangle$ and $\zeta \in \langle \sigma_{j+2}, \ldots, \sigma_{n-1} \rangle$. Then

$$\begin{aligned}
\tau^d_{j-1, j+1} \delta^{-1} \alpha = \zeta &\in \langle \sigma_1, \ldots, \sigma_\ell \rangle \cap \langle \sigma_{j+2}, \ldots, \sigma_{n-1} \rangle \\
&= (\langle \sigma_1, \ldots, \sigma_\ell \rangle \cap \Gamma_{n-1}) \cap \langle \sigma_{j+2}, \ldots, \sigma_{n-1} \rangle \\
&= \langle \sigma_1, \ldots, \sigma_\ell \rangle \cap (\Gamma_{n-1} \cap \langle \sigma_{j+2}, \ldots, \sigma_{n-1} \rangle) \\
&= \langle \sigma_1, \ldots, \sigma_\ell \rangle \cap \langle \sigma_{j+2}, \ldots, \sigma_{n-2} \rangle \\
&= \langle \sigma_{j+2}, \ldots, \sigma_\ell \rangle.
\end{aligned}$$

This means that, if we denote by Γ^*_i the subgroup of Γ_{n-1} given by $\{\tau_{j,k} : j, k \in \{0, \ldots, n-2\} \setminus \{i\}\}$ then $\alpha \in \Gamma^*_j$ for every $j \in I$ such that $j < \ell$. Our choice of α guarantees that $\alpha \in \Gamma^*_j$ for every $j \in I$ such that $j > \ell$. Since Γ_{n-1} satisfies the intersection condition, Corollary 3.1.41 implies that $\alpha \in \Gamma_{\overline{I}}$, and since it is also true that $\tau^c_{\ell, i_m+1}, \beta \in \Gamma_{\overline{I}}$, we may conclude that $\gamma \in \Gamma_{\overline{I}}$ as desired.

To finish the proof of (a) we let $I, J \subseteq \{0, \ldots, n-1\}$. Then

$$\begin{aligned}
&\langle \tau_{i,j} : i, j \in I \rangle \cap \langle \tau_{i,j} : i, j \in J \rangle \\
&= \Gamma_I \cap \Gamma_J = \Gamma_{\overline{\overline{I}}} \cap \Gamma_{\overline{\overline{J}}} = \left(\textstyle\bigcap_{i \in \overline{I}} \Gamma_i\right) \cap \left(\textstyle\bigcap_{i \in \overline{J}} \Gamma_i\right) \\
&= \textstyle\bigcap_{i \in \overline{I} \cup \overline{J}} \Gamma_i = \Gamma_{\overline{\overline{I \cap J}}} = \langle \tau_{i,j} : i, j \in I \cap J \rangle.
\end{aligned}$$

Item (b) follows from (a). Its proof is analogous to that of Proposition 3.1.35. ■

Exercise 3.1.61 Give an alternative proof of Proposition 3.1.60 (a) by assuming Proposition 3.1.60 (b) and following the next steps. With the hypotheses of Proposition 3.1.60 (b) define the relation $\leq$ as in (3.2) in the set of right cosets of the subgroups Γ_i.

1. Show that $\leq$ is a partial order.
2. Show that all sections of rank $n-1$ in that poset are $(n-1)$-polytopes.

3. Show that the poset is connected.
4. Use Proposition 2.1.44 and Theorem 3.1.53 to show that Γ satisfies the intersection condition.
5. Conclude that item (a) of Proposition 3.1.60 follows from item (b) of the same proposition.

Proposition 3.1.60 is a useful tool to verify that some groups satisfy the intersection property by computing only few intersections of subgroups. For example, if $n = 4$ then we can combine Propositions 3.1.58 and 3.1.60 (b) to conclude that a string even group $\Gamma = \langle \sigma_1, \sigma_2, \sigma_2 \rangle$ satisfying

$$\langle \sigma_1 \rangle \cap \langle \sigma_2 \rangle = \langle \sigma_2 \rangle \cap \langle \sigma_3 \rangle = \mathrm{id}, \tag{3.17}$$

$$\langle \sigma_1, \sigma_2 \rangle \cap \langle \sigma_2, \sigma_3 \rangle = \langle \sigma_2 \rangle, \tag{3.18}$$

also satisfies the intersection condition. This idea generalises by considering sets of generators of the form $\langle \sigma_i, \sigma_{i+1}, \dots, \sigma_j \rangle$ where there is no gap in the set of indices of the generators σ_k.

Proposition 3.1.62 *Let $n \geq 3$ and let $\Gamma = \langle \sigma_1, \dots, \sigma_{n-1} \rangle$ be a string even group. If, in addition,*

$$\langle \sigma_k \colon i_1 \leq k \leq i_2 \rangle \cap \langle \sigma_k \colon j_1 \leq k \leq j_2 \rangle = \langle \sigma_k \colon j_1 \leq k \leq i_2 \rangle \tag{3.19}$$

for every $i, j \in \{1, \dots, n-1\}$ such that $i_1 \leq i_2$, $j_1 \leq j_2$, $i_1 < j_1 \leq i_2 + 1$ and $i_2 < j_2$ then Γ also satisfies the intersection condition (3.10).

Proof We proceed by induction over n. When $n = 3$ the conditions on the parameters i_1, i_2, j_1 and j_2 give us only one intersection to check, namely when $i_1 = i_2 = 1$ and $j_1 = j_2 = 2$, and (3.19) reduces to (3.16).

Assume that $n \geq 4$ and that the statement holds if the rank is at most $n - 1$. By taking into account that (3.19) holds whenever $j_2 < n - 1$ (resp. $i_1 > 1$) we may use the induction hypothesis to conclude that $\langle \sigma_1, \dots, \sigma_{n-2} \rangle$ satisfies (3.10) (resp. that $\langle \sigma_2, \dots, \sigma_{n-1} \rangle$ satisfies (3.10)). The proposition now follows from Proposition 3.1.60 and from (3.19) for $i_1 = 1$, $i_2 = n - 2$, $j_1 = 2$ and $j_2 = n - 1$. ■

The criterion in Proposition 3.1.62 has obvious advantages over (3.10) when determining if a given group satisfies the intersection condition. On one hand, we only need to use generators σ_k, and not other general generators $\tau_{k,\ell}$. On the other hand, the number of intersections to be satisfied in the statement of Proposition 3.1.62 grows polynomially on the rank, whereas that required for (3.10) grows exponentially on the rank.

We next present an analogue for rotary polytopes of Lemma 3.1.37. It can be proven following the same lines as Lemma 3.1.37.

Lemma 3.1.63 *Let* $\Gamma = \langle \sigma_1, \ldots, \sigma_{n-1} \rangle$ *and* $\Gamma' = \langle \sigma'_1, \ldots, \sigma'_{n-1} \rangle$ *be string even groups and assume that* Γ' *satisfies the intersection condition* (3.10). *If there exists an epimorphism from* Γ *to* Γ' *mapping each* σ_j *to* σ'_j *with the property that the restriction to* $\Gamma_{n-1} = \langle \sigma_1, \ldots, \sigma_{n-2} \rangle$ *or to* $\Gamma_0 = \langle \sigma_2, \ldots, \sigma_{n-1} \rangle$ *is an isomorphism, then* Γ *also satisfies the intersection condition* (3.10).

Exercise 3.1.64 Prove Lemma 3.1.63.

In what follows we shall concentrate on the automorphism groups of chiral polytopes.

In Section 2.3.2 we defined the enantiomorph of a rooted chiral polytope $\mathcal{P}_{(\Phi)}$ as the rooted polytope $\mathcal{P}_{(\Phi^0)}$, in the understanding that the latter is isomorphic (as a rooted polytope) to $\mathcal{P}_{(\Phi^i)}$ for any i. Elsewhere the base flag has been omitted from the notation and the enantiomorph of a (rooted) chiral polytope $\mathcal{P}$ has been called $\overline{\mathcal{P}}$ or $\mathcal{P}^-$.

If $\sigma_1, \ldots, \sigma_{n-1}$ are the standard rotations of a chiral polytope $\mathcal{P}$ with respect to Φ then

$$\Phi^i \sigma_j = (\Phi \sigma_j)^i = \Phi^{j,j-1,i} = (\Phi^i)^{i,j,j-1,i}. \tag{3.20}$$

We claim that this flag is distinct from $(\Phi^i)^{j,j-1}$ if $j-2 \leq i \leq j+1$. Indeed, if $i \in \{j-1, j\}$ then $(\Phi^i)^{i,j,j-1,i} = (\Phi^i)^{j-1,j}$; on the other hand,

$$(\Phi^{j+1})^{j+1,j,j-1,j+1} = (\Phi^{j+1})^{j+1,j,j+1,j-1},$$
$$(\Phi^{j-2})^{j-2,j,j-1,j-2} = (\Phi^{j-2})^{j,j-2,j-1,j-2}.$$

In each case, if $(\Phi^i)^{i,j,j-1,i} = (\Phi^i)^{j,j-1}$, there exist a flag Ψ and an index k such that $\Psi^{(k,k+1)^2} = \Psi$. For example, if $(\Phi^{j+1})^{j+1,j,j-1,j+1} = (\Phi^{j+1})^{j,j-1}$ then $\Phi^{j,j+1,j-1} = \Phi^{j+1,j,j-1}$, implying that $\Phi^{j,j+1} = \Phi^{j+1,j}$ and therefore $\Phi^{(j,j+1)^2} = \Phi$ (so here $\Psi = \Phi$ and $k = j$). By denoting by ξ_i the ith standard rotation with respect to Ψ, we observe that $\Psi^{(k,k+1)^\ell} = \Psi \xi_{k+1}^{-\ell}$, implying that the order of ξ_{k+1} is 2, contradicting Lemma 2.3.16.

The previous analysis establishes that the set of standard rotations with respect to Φ is not the same as the set with respect to any of its adjacent flags. In the next result we exhibit the jth standard rotation with respect to Φ^i in terms of the standard rotations with respect to Φ.

Proposition 3.1.65 *Let $i \in \{0, \ldots, n-1\}$, let $\mathcal{P}$ be a chiral n-polytope, let Φ be a flag of $\mathcal{P}$, let σ_j be the jth standard rotation with respect to Φ, and let ξ_j be the jth standard rotation with respect to Φ^i. Then*

$$\xi_j = \begin{cases} \sigma_{i-1}\sigma_i^2 & \text{if } j = i-1, \\ \sigma_j^{-1} & \text{if } j \in \{i, i+1\}, \\ \sigma_{i+1}^2\sigma_{i+2} & \text{if } j = i+2, \\ \sigma_j & \text{otherwise.} \end{cases}$$

Proof The case when $j \notin \{i-1, i, i+1, i+2\}$ follows easily from (3.20). For the remaining cases, we recall that $\Phi^i\xi_j = (\Phi^i)^{j,j-1}$ and observe that

$$\begin{aligned}
(\Phi^i)^{i-1,i-2} &= \Phi^{i,i-1,i,i-2,i} = (\Phi\sigma_i)^{i,i-1,i-1,i-2,i} = \Phi^{i,i-1,i-1,i-2,i}\sigma_i \\
&= (\Phi\sigma_i)^{i-1,i-2,i}\sigma_i = \Phi^{i-1,i-2,i}\sigma_i^2 = (\Phi\sigma_{i-1})^i\sigma_i^2 = \Phi^i\sigma_{i-1}\sigma_i^2, \\
(\Phi^i)^{i,i-1} &= \Phi^{i-1} = (\Phi^{i-1,i})^i = (\Phi\sigma_i^{-1})^i = \Phi^i\sigma_i^{-1}, \\
(\Phi^i)^{i+1,i} &= (\Phi\sigma_{i+1}^{-1})^i = \Phi^i\sigma_{i+1}^{-1}, \\
(\Phi^i)^{i+2,i+1} &= \Phi^{i+2,i,i+1} = \Phi^{i+2,i+1,i+1,i,i+1,i,i} = (\Phi\sigma_{i+2})^{i+1,i,i+1,i,i} \\
&= \Phi^{i+1,i,i+1,i,i}\sigma_{i+2} \\
&= (\Phi\sigma_{i+1})^{i+1,i,i}\sigma_{i+2} = (\Phi^{i+1,i,i})\sigma_{i+1}\sigma_{i+2} \\
&= (\Phi\sigma_{i+1})^i\sigma_{i+1}\sigma_{i+2} = \Phi^i\sigma_{i+1}^2\sigma_{i+2}.
\end{aligned}$$

The proposition follows from the regularity of the action of $\Gamma(\mathcal{P})$ on the flags of $\mathcal{P}$. ■

It follows from the necessary and sufficient condition for regularity in Theorem 3.1.53 that if $\mathcal{P}$ is chiral then no set of defining relations for the standard rotations with respect to some base flag Φ is also satisfied by the standard rotations with respect to Φ^i. We leave an explicit example for this as an exercise.

Exercise 3.1.66 Choose a base flag Φ of $\{4,4\}_{(2,1)}$ and let σ_1, σ_2 be the standard rotations with respect to Φ. Show a relation satisfied by the generating set $\{\sigma_1, \sigma_2\}$ of $\Gamma(\{4,4\}_{(2,1)})$ that is not satisfied by the generating set $\{\sigma_1^{-1}, \sigma_2^{-1}\}$ consisting of the standard rotations with respect to Φ^1.

The choice of base flag of a chiral polytope $\mathcal{P}$ is relevant not only when determining the defining relations of $\Gamma^+(\mathcal{P})$ with respect to the standard rotations. The following proposition provides another example of a statement valid for one choice of base flag but not for the other.

Proposition 3.1.67 *Let $\mathcal{P}$ and $\mathcal{Q}$ be two rotary n-polytopes with base flags Φ and Φ', respectively. Let $\Gamma^+(\mathcal{P}) = \langle\sigma_1,\dots,\sigma_{n-1}\rangle$ and $\Gamma^+(\mathcal{Q}) = \langle\sigma'_1,\dots,\sigma'_{n-1}\rangle$. There is a group epimorphism $\eta\colon \Gamma^+(\mathcal{P}) \to \Gamma^+(\mathcal{Q})$ such that $\eta(\sigma_i) = \sigma'_i$ for every i if and only if there is a covering from $\mathcal{P}$ to $\mathcal{Q}$ that maps Φ to Φ'.*

Proof Assume first that $\eta(\sigma_i) = \sigma'_i$ for every i extends to a group epimorphism from $\Gamma^+(\mathcal{P})$ to $\Gamma^+(\mathcal{Q})$. We may define the function $\hat{\eta}\colon \mathcal{F}l(\mathcal{P}) \to \mathcal{F}l(\mathcal{Q})$ that maps $\Phi\gamma$ to $\Phi'\eta(\gamma)$ and $\Phi^0\gamma$ to $(\Phi')^0\eta(\gamma)$ for every $\gamma \in \Gamma^+(\mathcal{P})$. Let $\gamma \in \Gamma^+(\mathcal{P})$ and $i \in \{0,\dots,n-1\}$. Then

$$\begin{aligned}\hat{\eta}((\Phi\gamma)^i) &= \hat{\eta}(\Phi^i\gamma) = \hat{\eta}(\Phi^0\tau_{0,i}\gamma) = (\Phi')^0\eta(\tau_{0,i})\eta(\gamma)\\ &= (\Phi')^0\tau'_{0,i}\eta(\gamma) = (\Phi')^i\eta(\gamma) = (\Phi'\eta(\gamma))^i = (\hat{\eta}(\Phi\gamma))^i,\end{aligned}$$

where $\tau'_{0,i} = \eta(\tau_{0,i}) = \sigma'_1\cdots\sigma'_i$. This tells us that $\hat{\eta}$ maps i-adjacent flags to i-adjacent flags and hence it is a covering that maps Φ to Φ'.

Conversely, assume that there is a covering $\hat{\eta}$ from $\mathcal{P}$ to $\mathcal{Q}$ that maps Φ to Φ' and let η be the function that assigns σ'_j to each σ_j. In order to show that η can be extended to a group epimorphism we only need to verify that every time that $\sigma_{i_1}\cdots\sigma_{i_k} = \mathrm{id}$ the expression $\sigma'_{i_1}\cdots\sigma'_{i_k} = \mathrm{id}$ also holds. This becomes clear when we apply these expressions to the appropriate flags:

$$\begin{aligned}\Phi'\sigma'_{i_1}\cdots\sigma'_{i_k} &= (\Phi')^{i_k,i_k-1,\dots,i_1,i_1-1} = (\hat{\eta}(\Phi))^{i_k,i_k-1,\dots,i_1,i_1-1}\\ &= \hat{\eta}(\Phi^{i_k,i_k-1,\dots,i_1,i_1-1})\\ &= \hat{\eta}(\Phi\sigma_{i_1}\cdots\sigma_{i_k}) = \hat{\eta}(\Phi) = \Phi',\end{aligned}$$

and the result follows from Proposition 2.1.80. ■

In the statement of Proposition 3.1.67 if either $\mathcal{P}$ or $\mathcal{Q}$ is regular then the choices of Φ and Φ' are irrelevant.

Exercise 3.1.68 Let $\mathcal{P}$ and $\mathcal{Q}$ be rotary polytopes with at least one of them regular, and let $\Gamma^+(\mathcal{P}) = \langle\sigma_1,\dots,\sigma_{n-1}\rangle$ and $\Gamma^+(\mathcal{Q}) = \langle\sigma'_1,\dots,\sigma'_{n-1}\rangle$. Show that there is a group epimorphism $\eta\colon \Gamma^+(\mathcal{P}) \to \Gamma^+(\mathcal{Q})$ such that $\eta(\sigma_i) = \sigma'_i$ for every i if and only if $\mathcal{P}$ covers $\mathcal{Q}$.

In case $\mathcal{P}$ and $\mathcal{Q}$ are chiral the statement of Exercise 3.1.68 may fail. In that situation it is necessary to use the version for rooted polytopes in Proposition 3.1.67.

Example 3.1.69 Figure 3.5 illustrates a covering from $\{4,4\}_{(3,1)}$ to $\{4,4\}_{(2,1)}$, by indicating the two preimages of each flag of $\{4,4\}_{(2,1)}$.

An observant reader might notice that according to the description in Section 2.4.2, the quotient should be named $\{4,4\}_{(1,2)}$ instead of $\{4,4\}_{(2,1)}$, if we want full consistency with the geometric setting. But $\{4,4\}_{(2,1)} \cong \{4,4\}_{(1,2)}$ since they are enantiomorphs of each other, and for general chiral polytopes it is not obvious which rooted polytope to choose in a given context. Regardless of the choice of base flags, the quotient can always be indicated by the surjection of the sets of flags as in our case. The covering from $\{4,4\}_{(3,1)}$ to $\{4,4\}_{(1,2)}$ in a geometric fashion is shown in Figure 3.6(a), where the translation group with which the quotient is taken contains only id and the translation by $(1,2)$.

However, if we choose as base flags of $\{4,4\}_{(3,1)}$ and $\{4,4\}_{(2,1)}$ the ones in grey in Figure 3.6(b) and (c) then there is no surjection mapping standard rotations of $\{4,4\}_{(3,1)}$ to standard rotations of $\{4,4\}_{(2,1)}$, despite the fact that $\{4,4\}_{(3,1)}$ covers $\{4,4\}_{(2,1)}$. Indeed, if $\Gamma(\{4,4\}_{(3,1)}) = \langle\sigma_1,\sigma_2\rangle$ and $\Gamma(\{4,4\}_{(2,1)}) = \langle\sigma_1',\sigma_2'\rangle$ then $\sigma_1\sigma_2^{-1}(\sigma_2^{-1}\sigma_1)^3 = \text{id}$ but $\sigma_1'(\sigma_2')^{-1}((\sigma_2')^{-1}\sigma_1')^3$ acts like a one-step translation to the right.

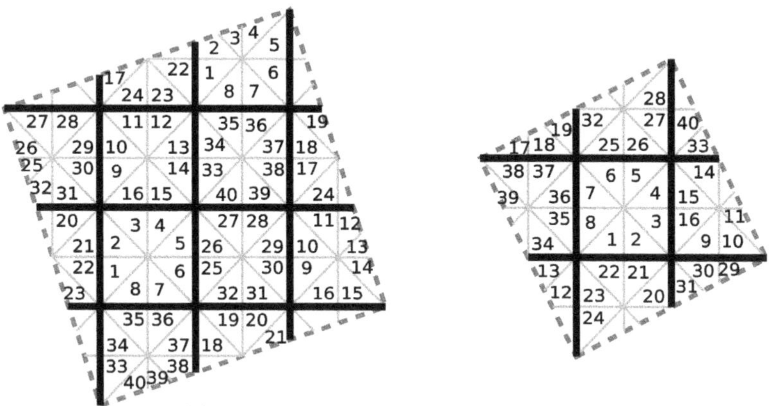

(a) (b)

Figure 3.5 A covering from (a) $\{4,4\}_{(3,1)}$ to (b) $\{4,4\}_{(2,1)}$.

When working with the dual $\mathcal{P}^\delta$ of a chiral polytope $\mathcal{P}$ we want to be able to express the ith standard rotation of $\mathcal{P}^\delta$ in terms of the standard rotations of $\mathcal{P}$. These expressions are dependent on the base flag chosen, as shown in the next result.

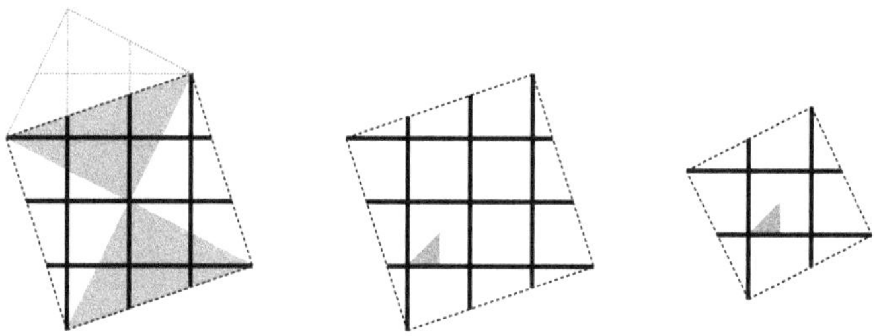

Figure 3.6 (a) The covering from $\{4,4\}_{(3,1)}$ to $\{4,4\}_{(1,2)}$; there is no covering mapping the grey flag of the polyhedron in (b) to the grey flag of the polyhedron in (c).

Proposition 3.1.70 *Let $\mathcal{P}$ be a chiral n-polytope and let σ_i be the ith standard rotation of $\mathcal{P}$ with respect to the base flag Φ. Then the ith standard rotation of the dual polytope $\mathcal{P}^\delta$ with respect to Φ is $\xi_i := \sigma_{n-i}^{-1}$; and with respect to Φ^0 is*

$$\zeta_i = \begin{cases} \sigma_{n-i}^{-1} & \text{if } i \leq n-3, \\ \sigma_2^{-1}\sigma_1^{-2} & \text{if } i = n-2, \\ \sigma_1 & \text{if } i = n-1. \end{cases}$$

Proof For convenience, we denote the flag Φ as Ψ, when thinking of it as a flag of $\mathcal{P}^\delta$. Then $\Phi^0 = \Psi^{n-1}$, and the proposition follows from the free action of the automorphism group on the flags together with the following equalities:

$$\Psi\xi_i = \Psi^{i,i-1} = \Phi^{n-i-1,n-i} = \Phi\sigma_{n-i}^{-1} = \Psi\sigma_{n-i}^{-1};$$

$$\Phi^0\zeta_i = \Psi^{n-1}\zeta_i = \Psi^{n-1,i,i-1} = \Psi^{n-1,i,i-1,n-1,n-1} = \Phi^{0,n-i-1,n-i,0,0};$$

$$\Phi^{0,n-i-1,n-i,0,0} = \begin{cases} \Phi^{n-i-1,n-i,0} = (\Phi\sigma_{n-i}^{-1})^0 = \Psi^{n-1}\sigma_{n-i}^{-1} & \text{if } i \leq n-3, \\ \Phi^{0,1,0,2,0} = (\Phi\sigma_2^{-1}\sigma_1^{-2})^0 = \Psi^{n-1}\sigma_2^{-1}\sigma_1^{-2} & \text{if } i = n-2, \\ \Phi^{1,0,0} = (\Phi\sigma_1)^0 = \Psi^{n-1}\sigma_1 & \text{if } i = n-1. \end{cases}$$

■

The choice of base flag Φ^0 in Proposition 3.1.70 is arbitrary. The expressions of the standard rotations of $\mathcal{P}^\delta$ with respect to the flag Φ^i can easily be obtained by translating the expressions in Proposition 3.1.70 with the help of Proposition 3.1.65. The details are left as an exercise.

Exercise 3.1.71 Let $\mathcal{P}$ be a chiral n-polytope with base flag Φ. For each $(i,j) \in \{1,\ldots,n-1\}^2$ determine an expression for the ith standard rotation of $\mathcal{P}^\delta$ with respect to the flag Φ^j in terms of the standard rotations of $\mathcal{P}$.

Proposition 3.1.70 provides a convenient condition on the standard rotations to determine if a chiral polytope is self-dual.

Corollary 3.1.72 *A chiral n-polytope $\mathcal{P}$ with standard rotations σ_i is properly self-dual if and only if there exists an isomorphism $\eta\colon \Gamma(\mathcal{P}) \to \Gamma(\mathcal{P})$ that maps σ_i to σ_{n-i}^{-1} for every $i \in \{1, \ldots, n-1\}$. It is improperly self-dual if and only if there exists an isomorphism $\eta\colon \Gamma(\mathcal{P}) \to \Gamma(\mathcal{P})$ such that $\eta(\sigma_{n-1}) = \sigma_1$, $\eta(\sigma_{n-2}) = \sigma_2^{-1}\sigma_1^{-2}$, and $\eta(\sigma_i) = \sigma_{n-i}^{-1}$ for every $i \in \{1, \ldots, n-3\}$.*

We saw in Proposition 2.3.25 that if $\mathcal{P}$ is an improperly self-dual chiral n-polytope then its extended group $\overline{\Gamma}(\mathcal{P})$ contains a duality δ_0 mapping a given flag Φ to Φ^0. We can compose this duality with the automorphism $\tau_{n-i-1,n-1}$ to obtain a duality $\delta_i = \tau_{n-i-1,n-1}\delta_0$ satisfying that

$$\Phi\delta_i = \Phi\tau_{n-i-1,n-1}\delta_0 = \Phi^{n-1,n-i-1}\delta_0 = (\Phi^{n-1}\delta_0)^i = (\Phi\delta_0)^{0,i} = \Phi^i$$

for any $i \in \{0, \ldots, n-1\}$ (see (2.5)). We can approach Proposition 3.1.65 in the special case of improperly self-dual polytopes by regarding the jth standard rotation ξ_j with respect to Φ^i as the conjugate $\delta_i^{-1}\sigma_{n-j}^{-1}\delta_i$ of the inverse of the $(n-j)$th standard rotation σ_{n-j} with respect to Φ. Indeed, in this case $\overline{\Gamma}(\mathcal{P})$ acts regularly on the flags of $\mathcal{P}$, and

$$\Phi^i\delta_i^{-1}\sigma_{n-j}^{-1}\delta_i = \Phi\sigma_{n-j}^{-1}\delta_i = \Phi^{n-j-1,n-j}\delta_i = (\Phi\delta_i)^{j,j-1} = \Phi^{i,j,j-1}. \tag{3.21}$$

The previous discussion provides an example of an application of conjugation by dualities. Next we present three results regarding such conjugations. For the first of them, recall from Proposition 2.3.25 that properly self-dual chiral polytopes have a duality that fixes a given flag, and that such a duality is involutory.

Proposition 3.1.73 *Let $\mathcal{P}$ be a properly self-dual chiral n-polytope, let σ_j be the jth standard rotation of $\mathcal{P}$ with respect to the base flag Φ, and let δ be the unique duality that fixes Φ. Then $\delta\sigma_j\delta = \sigma_{n-j}^{-1}$.*

Proof We observe that

$$\Phi\delta\sigma_j\delta = \Phi\sigma_j\delta = \Phi^{j,j-1}\delta = (\Phi\delta)^{n-j-1,n-j} = \Phi^{n-j-1,n-j} = \Phi\sigma_{n-j}^{-1}.$$

The proposition follows from the facts that $\delta\sigma_j\delta \in \Gamma(\mathcal{P})$ and that $\Gamma(\mathcal{P})$ acts regularly on each flag orbit. ■

Proposition 3.1.74 *Let $\mathcal{P}$ be an improperly self-dual chiral n-polytope, let σ_j be the jth standard rotation of $\mathcal{P}$ with respect to the base flag Φ, and let δ_i be the duality that maps Φ to Φ^i. Then*

$$\delta_i^{-1}\sigma_j\delta_i = \begin{cases} \sigma_{i+2}^{-1}\sigma_{i+1}^{-2} & \text{if } j = n-i-2, \\ \sigma_{n-j} & \text{if } j \in \{n-i-1, n-i\}, \\ \sigma_i^{-2}\sigma_{i-1}^{-1} & \text{if } j = n-i+1, \\ \sigma_{n-j}^{-1} & \text{otherwise.} \end{cases}$$

Proof From (3.21) we have that $\delta_i^{-1}\sigma_{n-j}^{-1}\delta_i = \xi_j$, where ξ_j is the jth standard rotation with respect to Φ^i. It follows that $\delta_i^{-1}\sigma_j\delta_i = \xi_{n-j}^{-1}$. The proposition now follows from Proposition 3.1.65 by rewriting ξ_{n-j}^{-1} in terms of the standard rotations with respect to Φ. For example, if $j = n - i - 2$ then

$$\xi_{n-j}^{-1} = \xi_{i+2}^{-1} = (\sigma_{i+1}^2\sigma_{i+2})^{-1} = \sigma_{i+2}^{-1}\sigma_{i+1}^{-2}.$$

■

Exercise 3.1.75 Prove Proposition 3.1.74 without using Proposition 3.1.65, by evaluating $\delta_i^{-1}\sigma_j\delta_i$ on the base flag Φ.

Corollary 3.1.76 *Let $\mathcal{P}$ be an improperly self-dual chiral* 4*-polytope, and let σ_j and δ_0 be as in Proposition 3.1.74. Then*

$$\delta_0^2 = \sigma_1\sigma_2\sigma_3, \tag{3.22}$$

$$\delta_0^{-1}\sigma_1\sigma_2\delta_0 = \sigma_1\sigma_2\sigma_3\sigma_1^{-1}. \tag{3.23}$$

Proof We first apply δ_0^2 to Φ to obtain

$$\Phi\delta_0^2 = \Phi^0\delta_0 = (\Phi\delta_0)^3 = \Phi^{0,3} = \Phi\sigma_1\sigma_2\sigma_3,$$

proving (3.22). Then we use Proposition 3.1.74 to write

$$\delta_0^{-1}\sigma_1\delta_0 = \sigma_3^{-1}, \quad \delta_0^{-1}\sigma_2\delta_0 = \sigma_2^{-1}\sigma_1^{-2}.$$

From here, (3.23) follows easily. ■

3.1.4 Applications of Theorem 3.1.53

Building chiral maps and polytopes has been harder than their regular counterparts. Few chiral maps were built from combinatorial methods, like those defined by Heffter and Sherk in [124, 253], respectively. Some chiral polytopes were constructed using geometric arguments, like the chiral maps on the torus in [13, 49] with techniques of Euclidean geometry, or the chiral 4-polytopes with toroidal facets in [207, 248] using hyperbolic geometry. These combinatorial and geometric methods had limitations.

- The constructed objects were forced to have rank 3 or 4.
- There was not much freedom in their Schläfli types.

By contrast, many constructions of chiral polytopes have used Theorem 3.1.53. In Propositions 3.1.84, 3.1.92 and 3.1.111 we include a small sample of different ways in which this theorem has been used.

Chiral Polyhedra with 40 Flags

Here we find all chiral polyhedra with precisely 40 flags. We start with some results about the structure of the automorphism groups or rotary polytopes.

Lemma 3.1.77 *Let $\mathcal{P}$ be a rotary polyhedron with standard rotations σ_1 and σ_2. If for some $i \in \{1,2\}$ the subgroup $\langle \sigma_i \rangle$ is normal in $\Gamma^+(\mathcal{P})$ then $\sigma_{3-i}^2 = \mathrm{id}$.*

Proof Let F_0 and F_2 be the vertex and 2-face of the base flag Φ. Then the conjugates $\gamma^{-1}\langle \sigma_1 \rangle \gamma$ and $\gamma^{-1}\langle \sigma_2 \rangle \gamma$ are the stabilisers of the 2-face $F_2\gamma$ and of the vertex $F_0\gamma$, respectively.

Assume that $\langle \sigma_1 \rangle$ is normal in $\Gamma(\mathcal{P})$, then σ_1 fixes all 2-faces of $\mathcal{P}$. On the other hand, σ_1 cyclically permutes all edges contained in F_2. It follows that there is a 2-face F_2' such that every edge of F_2 also belongs to F_2'. Furthermore, by the diamond condition, F_2' is distinct from F_2. Thus, the subgraph of the flag graph of $\mathcal{P}$ induced by all flags in F_2 and F_2' has all its vertices with degree 3. We can conclude that $\mathcal{P}$ has precisely two faces, and the order of σ_2 is 2. The case when $\langle \sigma_2 \rangle \triangleleft \Gamma(\mathcal{P})$ can be proven by a dual argument. ■

Combining Lemmas 3.1.77 and 2.3.16 we get the following results.

Corollary 3.1.78 *Let $\mathcal{P}$ be a rotary polyhedron with type $\{p,q\}$ for some $p,q \geq 3$. Then $\Gamma^+(\mathcal{P})$ is non-abelian.*

Corollary 3.1.79 *The automorphism group of every chiral n-polytope is non-abelian.*

Even though the automorphism groups of rotary polyhedra with type $\{4,4\}$ are non-abelian, their structure is not all that complicated, as shown by the next lemma.

Lemma 3.1.80 *Let $\mathcal{P}$ be a rotary polytope with type $\{4,4\}$, let σ_1 and σ_2 be the standard rotations of $\mathcal{P}$ and let $\Delta = \langle \sigma_1\sigma_2^{-1}, \sigma_1^{-1}\sigma_2 \rangle$. Then Δ is an abelian normal subgroup of $\Gamma^+(\mathcal{P})$. Furthermore,*

$$\Gamma^+(\mathcal{P}) \cong \Delta \rtimes \langle \sigma_1 \rangle \cong \Delta \rtimes \langle \sigma_2 \rangle.$$

Proof To show that $\sigma_1\sigma_2^{-1}$ commutes with $\sigma_1^{-1}\sigma_2$, we use that the orders of σ_1 and σ_2 are both 4, that $\sigma_1\sigma_2 = \sigma_2^{-1}\sigma_1^{-1}$, and that $\sigma_1^{-1}\sigma_2^{-1} = \sigma_2\sigma_1$. Indeed,

$$\sigma_1\sigma_2^{-1}\sigma_1^{-1}\sigma_2 = \sigma_1^2\sigma_2^2 = \sigma_1^{-2}\sigma_2^{-2} = \sigma_1^{-1}\sigma_2\sigma_1\sigma_2^{-1}.$$

The conjugates $\sigma_1^{-1}(\sigma_1\sigma_2^{-1})\sigma_1$ and $\sigma_2^{-1}(\sigma_1\sigma_2^{-1})\sigma_2$ both equal $\sigma_2^{-1}\sigma_1 = (\sigma_1^{-1}\sigma_2)^{-1} \in \Delta$, while $\sigma_1(\sigma_1^{-1}\sigma_2)\sigma_1^{-1}$ and $\sigma_2(\sigma_1^{-1}\sigma_2)\sigma_2^{-1}$ both equal $\sigma_2\sigma_1^{-1} = (\sigma_1\sigma_2^{-1})^{-1} \in \Delta$. This shows that $\Delta \triangleleft \Gamma^+(\mathcal{P})$.

The four cosets Δ, $\Delta\sigma_1$, $\Delta\sigma_1^2$ and $\Delta\sigma_1^3$ are distinct. To see this it suffices to show that $\Delta\sigma_1^2 \neq \Delta$. Assume to the contrary that $\sigma_1^2 \in \Delta$, then we use that $\sigma_2\sigma_1 = \sigma_1^{-1}\sigma_2^{-1}$ and that Δ is abelian to obtain

$$\sigma_1\sigma_2 = (\sigma_1^2)(\sigma_1^{-1}\sigma_2) = (\sigma_1^{-1}\sigma_2)(\sigma_1^2) = \sigma_1^{-1}\sigma_1^{-1}\sigma_2^{-1}\sigma_1 = \sigma_1^2\sigma_2^{-1}\sigma_1.$$

It follows that $\sigma_1^{-1}\sigma_2 = \sigma_2^{-1}\sigma_1 = (\sigma_1^{-1}\sigma_2)^{-1}$, and this forces $\sigma_1^{-1}\sigma_2$ to have order 2. Also $\sigma_1\sigma_2^{-1}$ must have order 2; this can be deduced by proceeding similarly, but studying the product $(\sigma_1^2)(\sigma_1\sigma_2^{-1}) = (\sigma_1\sigma_2^{-1})(\sigma_1^2)$ instead. Since Δ is abelian, its non-trivial elements must be $\sigma_1\sigma_2^{-1}$, $\sigma_1^{-1}\sigma_2$ and $\sigma_1\sigma_2^{-1}\sigma_1^{-1}\sigma_2 = \sigma_1^2\sigma_2^2$. Direct inspection shows that $\sigma_1^2 \in \Delta$ violates the intersection condition, which yields the desired contradiction.

In order to show that $\Gamma^+(\mathcal{P}) \cong \Delta \rtimes \langle\sigma_1\rangle$ it suffices to verify that $\Gamma^+(\mathcal{P}) = \Delta\langle\sigma_1\rangle$. With this in mind we shall prove that for each $k \in \{0,1,2,3\}$ the coset $\Delta\sigma_1^k\sigma_2$ equals $\Delta\sigma_1^{k+1}$, since this implies that these four are the only cosets of Δ in $\Gamma^+(\mathcal{P})$. Indeed,

$$\begin{aligned}
\Delta\sigma_2 &= \Delta\sigma_1\sigma_2^{-1}\sigma_2 = \Delta\sigma_1,\\
\Delta\sigma_1\sigma_2 &= \Delta\sigma_2^{-1}\sigma_1^{-1} = \Delta\sigma_1^{-1}\sigma_2\sigma_2^{-1}\sigma_1^{-1} = \Delta\sigma_1^2,\\
\Delta\sigma_1^2\sigma_2 &= \Delta\sigma_1\sigma_2^{-1}\sigma_1^{-1} = \Delta\sigma_1^{-1} = \Delta\sigma_1^3,\\
\Delta\sigma_1^3\sigma_2 &= \Delta\sigma_1^{-1}\sigma_2 = \Delta.
\end{aligned}$$

Similar arguments can be used to prove that $\Gamma^+(\mathcal{P}) \cong \Delta \rtimes \langle\sigma_2\rangle$. ■

Corollary 3.1.81 *If $\mathcal{P}$ is the universal polyhedron with type $\{4,4\}$ then $\Gamma^+(\mathcal{P}) \cong \mathbb{Z}^2 \rtimes \mathbb{Z}_4$. Furthermore, $\mathcal{P}$ is isomorphic to the regular tessellation of $\mathbb{E}^2$ by squares.*

Proof From Lemma 3.1.80 we know that $\langle\sigma_1\sigma_2^{-1}, \sigma_1^{-1}\sigma_2\rangle$ is abelian and that $\Gamma^+(\mathcal{P}) = \langle\sigma_1\sigma_2^{-1}, \sigma_1^{-1}\sigma_2\rangle \rtimes \langle\sigma_1\rangle$. It follows that $\Gamma^+(\mathcal{P})$ is isomorphic to a quotient of $\mathbb{Z}^2 \rtimes \mathbb{Z}_4$. On the other hand,

$$\sigma_1^{-1}(\sigma_1^{-1}\sigma_2)\sigma_1 = \sigma_1^{-2}\sigma_1^{-1}\sigma_2^{-1} = \sigma_1\sigma_2^{-1},$$

and so $\sigma_1\sigma_2^{-1}$ and $\sigma_1^{-1}\sigma_2$ have the same order. Since there are infinitely many rotary polytopes of type $\{4,4\}$ (see Theorem 2.4.8), this order must be ∞ and $\langle \sigma_1\sigma_2^{-1}, \sigma_1^{-1}\sigma_2 \rangle \cong \mathbb{Z}^2$.

The regular tessellation of $\mathbb{E}^2$ by squares has type $\{4,4\}$ and so it must be a quotient of the universal polyhedron with type $\{4,4\}$. For the reasons explained above, its rotation group with respect to the standard rotations σ_1' and σ_2' must also be $\langle \sigma_1'(\sigma_2')^{-1}, (\sigma_1')^{-1}\sigma_2' \rangle \rtimes \langle \sigma_1' \rangle \cong \mathbb{Z}^2 \rtimes \mathbb{Z}_4$. Now, $\sigma_1'(\sigma_2')^{-1}$ and $(\sigma_1')^{-1}\sigma_2'$ are one-step translations along the y- and x-axes, respectively. Hence there is an isomorphism from $\langle \sigma_1\sigma_2^{-1}, \sigma_1^{-1}\sigma_2 \rangle$ to $\langle \sigma_1'(\sigma_2')^{-1}, (\sigma_1')^{-1}\sigma_2' \rangle$ that maps $\sigma_1\sigma_2^{-1}$ to $\sigma_1'(\sigma_2')^{-1}$ and $\sigma_1^{-1}\sigma_2$ to $(\sigma_1')^{-1}\sigma_2'$. From here it is straightforward to prove that the rotation group of $\{4,4\}$ and the rotation group of the regular tessellation by squares have identical presentations. The isomorphism follows from the uniqueness in Theorem 3.1.53. ∎

If explicitly asked to, distinct researchers on polytopes theory would prove Lemma 3.1.80 and Corollary 3.1.81 in various distinct ways. Here we provided a proof using mostly group theory and Theorem 3.1.53, but mathematicians with more geometric taste may find alternative proofs relating automorphisms with isometries of Euclidean plane or of the torus. We next give an idea of one way in which this could be done.

Remark 3.1.82 The full symmetry group of the tessellation $\{4,4\}$ of $\mathbb{E}^2$ by squares is isomorphic to the Coxeter group $[4,4]$ (see [48, 140]). Using this fact it can be shown that the orientation-preserving symmetry group $G^+(\{4,4\})$ of $\{4,4\}$ admits the presentation

$$\langle \sigma_1, \sigma_2 \colon \sigma_1^4 = \sigma_2^4 = (\sigma_1\sigma_2)^2 = \mathrm{id} \rangle$$

with respect to the standard rotations σ_1 and σ_2. It follows that every rotary polyhedron with type $\{4,4\}$ is a quotient of the Euclidean tessellation by squares.

The translation subgroup T of $G^+(\{4,4\})$ is normal and has index 4; furthermore, any normal subgroup K of $G^+(\{4,4\})$ such that the orders of $K\sigma_1$ and $K\sigma_2$ in $G^+(\{4,4\})/K$ is 4 is a subgroup of T. This implies that all rotary polyhedra with type $\{4,4\}$ can be visualised as maps on the torus. If $\mathcal{P} = \{4,4\}/K$ is one such polyhedron, then T/K is a normal subgroup of index 4 of $\Gamma^+(\mathcal{P})$. The conclusions of Lemma 3.1.80 follow easily from the fact that $\sigma_1\sigma_2^{-1}$ and $\sigma_1^{-1}\sigma_2$ generate T.

Remark 3.1.83 When visualising $\{4,4\}_{(a,b)}$ as a map on the torus, the automorphisms $\sigma_1\sigma_2^{-1}$ and $\sigma_1^{-1}\sigma_2$ can still be understood as translations. Indeed, if Φ is the base flag then $\Phi\sigma_1\sigma_2^{-1} = \Phi^{1,0}\sigma_2^{-1} = \Phi^{1,2,1,0}$ and

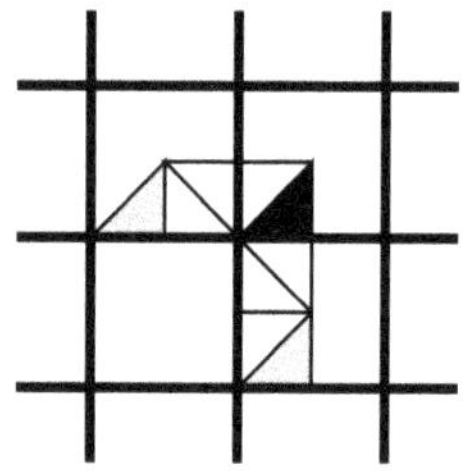

Figure 3.7 The flags $\Phi\sigma_1\sigma_2^{-1}$ and $\Phi\sigma_1^{-1}\sigma_2$.

$\Phi\sigma_1^{-1}\sigma_2 = \Phi^{0,1}\sigma_2 = \Phi^{2,1,0,1}$. By choosing Φ as the black flag in Figure 3.7, the flags $\Phi\sigma_1\sigma_2^{-1}$ and $\Phi\sigma_1^{-1}\sigma_2$ are the grey flags, and are translates of Φ by vectors $(-1,0)$ and $(0,-1)$, respectively.

Now we present our first application of Theorem 3.1.53. In the following proposition we find all rotary polytopes with a certain fixed number of flags. The statement does not show an algebraic flavour, but the proof will be group-theoretical in essence.

Proposition 3.1.84 *Up to isomorphism there is precisely one chiral polyhedron with 40 flags, namely* $\{4,4\}_{(2,1)}$.

Proof Let $\mathcal{P}$ be a chiral polyhedron with 40 flags and let σ_1 and σ_2 be the standard rotations with respect to some flag Φ. From the size of $\mathcal{F}l(\mathcal{P})$ we know that $|\Gamma(\mathcal{P})| = 20$ and therefore the orders of σ_1 and σ_2 must divide 20. Furthermore, Proposition 2.3.14 (c) and Lemma 2.3.16 tell us that these orders must be greater or equal to 3, and that $\mathcal{P}$ must have at least three vertices and at least three facets. Now we use that the vertices and facets are in a one-to-one correspondence with the cosets of $\langle\sigma_2\rangle$ and $\langle\sigma_1\rangle$, respectively, to deduce that each of the orders of σ_1 and σ_2 must be less than 7, and hence each of them must be either 4 or 5. Moreover, the third Sylow theorem implies that if a group of order 20 has a subgroup of order 5 then it must be normal. Thus, Lemma 3.1.77 implies that $\mathcal{P}$ has type $\{4,4\}$.

From Lemma 3.1.80 we know that the subgroup $\Delta = \langle\sigma_1\sigma_2^{-1}, \sigma_1^{-1}\sigma_2\rangle$ has order 5. The intersection property forces $\sigma_1\sigma_2^{-1}$ and $\sigma_1^{-1}\sigma_2$ to be non-trivial, and hence $\Delta = \langle\sigma_1\sigma_2^{-1}\rangle = \langle\sigma_1^{-1}\sigma_2\rangle$. In particular, $\langle\sigma_1^{-1}\sigma_2\rangle \cong \mathbb{Z}_5$ and $\sigma_1^{-1}\sigma_2 = (\sigma_1\sigma_2^{-1})^k$ for some $k \in \{1,2,3,4\}$. We can discard $k = 1$, since otherwise we would have that $\sigma_1^2 = \sigma_2^2$, contradicting the intersection condition. We can also discard $k = 4$ since that would imply that $\sigma_1^{-1}\sigma_2 = \sigma_2\sigma_1^{-1}$, contradicting Corollary 3.1.78. In the case that $k = 3$ we use Proposition 3.1.70 to imply that the standard rotations ξ_1 and ξ_2 (taken with respect to Φ^1) satisfy that $\xi_1\xi_2^{-1} = \sigma_1^{-1}\sigma_2 = (\sigma_1\sigma_2^{-1})^3 = (\xi_1^{-1}\xi_2)^3$. Since $\xi_1\xi_2^{-1} \in \Delta$, we have that $(\xi_1\xi_2^{-1})^2 = (\xi_1^{-1}\xi_2)^6 = \xi_1^{-1}\xi_2$, which tells us that the rooted polytope $\mathcal{P}_{(\Phi)}$ is the enantiomorph of a rooted polytope for which $k = 2$. Finally, it is routine to verify that the generators σ_1 and σ_2 of $\{4,4\}_{(2,1)}$ satisfy $\sigma_1^{-1}\sigma_2 = (\sigma_1\sigma_2^{-1})^2$, where the base flag Φ is chosen as a white flag in Figure 2.18. The discussion above implies that up to enantiomorphism $\Gamma^+(\{4,4\}_{(2,1)})$ is the only string even group $\langle\sigma_1,\sigma_2\rangle$ with order 20 satisfying the intersection condition (3.10), and hence $\{4,4\}_{(2,1)}$ is the only rotary polyhedron with 40 flags. ■

While proving Proposition 3.1.84 we only used chirality to discard that the order of σ_1 or of σ_2 could be 2. Therefore, the same arguments can be used to prove the following.

Corollary 3.1.85 *There are no orientably regular polyhedra with type $\{p,q\}$ for some $p,q \geq 3$ with 40 flags.*

The arguments in the proof of Proposition 3.1.84 are very specific to the particular number 40 of flags. If this number is increased, it may no longer be possible to specify only one possible Schläfli type for the polyhedron. Moreover, if the type is not $\{4,4\}$ then we may not find an analogue of Lemma 3.1.80 to provide conditions on the structure of the automorphism group. This suggests that it is not practical to use this method to try to find all chiral polytopes with M flags for large values of M without incorporating additional ideas.

There are other methods to search for rotary polytopes with a given number M of flags with the help of computer software. One of them consists of the following. Let Σ be the infinite group with presentation

$$\langle\sigma_1,\ldots,\sigma_{n-1}\colon (\sigma_i\cdots\sigma_j)^2 = \text{id} \quad \text{for every } i < j\rangle. \tag{3.24}$$

(This is the universal string even group.) Since the automorphism group of every chiral n-polytope $\mathcal{P}$ satisfies (3.7), the group $\Gamma^+(\mathcal{P})$ is the quotient of Σ by a normal subgroup Δ. When representing $\Gamma^+(\mathcal{P})$ with generators and relations, Δ is precisely the normal closure of the relators in the presentation with respect to the standard rotations of $\mathcal{P}$. Furthermore, $|\Gamma^+(\mathcal{P})|$ is the index of Δ in Σ. Hence, if $\mathcal{P}$ has M flags, the index of Δ in Σ must be $M/2$. So the first step is to find all normal subgroups of Σ of index $M/2$. For each such normal subgroup Δ we still need to verify that the quotient Σ/Δ is the rotation group of a rotary polytope. The images $\sigma_i\Delta$ of the generators σ_i under the quotient will satisfy (3.7), so the second step is to determine if these generators are non-trivial and also satisfy the intersection condition (3.10). The final step is to determine which quotients are rotation groups of regular polytopes and which ones are automorphism groups of chiral polytopes. This can be done by finding all subgroups Δ for which there is a group automorphism η satisfying that

$$\eta(\sigma_i \Delta) = \begin{cases} (\sigma_1 \Delta)^{-1} & \text{if } i = 1, \\ (\sigma_1 \Delta)^2 (\sigma_2 \Delta) & \text{if } i = 2, \\ \sigma_i \Delta & \text{if } i > 2, \end{cases}$$

which is equivalent to requiring $\mathcal{P}$ to be regular. This is a sketch of the procedure that was used to produce the atlas in [33, 34]. The three steps just mentioned are tedious and time-consuming; even for small values of the number M of flags it becomes necessary to count with the support of a computer software able to manipulate groups.

Exercise 3.1.86 With the help of specialised software on group theory find all normal subgroups of index 20 in the group Σ defined in (3.24). Use those normal subgroups to confirm that the only chiral polyhedron with 40 flags is $\{4,4\}_{(2,1)}$.

Infinitely Many Chiral 4-Polytopes

The next application of Theorem 3.1.53 uses some results on the cube and on toroidal polyhedra with type $\{4,4\}$. We start by giving a presentation of the rotation group of the cube in terms of generators and relations.

Lemma 3.1.87 *The rotation group of the cube is isomorphic to the symmetric group S_4 and admits the presentation*

$$\langle x, y : x^4 = y^3 = (xy)^2 = \text{id} \rangle. \tag{3.25}$$

Proof Let Δ be the rotation group of the cube, and let $\mathcal{X}$ be the set of four diagonals between opposite vertices of the cube. The half-turn about the line spanned by the midpoints of two opposite edges F_1 and F_1' of the cube swaps the two elements of $\mathcal{X}$ contained in the plane spanned by F_1 and F_1', while it fixes the other two elements in $\mathcal{X}$. It is instructive to stare at a cube and convince ourselves that all transpositions in the symmetric group $S_{\mathcal{X}}$ are attained by such half-turns. The only element in Δ fixing $\mathcal{X}$ pointwise is id, and this implies that $S_{\mathcal{X}}$ is isomorphic to a subgroup of Δ. Since $S_{\mathcal{X}}$ has 24 elements and the cube has 48 flags, Δ must be isomorphic to $S_{\mathcal{X}}$.

Let Γ be the group with presentation (3.25) and let σ_1 and σ_2 be the standard rotations of the cube. Then $\Delta = \langle \sigma_1, \sigma_2 \rangle$ satisfies $\sigma_1^4 = \sigma_2^3 = \text{id}$ as well

Table 3.1. *Cosets of the subgroup $\Lambda = \langle x \rangle$ of the rotation group of the cube (here x represents σ_1).*

	Λ	Λy	Λy^2	$\Lambda y^2 x$	$\Lambda y x^{-1}$	$\Lambda y x^{-1} y$
$\cdot x$	Λ	Λy^2	$\Lambda y^2 x$	$\Lambda y x^{-1}$	Λy	$\Lambda y x^{-1} y$
$\cdot y$	Λy	Λy^2	Λ	$\Lambda y x^{-1}$	$\Lambda y x^{-1} y$	$\Lambda y^2 x$

as $(\sigma_1\sigma_2)^2 = \mathrm{id}$ (see the second and fourth items of Proposition 2.2.16), and therefore it is a quotient of Γ. To prove the isomorphism it suffices to show that the order of Γ is less or equal to 24, which is the number of flags of the cube divided by 2. In order to do that we present in the first row of Table 3.1 the list of cosets of the group $\Lambda = \langle x \rangle$; the second and third rows indicate the coset obtained by multiplying the coset in the first row by x and y, respectively. For example, by using $(xy)^2 = \mathrm{id}$ (and hence $yx = x^{-1}y^{-1}$, $xy = y^{-1}x^{-1}$ and $yxy = x^{-1}$), we verify that

$$\begin{aligned}(\Lambda y x^{-1} y)x &= \Lambda (yx)x(xy)x = \Lambda x^{-1}y^{-1}xy^{-1}x^{-1}x\\ &= \Lambda y^{-1}xy^{-1} = \Lambda y^2 x y^2 = \Lambda y x^{-1} y.\end{aligned}$$

Since Λ has order 4 and at most 6 right cosets, the order of Γ is indeed 24. ∎

The next couple of results expand the description in Lemma 3.1.80 of rotary polyhedra with type $\{4,4\}$.

Lemma 3.1.88 *The rotation group $\Gamma^+(\{4,4\}_{(a,b)})$ admits the presentation*

$$\langle \sigma_1, \sigma_2 \colon \sigma_1^4 = \sigma_2^4 = (\sigma_1\sigma_2)^2 = (\sigma_1\sigma_2^{-1})^a(\sigma_1^{-1}\sigma_2)^b = \mathrm{id}\rangle \tag{3.26}$$

with respect to the standard rotations.

Proof The rotation group of every polyhedron with type $\{4,4\}$ satisfies that $\sigma_1^4 = \sigma_2^4 = (\sigma_1\sigma_2)^2 = \mathrm{id}$. With the help of Remark 3.1.83 it is easy to verify that the rotation group of $\{4,4\}_{(a,b)}$ also satisfies $(\sigma_1\sigma_2^{-1})^a(\sigma_1^{-1}\sigma_2)^b = \mathrm{id}$. Hence the group Λ with presentation (3.26) covers $\Gamma^+(\{4,4\}_{(a,b)})$, and therefore it has order at least $|\Gamma^+(\{4,4\}_{(a,b)})| = 4(a^2+b^2)$. (To compute the order of $\Gamma^+(\{4,4\}_{(a,b)})$ we count the number of flags of $\{4,4\}_{(a,b)}$ as 8 times the number a^2+b^2 of squares, and then divide by 2.)

To show that the group Λ with presentation (3.26) has order at most $4(a^2+b^2)$, we use that it is a quotient of $\Gamma^+(\{4,4\})$. Lemma 3.1.80 implies that $\Delta := \langle \sigma_1\sigma_2^{-1}, \sigma_1^{-1}\sigma_2\rangle$ is an abelian normal subgroup of Λ of index 4. The relation $(\sigma_1\sigma_2^{-1})^a(\sigma_1^{-1}\sigma_2)^b = \mathrm{id}$ is equivalent to

$$\begin{aligned}\mathrm{id} &= \sigma_1^{-1}(\sigma_1\sigma_2^{-1})^a(\sigma_1^{-1}\sigma_2)^b\sigma_1 = (\sigma_1^{-1}\sigma_1\sigma_2^{-1}\sigma_1)^a(\sigma_1^{-1}\sigma_1^{-1}\sigma_2\sigma_1)^b \\ &= (\sigma_2^{-1}\sigma_1)^a(\sigma_1^{-2}\sigma_1^{-1}\sigma_2^{-1})^b = (\sigma_2^{-1}\sigma_1)^a(\sigma_1\sigma_2^{-1})^b = (\sigma_2^{-1}\sigma_1)^a(\sigma_2\sigma_1^{-1})^{-b}.\end{aligned}$$

Inverting both sides and using that $\sigma_1\sigma_2^{-1}$ commutes with $\sigma_1^{-1}\sigma_2$ we obtain the equivalent relation $(\sigma_1\sigma_2^{-1})^b(\sigma_1^{-1}\sigma_2)^{-a} = \mathrm{id}$. It follows that Δ is covered by the quotient of $\mathbb{Z}^2$ by the group $\langle(a,b),(b,-a)\rangle$. This quotient has as many elements as points with integer coordinates in the square S of $\mathbb{E}^2$ with vertices $(0,0)$, (a,b), $(b,-a)$, $(a+b,b-a)$. With the help of the auxiliary tessellation by squares with vertices in the points $(\frac{1}{2},\frac{1}{2})+\mathbb{Z}^2$, we observe that the desired number equals the area a^2+b^2 of S. This proves that Λ and $\Gamma^+(\{4,4\}_{(a,b)})$ have the same order, and therefore they are isomorphic. ■

Corollary 3.1.89 *Let a, b be relatively prime, let $\mathcal{P}$ be a chiral polyhedron with type $\{4,4\}$ whose rotation group satisfies the relations in (3.26) in terms of the standard rotations σ_1, σ_2 of $\mathcal{P}$, and assume that the order of $\sigma_1\sigma_2^{-1}$ is a^2+b^2. Then $\mathcal{P}$ is isomorphic to $\{4,4\}_{(a,b)}$.*

Proof Lemma 3.1.88 implies that $\Gamma^+(\mathcal{P})$ is a quotient of $\Gamma^+(\{4,4\}_{(a,b)})$. We use Lemma 3.1.80 and the fact that the subgroup $\Delta = \langle\sigma_1\sigma_2^{-1},\sigma_1^{-1}\sigma_2\rangle$ of $\Gamma(\mathcal{P})$ contains $\langle\sigma_1\sigma_2^{-1}\rangle$ to imply that the order of $\langle\sigma_1\sigma_2^{-1}\rangle\langle\sigma_1\rangle$ is already $4(a^2+b^2)$, and so there is an isomorphism from $\Gamma^+(\mathcal{P})$ to the group Λ defined by the presentation (3.26). The uniqueness in Theorem 3.1.53 forces $\mathcal{P}$ to be isomorphic to $\{4,4\}_{(a,b)}$. ■

Remark 3.1.90 The order of $\sigma_1\sigma_2^{-1}$ in $\Gamma^+(\{4,4\}_{(a,b)})$ when a and b are relatively prime is indeed a^2+b^2. To verify this it suffices to recall that $(\sigma_1\sigma_2^{-1})^k$ acts like the translation by $(-k,0)$ and $(\sigma_1^{-1}\sigma_2)^k$ like the translation by $(0,-k)$ (here we took as base flag a flag in the position of the black one in Figure 3.7). From Section 2.4.1 we know that $\{4,4\}_{(a,b)}$ is the quotient of the plane tessellation $\{4,4\}$ by the group Λ of translations generated by (a,b) and $(b,-a)$. The smallest positive k for which $(\sigma_1\sigma_2^{-1})^k = \mathrm{id}$ is the smallest positive k for which $(k,0)\in\Lambda$. If we assume that

$$(k,0) = \mu(a,b) + \nu(b,-a) = (\mu a + \nu b, \mu b - \nu a)$$

then, since a and b are relatively prime, the second coordinate forces $\mu = az$ and $\nu = bz$ for some $z\in\mathbb{Z}^+$. The smallest value of $\mu a+\nu b$ is a^2+b^2, and it is attained when $z=1$. Therefore the order of $\sigma_1\sigma_2^{-1}$ is precisely a^2+b^2.

On the other hand, the order of $\sigma_1\sigma_2^{-1}$ in $\Gamma^+(\{4,4\}_{(4,2)})$ is 10, and not $4^2+2^2=20$. This is illustrated in Figure 3.8 through the action of $(\sigma_1\sigma_2^{-1})^k$ on the grey flag Φ; the label $k\in\{1,\dots,10\}$ indicates the square of the flag $\Phi(\sigma_1\sigma_2^{-1})^k$.

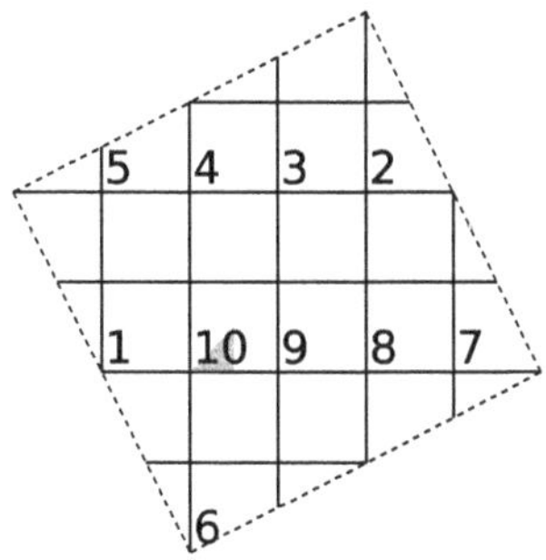

Figure 3.8 The order of $\sigma_1\sigma_2^{-1}$ is 10.

Exercise 3.1.91 Determine the order of $\sigma_1\sigma_2^{-1}$ in $\Gamma^+(\{4,4\}_{(a,b)})$ for arbitrary a,b. Does it depend on the choice of base flag?

Next we present our second application of Theorem 3.1.53. In the following proposition we choose an infinite set $\mathcal{Z}$ of chiral polyhedra, and for every polyhedron $\mathcal{P}$ in $\mathcal{Z}$ we show existence of a chiral 4-polytope whose facet is isomorphic to $\mathcal{P}$.

Proposition 3.1.92 *Let a,b be positive integers such that a^2+b^2 is an odd prime. Then there exists a chiral 4-polytope with type $\{4,4,3\}$ whose facets are isomorphic to $\{4,4\}_{(a,b)}$.*

Proof Let $p = a^2+b^2$, and let $PL_2(\mathbb{Z}_p)$ denote the quotient of the group of invertible 2×2 matrices with coefficients in $\mathbb{Z}_p$ by the subgroup $\{\lambda Id\colon \lambda \in \mathbb{Z}_p^*\}$ of scalar matrices. (Here $\mathbb{Z}_p^*$ denotes the multiplicative group of $\mathbb{Z}_p$.) From Lemma B.0.2 we know that there exists $i \in \mathbb{Z}_p$ such that $i^2 = -1$. For such a value of i we consider the (classes of) matrices

$$\sigma_1 = \begin{bmatrix} -i & 0 \\ 0 & 1 \end{bmatrix}, \quad \sigma_2 = \begin{bmatrix} -i & i \\ 0 & 1 \end{bmatrix}, \quad \sigma_3 = \begin{bmatrix} 1 & -1 \\ 1 & 0 \end{bmatrix}, \tag{3.27}$$

in $PL_2(\mathbb{Z}_p)$. Let $\Gamma = \langle\sigma_1,\sigma_2,\sigma_3\rangle$.

It is routine to verify that the orders of σ_1, σ_2 and σ_3 are 4, 4 and 3, respectively, and also that Γ satisfies (3.7).

In order to prove that Γ is the rotation group of a rotary 4-polytope with type $\{4,4,3\}$ we only need to show that it satisfies the intersection conditions (3.17) and (3.18). Condition (3.17) can be verified directly by listing the four elements in $\langle\sigma_1\rangle$, the four elements in $\langle\sigma_2\rangle$, and the three elements in $\langle\sigma_3\rangle$. To prove (3.18) we first note that σ_1 and σ_2 are classes of upper triangular matrices, implying that every element in the intersection $\Lambda := \langle\sigma_1,\sigma_2\rangle \cap \langle\sigma_2,\sigma_3\rangle$ must be a class of triangular matrices as well. Moreover, $\langle\sigma_2,\sigma_3\rangle$ satisfies the relations in (3.25), and Lemma 3.1.87 tells us that $\langle\sigma_2,\sigma_3\rangle$ is isomorphic to a quotient of S_4. Its order must be a multiple of 12, since it has elements of orders 3 and 4; but S_4 has no normal subgroup of order 2, which implies that $\langle\sigma_2,\sigma_3\rangle \cong S_4$. The intersection Λ contains $\langle\sigma_2\rangle \cong \mathbb{Z}_4$ and is contained on $\langle\sigma_2,\sigma_3\rangle \cong S_4$. It follows that the index of Λ in $\langle\sigma_2,\sigma_3\rangle$ is a divisor of 6. To show that the index is 6 it suffices to exhibit four distinct right cosets of Λ. Clearly $\sigma_3 \notin \Lambda$, since it is not an upper triangular matrix. It follows that Λ, $\Lambda\sigma_3$ and $\Lambda\sigma_3^2$ are three distinct cosets. Furthermore,

$$\sigma_3^2\sigma_2 = \begin{bmatrix} 0 & -1 \\ -i & i-1 \end{bmatrix}, \quad \sigma_3^2\sigma_2\sigma_3 = \begin{bmatrix} -1 & 0 \\ -1 & i \end{bmatrix}, \quad \sigma_3^2\sigma_2\sigma_3^2 = \begin{bmatrix} -1 & 1 \\ i-1 & 1 \end{bmatrix},$$

none of which is upper triangular. It follows that $\Lambda\sigma_3^2\sigma_2$ is a coset distinct to Λ, $\Lambda\sigma_3$ and $\Lambda\sigma_3^2$. We can conclude that $\Lambda = \langle\sigma_2\rangle$ and therefore Γ is the rotation group of a rotary 4-polytope $\mathcal{P}$.

Now we study the facets of $\mathcal{P}$ through its rotation group $\langle\sigma_1,\sigma_2\rangle$. An inductive procedure shows that

$$(\sigma_1\sigma_2^{-1})^b = \begin{bmatrix} 1 & -bi \\ 0 & 1 \end{bmatrix}, \quad (\sigma_1^{-1}\sigma_2)^a = \begin{bmatrix} 1 & -a \\ 0 & 1 \end{bmatrix},$$

and therefore

$$(\sigma_1\sigma_2^{-1})^b(\sigma_1^{-1}\sigma_2)^a = \begin{bmatrix} 1 & -(a+bi) \\ 0 & 1 \end{bmatrix}.$$

Lemma B.0.4 implies that $a+bi \equiv 0$ (modulo p), and Corollary 3.1.89 implies that the facets of $\mathcal{P}$ are isomorphic to $\{4,4\}_{(a,b)}$. We can conclude that $\mathcal{P}$ is chiral since it is rotary and has chiral facets. ■

Remark 3.1.93 The set of chiral polyhedra $\{4,4\}_{(a,b)}$ where a and b satisfy the hypothesis of Proposition 3.1.92 is infinite. This is a consequence of Proposition B.0.3 and Dirichlet's Theorem B.0.1. Here we use the particular case that states that there are infinitely many prime numbers congruent to 1 modulo 4.

Remark 3.1.94 Proposition 3.1.92 talks about a family of chiral toroidal polyhedra, each of whose members is isomorphic to the facets of a chiral 4-polytope. This is a particular case of the results in [248, Section 7] where more 4-polytopes with type $\{4,4,3\}$ are constructed from the matrices in (3.27). Moreover, further sections of [248] provide constructions of chiral 4-polytopes with types $\{4,4,4\}$, $\{6,3,3\}$, $\{3,6,3\}$ and $\{6,3,6\}$, all in the same spirit as that in

Proposition 3.1.92. (See [205] for further details of chiral extensions of chiral toroidal polyhedra.)

Remark 3.1.95 The matrices in (3.27), when considered with complex coefficients, have a geometric meaning as Möbius transformations of the Riemann sphere, and encode isometries of hyperbolic 3-space. These isometries constitute a generating set of the rotation group of the paracompact hyperbolic tessellation with type $\{4,4,3\}$ (see for example [144, Chapter 13]). The facets of this tessellation are all combinatorially equivalent to tessellations $\{4,4\}$ of the Euclidean plane; hence they are infinite, and the vertex set of each of them is contained in a horosphere. For more details about these geometric aspects see [248].

Rotary Polyhedra Whose Rotation Group is A_6

Our last application of Theorem 3.1.53 consists of finding all rotary polyhedra $\mathcal{P}$ such that $\Gamma^+(\mathcal{P})$ is isomorphic to the alternating group A_6. We shall see that all such rotary polyhedra are regular, and therefore there are no chiral polyhedra with automorphism group A_6.

We start by recalling some well-known facts about the alternating group A_6. Let $\mathcal{X}_6 := \{1,2,3,4,5,6\}$, then A_6 consists of all even permutations of the elements in $\mathcal{X}_6$. Therefore the cyclic structures of the elements in A_6 are

- double transpositions,
- 3-cycles,
- double 3-cycles,
- 4-cycle and transposition,
- 5-cycles.

The action of A_6 on $\mathcal{X}_6$ is transitive (for each $a,b \in \mathcal{X}_6$ there exists $\alpha \in A_6$ such that $a\alpha = b$) and primitive (there is no non-trivial partition of $\mathcal{X}_6$ invariant under the action of every element of A_6).

The automorphism group of A_6 contains all conjugations by elements in S_6 as a subgroup of index 2. Each of these conjugations acts on a given element of A_6 by preserving the cyclic structure while renaming the elements of $\mathcal{X}_6$. As a consequence, if $\alpha, \beta \in A_6$ have the same cyclic structure then there is an automorphism of A_6 mapping α to β.

The remaining automorphisms of A_6 map 3-cycles to products of disjoint 3-cycles, and vice versa. Each $\alpha \in A_6$ that is neither a 3-cycle nor the product of two disjoint 3-cycles is mapped to some element with the same cyclic structure. One way of describing such automorphisms is through the action of A_6 on two different substructures of the complete graph K_6: one of them is

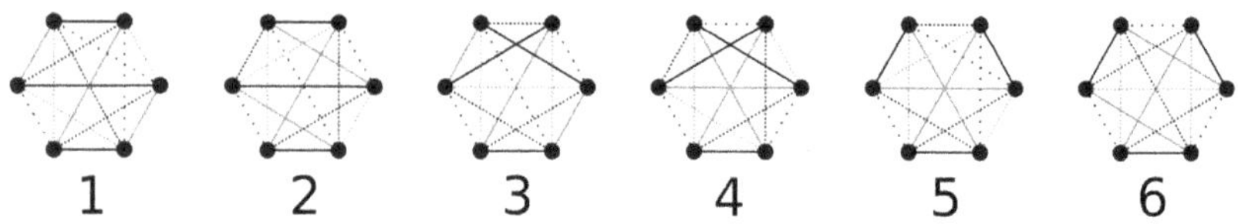

Figure 3.9 The six partitions of the edge set of K_6 into perfect matchings.

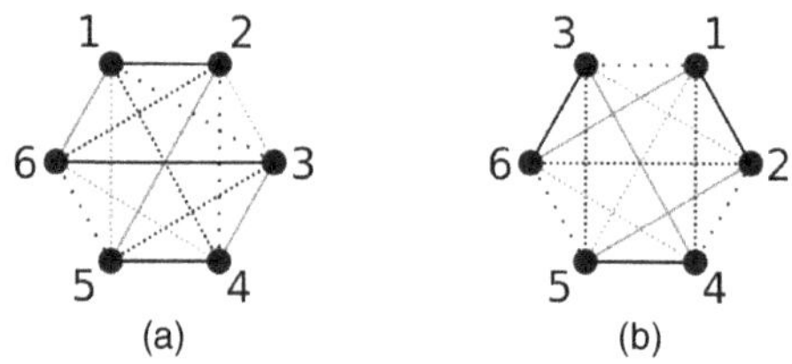

Figure 3.10 (a) A partition of the edge set of K_6; (b) its image under $(1,2,3)$.

the set of its six vertices, the other one is the set of the partitions of its edges into perfect matchings, illustrated in Figure 3.9. By labelling these two sets with the elements of $\mathcal{X}_6$ we get automorphisms of A_6 that do not preserve all cyclic structures.

For example, when applying the permutation $(1,2,3)$ to the vertices labelled as in Figure 3.10(a), carrying along the partition of edges, the partition numbered 1 in Figure 3.9 is mapped to the partition in Figure 3.10(b). Disregarding the labels of the vertices and considering only their positions, this is the same partition as that numbered 6 in Figure 3.9, since the colours are only used to distinguish the five perfect matchings in each partition. Continuing this way we obtain that the permutation $(1,2,3)$ on the vertices labelled as in Figure 3.10(a) corresponds to the permutation $(1,6,3)(2,5,4)$ on the partitions. Hence these particular labellings of vertices and partitions induce an automorphism of A_6 that maps $(1,2,3)$ to $(1,6,3)(2,5,4)$. Distinct labellings of the vertices induce distinct automorphisms of A_6.

Exercise 3.1.96 Let ς be the automorphism of A_6 just described; compute $\varsigma(\alpha)$ for all α in $\{(1,4,5),(2,4)(1,3,6,5),(1,2,3)(4,5,6)\}$.

To better illustrate generating sets of A_6 and their properties we shall use diagrams representing their actions on $\mathcal{X}_6$, explained in the following paragraphs. Perhaps the first appearance of similar diagrams occurred when the Cayley graphs were introduced [27]; there the groups act on their own

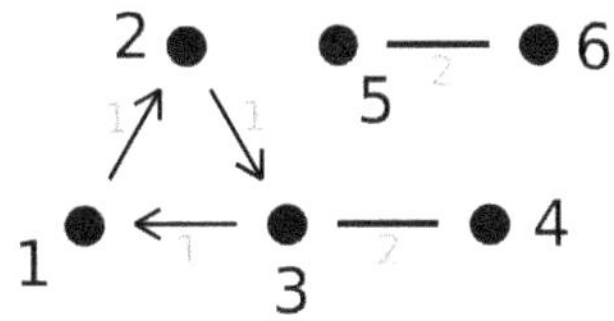

Figure 3.11 The Schreier diagram for $\alpha_1 = (1,2,3)$ and $\alpha_2 = (3,4)(5,6)$.

elements. Later this notion was generalised in [238] so that the action is on the cosets of any subgroup. There is a rather small jump to consider diagrams from actions of groups on arbitrary sets, without explicitly mentioning subgroups (and their cosets) involved in the action. Such diagrams have appeared in many distinct contexts. For example, in [40] adaptations of these diagrams were used to obtain results of the even subgroups of rotary maps of type $\{3,7\}$. More recently, in [230] these diagrams were used to study automorphism groups of two-orbit polytopes. Elsewhere similar diagrams have been used under more than one name; here we shall refer to them as *Schreier diagrams*.

Given a subset $\{\alpha_1, \dots, \alpha_k\}$ of A_6 we construct a digraph whose nodes are the elements of $\mathcal{X}_6$, where there is an arrow from a to b labelled i whenever $a\alpha_i = b$. If α_i contains the transposition (a,b) then the diagram contains symmetric arrows labelled i between a and b, and we may replace them with an undirected edge labelled i between those two nodes. Figure 3.11 illustrates the diagram for $\alpha_1 = (1,2,3)$ and $\alpha_2 = (3,4)(5,6)$.

Theorem 3.1.53 tells us that in order to find all rotary polyhedra $\mathcal{P}$ such that $\Gamma^+(\mathcal{P}) \cong A_6$ it suffices to find all pairs $\{\sigma_1, \sigma_2\} \subseteq A_6$ satisfying certain properties. Instead of this generating set, we shall look for all pairs $\{\sigma_1, \tau_{0,2}\}$ where $\tau_{0,2}$ is an involution. (Recall that $\tau_{0,2} = \sigma_1\sigma_2$.) In each such choice we need to determine if

- $\langle \sigma_1, \sigma_2 \rangle = \langle \sigma_1, \tau_{0,2} \rangle = A_6$,
- $\langle \sigma_1 \rangle \cap \langle \sigma_2 \rangle = \langle \sigma_1 \rangle \cap \langle \sigma_1^{-1} \tau_{0,2} \rangle = \{\mathrm{id}\}$.

There is no need to verify (3.7), since $(\sigma_1\sigma_2)^2 = \tau_{0,2}^2$ trivially holds.

Remark 3.1.97 If $\{\sigma_1, \tau_{0,2}\}$ and $\{\sigma_1', \tau_{0,2}'\}$ are two pairs of elements of A_6 satisfying the items above, and if there exists a group automorphism $\zeta\colon A_6 \to A_6$ such that $\zeta(\sigma_1) = \sigma_1'$ and $\zeta(\tau_{0,2}) = \tau_{0,2}'$ then the pairs $\{\sigma_1, \sigma_2 := \sigma_1^{-1}\tau_{0,2}\}$ and $\{\sigma_1', \sigma_2' := (\sigma_1')^{-1}\tau_{0,2}'\}$ induce isomorphic rotary polyhedra. The isomorphism of polyhedra is given by $\langle \sigma_i \rangle \gamma \mapsto \langle \sigma_i' \rangle (\zeta(\gamma))$ for $i \in \{1,2\}$, and $\langle \tau_{0,2} \rangle \gamma \mapsto \langle \tau_{0,2}' \rangle (\zeta(\gamma))$. (Recall that for rotary polyhedra $\Gamma_0 = \langle \sigma_2 \rangle$, $\Gamma_1 = \langle \tau_{0,2} \rangle$ and $\Gamma_2 = \langle \sigma_1 \rangle$.)

Remark 3.1.98 In what follows we shall look for candidate Schreier diagrams for the generating set $\{\sigma_1, \tau_{0,2}\}$. In view of Remark 3.1.97, the labels of the vertices on the diagram are redundant, since distinct labellings correspond to isomorphic choices of generators.

From the list of cyclic structures of the elements of A_6 we see that the only involutory elements are the double transpositions, and thus $\tau_{0,2} = (a,b)(c,d)$ for some $a,b,c,d \in \{1,\ldots,6\}$.

The automorphism σ_1 of $\mathcal{P}$ cannot be an involution, since otherwise $\langle \sigma_1, \tau_{0,2} \rangle$ would be dihedral. We can conclude that σ_1 is either a 3-cycle, a product of disjoint 3-cycles, a product of disjoint transposition and 4-cycle, or a 5-cycle. We next discard two of these possibilities.

Proposition 3.1.99 *Let $\mathcal{P}$ be a rotary polyhedron with $\Gamma^+(\mathcal{P}) = A_6$. Then neither σ_1 nor σ_2 is a* 3*-cycle.*

Proof Under the natural action of A_6 on $\mathcal{X}_6$, the subgroup of generated by a 3-cycle and a product of two disjoint transpositions must induce at least two orbits on $\mathcal{X}_6$ (see Figure 3.11). Since $\Gamma^+(\mathcal{P}) = \langle \sigma_1, \tau_{0,2} \rangle = \langle \tau_{0,2}, \sigma_2 \rangle$ and $\tau_{0,2}$ must be a product of two disjoint transpositions, neither σ_1 nor σ_2 is a 3-cycle. ■

Proposition 3.1.100 *Let $\mathcal{P}$ be a rotary polyhedron with $\Gamma^+(\mathcal{P}) = A_6$. Then neither σ_1 nor σ_2 is a product of disjoint* 3*-cycles.*

Proof Here we use that the products of two disjoint transpositions form an orbit of A_6 under its automorphism group, while the 3-cycles together with the products of disjoint 3-cycles form another orbit. If $\Gamma^+(\mathcal{P}) = \langle \sigma_1, \tau_{0,2} \rangle$ with σ_1 being the product of two disjoint 3-cycles and $\tau_{0,2}$ an involution, then there exists a group automorphism ζ of A_6 mapping σ_1 to a 3-cycle and $\tau_{0,2}$ to an involution. If, in addition, $\langle \sigma_1, \tau_{0,2} \rangle = A_6$ then Remark 3.1.97 and the fact that $A_6 = \langle \sigma_1, \tau_{0,2} \rangle = \langle \zeta(\sigma_1), \zeta(\tau_{0,2}) \rangle$ contradict Proposition 3.1.99. ■

The previous two propositions together with the list of cyclic structures of the elements in A_6 directly imply the following result.

Corollary 3.1.101 *The first entry of the type of a rotary polyhedron $\mathcal{P}$ with $\Gamma^+(\mathcal{P}) \cong A_6$ must be at least* 4.

When considering the cases where σ_1 is a 5-cycle or the product of a 4-cycle and a disjoint transposition, we shall make use of the next result.

Lemma 3.1.102 *Let $\mathcal{P}_1$ and $\mathcal{P}_2$ be two rotary polyhedra and let $\mathcal{D}_1$ and $\mathcal{D}_2$ be Schreier diagrams of $\Gamma^+(\mathcal{P}_1)$ and $\Gamma^+(\mathcal{P}_2)$, respectively, with respect to the* *generators σ_1 and $\tau_{0,2}$. If $\mathcal{D}_1$ can be obtained from $\mathcal{D}_2$ by reversing all arrows corresponding to σ_1, then $\mathcal{P}_1 \cong \mathcal{P}_2$.*

Proof Let Φ_1 be the base flag of $\mathcal{P}_1$, so that $\Phi_1\sigma_1 = \Phi_1^{1,0}$ and $\Phi_1\tau_{0,2} = \Phi_1^{0,2}$. When establishing that $\Gamma^+(\mathcal{P}_1) = \langle\sigma_1, \tau_{0,2}\rangle$ we are implicitly assuming $\mathcal{P}_1$ to be the rooted polyhedron $(\mathcal{P}_1)_{(\Phi_1)}$.

We shall show that the corresponding generators for $(\mathcal{P}_1)_{(\Phi_1^0)}$ are σ_1^{-1} and $\tau_{0,2}$. Indeed,

$$\Phi_1^0\sigma_1^{-1} = (\Phi_1\sigma_1^{-1})^0 = \Phi_1^{0,1,0} = (\Phi_1^0)^{1,0},$$
$$\Phi_1^0\tau_{0,2} = (\Phi_1\tau_{0,2})^0 = \Phi_1^{0,2,0} = (\Phi_1^0)^{0,2}.$$

In this situation the Schreier diagram of $\Gamma^+(\mathcal{P}_1)$ with the same vertex set and action as $\mathcal{D}_1$ but with respect to σ_1^{-1} and $\tau_{0,2}$ is $\mathcal{D}_2$. This establishes that as a rooted polyhedron $\mathcal{P}_2 = (\mathcal{P}_1)_{(\Phi_1^0)}$. In particular, if $\mathcal{P}_1$ is chiral then $\mathcal{P}_2$ is the enantiomorph of $\mathcal{P}_1$. In any case, $\mathcal{P}_1 \cong \mathcal{P}_2$. ■

Transitivity of the action of A_6 on $\mathcal{X}_6$ implies that the Schreier diagram of $\Gamma^+(\mathcal{P}) \cong A_6$ with respect to $\{\sigma_1, \tau_{0,2}\}$ must be connected. Therefore, if σ_1 is the product of a 4-cycle and a disjoint transposition then the Schreier diagram must contain the upper-left diagram in Figure 3.12, up to relabelling the vertices. Similarly, if σ_1 is a 5-cycle then the Schreier diagram must contain the lower-left diagram in Figure 3.12. In order to complete the diagram it only remains to determine the second transposition involved in $\tau_{0,2}$. Up to isomorphism and the reversal of all arrows, the only candidate Schreier diagrams for $\Gamma^+(\mathcal{P})$ are (a) to (h) in Figure 3.12. There, labels 1 and 2 on edges and arrows represent the actions of σ_1 and $\tau_{0,2}$, respectively; in each case, the added transposition for $\tau_{0,2}$ is represented by a dashed line.

With the following results we discard candidates (a), (c), (d), (f) and (g).

Proposition 3.1.103 *The permutations represented in Figure 3.12(a) and (c) do not generate A_6.*

Proof The action in these cases is imprimitive; in each of them the partition $\{\{1,3,5\},\{2,4,6\}\}$ is invariant under both permutations. ■

Proposition 3.1.104 *The permutations represented in Figure 3.12(d), (f) and (g) do not generate A_6.*

Proof In cases (d) and (g) we compute $\sigma_2 = \sigma_1^{-1}\tau_{0,2}$. In case (d) we have $\sigma_2 = (1,5,4)(2,3,6)$ while in case (g) we have $\sigma_2 = (1,2,3)(4,6,5)$.

The intersection condition is easily verified in these two cases, and therefore the resulting groups are the even subgroups of rotary polyhedra $\mathcal{P}_1$ and $\mathcal{P}_2$ with

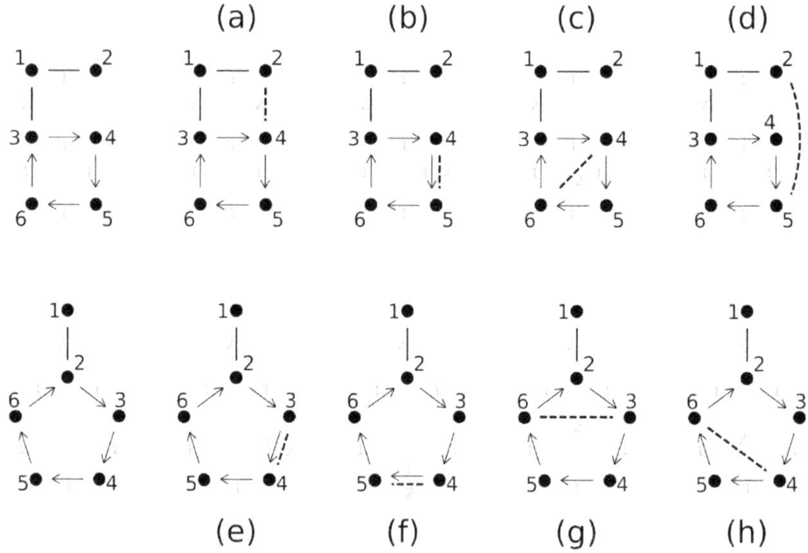

Figure 3.12 The diagrams (a) - (h) are all the candidate Schreier diagrams for polyhedra with even subgroup A_6.

types $\{4,3\}$ and $\{5,3\}$, respectively. But then $\mathcal{P}_1^\delta$ and $\mathcal{P}_2^\delta$ are rotary polyhedra with even subgroup isomorphic to A_6 and first entry of their types equal to 3, contradicting Corollary 3.1.101. In particular, the groups induced by diagrams (d) and (g) are proper subgroups of A_6.

It remains to show that diagram (f) induces a subgroup of that induced by diagram (g), implying that from diagram (f) we also get a proper subgroup of A_6. Indeed, if σ_1 and $\tau_{0,2}$ are as in diagram (g) then σ_1 and $\sigma_1^{-2}\tau_{0,2}\sigma_1\tau_{0,2}\sigma_1^{-1}\tau_{0,2}\sigma_1^2$ are the generators indicated in diagram (f). ■

In fact, the groups induced by the diagrams in Figure 3.12(d), (f) and (g) are even subgroups of familiar regular polyhedra.

Exercise 3.1.105 Determine the regular polyhedra whose even subgroups are those with Schreier diagrams in Figure 3.12(d), (f) and (g). Illustrate specific sets of six elements where the actions of σ_1 and $\tau_{0,2}$ are precisely those of the diagrams.

We are left with the diagrams in Figure 3.12(b), (e) and (h). In what follows we shall show that the groups induced by those diagrams are all A_6, and each of them corresponds to the even subgroup of a regular polyhedron.

Proposition 3.1.106 *The groups induced by the diagrams in Figure 3.12(b), (e) and (h) are all even subgroups of rotary polyhedra. Moreover, the polyhedron arising from diagram (b) is the dual of that arising from diagram (h).*

Proof In order to show that they correspond to even subgroups of rotary polyhedra it suffices to show that in each case $\langle\sigma_1\rangle \cap \langle\sigma_1^{-1}\tau_{0,2}\rangle = \{\mathrm{id}\}$ holds (see Theorem 3.1.53). For diagrams (b), (e) and (h) we have that $\sigma_1^{-1}\tau_{0,2}$ is $(1,2,3,6,4)$, $(1,2,6,5,3)$ and $(1,2,4,3)(5,6)$, respectively. The trivial intersection can easily be verified by computing all powers of σ_1 and of $\sigma_1^{-1}\tau_{0,2}$.

To show the duality between the polyhedra $\mathcal{P}_h$ and $\mathcal{P}_b$ arising from diagrams (h) and (b), respectively, let $\tau_{0,2} = (1,2)(4,6)$, $\sigma_1 = (2,3,4,5,6)$ be the generators indicated in diagram (h), and let $\tau'_{0,2} = (1,3)(4,5)$, $\sigma'_1 = (1,2)(3,4,5,6)$ be the generators indicated in diagram (b). Then $\Gamma^+(\mathcal{P}_h) = \langle\sigma_1,\sigma_2\rangle$ with $\sigma_2 := \sigma_1^{-1}\tau_{0,2} = (1,2,4,3)(5,6)$, and $\Gamma^+(\mathcal{P}_b) = \langle\sigma'_1,\sigma'_2\rangle$ with $\sigma'_2 := (\sigma'_1)^{-1}\tau'_{0,2} = (1,2,3,6,4)$. According to Proposition 3.1.70 we may take $\sigma_2^{-1} = (1,3,4,2)(5,6)$ and $\sigma_1^{-1} = (2,6,5,4,3)$ as standard generators of $\Gamma^+(\mathcal{P}_h^\delta)$. On the other hand, by defining $\xi = (1,5,2,4,3,6)$ we have that $\xi\sigma'_i\xi^{-1} = \sigma_{3-i}^{-1}$. Since conjugation by ξ represents an automorphism of A_6, we have an isomorphism between $\Gamma^+(\mathcal{P}_b)$ and $\Gamma^+(\mathcal{P}_h^\delta)$ that maps standard generators to standard generators. The uniqueness in Theorem 3.1.53 implies that $\mathcal{P}_b \cong \mathcal{P}_h^\delta$. ■

Proposition 3.1.107 *The groups induced by the diagrams in Figure 3.12(b), (e) and (h) are all the alternating groupA_6.*

Proof Let Γ_b, Γ_e and Γ_h be the groups induced by the diagrams in Figure 3.12(b), (e) and (h), respectively.

We first consider the group Γ_e. Here,

$$\alpha := \tau_{0,2}\sigma_1^2\tau_{0,2}\sigma_1^{-1}\tau_{0,2}\sigma_1 = (3,6,5) \in \Gamma_e.$$

Furthermore, $\beta := \sigma_1^2\alpha\sigma_1^{-2} = (3,6,4)$ and $\delta := \sigma_1\alpha\sigma_1^{-1} = (2,5,4)$ are also in Γ_e. Standard results imply that $\langle\alpha,\beta\rangle$ is the alternating group on $\{3,4,5,6\}$, and therefore that $\langle\alpha,\beta,\gamma\rangle$ is the alternating group on $\{2,3,4,5,6\}$. It follows that the stabiliser of 1 in Γ_e has at least $60 = |A_5|$ elements, and since Γ_e acts transitively on $\mathcal{X}_6$, Γ_e must have at least $6{\cdot}60 = 360$ elements. But $|A_6| = 360$, forcing $\Gamma_e = A_6$.

The analysis for the group Γ_h is analogous. In that case,

$$\alpha' := \tau_{0,2}\sigma_1^{-1}\tau_{0,2}\sigma_1^{-2}\tau_{0,2}\sigma_1^2 = (3,4,5)$$

Figure 3.13 Labellings of vertices that illustrate the automorphisms of A_6 (a) for the diagram in Figure 3.12(e), and (b) for that in Figure 3.12(h).

can be used instead of α. The result for Γ_b now follows from Proposition 3.1.106 and Remark 2.1.84. ∎

Proposition 3.1.108 *The groups induced by the diagrams in Figure 3.12(b), (e) and (h) are even subgroups of automorphism groups of regular polyhedra.*

Proof We shall use Corollary 3.1.54 by exhibiting the existence of group automorphisms η of A_6 such that $\eta(\sigma_1) = \sigma_1^{-1}$ and $\eta(\sigma_2) = \sigma_1^2\sigma_2$. With this in mind, we require η to map σ_1 to σ_1^{-1} while fixing $\tau_{0,2}$. Indeed, in that case

$$\eta(\sigma_2) = \eta(\sigma_1^{-1}\tau_{0,2}) = \eta(\sigma_1^{-1})\tau_{0,2} = \sigma_1\tau_{0,2} = \sigma_1^2\sigma_2.$$

Furthermore, in view of Proposition 3.1.106 it suffices to find such a group automorphism only for diagrams (e) and (h).

When considering diagram (e) the relevant permutations are $\sigma_1 = (2,3,4,5,6)$ and $\tau_{0,2} = (1,2)(3,4)$. In this case the labelling of the vertices in Figure 3.13(a) together with the labelling of the partitions in Figure 3.9 illustrate the desired automorphism of A_6.

In the case of diagram (h) the permutations are $\sigma_1 = (2,3,4,5,6)$ and $\tau_{0,2} = (1,2)(4,6)$. Here the labelling of the vertices in Figure 3.13(b) and the labelling of the partitions in Figure 3.9 illustrate the desired automorphism of A_6. ∎

The regular polyhedra addressed in Proposition 3.1.108 have been mentioned elsewhere. In particular, one can identify each such polyhedron $\mathcal{Q}$ in the atlas [116] and in the lists of maps [41]. To do this we may first look for orientable polyhedra (or maps) $\mathcal{P}$ having the type of $\mathcal{Q}$, and whose automorphism group has size 720 (and so the even subgroup has order $|A_6| = 360$). Then we must verify that the generators of $\Gamma^+(\mathcal{Q})$ satisfy the relations given for $\Gamma(\mathcal{P})$. When doing so we must take into account that the relations for $\Gamma(\mathcal{P})$ may not be given in terms of the generators σ_1 and $\tau_{0,2}$; in fact, since $\mathcal{P}$ is orientably regular, $\langle\sigma_1, \tau_{0,2}\rangle$ is a subgroup of index 2 in $\Gamma(\mathcal{P})$. Hence we must rewrite the defining relations for $\Gamma(\mathcal{P})$ in terms of σ_1 and $\tau_{0,2}$ (this is always possible; see Exercise 3.1.109). In the case where the relations are satisfied we can conclude that $\Gamma^+(\mathcal{Q})$ is a quotient of $\Gamma^+(\mathcal{P})$ and, since they have the same order, we can conclude that $\Gamma^+(\mathcal{Q}) \cong \Gamma^+(\mathcal{P})$. Theorem 3.1.53 can then be used to conclude that $\mathcal{Q} \cong \mathcal{P}$.

Exercise 3.1.109 Let $\mathcal{P}$ be an orientably regular polytope and let R be a set of defining relations of $\Gamma(\mathcal{P})$ in terms of the standard reflections. Show that every relation in R has an even number of factors ρ_i. Use this to specify a set of defining relations for $\Gamma^+(\mathcal{P})$ in terms of the standard rotations.

Exercise 3.1.110 Determine which polyhedra in [116] and which maps in [41] correspond to the regular polyhedra mentioned in Proposition 3.1.108.

The next proposition follows from Propositions 3.1.99, 3.1.100, 3.1.103, 3.1.104, 3.1.106 and 3.1.108.

Proposition 3.1.111 *There are no chiral polyhedra $\mathcal{P}$ with $\Gamma(\mathcal{P}) \cong A_6$.*

It is still possible to work by hand the details to determine all rotary n-polytopes whose even subgroup is isomorphic to A_6, for $n \geq 4$. Readers with knowledge of software specialised in groups are encouraged to use computational techniques to achieve the same goal.

Exercise 3.1.112 Determine all rotary n-polytopes whose even subgroup is isomorphic to A_6, for $n \geq 4$.

The method developed here to study rotary polyhedra with an even subgroup isomorphic to A_6 is limited in more than one sense. It relies on our knowledge of the alternating group; less studied groups would require a previous analysis of their properties, like for example their automorphism groups and their lists of involutions. Even when considering only alternating groups A_m, the difficulty increases rapidly with m since there are more conjugacy classes of involutions, and in general more conjugacy classes of permutations. This increases substantially the amount of cases to consider, particularly when looking for polytopes of a large rank.

In order to construct the atlas [116], computational techniques were used to analyse all groups in the database of small groups. (This database consisted at the time of all groups of order at most 2000, with the exceptions of those of orders 1024 and 1536.) The idea of using specific knowledge on groups was also used elsewhere; for example, in [162] Leemans and Vauthier determined all abstract regular polytopes whose automorphism groups are isomorphic to certain almost simple groups.

Later more efficient methods were used to obtain lists of maps and abstract polytopes where the sizes of the automorphism groups have some given upper bound (see [33, 34, 41]).

3.2 Connection Group

As a consequence of Proposition 2.1.79, $\Gamma(\mathcal{P})$ can be understood as a group of permutations on $\mathcal{F}l(\mathcal{P})$. Section 3.1 gives us an idea of the relevance of such a permutation group.

In Section 3.2.1 we study another relevant group of permutations of $\mathcal{F}l(\mathcal{P})$, called the *connection group*. The main source of this section is [191], which surveys previous knowledge in this topic. Then, in Section 3.2.2 we describe group actions that resemble that of the connection group; the main sources for this topic are [111, 113, 191]. Many of the results have been stated in other contexts (see [191] and the references within); some of those contexts seem more particular than abstract polytopes, like for example maps on surfaces, whereas some other are more general, like the combinatorial maps in [258].

3.2.1 The Permutation Group

Given an n-polytope $\mathcal{P}$, we define for $i \in \{0, \dots, n-1\}$ the permutation

$$\begin{array}{rccc} r_i \colon & \mathcal{F}l(\mathcal{P}) & \to & \mathcal{F}l(\mathcal{P}) \\ & \Phi & \mapsto & \Phi^i. \end{array} \tag{3.28}$$

The *connection group* of $\mathcal{P}$ is the group

$$\mathrm{Conn}(\mathcal{P}) = \langle r_0, \dots, r_{n-1} \rangle,$$

and its elements are called *connections*.

Example 3.2.1 Figure 3.14 illustrates a tetrahedron seen from above. Its four triangles are subdivided into their six flags; the flags of the bottom face are illustrated outside the large triangle. With this labelling of flags the generators of the connection group are

$$r_0 = (1,2)(3,4)(5,6)(7,8)(9,10)(11,12)(13,14)(15,16)(17,18)(19,20)(21,22)(23,24),$$
$$r_1 = (1,6)(2,3)(4,5)(7,12)(8,9)(10,11)(13,18)(14,15)(16,17)(19,24)(20,21)(22,23),$$
$$r_2 = (1,12)(2,11)(3,20)(4,19)(5,14)(6,13)(7,18)(8,17)(9,22)(10,21)(15,24)(16,23).$$

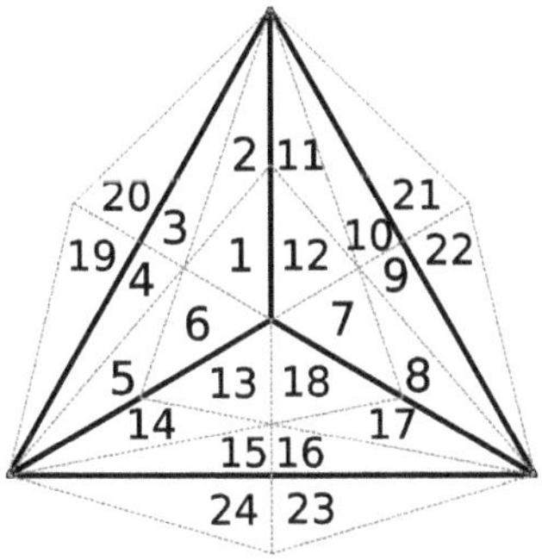

Figure 3.14 The flags of the tetrahedron.

The connections need not represent automorphisms. For instance, the generator r_1 in Example 3.2.1 maps flags 1 and 2 to 6 and 3, respectively. While flags 1 and 2 are 0-adjacent, flags 3 and 6 are not adjacent at all. From Lemma 2.1.77 we know that r_1 does not induce an automorphism of the tetrahedron.

Perhaps the origin of the connection group was in [102], where Grothendieck defines the connection group of an orientable map as the *cartographic group*, in terms of oriented half-edges, which yielded four generators instead of only three of them (see also [1]). This group is highly related to the *monodromy group* of a dessin, which corresponds to the monodromy group of certain meromorphic function from the underlying surface of the dessin to the sphere; this meromorphic function is called Belyĭ function (see for example [150]). The name 'monodromy group' was adopted for some permutation groups of substructures of a map (without requiring the possession of a complex structure, or even orientability) in numerous publications, for example [16, 206, 213].

When working with polytopes we deviate from the complex structure of Riemann surfaces; in particular we allow non-orientable order complexes and ranks higher than 3. For that reason we avoid the usage of the term 'monodromy', since for general polytopes it is no longer linked in a natural way to the monodromy group of complex analysis. Instead, we keep the term 'connection group' previously used for example in [273]. This name feels natural, since the connection group as a permutation group on the flags gives a recipe of how the flags are connected to manufacture the polytope (or map or maniplex).

We shall make $\mathrm{Conn}(\mathcal{P})$ act on the left. The following proposition justifies this choice.

Proposition 3.2.2 *Let $\mathcal{P}$ be a polytope. Then, for evey $\Phi \in \mathcal{F}l(\mathcal{P})$, every $\gamma \in \Gamma(\mathcal{P})$ and every $w \in \mathrm{Conn}(\mathcal{P})$,*

$$(w\Phi)\gamma = w(\Phi\gamma).$$

Proof Let $w = r_{i_1} \cdots r_{i_k} \in \mathrm{Conn}(\mathcal{P})$. We proceed by induction over k. When $k = 1$ we have directly from Lemma 2.1.77:,

$$(r_{i_1}\Phi)\gamma = (\Phi^{i_1})\gamma = (\Phi\gamma)^{i_1} = r_{i_1}(\Phi\gamma).$$

For general $k > 1$, assuming the result true for $k-1$ factors r_{i_j} we have

$$\begin{aligned}(r_{i_1}\cdots r_{i_k}\Phi)\gamma &= (r_{i_1}(r_{i_2}\cdots r_{i_k}\Phi))\gamma = ((r_{i_2}\cdots r_{i_k}\Phi)^{i_1})\gamma = ((r_{i_2}\cdots r_{i_k}\Phi)\gamma)^{i_1}\\ &= (r_{i_2}\cdots r_{i_k}(\Phi\gamma))^{i_1} = r_{i_1}(r_{i_2}\cdots r_{i_k}(\Phi\gamma)) = r_{i_1}\cdots r_{i_k}(\Phi\gamma).\end{aligned}$$

■

Remark 3.2.3 Proposition 3.2.2 can be alternatively proven from Proposition 2.1.78 by first noting that $r_{i_1} \cdots r_{i_k}\Phi = \Phi^{i_k,\ldots,i_1}$.

Exercise 3.2.4 Show that for every flag Φ and every $i_1, \ldots, i_k$, the expression $r_{i_1} \cdots r_{i_k}\Phi = \Phi^{i_k,\ldots,i_1}$ holds.

Now we present a tool to derive $\Gamma(\mathcal{P})$ from $\mathrm{Conn}(\mathcal{P})$.

Proposition 3.2.5 *Let $\mathcal{P}$ be a polytope. Under the understanding of $\Gamma(\mathcal{P})$ as a permutation group on the flags, $\Gamma(\mathcal{P})$ is the centraliser of* $\mathrm{Conn}(\mathcal{P})$.

Proof We denote by $C(\mathrm{Conn}(\mathcal{P}))$ the centraliser of $\mathrm{Conn}(\mathcal{P})$ in the symmetric group $S_{\mathcal{F}l(\mathcal{P})}$.

If we were to act $\Gamma(\mathcal{P})$ and $\mathrm{Conn}(\mathcal{P})$ on the same side, Proposition 3.2.2 implies that every element in $\Gamma(\mathcal{P})$ commutes with every element in $\mathrm{Conn}(\mathcal{P})$, and it follows that $\Gamma(\mathcal{P}) \subseteq C(\mathrm{Conn}(\mathcal{P}))$. Conversely, if $\gamma \in C(\mathrm{Conn}(\mathcal{P}))$ then the action of γ commutes with the action of r_i for every i. Lemma 2.1.79 implies that $\gamma \in \Gamma(\mathcal{P})$. ■

The roles of $\Gamma(\mathcal{P})$ and of $\mathrm{Conn}(\mathcal{P})$ in Proposition 3.2.5 cannot be reversed. In general the centraliser of $\Gamma(\mathcal{P})$ is larger than $\mathrm{Conn}(\mathcal{P})$.

Exercise 3.2.6 Find a polytope $\mathcal{P}$ for which $\mathrm{Conn}(\mathcal{P})$ is a proper subgroup of the centraliser of $\Gamma(\mathcal{P})$.

The connection group can be used as a tool to work with quotients of polytopes, and relates abstract polytopes with algebraic topology. Intuitively, we may think of a connection, when expressed as products of the generators r_i, as a family of paths on the flag graph (one path starting on each flag). If a connection w of $\mathcal{P}$ fixes Φ then the corresponding path that starts at Φ, is closed. A covering η from $\mathcal{P}$ to $\mathcal{Q}$ induces a mapping from the flag graph of $\mathcal{P}$ to the flag graph of $\mathcal{Q}$ that sends adjacent vertices by an edge labelled i to adjacent vertices by an edge labelled i. The image under this mapping of the path on the flag graph of $\mathcal{P}$ mentioned above must also be closed when the initial flag is $\eta(\Phi)$. Furthermore, products of connections correspond to concatenation of paths. Readers familiar with algebraic topology are encouraged to establish correspondences of concepts between that area and connections of polytopes. In any case, the following results can be proved directly.

Lemma 3.2.7 *Let $\mathcal{Q}_{(\Psi)}$ and $\mathcal{P}_{(\Phi)}$ be rooted n-polytopes with* $\mathrm{Conn}(\mathcal{Q}) = \langle r_0, \dots, r_{n-1}\rangle$ *and* $\mathrm{Conn}(\mathcal{P}) = \langle r'_0, \dots, r'_{n-1}\rangle$. *As rooted polytopes, $\mathcal{Q}_{(\Psi)}$ covers $\mathcal{P}_{(\Phi)}$ if and only if for every $i_1, \dots, i_k \in \{0, \dots, n-1\}$ such that $r_{i_1} \cdots r_{i_k}\Psi = \Psi$ we have that $r'_{i_1} \cdots r'_{i_k}\Phi = \Phi$.*

Proof If η is a covering from $\mathcal{Q}_{(\Psi)}$ to $\mathcal{P}_{(\Phi)}$ and $r_{i_1} \cdots r_{i_k}\Psi = \Psi$ then

$$\begin{aligned}\Phi = \eta(\Psi) = \eta(r_{i_1} \cdots r_{i_k}\Psi) = \eta(\Psi^{i_k,\dots,i_1}) = (\eta(\Psi))^{i_k,\dots,i_1} = \Phi^{i_k,\dots,i_1}\\ = r_{i_1} \cdots r_{i_k}\Phi.\end{aligned}$$

For the converse we intend to define $\eta\colon \mathcal{F}l(\mathcal{Q}) \to \mathcal{F}l(\mathcal{P})$ by mapping the flag $r_{j_1} \cdots r_{j_m}\Psi$ to $r'_{j_1} \cdots r'_{j_m}\Phi$ for every $m \in \mathbb{Z}^+$ and every $j_1, \dots, j_m \in \{0, \dots, n-1\}$. For η to be well defined we require that whenever $r_{i_1} \cdots r_{i_k}\Psi = r_{j_1} \cdots r_{j_m}\Psi$ then $r'_{i_1} \cdots r'_{i_k}\Phi = r'_{j_1} \cdots r'_{j_m}\Phi$. However, if $r_{i_1} \cdots r_{i_k}\Psi = r_{j_1} \cdots r_{j_m}\Psi$ then it must be true that $r_{j_m} \cdots r_{j_1} r_{i_1} \cdots r_{i_k}\Psi = \Psi$; by our hypothesis $r'_{j_m} \cdots r'_{j_1} r'_{i_1} \cdots r'_{i_k}\Phi = \Phi$, and we reach the desired conclusion. Furthermore, for every $j \in \{0, \dots, n-1\}$ and $x \in \mathrm{Conn}(\mathcal{Q})$, we have that

$$\eta((x\Psi)^j) = \eta(r_j x\Psi) = r'_j \eta(x\Psi) = (\eta(x\Psi))^j,$$

and therefore η is a covering. ■

Proposition 3.2.8 *Let η be a covering from $\mathcal{P}$ to $\mathcal{Q}$, let* $\mathrm{Conn}(\mathcal{P}) = \langle r_0, \dots, r_n\rangle$ *and* $\mathrm{Conn}(\mathcal{Q}) = \langle r'_0, \dots, r'_n\rangle$. *Then there is a group epimorphism*

$$\hat{\eta}\colon \mathrm{Conn}(\mathcal{P}) \to \mathrm{Conn}(\mathcal{Q})$$

that maps r_i to r'_i for every i. Furthermore, if $\Phi \in \mathcal{F}l(\mathcal{P})$ then

$$\hat{\eta}(\mathrm{Stab}_{\mathrm{Conn}(\mathcal{P})}(\Phi)) \subseteq \mathrm{Stab}_{\mathrm{Conn}(\mathcal{Q})}(\eta(\Phi)).$$

Proof Recall that a function ζ from a generating set $\mathcal{X}$ of a group Γ to some subset of a group Δ can be extended to a group homomorphism if and only if every relation of Γ in terms of $\mathcal{X}$ translates to a relation of Δ in $\zeta(\mathcal{X})$. Here we are using $\zeta : \{r_0, \dots, r_n\} \to \mathrm{Conn}(\mathcal{Q})$ defined as $\zeta(r_i) = r'_i$. We need to show that if $r_{i_1} \cdots r_{i_k} = \mathrm{id}_{\mathrm{Conn}(\mathcal{P})}$ then $r'_{i_1} \cdots r'_{i_k} = \mathrm{id}_{\mathrm{Conn}(\mathcal{Q})}$. For any given flag Ψ of $\mathcal{Q}$ we have that there exists $\Phi \in \mathcal{F}l(\mathcal{P})$ such that $\eta(\Phi) = \Psi$; using that η is a covering we deduce that

$$r'_{i_1} \cdots r'_{i_k} \Psi = r'_{i_1} \cdots r'_{i_k}(\eta(\Phi)) = \eta(r_{i_1} \cdots r_{i_k} \Phi) = \eta(\Phi) = \Psi. \qquad (3.29)$$

It follows that $r'_{i_1} \cdots r'_{i_k}$ is the trivial permutation on $\mathcal{F}l(\mathcal{Q})$ and ζ extends to the desired group homomorphism $\hat{\eta}$. Moreover, $\hat{\eta}$ must be an epimorphism since its image contains the standard generating set of $\mathrm{Conn}(\mathcal{Q})$.

The last part of the statement follows directly from (3.29). ■

Corollary 3.2.9 *Let Φ and Ψ be two flags of $\mathcal{P}$ in the same orbit under $\Gamma(\mathcal{P})$. If $w \in \mathrm{Conn}(\mathcal{P})$ is such that $w\Phi = \Phi$ then $w\Psi = \Psi$.*

Proof Isomorphisms of polytopes are special cases of quotients. In particular, an automorphism is a quotient of a polytope into itself. Proposition 3.2.8 implies that the identity automorphism in $\mathrm{Conn}(\mathcal{P})$ maps $\mathrm{Stab}(\Phi)$ into $\mathrm{Stab}(\Psi)$. ■

Exercise 3.2.10 Show that a chiral polytope is not a quotient of any non-orientable polytope.

The following result is the converse of Proposition 3.2.8.

Proposition 3.2.11 *Let $\mathcal{P}$ and $\mathcal{Q}$ be n-polytopes with $\mathrm{Conn}(\mathcal{P}) = \langle r_0, \dots, r_{n-1}\rangle$ and $\mathrm{Conn}(\mathcal{Q}) = \langle r'_0, \dots, r'_{n-1}\rangle$. Let $\Phi \in \mathcal{F}l(\mathcal{P})$ and $\Psi \in \mathcal{F}l(\mathcal{Q})$. If there is a group epimorphism $\hat{\eta}\colon \mathrm{Conn}(\mathcal{P}) \to \mathrm{Conn}(\mathcal{Q})$ mapping r_i to r'_i for all i with the property that $\hat{\eta}(\mathrm{Stab}_{\mathrm{Conn}(\mathcal{P})}(\Phi)) \subseteq \mathrm{Stab}_{\mathrm{Conn}(\mathcal{Q})}(\Psi)$, then there is a covering η from $\mathcal{P}$ to $\mathcal{Q}$ that maps Φ to Ψ.*

Proof We intend to define η by using that $\mathrm{Conn}(\mathcal{P})$ acts transitively on $\mathcal{F}l(\mathcal{P})$ as follows. An arbitrary flag of $\mathcal{P}$ is of the form $w\Phi$ for some $w \in \mathrm{Conn}(\mathcal{P})$, and we want η to map it to $\hat{\eta}(w)\Psi$. For this to make sense we need that if $w\Phi = w'\Phi$ then $\eta(w\Phi) = \eta(w'\Phi)$. Indeed, if $w\Phi = w'\Phi$ then $(w')^{-1}w \in \mathrm{Stab}_{\mathrm{Conn}(\mathcal{P})}(\Phi)$ and by hypothesis $(\hat{\eta}(w'))^{-1}\hat{\eta}(w) = \hat{\eta}((w')^{-1}w) \in \mathrm{Stab}_{\mathrm{Conn}(\mathcal{Q})}(\Psi)$. It follows that

$$\eta(w'\Phi) = \hat{\eta}(w')\Psi = \hat{\eta}(w)\Psi = \eta(w\Phi).$$

It remains to show that the function η just defined is a covering. Given $i \in \{0, \ldots, n-1\}$ and $w \in \mathrm{Conn}(\mathcal{P})$,

$$\begin{aligned}\eta((w\Phi)^i) &= \eta(r_i w\Phi) = \hat{\eta}(r_i w)\Phi = \hat{\eta}(r_i)\hat{\eta}(w)\Phi = r_i'\hat{\eta}(w)\Phi \\ &= r_i'(\eta(w\Phi)) = (\eta(w\Phi))^i\end{aligned}$$

and the proposition follows. ∎

In general the connection and automorphism groups have very different properties. We know from the second item of Proposition 2.1.80 that $\Gamma(\mathcal{P})$ acts transitive on the flags if and only if $\mathcal{P}$ is regular. On the other hand, the strong flag-connectivity of every polytope implies the following remark.

Remark 3.2.12 The connection group of every polytope acts transitively on the flags.

Corollary 3.2.13 *For any given polytope $\mathcal{P}$,*

$$|\Gamma(\mathcal{P})| \leq |\mathcal{F}l(\mathcal{P})| \leq |\mathrm{Conn}(\mathcal{P})|.$$

Actually, it is not easy to determine the structure as an abstract group of $\mathrm{Conn}(\mathcal{P})$ for general polytopes $\mathcal{P}$. Even determining the size of this group is difficult. This is not the case, however, if $\mathcal{P}$ is regular.

Theorem 3.2.14 *Let $\mathcal{P}$ be a regular polytope. Then there is an isomorphism from* $\mathrm{Conn}(\mathcal{P})$ *to* $\Gamma(\mathcal{P})$ *that for every i maps the generator r_i defined in* (3.28) *to the standard reflection ρ_i.*

Proof Let Φ be the base flag of $\mathcal{P}$. We define $\eta\colon \mathrm{Conn}(\mathcal{P}) \to \Gamma(\mathcal{P})$ so that for each $w \in \mathrm{Conn}(\mathcal{P})$, $\eta(w)$ is the only automorphism mapping Φ to $w\Phi$. From the definition of the generators of both groups, $\eta(r_i) = \rho_i$ for every i.

Given $w, t \in \mathrm{Conn}(\mathcal{P})$ we verify that

$$wt\Phi = w(\Phi(\eta(t))) = (w\Phi)\eta(t) = \Phi\eta(w)\eta(t).$$

This proves that $\eta(wt) = \eta(w)\eta(t)$ and hence η is a group homomorphism. Furthermore, since the standard generating set of $\Gamma(\mathcal{P})$ is contained in the image, η is surjective.

It only remains to show that η is one-to-one. Assume that for some $w \in \mathrm{Conn}(\mathcal{P})$ we have that $\eta(w) = \mathrm{id}$. This means that $w\Phi = \Phi$. Given $\Psi \in \mathcal{F}l(\mathcal{P})$, there exists $\gamma \in \Gamma(\mathcal{P})$ such that $\Psi = \Phi\gamma$. Then

$$w\Psi = w(\Phi\gamma) = (w\Phi)\gamma = \Phi\gamma = \Psi,$$

and we can conclude that $w = \mathrm{id}$. ∎

The next proposition is in some way a converse of Theorem 3.2.14.

Proposition 3.2.15 *A polytope $\mathcal{P}$ is regular if and only if* $\mathrm{Conn}(\mathcal{P})$ *acts regularly on* $\mathcal{F}l(\mathcal{P})$.

Proof Assume first that $\mathcal{P}$ is regular. We already know that $\mathrm{Conn}(\mathcal{P})$ acts transitively on $\mathcal{F}l(\mathcal{P})$. Now assume that $w, t \in \mathrm{Conn}(\mathcal{P})$ are such that $w\Phi = t\Phi$ where Φ is the base flag of $\mathcal{P}$. Then $t^{-1}w\Phi = \Phi$, and the homomorphism η defined in the proof of Theorem 3.2.14 maps $t^{-1}w \in \mathrm{Conn}(\mathcal{P})$ to $\mathrm{id} \in \Gamma(\mathcal{P})$. Since η is an isomorphism, we conclude that $t = w$, and the action of $\mathrm{Conn}(\mathcal{P})$ on $\mathcal{F}l(\mathcal{P})$ is regular.

Now assume that $\mathcal{P}$ is an n-polytope satisfying that $\mathrm{Conn}(\mathcal{P})$ acts regularly on $\mathcal{F}l(\mathcal{P})$. For $i \in \{0, \ldots, n-1\}$ and for a fixed flag Φ we define $\hat{\rho}_i : \mathcal{F}l(\mathcal{P}) \to \mathcal{F}l(\mathcal{P})$ so that $(w\Phi)\hat{\rho}_i = wr_i\Phi$ for every $w \in \mathrm{Conn}(\mathcal{P})$. Regularity of the action of $\mathrm{Conn}(\mathcal{P})$ implies that $\hat{\rho}_i$ is a well-defined bijection; furthermore, $\Phi\hat{\rho}_i = r_i\Phi = \Phi^i$. Next we show that $\hat{\rho}_i$ is an automorphism; we shall do this by using Lemma 2.1.79. Indeed, for $j \in \{0, \ldots, n-1\}$, if $\Psi \in \mathcal{F}l(\mathcal{P})$ and $w \in \mathrm{Conn}(\mathcal{P})$ are such that $\Psi = w\Phi$ then

$$(\Psi^j)\hat{\rho}_i = (w\Phi)^j\hat{\rho}_i = (r_jw\Phi)\hat{\rho}_i = r_jwr_i\Phi = r_j(\Psi\hat{\rho}_i) = (\Psi\hat{\rho}_i)^j.$$

Using Theorem 2.2.9 we conclude that $\mathcal{P}$ is regular. ■

Proposition 3.2.15 has the following consequence.

Corollary 3.2.16 *A finite polytope $\mathcal{P}$ is regular if and only if* $\Gamma(\mathcal{P}) \cong \mathrm{Conn}(\mathcal{P})$.

Proof If $\mathcal{P}$ is regular, Theorem 3.2.14 guarantees the existence of the isomorphism. Conversely, if $\Gamma(\mathcal{P}) \cong \mathrm{Conn}(\mathcal{P})$ then the action of $\mathrm{Conn}(\mathcal{P})$ on $\mathcal{F}l(\mathcal{P})$ must be regular, since it is transitive and $|\mathcal{F}l(\mathcal{P})| = |\Gamma(\mathcal{P})| = |\mathrm{Conn}(\mathcal{P})|$. Under this condition, we know from Proposition 3.2.15 that $\mathcal{P}$ must be regular. ■

The finiteness hypothesis is necessary in the proof of Corollary 3.2.16 to guarantee regularity of the action of $\mathrm{Conn}(\mathcal{P})$ on $\mathcal{F}l(\mathcal{P})$ when assuming $\Gamma(\mathcal{P}) \cong \mathrm{Conn}(\mathcal{P})$. This argument cannot be used if $\mathcal{P}$ is infinite, since then $\mathrm{Conn}(\mathcal{P})$ is infinite and it is possible for an infinite group to act in two different ways on the same set, so that one of these actions is regular and the other is still transitive but not regular.

Problem 11 Determine whether there exists a non-regular infinite polytope $\mathcal{P}$ such that $\Gamma(\mathcal{P}) \cong \mathrm{Conn}(\mathcal{P})$.

As a consequence of Theorem 3.2.14 we have that if Φ is the base flag of $\mathcal{P}$ then

$$r_{i_1} \cdots r_{i_k}\Phi = \Phi\rho_{i_1} \cdots \rho_{i_k}.$$

This can be used to illustrate $\mathcal{P}$ as a map in case it has rank 3, and we have available a presentation of $\Gamma(\mathcal{P})$ in terms of generators and relations on the standard reflections. The main idea is to start with a collection of faces determining a connected region of the plane and use connections that stabilise a flag to impose identifications on that region.

As an example we take the polyhedron $\mathcal{P} = \{3,8\}{*}96$ in [116]. One of the presentations given for $\Gamma(\mathcal{P})$ is

$$\left\langle \rho_0, \rho_1, \rho_2 \colon \rho_0^2 = \rho_1^2 = \rho_2^2 = (\rho_0\rho_2)^2 = (\rho_0\rho_1)^3 = (\rho_2\rho_0\rho_1\rho_2\rho_1\rho_2\rho_0\rho_1)^2 \right\rangle. \tag{3.30}$$

The name of the polyhedron indicates that it has type $\{3, 8\}$, although that is not obvious from the relations. It also indicates that it has 96 flags. It would still be possible to derive a drawing of $\mathcal{P}$ as a map without assuming any of these two facts, but it becomes lengthier. We will proceed by assuming that the relation $(\rho_1\rho_2)^8$ also holds in $\Gamma(\mathcal{P})$. It is a good exercise to derive this last relation from the defining relations of $\Gamma(\mathcal{P})$.

Exercise 3.2.17 Show that the relation $(\rho_1\rho_2)^8$ can be derived from those in (3.30).

We begin by placing eight triangles around a vertex like in Figure 3.15(a), without making any assumption on the second triangle of the edges that in that figure belong to only one triangle. We use the isomorphism in Theorem 3.2.14 to imply that

$$\Phi = \Phi(\rho_2\rho_0\rho_1\rho_2\rho_1\rho_2\rho_0\rho_1)^2 = (r_2r_0r_1r_2r_1r_2r_0r_1)^2\Phi \tag{3.31}$$

for every flag Φ. Starting from the flag Φ_0 shown in grey, we define $\Phi_1 = r_1\Phi$, $\Phi_2 = r_0r_1\Phi, \ldots, \Phi_{16} = (r_2r_0r_1r_2r_1r_2r_0r_1)^2\Phi$. Then we trace the sequence of flags $\Phi_0, \Phi_1, \ldots, \Phi_{16}$, marked as dots in Figure 3.15(b). From (3.31) we know that the two grey flags in Figure 3.15(b) are the same; consequently, the two edges of these grey flags can be identified in the orientable way. For the sake of the drawing we instead identify the edges of $r_1r_0r_2\Phi_0$ and $\Phi_{13} = r_1r_0r_2\Phi_{16}$, as shown in Figure 3.15(c).

Next we consider as base flag Ψ_0 the one marked in grey in Figure 3.16(a). The identification illustrated in Figure 3.15 tells us that

$$r_0(r_1r_2)^3r_1r_0(r_1r_2)^3r_1\Psi_0 = \Psi_0.$$

Corollary 3.2.9 can be used on $\Phi_0^2 = \Phi_0\rho_2$ to force the map to have also the extra identification shown in Figure 3.16(b). Furthermore, the two identifications obtained so far can be combined to obtain a single identification

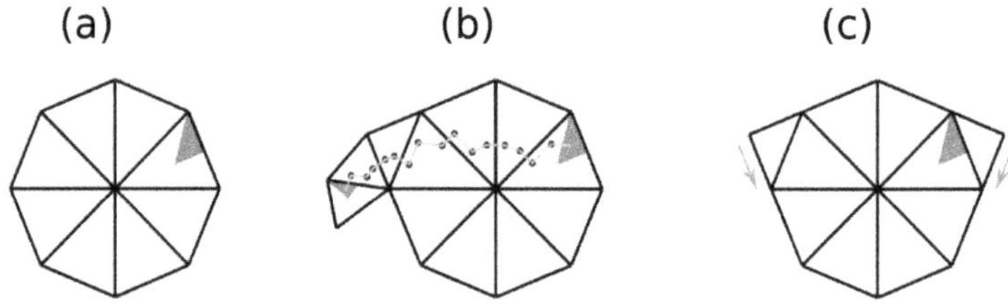

Figure 3.15 (a) A flag Φ_0 of the map; (b) a flag equivalent to Φ_0; (c) the induced identification.

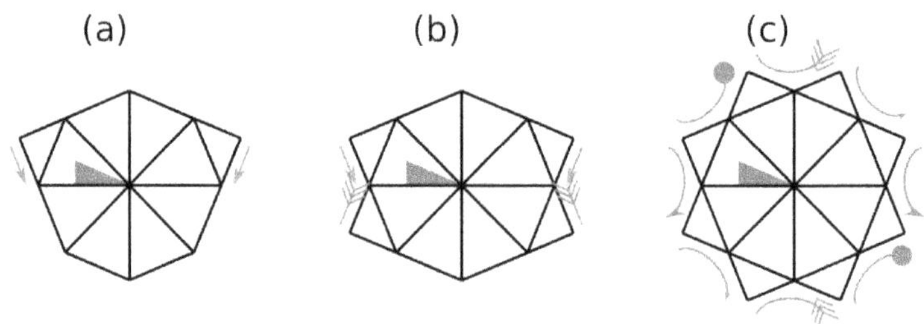

Figure 3.16 (a) A flag Ψ_0 of the map; (b) a second identification of the map; (c) all identifications of the map.

that completes the eight triangles around the vertex that appears at the left (and at the right) of the drawing. Finally, applying again Corollary 3.2.9 on Φ_0, Φ_0^2 and their images under $\langle\sigma_2\rangle$, results in further identifications of the map as illustrated in Figure 3.16(c).

We claim that Figure 3.16(c) is a representation of the map $\mathcal{M}$ corresponding to the polyhedron {3,8}*96. The standard way to verify this is by determining the existence of the standard reflections ρ_0, ρ_1 and ρ_2, and by checking the validity on them of the defining relations of $\Gamma(\{3,8\}*96)$. Instead of dealing with automorphisms, we may corroborate that $(r_2r_0r_1r_2r_1r_2r_0r_1)^2\Psi = \Psi$ holds for every flag Ψ. In order to do this it suffices to consider only the flag Φ_0 in grey in Figure 3.15 together with the flags Φ_0^1, $\Phi_0^{1,0}$, Φ_0^2, $\Phi_0^{2,1}$ and $\Phi_0^{2,1,0}$, due to the obvious action of the dihedral group D_8 as an automorphism group. Clearly,

$$r_i^2\Psi = (r_0r_2)^2\Psi = (r_0r_1)^3\Psi = \Psi$$

holds for all i, telling us that the connection group of the map in Figure 3.16(c) is a quotient of $\mathrm{Conn}(\mathcal{M})$ that maps standard generators to standard generators. From Proposition 3.2.11 together with the defining relations of $\Gamma(\mathcal{P})$, we know that our map is indeed a quotient of $\mathcal{M}$. Since our map was obtained only by implementing identifications imposed by the defining relations of $\Gamma(\mathcal{P})$, it must be isomorphic to $\mathcal{M}$. Alternatively, we may use that $\Gamma(\mathcal{P})$ has order 96 and that our drawing has 16 triangles to conclude that no further identification is needed.

We were able to find a drawing of the map corresponding to the polyhedron {3,8}*96 in a small number of steps for more than one reason. On one hand, the number of flags of {3,8}*96 is small. On the other hand, there is only one additional relation to those given by the type, and it involves relatively few factors in the standard generating set. Larger maps may require us to repeat the steps we made here many more times, and perhaps also to develop some intuition on where to draw each identification given that in general there are many places in the picture where this can be done. One should expect that if the polyhedron has many flags then it will be harder to find a 'nice' picture.

Exercise 3.2.18 Draw a map for each of the following polyhedra.

1. {6,4}*48a of [116], whose automorphism group has defining relations
$$\rho_0^2 = \rho_1^2 = \rho_2^2 = (\rho_0\rho_2)^2 = (\rho_0\rho_1\rho_2\rho_1)^2 = (\rho_1\rho_2)^4 = (\rho_0\rho_1)^6 = \mathrm{id}.$$
2. {4,6}*72 of [116], whose automorphism group has defining relations
$$\rho_0^2 = \rho_1^2 = \rho_2^2 = (\rho_0\rho_2)^2 = (\rho_0\rho_1)^4 = (\rho_2\rho_0\rho_1)^4 = (\rho_1\rho_2)^6$$
$$= \rho_1\rho_2\rho_1\rho_0\rho_1\rho_2\rho_0\rho_1\rho_0\rho_1\rho_2\rho_1\rho_0 = \mathrm{id}.$$

The connection group of an n-polytope $\mathcal{P}$ also has an *even subgroup*, denoted by $\mathrm{Conn}^+(\mathcal{P})$, consisting of all elements that can be expressed as a product of an even number of factors r_i. Next we show similarities between the behaviour of the even subgroup of a rotary polytope with the behaviour of the even subgroup of the connection group.

Proposition 3.2.19 *Let $\mathcal{P}$ be an n-polytope, let* $\mathrm{Conn}(\mathcal{P}) = \langle r_0, \ldots, r_{n-1}\rangle$, *and for $i \in \{1, \ldots, n-1\}$ let $s_i = r_{i-1}r_i$. Then* $\mathrm{Conn}^+(\mathcal{P}) = \langle s_1, \ldots, s_{n-1}\rangle$.

Proof It is clear that $\langle s_1, \ldots, s_{n-1}\rangle \subseteq \mathrm{Conn}^+(\mathcal{P})$. For the other inclusion it suffices to observe that $r_i r_j \in \langle s_1, \ldots, s_{n-1}\rangle$ for every $i, j \in \{0, \ldots, n-1\}$, since every element in $\mathrm{Conn}^+(\mathcal{P})$ can be expressed as a product of an even number of factors r_k. Indeed, if $i < j$ then $r_i r_j = s_{i+1} \cdots s_j$ while if $i > j$ then $r_i r_j = s_i^{-1} \cdots s_{j+1}^{-1}$. ■

Proposition 3.2.20 *Let $\mathcal{P}$ be a polytope. Then the index of* $\mathrm{Conn}^+(\mathcal{P})$ *in* $\mathrm{Conn}(\mathcal{P})$ *is at most* 2. *Forthermore, if $\mathcal{P}$ is orientable then the index is* 2.

Proof Given $w \in \mathrm{Conn}(\mathcal{P})$, at least one of w and wr_0 belongs to $\mathrm{Conn}^+(\mathcal{P})$. It follows that if $\mathrm{Conn}^+(\mathcal{P})$ is not equal to $\mathrm{Conn}(\mathcal{P})$ then it has only two right cosets: $\mathrm{Conn}^+(\mathcal{P})$ and $\mathrm{Conn}^+(\mathcal{P})r_0$. In that situation $\mathrm{Conn}^+(\mathcal{P})$ is an index-2 subgroup.

Assume now that $\mathcal{P}$ is orientable. Then $\mathcal{F}l(\mathcal{P})$ admits a bicolouring such that for each flag Φ and each $i \in \{0, \dots, n-1\}$, Φ and Φ^i have distinct colours. If $w \in \mathrm{Conn}^+(\mathcal{P})$ then $w = r_{i_1} \cdots r_{i_{2k}}$ for some non-negative integer k; it follows that the flags $w\Phi = r_{i_1} \cdots r_{i_{2k}} \Phi = \Phi^{i_{2k}, \dots, i_1}$ and Φ have the same colour. This implies that $r_0 \notin \mathrm{Conn}^+(\mathcal{P})$, and therefore $\mathrm{Conn}^+(\mathcal{P})$ has index 2 in $\mathrm{Conn}(\mathcal{P})$. ■

Some word of caution is in place here. The second part of Proposition 3.2.20 is an implication in only one direction; the converse is not true in general.

Exercise 3.2.21 Find a non-orientable polyhedron $\mathcal{P}$ such that the index of $\mathrm{Conn}^+(\mathcal{P})$ in $\mathrm{Conn}(\mathcal{P})$ is 2.

For orientable rotary polytopes $\mathcal{P}$ the generators of the even subgroups of $\Gamma(\mathcal{P})$ and of $\mathrm{Conn}(\mathcal{P})$ have similar actions on the flags. The next result follows directly from the definitions of σ_i and of s_i.

Remark 3.2.22 Let $\mathcal{P}$ be an orientable rotary polytope with base flag Φ. Let $\Gamma^+(\mathcal{P}) = \langle \sigma_1, \dots, \sigma_{n-1} \rangle$ and $\mathrm{Conn}^+(\mathcal{P}) = \langle s_1, \dots, s_{n-1} \rangle$. Then $s_{i_1} \cdots s_{i_k} \Phi = \Phi \sigma_{i_1} \cdots \sigma_{i_k}$.

When the rotation subgroup of an orientable rotary polyhedron $\mathcal{P}$ is presented through generators and relations in terms of the standard rotations we may make a picture of the map corresponding to $\mathcal{P}$ in a similar way as we did for regular polyhedra when their automorphism groups are given in terms of the standard reflections. There are some important aspects to keep in mind; they are explained next.

If Φ is the base flag then $\Phi\sigma_i = \Phi^{i,i-1} = r_{i-1} r_i \Phi$. This is used to derive connections that fix the base flag from the defining relations of $\Gamma(\mathcal{P})$.

In the case when $\mathcal{P}$ is chiral then the connections arising from defining relations stabilise all flags in the flag orbit of the base flag, but may not stabilise the flags in the other flag orbit. Hence, when deriving identifications of the map such connections should only be applied to flags in the orbit of the base flag.

In what follows we shall derive a picture of the map C7.2 of [41], whose rotation group has defining relations

$$(\sigma_1 \sigma_2)^2 = \sigma_1^{-7} = \sigma_2 \sigma_1^{-3} \sigma_2^2 \sigma_1^{-1} = \mathrm{id}.$$

(This is the map that Edmonds found as the first example of a chiral map on a surface of genus greater than 1.) The notation C7.2 indicates that it is chiral and lies on an orientable surface of genus 7. Furthermore, the list of maps also indicates that its type is $\{7,7\}$ and that $\Gamma(\mathcal{P})$ has order 56. From here we can deduce that the map has 112 flags and hence it has 8 heptagonal faces.

We start our procedure by drawing one central heptagon together with another seven heptagons, each sharing an edge with the central one (see Figure 3.17(a)). The remaining six edges of each of these seven heptagons are not assumed yet to be in other particular heptagons. We choose the grey flag as our base flag Φ, and rewrite the action on Φ of the only relation not obtained from the type as

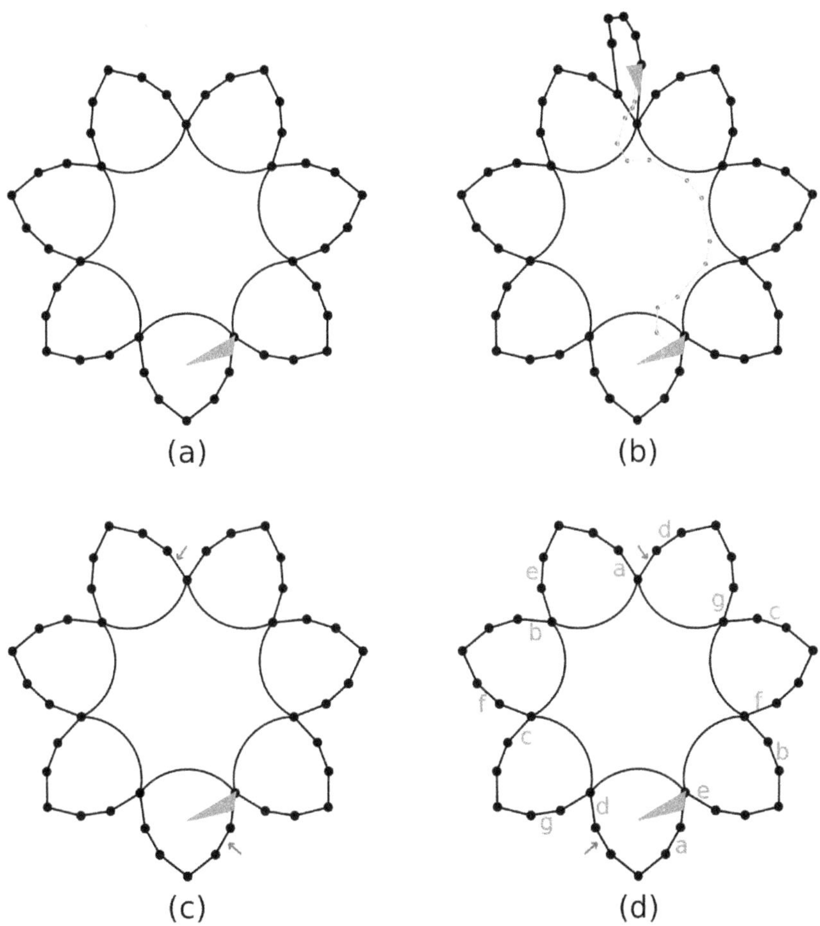

Figure 3.17 (a) Eight faces and a flag Φ of the map; (b) a flag equivalent to Φ; (c) the induced identification; (d) other identifications obtained in an analogous way.

$$r_1r_2(r_1r_0)^3(r_1r_2)^2r_1r_0\Phi = \Phi\sigma_2\sigma_1^{-3}\sigma_2^2\sigma_1^{-1} = \Phi.$$

For the sake of the drawing we instead consider the inverse connection:

$$r_0r_1(r_2r_1)^2(r_0r_1)^3r_2r_1\Phi = \Phi. \tag{3.32}$$

Next we trace the path from Φ to $r_0r_1(r_2r_1)^2(r_0r_1)^3r_2r_1\Phi$ induced by the connections in the last expression to obtain the identification on Figure 3.17(b).

That identification is equivalent to identifying orientably the two edges pointed at by arrows in Figure 3.17(c). From the fact that the 7-fold rotation α about the centre of the diagram must be an automorphism of the map, we use Corollary 3.2.9 on the flags $\Phi\langle\alpha\rangle$ to obtain the identifications with labels a to g in Figure 3.17(d).

Applying the same idea on $\Phi^{1,0,1}$ would force an identification of the edges pointed at by arrows in Figure 3.17(d). However, Φ and $\Phi^{1,0,1}$ belong to distinct flag orbits and we can no longer apply Corollary 3.2.9. We cannot assume that the flag

$$r_0r_1(r_2r_1)^2(r_0r_1)^3r_2r_1\Phi^{1,0,1} = r_0r_1(r_2r_1)^2(r_0r_1)^3r_2r_0r_1\Phi$$

equals $\Phi^{1,0,1}$, since the automorphism γ mapping $r_1r_0r_1r_0r_1(r_2r_1)^2(r_0r_1)^3$ $r_2r_0r_1\Phi$ to Φ may not be trivial. Later it shall become clear that, indeed, those two edges will not be identified to each other.

To derive further identifications in the drawing of the map we now start with the flag $\Phi^{0,1,0,1}$ which belongs to the orbit of Φ. This is the grey flag in Figure 3.18(a). Corollary 3.2.9 tells us that (3.32) holds when Φ is replaced by $\Phi^{0,1,0,1}$. The path that starts at $\Phi^{0,1,0,1}$ and is determined by the connections in (3.32) uses identifications a and e, and is shown in Figure 3.18(a). This induces the orientable identification of the edges of the flags $\Phi^{(0,1)^3}$ and $\Phi^{1,2,1,2,1,0,1}$, labelled as h in Figure 3.18(b). As above, Corollary 3.2.9 applied to the images of $\Phi^{0,1,0,1}$ under the 7-fold rotation $\langle\alpha\rangle$ forces identifications i to p in Figure 3.18(b).

We still have edges that belong to only one heptagon. To determine more identifications between pairs of these edges we now consider the flag $\Phi^{1,0}$. This flag is shown in grey in Figure 3.18(c), along with the corresponding path determined by the connections in (3.32). The same arguments as above tell us that the two edges labelled q in Figure 3.18(d) must be identified. A last application of Corollary 3.2.9 on the flags $\Phi^{1,0}\langle\alpha\rangle$ yields identifications r to w in Figure 3.18(d). Now that every edge belongs to two faces, we finish our drawing of the map C7.2.

After finishing the diagram of the map it is worth verifying that the result is indeed a chiral map; this is a redundancy that helps detecting potential mistakes in the process.

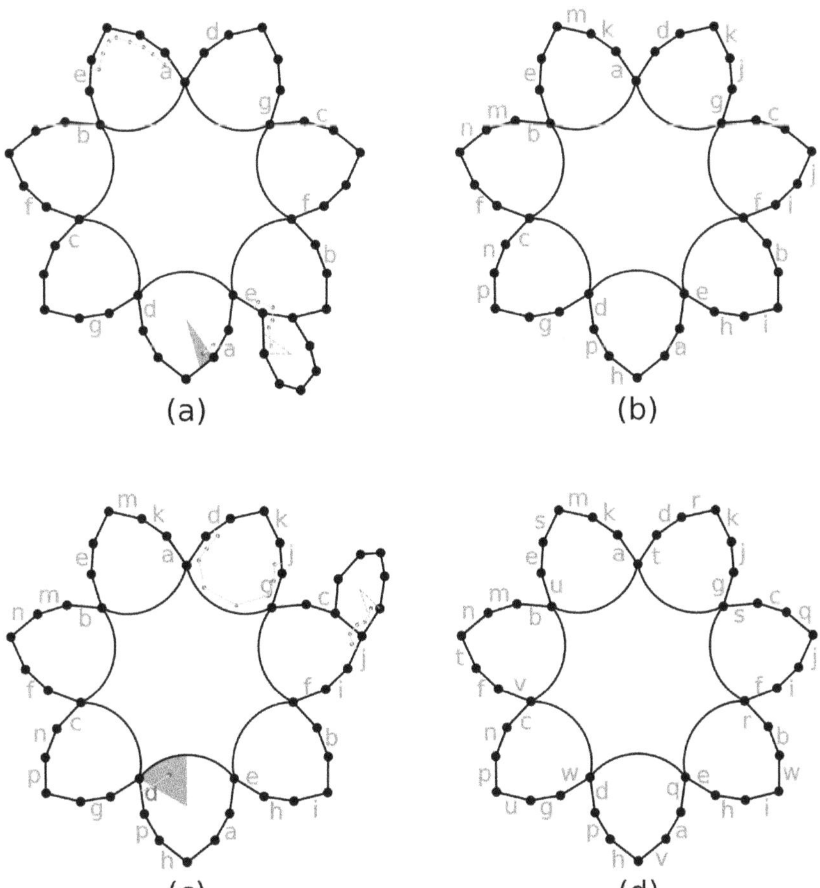

Figure 3.18 (a) The flag $\Phi^{0,1,0,1}$ and an equivalent flag; (b) a set of new identifications of the map; (c) the flag $\Phi^{0,1,0,1}$ and an equivalent flag; (d) all remaining identifications of the map.

Exercise 3.2.23 Show that the map in Figure 3.18(d) represents a polyhedron.

Exercise 3.2.24 Choose a base flag of the map in Figure 3.18(d) and show that the automorphism group contains the standard rotations with respect to that flag.

Remark 3.2.25 A description of Edmond's map can be found in [53, Section 21.3], and can be used to corroborate that the map corresponds to that illustrated in Figure 3.18.

Exercise 3.2.26 Draw the following chiral maps.

1. C7.1 of [41] with type $\{6,9\}$, whose automorphism group has defining relations

$$\sigma_1^6 = (\sigma_1\sigma_2)^2 = (\sigma_1\sigma_2^{-2})^2 = \sigma_2^{-1}\sigma_1^{-3}\sigma_2\sigma_1^{-1}\sigma_2^{-1} = \mathrm{id}.$$

2. C10.2 of [41], whose automorphism group has defining relations

$$\sigma_1^4 = \sigma_2^8 = (\sigma_1\sigma_2)^2 = \sigma_1\sigma_2^{-3}\sigma_1^2\sigma_2^{-2}\sigma_1\sigma_2^{-1} = \mathrm{id}.$$

The next result is a consequence of Proposition 3.1.67 in the context of quotients by automorphisms (see Definition 3.1.26).

Proposition 3.2.27 *Let $\mathcal{P}$ and $\mathcal{Q}$ be orientable rotary n-polytopes. If $\mathcal{Q}$ is a quotient of $\mathcal{P}$ then it is the quotient by automorphisms $\mathcal{Q} \cong \mathcal{P}/N$ for some subgroup $N \triangleleft \Gamma^+(\mathcal{P})$ not containing any of the standard rotations, and such that $\Gamma^+(\mathcal{P})/N$ satisfies the intersection property* (3.10).

Proof Proposition 3.1.67 tells us that there exists a surjective group homomorphism $\eta\colon \Gamma^+(\mathcal{P}) \to \Gamma^+(\mathcal{Q})$ mapping the ith standard rotation to the ith standard rotation (for a suitable choice of base flag). Then $N := \ker(\eta) \triangleleft \Gamma^+(\mathcal{P})$, and since $\Gamma^+(\mathcal{Q}) \cong \Gamma^+(\mathcal{P})/N$, this group must satisfy the intersection condition, and N must not contain any standard rotation.

In order to show that $\mathcal{Q} \cong \mathcal{P}/N$, let $\Gamma^+(\mathcal{P}) = \langle\sigma_1,\ldots,\sigma_{n-1}\rangle$ and $\Gamma^+(\mathcal{Q}) = \langle\hat{\sigma}_1,\ldots,\hat{\sigma}_{n-1}\rangle$. From the previous discussion we may rewrite $\hat{\sigma}_i$ as $N\sigma_i$. Then the i-faces of both $\mathcal{Q}$ and $\mathcal{P}/N$ are represented by the same cosets in the coset representation $(\langle N\sigma_1,\ldots,N\sigma_{n-1}\rangle,\{\Gamma_0,\ldots,\Gamma_{n-1}\})$. The uniqueness in Theorem 3.1.53 finishes the proof. ■

As a consequence of Theorem 3.2.14 we know that if $\mathcal{P}$ is a regular polytope then the size and structure of $\mathrm{Conn}(\mathcal{P})$ can be easily obtained from those of $\Gamma(\mathcal{P})$. The situation when $\mathcal{P}$ is chiral is very different. This will not be justified immediately. We shall see in Section 3.3 that in that case the size of $\mathrm{Conn}(\mathcal{P})$ may be very large in comparison to that of $\Gamma(\mathcal{P})$. Proposition 4.4.18 in Chapter 4 is the main structural result relating $\mathrm{Conn}(\mathcal{P})$ to $\Gamma(\mathcal{P})$ when $\mathcal{P}$ is chiral. Here we describe the action of $\mathrm{Conn}^+(\mathcal{P})$ on only one flag orbit.

Proposition 3.2.28 *Let $\mathcal{P}$ be a chiral polytope and let $\mathcal{X}$ be one flag orbit under $\Gamma(\mathcal{P})$. If $\Phi \in \mathcal{X}$ and $x, y \in \mathrm{Conn}^+(\mathcal{P})$ are such that $x\Phi = y\Phi$ then $x\Psi = y\Psi$ for every $\Psi \in \mathcal{X}$.*

Proof Since $\Psi \in \mathcal{X}$ it can be written as $\Psi = \Phi\gamma$ for some $\gamma \in \Gamma(\mathcal{P})$. Then

$$x\Psi = x\Phi\gamma = y\Phi\gamma = y\Psi.$$

■

Proposition 3.2.29 *Let $\mathcal{P}$ be a chiral polytope with base flag Φ_0, let $\Gamma(\mathcal{P}) \cong \langle\sigma_1, \dots, \sigma_{n-1}\rangle$ and $\mathrm{Conn}(\mathcal{P}) \cong \langle r_0, \dots, r_{n-1}\rangle$, and let $s_i := r_{i-1}r_i$ for $i \in \{1, \dots, n-1\}$. Then there is a surjective group homomorphism from $\langle s_1, \dots, s_{n-1}\rangle$ to $\Gamma(\mathcal{P})$ mapping s_i to σ_i for all i, whose kernel is* $\mathrm{Stab}_{\mathrm{Conn}(\mathcal{P})}(\Phi_0)$.

Proof For any given $x \in \mathrm{Conn}^+(\mathcal{P})$ let γ_x denote the unique automorphism such that $x\Phi_0 = \Phi_0\gamma_x$. It is routine to show that the mapping $x \mapsto \gamma_x$ is the desired surjective group homomorphism from $\mathrm{Conn}^+(\mathcal{P})$ to $\Gamma(\mathcal{P})$. It follows from Proposition 3.2.28 that the kernel of this automorphism is precisely $\mathrm{Stab}_{\mathrm{Conn}(\mathcal{P})}(\Phi_0)$. ■

Remark 3.2.30 Proposition 3.2.28 tells us that $\mathrm{Conn}^+(\mathcal{P})$ induces relatively few permutations when restricted to each flag orbit of a chiral polytope $\mathcal{P}$ (one for each flag in the orbit). Still, Proposition 3.2.20 states that the index of $\mathrm{Conn}^+(\mathcal{P})$ in $\mathrm{Conn}(\mathcal{P})$ is 2, and in principle $\mathrm{Conn}(\mathcal{P})$ can be very large with respect to the size of each flag orbit. We shall confirm in Sections 3.3 and 4.4 that the stabiliser of a flag Φ under $\mathrm{Conn}^+(\mathcal{P})$ acts non-trivially on the flag orbit of Φ^0, corroborating Proposition 3.2.15.

3.2.2 Flag Actions

The connection group of a polytope $\mathcal{P}$ is one of many groups that have a similar action on $\mathcal{P}$, where some elements map every flag to its i-adjacent flag.

Definition 3.2.31 We say that an n-polytope $\mathcal{P}$ admits a *flag action* from a group $W = \langle w_0, \dots, w_{n-1}\rangle$ whenever there is an action of W on $\mathcal{F}l(\mathcal{P})$ where $w_i\Phi = \Phi^i$ for every $\Phi \in \mathcal{F}l(\mathcal{P})$ and every $i \in \{0, \dots, n-1\}$.

Flag actions have the following purely group-theoretical definition. The proof of the equivalence is straightforward.

Remark 3.2.32 An n-polytope admits a flag action from the sggi $W = \langle w_0, \dots, w_{n-1}\rangle$ if and only if there is an epimorphism from W to $\mathrm{Conn}(\mathcal{P})$ that maps w_i to r_i for every i.

Perhaps the first mention of flag actions on polytopes was in [111, Section 3]. Here we follow the spirit of that article and consider only flag actions of sggi, since no gain is apparent for our purposes by allowing flag actions of more general groups.

Example 3.2.33 Every n-polytope $\mathcal{P}$ admits a flag action from the universal string Coxeter group $[\infty, \dots, \infty]$ (with $n-1$ entries ∞). In order to verify this, it suffices to ensure that the desired action of every defining relation of $[\infty, \dots, \infty]$ on every flag of $\mathcal{P}$ is trivial. Let $[\infty, \dots, \infty] = \langle w_0, \dots, w_{n-1}\rangle$ and let $\Phi \in \mathcal{F}l(\mathcal{P})$. Then the only defining relations for $[\infty, \dots, \infty]$ are $(w_i w_j)^2$ for every $i, j \in \{0, \dots, n-1\}$ with the property that $|i-j| \geq 2$. In that case, the desired action on Φ is

$$(w_i w_j)^2 \Phi = w_i w_j w_i \Phi^j = w_i w_j \Phi^{j,i} = w_i \Phi^{j,i,j} = \Phi^{j,i,j,i},$$

which indeed equals Φ whenever $|i-j| \geq 2$. We can conclude that $\mathcal{P}$ admits a flag action from $[\infty, \dots, \infty]$.

Example 3.2.34 Every equivelar polytope $\mathcal{P}$ with type $\{p_1, \dots, p_{n-1}\}$ admits a flag action from the universal string Coxeter group $[p_1, \dots, p_{n-1}]$. The argumentation follows the lines of Example 3.2.33. It is only necessary to add that the relation $(w_{i-1} w_i)^{p_i}$ is satisifed since the ith entry of the type of $\mathcal{P}$ is precisely p_i.

The flag action is a generalisation of the connection group also in the sense of the next two result, analogous (and with analogous justifications) to Remark 3.2.12 and Proposition 3.2.2.

Remark 3.2.35 Every flag action on a polytope $\mathcal{P}$ is transitive on $\mathcal{F}l(\mathcal{P})$.

Proposition 3.2.36 *Let W be an sggi with a flag action on a polytope $\mathcal{P}$. Then $(x\Phi)\gamma = x(\Phi\gamma)$ for every $\Phi \in \mathcal{F}l(\mathcal{P})$, $\gamma \in \Gamma(\mathcal{P})$ and $x \in W$.*

The next proposition follows directly from Remark 3.2.32 and Proposition 3.2.8.

Proposition 3.2.37 *If $\mathcal{P}$ covers $\mathcal{Q}$ then $\mathcal{Q}$ admits a flag action from* $\mathrm{Conn}(\mathcal{P})$.

The previous result and Theorem 3.2.14 imply the next proposition.

Proposition 3.2.38 *Let $\mathcal{P}$ be a regular polytope with $\Gamma(\mathcal{P}) \cong \langle \rho_0, \ldots, \rho_{n-1} \rangle$ and let $\mathcal{Q}$ be a quotient of $\mathcal{P}$ with* $\mathrm{Conn}(\mathcal{Q}) \cong \langle r_0, \ldots, r_{n-1} \rangle$. *Then there is a surjective group homomorphism from $\Gamma(\mathcal{P})$ to* $\mathrm{Conn}(\mathcal{Q})$ *mapping ρ_i to r_i for every i.*

As a converse of Proposition 3.2.37 for the particular case when $\mathcal{P}$ is regular, we have the following result.

Proposition 3.2.39 *If $\mathcal{Q}$ admits a flag action from* $\mathrm{Conn}(\mathcal{P})$ *and $\mathcal{P}$ is regular then $\mathcal{P}$ covers $\mathcal{Q}$.*

Proof Let Φ be the base flag of $\mathcal{P}$. Remark 3.2.32 tells us that there is a group epimorphism $\hat{\eta}$ from $\mathrm{Conn}(\mathcal{P})$ to $\mathrm{Conn}(\mathcal{Q})$. Furthermore, Proposition 3.2.15 tells us that $\mathrm{Stab}_{\mathrm{Conn}(\mathcal{P})}(\Phi) = \mathrm{id}$, and so $\hat{\eta}(\mathrm{Stab}_{\mathrm{Conn}(\mathcal{P})}(\Phi)) \subseteq \mathrm{Stab}_{\mathrm{Conn}(\mathcal{Q})}(\Psi)$ for every $\Psi \in \mathcal{F}l(\mathcal{Q})$. From Proposition 3.2.11 we can conclude that for each flag Ψ of $\mathcal{Q}$ there is a covering from $\mathcal{P}$ to $\mathcal{Q}$ mapping Φ to Ψ. ■

The last proposition need not hold if the hypothesis of $\mathcal{P}$ being regular is removed.

Example 3.2.40 Let $\mathcal{P}$ be the toroidal polyhedron with four hexagons and eight triangles in Figure 3.19, and let $\mathcal{Q}$ be the regular polyhedron constructed from the hexagon as in Exercise 2.1.61 (see also Exercise 2.2.4). Then $\mathcal{Q}$ has only two hexagonal facets, both sharing the six vertices and the six edges. Furthermore,

$$\Gamma(\mathcal{Q}) \cong \langle \rho_0, \rho_1 \rangle \times \langle \rho_2 \rangle \cong D_6 \times \mathbb{Z}_2,$$

and Theorem 3.2.14 implies that $\mathrm{Conn}(\mathcal{Q}) \cong D_6 \times \mathbb{Z}_2$ as well.

We claim that there is a group homomorphism ζ from $\mathrm{Conn}(\mathcal{P})$ to $\mathrm{Conn}(\mathcal{Q})$ that maps standard generators to standard generators. In general it is not easy to verify such a claim. Here the connection group of $\mathcal{Q}$ is small enough to do explicit computations, and they are left as an exercise.

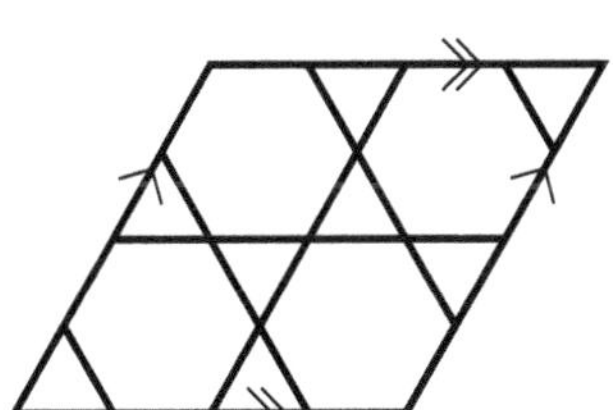

Figure 3.19 A toroidal polyhedron whose connection group has a flag action on a polyhedron with type $\{6, 2\}$.

The existence of ζ guarantees the flag action of $\mathrm{Conn}(\mathcal{P})$ on $\mathcal{Q}$. However, $\mathcal{P}$ does not cover $\mathcal{Q}$ since $\mathcal{P}$ has some triangular facets, each of which should cover a facet of $\mathcal{Q}$; since the facets of the latter are hexagons, no such covering is possible.

Exercise 3.2.41 Let $\mathcal{P}$ and $\mathcal{Q}$ be as in Example 3.2.40.

(a) Number the flags of $\mathcal{P}$ and find the explicit permutations corresponding to r_0, r_1, r_2 and $(r_1 r_2)^2$.
(b) Show that the only conjugates of $(r_1 r_2)^2$ are itself, $r_0(r_1 r_2)^2 r_0$ and $r_1 r_0 (r_1 r_2)^2 r_0 r_1$; also determine $\langle (r_1 r_2)^2, r_0(r_1 r_2)^2 r_0, r_1 r_0 (r_1 r_2)^2 r_0 r_1 \rangle$, the normal closure of $(r_1 r_2)^2$ in $\mathrm{Conn}(\mathcal{P})$.
(c) Use that $\mathrm{Conn}(\mathcal{Q}) \cong \Gamma(\mathcal{Q})$ admits the presentation

$$\left\langle \hat{r}_0, \hat{r}_1, \hat{r}_2 : \hat{r}_0^2 = \hat{r}_1^2 = \hat{r}_2^2 = (\hat{r}_0\hat{r}_1)^6 = (\hat{r}_0\hat{r}_2)^2 = (\hat{r}_1\hat{r}_2)^2 = \mathrm{id} \right\rangle$$

to conclude that there is a group homomorphism from $\mathrm{Conn}(\mathcal{P})$ to $\mathrm{Conn}(\mathcal{Q})$ mapping r_i to $\hat{r}_i$ for every i.
(d) Compute the order of $\mathrm{Conn}(\mathcal{P})$.

Universal polytopes of a given type satisfy the following universal property, which is a generalisation of Proposition 3.1.25.

Corollary 3.2.42 *If $\mathcal{P}$ is an equivelar polytope with type $\{p_1, \ldots, p_{n-1}\}$ then it is a quotient of the universal polytope $\{p_1, \ldots, p_{n-1}\}$.*

Proof The universal polytope $\{p_1, \ldots, p_{n-1}\}$ is regular, and the type of $\mathcal{P}$ guarantees that $\mathrm{Conn}(\mathcal{P})$ satisfies all defining relations of $[p_1, \ldots, p_{n-1}] = \Gamma(\{p_1, \ldots, p_{n-1}\})$. The result follows from Proposition 3.2.39. ∎

An sggi W with a flag action on a polytope $\mathcal{P}$ carries all combinatorial information of $\mathcal{P}$. We may associate a given flag Φ to its stabiliser $\mathrm{Stab}_W(\Phi)$, and then an arbitrary flag Ψ is associated to the coset $w\mathrm{Stab}_W(\Phi)$, where $w \in W$ is such that $w\Phi = \Psi$; this is precisely the subset consisting of all elements of W that map Φ to Ψ. In this way, the action of an element $x \in W$ on the flag corresponding to $y\mathrm{Stab}_W(\Phi)$ is precisely the flag corresponding to $xy\mathrm{Stab}_W(\Phi)$. In particular, if w_i is the ith generator of W then $w_i(y\mathrm{Stab}_W(\Phi)) = (w_i y)\mathrm{Stab}_W(\Phi)$ corresponds to the i-adjacent flag of that corresponding to $y\mathrm{Stab}_W(\Phi)$. Once we have a description of the flags and their adjacencies, we have a description of the flag graph, and from there we can rebuild $\mathcal{P}$ itself (see Proposition 2.1.46).

From the previous discussion it is natural to define an abstract polytope as a quotient as follows.

Definition 3.2.43 A *quotient presentation* of a polytope $\mathcal{P}$ is a quotient $\mathcal{P}(W)/N$ where W is an sggi with a flag action on $\mathcal{P}$ and N is the stabiliser in W of some flag of $\mathcal{P}$.

Remark 3.2.44 If W is not a string C-group then $\mathcal{P}(W)$ is not a polytope, but it is still possible to construct the coset structure $\mathcal{P}(W)$ as in Section 3.1.1. In many instances the quotient presentation is used to take advantage of our knowledge of the group W, rather than of the actual structure $\mathcal{P}(W)/N$. As we shall see, much can be said of the polytope $\mathcal{P}$ even when W is not a string C-group. In any case, W must act regularly on the flags of $\mathcal{P}(W)$.

Remark 3.2.45 When W is an sggi and we express a combinatorial structure (in particular a polytope) as $\mathcal{P}(W)$, we are thinking of W as the automorphism group of $\mathcal{P}(W)$, as explained in Sections 3.1.1 and 3.1.2. Such a structure must be regular, and consequently the automorphism and connection groups are isomorphic (see Theorem 3.2.14). In view of this, the definition of flag action would work well from $\mathrm{Conn}(\mathcal{P}(W))$ instead of W. For the sake of consistency with [111, 113], we keep the convention that the automorphism group of $\mathcal{P}(W)$ is the one that has the flag action.

Definition 3.2.43 suggests that a polytope $\mathcal{P}$ may have many quotient presentations, by considering stabilisers N of distinct flags of $\mathcal{P}$.

Proposition 3.2.46 *Let $\mathcal{P}$ be a k-orbit polytope that admits a flag action from the sggi W. Then $\mathcal{P}$ has precisely k distinct quotient presentations of the form $\mathcal{P}(W)/N$.*

Proof Let $w_0, \ldots, w_{n-1}$ be the generators of W. As noted in the discussion preceding Definition 3.2.43, if $\Phi \in \mathcal{F}l(\mathcal{P})$ and $N = \mathrm{Stab}_W(\Phi)$ then $\mathcal{P}$ can be recovered explicitly by considering its flags as the left cosets of N, where xN and yN are i-adjacent if and only if $xN = w_i yN$. From this, together with Lemma 2.1.79, we can conclude that if N is also the stabiliser of some flag Ψ then there is an automorphism γ mapping Φ to Ψ, given as a permutation of flags by $(x\Phi)\gamma := x\Psi$ for every $x \in W$, since $x\Phi$ is associated to xN when associating N to Φ, while xN is associated to $x\Psi$ when associating N to Ψ. Hence, $\mathcal{P}$ has at least k quotient presentations, since flags in distinct flag orbits have distinct stabilisers.

On the other hand, if $\Psi = \Phi\gamma$ for some $\gamma \in \Gamma(\mathcal{P})$ then Proposition 3.2.36 tells us that $x \in W$ is such that $x\Phi = \Phi$ if and only if

$$\Psi = \Phi\gamma = (x\Phi)\gamma = x(\Phi\gamma) = x\Psi,$$

and therefore $\mathrm{Stab}_W(\Psi) = N$. It follows that $\mathcal{P}$ exactly k quotient presentations, one for each flag orbit. ■

As a consequence of the previous proof we have the following corollary.

Corollary 3.2.47 *Let W be an sggi with a flag action on a polytope $\mathcal{P}$. Two flags of $\mathcal{P}$ have the same stabiliser under the action of W if and only if the flags belong to the same orbit under $\Gamma(\mathcal{P})$.*

Even when a polytope has more than one quotient presentation with respect to the sggi W, the corresponding flag stabilisers are related as follows.

Lemma 3.2.48 *Let W be an sggi with a flag action on a polytope $\mathcal{P}$. Let $\Phi, \Psi \in \mathcal{F}l(\mathcal{P})$ and let $x \in W$ such that $\Psi = x\Phi$. Then*

$$\mathrm{Stab}_W(\Psi) = x\mathrm{Stab}_W(\Phi)x^{-1}.$$

Proof If $y \in \mathrm{Stab}_W(\Phi)$ then $\Phi = y\Phi$. Then

$$xyx^{-1}\Psi = xy\Phi = x\Phi = \Psi.$$

It follows that $x\mathrm{Stab}_W(\Phi)x^{-1} \subseteq \mathrm{Stab}_W(\Psi)$. The other inclusion is analogous, by rewriting $\Phi = x^{-1}\Psi$, interchanging the roles of Φ and Ψ, and the roles of x and x^{-1}. ■

Corollary 3.2.49 *All flag stabilisers of a flag action are conjugates.*

Example 3.2.50 Let $\mathcal{P}$ and $\mathcal{R}$ be the toroidal polyhedra in Figure 3.20(a) and (b), respectively, and let Φ be the grey flag of $\mathcal{P}$. We saw in Example 2.2.22 that $\mathcal{R}$ is regular, that $\mathcal{P}$ is a two-orbit polyhedron, and that Φ and Φ^1 belong to distinct orbits.

We claim that $\mathcal{R}$ covers $\mathcal{P}$; any two flags (resp. vertices, edges, squares) of $\mathcal{R}$ that differ by the translation by a vector in $\langle(3,0),(0,2)\rangle$ are identified to obtain $\mathcal{P}$. When doing so, i-adjacencies are preserved for $i \in \{0,1,2\}$. Figure 3.20(c) shows six fundamental regions under the translation group just mentioned.

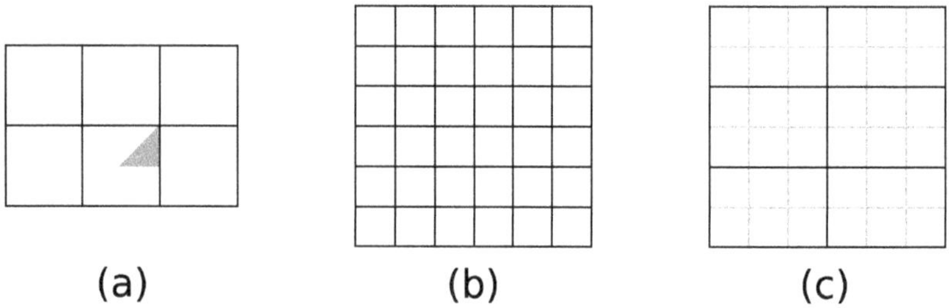

Figure 3.20 (a) The polyhedron $\mathcal{P}$; (b) the polyhedron $\mathcal{R}$; in (c) we represent that $\mathcal{R}$ covers $\mathcal{P}$.

We know from Proposition 3.2.37 that $\mathcal{P}$ admits a flag action from $\mathrm{Conn}(\mathcal{R})$. Let $\mathrm{Conn}(\mathcal{R}) = \langle r_0, \ldots, r_{n-1} \rangle$, let $N = \mathrm{Stab}_{\mathrm{Conn}(\mathcal{R})}(\Phi)$ and let $N' = \mathrm{Stab}_{\mathrm{Conn}(\mathcal{R})}(\Phi^1)$. Then $\mathcal{P}$ admits the quotient presentations $\mathcal{R}/N$ and $\mathcal{R}/N'$. In what follows we describe N and N'.

Because of Remark 3.2.35, N must have six elements (the number of elements of $\mathrm{Conn}(\mathcal{R})$ over the number of flags of $\mathcal{P}$). Let $a = (r_0 r_1 r_2 r_1)^2$ and $b = (r_2 r_1 r_0 r_1)^3$. Direct computations show that $a, b \in N$, that $a^3 = b^2 = \mathrm{id}$, and that $ab = ba$. Hence N is a cyclic group of order 6 generated by ab. From Lemma 3.2.48 we know that $N' = r_1 N r_1$, and therefore it is also a cyclic group of order 6; it is generated by $(r_1 r_0 r_1 r_2)^2$ and $(r_1 r_2 r_1 r_0)^3$.

In view of the previous discussion, we can think of a quotient presentation $\mathcal{P}/N$ as a presentation of a rooted polytope $\mathcal{Q}_{(\Phi)}$, where the flag Φ has N as its stabiliser. Given any two choices Φ, Φ' of such flags in the same orbit under $\Gamma(\mathcal{Q})$ there is an isomorphism of rooted polytopes taking one choice to the other.

Whenever the size and structure of the connection group of a certain polytope $\mathrm{Conn}(\mathcal{P})$ are well understood, it may be convenient to use this group for quotient presentations of $\mathcal{P}$, since $\mathrm{Conn}(\mathcal{P})$ is the smallest sggi with a flag action on $\mathcal{P}$ (in the sense that any other such sggi covers $\mathrm{Conn}(\mathcal{P})$). However, the connection group of distinct polytopes may have very distinct properties, making connection groups hard to work with when studying general properties of polytopes and their flag actions.

In view of Example 3.2.33 and Proposition 3.2.39, every n-polytope is a quotient of the universal polytope $\{\infty, \ldots, \infty\} = \mathcal{P}([\infty, \ldots, \infty])$. Given our knowledge of the group $[\infty, \ldots, \infty]$, it is sometimes easier to use quotient presentations with respect to this group instead of any other one, with the advantage that the obtained results will be valid for every polytope. This choice was taken for example in [214] in the context of maps on surfaces.

Quotient presentations are also a useful tool to work with quotients of polytopes, as shown by the following results.

Proposition 3.2.51 *Let $\mathcal{Q}$ and $\mathcal{R}$ be polytopes admitting a flag action from an sggi Γ. Then $\mathcal{R}$ is a quotient of $\mathcal{Q}$ if and only if there exist $N, M \leq \Gamma$ such that $\mathcal{Q}$ and $\mathcal{R}$ have respectively quotient presentations $\mathcal{P}(\Gamma)/N$ and $\mathcal{P}(\Gamma)/M$, and $N \leq M$. In those cases the degree of the covering is the index of N in M.*

Proof Let $\Phi \in \mathcal{F}l(\mathcal{Q})$, let $N = \mathrm{Stab}_\Gamma(\Phi)$ and let $\Gamma = \langle r_0, \ldots, r_{n-1} \rangle$.

If η is a covering from $\mathcal{Q}$ to $\mathcal{R}$ and $x := r_{i_1} \cdots r_{i_k} \in N$ then

$$\eta(\Phi) = \eta(x\Phi) = \eta(\Phi^{i_k,\ldots,i_1}) = (\eta(\Phi))^{i_k,\ldots,i_1} = x\eta(\Phi). \tag{3.33}$$

Hence $N \leq \mathrm{Stab}_\Gamma(\eta(\Phi))$, allowing us to set $M := \mathrm{Stab}_\Gamma(\eta(\Phi))$.

Conversely, if $N \leq M$ we may use the bijection between the flags and the left cosets of the stabiliser of some flag to define the mapping $\eta \colon \mathcal{F}l(\mathcal{P}) \to \mathcal{F}l(\mathcal{Q})$ as $xN \mapsto xM$. This is a covering since $\eta((xN)^i) = \eta(r_i xN) = r_i xM = (xM)^i$ for every i.

To determine the degree of the covering we only note that if a denotes the index of N in M then M is the disjoint union of a left cosets of N. The image of any such left coset under η is M. We conclude that the degree is a. ■

Remark 3.2.52 From (3.33) we can conclude that flag actions commute with coverings. This is convenient in the situation when $\mathcal{Q}$ and $\mathcal{R}$ admit a flag action from the same sggi Γ (for instance from $[\infty, \ldots, \infty]$). In such a scenario we do not need to define notation for $\mathrm{Conn}(\mathcal{Q})$ and $\mathrm{Conn}(\mathcal{R})$, and instead we may use the elements of Γ. In other words, if $\Phi \in \mathcal{F}l(\mathcal{Q})$ and $\Psi \in \mathcal{F}l(\mathcal{R})$ then it makes sense to speak of $r_i\Phi$ and $r_i\Psi$ in the understanding that r_i is the ith generator of Γ.

Remark 3.2.53 Proposition 3.2.51 does not imply that if $\mathcal{Q} = \mathcal{P}(\Gamma)/N$, $\mathcal{R} = \mathcal{P}(\Gamma)/M$ and $\mathcal{Q}$ covers $\mathcal{R}$ then $N \leq M$. The groups N and M play the role of flags Φ and Ψ of $\mathcal{P}$ and $\mathcal{Q}$, respectively, and so the covering must be such that it maps Φ to Ψ (compare with Proposition 3.1.67).

For example, the toroidal polyhedron $\mathcal{Q}$ in Figure 3.21(a) covers the toroidal polyhedron $\mathcal{R}$ in Figure 3.21(b) through the symmetry group of order 2 generated by the translation by the vector $(0, 2)$. However, as it will become apparent, any covering from $\mathcal{Q}$ to $\mathcal{R}$ must map the grey flag Φ in

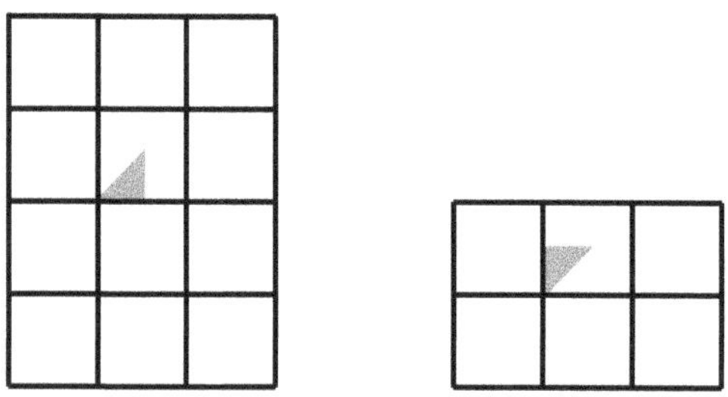

(a) (b)

Figure 3.21 The polyhedron in (a) covers that in (b), but not as rooted polytopes.

Figure 3.21(a) to a flag containing a horizontal edge in Figure 3.21(b). If Ψ is the grey flag in Figure 3.21(b) then $(r_0r_1r_2r_1)^3 \in \mathrm{Stab}_{[\infty,\infty]}(\Phi)$ and $(r_0r_1r_2r_1)^3 \notin \mathrm{Stab}_{[\infty,\infty]}(\Psi)$. It follows that if $\{\infty,\infty\}/N$ and $\{\infty,\infty\}/M$ are quotient presentations of the rooted polytopes $\mathcal{Q}_{(\Phi)}$ and $\mathcal{R}_{(\Psi)}$, respectively, then N is not a subgroup of M.

The following result describes the connection group of a polytope in terms of a given flag action on it. Recall that the *core* $\mathrm{Core}_\Gamma(N)$ of a subgroup N of a group Γ is the largest normal subgroup of Γ contained in N, and that it may be computed as the intersection of all conjugates of N by elements in Γ.

Proposition 3.2.54 *Let $\mathcal{Q} = \mathcal{P}(\Gamma)/N$ for some sggi Γ and some $N \leq \Gamma$. Then* $\mathrm{Conn}(\mathcal{Q}) \cong \Gamma/\mathrm{Core}_\Gamma(N)$.

Proof From the definition of flag action we know that there exists a group homomorphism $\eta\colon \Gamma \to \mathrm{Conn}(\mathcal{Q})$ that maps generators to generators. It suffices to show that the kernel $\ker(\eta)$ is precisely $\mathrm{Core}_\Gamma(N)$.

An element $\gamma \in \ker(\eta)$ acts like id on every flag of $\mathcal{Q}$, and hence $\gamma \in N$. Furthermore, Lemma 3.2.48 and Corollary 3.2.49 imply that $\gamma \in xNx^{-1}$ for every $x \in \Gamma$. From here we conclude that $\ker(\eta) \leq \mathrm{Core}_\Gamma(N)$. Conversely, if $\gamma \in \mathrm{Core}_\Gamma(n)$ then $\gamma \in xNx^{-1}$ for every $x \in \Gamma$, and Lemma 3.2.48 forces $\gamma \in \ker(\eta)$. ∎

We turn our attention to describing $\Gamma(\mathcal{Q})$ from a quotient presentation for $\mathcal{Q}$.

Lemma 3.2.55 *Let $\mathcal{Q} = \mathcal{P}(W)/N$ be the quotient presentation for $\mathcal{Q}$ given by the stabiliser N of a flag Φ, and let $x \in W$. There is an automorphism $\gamma \in \Gamma(\mathcal{Q})$ mapping Φ to $x\Phi$ if and only if $x \in$* $\mathrm{Norm}(N)$.

Proof By definition, $x \in \mathrm{Norm}(N)$ if and only if $xNx^{-1} = N$. Now we use Corollary 3.2.47 and Lemma 3.2.48 to conclude that there exists $\gamma \in \Gamma(\mathcal{Q})$ such that $\Phi\gamma = x\Phi$ if and only if the stabiliser xNx^{-1} of $x\Phi$ equals the stabiliser N of Φ. ∎

Example 3.2.56 Let $\mathcal{R}$ be the toroidal polyhedron in Figure 3.21(b) and let Φ be the grey flag. Then Φ^1 and Φ belong to distinct orbits. If $N = \mathrm{Stab}(\Phi)$ then we can verify directly that $r_1 \notin \mathrm{Norm}(N)$. Indeed, $(r_0r_1r_2r_1)^2 \in N$, but $r_1(r_0r_1r_2r_1)^2r_1$ does not fix Φ and hence it is not in N.

Lemma 3.2.55 is the key to show a way to describe $\Gamma(\mathcal{Q})$ given a quotient presentation of $\mathcal{Q}$.

Theorem 3.2.57 *If $\mathcal{P}(W)/N$ is a quotient presentation of $\mathcal{Q}$ then* $\Gamma(\mathcal{Q}) \cong \mathrm{Norm}_W(N)/N$.

Proof Given a base flag Φ, we know from Lemma 3.2.55 that we can define a function $\mu\colon \mathrm{Norm}(N) \to \Gamma(\mathcal{Q})$ that maps $x \in \mathrm{Norm}(N)$ to the unique automorphism γ_x that maps Φ to $x\Phi$. Under this setting, if $x, y \in \mathrm{Norm}(N)$ then

$$\Phi\mu(x)\mu(y) = \Phi\gamma_x\gamma_y = x\Phi\gamma_y = xy\Phi = \Phi\gamma_{xy} = \Phi\mu(xy).$$

We can conclude that μ is a group homomorphism. Clearly the kernel of μ is N, and the theorem follows. ■

Corollary 3.2.58 *Let $\mathcal{P}(\Gamma)/N$ be a quotient presentation of $\mathcal{Q}$. Then $\mathcal{Q}$ is rotary if and only if $N \triangleleft \Gamma^+$; it is regular if and only if $N \triangleleft \Gamma$; and it is chiral if and only if $N \triangleleft \Gamma^+$ but N is not normal in Γ.*

Proof Let $\Phi \in \mathcal{F}l(\mathcal{Q})$. By definition, $\mathcal{Q}$ is regular if and only for every $x \in \Gamma$ there exists $\gamma_x \in \Gamma(\mathcal{Q})$ such that $x\Phi = \Phi\gamma_x$. Then Lemma 3.2.55 implies that $\mathcal{Q}$ is regular if and only if $N \triangleleft \Gamma$.

If $\mathcal{Q}$ is rotary then we use the same argument as above with $x \in \Gamma^+$ to conclude that $N \triangleleft \Gamma^+$. Conversely, if $N \triangleleft \Gamma^+$ then we may proceed as in the proof of Theorem 3.2.57 to conclude that $\mathcal{Q}$ is rotary. The assertion regarding chirality follows from those about regular and rotary polytopes. ■

3.2.3 Petrie Paths

We conclude this section by relating the Petrie paths with connections. We start by finding a Petrie path in terms of the connection group.

Lemma 3.2.59 *Let $n \geq 3$, let $\mathcal{P}$ be an n-polytope and let Φ be a flag of $\mathcal{P}$. Then there is a Petrie path π_Φ of $\mathcal{P}$ such that the ith vertex and the ith edge of π_Φ are the vertex and edge in $(r_{n-1}\cdots r_0)^i\Phi$. Furthermore, every Petrie path π is the path π_Φ for some flag Φ.*

Proof Let Φ_0 be some fixed flag of $\mathcal{P}$, and for every j let F_j be its j-face.

Let $v_1 := F_0$, $e_1 := F_1$, and for $i \geq 2$ let v_i and e_i be respectively the vertex and edge in $(r_{n-1}\cdots r_0)^{i-1}\Phi_0$. We start by noting that for every ℓ and i, the ℓ-face of $(r_{n-1}\cdots r_0)^i\Phi_0$ is contained in the $(\ell+1)$-face of $(r_{n-1}\cdots r_0)^{i-1}\Phi_0$. It follows that $v_{i+1} \leq e_i$ for any given i, and that for every $\ell \in \{1,\ldots,n-1\}$ the edge e_i is contained in the ℓ-face of $(r_{n-1}r_{n-2}\cdots r_0)^{i-\ell}\Phi_0$. From this we conclude on one hand that the ordered edge set $(\ldots,e_1,e_2,e_3,\ldots)$ induces a path π_Φ, and on the other, that for $j \in \{1,\ldots,n-1\}$ the face F_j contains the edges $e_1,\ldots,e_j$.

Now we show that π_Φ is a Petrie path. Under the assumption that the face F_j plays the role of the face $F_{j,1}$ for $j \in \{2, \ldots, n-1\}$, we just showed that all these faces satisfy the first item of Definition 2.1.66. With the same idea, for every i we define $F_{k,i}$ to be the k-face of $(r_{n-1} \cdots r_0)^{i-1}\Phi_0$ and the first item still holds. For the remaining items of Definition 2.1.66 it suffices to show that for every k and i the faces $F_{k,i}$ and $F_{k,i+1}$ are distinct, but this follows from the diamond condition on $\mathcal{P}$, since the k-faces of $r_{k-1} \cdots r_0(r_{n-1} \cdots r_0)^{i-1}\Phi_0$ and $r_k \cdots r_0(r_{n-1} \cdots r_0)^{i-1}\Phi_0$ are distinct.

Finally, if π is a Petrie path then there exist faces $F_{j,k}$ as in Definition 2.1.66. Let Φ be the flag consisting of the vertex between e_0 and e_1, the edge e_1 and the faces $F_{j,1}$ for $j \in \{2, \ldots, n-1\}$. To verify that $\pi = \pi_\Phi$ it suffices to show that $(r_{n-1} \cdots r_0)^i \Phi$ consists of the vertex between e_{i-1} and e_i together with the edge e_i and the faces $F_{j,i}$ for $j \in \{2, \ldots, n-1\}$. The details are left as an exercise. ■

Exercise 3.2.60 Let π be a Petrie path with edges $\ldots e_{-1}, e_0, e_1, \ldots$; for every j and k let $F_{j,k}$ be as in Definition 2.1.66, and let Φ be the flag consisting of the vertex between e_0 and e_1, the edge e_1 and the faces $F_{j,1}$ for $j \in \{2, \ldots, n-1\}$. Show that for every $i \in \mathbb{Z}$ the flag $(r_{n-1} \cdots r_0)^i \Phi$ consists of the vertex between e_{i-1} and e_i together with the edge e_i and the faces $F_{j,i}$ for $j \in \{2, \ldots, n-1\}$.

Corollary 3.2.61 *Let $n \geq 3$ and let $\mathcal{P}$ be an n-polytope whose Petrie paths have all the same length m. Then m is the order of the connection $r_{n-1} \cdots r_0$ (or equivalently, of the connection $r_0 \cdots r_{n-1}$).*

Corollary 3.2.62 *Let $n \geq 3$ and let $\mathcal{P}$ be a regular n-polytope with $\Gamma(\mathcal{P}) = \langle \rho_0, \ldots, \rho_{n-1} \rangle$. Then the size of the Petrie paths of $\mathcal{P}$ is the order of $\rho_0 \cdots \rho_{n-1}$. Furthermore, $\rho_{n-1} \cdots \rho_0$ and $\rho_0 \cdots \rho_{n-1}$ act like one-step rotations on some Petrie path of $\mathcal{P}$.*

Proof If Φ_0 is the base flag of $\mathcal{P}$ then $\Phi_0(\rho_{n-1} \cdots \rho_0)^k = (r_{n-1} \cdots r_0)^k \Phi_0$. The fact that $\rho_{n-1} \cdots \rho_0$ acts like a one-step rotation on some Petrie path π follows from the discussion in the proof of Lemma 3.2.59. Since $\rho_0 \cdots \rho_{n-1} = (\rho_{n-1} \cdots \rho_0)^{-1}$, the automorphism $\rho_0 \cdots \rho_{n-1}$ must act like a one-step rotation on π, but in the opposite direction. ■

Exercise 3.2.63 Show that the Petrie paths of a regular polytope $\mathcal{P}$ and of its dual have the same lengths.

In what follows we shall identify the flags of a Petrie path π of $\mathcal{P}$ with some specific flags of $\mathcal{P}$. This will help us determine the automorphisms of π that are restrictions of automorphisms of $\mathcal{P}$ that preserve π. This idea is illustrated in the next example.

Example 3.2.64 Let $\mathcal{P}$ be the tiling $\{4,4\}$ of squares of the Euclidean plane $\mathbb{E}^2$ with vertex set given by all points with integer coordinates, and let Φ be the flag consisting of the point $F_0 = (0,0)$, the edge F_1 between F_0 and $(1,0)$, and the square F_2 determined by the edge F_1 and the point $(0,1)$. Then the Petrie path π_Φ is the one illustrated in Figure 3.22, where Φ is represented as the dark grey flag.

We let $\Psi := r_0 r_2 \Phi$; this flag contains vertex $(1,0)$, the edge F_1, together with the square determined by F_1 and the point $(0,-1)$ (this is the light grey flag in Figure 3.22). It is a good exercise to verify step by step that the Petrie path π_Ψ consists of the same vertices and edges as π_Φ near F_0. The geometric representation can be used to see that the rotation about the midpoint of F_1 preserves π_Φ while mapping Φ to Ψ.

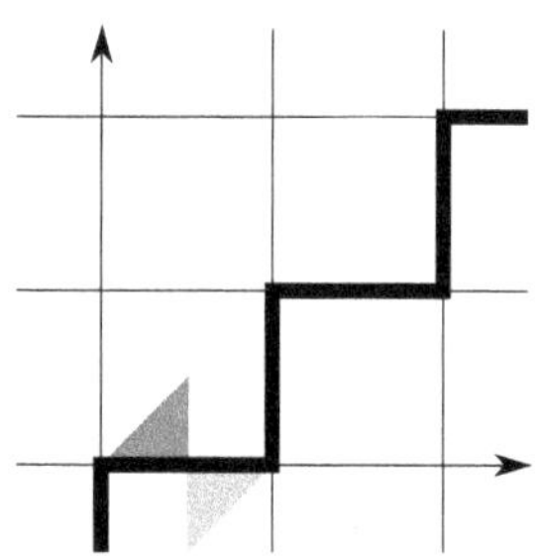

Figure 3.22 The Petrie paths π_Φ and π_Ψ induced by the grey flags Φ and Ψ coincide.

In order to do the identification of flags of $\mathcal{P}$ and of its Petrial we shall study the following connections. For a given n-polytope $\mathcal{P}$ with $\mathrm{Conn}(\mathcal{P}) = \langle r_0, \ldots, r_{n-1} \rangle$ and for every $0 \leq k \leq \ell < n$, we define the connection $z_{k,\ell}$ recursively as follows. For $\ell \in \{0, \ldots, n-1\}$ we define $z_{\ell,\ell} = r_\ell$, and for $k < \ell$ we define

$$z_{k,\ell} = z_{k+1,\ell} r_k r_{k+1} \cdots r_\ell. \tag{3.34}$$

For example, $z_{1,2} = r_2 r_1 r_2$ and $z_{0,2} = r_2 r_1 r_2 r_0 r_1 r_2$.

Lemma 3.2.65 *Let $k \leq \ell$ and let $z_{k,\ell}$ be defined as above.*

- *If $k < \ell$ then $r_\ell r_{\ell-1} \cdots r_k z_{k+1,\ell} = z_{k,\ell}$.*
- *For every $0 \leq k \leq \ell < n$ the element $z_{k,\ell}$ is an involution.*
- *Conjugation by $z_{k+1,\ell}$ inverts the element $r_\ell r_{\ell-1} \cdots r_k$.*

Proof To prove the first item we proceed by induction over $m := \ell - k$. If $m = 1$ then $r_\ell r_{\ell-1} z_{\ell,\ell} = r_\ell r_{\ell-1} r_\ell = z_{k,\ell}$. For $m \geq 2$ and assuming the hypothesis valid for values at most $m - 1$, we have that

$$\begin{aligned} r_\ell r_{\ell-1} \cdots r_k z_{k+1,\ell} &= r_\ell \cdots r_k z_{k+2,\ell} r_{k+1} \cdots r_\ell \\ &= r_\ell \cdots r_{k+1} z_{k+2,\ell} r_k \cdots r_\ell \\ &= z_{k+1,\ell} r_k \cdots r_\ell = z_{k,\ell}. \end{aligned}$$

The second item follows from the first one through a backward induction process. Indeed, it holds for $k = \ell$; and if $k < \ell$ then

$$z_{k,\ell}^2 = (z_{k+1,\ell} r_k r_{k+1} \cdots r_\ell)(r_\ell r_{\ell-1} \cdots r_k z_{k+1,\ell}) = z_{k+1,\ell}^2 = \mathrm{id}.$$

The third item also follows from the first one and the definition of $z_{k,\ell}$. ■

Exercise 3.2.66 Let $\mathcal{P}$ be an n-polytope for some $n \geq 3$, and let $z_{k,\ell}$ be as in (3.34). Show that the number of factors r_i in the definition of $z_{2,\ell}$ is even if and only if $n \equiv a \pmod 4$ for $a \in \{1, 2\}$.

Next we relate the connections $z_{k,\ell}$ with the Petrie walk π_Φ.

Lemma 3.2.67 *Let $\mathcal{P}$, Φ and π_Φ be as in Lemma 3.2.59, let $z_{k,\ell}$ be as in* (3.34)*, and let $\Psi = r_0 z_{2,n-1} \Phi$. Then the vertex of the flag $(r_{n-1} \cdots r_0)^i \Phi$ is the vertex of the flag $(r_{n-1} \cdots r_0)^{1-i} \Psi$, and the edge of the flag $(r_{n-1} \cdots r_0)^i \Phi$ is the edge of the flag $(r_{n-1} \cdots r_0)^{-i} \Psi$.*

Proof The flag $(r_{n-1} \cdots r_0)^j \Psi$ can be rewritten as $(r_{n-1} \cdots r_0)^j r_0 z_{2,n-1} \Phi$. We may use the first item of Lemma 3.2.65 for $k \in \{0, 1\}$ and $\ell = n - 1$ together with the third item of the same lemma to manipulate the connection applied to Φ as follows:

$$\begin{aligned} (r_{n-1} \cdots r_0)^j r_0 z_{2,n-1} &= (r_{n-1} \cdots r_0)^{j-1} r_{n-1} \cdots r_1 z_{2,n-1} \\ &= (r_{n-1} \cdots r_0)^{j-1} z_{1,n-1} \\ &= z_{1,n-1} (r_0 \cdots r_{n-1})^{j-1} = z_{2,n-1} r_0 (r_0 \cdots r_{n-1})^j \\ &= r_0 z_{2,n-1} (r_{n-1} \cdots r_0)^{-j}. \end{aligned}$$

We have just shown that $(r_{n-1} \cdots r_0)^j \Psi = r_0 z_{2,n-1} (r_{n-1} \cdots r_0)^{-j} \Phi$. By replacing j by $-i$, the last expression shows that the edges of $(r_{n-1} \cdots r_0)^i \Phi$ and $(r_{n-1} \cdots r_0)^{-i} \Psi$ are the same, since $r_0 z_{2,n-1}$ has no factor r_1.

Now we rewrite $r_0 z_{2,n-1}(r_{n-1}\cdots r_0)^{-j}$ as $z_{2,n-1}r_1\cdots r_{n-1}(r_{n-1}\cdots r_0)^{-j+1}$ to conclude that $(r_{n-1}\cdots r_0)^j\Psi = z_{1,n-1}(r_{n-1}\cdots r_0)^{1-j}\Phi$. The replacement of j by $1-i$ shows that the vertices of $(r_{n-1}\cdots r_0)^i\Phi$ and $(r_{n-1}\cdots r_0)^{1-i}\Psi$ are the same, since $z_{1,n-1}$ has no factor r_0. ■

The main idea of Lemmas 3.2.59 and 3.2.67 is that the flags Φ and $r_0 z_{2,n-1}\Phi$ behave like 0-adjacent flags on π_Φ while $r_0 z_{2,n-1}\Phi$ and $r_{n-1}\cdots r_0\Phi$ behave as 1-adjacent flags on π_Φ. We may identify the flags of π_Φ with the orbit of $\{\Phi, r_0 z_{2,n-1}\Phi\}$ under $\langle r_{n-1}\cdots r_0\rangle$. If we denote by w_0 and w_1 the standard generators of $\text{Conn}(\pi_\Phi)$ then w_0 and w_1 correspond to $r_0 z_{2,n-1}$ and

$$(r_{n-1}\cdots r_0)(r_0 z_{2,n-1}) = r_{n-1}\cdots r_1 z_{2,n-1} = z_{1,n-1},$$

respectively (see the first item of Lemma 3.2.65).

Exercise 3.2.68 Let $\mathcal{P}$ be the tessellation of $\mathbb{E}^3$ by cubes with vertex set in $\mathbb{Z}^3$. Let Φ be the flag consisting of the point $F_0 = (0,0,0)$, the edge F_1 from F_0 to $(1,0,0)$, the face F_2 containing F_1 and the point $(0,1,0)$, and the cube F_3 containing F_2 and the point $(0,0,1)$.

- Find the Petrie path π_Φ as defined in Lemma 3.2.59 with an explicit list of vertices, edges, squares and cubes.
- Let $\Psi_1 = r_0 z_{2,3}\Phi$ and $\Psi_2 = z_{1,3}\Psi$. Corroborate that the Petrie paths π_{Ψ_1} and π_{Ψ_2} have the same sets of vertices, edges, squares and cubes as π_Φ.

We know from Proposition 2.2.6 that if $\mathcal{P}$ is regular then $\Gamma(\mathcal{P})$ acts transitively on the set of Petrie paths of $\mathcal{P}$. We are now ready to show that this is sometimes true for chiral polytopes as well.

Proposition 3.2.69 *Let $\mathcal{P}$ be a chiral n-polytope for some n not divisible by 4. Then $\Gamma(\mathcal{P})$ acts transitively on the set of Petrie paths of $\mathcal{P}$.*

Proof Let π_1 and π_2 be two Petrie paths of $\mathcal{P}$. From Lemma 3.2.59 we know that there exist flags Φ and Ψ such that $\pi_1 = \pi_\Phi$ and $\pi_2 = \pi_\Psi$. If $\Psi = \Phi\gamma$ for some $\gamma \in \Gamma(\mathcal{P})$ then we can conclude from Lemma 3.2.59 that $\pi_2 = \pi_1\gamma$. In the remainder of the proof assume that Φ and Ψ are in distinct orbits.

Assume first that n is odd. Then Ψ and $r_{n-1}\cdots r_0\Psi$ belong to distinct flag orbits, and therefore Φ is in the orbit of $r_{n-1}\cdots r_0\Psi$. From Lemma 3.2.59 we know that $\pi_2 = \pi_{r_{n-1}\cdots r_0\Psi}$, and we may conclude with the argument in the previous paragraph that $\pi_2 = \pi_1\gamma$ for some $\gamma \in \Gamma(\mathcal{P})$.

Now, if $n \equiv 2$ (modulo 4) then $z_{2,n-1}$ has an even number of factors r_i (see Exercise 3.2.66). It follows that Φ and $r_0 z_{2,n-1}\Psi$ are in the same orbit, and on the other hand, we know from Lemma 3.2.67 that $\pi_\Psi = \pi_{r_0 z_{2,n-1}\Psi}$. Once again, the argument in the first paragraph of the proof shows that $\pi_2 = \pi_1 \gamma$ for some $\gamma \in \Gamma(\mathcal{P})$. ■

Corollary 3.2.70 *Let $\mathcal{P}$ be a chiral polytope of rank n not divisible by 4. Then the lengths of the left Petrie paths and the right Petrie paths are equal.*

The conclusion of Proposition 3.2.69 needs not hold if n is divisible by 4. Indeed, some rotary 4-polytopes discussed in [54, 230, 248] were verified to be chiral by noting that they have Petrie paths of distinct sizes. The integers that are lengths of Petrie paths of chiral 4-polytopes have not been studied in depth. We next state problems 11 and 12 in [220].

Problem 12 Determine the pairs $\{a,b\} \subseteq \mathbb{Z} \cup \{\infty\}$ such that there exists a chiral 4-polytope with Petrie paths of sizes a and b.

Problem 13 Determine all $a \in \mathbb{Z} \cup \{\infty\}$ such that there is a chiral 4-polytope whose Petrie paths all have size a.

Given our current little knowledge about chiral polytopes of ranks higher than 4, we state a more general problem for ranks $n = 4m$ for $m > 1$ (see problem 10 in [220]).

Problem 14 For each $m \geq 2$ find an example of a chiral $4m$-polytope whose Petrie paths have distinct sizes, and an example of a chiral $4m$-polytope whose Petrie paths are all of the same size. Else, show the corresponding non-existence.

The results about Petrie paths in this section are useful not only for chiral polytopes, but can be applied to more general classes of polytopes.

Exercise 3.2.71 Determine sufficient conditions for $n \geq 3$ and for $I \subset \{0, \ldots, n-1\}$ that guarantee that if $\mathcal{P}$ is a two-orbit n-polytope in class 2_I then $\Gamma(\mathcal{P})$ acts transitively on the Petrie paths.

3.3 Chirality Group

Unlike the automorphism and the connection groups, the chirality group cannot be defined for every polytope. It was originally introduced in [16] in the context of orientable rotary maps and hypermaps (*hypermaps* are generalisations of

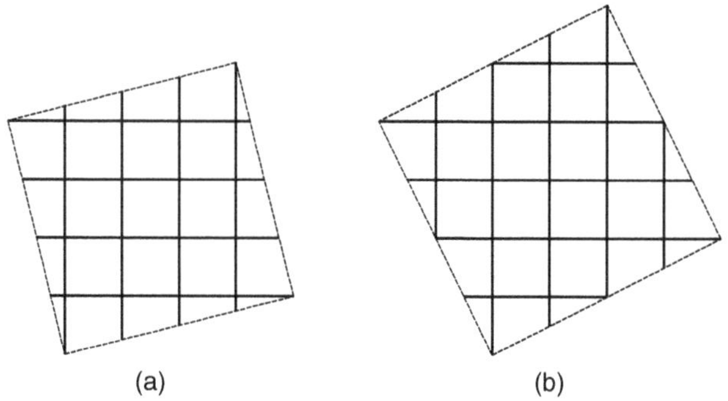

Figure 3.23 Toroidal polyhedra (a) $\{4,4\}_{(4,1)}$ and (b) $\{4,4\}_{(4,2)}$.

maps on surfaces where edges need not belong to only two faces and need not contain only two vertices, see [47]) with the idea of providing every such object of a number and a group that indicate how far it is from being regular. The chirality groups of the maps on surfaces of low genera were found in [15], and later the notion was extended to rotary orientable abstract polytopes in [78].

Let us consider the toroidal polyhedra $\{4,4\}_{(4,1)}$ and $\{4,4\}_{(4,2)}$, illustrated in Figure 3.23. They are both chiral, and without an appropriate definition, at first sight they seem to be equally chiral (see Theorem 2.4.8). However, the toroidal polyhedron $\{4,4\}_{(4,2)}$ covers the regular polyhedron $\{4,4\}_{(2,0)}$, and therefore it can be understood as the union of five copies of this regular map as shown in Figure 3.24(b). On the other hand, the largest regular structure that is a quotient of $\{4,4\}_{(4,1)}$ is the toroidal non-polytopal map consistint of only one square with its opposite edges identified orientably (see Figure 3.24(a)). In other words the minimum numbers of regular pieces of rank 3 in which we may decompose $\{4,4\}_{(4,2)}$ and $\{4,4\}_{(4,1)}$ are 5 and 17, respectively. Hence, we may associate the polyhedra $\{4,4\}_{(4,2)}$ and $\{4,4\}_{(4,1)}$ to the indices 5 and 17, in the understanding that the larger the index, the less regular the polytope is. It is with this idea in mind that $\{4,4\}_{(4,1)}$ was regarded in [16] as 'more chiral' than $\{4,4\}_{(4,2)}$.

Exercise 3.3.1 Show that the maniplexes $\{4,4\}_{(1,0)}$ and $\{4,4\}_{(2,0)}$ are the largest regular quotients of the polyhedra $\{4,4\}_{(4,1)}$ and $\{4,4\}_{(4,2)}$, respectively, by using the techniques in Sections 2.4.2 and 3.2.2.

The previous discussion suggests the following definition.

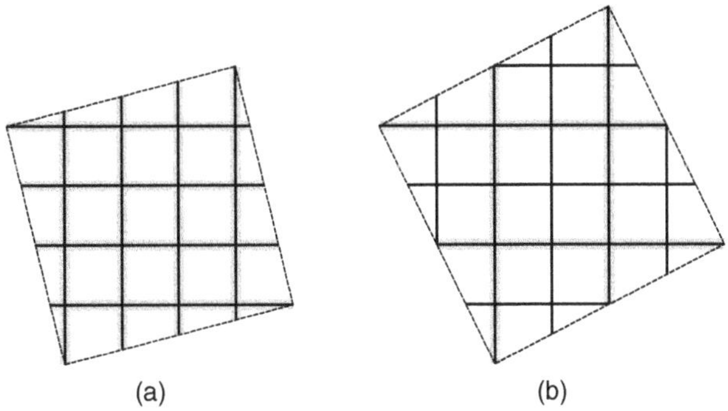

Figure 3.24 Regular quotients of the toroidal polyhedra (a) $\{4,4\}_{(4,1)}$ and (b) $\{4,4\}_{(4,2)}$.

Definition 3.3.2 The *irregularity index by quotients* of a polytope $\mathcal{P}$ is the minimum degree of the coverings from $\mathcal{P}$ to its regular quotients.

Remark 3.3.3 As illustrated in Figure 3.24(a), the regular quotient of $\mathcal{P}$ in Definition 3.3.2 may not be a polytope. In that case it is isomorphic to the map $\{4,4\}_{(1,0)}$, whose edges have only one vertex and therefore it does not satisfy the diamond condition. This quotient will be formalised in Lemma 3.3.6. We shall see later that this quotient may even fail to be a maniplex.

It is natural to investigate not only regular quotients of $\mathcal{P}$, but also regular covers of $\mathcal{P}$. The polyhedra $\{4,4\}_{(4,1)}$ and $\{4,4\}_{(4,2)}$ are covered by $\{4,4\}_{(17,0)}$ and $\{4,4\}_{(10,0)}$, respectively, as illustrated in Figure 3.25. Furthermore, these are their smallest regular covers.

Exercise 3.3.4 Show that the maniplexes $\{4,4\}_{(17,0)}$ and $\{4,4\}_{(10,0)}$ are the smallest regular polyhedra with type $\{4,4\}$ that cover the polyhedra $\{4,4\}_{(4,1)}$ and $\{4,4\}_{(4,2)}$, respectively, by using the techniques in Sections 2.4.2 and 3.2.2.

Definition 3.3.5 The *irregularity index by covers* of a polytope $\mathcal{P}$ is the minimum degree of a covering from a regular maniplex to $\mathcal{P}$.

More details about the smallest regular maniplexes that cover a given polytope are presented in Section 4.3.

When computing the irregularity index by covers of $\{4,4\}_{(4,1)}$ and $\{4,4\}_{(4,2)}$ we observe that again these numbers are 17 and 5, respectively. This suggests that the irregularity indexes by quotients and by covers of a given polytope are

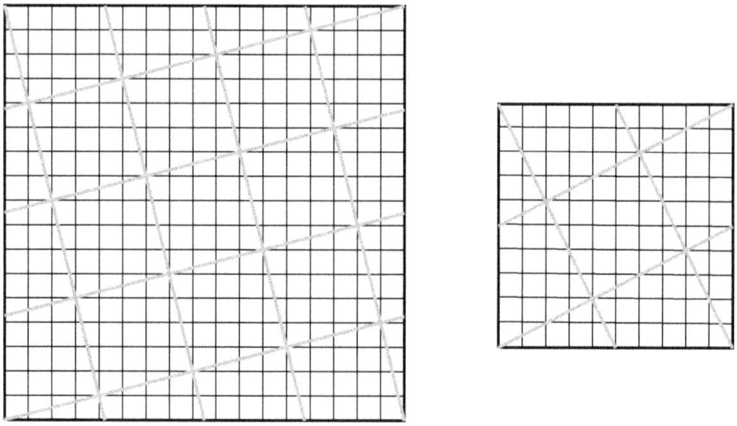

(a) (b)

Figure 3.25 Regular covers of the toroidal polyhedra (a) $\{4,4\}_{(4,1)}$ and (b) $\{4,4\}_{(4,2)}$.

equal. Next we shall show that this is always the case, and also that for a fixed polytope $\mathcal{P}$, these two indices are the cardinality of a certain relevant group. In order to do this we make use of Example 3.2.33 to guarantee the existence of a quotient presentation of $\mathcal{P}$ with respect to the universal polytope $[\infty, \dots, \infty]$.

Lemma 3.3.6 *Let $\mathcal{P}$ be a rotary n-polytope, let $\rho_0, \dots, \rho_{n-1}$ denote the standard generators of $[\infty, \dots, \infty]$, and let $N \leq [\infty, \dots, \infty]$ such that $\{\infty, \dots, \infty\}/N$ is a quotient presentation of $\mathcal{P}$ in terms of the universal polytope $\{\infty, \dots, \infty\}$. Then*

1. *the largest regular structure that is a quotient of $\mathcal{P}$ can be described through the quotient presentation $\{\infty, \dots, \infty\}/(N \cdot \rho_0 N \rho_0)$;*
2. *the smallest regular maniplex that covers $\mathcal{P}$ has $\{\infty, \dots, \infty\}/(N \cap \rho_0 N \rho_0)$ as a quotient presentation.*

Proof Recall from Proposition 3.1.30 that a structure $\{\infty, \dots, \infty\}/M$ is regular (in the sense that the coset representation is flag-transitive) if and only if $M \triangleleft [\infty, \dots, \infty]$. It follows that if $\mathcal{P}$ is regular then $N = N \cdot \rho_0 N \rho_0 = N \cap \rho_0 N \rho_0$, and the two items hold, since $\mathcal{P}$ is at the same time the largest regular quotient and the smallest regular cover of itself.

Example 3.2.33 and Proposition 3.2.39 imply that every polytope is a quotient of $\{\infty, \dots, \infty\}$. Furthermore, Proposition 3.2.51 implies that if $\mathcal{P}$ covers $\mathcal{Q}$ (resp. is covered by $\mathcal{Q}$) then there exists a quotient presentation $\{\infty, \dots, \infty\}/M$ of $\mathcal{Q}$ such that $N \leq M$ (resp. $M \leq N$).

Now, if $\mathcal{P}$ is chiral, we know from Corollary 3.2.58 that $N \triangleleft [\infty, \ldots, \infty]^+$, and from Proposition 3.2.46, Corollary 3.2.47 and Lemma 3.2.48 that the only quotient presentations of $\mathcal{P}$ with respect to $\{\infty, \ldots, \infty\}$ are $\{\infty, \ldots, \infty\}/N$ and $\{\infty, \ldots, \infty\}/\rho_0 N \rho_0$.

If $\mathcal{Q}$ is a regular structure covered by $\mathcal{P}$ then there is a quotient presentation $\{\infty, \ldots, \infty\}/M$ of $\mathcal{Q}$ such that $M \triangleleft [\infty, \ldots, \infty]$ and either $N \leq M$ or $\rho_0 N \rho_0 \leq M$. In any case, a normal subgroup of $[\infty, \ldots, \infty]$ that contains N (or $\rho_0 N \rho_0$) must contain all its conjugates as well, implying that $N \cdot \rho_0 N \rho_0 \leq M$. To conclude the first item it suffices to show that $N \cdot \rho_0 N \rho_0 \triangleleft [\infty, \ldots, \infty]$. To this end we first note that the fact that N has only two conjugates in $[\infty, \ldots, \infty]$ implies that $\rho_0 N \rho_0 \triangleleft [\infty, \ldots, \infty]^+$. This in turn implies that for any given $n \in [\infty, \ldots, \infty]^+$, we have that

$$n^{-1} N \rho_0 N \rho_0 n = N \cdot \rho_0 N \rho_0. \tag{3.35}$$

Furthermore, for any $n_1, n_2 \in N$,

$$\rho_0(n_1 \rho_0 n_2 \rho_0)\rho_0 = n_2\big[n_2^{-1}(\rho_0 n_1 \rho_0) n_2\big] \in N \cdot \rho_0 N \rho_0, \tag{3.36}$$

and the normality of $N \cdot \rho_0 N \rho_0$ in $[\infty, \ldots, \infty]$ follows.

Similarly, if $\mathcal{Q}$ is a regular maniplex that covers $\mathcal{P}$ then it has a quotient presentation $\{\infty, \ldots, \infty\}/M$ such that $M \triangleleft [\infty, \ldots, \infty]$ and either $M \leq N$ or $M \leq \rho_0 N \rho_0$. But any normal subgroup contained in N (or $\rho_0 N \rho_0$) must be contained in all its conjugates. The second item follows from the fact that $N \cap \rho_0 N \rho_0 \triangleleft [\infty, \ldots, \infty]$. ∎

Here we recall that both N and $\rho_0 N \rho_0$ are normal in $[\infty, \ldots, \infty]^+$. This allows us to use the second isomorphism theorem for groups to observe that

$$(N \cdot \rho_0 N \rho_0)/N \cong \rho_0 N \rho_0/(N \cap \rho_0 N \rho_0) \cong N/(N \cap \rho_0 N \rho_0).$$

This gives us the final ingredient to define the chirality index and the chirality group of $\mathcal{P}$.

Definition 3.3.7 Let $[\infty, \ldots, \infty] = \langle \rho_0, \ldots, \rho_{n-1} \rangle$ and let $\{\infty, \ldots, \infty\}/N$ be a quotient presentation of a rotary polytope $\mathcal{P}$. We define the *chirality group* of $\mathcal{P}$ as

$$\chi(\mathcal{P}) := (N \cdot \rho_0 N \rho_0)/N \cong N/(N \cap \rho_0 N \rho_0),$$

and its *chirality index* $\kappa(\mathcal{P})$ as the order of $\chi(\mathcal{P})$.

Remark 3.3.8 In Definition 3.3.7 and the discussion preceding it we expressed $\mathcal{P}$ through the quotient presentation $\{\infty, \ldots, \infty\}/N$. The polytope $\{\infty, \ldots, \infty\}$ plays no role, and can be substituted by $\mathcal{P}(\Gamma)$ for any sggi Γ having a flag action on $\mathcal{P}$. Indeed, any such Γ can be expressed as

$[\infty, \ldots, \infty]/M$ for some $M \triangleleft [\infty, \ldots, \infty]$ (Proposition 3.1.30 applies for the coset representation of any sggi, and not only for polytopes). Furthermore, $M \leq N$. We may choose $N' = N/M$ and then $\mathcal{P}$ admits the quotient presentation $\mathcal{P}(\Gamma)/N'$. If $\Gamma = \langle \rho'_0, \ldots, \rho'_{n-1} \rangle$ then

$$(N' \cdot \rho'_0 N' \rho'_0)/N' = [(N/M) \cdot \rho'_0 (N/M) \rho'_0]/(N/M) \cong (N \cdot \rho_0 N \rho_0)/N.$$

Example 3.3.9 Let $\mathcal{P} = \{4,4\}_{(3,1)}$ and let $\mathcal{R}$ be the tiling $\{4,4\}$ of $\mathbb{E}^2$ by unit squares with vertex set in $\mathbb{Z}^2$. We know from Section 2.4.1 or Corollary 3.1.81 that $\{4,4\}_{(3,1)}$ is a quotient of the tessellation $\{4,4\}$ by squares of $\mathbb{E}^2$. The stabiliser $N = \mathrm{Stab}_{\mathrm{Conn}(\{4,4\})}(\Phi)$ of a flag Φ of $\{4,4\}_{(3,1)}$ corresponds to the set of connections mapping the corresponding flag $\hat{\Phi}$ of $\{4,4\}$ to all other preimages of Φ under the covering (see Remark 3.2.45). Since $\{4,4\}$ is regular, the choice of $\hat{\Phi}$ is arbitrary and we choose the grey flag in Figure 3.26. Furthermore, $\mathrm{Conn}(\{4,4\}) \cong \Gamma(\{4,4\})$, and so there is a unique connection mapping the flag $\hat{\Phi}$ to each of the white flags in Figure 3.26(a). Then $\rho_0 N \rho_0$ corresponds to the set of connections of $\{4,4\}$ mapping the gray flag to the white flags in Figure 3.26(b).

Direct computations show that the groups are

$$\begin{aligned} N &= \left\langle (\rho_0\rho_1\rho_2\rho_1)^3 \rho_1\rho_0\rho_1\rho_2, (\rho_0\rho_1\rho_2\rho_1)^{-1} (\rho_1\rho_0\rho_1\rho_2)^3 \right\rangle \\ &= \left\langle (\sigma_1\sigma_2^{-1})^3 \sigma_1^{-1}\sigma_2, (\sigma_1\sigma_2^{-1})^{-1} (\sigma_1^{-1}\sigma_2)^3 \right\rangle. \end{aligned}$$

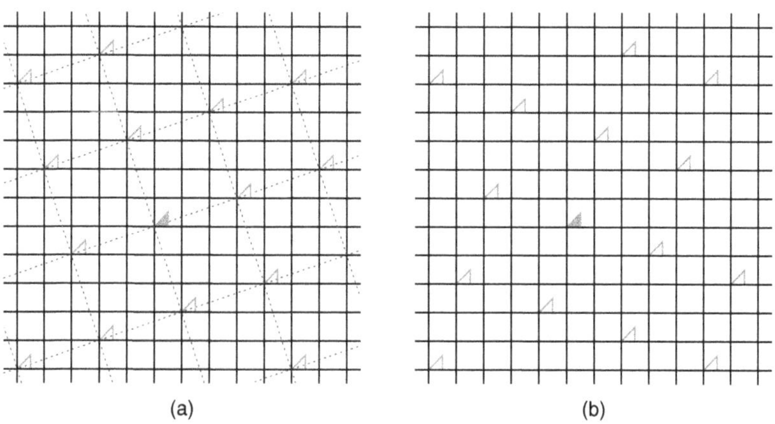

Figure 3.26 Graphic illustration of the groups (a) N and (b) $\rho_0 N \rho_0$.

$$\rho_0 N \rho_0 = \left\langle (\rho_0\rho_1\rho_2\rho_1)^3(\rho_1\rho_0\rho_1\rho_2)^{-1}, \rho_0\rho_1\rho_2\rho_1(\rho_1\rho_0\rho_1\rho_2)^3 \right\rangle$$
$$= \left\langle (\sigma_1\sigma_2^{-1})^3(\sigma_1^{-1}\sigma_2)^{-1}, \sigma_1\sigma_2^{-1}(\sigma_1^{-1}\sigma_2)^3 \right\rangle.$$

We know from Lemma 3.1.80 that $\langle \sigma_1\sigma_2^{-1}, \sigma_1^{-1}\sigma_2 \rangle \cong \mathbb{Z}^2$, and this allows us to use standard computations to determine that

$$N \cap \rho_0 N \rho_0 \cong \langle (3,1), (-1,3) \rangle \cap \langle (3,-1), (1,3) \rangle = \langle (5,5), (-5,5) \rangle,$$

and therefore

$$\chi(\{4,4\}_{(3,1)}) = N/(N \cap \rho_0 N \rho_0) \cong \langle (3,1), (-1,3) \rangle / \langle (5,5), (-5,5) \rangle \cong \mathbb{Z}^5.$$

We conclude that $\kappa(\{4,4\}_{(3,1)}) = 5$.

In view of Remark 3.3.8 we may compute the core of the stabiliser in an sggi Γ of a flag of a rotary polytope, regardless of the group Γ.

Lemma 3.3.10 *Let $\Gamma = \langle \rho_0, \ldots, \rho_{n-1} \rangle$ be an orientable sggi and let $N \leq \Gamma$ be such that* $\mathrm{Norm}(N) = \Gamma^+$. *Then* $\mathrm{Core}_\Gamma(N) = N \cap \rho_0 N \rho_0$.

Proof This follows from the arguments in the proof of Lemma 3.3.6, since (3.35) and (3.36) hold for any $\Gamma = \langle \rho_0, \ldots, \rho_{n-1} \rangle$ and any $N \triangleleft \Gamma^+$. ■

Next we show that this definition of the chirality index is what we would expect it to be.

Proposition 3.3.11 *The irregularity indices by quotients and by covers of any rotary polytope $\mathcal{P}$ are both equal to the chirality index of $\mathcal{P}$.*

Proof The three indices are 1 if $\mathcal{P}$ is regular. For the reminder of the proof assume that $\mathcal{P}$ is chiral.

Let $N \leq [\infty, \ldots, \infty]$ such that $\mathcal{P}$ has quotient presentation $\{\infty, \ldots, \infty\}/N$. Then $\mathrm{Norm}(N) = [\infty, \ldots, \infty]^+$ and $\Gamma(\mathcal{P}) \cong [\infty, \ldots, \infty]^+/N$ (see Theorem 3.2.57). From Lemma 3.3.6 we know that the largest regular structure $\mathcal{Q}$ that is a quotient of $\mathcal{P}$ and the smallest regular maniplex $\mathcal{R}$ that covers $\mathcal{P}$ have quotient presentations $\{\infty, \ldots, \infty\}/(N \cdot \rho_0 N \rho_0)$ and $\{\infty, \ldots, \infty\}/(N \cap \rho_0 N \rho_0)$, respectively. It follows from Proposition 3.2.51 that the irregularity indices by quotients and by covers are the indices of N in $N \cdot (\rho_0 N \rho_0)$ and of $N \cap (\rho_0 N \rho_0)$ in N, respectively.

On the other hand, $\chi(\mathcal{P})$ is the index of N in $N \cdot \rho_0 N \rho_0$, or equivalently the index of $N \cap \rho_0 N \rho_0$ in N. ■

Corollary 3.3.12 *A rotary polytope $\mathcal{P}$ is regular if and only if $\chi(\mathcal{P}) = \{\mathrm{id}\}$.*

Example 3.3.13 In order to find $\chi(\{4,4\}_{(2,1)})$ it is not necessary to determine explicitly the stabiliser N of a flag under some flag action. Instead, we shall show below that it is a quotient of $\{4,4\}_{(5,0)}$. Since $\{4,4\}_{(5,0)}$ has 25 squares and $\{4,4\}_{(2,1)}$ has 5 squares, the degree of the covering will be 5. Next we use that the only positive divisors of 5 are 1 and 5 to conclude that the irregularity index by covers of $\{4,4\}_{(2,1)}$ is one of these two numbers. But $\{4,4\}_{(2,1)}$ is not regular, forcing the irregularity index by covers, and hence the chirality index, of $\{4,4\}_{(2,1)}$ to be 5. Being $\mathbb{Z}_5$ the only group of order 5, the only possibility for $\chi(\{4,4\}_{(2,1)})$ is to be isomorphic to $\mathbb{Z}_5$.

Now we show that $\{4,4\}_{(2,1)}$ is a quotient of $\{4,4\}_{(5,0)}$. From Lemma 3.1.88 we know that

$$\Gamma(\{4,4\}_{(2,1)}) = \left\langle \sigma_1, \sigma_2 \colon \sigma_1^4 = \sigma_2^4 = (\sigma_1\sigma_2)^2 = (\sigma_1\sigma_2^{-1})^2\sigma_1^{-1}\sigma_2 = \mathrm{id} \right\rangle,$$
$$\Gamma(\{4,4\}_{(5,0)}) = \left\langle \hat{\sigma_1}, \hat{\sigma_2} \colon \hat{\sigma_1}^4 = \hat{\sigma_2}^4 = (\hat{\sigma_1}\hat{\sigma_2})^2 = (\hat{\sigma_1}\hat{\sigma_2}^{-1})^5 = \mathrm{id} \right\rangle.$$

In order to use Proposition 3.1.67 we only need to show that $(\sigma_1\sigma_2^{-1})^5 = \mathrm{id}$ by using the defining relations of $\Gamma(\{4,4\}_{(2,1)})$; in particular, $(\sigma_1\sigma_2^{-1})^2\sigma_1^{-1}\sigma_2$ and all its conjugates equal id. We also shall use that $\sigma_1\sigma_2^{-1}$ and $\sigma_1^{-1}\sigma_2$ commute (see Lemma 3.1.80).

$$\begin{aligned}
\mathrm{id} &= [(\sigma_1\sigma_2^{-1})^2\sigma_1^{-1}\sigma_2]^2\sigma_1^{-1}[(\sigma_1\sigma_2^{-1})^2\sigma_1^{-1}\sigma_2]\sigma_1\\
&= (\sigma_1\sigma_2^{-1})^4(\sigma_1^{-1}\sigma_2)^2\sigma_1^{-1}(\sigma_1\sigma_2^{-1})^2\sigma_1\sigma_1^{-1}(\sigma_1^{-1}\sigma_2)\sigma_1\\
&= (\sigma_1\sigma_2^{-1})^4(\sigma_1^{-1}\sigma_2)^2(\sigma_2^{-1}\sigma_1)^2\sigma_1^{-2}\sigma_1^{-1}\sigma_2^{-1}\\
&= (\sigma_1\sigma_2^{-1})^4(\sigma_1^{-1}\sigma_2)^2(\sigma_1^{-1}\sigma_2)^{-2}\sigma_1^{-3}\sigma_2^{-1} = (\sigma_1\sigma_2^{-1})^5.
\end{aligned}$$

Despite the fact that the chirality group is defined in terms of a flag action, and hence in terms of the connection group, it is also related to the automorphism group.

Proposition 3.3.14 *Let $\mathcal{P}$ be a chiral polytope. Then $\chi(\mathcal{P})$ is isomorphic to a normal subgroup of $\Gamma(\mathcal{P})$.*

Proof Let $N \leq [\infty, \ldots, \infty]$ such that $\{\infty, \ldots, \infty\}/N$ is a quotient presentation for $\mathcal{P}$. Then $N \triangleleft [\infty, \ldots, \infty]^+$ and $N = \mathrm{Stab}_{[\infty,\ldots,\infty]}(\Phi)$ for some $\Phi \in \mathcal{F}l(\mathcal{P})$.

Recall from Theorem 3.2.57 and Corollary 3.2.58 that

$$\Gamma(\mathcal{P}) \cong \mathrm{Norm}_{[\infty,\ldots,\infty]}(N)/N = [\infty, \ldots, \infty]^+/N.$$

Then

$$\chi(\mathcal{P}) = (N \cdot \rho_0 N \rho_0)/N \leq [\infty, \ldots, \infty]^+/N \cong \Gamma(\mathcal{P}).$$

It follows from Lemma 3.3.10 that $\chi(\mathcal{P})$ is in fact a normal subgroup of $[\infty, \ldots, \infty]^+/N$, and the proposition follows. ∎

Actually, the chirality group can be defined in terms of the automorphism group; the chirality group was defined in [78] in that way.

Proposition 3.3.15 *Let $\mathcal{P}$ be a rotary polytope with $\Gamma^+(\mathcal{P}) = \langle \sigma_1, \ldots, \sigma_{n-1} \rangle$, and let $\mathcal{N} \leq [\infty, \ldots, \infty]^+$ such that $\Gamma^+(\mathcal{P}) \cong [\infty, \ldots, \infty]^+/\mathcal{N}$. Then*

$$\chi(\mathcal{P}) \cong (\mathcal{N} \cdot \rho_0 \mathcal{N} \rho_0)/\mathcal{N} \cong \rho_0 \mathcal{N} \rho_0/(\mathcal{N} \cap \rho_0 \mathcal{N} \rho_0) \cong \mathcal{N}/(\mathcal{N} \cap \rho_0 \mathcal{N} \rho_0),$$

where ρ_0 is the first generator of $[\infty, \ldots, \infty]$.

Proof Let Φ be the base flag of $(\mathcal{P})$ and let $N = \mathrm{Stab}_{[\infty,\ldots,\infty]}(\Phi)$. We note that we are considering two distinct actions of the group $[\infty, \ldots, \infty]^+$ on $\mathcal{F}l(\mathcal{P})$, so that in this context $\sigma_i \Phi$ is defined as $r_0 r_1 \Phi$. Then Remark 3.2.22 tells us that $N = \mathcal{N}$. It follows that

$$\begin{aligned}\chi(\mathcal{P}) &= (N \cdot \rho_0 N \rho_0)/N \cong \rho_0 N \rho_0/(N \cap \rho_0 N \rho_0) \cong N/(N \cap \rho_0 N \rho_0) \\ &\cong (\mathcal{N} \cdot \rho_0 \mathcal{N} \rho_0)/\mathcal{N} \cong \rho_0 \mathcal{N} \rho_0/(\mathcal{N} \cap \rho_0 \mathcal{N} \rho_0) \cong \mathcal{N}/(\mathcal{N} \cap \rho_0 \mathcal{N} \rho_0).\end{aligned}$$

∎

Exercise 3.3.16 Let $\mathcal{P} = \{4,4\}_{(a,b)}$ with $m := a^2 + b^2$ a prime number. Show that $\chi(\mathcal{P}) \cong \mathbb{Z}_m$.

From Proposition 3.3.15 we deduce the following method to determine the chirality group from a presentation of the automorphism group. This method was used in [15] to determine the chirality groups of all chiral polyhedra that can be seen as maps on surfaces of genus at most 15, including all chiral toroidal polyhedra.

Corollary 3.3.17 *Let $\mathcal{P}$ be a chiral polytope, let $\sigma_1, \ldots, \sigma_{n-1}$ be the standard generators of $[\infty, \ldots, \infty]^+$, let $\mathcal{N} \leq [\infty, \ldots, \infty]^+$ such that $\Gamma(\mathcal{P}) = [\infty, \ldots, \infty]^+/\mathcal{N}$, and let $R = \{h_i(\sigma_1, \ldots, \sigma_{n-1})\}_{i \in I}$ be such that $\mathcal{N}$ is the normal closure of R. Then $\chi(\mathcal{P})$ is isomorphic to the normal closure in $[\infty, \ldots, \infty]^+/\mathcal{N}$ of the set $R' = \{h_i(\sigma_1^{-1}, \sigma_1^2\sigma_2, \sigma_3, \sigma_4, \ldots, \sigma_{n-1}) \cdot \mathcal{N}\}_{i \in I}$.*

Proof From Proposition 3.3.15 we know that $\chi(\mathcal{P}) \cong (\mathcal{N} \cdot \rho_0 \mathcal{N} \rho_0)/\mathcal{N}$. By taking $i = 0$ in Proposition 3.1.65 we observe that $\rho_0 \mathcal{N} \rho_0$ is the normal closure of $\{h_i(\sigma_1^{-1}, \sigma_1^2\sigma_2, \sigma_3, \sigma_4, \ldots, \sigma_{n-1})\}_{i \in I}$, and the corollary follows. ∎

Example 3.3.18 Lemma 3.1.88 tells us that

$$\Gamma(\{4,4\}_{(5,3)}) = \left\langle \sigma_1, \sigma_2 \colon \sigma_1^4 = \sigma_2^4 = (\sigma_1\sigma_2)^2 = (\sigma_1\sigma_2^{-1})^5(\sigma_1^{-1}\sigma_2)^3 = \mathrm{id} \right\rangle.$$

Furthermore,

$$\begin{aligned} \mathrm{id} &= \sigma_1^{-1}[(\sigma_1\sigma_2^{-1})^5(\sigma_1^{-1}\sigma_2)^3]\sigma_1 = (\sigma_2^{-1}\sigma_1)^5(\sigma_1^{-2}\sigma_2\sigma_1)^3 \\ &= (\sigma_1^{-1}\sigma_2)^{-5}(\sigma_1\sigma_2^{-1})^3, \end{aligned}$$

and hence we may use Lemma 3.1.80 to imply that

$$\mathrm{id} = [(\sigma_1\sigma_2^{-1})^5(\sigma_1^{-1}\sigma_2)^3][(\sigma_1^{-1}\sigma_2)^{-5}(\sigma_1\sigma_2^{-1})^3]^4 = (\sigma_1\sigma_2^{-1})^{17}(\sigma_1^{-1}\sigma_2)^{-17}.$$

We continue by noting that

$$\begin{aligned} \mathrm{id} &= (\sigma_1\sigma_2^{-1})^{17}(\sigma_1^{-1}\sigma_2)^{-17}[\sigma_1^{-1}(\sigma_1\sigma_2^{-1})^{17}(\sigma_1^{-1}\sigma_2)^{-17}\sigma_1] \\ &= (\sigma_1\sigma_2^{-1})^{17}(\sigma_1^{-1}\sigma_2)^{-17}(\sigma_2^{-1}\sigma_1)^{17}(\sigma_1^{-2}\sigma_2\sigma_1)^{-17} \\ &= (\sigma_1\sigma_2^{-1})^{17}(\sigma_1^{-1}\sigma_2)^{-17}(\sigma_1^{-1}\sigma_2)^{-17}(\sigma_1\sigma_2^{-1})^{-17} = (\sigma_1^{-1}\sigma_2)^{34}, \end{aligned}$$

which tells us that the order of $\sigma_1^{-1}\sigma_2$ is either 1, 2, 17 or 34.

If $[\infty,\infty] = \langle \hat{\sigma}_1, \hat{\sigma}_2 \rangle$ then the group $\mathcal{N}$ in Corollary 3.3.17 is the normal closure in $[\infty,\infty]$ of $R = \{\hat{\sigma}_1^4, \hat{\sigma}_2^4, (\hat{\sigma}_1\hat{\sigma}_2)^2, (\hat{\sigma}_1\hat{\sigma}_2^{-1})^5(\hat{\sigma}_1^{-1}\hat{\sigma}_2)^3\}$. The first three elements of the set R' are σ_1^{-4}, $(\sigma_1^2\sigma_2)^4 = \sigma_1\sigma_2^{-4}\sigma_1^{-1}$ and $(\sigma_1\sigma_2)^2$, all of them corresponding to the identity element in $\Gamma(\{4,4\}_{(5,3)})$. Hence, in order to find $\chi(\{4,4\}_{(5,3)})$, we only need to compute the normal closure in $\Gamma(\{4,4\}_{(5,3)})$ of

$$(\sigma_1^{-1}\sigma_2^{-1}\sigma_1^{-2})^5(\sigma_1^3\sigma_2)^3 = (\sigma_2\sigma_1^{-1})^5(\sigma_1^{-1}\sigma_2)^3 = (\sigma_1\sigma_2^{-1})^{-5}(\sigma_1^{-1}\sigma_2)^3,$$

which equals $(\sigma_1^{-1}\sigma_2)^6$ given the defining relations of $\Gamma(\{4,4\}_{(5,3)})$. To determine this normal closure we use Lemma 3.1.80 to imply that it is generated by $\alpha := (\sigma_1^{-1}\sigma_2)^6$ and its conjugates by σ_1, σ_1^{-1} and σ_1^2. We start by rewriting the latter conjugate:

$$\sigma_1^2\alpha\sigma_1^2 = \sigma_1^2(\sigma_1^{-1}\sigma_2)^6\sigma_1^{-2} = (\sigma_1\sigma_2\sigma_1^{-2})^6 = (\sigma_2^{-1}\sigma_1^{-3})^6 = (\sigma_2^{-1}\sigma_1)^6 = \alpha^{-1}.$$

Now, $\sigma_1\alpha\sigma_1^{-1} = (\sigma_2\sigma_1^{-1})^6 = (\sigma_1\sigma_2^{-1})^{-6}$, but from our previous analysis we know that

$$(\sigma_1\sigma_2^{-1})^{-6} = (\sigma_1\sigma_2^{-1})^{-6}[(\sigma_1^{-1}\sigma_2)^{-5}(\sigma_1\sigma_2^{-1})^3]^2 = (\sigma_1^{-1}\sigma_2)^{-10} = \alpha^{-13}.$$

(Here we used that $78 = 6 \cdot 13 \equiv 10$ modulo 17, and also modulo 2.) Furthermore,

$$\sigma_1^{-1}\alpha\sigma_1 = (\sigma_1^{-2}\sigma_2\sigma_1)^6 = (\sigma_1^{-3}\sigma_2^{-1})^6 = [(\sigma_1\sigma_2^{-1})^{-6}]^{-1} = \alpha^{13}.$$

We can conclude that the normal closure of α is merely $\langle \alpha \rangle$.

Since the order of $\sigma_1^{-1}\sigma_2$ is either 1, 2, 17 or 34, the order of α must be either 1 or 17. However, it cannot be 1 since otherwise $\chi(\{4,4\}_{(5,3)})$ would be trivial, forcing $\{4,4\}_{(5,3)}$ to be regular. Therefore, $\chi(\{4,4\}_{(5,3)}) \cong \langle\alpha\rangle \cong \mathbb{Z}_{17}$, and $\kappa(\{4,4\}_{(5,3)}) = 17$.

The argumentation in Example 3.3.18 was based purely on group theoretical methods. Some parts of it can be simplified by the usage of the theory developed in Section 2.4.2.

Exercise 3.3.19 Provide a proof of the fact that $\chi(\{4,4\}_{(5,3)}) \cong \mathbb{Z}_{17}$ by using the description of $\{4,4\}_{(5,3)}$ in Section 2.4.2.

So far we have only encountered cyclic groups as chirality groups of chiral polytopes. In fact, it was established in [15, Theorems 3 and 4] that the chirality group of every chiral toroidal map is cyclic. We shall see shortly that some non-cyclic groups are also the chirality groups of polytopes.

In [16, Theorems 20 and 26] it was proven that every finite abelian group and every affine group $AGL(q)$ with q an odd power of 2 is the chirality group of a chiral hypermap. The proofs of these theorems rely on the existence of two generators of certain groups Γ satisfying a series of requirements so that they are the standard rotations of chiral hypermaps $\mathcal{M}$ with $\Gamma(\mathcal{M}) = \Gamma$ and whose chirality groups are the desired groups. Since not every hypermap is a map and not every map is a polyhedron, it is not yet proven that those groups are all chirality groups of polytopes.

When looking on the other direction, corollaries 23 and 24 of [16] state that for $n > 2$ the dihedral group D_n (with $2n$ elements) and the symmetric group S_n are not the chirality group of any chiral hypermap. This certainly implies that those groups are not the chirality groups of chiral polyhedra. However, the proofs rely on the fact that the automorphism groups of hypermaps have two standard generators. It is not clear that these arguments extend to polytopes of rank $n \geq 4$.

The research developed in [16] suggests to look for necessary conditions and sufficient conditions for groups to be the chirality groups of chiral polytopes.

Problem 15 Determine sufficient conditions for a group G to be the chirality group of a chiral polytope.

Problem 16 Determine necessary conditions for a group G to be the chirality group of a chiral polytope.

From the definition of chirality index we know that $\kappa(\mathcal{P}) \geq 1$ for every rotary polytope $\mathcal{P}$. Corollary 3.3.12 indicates that this lower bound is attained if and only if $\mathcal{P}$ is regular. On the other hand, Proposition 3.3.14 gives us the upper bound $\kappa(\mathcal{P}) \leq |\Gamma(\mathcal{P})|$. Some attention has been given to those polytopes that attain the upper bound.

Definition 3.3.20 We say that a finite polytope $\mathcal{P}$ is *totally chiral* if it is chiral and $\chi(\mathcal{P}) \cong \Gamma(\mathcal{P})$.

Explicit computation of the chirality groups of chiral polytopes is hard; nevertheless, there is a natural way to find totally chiral polytopes. If $\mathcal{P}$ is chiral then from Corollary 3.3.12 we know that $\kappa(\mathcal{P}) > 1$. If on top of that $\Gamma(\mathcal{P})$ is a simple group, we know from Proposition 3.3.14 that $\chi(\mathcal{P}) \cong \Gamma(\mathcal{P})$, forcing $\mathcal{P}$ to be totally chiral.

Example 3.3.21 Let $\sigma_1 = (1,6,8,7,5,3,2)$, $\sigma_2 = (1,2,3,4,5)(6,7,8)$, both elements of the alternating group A_8. In what follows we shall prove that $\langle \sigma_1, \sigma_2 \rangle$ is the automorphism group of a chiral polyhedron $\mathcal{P}$ with $\Gamma(\mathcal{P}) \cong A_8$. Since A_n is simple for every $n \geq 5$, our polytope $\mathcal{P}$ must be totally chiral.

We first show that $\langle \sigma_1, \sigma_2 \rangle = A_8$. In order to do that it suffices to show that it contain the 3-cycles $(k,7,8)$ for every $k \in \{1,\dots,6\}$. It can be verified directly that $(6,7,8) = \sigma_2^{10}$, while $(5,7,8) = \sigma_1^{-1}\sigma_2^{10}\sigma_1$. Since $\sigma_2^3 = (1,4,2,5,3)$, the 3-cycles $(k,7,8)$ with $k < 5$ are obtained by conjugating $(5,7,8)$ by powers of σ_2^3.

Next we make sure that $\langle \sigma_1, \sigma_2 \rangle$ satisfies (3.7) as well as the intersection condition (3.16). For the former it suffices to note that $\sigma_1\sigma_2 = (1,7)(4,5)$, while for the latter it is enough to observe that all non-trivial powers of σ_1 fix 4 but do not fix 1, whereas the only non-trivial powers of σ_2 that fix 4 are σ_2^5 and σ_2^{10}, both fixing also 1. We conclude from Theorem 3.1.53 and Proposition 3.1.58 that $\langle \sigma_1, \sigma_2 \rangle$ is the automorphism group of a rotary polyhedron $\mathcal{P}$.

Finally, to show that $\mathcal{P}$ is chiral we proceed by contradiction. Suppose that $\mathcal{P}$ is regular and $i_1, \dots, i_k \in \{1,2\}$, then Corollary 3.1.55 tells us that there is an automorphism of Γ mapping $\sigma_{i_1} \cdots \sigma_{i_k}$ to $\sigma_{i_1}^{-1} \cdots \sigma_{i_k}^{-1}$; in particular those two elements must have the same order. Direct computation shows that

$$\sigma_1\sigma_2\sigma_1^4\sigma_2^{-1}\sigma_1^3\sigma_2^3\sigma_1^{-1}\sigma_2^3 = (1,2,3,7)(4,5),$$
$$\sigma_1^{-1}\sigma_2^{-1}\sigma_1^{-4}\sigma_2\sigma_1^{-3}\sigma_2^{-3}\sigma_1\sigma_2^{-3} = (2,5,3,8,6),$$

with orders 4 and 5, respectively. The contradiction arises from the assumption of regularity.

The previous example was taken from [230, Figure 8], where it is also mentioned that there is no chiral polyhedron with symmetry group A_5, A_6 or A_7 (see also [156]). Moreover, in [16, Section 5] it is mentioned that there are no totally chiral maps on orientable surfaces of genera at most 101; in particular, totally chiral polyhedra when realised as maps on surfaces must have a high genus.

The results in [16, Section 5] stating that some finite simple groups are the chirality groups of chiral hypermaps do not imply directly that they are the chirality groups of polytopes, since the hypermaps constructed may not be maps, and in the case where they are maps they may not correspond to polyhedra. Instead, it was shown in [156] that every finite non-abelian simple group, with the exceptions of $PSL_2(q)$, $PSL_3(q)$, $PSU_3(q)$ and A_7, is the automorphism group of a chiral polyhedron. In particular, each such polyhedron is totally chiral. The list of exceptions becomes smaller when we ask which finite non-abelian simple groups are the automorphism group of a chiral polytope (of any rank). Up to now, that list has not been completely settled, but substantial progress has been made in [157, 161].

Problem 17 Determine which finite simple groups are the automorphism groups of chiral polytopes.

To date there is no published example of a totally chiral polytope whose chirality group is not simple. The results in [16, Section 6] and [78, Section 9] are highly related to the concepts we shall develop in Section 4.4, and they suggest that there may be some such totally chiral polytopes.

Problem 18 Determine which finite non-simple groups are the chirality groups of totally chiral polytopes.

Definition 3.3.20 is not convenient for infinite polytopes, since an isomorphism of infinite chirality and automorphism goups may not reflect the core of the concept of total chirality. It is more appropriate to use the groups involved in Definition 3.3.7 as follows.

Definition 3.3.22 Let $\mathcal{P}$ be a chiral polytope admiting the quotient presentation $\{\infty, \ldots, \infty\}/N$. We say that $\mathcal{P}$ is *totally chiral* if $N \cdot \rho_0 N \rho_0 = \{\infty, \ldots, \infty\}^+$.

In view of Theorem 3.2.57, Definitions 3.3.20 and 3.3.22 are equivalent for finite polytopes. In the general case Definition 3.3.22 states that the largest regular quotient of $\mathcal{P}$ (with quotient presentation $\{\infty, \ldots, \infty\}/N \cdot \rho_0 N \rho_0$) is the smallest possible orientable structure, namely two flags with the property that i-adjacency changes flags for every i.

Exercise 3.3.23 Let $\mathcal{P}$ be a finite chiral polytope and let Φ be a flag of $\mathcal{P}$. Show that $\mathcal{P}$ is totally chiral if and only if $\mathrm{Stab}_{\mathrm{Conn}(\mathcal{P})}(\Phi) \cong \Gamma(\mathcal{P})$.

In previous works there is little mention of infinite chiral polytopes (see [228, 230, 245, 249]). No systematic study of their chirality groups has been carried out, which suggests the following problems.

Problem 19 Determine whether there are infinite chiral polytopes with finite chirality index, and whether there are infinite chiral polytopes with infinite chirality index.

Problem 20 Are there infinite totally chiral polytopes?

Other classes of non-regular polytopes may admit a notion equivalent to the chirality group. Most of the results in this section generalise naturally to the remaining classes of two-orbit polytopes, making it possible to define an *irregularity group* with the same idea as the chirality group of chiral polytopes. However, difficulties arise when trying to define the irregularity group for k-orbit polytopes with $k \geq 3$.

As an example we take $\mathcal{P}$ to be the prism over an 8-gon. It is a three-orbit polyhedron with 96 flags. All flags containing the top and bottom faces are in the same orbit (light grey in Figure 3.27), while the flags containing the squares split in two orbits depending on whether the edge divides two squares or not (dark grey and white in Figure 3.27, respectively). There are 16 automorphisms of $\mathcal{P}$ preserving each of the top and bottom faces; they are the 8 rotations about the 8-fold axis ℓ together with 8 reflections about planes containing ℓ. The remaining 16 automorphisms are obtained by composing the previous 16 with the plane reflection that interchanges the top and bottom faces. The irregularity index by quotients of $\mathcal{P}$ (see Definition 3.3.2) must be given by the quotient to

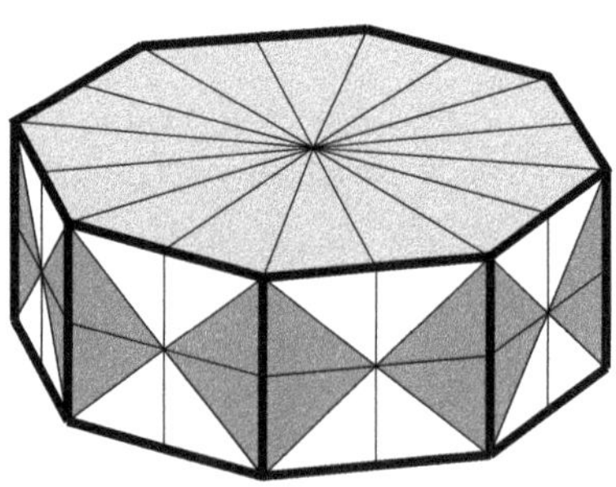

Figure 3.27 The three flag orbits of a prism over an octagon.

the cube, since the corresponding covering has degree 2 (see Example 2.1.30). On the other hand, the irregularity index by covers of $\mathcal{P}$ (see Definition 3.3.5) is 4, since it was proven in [112, Theorem 12] that the smallest regular cover of $\mathcal{P}$ has 384 flags. It follows that the irregularity indexes by quotients and by covers do not coincide in this small case; it is to be expected that the proportion of polytopes in which they are the same is very small. The determination of a reasonable definition of the chirality group of k-orbit polytopes (if any) is stated as problem 7 in [72].

We conclude this section by recalling that the chirality group and index measure how far a rotary polytope is from being regular. In [63, Chapter 3] this idea is taken to other notions, for example finding a group and index that help determine how far away a rotary polytope is from being self-dual.

4

Polytopes Constructed from Other Polytopes

It has not been easy to construct examples of chiral polytopes, particularly in ranks 5 and higher. A common way to construct polytopes has been to use specific procedures to derive them from other previously known polytopes. Here we do not intend to give an extensive theory on relationships between distinct classes of polytopes. Instead, we shall concentrate on some procedures that under the right hypotheses result in constructions for chiral polytopes.

Some known operations on polytopes are discussed in Section 4.1. In the past they have been mostly applied to regular polytopes; here we provide the theory of these operations with emphasis on the cases where the resulting polytopes are chiral. The notion of an abstract polytope embedded in another abstract polytope is developed in Section 4.2 for the chiral case. Quotients and covers of chiral polytopes are addressed in Section 4.3. Finally, Section 4.4 discusses the mix and comix of orientable rotary polytopes.

4.1 Operations

The term 'operation' on maps or polytopes has been used in more than one context (see for example [74, 134, 145] and [188, Chapter 7]). In all cases the idea is to use a known combinatorial structure to obtain a new one. Here we present only a few operations that can be used to obtain chiral polytopes from other polytopes. Moreover, these operations can be applied to general maniplexes, but sometimes when applied to polytopes the result is a maniplex that is not polytopal. This justifies our choice to define most of them for maniplexes, and therefore the results are also valid for polytopes when applicable.

4.1.1 Dual Operation

Mapping a maniplex $\mathcal{M}$ to its dual $\mathcal{M}^\delta$ is an easy example of an operation. As mentioned in Section 2.3, the result of the dual operation applied to a chiral polytope is again a chiral polytope; the same holds for maniplexes. Furthermore, $(\mathcal{M}^\delta)^\delta = \mathcal{M}$. For example, $\{3,6\}^\delta_{(a,b)} = \{6,3\}_{(a,b)}$.

We also recall that if $\mathcal{M}$ has rank n and two flags of $\mathcal{M}$ are i-adjacent then they are $(n-i-1)$-adjacent as flags of $\mathcal{M}^\delta$. From here we see that if $\mathcal{M}$ is rotary and σ_i is the ith standard rotation of $\mathcal{M}$ with respect to the base flag Φ then

$$\Phi\sigma_i^{-1} = \Phi^{i-1,i} = \Phi^{n-(n-i)-1, n-(n-i-1)-1}.$$

We can conclude that σ_i^{-1} is the $(n-i)$th standard rotation of $\mathcal{M}^\delta$ with respect to the same flag Φ.

4.1.2 Petrie Operation

Next we recall an operation that has been widely studied for maps on surfaces (see for example [269]). The generalisation to n-polytopes has also been addressed (see for example [135, Section 3]). Here we define its natural generalisation to n-maniplexes.

Let $\mathcal{M}$ be an n-maniplex and let $r_0, \ldots, r_{n-1}$ be the standard generators of $\mathrm{Conn}(\mathcal{M})$. The *Petrial* $\mathcal{M}^\pi$ of $\mathcal{M}$ (also called *Petrie* or *Petrie dual*) is the combinatorial structure that has the same set of flags as $\mathcal{M}$, but whose connection group has as standard generators $r'_0, \ldots, r'_{n-1}$, where

$$r'_i = \begin{cases} r_i & \text{if } i \neq n-3, \\ r_{n-3}r_{n-1} & \text{if } i = n-3. \end{cases}$$

We say that the *Petrie operation* maps $\mathcal{M}$ to $\mathcal{M}^\pi$.

Example 4.1.1 The Petrial of the tetrahedron is the hemi-cube. One way to see this is by rearranging the flags of the tetrahedron as in Figure 4.1. We note that 1- and 2-adjacencies are the same in both polyhedra, whereas two flags are 0-adjacent in the tetrahedron if and only if one is obtained from the other in the hemi-cube by the connection r_0r_2.

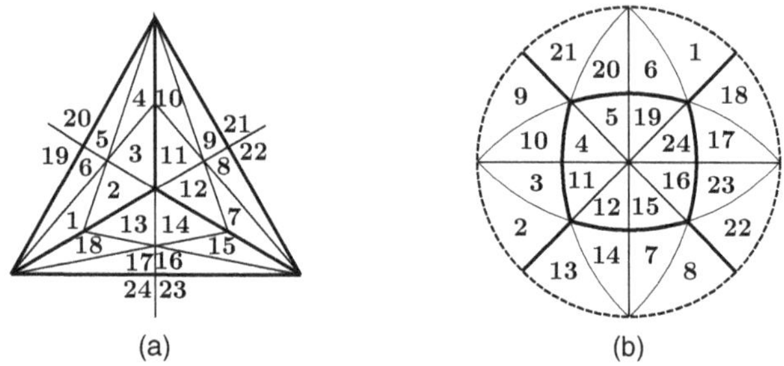

Figure 4.1 (a) Labels of the flags of the tetrahedron; (b) labels of the flags of its dual, the hemi-cube.

Exercise 4.1.2 Show that the Petrial of the cube is isomorphic to the toroidal polyhedron $\{6,3\}_{(2,0)}$.

Exercise 4.1.3 Show that the Petrial of a maniplex is a premaniplex.

Exercise 4.1.4 Show that if $\mathcal{P}$ is an orientable maniplex or a polytope then its Petrial is a maniplex.

Even if $\mathcal{M}$ is a polytope, $\mathcal{P}^{\pi}$ may not be polytopal. Indeed, $\{4,4\}^{\pi}_{(2,1)}$ contains only two decagonal 2-faces, and both contain all the edges; on the other hand, each vertex has degree 4, contradicting the diamond condition in that there are four edges containing a given vertex that are also contained in a given face. The following result provides many more examples in which the Petrial of a polytope is not a polytope (see [176, Theorem 5.7] for a more general version for the regular case).

Proposition 4.1.5 *Let $n \geq 4$ and let $\mathcal{P}$ be an equivelar n-polytope with type $\{p_1, \ldots, p_{n-1}\}$ such that p_{n-3} is odd. Then $\mathcal{P}^{\pi}$ is not polytopal.*

Proof Let $r_0, \ldots, r_{n-1}$ and $r'_0, \ldots, r'_{n-1}$ be the standard generators of $\mathrm{Conn}(\mathcal{P})$ and of $\mathrm{Conn}(\mathcal{P}^{\pi})$, respectively, and let $\Phi \in \mathcal{F}l(\mathcal{P})$. Then,

$$(r'_{n-4}r'_{n-3})^{p_{n-3}}\Phi = (r_{n-4}r_{n-3}r_{n-1})^{p_{n-3}}\Phi = (r_{n-4}r_{n-3})^{p_{n-3}}r_{n-1}^{p_{n-3}}\Phi$$
$$= r_{n-1}\Phi = r'_{n-1}\Phi.$$

It follows that the flag graph of $\mathcal{P}^{\pi}$ does not satisfy the path intersection property, since there is a path from Φ to $(r'_{n-4}r'_{n-3})^{p_{n-3}}\Phi$ using edges with labels $n-4$ and $n-3$, and a path consisting of only one edge with label $n-1$. Theorem 2.1.57 shows that $\mathcal{P}^{\pi}$ is not polytopal. ∎

Regardless of whether $\mathcal{M}$ and $\mathcal{M}^\pi$ are polytopes, it is true that $(\mathcal{M}^\pi)^\pi = \mathcal{M}$. The following results highlight similarities between a maniplex $\mathcal{M}$ and $\mathcal{M}^\pi$.

Proposition 4.1.6 *The $(n-2)$-skeleta of $\mathcal{M}$ and $\mathcal{M}^\pi$ are isomorphic.*

Proof Recall that the j-faces of a maniplex are the connected components after the removal of all edges labelled j. Since $\langle r_{n-3}, r_{n-1}\rangle = \langle r_{n-3}r_{n-1}, r_{n-1}\rangle$, the set of j-faces and their incidences of $\mathcal{M}$ and of $\mathcal{M}^\pi$ can be understood to be the same for all $j \le n-2$, since the connected components of $\mathcal{M}$ and of $\mathcal{M}^\pi$ after removing all edges labelled j have the same sets of flags. ■

Proposition 4.1.7 *Given any maniplex $\mathcal{M}$, $\Gamma(\mathcal{M}) \cong \Gamma(\mathcal{M}^\pi)$.*

Proof Let n be the rank of $\mathcal{M}$, let $\mathrm{Conn}(\mathcal{M}) = \langle r_0, \ldots, r_{n-1}\rangle$, let $\mathrm{Conn}(\mathcal{M}^\pi) = \langle r'_0, \ldots, r'_{n-1}\rangle$ and let $\gamma \in \Gamma(\mathcal{M})$. Then γ induces a permutation of $\mathcal{F}l(\mathcal{M}^\pi)$ in which $\Phi = \Psi^j$ for any given j if and only if $\Phi\gamma = (\Psi\gamma)^j$. Furthermore, if $\Phi = r'_{n-3}\Psi$ then $\Phi = r_{n-3}r_{n-1}\Psi$. This implies that

$$\Phi\gamma = (r_{n-3}r_{n-1}\Psi)\gamma = r_{n-3}r_{n-1}(\Psi\gamma) = r'_{n-3}(\Psi\gamma),$$

and the flags $\Phi\gamma$ and $\Psi\gamma$ are $(n-3)$-adjacent in $\mathcal{M}^\pi$. The proposition now follows from Proposition 2.1.27. ■

Proposition 4.1.7 implies that the Petrie operation permutes the classes of two-orbit polytopes. In particular, if $\mathcal{M}$ is chiral then two flags that are j-adjacent for $j \ne n-3$ are in distinct orbits also when seen as flags of $\mathcal{M}^\pi$, while those that are $(n-3)$-adjacent are in the same orbit. It follows that $\mathcal{M}^\pi$ is in class $2_{\{n-3\}}$.

Exercise 4.1.8 Let $I \subset \{0, \ldots, n-1\}$ and let $\mathcal{M}$ be an n-maniplex in class 2_I. Determine the class of $\mathcal{M}^\pi$.

If $\mathcal{M}$ is a 3-maniplex then the 2-faces of $\mathcal{M}^\pi$ are the Petrie paths of $\mathcal{M}$, and the Petrie paths of $\mathcal{M}^\pi$ correspond to the 2-faces of $\mathcal{M}$ (see Corollary 3.2.61). If $n \ge 4$ and $\mathcal{M}$ is an n-maniplex then $\mathcal{M}^\pi$ can be understood as the poset with the same $(n-2)$-skeleton, but with each vertex-figure replaced by its Petrial. Direct methods show that if $n \ge 4$, $\mathcal{M}$ is a chiral n-maniplex with type $\{p_1, \ldots, p_{n-1}\}$ and q is the length of all Petrie paths of the cofaces to the $(n-4)$-faces then the type of $\mathcal{M}^\pi$ is $\{p_1, \ldots, p_{n-3}, r, p_{n-1}\}$, where

$$r = \begin{cases} q & \text{if } q \text{ is even,} \\ 2q & \text{if } q \text{ is odd.} \end{cases}$$

Exercise 4.1.9 Determine the isomorphism type of the facets of the Petrial of the 4-cube $\{4,3,3\}$. How many of them are there?

Exercise 4.1.10 Give necessary and sufficient conditions for a and b to guarantee that $\{4,4\}^{\pi}_{(a,b)}$ is a polyhedron.

4.1.3 Facetting Operation

Let $k \in \mathbb{Z}^+$ be less than the degree of every vertex of a map $\mathcal{M}$. A *k-hole* of $\mathcal{M}$ is a closed or infinite walk with edge set $\{e_1, \dots, e_m\}$ or $\{\dots, e_{-1}, e_0, e_1, \dots\}$ in the 1-skeleton of $\mathcal{M}$ constructed as follows. For some flag Φ contained in e_1 and for $s \in \mathbb{Z}$, let Φ_s be the flag $[r_0r_1(r_2r_1)^{k-1}]^{s-1}\Phi$. We define e_s to be the edge in Φ_s. If there exists $m \neq 0$ with the property that $[r_0r_1(r_2r_1)^{k-1}]^m\Phi_s = \Phi$ then the hole is finite, otherwise it is infinite. Informally, the k-holes are constructed by traversing an edge, then choosing the kth edge on the right (according to some local orientation) as the next edge to traverse (keeping the local orientation), and repeating this process (see Figure 4.2). For convenience, we shall refer to the 2-holes simply by holes.

The *k-facetting* operation is defined only for maps (3-maniplexes) and is constructed by setting the k-holes as the new faces while keeping the vertex and edge sets. Let $\mathcal{M}$ be a 3-maniplex with $\mathrm{Conn}(\mathcal{M}) = \langle r_0, r_1, r_2\rangle$. We consider the set of flags of $\mathcal{M}$ with the same rules of 0- and 2-adjacency, with two flags Φ and Ψ being 1-adjacent if and only if $\Psi = r_1(r_2r_1)^{k-1}\Phi$. The resulting structure may not be flag-connected; $\mathcal{M}^{\varphi_k}$ is defined as the flag-connected component containing the base edge. Since we assume that the degree of every vertex is strictly less than k we have that the kth facetting operation applied to a polyhedron is a maniplex.

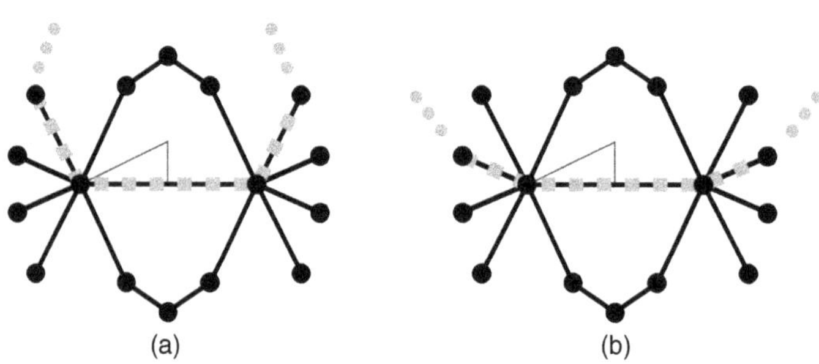

Figure 4.2 Portions of a 2-hole (a) and of a 3-hole (b) of a map; the base flag is indicated in each case.

Remark 4.1.11 If $\mathcal{M}$ is a 3-maniplex then $\mathcal{M}^{\varphi_1} = \mathcal{M}$.

Exercise 4.1.12 Show that the Petrie operation and the k-facetting operation commute; that is, $(\mathcal{M}^{\pi})^{\varphi_k} = (\mathcal{M}^{\varphi_k})^{\pi}$ for every 3-maniplex $\mathcal{M}$.

Example 4.1.13 When applying the 2-facetting operation to the icosahedron $\{3,5\}$, we obtain a polyhedron whose faces are pentagons, five of them around every vertex. The outcome is the great dodecahedron (see [55, Chapter 6] and [188, Section 7E]). This is an example where the automorphism group and the numbers of vertices and edges are preserved after the application of the facetting operation.

Example 4.1.14 The k-facetting operation sometimes reduces the vertex and edge sets. For example, the 2-facetting operation applied to the regular tessellation $\{3,6\}$ by triangles of the Euclidean plane yields the tessellation $\{6,3\}$ by hexagons. The centres of these hexagons are vertices of the original tessellation by triangles, and therefore the vertex set of $\{3,6\}^{\varphi_2}$ is a proper subset of the vertex set of $\{3,6\}$. The edge set and the symmetry group of $\{3,6\}^{\varphi_2}$ are also proper subsets of those of $\{3,6\}$. This is an instance where in the construction of the facetting operation we need to consider only one flag-connected component among several (among three in this case).

The combinatorial structure arising from the facetting operation applied to a polyhedron is not always a polyhedron. For instance, $(\{3,6\}_{(1,1)})^{\varphi_2}$ has only one hexagonal face and it has only three edges, each occurring twice around the face. Figure 4.3 on the left shows the polyhedron $\{3,6\}_{(1,1)}$ and its base

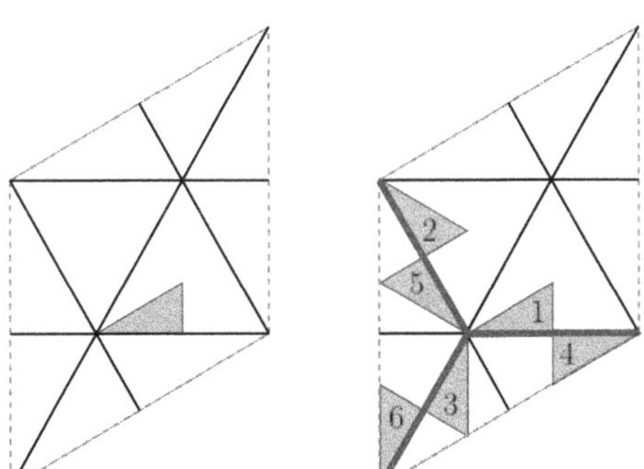

Figure 4.3 The operation φ_2 applied to $\{3,6\}_{(1,1)}$.

flag; on the right each flag Φ_i is indicated just by i, and hence the hole has length 6, but only 3 edges. This is not a polygon, and therefore $(\{3,6\}_{(1,1)})^{\varphi_2}$ is not a polyhedron.

Given a 3-maniplex $\mathcal{M}$ with all its vertices of degree q, the vertices and edges incident to the face containing the flag Φ of $\mathcal{M}^{\varphi_k}$ are the same as the ones incident to the face containing the flag Φ^2 of $\mathcal{M}^{\varphi_{q-k}}$. Indeed, if $(r_1r_2)^q = \mathrm{id}$ then

$$\begin{aligned}[r_0r_1(r_2r_1)^{q-k-1}]^{s-1}\Phi^2 &= [r_2r_0r_2r_1(r_2r_1)^{q-k-1}]^{s-1}r_2\Phi \\ &= r_2[r_0(r_2r_1)^{q-k}r_2]^{s-1}\Phi \\ &= r_2[r_0r_1(r_2r_1)^{k-1}]^{s-1}\Phi.\end{aligned}$$

It follows that with respect to suitable flags, $\mathcal{M}^{\varphi_k}$ and $\mathcal{M}^{\varphi_{q-k}}$ have the same vertex, edge and 2-face sets, as well as the same incidence rule; hence φ_k and φ_{q-k} are regarded as the same operation.

Whenever $\mathcal{M}$ has all its vertices of degree $2k$ then $\mathcal{M}^{\varphi_k}$ is just the regular polyhedron $\{r,2\}$, where r is the length of the k-hole induced by the base flag. Henceforth we may assume that $k < q/2$.

Proposition 4.1.15 *The composition of facetting operations follows the rule* $(\mathcal{M}^{\varphi_k})^{\varphi_l} = \mathcal{M}^{\varphi_{kl}}$.

Proof Let r_0, r_1, r_2 be the standard generators of $\mathrm{Conn}(\mathcal{M})$. From the definition of facetting we know that two flags Φ and Ψ of $(\mathcal{M}^{\varphi_k})^{\varphi_l}$ are 0- or 2-adjacent if and only if they are respectively 0- or 2-adjacent as flags of $\mathcal{M}$, and that they are 1-adjacent if and only if

$$\Psi = r_1(r_2r_1)^{k-1}[r_2r_1(r_2r_1)^{k-1}]^{l-1}\Phi = r_1(r_2r_1)^{kl-1}\Phi.$$

But these are precisely the adjacency rules for $\mathcal{M}^{\varphi_{kl}}$. ■

As a consequence of Proposition 4.1.15, if $\mathcal{M}$ has type $\{p,q\}$ and k is coprime with q then $\varphi_k^{-1} = \varphi_l$, where l is the multiplicative inverse of k in $\mathbb{Z}_q$. Now suppose that k and q have greatest common divisor $m \geq 2$. In that situation the vertices of $\mathcal{M}^{\varphi_k}$ have valency q/m. In some cases this implies that only part of the edges of $\mathcal{M}$ are also edges of $\mathcal{M}^{\varphi_k}$ (see Example 4.1.14). In some others, each vertex of $\mathcal{M}$ splits into at least two vertices of $\mathcal{M}^{\varphi_k}$.

Example 4.1.16 Let $\mathcal{P} = \{3,6\}_{(4,0)}$. Then every 2-face of $\mathcal{P}^{\varphi_2}$ is a hexagon that has as its centre a vertex of $\mathcal{P}$, as the one with thick edges corresponding to the base flag in Figure 4.4(a). We shall label the 16 vertices of $\mathcal{P}$ with $\{(x,y)\colon x,y \in \mathbb{Z}_4\}$, which correspond to coordinates with respect to the basis indicated in Section 2.4.2 (the vertices $(0,0)$, $(1,0)$ and $(0,1)$ are indicated in Figure 4.4(a)). Then the hexagon centred at

(x_0, y_0) shares an edge with the hexagons with centres at $(x_0, y_0) + (z, w)$ for $(z, w) \in \{\pm(1,1), \pm(-1,2), \pm(-2,1)\}$ (see Figure 4.4(b)). It follows that if $\mathcal{P}^{\varphi_2}$ has the hexagon centred at $(0,0)$ as a 2-face then it also has all hexagons centred at points in $\langle(1,1),(-1,2),(-2,1)\rangle$. But the subgroup of $\mathbb{Z}_4^2$ generated by $\{(1,1),(-1,2),(-2,1)\}$ is the entire $\mathbb{Z}_4^2$. This implies that $\mathcal{P}^{\varphi_2}$ has 16 hexagons. By using the basis $\{(1,1),(-1,2)\}$ for the hexagonal tiling we can see that $\mathcal{P}^{\varphi_2}$ is precisely $\{6,3\}_{(4,0)}$. It has indeed 16 hexagons, but it has 32 vertices contrasting with the 16 vertices of $\mathcal{P}$. This is one instance in which every vertex of $\mathcal{P}$ splits into two vertices of $\mathcal{P}^{\varphi_2}$.

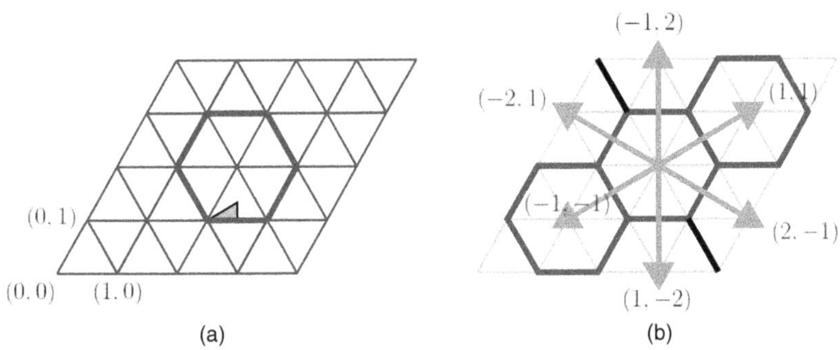

Figure 4.4 (a) A hexagon of $\{3,6\}^{\phi_2}_{(4,0)}$; (b) the six neighboring hexagons.

Exercise 4.1.17 Show that $(\{3,6\}_{(a,b)})^{\varphi_2}$ is a polyhedron for all $b \geq 1$ and $a > b$. Which polyhedron is it? Give necessary and sufficient conditions for a and b so that each vertex of $\{3,6\}_{(a,b)}$ splits into more than one vertex of $(\{3,6\}_{(a,b)})^{\varphi_2}$.

Next we study the automorphisms of $\mathcal{M}^{\varphi_k}$. Let $\mathcal{M}$ be a maniplex with $\mathrm{Conn}(\mathcal{M}) = \langle r_0, r_1, r_2\rangle$, let $\Phi \in \mathcal{F}l(\mathcal{M}^{\varphi_k})$, and let $\gamma \in \Gamma(\mathcal{M})$ be such that $\Phi\gamma \in \mathcal{F}l(\mathcal{M}^{\varphi_k})$. From Proposition 3.2.2 we know that $(r_0\Phi)\gamma$ and $(r_2\Phi)\gamma$ are respectively 0- and 2-adjacent to $\Phi\gamma$ in both, $\mathcal{M}$ and $\mathcal{M}^{\varphi_k}$. The 1-adjacent flag of Φ in $\mathcal{M}^{\varphi_k}$ is $r_1(r_2r_1)^{k-1}\Phi$, and Proposition 3.2.2 also implies that $(r_1(r_2r_1)^{k-1}\Phi)\gamma = r_1(r_2r_1)^{k-1}(\Phi\gamma)$, which is 1-adjacent to $\Phi\gamma$ in $\mathcal{M}^{\varphi_k}$. It follows that $\gamma \in \Gamma(\mathcal{M}^{\varphi_k})$. This proves the following result.

Proposition 4.1.18 *Let $\mathcal{M}$ be a 3-maniplex and let $\Delta \leq \Gamma(\mathcal{M})$ be the subgroup of all automorphisms that map each flag of $\mathcal{M}^{\varphi_k}$ to a flag of $\mathcal{M}^{\varphi_k}$. Then Δ is a subgroup of $\Gamma(\mathcal{M}^{\varphi_k})$.*

Corollary 4.1.19 *Let $\mathcal{M}$ be an s-orbit* 3*-maniplex. Then for each k, $\mathcal{M}^{\varphi_k}$ has at most s flag orbits.*

In particular, if $\mathcal{M}$ is a two-orbit maniplex then $\mathcal{M}^{\varphi_k}$ has at most two orbits on flags. Moreover, in the case when $\mathcal{M}$ and $\mathcal{M}^{\varphi_k}$ have two orbits on flags we have that Φ and Φ^0 are in distinct orbits under $\Gamma(\mathcal{M})$ if and only if they are in distinct orbits under $\Gamma(\mathcal{M}^{\varphi_k})$. The same statement holds for Φ and Φ^2, and also for Φ and $r_1(r_2r_1)^{k-1}\Phi$. In the particular case $k = 2$, if $\mathcal{M}$ is a two-orbit polyhedron in class 2 or 2_1 then Φ is in a distinct orbit as Φ^0, Φ^2 and $r_1r_2r_1\Phi$. We have proven the following.

Theorem 4.1.20 *Let $\mathcal{M}$ be a chiral* 3*-maniplex or a maniplex in class* 2_1*. Then $\mathcal{M}^{\varphi_2}$ is rotary.*

Exercise 4.1.21 Let $\mathcal{M}$ be a chiral 3-maniplex such that $\mathcal{M}^{\varphi_2}$ is chiral. Express the standard rotations of $\mathcal{M}^{\varphi_2}$ in terms of those of $\mathcal{M}$.

Exercise 4.1.22 Let $k \geq 3$ and let $\mathcal{M}$ be a two-orbit 3-maniplex such that $\mathcal{M}^{\varphi_k}$ is chiral. What can be said about the class of $\mathcal{M}$?

Exercise 4.1.23 Find a polyhedron $\mathcal{P}$ with $s \geq 3$ flag orbits and a flag Φ of $\mathcal{P}$ such that $\mathcal{P}^{\varphi_2}$ with base flag Φ is regular. Find a polyhedron $\mathcal{Q}$ with $t \geq 3$ flag orbits and a flag Ψ of $\mathcal{Q}$ such that $\mathcal{Q}^{\varphi_2}$ with base flag Ψ is chiral.

4.1.4 Halving Operation

The *halving* operation is defined only for 3-maniplexes whose 2-faces are all 4-gons. It consists of replacing the original edges by the diagonals of the squares and considering the vertex-figures as 2-faces, as explained below.

Let $\mathcal{M}$ be a maniplex with 4-gonal 2-faces and let $\mathrm{Conn}(\mathcal{M}) = \langle r_0, r_1, r_2\rangle$. Then the halving operation maps $\mathcal{M}$ to $\mathcal{M}^{\eta}$, where the flag set of $\mathcal{M}^{\eta}$ is constructed from the flags of $\mathcal{M}$ by establishing that two flags Φ and Ψ are 0-adjacent (resp. 1- and 2-adjacent) if and only if $\Psi = r_0r_1r_0\Phi$ (resp. $\Psi = r_2\Phi$ and $\Psi = r_1\Phi$). Then $\mathcal{M}^{\eta}$ is the flag-connected component containing the base flag.

Exercise 4.1.24 Show that $\mathcal{M}^{\eta}$ is a maniplex.

Example 4.1.25 When applying the halving operation to the cube we obtain one of its inscribed tetrahedra (see Figure 4.5). The 1- and 2-adjacencies are interchanged, whereas the 0-adjacency relates opposite vertices of a square instead of adjacent vertices of the cube.

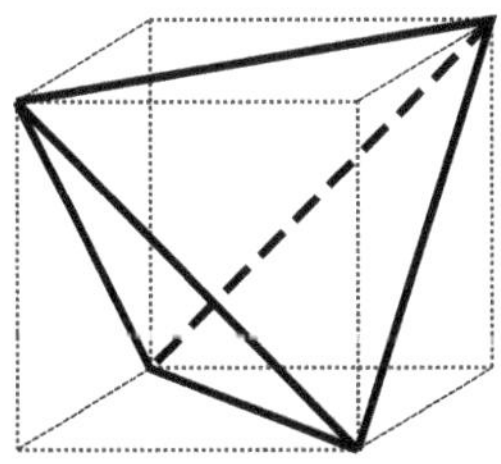

Figure 4.5 A tetrahedron $\{4,3\}^{\eta}$ inscribed on a cube $\{4,3\}$.

In the majority of the cases, the halving operation applied to a polyhedron yields a polyhedron. However, the halving operation may break the strong flag-connectivity. For example, $(\{4,4\}_{(2,0)})^{\eta} = \{4,4\}_{(1,1)}$.

Exercise 4.1.26 Determine the parameters a and b for which $(\{4,4\}_{(a,b)})^{\eta}$ is a polyhedron. Which polyhedron is it?

It follows from the definition of the adjacencies that the vertex set of $\mathcal{M}^{\eta}$ may be understood as a subset of the vertex set of $\mathcal{M}$, since the set of flags of a vertex of $\mathcal{M}^{\eta}$ is also the set of flags of a vertex of $\mathcal{M}$. Furthermore, there is an edge between two vertices of $\mathcal{M}^{\eta}$ if and only if they have a neighbour in common in one 2-face of $\mathcal{M}$. We shall abuse notation and refer by *vertices of* $\mathcal{M}^{\eta}$ to the set of vertices of $\mathcal{M}$ such that the flags contained in them are also flags of $\mathcal{M}^{\eta}$.

If the 1-skeleton of $\mathcal{M}$ is bipartite then the edges of $\mathcal{M}^{\eta}$ join vertices in the same part and $\mathcal{M}^{\eta}$ has as vertices only one part of the bipartition. Since every vertex of $\mathcal{M}^{\eta}$ contains the same number of flags as a vertex of $\mathcal{M}$, it follows that if the 1-skeleton of $\mathcal{M}$ is bipartite then $\mathcal{M}^{\eta}$ has half as many flags as $\mathcal{M}$. On the other hand, if the 1-skeleton of $\mathcal{M}$ is not bipartite then $\mathcal{M}$ and $\mathcal{M}^{\eta}$ have the same number of vertices, and hence, they also have the same number of flags. The details are left as an exercise.

Exercise 4.1.27 Show that if $\mathcal{M}$ is a 3-maniplex whose 1-skeleton is not bipartite then $\mathcal{M}$ and $\mathcal{M}^{\eta}$ have the same set of vertices.

The automorphism group of $\mathcal{M}^\eta$ is highly related to $\Gamma(\mathcal{M})$, as mentioned in the next proposition. Its proof is similar to those of Propositions 4.1.7 and 4.1.18, and it is left as an exercise.

Proposition 4.1.28 *Let $\mathcal{M}$ be a 3-maniplex and let Δ be the subgroup of $\Gamma(\mathcal{M})$ that maps flags of $\mathcal{M}^\eta$ to flags of $\mathcal{M}^\eta$. Then Δ is isomorphic to $\Gamma(\mathcal{M}^\eta)$.*

Exercise 4.1.29 Prove Proposition 4.1.28.

Corollary 4.1.30 *Let $\mathcal{M}$ be a 3-maniplex with 4-gonal 2-faces whose 1-skeleton is not bipartite. Then $\Gamma(\mathcal{M}^\eta) \cong \Gamma(\mathcal{M})$.*

In the same way that the vertices of $\mathcal{M}^\eta$ correspond to vertices of $\mathcal{M}$, each 2-face of $\mathcal{M}^\eta$ corresponds to a vertex-figure (and hence to a vertex) of $\mathcal{M}$. To see this let r_0, r_1, r_2 and r_0', r_1', r_2' be the standard generators of $\mathrm{Conn}(\mathcal{M})$ and of $\mathrm{Conn}(\mathcal{M}^\eta)$, respectively. The even subgroup of the connection group of a 2-face of $\mathcal{M}^\eta$ is generated by $r_0'r_1' = r_0r_1r_2r_0$, which is a conjugate in $\mathcal{M}$ of r_1r_2. That is, the vertices of the face F in some flag Φ are precisely the vertices of the flags $r_0(r_1r_2)^s r_0\Phi$, which are the neighbours of the vertex of $\mathcal{M}$ in the flag $r_0\Phi$ (see Figure 4.6). Therefore, if $\mathcal{M}$ has type $\{4,q\}$ then $\mathcal{M}^\eta$ is equivelar with type $\{q,q\}$.

If the 1-skeleton of a 3-maniplex $\mathcal{M}$ is bipartite then $\Gamma(\mathcal{M}^\eta)$ may be either isomorphic to $\Gamma(\mathcal{M})$, or to an index-2 subgroup of it, depending on whether there is an automorphism of $\mathcal{M}$ mapping a vertex of $\mathcal{M}^\eta$ to a vertex not in $\mathcal{M}^\eta$.

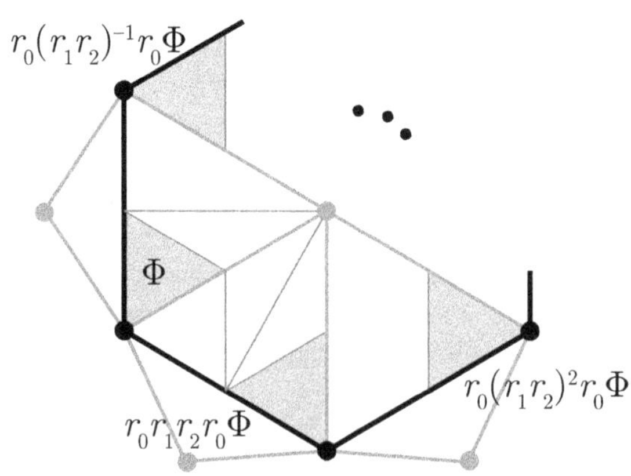

Figure 4.6 The maniplex $\mathcal{M}$ with 4-gonal 2-faces in grey, the maniplex $\mathcal{M}^\eta$ in black, and some flags $r_0(r_1r_2)^s r_0\Phi$ of $\mathcal{M}$ indicated in grey.

For example, if $\mathcal{M}$ is regular with 4-gonal 2-faces and bipartite 1-skeleton then $\Gamma(\mathcal{M}^\eta)$ is an index-2 subgroup of $\Gamma(\mathcal{M})$. On the other hand, if $\mathcal{P}$ is the rhombic dodecahedron (see the polyhedron in thick lines in Figure 2.32, or [55, Section 2.7]), then $\mathcal{P}^\eta$ is either a cube or an octahedron, depending on the choice of base flag. In either case, $\Gamma(\mathcal{P}) \cong \Gamma(\mathcal{P}^\eta)$. More generally, if $\mathcal{M}$ is a two-orbit 3-maniplex in class 2_{12} with 4-gonal 2-faces (like the rhombic dodecahedron) then $\mathcal{M}^\eta$ is regular for any choice of base flag (there are essentially two distinct choices).

Since all two-orbit 3-maniplexes which are not in class 2_{12} are vertex-transitive, if $\mathcal{M}$ is a two-orbit 3-maniplex not in class 2_{12} with 4-gonal 2-faces then $\mathcal{M}^\eta$ is also a two-orbit 3-maniplex. The adjacency rules in the construction of $\mathcal{M}^\eta$ directly imply the following result.

Proposition 4.1.31 *Let $\mathcal{M}$ be a two-orbit 3-maniplex with 4-gonal 2-faces. If $\mathcal{M}$ is in class 2_2 or 2_{02} (resp. 2 or 2_0, and 2_1 or 2_{01}) then $\mathcal{M}^\eta$ is in class 2_1 (resp. 2, and 2_{02}).*

4.1.5 The Petrie–Coxeter Polyhedron of $\mathcal{P}$

There are several ways to describe the polyhedron $\{4,6|4\}$ discovered by Petrie. One of them admits a generalisation to a construction of abstract polyhedra from abstract 4-polytopes.

We start with the regular tessellation $\{4,3,4\}$ of the Euclidean space by cubes with edge length 2. Then we replace each cube by a smaller copy of itself with edge length 1, whose centre and directions of its edges coincide with the ones of the original cube. This disconnects the structure, leaving only a set $\mathcal{S}$ of isolated cubes. Every 2-face f of $\{4,3,4\}$ is now represented by two (parallel) 2-faces f_1 and f_2 that belong to distinct cubes of $\mathcal{S}$. Hence each edge of f is represented by an edge of f_1 and an edge of f_2. The four vertices of these two edges are the vertices of a square (this is the white horizontal square in Figure 4.7). We add this square along with the two new edges for all choices of 2-face f and each edge of f.

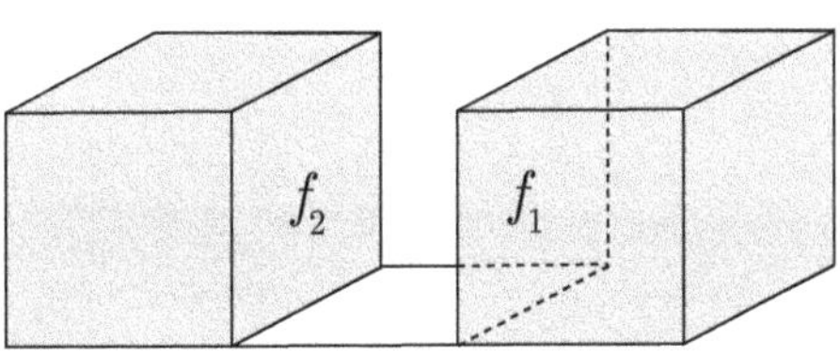

Figure 4.7 The new square from an edge of the parallel squares f_1 and f_2 of two cubes in $\mathcal{S}$.

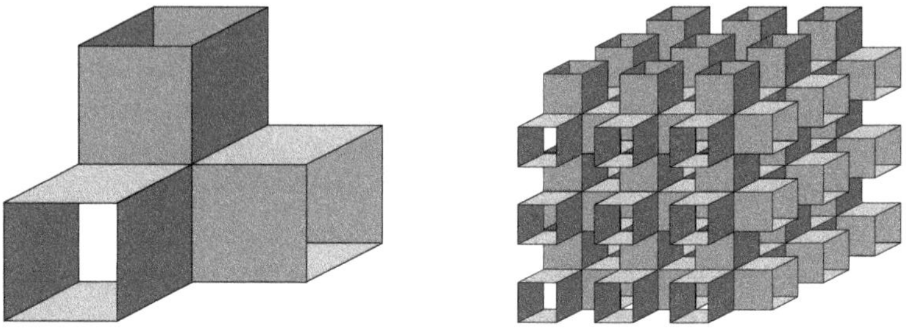

(a) (b)

Figure 4.8 The polyhedron $\{4,6|4\}$.

Each edge of the cubes in $\mathcal{S}$ is now surrounded by four squares, whereas each edge that does not belong to a cube in $\mathcal{S}$ belongs to precisely two squares. Finally we remove the 2-faces of the cubes in $\mathcal{S}$; in doing so, removing two squares around every edge of the cubes in $\mathcal{S}$. The resulting structure has square 2-faces, six at every vertex, and is precisely the polyhedron $\{4,6|4\}$. Figure 4.8(a) shows a portion of $\{4,6|4\}$ containing all squares surrounding a vertex; a wider region of the same polyhedron can be seen in Figure 4.8(b).

Exercise 4.1.32 Show that the facets of the Petrial $\{4,3,4\}^{\pi}$ of the tessellation of the Euclidean 3-space by cubes are isomorphic to the Petrie–Coxeter polyhedron $\{4,6|4\}$ (see [61, 104] for further description of this polyhedron). How many facets does $\{4,3,4\}^{\pi}$ have?

The topological way to extend this construction to the Petrie–Coxeter polyhedron $\mathcal{Q}$ of an arbitrary abstract 4-polytope $\mathcal{P}$ is described in [166], where the emphasis is taken on the case when $\mathcal{P}$ is regular and self-dual (see also [61]). There, the polyhedron $\mathcal{Q}$ is also called the *skew polyhedron related to* $\mathcal{P}$ due to the topological local structure of its embedding in the order complex of $\mathcal{P}$. The corresponding combinatorial construction is as follows.

Definition 4.1.33 Given a 4-polytope $\mathcal{P}$ with set of i-faces $\mathcal{P}_i$, its *Petrie–Coxeter polyhedron* $\mathcal{Q}$ has vertex set

$$\mathcal{Q}_0 = \{(v,c) \colon v \leq c, v \in \mathcal{P}_0, c \in \mathcal{P}_3\},$$

2-face set

$$\mathcal{Q}_1 = \{(e,f) \colon e \leq f, e \in \mathcal{P}_1, f \in \mathcal{P}_2\},$$

and edge set $\mathcal{Q}_2 = E_1 \cup E_2$, where

$$E_1 = \{(e,c) \colon e \leq c, e \in \mathcal{P}_1, c \in \mathcal{P}_3\}, \quad E_2 = \{(v,f) \colon v \leq f, v \in \mathcal{P}_0, f \in \mathcal{P}_2\}.$$

An edge $(e,c) \in E_1$ contains the two vertices (v,c) with $v \leq e$, and is contained in the two 2-faces (e,f) with $e \leq f \leq c$. On the other hand, an edge $(v,f) \in E_2$ contains the two vertices (v,c) with $f \leq c$, and is contained in the two faces (e,f) with $v \leq e \leq f$. As a consequence, a vertex (v,c) is contained in a 2-face (e,f) if and only if $v \leq e \leq f \leq c$.

In the construction of $\{4,6|4\}$, E_1 represents all edges in the cubes in $\mathcal{S}$ while E_2 represents the edges joining vertices of distinct cubes of $\mathcal{S}$. It is an instructive exercise to find a natural choice of a vertex and a square of $\{4,6|4\}$ for each pair of vertex and facet and of edge and square of $\{4,3,4\}$, respectively.

Proposition 4.1.34 *The structure $\mathcal{Q}$ just constructed is a polyhedron with 4-gonal faces.*

Proof It is straightforward that all flags have precisely three elements, and the diamond condition follows from the incidences in the definition.

Any 2-face (e,f) of $\mathcal{Q}$ is a 4-gon whose four vertices are $\{(v,c) \colon v \leq e, f \leq c\}$. Two opposite edges are in E_1 and the other two are in E_2. As posets the 2-faces are connected since their four vertices are distinct. We claim that the vertex-figure at vertex (v,c) is a $2q$-gon, where q is the number of edges containing v in c. In fact, if the vertex-figure at v in c is the q-cycle $(e_1,e_2,\dots,e_q,e_1)$ then that of (v,c) in $\mathcal{Q}$ is the subdivision of $((e_1,c),(e_2,c),\dots,(e_q,c),(e_1,c))$ obtained by inserting between (e_i,c) and (e_{i+1},c) the node (v,f_i), where f_i is the 2-face of $\mathcal{P}$ that contains e_i and e_{i+1}. In this way (v,c) is incident to q edges $(e_i,c) \in E_1$ and to q edges (v,f_i) in E_2. The connectedness of the vertex-figures follows from the connectedness of the vertex-figures of the facets of $\mathcal{P}$. By Proposition 2.1.44 it only remains to prove that $\mathcal{Q}$ is flag-connected.

Since the edge of a flag Φ is in E_1 if and only if the edge of Φ^1 is in E_2, we only need to prove that if Φ and Ψ are flags of $\mathcal{Q}$ with edges in E_1 then there is a sequence of adjacent flags between them. If $\Phi = [(v,c),(e,c),(e,f)]$ and $\Psi = [(v',c'),(e',c'),(e',f')]$ we construct the flags $\hat{\Phi} = (v,e,f,c)$ and $\hat{\Psi} = (v',e',f',c')$ of $\mathcal{P}$. Since $\mathcal{P}$ is flag-connected, there exist $i_1,\dots,i_k$ such that $\hat{\Psi} = \hat{\Phi}^{i_1,\dots,i_k}$. It then follows from the definition of the incidences of $\mathcal{Q}$ that $\Psi = \Phi^{w_1,\dots,w_k}$, where

$$w_j = \begin{cases} 0 & \text{if } i_j = 0, \\ 1,2,1 & \text{if } i_j = 1, \\ 2 & \text{if } i_j = 2, \\ 1,0,1 & \text{if } i_j = 3. \end{cases}$$

■

Exercise 4.1.35 Show that a 4-polytope $\mathcal{P}$ is non-orientable if and only if its Petrie–Coxeter polyhedron is non-orientable.

The topological way to visualise the Petrie–Coxeter polyhedron of a 4-polytope given in [166] suggests a strong relationship between this construction and duality. This is clear from the combinatorial definition above.

Remark 4.1.36 The Petrie–Coxeter polyhedra of the 4-polytopes $\mathcal{P}$ and $\mathcal{P}^{\delta}$ are isomorphic.

If $\mathcal{P}$ is a 4-polytope then its Petrie–Coxeter polyhedron $\mathcal{Q}$ has twice as many flags as $\mathcal{P}$. In fact, for $i \in \{1,2\}$ there is a bijection between the flags of $\mathcal{P}$ and the flags of $\mathcal{Q}$ containing edges in E_i. We shall use this fact to find the automorphism group of $\mathcal{Q}$.

Theorem 4.1.37 *Let $\mathcal{P}$ be a 4-polytope and let $\mathcal{Q}$ be its Petrie–Coxeter polyhedron. Then $\Gamma(\mathcal{Q})$ is isomorphic to $\overline{\Gamma}(\mathcal{P})$, the extended group of $\mathcal{P}$.*

Proof Since the Petrie–Coxeter polyhedron $\mathcal{Q}$ of a 4-polytope $\mathcal{P}$ is constructed from the incidences of $\mathcal{P}$, any automorphism γ of $\mathcal{P}$ induces an automorphism α_γ of $\mathcal{Q}$ defined on the flags by

$$[(v,c),(e,c),(e,f)]\alpha_\gamma = [(v\gamma,c\gamma),(e\gamma,c\gamma),(e\gamma,f\gamma)],$$
$$[(v,c),(v,f),(e,f)]\alpha_\gamma = [(v\gamma,c\gamma),(v\gamma,f\gamma),(e\gamma,f\gamma)].$$

On the other hand, any automorphism α of $\mathcal{Q}$ permuting the flags with edges in E_1 induces a bijection γ_α of the flags of $\mathcal{P}$ defined as follows. Given a flag $\Phi = [v,e,f,c]$ of $\mathcal{P}$ we define $\Phi\gamma_\alpha := [v',e',f',c']$, where $[(v,c),(e,c),(e,f)]\alpha = [(v',c'),(e',c'),(e',f')]$. From the definition of the adjacencies of $\mathcal{Q}$ we know that γ_α maps 0-adjacent and 2-adjacent flags to 0-adjacent and 2-adjacent flags, respectively. Furthermore, 1-adjacent (resp. 3-adjacent) flags of $\mathcal{P}$ correspond to pairs of flags Φ and Ψ of $\mathcal{Q}$ with $\Psi = \Phi^{1,2,1}$ (resp. $\Psi = \Phi^{1,0,1}$). It follows from Proposition 3.2.2 that γ_α preserves i-adjacencies of flags for all $i \in \{0,1,2,3\}$, and therefore any automorphism α of $\mathcal{Q}$ preserving E_1 induces an automorphism γ_α of $\mathcal{P}$. Clearly $\alpha_{(\gamma_\alpha)} = \alpha$ and $\gamma_{(\alpha_\gamma)} = \gamma$, and hence this establishes a bijection between $\Gamma(\mathcal{P})$ and the set of automorphisms of $\mathcal{Q}$ permuting the flags with edges in E_1. A standard procedure shows that this bijection is in fact a group isomorphism.

To determine the full automorphism group of $\mathcal{Q}$ it only remains to know if there exist automorphisms mapping a flag with its edge in E_1 to a flag with its edge in E_2. This certainly happens if $\mathcal{P}$ has a duality δ, in which case we can define an automorphism α_δ of $\mathcal{Q}$ mapping each flag $\Phi = [(v,c),(e,c),(e,f)]$

of $\mathcal{Q}$ with its edge in E_1 to $\Phi\alpha_\delta := [(c\delta, v\delta), (c\delta, e\delta), (f\delta, e\delta)]$, whose edge is in E_2; and mapping each flag $\Psi = [(v,c), (v, f), (e, f)]$ of $\mathcal{Q}$ with its edge in E_2 to $\Psi\alpha_\delta := [(c\delta, v\delta), (f\delta, v\delta), (f\delta, e\delta)]$, whose edge is in E_1. It is straightforward that $(\Psi^i)\alpha_\delta = (\Phi\alpha_\delta)^i$ for $i \in \{0, 1, 2\}$, and that $\Phi(\alpha_{\gamma_1}\alpha_{\gamma_2}) = \Phi\alpha_{\gamma_1\gamma_2}$ for any flag Φ of $\mathcal{Q}$ and any $\gamma_1, \gamma_2 \in \overline{\Gamma}(\mathcal{P})$. We conclude that $\overline{\Gamma}(\mathcal{P})$ is isomorphic to a subgroup of $\Gamma(\mathcal{Q})$.

Suppose now that $\mathcal{Q}$ has an automorphism γ mapping a flag containing an edge in E_1 to a flag with an edge in E_2. Then, since $\mathcal{Q}$ is flag-connected and all 2-faces of $\mathcal{Q}$ alternate edges in E_1 and E_2, γ maps all flags containing edges in E_1 to flags containing edges in E_2 and vice versa. We intend to prove that there exists a duality δ of $\mathcal{P}$ such that $\gamma = \alpha_\delta$ as defined above. To do so, we make use of Proposition 2.1.27 to show that $\mathcal{P}$ and $\mathcal{P}^\delta$ are isomorphic. We define a bijection η from $\mathcal{F}l(\mathcal{P})$ to $\mathcal{F}l(\mathcal{P})^\delta$ in the following way. For every flag $\Phi = (v_1, e_1, f_1, c_1)$ of $\mathcal{P}$ we construct the flag $\overline{\Phi} = [(v_1, c_1), (e_1, c_1), (e_1, f_1)]$ of $\mathcal{Q}$ whose edge is in E_1. Then $\Phi\eta := (c_2, f_2, e_2, v_2)$, where $\overline{\Phi}\gamma = [(v_2, c_2), (v_2, f_2), (e_2, f_2)]$.

To show that η maps i-adjacent flags to i-adjacent flags, we first note that

$$\begin{aligned}
\overline{\Phi^0} &= [(v_1', c_1), (e_1, c_1), (e_1, f_1)],\\
\overline{\Phi^1} &= [(v_1, c_1), (e_1', c_1), (e_1', f_1)],\\
\overline{\Phi^2} &= [(v_1, c_1), (e_1, c_1), (e_1, f_1')],\\
\overline{\Phi^3} &= [(v_1, c_1'), (e_1, c_1'), (e_1, f_1)],
\end{aligned}$$

where v_1', e_1', f_1', c_1' are respectively the vertex, edge, 2-face and 3-face completing the respective diamonds. It then follows from Proposition 3.2.2 that

$$\begin{aligned}
&[(v_1', c_1), (e_1, c_1), (e_1, f_1)]\gamma = [(v_2, c_2'), (v_2, f_2), (e_2, f_2)],\\
&[(v_1, c_1), (e_1', c_1), (e_1', f_1)]\gamma = (\overline{\Phi})^{1,2,1}\gamma = [(v_2, c_2), (v_2, f_2'), (e_2, f_2')],\\
&[(v_1, c_1), (e_1, c_1), (e_1, f_1')]\gamma = [(v_2, c_2), (v_2, f_2), (e_2', f_2)],\\
&[(v_1, c_1'), (e_1, c_1'), (e_1, f_1)]\gamma = (\overline{\Phi})^{1,0,1}\gamma = [(v_2', c_2), (v_2', f_2), (e_2, f_2)],
\end{aligned}$$

and therefore,

$$\begin{aligned}
(\Phi^0)\eta &= (c_2', f_2, e_2, v_2) = (\Phi\eta)^0,\\
(\Phi^1)\eta &= (c_2, f_2', e_2, v_2) = (\Phi\eta)^1,\\
(\Phi^2)\eta &= (c_2, f_2, e_2', v_2) = (\Phi\eta)^2,\\
(\Phi^3)\eta &= (c_2, f_2, e_2, v_2') = (\Phi\eta)^3,
\end{aligned}$$

where adjacencies on the left are taken in $\mathcal{P}$ and those on the right are taken in $\mathcal{P}^{\delta}$. This shows that η is the desired duality, and we can conclude that $\Gamma(\mathcal{Q}) \cong \overline{\Gamma}(\mathcal{P})$. ■

Theorem 4.1.37 implies that the Petrie–Coxeter polyhedron of a self-dual regular 4-polytope is a regular polyhedron. Details about the structure of its automorphism group can be found in [166].

If $\mathcal{P}$ is a regular but not self-dual 4-polytope, or if $\mathcal{P}$ is a self-dual two-orbit 4-polytope, then its Petrie–Coxeter polyhedron has two orbits on flags. We elaborate on this fact in the next result.

Proposition 4.1.38 *Let $\mathcal{P}$ be a 4-polytope and $\mathcal{Q}$ its Petrie–Coxeter polyhedron.*

(a) $\mathcal{Q}$ has two orbits on flags and belongs to class 2_{02} if and only if $\mathcal{P}$ is regular but not self-dual.
(b) Table 4.1 indicates necessary and sufficient conditions for $\mathcal{P}$ so that $\mathcal{Q}$ is in class 2_I for $I \neq \{0,2\}$. $\mathcal{P}$ must be a two-orbit 4-polytope in the class given in the second row of the column labelled 2_I. A letter p (resp. i) indicates that $\mathcal{P}$ is properly (resp. improperly) self-dual.

Proof All two-orbit self-dual 4-polytopes are either in class 2, in class 2_{03}, or in class 2_{12}. Therefore only the Petrie–Coxeter polyhedra of regular 4-polytopes that are not self-dual, and of self-dual 4-polytopes in these three classes have two orbits on flags.

From the proof of Theorem 4.1.37 we know that a pair of 0-adjacent flags (v,e,f,c) and (v',e,f,c) of a 4-polytope $\mathcal{P}$ are in the same orbit if and only if the pair of 0-adjacent flags $[(v,c),(e,c),(e,f)]$ and $[(v',c),(e,c),(e,f)]$ in E_1 of the Petrie–Coxeter polyhedron $\mathcal{Q}$ of $\mathcal{P}$ are also in the same orbit. Therefore, if $\mathcal{P}$ is a 4-polytope whose Petrie–Coxeter polyhedron $\mathcal{Q}$ has two orbits on flags, then 0-adjacent flags of $\mathcal{P}$ belong to the same orbit if and only if 0-adjacent flags of $\mathcal{Q}$ belong to the same orbit. Similarly, 2-adjacent flags of $\mathcal{P}$ belong to the same orbit if and only if 2-adjacent flags of $\mathcal{Q}$ belong to the same orbit.

Table 4.1. *Classes of two-orbit polyhedra from self-dual two-orbit 4-polytopes.*

$\mathcal{Q}$	2_{01}	2_{12}	2_0	2_1	2_2	2
$\mathcal{P}$	$p, 2_{03}$	$p, 2_{12}$	$i, 2_{03}$	$p, 2$	$i, 2_{12}$	$i, 2$

A 4-polytope $\mathcal{P}$ is properly self-dual if and only if there is a duality δ mapping some flag (v,e,f,c) of $\mathcal{P}$ to flag (c,f,e,v) of $\mathcal{P}^\delta$. This occurs if and only if the flag $[(v,c),(e,c),(e,f)]$ of the Petrie–Coxeter polyhedron of $\mathcal{P}$ is in the same flag orbit as its 1-adjacent flag $[(v,c),(v,f),(e,f)]$.

The previous discussion shows that if $\mathcal{Q}$ is the Petrie–Coxeter polyhedron of a 4-polytope $\mathcal{P}$ and $\mathcal{Q}$ has two flag orbits then

- $\mathcal{P}$ is regular and not self-dual if and only if $\mathcal{Q}$ is in class 2_{02},
- $\mathcal{P}$ is two-orbit in class 2_I and properly self-dual if and only if $\mathcal{Q}$ is in class $2_{(I\cap\{0,2\})\cup\{1\}}$,
- $\mathcal{P}$ is two-orbit in class 2_I and improperly self-dual if and only if $\mathcal{Q}$ is in class $2_{I\cap\{0,2\}}$.

■

Thus, the Petrie–Coxeter polyhedron $\mathcal{Q}$ of an improperly self-dual chiral 4-polytope $\mathcal{P}$ is chiral. These polytopes are considered in [137] from an algebraic point of view, where the generators of $\Gamma(\mathcal{Q})$ are obtained from elements in $\overline{\Gamma}(\mathcal{P})$ by means of a twisting operation in the sense of [188, Chapter 8]. We next present the main theorem about chiral Petrie–Coxeter polyhedra, where we derive the generators σ_1 and σ_2 of $\mathcal{Q}$ described in [137] in terms of the above definition and results.

Theorem 4.1.39 *Let $\mathcal{P}$ be an improperly self-dual chiral 4-polytope with type $\{p,q,p\}$, let δ be the duality that maps the base flag Φ of $\mathcal{P}$ to Φ^0, and let $\tau \in \Gamma(\mathcal{P})$ such that $\Phi\tau = \Phi^{0,2}$. Then the Petrie–Coxeter polyhedron $\mathcal{Q}$ of $\mathcal{P}$ is chiral with type $\{4,2q\}$ and holes with size p. Furthermore, if $\Gamma(\mathcal{Q}) = \langle\sigma_1,\sigma_2\rangle$ then there is a group isomorphism from $\overline{\Gamma}(\mathcal{P})$ to $\Gamma(\mathcal{Q})$ mapping δ to σ_1^{-1} and $\delta\tau$ to σ_2.*

Proof We know from Proposition 4.1.34 that $\mathcal{Q}$ is a polyhedron with 4-gonal faces. The fact that the 2-faces are 4-gons can also be seen as a consequence of Proposition 2.3.25 (b).

Theorem 4.1.37 states that there is a group isomorphism g from $\overline{\Gamma}(\mathcal{P})$ to $\Gamma(\mathcal{Q})$. This isomorphism maps an automorphism or duality γ of $\mathcal{P}$ to α_γ, as in the proof of that theorem.

Let $\Phi = [(v,c),(e,c),(e,f)]$ be the base flag of $\mathcal{Q}$, let v' be the vertex of e other than v, and let f' be the 2-face containing e and contained in c other than f. Then

$$\Phi^{0,2} = [(v',c),(e,c),(e,f')] = [(v\tau,c\tau),(e\tau,c\tau),(e\tau,f\tau)] = \Phi\alpha_\tau,$$

and therefore $\alpha_\tau = \sigma_1\sigma_2$. On the other hand,

$$\Phi^{0,1} = [(v',c),(v',f),(e,f)] = [(c\delta,v\delta),(c\delta,e\delta),(f\delta,e\delta)] = \Phi\alpha_\delta,$$

and we conclude that $\alpha_\delta = \sigma_1^{-1}$. It follows that $\alpha_{\delta\tau} = \alpha_\delta\alpha_\tau = \sigma_1^{-1}\sigma_1\sigma_2 = \sigma_2$, providing the desired isomorphism.

From (3.22) and (3.23) we know that if η_1, η_2, η_3 are the standard rotations of $\mathcal{P}$ then

$$g^{-1}(\sigma_2^{-2}) = (\tau\delta^{-1})^2 = \tau\delta^{-1}\tau\delta\delta^2 = \eta_1\eta_2\eta_1\eta_2\eta_3\eta_1^{-1}\eta_1\eta_2\eta_3 = \eta_3\eta_2\eta_3 = \eta_2^{-1},$$

which has order q. Moreover, $\tau\delta^{-1}$ is a duality and must have even order. Hence $\mathcal{Q}$ has type $\{4, 2q\}$. Another argument for this was given in the proof of Proposition 4.1.34.

The size of the holes is the order of $\sigma_1\sigma_2^{-1}$, since $\Phi\sigma_1\sigma_2^{-1} = \Phi^{1,2,1,0} = r_0r_1r_2r_1\Phi$. We know from (3.22) and (3.23) that

$$g^{-1}(\sigma_1\sigma_2^{-1}) = \delta^{-1}\tau\delta^{-1} = \delta^{-1}\tau\delta\delta^2 = \eta_1\eta_2\eta_3\eta_1^{-1}\eta_1\eta_2\eta_3 = \eta_1.$$

Hence the holes have length p. ■

Remark 4.1.40 In [137] the isomorphism is given so that δ is mapped to σ_1 and $\delta\tau$ to σ_2^{-1} (or rather, $\tau\delta^{-1}$ to σ_2). The rooted chiral polytope obtained as the image of this isomorphism from $\overline{\Gamma}(\mathcal{P})$ to $\Gamma(\mathcal{Q})$ is the enantiomorph of the one given in Theorem 4.1.39.

Remark 4.1.41 The Petrie–Coxeter polyhedron of a 4-polytope is a special instance of the constructions of polytopes developed in [225]. With such constructions it is possible to construct polytopes of each rank $n \geq 3$. However, in order for the resulting polytopes to be regular the starting polytopes must meet very restrictive conditions, and for ranks $n \geq 5$ it is not known if there are any polytopes that meet them. It is expected that if there are chiral n-polytopes with $n \geq 4$ arising from such constructions their study will be even harder than that when the resulting polytopes are regular.

4.1.6 Chiral Polytopes from Regular 4-Polytopes

The previous operations provide several ways to construct chiral polyhedra from 4-polytopes. For instance, if $\mathcal{P}$ is an improperly self-dual two-orbit 4-polytope in class 2_{03} then its Petrie–Coxeter polyhedron $\mathcal{Q}$ is in class 2_0, and $\mathcal{Q}^\eta$ and $\mathcal{Q}^\pi$ are chiral (although they may fail to be polyhedra). We know many more constructions for regular 4-polytopes than for two-orbit 4-polytopes, which shows the relevance of the constructions in this section.

Proposition 4.1.42 *Let $\mathcal{P}$ be a regular 4-polytope which is not self-dual, let $\mathcal{Q}$ be its Petrie–Coxeter polyhedron, and let $\mathcal{R} = (\mathcal{Q}^\eta)^{\varphi_2}$. Then $\mathcal{R}$ is a rotary maniplex. In this situation, if $\rho_0, \rho_1, \rho_2, \rho_3$ are the standard reflections of $\mathcal{P}$*

then there is an injective group homomorphism from $\Gamma^+(\mathcal{R})$ to $\Gamma(\mathcal{P})$ mapping the standard rotations to $\rho_3\rho_0\rho_1\rho_2$ and $\rho_2\rho_1$.

Proof Propositions 4.1.31, 4.1.38 and Theorem 4.1.20 imply that $\mathcal{R}$ is rotary.

Let Φ_1 be the base flag of $\mathcal{R}$ and let $\Gamma^+(\mathcal{R}) = \langle\sigma_1, \sigma_2\rangle$, so that $\Phi_1\sigma_1 = \Phi_1^{1,0}$ and $\Phi_1\sigma_2 = \Phi_1^{2,1}$. Let Φ_2 be the base flag of $\mathcal{Q}^\eta$ and let Φ_3 be the base flag of $\mathcal{Q}$. Then the flags $\Phi_1^{1,0}$ and $\Phi_1^{2,1}$ correspond in $\mathcal{Q}^\eta$ to the flags $\Phi_2^{1,2,1,0}$ and $\Phi_2^{2,1,2,1}$, respectively. Furthermore, $\Phi_2^{1,2,1,0}$ and $\Phi_2^{2,1,2,1}$ correspond in $\mathcal{Q}$ to the flags $\Phi_3^{2,1,2,0,1,0}$ and $\Phi_3^{1,2,1,2}$, respectively.

We may assume that the edge of Φ_3 is in E_1, since otherwise we consider instead as the base flag $\Phi_1' = \Phi_1^2$. Then $\Phi_2' = \Phi_2^2$ and $\Phi_3' = \Phi_3^1$, the latter having its edge in E_1. Let $v, e, f, c \in \mathcal{P}$ such that $\Phi_3 = [(v,c), (e,c), (e,f)]$.

Let Φ_4 be the flag (v, e, f, c) of $\mathcal{P}$, corresponding to Φ_3 when constructing $\mathcal{Q}$ as the Petrie–Coxeter polyhedron. Then the flags $\Phi_3^{2,1,2,0,1,0} = \Phi_3^{2,1,2,1,1,0,1,0}$ and $\Phi_3^{1,2,1,2}$ correspond respectively to $\Phi_4^{2,1,3,0} = \Phi_4\rho_0\rho_3\rho_1\rho_2$ and $\Phi_4^{1,2} = \Phi_4\rho_2\rho_1$ (see the proof of Proposition 4.1.34).

Let $i_\mathcal{R}$ be the natural inclusion of the orbit under $\Gamma^+(\mathcal{R})$ of the base flag Φ_1 of $\mathcal{R}$ into the flag set of $\mathcal{P}$. That is, a flag Φ of $\mathcal{R} = (\mathcal{Q}^\eta)^{\varphi_\in}$ is first mapped to itself seen as a flag of $\mathcal{Q}^\eta$, then again to itself as a flag of $\mathcal{Q}$, and flags $[(v,c), (e,c), (e,f)]$ and $[(v,c), (v,f), (e,f)]$ are finally mapped to flag (v, e, f, c) of $\mathcal{P}$. In particular, $(\Phi_1)i_\mathcal{R} = \Phi_4$. The previous discussion shows that $i_\mathcal{R}\rho_3\rho_0\rho_1\rho_2 = \sigma_1 i_\mathcal{R}$ and $i_\mathcal{R}\rho_2\rho_1 = \sigma_2 i_\mathcal{R}$, establishing the desired group homomorphism. ■

Proposition 4.1.42 suggests that we can construct $\mathcal{R}$ in a purely group-theoretical way. Given the automorphism group $\Gamma = \langle\rho_0, \rho_1, \rho_2, \rho_3\rangle$ of a regular 4-polytope which is not self-dual, we define σ_1 and σ_2 as $\rho_3\rho_0\rho_1\rho_2$ and $\rho_2\rho_1$, respectively. If the group $\langle\sigma_1, \sigma_2\rangle$ satisfies the intersection condition (3.1.58) then it is the rotation group of a rotary polyhedron. This is then a hybrid form of mixing operation in the sense of [188, Chapter 7]. We next give an example when the resulting polyhedron is regular and one when it is chiral.

Example 4.1.43 Let $\mathcal{P}$ be the *hemi-4-cube* $\{4, 3, 3\}_4$ (this is the 4-polytope obtained from the 4-cube by identifying antipodal j-faces for each j), let ρ_0, ρ_1, ρ_2, ρ_3 be the standard reflections of $\mathcal{P}$, and let $\mathcal{Q}$ be its Petrie–Coxeter polyhedron. Then the elements $\sigma_1 := \rho_3\rho_0\rho_1\rho_2$ and $\sigma_2 := \rho_2\rho_1$ have orders 4 and 3, respectively. In fact, $(\mathcal{Q}^\eta)^{\varphi_2}$ is isomorphic to the cube; in particular it is regular. The details are left as an exercise.

Exercise 4.1.44 Let $\mathcal{Q}$ be the Petrie–Coxeter polyhedron of the hemi-4-cube $\{4,3,3\}_4$. Show that $(\mathcal{Q}^{\eta})^{\varphi_2}$ is isomorphic to the cube.

Example 4.1.45 We denote by $\{5,3,3\}$ the 120-cell in $\mathbb{E}^4$, a convex 4-polytope with 120 dodecahedral facets whose vertex-figures are tetrahedra (see [55, Chapter VII]). As an abstract polytope it is universal with type $\{5,3,3\}$.

Let $\mathcal{P} = \{5,3,3\}^{\pi}$, that is, the Petrial of the 120-cell. We know from Proposition 4.1.5 that $\mathcal{P}$ is not polytopal, but we may still try to construct a chiral polyhedron from the group-theoretical construction suggested above. Let $\Gamma(\{5,3,3\}) = \langle \rho_0, \rho_1, \rho_2, \rho_3 \rangle$, and then $\Gamma(\mathcal{P}) = \langle \rho_0, \rho_1', \rho_2, \rho_3 \rangle$, with $\rho_1' = \rho_1\rho_3$. Then $\sigma_1 := \rho_3\rho_0\rho_1'\rho_2 = \rho_0\rho_1\rho_2$ and $\sigma_2 := \rho_2\rho_1' = \rho_2\rho_1\rho_3$. Since σ_1 and σ_2 are conjugates of 1-step rotations about Petrie paths on the base facet and vertex-figure of $\{5,3,3\}$, the intersection condition in $\Gamma := \langle \sigma_1, \sigma_2 \rangle$ holds. (The Petrie paths of a dodecahedron do not repeat vertices, and hence they are polygons.) Therefore Γ is the automorphism group of a rotary polyhedron with type $\{10,4\}$. (We recall that the sizes of the Petrie polygons of the dodecahedron and the tetrahedron are respectively 10 and 4.)

Let $\gamma_1 := \sigma_1^{-1}\sigma_2\sigma_1^5\sigma_2^2$ and $\gamma_2 := \sigma_1\sigma_2^{-1}\sigma_1^5\sigma_2^2$; one is obtained from the other by replacing each factor σ_i with σ_i^{-1}. The orders of γ_1 and γ_2 are not the same. To see this, note that

$$\begin{aligned}\gamma_1 &= \rho_2\rho_1\rho_0\rho_2\rho_1\rho_3(\rho_0\rho_1\rho_2)^5(\rho_3\rho_1\rho_2)^2\\ &= \rho_2\rho_1\rho_2\rho_3\left[\rho_0\rho_1(\rho_0\rho_1\rho_2)^5\rho_1\rho_2\right]\rho_3\rho_2\rho_1\rho_2,\end{aligned}$$

which is an involution since $(\rho_0\rho_1\rho_2)^5$ is central in the symmetry group $\langle \rho_0, \rho_1, \rho_2 \rangle$ of the dodecahedron. On the other hand, $\gamma_2 = \rho_0\rho_1\rho_2\rho_3\rho_1\rho_2(\rho_0\rho_1\rho_2)^5(\rho_2\rho_1\rho_3)^2$ is not an involution, since otherwise $(\rho_0\rho_1\rho_2)^5$ would be forced to commute with $(\rho_2\rho_1\rho_3)^2\rho_0\rho_1\rho_2\rho_3\rho_1\rho_2$, and hence also with $\rho_3\rho_2\rho_1\rho_0\rho_1\rho_2\rho_3$. However, from the geometry of $\{5,3,3\}$ it is easy to see that the images of the base vertex under $(\rho_0\rho_1\rho_2)^5\rho_3\rho_2\rho_1\rho_0\rho_1\rho_2\rho_3$ and $\rho_3\rho_2\rho_1\rho_0\rho_1\rho_2\rho_3(\rho_0\rho_1\rho_2)^5$ do not belong to a common dodecahedron, and hence those group elements are not the same. Alternatively, an algebra computer software can easily verify that $\sigma_1\sigma_2^{-1}\sigma_1^5\sigma_2^2$ has order 6. As a result we have that $(\mathcal{Q}^{\eta})^{\varphi_2}$ is chiral.

We can also construct chiral 4-polytopes from some regular 4-polytopes $\mathcal{P}$ with $\Gamma(\mathcal{P}) = \langle \rho_0, \rho_1, \rho_2, \rho_3 \rangle$, by defining $\sigma_1 := \rho_3\rho_0\rho_1\rho_2$, $\sigma_2 := \rho_2\rho_1$ and $\sigma_3 := \rho_1\rho_0$. Relations (3.7) hold, and so $\langle \sigma_1, \sigma_2, \sigma_3 \rangle$ is the rotation group of an orientable rotary 4-polytope if and only if σ_1, σ_2 and σ_3 are non-trivial and the intersection condition (3.10) holds. In any case, $\langle \sigma_1, \sigma_2, \sigma_3 \rangle = \Gamma^+(\mathcal{P})$.

Example 4.1.46 Let $\mathcal{P}$ be the 600-cell $\{3,3,5\}$ (see [55, Chapter VII]). It has 600 icosahedral facets and its vertex-figures are icosahedra. It is the universal polytope with type $\{3,3,5\}$.

Let ρ_0, ρ_1, ρ_2, ρ_3 be the standard reflections of $\mathcal{P}$ and let $\sigma_1 := \rho_3\rho_0\rho_1\rho_2$, $\sigma_2 := \rho_2\rho_1$ and $\sigma_3 := \rho_1\rho_0$. Then $\langle \sigma_2, \sigma_3 \rangle$ is isomorphic to the rotation group of the tetrahedral base facet of $\mathcal{P}$.

To prove the intersection condition of $\langle \sigma_1, \sigma_2, \sigma_3 \rangle$, we make use of the dual version of Proposition 3.1.34, that is, we need to show that $\langle \sigma_1 \rangle \cap \langle \sigma_2, \sigma_3 \rangle = \{\mathrm{id}\}$ and that $\langle \sigma_1, \sigma_2 \rangle \cap \langle \sigma_2, \sigma_3 \rangle = \langle \sigma_2 \rangle$. This can be easily done with the help of a computer software.

Alternatively, σ_1 is a conjugate of a one-step rotation along a Petrie polygon of $\{3,3,5\}$ and $\langle \sigma_2, \sigma_3 \rangle$ is contained in the vertex stabiliser. Since the Petrie polygons of $\{3,3,5\}$ do not repeat vertices, $\langle \sigma_1 \rangle \cap \langle \sigma_2, \sigma_3 \rangle = \{\mathrm{id}\}$.

A tedious calculation shows that $\langle \sigma_1, \sigma_2 \rangle$ has size 1,440, and therefore it is a proper subgroup of $\langle \sigma_1, \sigma_2, \sigma_3 \rangle = \Gamma^+(\{3,3,5\})$, which has order 7,200. In particular, $\langle \sigma_1, \sigma_2 \rangle$ does not contain σ_3. Since $\langle \sigma_2 \rangle \subseteq \langle \sigma_1, \sigma_2 \rangle \cap \langle \sigma_2, \sigma_3 \rangle$, and $\langle \sigma_2 \rangle$ is maximal in $\langle \sigma_2, \sigma_3 \rangle \cong A_4$, the intersection of these two subgroups must be $\langle \sigma_2 \rangle$.

Finally, to show that the resulting 4-polytope is chiral it suffices to note that $\sigma_1^{-1}\sigma_3 = \rho_2\rho_3\rho_0\rho_1$ is a conjugate of a one-step rotation along a Petrie polygon and has order 30 (see Corollary 3.2.62 and [55, Table 1]), while $\sigma_1\sigma_3 = \rho_0\rho_1\rho_3\rho_2\rho_1\rho_0$ is a conjugate of $\rho_3\rho_2$, and therefore has order 5.

So far, we do not know operations on regular polytopes that produce chiral polytopes of ranks higher than 4.

Problem 21 Find operations on regular polytopes that produce chiral n-polytopes from regular polytopes for $n \geq 5$.

4.2 Embeddings

Several convex polytopes are known to allow embeddings in other convex polytopes. For example, the regular convex p-gon admits a natural embedding

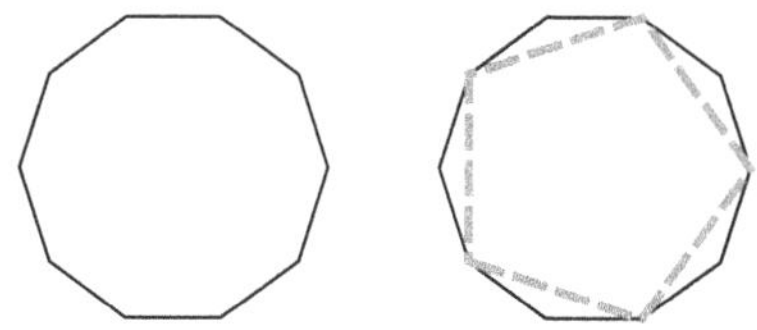

Figure 4.9 The pentagon embedded in the decagon.

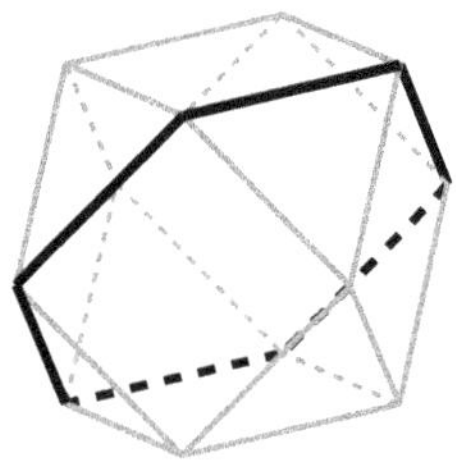

Figure 4.10 The hexagon embedded in the cuboctahedron.

in the regular convex $2p$-gon by considering as a vertex set only a subset of p mutually non-adjacent vertices (the case $p = 5$ is illustrated in Figure 4.9). In the cube with the vertex set $\{(\pm 1, \pm 1, \pm 1)\}$, the set of vertices with an even number of positive coordinates and the set of vertices with an odd number of positive coordinates are the vertex sets of two embedded tetrahedra (see Figure 4.5).

In the previous examples the symmetry group of the embedded polytope is contained in the symmetry group of the bigger polytope, and in each case both polytopes have the same rank. These conditions need not be satisfied in general. For example, the dodecahedron $\{5,3\}$ has five inscribed cubes $\{4,3\}$; however, the dodecahedron admits no 4-fold symmetries and therefore $\Gamma(\{4,3\})$ is not contained in $\Gamma(\{5,3\})$ (see [55, Section 3.6]). On the other hand, a regular convex hexagon can be embedded as an equatorial cycle of a cuboctahedron (see Figure 4.10), where the reflections of the hexagon fixing midpoints of edges are realised by plane reflections of the cuboctahedron, and the reflections of the hexagon fixing vertices are realised by half-turns of the cuboctahedron (see [55, Section 2.3]).

The idea of embeddings of convex polytopes naturally extends to abstract polytopes, as follows. An *embedding* of a polytope $\mathcal{Q}$ in a polytope $\mathcal{P}$ is an injection of the vertex set of $\mathcal{Q}$ into the vertex set of $\mathcal{P}$. When there is no possibility of confusion we shall abuse notation and refer to the image of the embedding as the set of vertices of $\mathcal{Q}$.

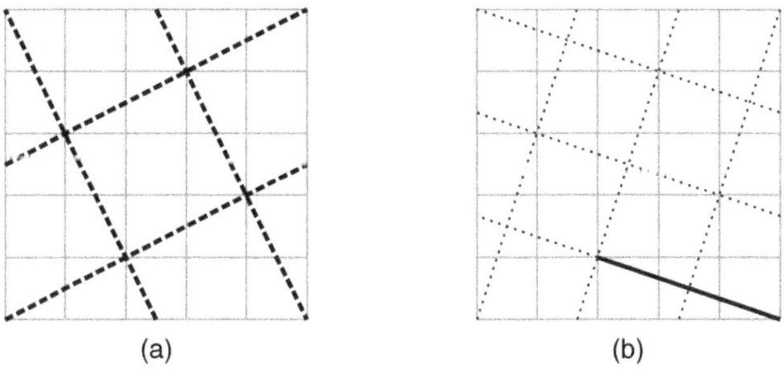

Figure 4.11 The polyhedron $\{4,4\}_{(2,1)}$ embedded in two ways: (a) inducing a subdivision of the torus into five squares and (b) inducing a double cover of the torus into itself.

We say that the embedding ϵ of $\mathcal{Q}$ in $\mathcal{P}$ is *symmetric* whenever there is an isomorphism ϕ from $\Gamma(\mathcal{Q})$ to a subgroup Δ of $\Gamma(\mathcal{P})$ with the properties that Δ preserves the vertex set of $\mathcal{Q}$ and $\epsilon(v\gamma) = (\epsilon(v))\phi(\gamma)$ for every vertex v of $\mathcal{Q}$ and every $\gamma \in \Gamma(\mathcal{Q})$. In other words, there is a subgroup of $\Gamma(\mathcal{P})$ acting on the vertex set of $\mathcal{Q}$ in the same way as $\Gamma(\mathcal{Q})$ does. In this sense, the tetrahedron admits a symmetric embedding in the cube, whereas the cube does not admit a symmetric embedding in the dodecahedron.

Such a general definition provides no intuition on whether the faces of ranks at least 1 have a representation in the poset of $\mathcal{P}$. However, if $\mathcal{Q}$ is edge-transitive and the embedding is symmetric we can represent the 1-skeleton in the order complex of $\mathcal{P}$ as follows. We may define in the order complex of $\mathcal{P}$ an arc e between two vertices which are adjacent in $\mathcal{Q}$ and then all topological edges are the images of e under ϕ.

To better illustrate this notion, consider $\mathcal{P} := \{4,4\}_{(5,0)}$ with vertex set $\{(x,y)\colon x,y \in \mathbb{Z}_5\}$. Let $\mathcal{Q} := \{4,4\}_{(2,1)}$ and embed its five vertices on the set $\{(0,0),(2,1),(4,2),(1,3),(3,4)\}$ in such a way that the preimage of $[(2,1),$ $(4,2),$ $(3,4),$ $(1,3)]$ in that order constitutes a square of $\mathcal{Q} := \{4,4\}_{(2,1)}$. It is clear that the rotation by $\pi/2$ of $\mathcal{P}$ around the centre of the square with vertices $(2,2),(2,3),(3,2),(3,3)$ fixes vertex $(0,0)$ and cyclically permutes $(2,1)$, $(4,2)$, $(3,4)$ and $(1,3)$. On the other hand, easy calculations show that the rotation by $\pi/2$ around vertex $(2,1)$ cyclically permutes $(4,2)$, $(1,3)$, $(0,0)$ and $(3,4)$. These two rotations belong to $\Gamma(\mathcal{P})$ and act precisely like the generators σ_1 and σ_2 of $\mathcal{Q}$. We may conclude that we have a symmetric embedding of $\mathcal{Q}$ in $\mathcal{P}$.

Now, the order complex of $\mathcal{P}$ is a subdivision of the torus. We may embed the edge between $(0,0)$ and $(2,1)$ as the diagonal of the rectangle of area 2 whose other two vertices are $(0,1)$ and $(2,0)$. Then the remaining edges are the ones shown in Figure 4.11(a). This embedding induces a subdivision of the torus into squares, forming a map isomorphic to $\{4,4\}_{(2,1)}$ in the order complex of $\{4,4\}_{(5,0)}$.

In general, an embedding of $\mathcal{Q}$ in $\mathcal{P}$ does not induce another tessellation of the order complex of $\mathcal{P}$. In the previous example we may embed the edge between vertices $(0,0)$ and $(2,1)$ as the bold line segment shown in Figure 4.11(b). When doing so, the resulting squares no longer subdivide the torus. In fact, the embedding suggests a double cover from the order complex of $\{4,4\}_{(2,1)}$ to the order complex of $\{4,4\}_{(5,0)}$, since every point of the map $\{4,4\}_{(5,0)}$ that is not in an edge of $\{4,4\}_{(2,1)}$ belongs to two squares of $\{4,4\}_{(2,1)}$.

A clearer idea of how far an embedding in $\mathcal{P}$ can be from inducing a tessellation on the order complex of $\mathcal{P}$ is given by the embedding of a triangle on the rank 4 regular toroidal polytope $\{4,3,4\}_{(3,0,0)}$ along a straight line (say, the X-axis).

Symmetric embeddings of regular or chiral polytopes can equivalently be defined in terms of their automorphism groups. Assume that for some orientable rotary polytopes $\mathcal{P}$ and $\mathcal{Q}$ there is an injective group homomorphism ϕ from $\Gamma^+(\mathcal{Q})$ to $\Gamma^+(\mathcal{P})$. We say that ϕ induces a symmetric embedding ϵ of $\mathcal{Q}$ on $\mathcal{P}$ whenever ϕ maps the automorphism group of the base vertex-figure of $\mathcal{Q}$ into the automorphism group of the base vertex-figure of $\mathcal{P}$, and the preimage under ϕ of the automorphism group of the base vertex-figure of $\mathcal{P}$ is the automorphism group of the base vertex-figure of $\mathcal{Q}$. From this point of view, ϵ maps the base vertex of $\mathcal{Q}$, represented by $\langle\sigma_2,\ldots,\sigma_{n-1}\rangle$, to the base vertex v_0 of $\mathcal{P}$; any other vertex $\langle\sigma_2,\ldots,\sigma_{n-1}\rangle\gamma$ is mapped into the vertex $v_0\phi(\gamma)$ of $\mathcal{P}$.

To illustrate this we return to the example above where $\mathcal{P}\cong\{4,4\}_{(5,0)}$ and $\mathcal{Q}\cong\{4,4\}_{(2,1)}$. We recall that $\Gamma(\mathcal{P})\cong D_4\ltimes(\mathbb{Z}_5\times\mathbb{Z}_5)$ where the factor D_4 corresponds to $\langle\rho_1,\rho_2\rangle$, and its action on $\mathbb{Z}_5\times\mathbb{Z}_5$ is given by $(x,y)\rho_1=(y,x)$ and $(x,y)\rho_2=(x,-y)$. In particular, $(x,y)\sigma_2=(y,-x)$. On the other hand, $\Gamma(\mathcal{Q})\cong\mathbb{Z}_4\ltimes\mathbb{Z}_5$ where $\mathbb{Z}_4$ corresponds to $\langle\sigma_2\rangle$ and $\mathbb{Z}_5$ may be understood as the subgroup $\langle(2,1)\rangle$ of $\mathbb{Z}_5\times\mathbb{Z}_5$. With this setting, $\mathbb{Z}_4$ acts on $\mathbb{Z}_5$ by $(x,y)\sigma_2=(y,-x)$. Hence there is an injective group homomorphism from $\Gamma(\mathcal{Q})$ to $\Gamma(\mathcal{P})$, where the rotation group of the vertex-figure of $\mathcal{Q}$ is mapped to that of $\mathcal{P}$.

The algebraic version of the definition of symmetric embedding above includes the case when $\mathcal{P}$ and $\mathcal{Q}$ are orientably regular. On the other hand, if $\mathcal{P}$ and $\mathcal{Q}$ have isomorphic vertex-figures, $\mathcal{Q}$ is orientably regular and $\mathcal{P}$ is chiral then the embedding as defined above cannot be symmetric, since the reflections of $\mathcal{Q}$ will not be the restriction of any automorphism of $\mathcal{P}$.

Exercise 4.2.1 Find a toroidal regular polyhedron $\mathcal{Q}$ with type $\{4,4\}$ that admits a non-symmetric embedding in a chiral polyhedron $\mathcal{P}$ of the same type with the following properties.

- There is an injective group homomorphism from $\Gamma^+(\mathcal{Q})$ to $\Gamma^+(\mathcal{P})$ mapping the rotation subgroup of the stabiliser of the base vertex of $\mathcal{Q}$ to the stabiliser of the base vertex of $\mathcal{P}$.
- There is no automorphism of $\mathcal{P}$ whose restriction to the vertices of $\mathcal{Q}$ acts like a reflection of $\mathcal{Q}$.

An example of a symmetric embedding of a regular polytope on a chiral polytope is the pentagon embedded on $\{4,4\}_{(2,1)}$, in such a way that any two consecutive edges of the pentagon are opposite edges around a vertex of $\{4,4\}_{(2,1)}$. (That is, the pentagon is embedded as a hole of $\{4,4\}_{(2,1)}$.) Finding such a symmetric embedding of a regular n-polytope on a chiral n-polytope with the same rank is a more challenging task. In that case, n must be at least 4.

Problem 22 Is there a symmetric embedding of a regular n-polytope in a chiral n-polytope for some n?

Exercise 4.2.2 Prove that the previous problem has a negative answer for $n = 3$.

Problem 23 Is there a symmetric embedding of a chiral polytope in a non-orientably regular polytope?

Problem 24 Is there a symmetric embedding of a non-orientably regular polytope in a chiral polytope?

In [11] it is shown that the toroidal polyhedron $\{4,4\}_{(a,b)}$ admits a symmetric embedding in $\{4,4\}_{(a^2+b^2,0)}$, for all a, b such that $ab(b-a) \neq 0$. The embedding is obtained by applying on the right the linear transformation g with matrix

$$\begin{pmatrix} a & -b \\ b & a \end{pmatrix}$$

to the square fundamental region $X_1 \subseteq \mathbb{R}^2$ determined by the vectors (a,b) and $(-b,a)$, corresponding to $\{4,4\}_{(a,b)}$. The vertices of $\{4,4\}_{(a,b)}$ are mapped via g into points with integer coordinates in the resulting square $X_2 := X_1 g$, defined by the vectors $(a^2+b^2,0)$ and $(0,a^2+b^2)$. The transformation g is the composition of an isometry and an homotety, implying that the squares of $\{4,4\}_{(a,b)}$ are mapped into (bigger) squares in X_2 whose sides may no longer be parallel to the coordinate axes. Under this transformation, the centre of the square with coordinates $(0,0)$, $(1,0)$, $(1,1)$ and $(0,1)$ is mapped to

$\left(\frac{1}{2}(a+b), \frac{1}{2}(a-b)\right)$, whose coordinates are either both integers or both halves of odd integers. This is true in general; the centre of a square of $\{4,4\}_{(a,b)}$ is mapped to either a point of X_2 with integer coordinates, or to a point of X_2 whose coordinates are halves of odd integers. We recall that the region X_2 corresponds to the regular polyhedron $\{4,4\}_{(a^2+b^2,0)}$, and therefore we may view g as an inclusion of the vertex set of $\{4,4\}_{(a,b)}$ in the vertex set of $\{4,4\}_{(a^2+b^2,0)}$. Under this understanding, there are automorphisms of $\{4,4\}_{(a^2+b^2,0)}$ that act precisely like the rotations around vertices and centres of squares of $\{4,4\}_{(a,b)}$, implying that the embedding is symmetric.

Chiral toroidal polyhedra with type $\{3,6\}$ can be treated in a similar way. A fundamental region for the map $\{3,6\}_{(a,b)}$ is transformed into a fundamental region of the regular map $\{3,6\}_{(a^2+ab+b^2,0)}$ by the linear transformation h with matrix

$$\begin{pmatrix} a+b & -b \\ b & a \end{pmatrix}$$

with respect to the basis $\{(1,0),(1/2,\sqrt{3}/2)\}$. (We recall that the vectors generating $\{3,6\}_{(a,b)}$ are (a,b) and $(-b,a+b)$.) All vertices of $\{3,6\}_{(a,b)}$ have integer coordinates, and hence they are mapped into vertices of $\{3,6\}_{(a^2+ab+b^2,0)}$. In a similar fashion to the case with type $\{4,4\}$, the centres of triangles are located in points $(x/3, y/3)$ with $x \equiv y$ (modulo 3), and hence they are mapped either to vertices or to centres of triangles. Since h is the composition of an isometry with a homotety, all faces of $\{3,6\}_{(a,b)}$ are mapped to equilateral triangles embedded in $\{3,6\}_{(a^2+ab+b^2,0)}$. Hence this embedding of $\{3,6\}_{(a,b)}$ on $\{3,6\}_{(a^2+ab+b^2,0)}$ is symmetric. Figure 4.12 shows the embedding of $\{3,6\}_{(1,2)}$ in $\{3,6\}_{(7,0)}$.

Finally, chiral toroidal polyhedra with type $\{6,3\}$ require more care. The standard argument of considering the dual polyhedron with type $\{3,6\}$, embedding it on a polyhedron $\mathcal{P}$ and embedding the original polyhedron on the dual of $\mathcal{P}$ could fail in that vertices may be mapped into centres of faces. Other ideas

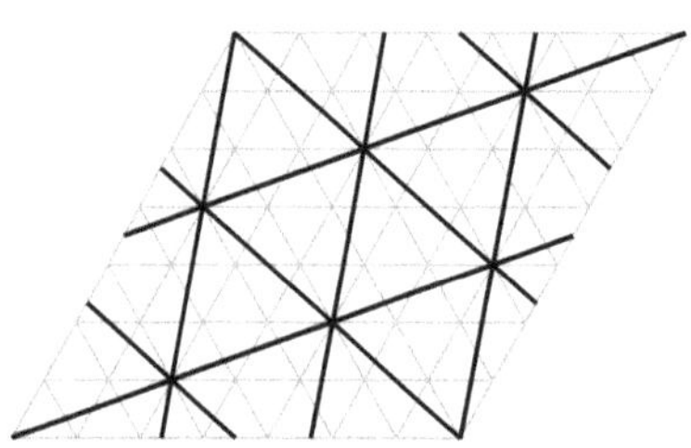

Figure 4.12 The polyhedron $\{3,6\}_{(1,2)}$ embedded in $\{3,6\}_{(7,0)}$.

of embeddings require some care in that the embedding will not be symmetric if centres of hexagons are mapped to vertices.

Exercise 4.2.3 Show that the chiral polyhedron $\{6,3\}_{(a,b)}$ can be embedded symmetrically in $\{6,3\}_{(a^2+ab+b^2,0)}$ whenever 3 does not divide $a+2b$. Is there a rotary polyhedron $\{6,3\}_{(a,b)}$ with $a + 2b \equiv 0$ (modulo 3) that admits a symmetric embedding in $\{6,3\}_{(a^2+ab+b^2)}$?

Exercise 4.2.4 Show that for every $a > b > 0$ the chiral polyhedron $\{6,3\}_{(a,b)}$ can be embedded (perhaps in a non-symmetric way) on $\{6,3\}_{(3(a^2+ab+b^2),0)}$ in such a way that there are automorphisms of $\{6,3\}_{(3(a^2+ab+b^2),0)}$ corresponding to all rotations centred at vertices of $\{6,3\}_{(a,b)}$.

Among non-symmetric embeddings, it is of interest when a regular polytope $\mathcal{Q}$ is embedded *chirally* in some other polytope $\mathcal{P}$. With a chiral embedding we mean that there is an isomorphism ϕ from $\Gamma^+(\mathcal{Q})$ to a subgroup Δ of $\Gamma(\mathcal{P})$, with the extra conditions that the rotation group of the base vertex-figure of $\mathcal{Q}$ is mapped by ϕ to a subgroup of the rotation group of the base vertex-figure of $\mathcal{P}$, and the preimage under ϕ of the base vertex-figure of $\mathcal{P}$ is the rotation group of the base vertex-figure of $\mathcal{Q}$. We further require that ϕ does not extend to an isomorphism satisfying those conditions from $\Gamma(\mathcal{Q})$ to a subgroup of $\Gamma(\mathcal{P})$.

It is not hard to find chiral embeddings of regular polytopes on chiral polytopes. For example, the regular toroidal polyhedron $\{4,4\}_{(2,0)}$ can be embedded chirally in $\{4,4\}_{(4,2)}$ (see Figure 4.13). A more challenging problem is to find chiral embeddings of regular polytopes in regular polytopes.

Problem 25 Are there chiral embeddings of regular n-polytopes in other regular n-polytopes?

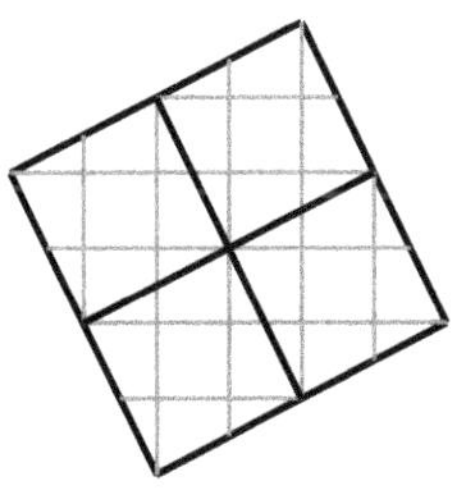

Figure 4.13 The polyhedron $\{4,4\}_{(2,0)}$ embedded chirally in $\{4,4\}_{(4,2)}$.

4.3 Quotients and Covers

A natural way to find chiral polytopes is to consider particular quotients of other regular or chiral polytopes. The first finite chiral polytopes of rank 5 and the polytopes in the atlas of chiral polytopes with small regular covers were found in this way (see [38, 118]).

Let $\mathcal{P}$ be a regular polytope with rotation group $\Gamma^+(\mathcal{P}) = \langle \sigma_1, \dots, \sigma_{n-1} \rangle$. We recall from Propositions 3.1.67 and 3.2.27 that $\mathcal{Q}$ is an orientable rotary quotient of $\mathcal{P}$ if and only if $\mathcal{P}$ is orientable and there is a normal subgroup N of $\Gamma^+(\mathcal{P})$ satisfying

- $\mathcal{Q} \cong \mathcal{P}/N$,
- $\sigma_i \notin N$ for all i,
- $\Gamma^+(\mathcal{P})/N$ satisfies the intersection property (3.10) with respect to the generators $N\sigma_i$.

Furthermore, $\mathcal{Q}$ is chiral if and only if N is not normal in $\Gamma(\mathcal{P})$.

Hence, to determine all chiral quotients of a given orientably regular polytope $\mathcal{P}$ we first need to find all conjugacy classes of normal subgroups of $\Gamma^+(\mathcal{P})$ not containing any generator σ_i, then discard those that are normal in $\Gamma(\mathcal{P})$, and finally determine for which of them the intersection property (3.10) is satisfied. As mentioned in the previous chapter, this was a key idea in the determination of the list of chiral maps in [41].

Example 4.3.1 The cube $\{4,3\}$ has no chiral quotients. To see this, we recall from Lemma 3.1.87 that $\Gamma^+(\{4,3\})$ is isomorphic to the symmetric group on the main diagonals of the cube. The only possible choice of nontrivial normal subgroup of $\Gamma^+(\{4,3\})$ is the subgroup isomorphic to A_4, but the quotient would have only two elements. Clearly it cannot be the rotation group of a chiral 3-polytope.

Example 4.3.2 The toroidal polyhedron $\{4,4\}_{(2,1)}$ is a quotient of $\{4,4\}_{(5,0)}$, where $N \cong \mathbb{Z}_5$ is the group generated precisely by the translation by the vector $(2,1)$ (the algebraic details are left as an exercise). Figure 4.11(a) shows the map $\{4,4\}_{(5,0)}$ divided into five dashed squares, each of which is subdivided as in Figure 1.2. Each of the five dashed squares is a fundamental region for the action of N on the map $\{4,4\}_{(5,0)}$.

Exercise 4.3.3 Show that the group N in Example 4.3.2 is normal in $\Gamma^+(\{4,4\}_{(5,0)})$, but not in $\Gamma(\{4,4\}_{(5,0)})$.

Recall from Section 3.2 that every n-polytope $\mathcal{Q}$ is a quotient of $\mathcal{U}_n$. Sometimes it is convenient to express $\mathcal{Q}$ as a quotient of a smaller polytope. In particular, if $\mathcal{Q}$ is finite, we may ask its cover to be finite as well. In what follows we explore chiral polytopes having a smallest regular cover. The question about existence of finite regular covers of finite chiral polytopes is answered in the next section.

Definition 4.3.4 We say that a regular polytope $\mathcal{P}$ is the *smallest regular cover* of another polytope $\mathcal{Q}$ whenever

- $\mathcal{Q}$ is a quotient of $\mathcal{P}$,
- if $\mathcal{R}$ is a regular cover of $\mathcal{Q}$ then $\mathcal{R}$ covers $\mathcal{P}$.

The previous definition does not ask any particular requirements of symmetry for $\mathcal{Q}$. In particular, if $\mathcal{Q}$ is regular then it is clearly its own smallest regular cover. Not every abstract polytope has a smallest regular cover. For example, the tomotope $\mathcal{T}$ described in [203] satisfies that there is an infinite set $\{\mathcal{P}_i\}_{i \in \mathbb{N}}$ of regular covers with the properties that if $i \neq j$ then $\mathcal{P}_i$ does not cover $\mathcal{P}_j$, and every regular cover of $\mathcal{T}$ covers at least one $\mathcal{P}_i$.

The smallest regular cover of many polytopes can be constructed from their connection groups.

Proposition 4.3.5 *Let $\mathcal{Q}$ be a polytope and suppose that* $\mathrm{Conn}(\mathcal{Q})$ *is a string C-group. Then $\mathcal{Q}$ has a smallest regular cover $\mathcal{P}$ with* $\Gamma(\mathcal{P}) \cong \mathrm{Conn}(\mathcal{Q})$.

Proof Let $r_0, \ldots, r_{n-1}$ be the standard generators of $\mathrm{Conn}(\mathcal{Q})$. Theorem 3.1.20 tells us that there exists a regular polytope $\mathcal{P}$ whose standard reflections are precisely $r_0, \ldots, r_{n-1}$. Furthermore, $\mathcal{Q}$ admits the natural flag action from $\Gamma(\mathcal{P})$. It follows from Proposition 3.2.39 that $\mathcal{P}$ covers $\mathcal{Q}$.

Assume that $\mathcal{R}$ is a regular polytope that covers $\mathcal{Q}$. Then by Proposition 3.2.38 there is a group epimorphism from $\Gamma(\mathcal{R})$ to $\Gamma(\mathcal{P})$ that maps standard reflections to standard reflections. But then Proposition 3.2.39, combined with Theorem 3.2.14, implies that $\mathcal{R}$ covers $\mathcal{P}$. Therefore $\mathcal{P}$ is the smallest regular cover of $\mathcal{Q}$. ■

Remark 4.3.6 In case that $\mathrm{Conn}(\mathcal{Q})$ is not a string C-group then it is still the automorphism group of a regular maniplex that covers $\mathcal{Q}$ and that is covered by any other regular maniplex that covers $\mathcal{Q}$. The verification of this fact follows the same lines as in the case where $\mathcal{Q}$ is a polytope.

Here we are only interested in the case when $\mathcal{Q}$ is chiral. For more details of smallest regular covers in the general case, the reader is referred to [191, Sections 3 and 6].

Example 4.3.7 It is possible to verify directly that $\mathrm{Conn}(\{4,4\}_{(2,1)}) \cong \Gamma(\{4,4\}_{(5,0)})$. An alternative way to determine that $\{4,4\}_{(5,0)}$ is the smallest regular cover of $\{4,4\}_{(2,1)}$ consists of noting that in Figure 4.11(a), $\{4,4\}_{(5,0)}$ is divided into five regions, each of which becomes isomorphic to $\{4,4\}_{(2,1)}$ when acted on by the group generated by the translation by the vector $(2,1)$. Moreover, $\{4,4\}_{(5,0)}$ has 25 squares and therefore it is a 5-fold cover of $\{4,4\}_{(2,1)}$, which has five squares.

Assume, on the contrary, that $\mathrm{Conn}(\{4,4\}_{(2,1)})$ is not isomorphic to $\Gamma(\{4,4\}_{(5,0)})$. Then, by Proposition 3.2.38, it is a k_1-fold quotient of $\Gamma(\{4,4\}_{(5,0)})$ for some $k_1 > 1$. In particular, the even subgroup of $\mathrm{Conn}(\{4,4\}_{(2,1)})$ has degree k_1 as a quotient of $\Gamma^+(\{4,4\}_{(5,0)})$. On the other hand, Proposition 3.2.29 implies that $\Gamma^+(\{4,4\}_{(2,1)})$ has degree k_2 as a quotient of the even subgroup of $\mathrm{Conn}(\{4,4\}_{(2,1)})$ for some $k_2 > 1$. Hence $\{4,4\}_{(5,0)}$ is a k_1k_2-fold cover of $\{4,4\}_{(2,1)}$ and $k_1k_2 = 5$, a contradiction.

Not every chiral polytope has a string C-group as its connection group. The chiral 5-polytope described in Exercises 2.1.21, 2.4.13 and 2.4.14 is isomorphic to the polytope with type $\{3,4,4,3\}$ in [38, Section 5], where an alternative description is given. One can show that these two polytopes are isomorphic by expressing their automorphism groups as permutation groups on the vertices or on the facets. The permutation representation given in [38] is

$$\begin{aligned} \sigma_1 &= (1,2,3), \\ \sigma_2 &= (1,3,2,4), \\ \sigma_3 &= (1,5,4,3), \\ \sigma_4 &= (1,2,3)(4,6,5). \end{aligned} \tag{4.1}$$

It is not hard to verify that these generators indeed satisfy (3.7) and the intersection condition (3.10).

Exercise 4.3.8 Prove that for some labelling in the set $\{1,2,3,4,5,6\}$ of the vertices of the 5-polytope $\mathcal{P}$ described in Exercise 2.1.21, the corresponding permutation representation of the generators σ_i of $\Gamma(\mathcal{P})$ is the one in (4.1).

Let the connection group of this 5-polytope be $\mathcal{M} := \langle r_0, r_1, r_2, r_3, r_4\rangle$. Thanks to Proposition 3.2.2, in order to describe an element of $\mathcal{M}$ we only need to show the image of a flag on each orbit. It is tedious but straightforward to verify that the elements

$$(r_3r_2(r_0r_1r_2)^2r_3r_2r_1r_0r_1r_2)^2,$$
$$r_3(r_1r_2(r_4r_3r_2)^2r_1r_2r_3r_4r_3r_2)^2r_3((r_3r_2r_1)^2r_2r_3r_2r_1)^2,$$

act like id on all flags of one flag orbit, and like $(r_1r_2)^2$ on all flags of the other flag orbit. Hence the permutation w with this action belongs to $\langle r_0, r_1, r_2, r_3\rangle \cap \langle r_1, r_2, r_3, r_4\rangle$. On the other hand, for any flag Ψ of $\{4,4\}_{(2,1)}$, the set of flags consisting of Ψ, Ψ^0 and their translates form a block of imprimitivity under the action of $\mathrm{Conn}(\{4,4\}_{(2,1)})$ (a *block of imprimitivity* of the action of a group Γ is a part of a non-trivial partition preserved by Γ). Under this setting, a flag Φ of the 5-polytope and $\Phi^{1,2,1,2}$ are in different blocks of imprimitivity when considering the action (on the section F_4/F_0 between a 4-face F_4 and a vertex F_0) of $\langle r_1, r_2, r_3\rangle \cong \mathrm{Conn}(\{4,4\}_{(2,1)})$. Hence there is no element in $\langle r_1, r_2, r_3\rangle$ acting like id on Φ and like $(r_1r_2)^2$ on Φ^1. Therefore $\mathcal{M}$ is not a string C-group. In Section 4.4 we give another proof of the failure of the intersection condition for $\mathcal{M}$, without explicitly exhibiting an element in $(\langle r_0, r_1, r_2, r_3\rangle \cap \langle r_1, r_2, r_3, r_4\rangle) \setminus \langle r_1, r_2, r_3\rangle$ (see Example 4.4.27).

Many chiral polytopes do have string C-groups as connection groups, as shown by the next proposition.

Proposition 4.3.9 *Let $\mathcal{P}$ be a polytope all of whose facets (or vertex-figures) are regular and isomorphic. Then* $\mathrm{Conn}(\mathcal{P})$ *is a string C-group.*

Proof If $\mathcal{P}$ has regular vertex-figures then its dual $\mathcal{P}^\delta$ has regular facets. Since $\mathrm{Conn}(\mathcal{P})$ is a string C-group if and only if $\mathrm{Conn}(\mathcal{P}^\delta)$ is a string C-group, we may assume that $\mathcal{P}$ has regular facets.

Let $\mathrm{Conn}(\mathcal{P}) = \langle r_0, \dots, r_{n-1}\rangle$ and suppose that the facets of $\mathcal{P}$ are isomorphic to some regular $(n-1)$-polytope $\mathcal{Q}$.

We first show that $\langle r_0, \dots, r_{n-2}\rangle$ is a string C-group. The orbit under $\langle r_0, \dots, r_{n-2}\rangle$ of a flag containing a given facet F_{n-1} consists precisely of the set of flags containing F_{n-1}. Since the facets of $\mathcal{P}$ are regular and isomorphic to $\mathcal{Q}$, Corollary 3.2.9 implies that a word on the generators r_i for $i \in \{0, \dots, n-2\}$ acts like id on a flag of $\mathcal{P}$ if and only if it acts like id in all flags of $\mathcal{P}$. Hence $\langle r_0, \dots, r_{n-2}\rangle$ acts regularly on the set of flags containing F_{n-1}, and this action is precisely that of the connection group of F_{n-1}. It follows from Theorem 3.2.14 that $\langle r_0, \dots, r_{n-2}\rangle \cong \Gamma(\mathcal{Q})$, and therefore it is a string C-group.

Let $j \in \{1, \dots, n-1\}$, let Φ be an arbitrary flag containing the facet F_{n-1} and let $w \in \langle r_0, \dots, r_{n-2}\rangle \cap \langle r_j, \dots, r_{n-1}\rangle$. Then the flag $w\Phi$ contains F_{n-1}, and also contains the i-face of Φ for every $i \in \{0, \dots, j-1\}$. Since $\mathcal{P}$ is strongly flag-connected, there exists an element $w' \in \langle r_j, \dots, r_{n-2}\rangle$ mapping Φ to $w\Phi$. But the action of $\langle r_0, \dots, r_{n-2}\rangle$ on the set of flags containing F_{n-1} is regular, implying that $w = w' \in \langle r_j, \dots, r_{n-2}\rangle$. It now follows from Proposition 3.1.34 that $\mathrm{Conn}(\mathcal{P})$ is a string C-group. ■

Since every polygon is combinatorially regular, the next corollary follows (see [191, Proposition 6.1] for a more general version).

Corollary 4.3.10 *If $\mathcal{P}$ is a chiral polyhedron then* Conn($\mathcal{P}$) *is a string C-group.*

In view of Proposition 4.3.9, if $\mathcal{P}$ is a chiral polytope whose connection group is not a string C-group then $\mathcal{P}$ has chiral facets and chiral vertex-figures. We next present an alternative way to express the smallest regular cover of a chiral polytope (compare with Lemma 3.3.6).

Proposition 4.3.11 *Let $\mathcal{P}$ be a chiral n-polytope, let $\mathcal{U}$ be the universal polytope with type $\{\infty, \dots, \infty\}$ of rank n, let $\rho_0, \dots, \rho_{n-1}$ be the standard reflections of $\Gamma(\mathcal{U})$, and assume that $\mathcal{P}$ can be expressed as the quotient by automorphisms $\mathcal{U}/\mathcal{N}$ for some $\mathcal{N} \leq \Gamma(\mathcal{U})$. Then the smallest regular maniplex that covers $\mathcal{P}$ is $\mathcal{U}/(\mathcal{N} \cap \rho_0\mathcal{N}\rho_0)$.*

Proof Let Φ be the base flag and let $\hat{\sigma}_1, \dots, \hat{\sigma}_{n-1}$ be the standard rotations of $\mathcal{P}$. For each $i \in \{1, \dots, n-1\}$ let $\sigma_i = \rho_{i-1}\rho_i$ and let $N \leq \Gamma^+(\mathcal{U}_n)$ be such that $\mathcal{P}$ admits the quotient presentation $\mathcal{U}/N$.

We know that $\sigma_i \Phi = \Phi^{i,i-1} = \Phi\hat{\sigma}_i$ (recall that σ_i acts on the left like a connection); hence $\sigma_{j_1} \cdots \sigma_{j_k}\Phi = \Phi\hat{\sigma}_{j_1} \cdots \hat{\sigma}_{j_k}$ for every $j_1, \dots, j_k \in \{0, \dots, n-1\}$.

Now, $\sigma_{j_1} \cdots \sigma_{j_k}\Phi = \Phi$ if and only if $\sigma_{j_1} \cdots \sigma_{j_k} \in N$, and on the other hand, this happens if and only if $\Phi = \Phi\hat{\sigma}_{j_1} \cdots \hat{\sigma}_{j_k}$, which is equivalent to $\sigma_{j_1} \cdots \sigma_{j_k} \in \mathcal{N}$. This forces $N = \mathcal{N}$.

We know from Lemma 3.3.6 that the smallest regular maniplex that covers $\mathcal{P}$ is $\mathcal{U}/(N \cap \rho_0 N \rho_0)$. Since $\mathcal{U}/(N \cap \rho_0 N \rho_0)$ is a regular maniplex, the arguments above show that $N \cap \rho_0 N \rho_0$ is the stabiliser of the base flag through the flag action, but also it is the subgroup of $\Gamma(\mathcal{U})$ such that this maniplex is the quotient by automorphisms $\mathcal{U}/(N \cap \rho_0 N \rho_0)$ and the result follows from the fact that $N = \mathcal{N}$. ∎

The following characterisation in terms of the chirality group of chiral polytopes whose connection groups are string C-groups first appeared in [68]. Here we understand the chirality group of a polytope as a subgroup of its automorphism group as described in Proposition 3.3.15.

Proposition 4.3.12 *Let $\mathcal{P}$ be a chiral polytope, let $\mathcal{Q}$ be the base facet $F_{n-1}/\varnothing$, let $\mathcal{R}$ be the vertex-figure $\mathcal{P}/F_0$ at the base vertex F_0, and let $\mathcal{M}$ be the* medial section *F_{n-1}/F_0. The connection group* Conn($\mathcal{P}$) *is a string C-group if and only if $\chi(\mathcal{Q}) \cap \chi(\mathcal{R}) \leq \chi(\mathcal{M})$.*

Proof Let $\mathcal{U}_n$ denote the universal n-polytope of type $\{\infty, \dots, \infty\}$, let W_n be its automorphism group, and let $\rho_0, \dots, \rho_{n-1}$ be its standard reflections. Let $\mathcal{N} \leq W_n$ be such that $\mathcal{P}$ is the quotient by automorphsms $\mathcal{U}_n/\mathcal{N}$. Then $\Gamma(\mathcal{P}) \cong W_n/\mathcal{N}$.

Now we describe $\mathcal{Q}$, $\mathcal{R}$ and $\mathcal{M}$ as quotients by automorphisms. Let $W_{0,n-2}$, $W_{1,n-1}$ and $W_{1,n-2}$ denote the subgroups $\langle \rho_0, \dots, \rho_{n-2} \rangle$, $\langle \rho_1, \dots, \rho_{n-1} \rangle$ and $\langle \rho_1, \dots, \rho_{n-2} \rangle$, respectively. Then the only elements of $W_{0,n-2}$ that fix the base flag of $\mathcal{Q}$ (that is, the restriction to the facet of the base flag of $\mathcal{P}$) are those in $\mathcal{N} \cap W_{0,n-2}$. It follows that $\mathcal{Q}$ is isomorphic to the quotient $\mathcal{U}_{n-1}/(\mathcal{N} \cap W_{0,n-2})$. Similarly, $\mathcal{R}$ and $\mathcal{M}$ are isomorphic to the quotients $\mathcal{U}_{n-1}/(\mathcal{N} \cap W_{1,n-1})$ and $\mathcal{U}_{n-2}/(\mathcal{N} \cap W_{1,n-2})$, respectively.

From Proposition 3.3.15 we know that

$$\chi(\mathcal{Q}) = [(\rho_0 \mathcal{N} \rho_0) \cap W_{0,n-2}]/(\mathcal{N} \cap \rho_1 \mathcal{N} \rho_1 \cap W_{0,n-2}),$$
$$\chi(\mathcal{R}) = [(\rho_0 \mathcal{N} \rho_0) \cap W_{1,n-1}]/(\mathcal{N} \cap \rho_1 \mathcal{N} \rho_1 \cap W_{1,n-1}),$$
$$\chi(\mathcal{M}) = [(\rho_0 \mathcal{N} \rho_0) \cap W_{1,n-2}]/(\mathcal{N} \cap \rho_1 \mathcal{N} \rho_1 \cap W_{1,n-2}).$$

Each of these groups admits a natural embedding into $W_n/(\mathcal{N} \cap \rho_1 \mathcal{N})$ by simply mapping $w \cdot (\mathcal{N} \cap \rho_1 \mathcal{N} \rho_1 \cap W_{a,b})$ to $w \cdot (\mathcal{N} \cap \rho_1 \mathcal{N} \rho_1)$.

The smallest regular maniplex that covers $\mathcal{P}$ is the quotient $\mathcal{U}_n/(\mathcal{N} \cap \rho_1 \mathcal{N} \rho_1)$ (see Proposition 4.3.11), and its automorphism group is $W_n/(\mathcal{N} \cap \rho_1 \mathcal{N} \rho_1)$.

Proposition 2.3.8 forces $\mathcal{Q}$ to have regular facets and $\mathcal{R}$ to have regular vertex-figures. Then Proposition 4.3.9 implies that their connection groups $W_{0,n-2}/(\mathcal{N} \cap \rho_1 \mathcal{N} \rho_1 \cap W_{0,n-2})$ and $W_{1,n-1}/(\mathcal{N} \cap \rho_1 \mathcal{N} \rho_1 \cap W_{1,N-1})$ are string C-groups. From Proposition 3.1.34 we know that if $W_n/(\mathcal{N} \cap \rho_1 \mathcal{N} \rho_1)$ is not a string C-group it is because $W_{0,n-2} \cdot \mathcal{N} \cap \rho_1 \mathcal{N} \rho_1$ and $W_{1,n-1} \cdot \mathcal{N} \cap \rho_1 \mathcal{N} \rho_1$ intersect in a subgroup that contains $W_{1,n-2} \cdot \mathcal{N} \cap \rho_1 \mathcal{N} \rho_1$ properly.

In view of the previous discussion, the following list of statements are equivalent.

- $\chi(\mathcal{Q}) \cap \chi(\mathcal{R})$ is not contained in $\chi(\mathcal{M})$.
- There exist $w_0 \in \rho_1 \mathcal{N} \rho_1 \cap W_{0,n-2}$ and $w_1 \in \rho_1 \mathcal{N} \rho_1 \cap W_{1,n-1}$ such that $w_0 \cdot \mathcal{N} \cap \rho_1 \mathcal{N} \rho_1 = w_1 \cdot \mathcal{N} \cap \rho_1 \mathcal{N} \rho_1$, and such that this coset cannot be written as $\tilde{w} \cdot \mathcal{N} \cap \rho_1 \mathcal{N} \rho_1$ for any $\tilde{w} \in \rho_1 \mathcal{N} \rho_1 \cap W_{1,n-2}$.
- The coset $w_0 \cdot \mathcal{N} \cap \rho_1 \mathcal{N} \rho_1$ is an element of $[W_{0,n-2} \cdot (\mathcal{N} \cap \rho_1 \mathcal{N} \rho_1)]/(\mathcal{N} \cap \rho_1 \mathcal{N} \rho_1)$, also of $[W_{1,n-1} \cdot (\mathcal{N} \cap \rho_1 \mathcal{N} \rho_1)]/(\mathcal{N} \cap \rho_1 \mathcal{N} \rho_1)$, but not of $[W_{1,n-2} \cdot (\mathcal{N} \cap \rho_1 \mathcal{N} \rho_1)]/(\mathcal{N} \cap \rho_1 \mathcal{N} \rho_1)$.
- The connection group $W/(\mathcal{N} \cap \rho_1 \mathcal{N} \rho_1)$ of the smallest maniplex that covers $\mathcal{P}$ does not satisfy the intersection condition. ∎

Proposition 4.3.12 has several consequences, some of which are listed in the following.

Corollary 4.3.13 *Let $\mathcal{P}$ be a chiral* 4*-polytope with facets and vertex figures isomorphic to polyhedra $\mathcal{Q}$ and $\mathcal{R}$, respectively, and with standard rotations σ_1, σ_2 and σ_3. The connection group* $\mathrm{Conn}(\mathcal{P})$ *is a string C-group if and only if $\chi(\mathcal{Q}) \cap \chi(\mathcal{R})$ contains no non-trivial power of σ_2.*

Proof We shall denote the standard rotations of $\mathcal{P}$ by σ_1, σ_2, σ_3, and follow the notation of Proposition 4.3.12. Under these hypotheses the medial section is a polygon, and hence it is regular; it follows that $\chi(\mathcal{M}) = \{\mathrm{id}\}$. On the other hand we know that $\chi(\mathcal{Q}) \subseteq \langle\sigma_1,\sigma_2\rangle$ while $\chi(\mathcal{R}) \subseteq \langle\sigma_2,\sigma_3\rangle$. From this we can conclude that $\chi(\mathcal{Q}) \cap \chi(\mathcal{R}) \subseteq \langle\sigma_1,\sigma_2\rangle \cap \langle\sigma_2,\sigma_3\rangle = \langle\sigma_2\rangle$; this group is contained in $\chi(\mathcal{M})$ if and only if it contains no non-trivial power of σ_2. ■

Corollary 4.3.14 *Let $\mathcal{P}$ be a chiral polytope with totally chiral facets and totally chiral vertex-figures, and let F_i be the i-face of the base flag. The connection group* $\mathrm{Conn}(\mathcal{P})$ *is a string C-group if and only if the medial section F_{n-1}/F_0 is also totally chiral.*

Proof Let $\sigma_1,\ldots,\sigma_{n-1}$ be the standard rotations of $\mathcal{P}$. Since $\mathcal{Q} := F_{n-1}/\varnothing$ and $\mathcal{R} := \mathcal{P}/F_0$ are totally chiral, $\chi(\mathcal{Q}) = \langle\sigma_1,\ldots,\sigma_{n-2}\rangle$ and $\chi(\mathcal{R}) = \langle\sigma_2,\ldots,\sigma_{n-1}\rangle$. The intersection condition in $\Gamma(\mathcal{P})$ forces $\chi(\mathcal{Q}) \cap \chi(\mathcal{R}) = \langle\sigma_2,\ldots,\sigma_{n-2}\rangle$, which is contained in $\chi(F_{n-1}/F_0)$ if and only if F_{n-1}/F_0 is totally chiral. ■

Corollary 4.3.15 *Let $\mathcal{P}$ be a chiral polytope, let F_i be the i-face of the base flag, and let $\Gamma_0 \leq \Gamma(\mathcal{P})$ denote the stabiliser of F_0. If $\chi(F_{n-1}/\varnothing) \cap \Gamma_0 = \{\mathrm{id}\}$ then* $\mathrm{Conn}(\mathcal{P})$ *is a string C-group.*

Proof If $\mathcal{Q} := F_{n-1}/\varnothing$ and $\mathcal{R} = \mathcal{P}/F_0$ then $\chi(\mathcal{Q}) \cap \Gamma_0 = \{\mathrm{id}\}$ and $\chi(\mathcal{R}) \subseteq \Gamma_0$. Hence, $\chi(\mathcal{Q}) \cap \chi(\mathcal{R}) = \{\mathrm{id}\}$, which is always contained in the chirality group of the medial section. ■

Example 4.3.16 We found in Example 3.3.18 that $\chi(\{4,4\}_{(5,3)})$ is cyclic generated by one translation. It follows that if $\mathcal{P}$ is a chiral polytope with base facet $\mathcal{Q}$ isomorphic to $\{4,4\}_{(5,3)}$, and $\Gamma_0 \leq \Gamma(\mathcal{P})$ is the stabiliser of the base vertex, then $\chi(\mathcal{Q}) \cap \Gamma_0 = \{\mathrm{id}\}$. Corollary 4.3.15 tells us that $\mathrm{Conn}(\mathcal{P})$ is a string C-group.

Remark 4.3.17 The chirality group of every rotary toroidal polyhedron is in fact cyclic generated by a translation, as shown in [15, Theorems 3 and 4].

The arguments in Example 4.3.16 show that the connection group of every rotary 4-polytope with toroidal facets (or vertex-figures) is a string C-group.

The previous results shed some light on whether the connection group of a given chiral polytope with chiral facets and vertex-figures is a string C-group. Given the lack of examples of such polytopes in higher ranks we have little basis to expect which of the two possible outcomes is predominant.

Problem 26 For each $n \geq 4$ let $f_n(x)$ (resp. $g_n(x)$) determine the number of chiral n-polytopes with chiral facets and chiral vertex-figures that have at most x flags, and whose connection group is a string C-group (resp. is not a string C-group). For each n, is either of

$$\lim_{x\to\infty} \frac{f_n(x)}{g_n(x)}, \qquad \lim_{x\to\infty} \frac{g_n(x)}{f_n(x)}$$

equal to 0? If not, does it converge?

In several publications, smallest regular covers were called *minimal regular covers* (see for example [5, 112, 189, 231]). In all these works only rank-3 polytopes were considered and, by Corollary 4.3.10, they all have smallest regular covers. Here we keep the term 'minimal regular cover' of $\mathcal{P}$ to denote regular covers that do not cover any other proper regular cover of $\mathcal{P}$.

The smallest regular cover of chiral polytopes has appeared several times in the literature in different contexts. In [270] it is defined for chiral maps in terms of the parallel product (see also Section 4.4). The smallest regular cover of a chiral n-polytope $\mathcal{P}$ is denoted by $\mathcal{P}^W$ in [78], as long as $\mathrm{Conn}(\mathcal{P})$ satisfies the intersection condition; in that work it is constructed in terms of the action of the universal Coxeter group on $\mathcal{F}l(\mathcal{P})$ (see also [64] for an equivalent definition in terms of the mix of chiral polytopes). Finally, the smallest regular cover of chiral polytopes with regular facets is called *mixed regular cover* in [218], and is defined in terms of permutation representations.

Little is known about minimal regular covers of chiral polytopes whose connection group is not a sting C-group. A chiral 4-polytope with at least two minimal regular covers was presented in [68, Section 4].

Problem 27 Determine whether there exists a chiral polytope whose connection group is not a string C-group but it still has a unique minimal regular (polytopal) cover.

Problem 28 Determine whether there exists a chiral polytope with infinitely many minimal regular covers.

Instead of looking for regular covers of chiral polytopes we may study chiral covers of regular polytopes. For example, $\{4,4\}_{(4,2)}$ covers $\{4,4\}_{(2,0)}$, since the group generated by the translations in $\mathbb{E}^3$ by the vectors $(4,2)$ and $(-2,4)$ is contained on the group generated by the translations $(2,0)$ and $(0,2)$ (see Propostion 3.2.51).

Not every regular polytope admits a chiral cover.

Example 4.3.18 The universal n-polytope $\mathcal{U}_n$ with type $\{\infty,\dots,\infty\}$ admits no chiral cover. Assume to the contrary that some n-polytope $\mathcal{P}$ with $\Gamma(\mathcal{P}) = \langle\bar{\sigma}_1,\dots,\bar{\sigma}_{n-1}\rangle$ is such a cover and let $\Gamma^+(\mathcal{U}_n) = \langle\sigma_1,\dots,\sigma_{n-1}\rangle$. Since $\mathcal{P}$ is not universal then some non-trivial relation in terms of the $\bar{\sigma}_i$ must hold. But then the analogous relation in terms of the σ_i also holds, contradicting the universality of $\mathcal{U}_n$.

This topic has not been explored deeply, which motivates the following natural question.

Problem 29 For $n \geq 3$, is $\mathcal{U}_n$ the only regular polytope of rank n that does not admit a chiral cover?

Partial answers for the cases of finite polytopes and rank-3 polytopes are also of great interest.

The connection group plays a central role in determining that some chiral polytopes have smallest regular covers. Is there an analog for smallest chiral covers of regular polytopes?

Problem 30 Is there a regular polytope that has a smallest chiral cover?

4.4 Mixing

We recall that a rooted polytope is a polytope with a specified base flag. An analogous adjective can similarly be applied to maps. The parallel product $\mathcal{M}_1||\mathcal{M}_2$ of two rooted maps $\mathcal{M}_1$ and $\mathcal{M}_2$ with base flags Φ_1 and Φ_2 was defined in [270] as follows. The set of flags $\mathcal{F}l(\mathcal{M}_1||\mathcal{M}_2)$ is contained in the cartesian product $\mathcal{F}l(\mathcal{M}_1) \times \mathcal{F}l(\mathcal{M}_2)$. A flag (Ψ_1,Ψ_2) is i-adjacent to a flag (Υ_1,Υ_2) whenever $\Upsilon_j^i = \Psi_j$ for $j \in \{1,2\}$. The resulting combinatorial structure may be disconnected; in any case we define $\mathcal{M}_1||\mathcal{M}_2$ to be just the connected component of the flag (Φ_1,Φ_2).

We may extend this idea and define the *parallel product* of two maniplexes $\mathcal{P}$ and $\mathcal{Q}$ rooted on base flags Φ and Ψ as the connected component of the flag (Φ, Ψ) when considering $\mathcal{F}l(\mathcal{P}) \times \mathcal{F}l(\mathcal{Q})$, with two elements being i-adjacent if and only if for $j \in \{1,2\}$ their jth entries are i-adjacent. The resulting structure is denoted by $\mathcal{P}\diamondsuit\mathcal{Q}$, and in the instances of regular and chiral polytopes it is also called the *mix* of $\mathcal{P}$ and $\mathcal{Q}$. Here we extend this terminology to polytopes with no restriction on their symmetry. We emphasise that the mix is not an operation in the category of maniplexes; it is an operation of rooted maniplexes. With this in mind, we shall denote it by $\mathcal{P}_{(\Phi)}\diamondsuit\mathcal{Q}_{(\Psi)}$ unless the choices of base flags Φ and Ψ are irrelevant.

Remark 4.4.1 The mix of maniplexes is commutative and associative.

Example 4.4.2 With the previous definition, the mix of a square and a hexagon with flags labelled as in Figure 4.14, both rooted on flag 1, is the dodecagon with flags in $\{1, \ldots, 8\} \times \{1, \ldots, 12\}$ with the property that the entries are congruent modulo 4. This is only one of the four connected components when considering the entire set $\{1, \ldots, 8\} \times \{1, \ldots, 12\}$. The remaining three components also yield dodecagons.

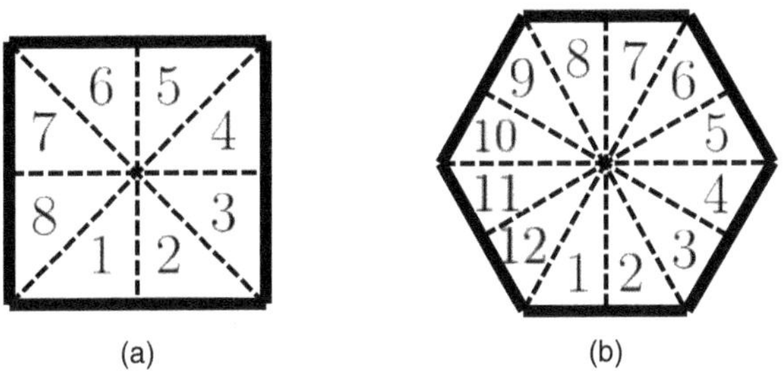

Figure 4.14 Square (a) and hexagon (b) with flag labels.

Proposition 4.4.3 *Let $\mathcal{P}$ and $\mathcal{Q}$ be two maniplexes and assume that $\mathcal{P}$ is regular. Then $\mathcal{P}_{(\Phi_1)}\diamondsuit\mathcal{Q}_{(\Psi)} \cong \mathcal{P}_{(\Phi_2)}\diamondsuit\mathcal{Q}_{(\Psi)}$, for all flags Φ_1 and Φ_2 of $\mathcal{P}$ and every flag Ψ of $\mathcal{Q}$.*

Proof Let $\gamma \in \Gamma(\mathcal{P})$ be such that $\Phi_1\gamma = \Phi_2$. We define a bijection $\hat{\gamma}$ between $\mathcal{F}l(\mathcal{P}_{(\Phi_1)}\diamondsuit\mathcal{Q}_{(\Psi)})$ and $\mathcal{F}l(\mathcal{P}_{(\Phi_2)}\diamondsuit\mathcal{Q}_{(\Psi)})$ given by $(\Phi, \Upsilon) \mapsto (\Phi\gamma, \Upsilon)$ for every flag Φ of $\mathcal{P}$ and every flag Υ of $\mathcal{Q}$. For every i we have that

$$(\Phi, \Upsilon)^i\hat{\gamma} = (\Phi^i, \Upsilon^i)\hat{\gamma} = (\Phi^i\gamma, \Upsilon^i) = (\Phi\gamma, \Upsilon)^i = [(\Phi, \Upsilon)\hat{\gamma}]^i,$$

proving at once that $\hat{\gamma}$ maps flags of $\mathcal{P}_{(\Phi_1)}\diamondsuit\mathcal{Q}_{(\Psi)}$ to flags of $\mathcal{P}_{(\Phi_2)}\diamondsuit\mathcal{Q}_{(\Psi)}$, and the desired isomorphism. ■

As a consequence of Proposition 4.4.3, if $\mathcal{P}$ is regular we write $\mathcal{P}\Diamond\mathcal{Q}_{(\Psi)}$ instead of $\mathcal{P}_{(\Phi)}\Diamond\mathcal{Q}_{(\Psi)}$.

Example 4.4.4 Let $\mathcal{P}$ be a non-orientable n-maniplex with base flag Φ_0, and let $\mathcal{D}$ be the n-premaniplex consisting of only two flags Ψ_0 and Ψ_1, each i-adjacent to the other for every $i \in \{0,\dots,n-1\}$. For each $i \in \{0,\dots,n-1\}$, each flag (Φ,Ψ_j) is i-adjacent in $\mathcal{P}_{(\Phi_0)}\Diamond\mathcal{D}_{(\Psi_0)}$ to some flag (Φ',Ψ_{1-j}), and hence $\mathcal{P}_{(\Phi_0)}\Diamond\mathcal{D}_{(\Psi_0)}$ is orientable. The projection of $\mathcal{P}_{(\Phi_0)}\Diamond\mathcal{D}_{(\Psi_0)}$ to the first coordinate is a covering to $\mathcal{P}$, implying that $\mathcal{P}$ is a quotient of $\mathcal{P}_{(\Phi_0)}\Diamond\mathcal{D}_{(\Psi_0)}$. Since $\mathcal{P}$ is non-orientable and $\mathcal{P}_{(\Phi_0)}\Diamond\mathcal{D}_{(\Psi_0)}$ is orientable, the degree of the covering is at least 2. But the number of flags when constructing $\mathcal{P}_{(\Phi_0)}\Diamond\mathcal{D}_{(\Psi_0)}$ is at most twice that of $\mathcal{P}$, forcing the degree of the covering to be at most 2. Hence $\mathcal{P}_{(\Phi_0)}\Diamond\mathcal{D}_{(\Psi_0)}$ is a double cover of $\mathcal{P}$. This allows us to define $\mathcal{P}_{(\Phi_0)}\Diamond\mathcal{D}_{(\Psi_0)}$ as the *orientable double cover* of $\mathcal{P}$. This way of defining the orientable double cover does not depend on the choice of base flags; the details are left as an exercise.

Exercise 4.4.5 With $\mathcal{P}$ and $\mathcal{D}$ as in Example 4.4.4, show that the mix $\mathcal{P}\Diamond\mathcal{D}$ does not depend on the base flags of $\mathcal{P}$ and $\mathcal{D}$.

With this definition of $\mathcal{D}$, we may generalise the corollary after theorem 1 in [270] to the following.

Proposition 4.4.6 *A maniplex $\mathcal{P}$ is orientable if and only if $\mathcal{P}\Diamond\mathcal{D} \cong \mathcal{P}$.*

Proof We first assume that $\mathcal{P}$ is orientable and let X_1 and X_2 be the parts of the bipartition of the flag graph, with the base flag $\Psi_1 \in X_1$. Let Φ_1, Φ_2 be the flags of $\mathcal{D}$, with Φ_1 being the base flag. Then a flag (Ψ,Φ_1) of $\mathcal{P}\Diamond\mathcal{D}$ with $\Psi \in X_1$ is adjacent only to flags (Υ,Φ_2) with $\Upsilon \in X_2$, and vice versa. It follows that the connected component in $\mathcal{P}\Diamond\mathcal{D}$ of the base flag (Ψ_1,Φ_1) contains precisely the flag set

$$\{(\Psi,\Phi_1)\colon \Psi \in X_1\} \cup \{(\Upsilon,\Phi_2)\colon \Upsilon \in X_2\}.$$

The projection to the first entry gives the desired isomorphism.

The converse follows from the discussion in Example 4.4.4. ■

From Proposition 4.4.6 it is clear that if $\mathcal{P}$ is non-orientably regular and $\mathcal{Q}_{(\Psi)}$ is chiral then

$$\mathcal{P}\Diamond\mathcal{Q}_{(\Psi)} \cong \mathcal{P}\Diamond(\mathcal{Q}_{(\Psi)}\Diamond\mathcal{D}) \cong (\mathcal{P}\Diamond\mathcal{D})\Diamond\mathcal{Q}_{(\Psi)},$$

and so the mix $\mathcal{P}\Diamond\mathcal{Q}_{(\Psi)}$ is just the mix of $\mathcal{Q}_{(\Psi)}$ with the orientable double cover of $\mathcal{P}$. For our purposes we may assume that whenever we mix a regular polytope $\mathcal{P}$ with a rooted chiral polytope $\mathcal{Q}_{(\Psi)}$ then $\mathcal{P}$ is orientably regular.

As Example 4.4.4 shows, the mix is a construction that does not require $\mathcal{P}$ and $\mathcal{Q}$ to satisfy the diamond condition or the strong flag-connectivity, or even to be maniplexes; it can be applied to more general structures. This suggests that $\mathcal{P}\Diamond\mathcal{Q}$ may not be always a polytope. Even when $\mathcal{P}$ and $\mathcal{Q}$ are regular polytopes, $\mathcal{P}\Diamond\mathcal{Q}$ need not be a polytope; an example of this can be found in [188, Section 7A]. If $\mathcal{P}$ is not orientable then even $\mathcal{P}\Diamond\mathcal{D}$ may fail to be a polytope, with $\mathcal{D}$ defined as in Example 4.4.4.

Exercise 4.4.7 Prove that if a 4-polytope $\mathcal{P}$ has non-orientable facets and non-orientable vertex-figures then $\mathcal{P}\Diamond\mathcal{D}$ is not polytopal.

On many occasions it is convenient to express the mix of polytopes in terms of quotient presentations.

Proposition 4.4.8 *Let $\mathcal{P}$ be a regular polytope and let $M, N \leq \Gamma(\mathcal{P})$ such that the rooted polytopes $\mathcal{Q}_{(\Phi)}$ and $\mathcal{R}_{(\Psi)}$ have quotient presentations $\mathcal{Q} = \mathcal{P}/M$ and $\mathcal{R} = \mathcal{P}/N$. Then $\mathcal{Q}_{(\Phi)}\Diamond\mathcal{R}_{(\Psi)}$ has quotient presentation $\mathcal{P}/(M \cap N)$.*

Proof Recall that M and N are the stabilisers of the base flags Φ and Ψ of $\mathcal{Q}$ and $\mathcal{R}$, respectively. The flag action of $\gamma \in \Gamma(\mathcal{P})$ stabilises the pair (Φ, Ψ) if and only if γ stabilises Φ as a flag of $\mathcal{Q}$, and Ψ as a flag of $\mathcal{R}$, and this happens if and only if $\gamma \in M \cap N$. ■

Exercise 4.4.9 Give an alternative proof of Proposition 4.4.6 using quotient presentations.

The following proposition provides a universal property of the mix in the category of premaniplexes.

Proposition 4.4.10 *The mix $\mathcal{P}_{(\Phi)}\Diamond\mathcal{Q}_{(\Psi)}$ covers both $\mathcal{P}_{(\Phi)}$ and $\mathcal{Q}_{(\Psi)}$. Furthermore, if $\mathcal{R}_{(\Upsilon)}$ covers both $\mathcal{P}_{(\Phi)}$ and $\mathcal{Q}_{(\Psi)}$ then $\mathcal{R}_{(\Upsilon)}$ covers $\mathcal{P}_{(\Phi)}\Diamond\mathcal{Q}_{(\Psi)}$.*

Proof Let $\mathcal{U}/M$ and $\mathcal{U}/N$ be quotient presentations of $\mathcal{P}_{(\Phi)}$ and $\mathcal{Q}_{(\Psi)}$, respectively. Proposition 4.4.8 tells us that $\mathcal{P}_{(\Phi)}\Diamond\mathcal{Q}_{(\Psi)}$ admits the quotient presentation $\mathcal{U}/(M \cap N)$. Now both conclusions of the statement follow from Proposition 3.2.51 (see also Remark 3.2.53). ■

Corollary 4.4.11 *For every rooted polytope $\mathcal{P}_{(\Phi)}$ we have that $\mathcal{P}_{(\Phi)} \Diamond \mathcal{P}_{(\Phi)}$ is isomorphic to $\mathcal{P}_{(\Phi)}$. Moreover, if $\mathcal{Q}_{(\Psi)}$ is a quotient of $\mathcal{P}_{(\Phi)}$ then $\mathcal{P}_{(\Phi)} \Diamond \mathcal{Q}_{(\Psi)}$ is isomorphic to $\mathcal{P}_{(\Phi)}$.*

Theorem 3.1.53 establishes when a group is the rotation group of an orientable rotary polytope. If Γ is the automorphism group of a non-orientably regular polytope $\mathcal{P}$ then it still satisfies the hypotheses of that theorem when taking the standard rotations as generators. Therefore it is possible to construct an orientably regular polytope $\mathcal{Q}$ such that $\Gamma^+(\mathcal{Q}) = \Gamma$. Since $\mathcal{Q}$ is orientable, it must cover the structure $\mathcal{D}$ defined in Example 4.4.4. As we shall see, $\mathcal{Q}$ also covers $\mathcal{P}$. Orientability of $\mathcal{Q}$ implies that if Φ is a flag of $\mathcal{Q}$ and $w \in \mathrm{Conn}(\mathcal{Q})$ such that $w\Phi = \Phi$ then w can be written as a word on the set $\{r_0, \dots, r_{n-1}\}$ of some even length. Then there exist $i_1, \dots, i_\ell \in \{1, \dots, n-1\}$ such that $w\Phi = \Phi\sigma_{i_1} \cdots \sigma_{i_\ell}$, as in the proof of Proposition 2.2.14. If Ψ is a flag of $\mathcal{P}$ and w' denotes the connection of $\mathcal{P}$ obtained as the same word as w on the set $\{r_0, \dots, r_{n-1}\}$, then $w'\Psi = \Psi\sigma_{i_1} \cdots \sigma_{i_\ell} = \Psi$, since $\Gamma^+(\mathcal{P}) = \Gamma^+(\mathcal{Q})$. Using Lemma 3.2.7 we conclude that $\mathcal{Q}$ also covers $\mathcal{P}$. Then Proposition 4.4.10 implies that $\mathcal{Q}$ covers the orientable double cover of $\mathcal{P}$. But $\mathcal{Q}$ is a double cover of $\mathcal{P}$ itself, since $\mathcal{Q}$ has two flag orbits under Γ. It follows that $\mathcal{Q} \cong \mathcal{P} \Diamond \mathcal{D}$. This provides an alternative way to obtain the orientable double cover of any non-orientably regular polytope.

In general it is hard to tell the symmetry type of $\mathcal{P}_{(\Phi)} \Diamond \mathcal{Q}_{(\Psi)}$ in terms of the symmetry types of $\mathcal{P}$ and $\mathcal{Q}$. We next present a result for rotary polytopes in this direction.

Proposition 4.4.12 *(a) If $\mathcal{P}$ and $\mathcal{Q}$ are regular maniplexes then $\mathcal{P} \Diamond \mathcal{Q}$ is also regular.*

(b) If $\mathcal{P}$ and $\mathcal{Q}$ are orientable rotary maniplexes then $\mathcal{P}_{(\Phi)} \Diamond \mathcal{Q}_{(\Psi)}$ is also orientable rotary for every choice of base flags Φ and Ψ.

Proof Let $\mathcal{P}_{(\Phi)} = \mathcal{U}/N$ and $\mathcal{Q}_{(\Psi)} = \mathcal{U}/M$ be quotient presentations.

If $\mathcal{P}$ and $\mathcal{Q}$ are regular then $N, M \lhd \Gamma(\mathcal{U})$ and therefore we also have that $N \cap M \lhd \mathcal{U}$. Hence $\mathcal{P} \Diamond \mathcal{Q} \cong \mathcal{U}_n/(M \cap N)$ is also regular.

On the other hand, if $\mathcal{P}$ and $\mathcal{Q}$ are orientable rotary then $N, M \lhd \Gamma^+(\mathcal{U})$ and therefore we also have that $N \cap M \lhd \Gamma^+(\mathcal{U})$. Hence $\mathcal{P}_{(\Phi)} \Diamond \mathcal{Q}_{(\Psi)} \cong \mathcal{U}/(M \cap N)$ is also orientable rotary. ∎

The mix of two chiral polytopes may be either regular or chiral. For example, if $\mathcal{Q}$ is chiral then we have that $\mathcal{Q}_{(\Phi)} \Diamond \mathcal{Q}_{(\Phi)} \cong \mathcal{Q}$ is also chiral, for every choice of base flag Φ. On the other hand, the mix of two chiral polytopes may be regular. An example of this can be constructed from $\mathcal{Q} = \{4,4\}_{(2,1)}$ as follows. Given a base flag Φ, the rooted polytope $\mathcal{Q}_{(\Phi)}$ admits the quotient presentation $\cong \{4,4\}/N$ for some $N \lhd [4,4]^+$ which is not normal in $[4,4]$. In this situation, $\mathrm{Core}_{[4,4]}(N) = N \cap \rho_0 N \rho_0$. Therefore,

$$\begin{aligned}\mathcal{Q}_{(\Phi)} \Diamond \mathcal{Q}_{(\Phi^0)} &= (\{4,4\}/N)\Diamond(\{4,4\}/\rho_0 N \rho_0) = \{4,4\}/(N \cap \rho_0 N \rho_0) \\ &= \{4,4\}/\mathrm{Core}_{[4,4]}(N),\end{aligned}$$

which is regular.

Exercise 4.4.13 Find a regular polytope $\mathcal{P}$, a chiral polytope $\mathcal{Q}$ and a flag Φ of $\mathcal{Q}$ such that

(a) $\mathcal{P} \Diamond \mathcal{Q}_{(\Phi)}$ is regular,
(b) $\mathcal{P} \Diamond \mathcal{Q}_{(\Phi)}$ is chiral.

When mixing an orientably regular polytope with a chiral polytope, the two choices of chirality forms for the latter yield isomorphic results.

Proposition 4.4.14 *Let $\mathcal{P}$ be a chiral polytope, let Φ be a flag of $\mathcal{P}$ and let $\mathcal{Q}$ be an orientably regular polytope. Then $\mathcal{P}_{(\Phi)} \Diamond \mathcal{Q}$ is isomorphic to $\mathcal{P}_{(\Phi^0)} \Diamond \mathcal{Q}$.*

Proof Let Ψ be the base flag of $\mathcal{Q}$, so that $\mathcal{P}_{(\Phi)} \Diamond \mathcal{Q}$ is rooted on (Φ, Ψ). Then Proposition 4.4.3 implies that

$$\mathcal{P}_{(\Phi)} \Diamond \mathcal{Q}_{(\Psi)} = \mathcal{P}_{(\Phi^0)} \Diamond \mathcal{Q}_{(\Psi^0)} \cong \mathcal{P}_{(\Phi^0)} \Diamond \mathcal{Q}_{(\Psi)}.$$

■

The statement of Proposition 4.4.14 may not hold if both $\mathcal{P}$ and $\mathcal{Q}$ are chiral. The simplest example of this is when $\mathcal{P} = \mathcal{Q}$, that is, $\mathcal{P}_{(\Phi)} \Diamond \mathcal{P}_{(\Phi)} \cong \mathcal{P}_{(\Phi)} \neq \mathcal{P}_{(\Phi)} \Diamond \mathcal{P}_{(\Phi^0)}$.

If $\mathcal{P}$ and $\mathcal{Q}$ are orientable rotary then their smallest regular covers are related to the regularity or chirality of $\mathcal{P} \Diamond \mathcal{Q}$.

Lemma 4.4.15 *Let $\mathcal{P}_{(\Phi)}$ be a rooted orientable rotary polytope and assume that $\mathcal{Q}_{(\Psi)}$ is a rooted chiral polytope of the same rank. Then $\mathcal{P}_{(\Phi)} \Diamond \mathcal{Q}_{(\Psi)}$ is regular if and only if it covers the smallest regular covers of $\mathcal{P}$ and of $\mathcal{Q}$.*

Proof Assume that $\mathcal{P}_{(\Phi)} = \mathcal{U}/N$ and $\mathcal{Q}_{(\Psi)} = \mathcal{U}/M$ are quotient presentations, and let $\mathcal{Q}'$ be the smallest regular cover of $\mathcal{Q}$. Then $\mathcal{P}_{(\Phi)} \Diamond \mathcal{Q}_{(\Psi)}$ and $\mathcal{Q}'$ have quotient presentations $\mathcal{U}/(N \cap M)$ and $\mathcal{U}/(M \cap \rho_0 M \rho_0)$, respectively, where ρ_0 is the first standard reflection of $\Gamma(\mathcal{U})$.

Assume first that $\mathcal{P}_{(\Phi)} \Diamond \mathcal{Q}_{(\Psi)}$ is regular, that is, $N \cap M \lhd \Gamma(\mathcal{U})$. Since $\mathrm{Core}_{\Gamma(\mathcal{U})}(M) = M \cap \rho_0 M \rho_0$, we have that $N \cap M \subseteq M \cap \rho_0 M \rho_0$. Then Proposition 3.2.51 tells us that $\mathcal{P}_{(\Phi)} \Diamond \mathcal{Q}_{(\Psi)} \cong \mathcal{U}/(N \cap M)$ covers

$\mathcal{U}/(M \cap \rho_0 M \rho_0) \cong \mathcal{Q}'$. A similar argument shows that $\mathcal{P}_{(\Phi)} \Diamond \mathcal{Q}_{(\Psi)}$ covers the smallest regular cover of $\mathcal{P}$.

Conversely, if $\mathcal{P}_{(\Phi)} \Diamond \mathcal{Q}_{(\Psi)}$ covers both $\mathcal{Q}'$ and the smallest regular cover of $\mathcal{P}$ then $N \cap M \subseteq N \cap (\rho_0 N \rho_0)$ and $N \cap M \subseteq M \cap (\rho_0 M \rho_0)$. It follows that

$$N \cap M \subseteq [N \cap (\rho_0 N \rho_0)] \cap [M \cap (\rho_0 M \rho_0)] \subseteq N \cap M.$$

Hence $N \cap M$ is normal since it is the intersection of two normal subgroups of $\Gamma(\mathcal{U})$, implying that $\mathcal{P}_{(\Phi)} \Diamond \mathcal{Q}_{(\Psi)} = \mathcal{U}/(N \cap M)$ is regular. ■

The definition of mix given above may be hard to use in practice. For rotary polytopes the mix can be equivalently defined in terms of their automorphism groups, which provides an extra tool for direct calculations.

Given two groups $\Delta = \langle \delta_1, \ldots, \delta_k \rangle$ and $\Lambda = \langle \lambda_1, \ldots, \lambda_k \rangle$ with prescribed generators, their *mix* is defined as $\Delta \Diamond \Lambda := \langle \eta_1, \ldots, \eta_k \rangle \subseteq \Delta \times \Lambda$, where $\eta_i := (\delta_i, \lambda_i)$ for every i. It is routine to show that $\Delta \Diamond \Lambda$ covers both Δ and Λ by mapping η_i to δ_i and to λ_i, respectively. Furthermore, if some group $\Gamma = \langle \gamma_1, \ldots, \gamma_k \rangle$ covers both Δ and Λ by mapping respectively γ_i to δ_i and to λ_i, then there is a surjective group homomorphism from Γ to $\Delta \Diamond \Lambda$ mapping γ_i to η_i.

We could have defined the mix $\mathcal{P}_{(\Phi)} \Diamond \mathcal{Q}_{(\Psi)}$ of two rooted regular polytopes $\mathcal{P}_{(\Phi)}$ and $\mathcal{Q}_{(\Psi)}$ as the combinatorial structure arising from the sggi $\Gamma(\mathcal{P}) \Diamond \Gamma(\mathcal{Q})$ as in Section 3.1, with prescribed generators given by the abstract reflections taken with respect to Φ and to Ψ (see [188, Section 5A]). Similarly, if $\mathcal{P}_{(\Phi)}$ and $\mathcal{Q}_{(\Psi)}$ are rooted orientable rotary polytopes then their mix could have been defined as the combinatorial structure arising from $\Gamma^+(\mathcal{P}) \Diamond \Gamma^+(\mathcal{Q})$ as in Section 3.1, with prescribed generators given by the abstract rotations taken with respect to Φ and to Ψ. The next proposition shows that these definitions are equivalent to the corresponding particular cases of our original definition of mix. For convenience we use the notation

$$\Gamma(\mathcal{P}_{(\Phi)}) = \langle \sigma_1, \ldots, \sigma_{n-1} \rangle \tag{4.2}$$

to indicate that the standard rotations are taken with respect to the flag Φ.

Proposition 4.4.16 *(a) If $\mathcal{P}$ and $\mathcal{Q}$ are regular maniplexes then* $\Gamma(\mathcal{P} \Diamond \mathcal{Q}) \cong \Gamma(\mathcal{P}) \Diamond \Gamma(\mathcal{Q})$.

(b) If $\mathcal{P}_{(\Phi)}$ and $\mathcal{Q}_{(\Psi)}$ are rooted orientable rotary maniplexes then $\Gamma^+(\mathcal{P}_{(\Phi)} \Diamond \mathcal{Q}_{(\Psi)}) \cong \Gamma^+(\mathcal{P}_{(\Phi)}) \Diamond \Gamma^+(\mathcal{Q}_{(\Psi)})$.

Proof Assume first that $\mathcal{P}$ and $\mathcal{Q}$ are regular. For each i we proceed in both coordinates as in the proof of Proposition 4.4.3 to conclude that there is an isomorphism $\hat{\rho}_i$ between $\Gamma(\mathcal{P}_{(\Phi)} \Diamond \mathcal{Q}_{(\Psi)})$ and $\Gamma(\mathcal{P}_{(\Phi^i)} \Diamond \mathcal{Q}_{(\Psi^i)})$ that maps

(Φ, Ψ) to (Φ^i, Ψ^i). But $(\Phi, \Psi)^i = (\Phi^i, \Psi^i)$, implying that $\Gamma(\mathcal{P}_{(\Phi)} \Diamond \mathcal{Q}_{(\Psi)}) = \Gamma(\mathcal{P}_{(\Phi^i)} \Diamond \mathcal{Q}_{(\Psi^i)})$ and $\hat{\rho}_i$ is the ith standard reflection of $\mathcal{P} \Diamond \mathcal{Q}$. On the other hand, $\hat{\rho}_i$ is the pair consisting of the ith standard reflections of $\mathcal{P}$ and $\mathcal{Q}$. This proves the first item. The second item is analogous, using standard rotations instead of standard reflections. ■

This group-theoretical point of view simplifies the process of determining the mix of a square and a hexagon (see Example 4.4.2). If the automorphism group of the square is $\langle \rho_0, \rho_1 \rangle$ and the one of the hexagon is $\langle \rho_0', \rho_1' \rangle$ then the automorphism group of the mix is $\langle (\rho_0, \rho_0'), (\rho_1, \rho_1') \rangle$. Since both generators are involutions, the automorphism group is dihedral with twice as many elements as the order of the product $(\rho_0\rho_1, \rho_0'\rho_1')$. Since the order of $\rho_0\rho_1$ is 4, and the order of $\rho_0'\rho_1'$ is 6 it follows that the order of $(\rho_0\rho_1, \rho_0'\rho_1')$ is 12 and the mix is a dodecagon. Alternatively, since both polygons are regular, we know that the mix is regular and it suffices to consider the mix of the rotation subgroups of these polygons. This is easily verified to be a cyclic group of order 12, implying that the mix is indeed a dodecagon.

Exercise 4.4.17 Determine which toroidal polyhedra are $\{4,4\}_{(2,0)} \Diamond \{4,4\}_{(3,0)}$ and $\{4,4\}_{(2,0)} \Diamond \{4,4\}_{(2,1)}$.

As a corollary of Proposition 4.3.9, every chiral polytope $\mathcal{Q}$ with regular facets or vertex-figures has a polytopal smallest regular cover, and it is precisely the regular polytope whose automorphism group is isomorphic to $\mathrm{Conn}(\mathcal{Q})$. This highlights the relevance of determining $\mathrm{Conn}(\mathcal{Q})$. The connection group of a chiral maniplex has an expression in terms of the mix, as shown in the next proposition (see [191, Theorem 7.2] for an alternative proof).

Proposition 4.4.18 *Let $\mathcal{Q}_{(\Phi)}$ be a rooted chiral polytope with $\Gamma(\mathcal{Q}_{(\Phi)}) = \langle \sigma_1, \ldots, \sigma_{n-1} \rangle$. Then*

(a) $\mathrm{Conn}^+(\mathcal{Q}) \cong \Gamma(\mathcal{Q}_{(\Phi)}) \Diamond \Gamma(\mathcal{Q}_{(\Phi^0)})$;
(b) $\mathrm{Conn}(\mathcal{Q}) \cong (\Gamma(\mathcal{Q}_{(\Phi)}) \Diamond \Gamma(\mathcal{Q}_{(\Phi^0)})) \rtimes \mathbb{Z}_2$.

Proof Let $\mathcal{Q}_{(\Phi)} = \mathcal{U}/N$ be a quotient presentation and let ρ_0 be the first standard generator of $\Gamma(\mathcal{U})$. Then $\mathcal{Q}_{(\Phi^0)} = \mathcal{U}/(\rho_0 N \rho_0)$. Hence $\mathcal{Q}_{(\Phi)} \Diamond \mathcal{Q}_{(\Phi^0)} = \mathcal{U}/(N \cap \rho_0 N \rho_0)$. Lemma 3.3.10 then implies that $\mathcal{Q}_{(\Phi)} \Diamond \mathcal{Q}_{(\Phi^0)} = \mathcal{U}/\mathrm{Core}_{\Gamma(\mathcal{U})}(N)$. Finally, from Propositions 3.2.54 and 4.4.16 we know that

$$\Gamma^+(\mathcal{Q}_{(\Phi)}) \Diamond \Gamma^+(\mathcal{Q}_{(\Phi^0)}) = \Gamma^+(\mathcal{Q}_{(\Phi)} \Diamond \mathcal{Q}_{(\Phi^0)}) \cong \Gamma^+(\mathcal{U})/\mathrm{Core}_{\Gamma(\mathcal{U})}(N)$$
$$\cong \mathrm{Conn}^+(\mathcal{Q}).$$

Since $\mathrm{Conn}^+(\mathcal{Q})$ has index 2 in $\mathrm{Conn}(\mathcal{Q})$ and each standard generator of $\mathrm{Conn}(\mathcal{Q})$ is an involution not contained in $\mathrm{Conn}^+(\mathcal{Q})$, we have that $\mathrm{Conn}(\mathcal{Q}) \cong \mathrm{Conn}^+(\mathcal{Q}) \rtimes \mathbb{Z}_2$. ■

It is possible to use Proposition 4.4.18 to determine the chirality index of chiral polytopes.

Corollary 4.4.19 *The chirality index of a finite chiral polytope $\mathcal{P}$ can be computed as*

$$\kappa(\mathcal{P}) = |\mathrm{Conn}^+(\mathcal{P})|/|\Gamma^+(\mathcal{P})|.$$

Proof Let $\mathcal{P}_{(\Phi)} = \mathcal{U}/N$ be a quotient presentation. Then the chirality index of $\mathcal{P}$ is

$$|N/(N \cap \rho_0 N \rho_0)| = \left|\frac{\Gamma^+(\mathcal{U})/(N \cap \rho_0 N \rho_0)}{\Gamma^+(\mathcal{U})/N}\right| = \left|\frac{\Gamma^+(\mathcal{P}_{(\Phi)} \Diamond \mathcal{P}_{(\Phi^0)})}{\Gamma^+(\mathcal{P})}\right|$$
$$= \left|\frac{\mathrm{Conn}^+(\mathcal{P})}{\Gamma^+(\mathcal{P})}\right|.$$

■

The mix of polytopes is useful not only to give an alternative description of the connection group of chiral polytopes, it also gives the following characterisation of the chirality group.

Lemma 4.4.20 *Let $\mathcal{Q}_{(\Phi)}$ be a rooted chiral polytope. Then $\chi(\mathcal{Q})$ is isomorphic to the kernel of the natural projection from $\Gamma(\mathcal{Q}_{(\Phi)}) \Diamond \Gamma(\mathcal{Q}_{(\Phi^0)})$ to $\Gamma(\mathcal{Q}_{(\Phi)})$. Moreover, $\Gamma(\mathcal{Q}_{(\Phi)}) \Diamond \Gamma(\mathcal{Q}_{(\Phi^0)})$ contains a subgroup isomorphic to $\chi(\mathcal{Q}) \times \chi(\mathcal{Q})$.*

Proof Let $\mathcal{Q} = \mathcal{U}/N$ be a quotient presentation. Then $\mathcal{Q}_{(\Phi^0)}$ and $\mathcal{Q}_{(\Phi)} \Diamond \mathcal{Q}_{(\Phi^0)}$ admit the quotient presentations $\mathcal{U}/\rho_0 N \rho_0$ and $\mathcal{U}/(N \cap \rho_0 N \rho_0)$, respectively. Furthermore, $\Gamma(\mathcal{Q}) \cong \Gamma^+(\mathcal{U})/N$ and $\Gamma(\mathcal{Q}_{(\Phi)} \Diamond \mathcal{Q}_{(\Phi^0)}) \cong \Gamma^+(\mathcal{U})/(N \cap \rho_0 N \rho_0)$.

Then the natural projection η_1 from $\Gamma(\mathcal{Q}_{(\Phi)}) \Diamond \Gamma(\mathcal{Q}_{(\Phi^0)})$ to $\Gamma(\mathcal{Q}_{(\Phi)})$ maps an element $w \cdot (N \cap \rho_0 N \rho_0)$ to $w \cdot N$. It follows that the kernel of this projection is $N/(N \cap \rho_0 N \rho_0) \cong \chi(\mathcal{Q})$.

A similar argument shows that $\chi(\mathcal{Q})$ is isomorphic to the kernel of the natural projection η_2 from $\Gamma(\mathcal{Q}_{(\Phi)}) \Diamond \Gamma(\mathcal{Q}_{(\Phi^0)})$ to $\Gamma(\mathcal{Q}_{(\Phi^0)})$. Since $\ker(\eta_1) \cap \ker(\eta_2)$ is trivial we conclude that $\chi(\mathcal{Q}) \times \chi(\mathcal{Q}) \cong \ker(\eta_1) \times \ker(\eta_2) \subseteq \Gamma(\mathcal{Q}_{(\Phi)}) \Diamond \Gamma(\mathcal{Q}_{(\Phi^0)})$. ■

The next exercise provides another interpretation to Proposition 4.3.12 in view of Proposition 4.4.18 and Lemma 4.4.20.

Exercise 4.4.21 Use Proposition 4.4.18 to write the even connection group of each chiral polytope $\mathcal{P}$ as $\mathrm{Conn}^+(\mathcal{P}) = \Gamma(\mathcal{P}_{(\Phi)})\Diamond\Gamma(\mathcal{P}_{(\Phi^1)})$ for some chosen flag Φ, and use Lemma 4.4.20 to express $\chi(\mathcal{P})$ as the set of elements (γ, id) that belong to $\mathrm{Conn}^+(\mathcal{P})$. Reinterpret with respect to this setting the intersection condition for $\mathrm{Conn}^+(\mathcal{P})$ and the condition $\chi(\mathcal{Q}) \cap \chi(\mathcal{R}) \subseteq \chi(\mathcal{M})$ given in Proposition 4.3.12. Use this to give an alternative proof of Proposition 4.3.12.

From Proposition 4.4.16 we know that if $\mathcal{P}_{(\Phi)}$ and $\mathcal{Q}_{(\Psi)}$ are orientable rotary polytopes then $\mathcal{P}_{(\Phi)}\Diamond\mathcal{Q}_{(\Psi)}$ is an orientable rotary polytope if and only if $\Gamma^+(\mathcal{P}_{(\Phi)})\Diamond\Gamma^+(\mathcal{Q}_{(\Psi)})$ satisfies the intersection condition (3.10). As we mentioned before, the mix of rotary polytopes is not always a polytope; not even if both of them are regular. We next present sufficient conditions for the mix of orientable rotary polytopes $\mathcal{P}$ and $\mathcal{Q}$ to be a polytope.

Theorem 4.4.22 *Let $\mathcal{P}$ and $\mathcal{Q}$ be orientable rotary polyhedra. Then $\mathcal{P}\Diamond\mathcal{Q}$ is an orientable rotary polyhedron regardless of the choice of base flags.*

Proof Let $\Gamma^+(\mathcal{P}) = \langle\sigma_1, \sigma_2\rangle$, $\Gamma^+(\mathcal{Q}) = \langle\sigma_1', \sigma_2'\rangle$, and let $\nu_i = (\sigma_i, \sigma_i')$. Let $\mu \in \langle\nu_1\rangle \cap \langle\nu_2\rangle$. Then $\mu = (\sigma_1^k, (\sigma_1')^k) = (\sigma_2^m, (\sigma_2')^m)$ for some k and m. Because of the intersection condition on $\mathcal{P}$ and $\mathcal{Q}$, $\mu = (\mathrm{id}, \mathrm{id})$. Hence $\langle\nu_1\rangle \cap \langle\nu_2\rangle = \{\mathrm{id}\}$ and the intersection condition holds. ■

Theorem 4.4.23 *Let $\mathcal{P}_{(\Phi)}$ and $\mathcal{Q}_{(\Psi)}$ be rooted orientable rotary n-polytopes with facets isomorphic to the $(n-1)$-polytopes $\mathcal{P}'$ and $\mathcal{Q}'$, respectively. Let Φ' and Ψ' be the restrictions of the flags Φ and Ψ to the base facet. Assume that $\mathcal{P}'_{(\Phi')}$ covers $\mathcal{Q}'_{(\Psi')}$. Then $\mathcal{P}_{(\Phi)}\Diamond\mathcal{Q}_{(\Psi)}$ is also a polytope.*

Proof Let $\Gamma^+ = \Gamma^+(\mathcal{P}_{(\Phi)}\Diamond\mathcal{Q}_{(\Psi)}) = \langle\eta_1, \ldots, \eta_{n-1}\rangle$ and $\Gamma^+(\mathcal{P}_{(\Phi)}) = \langle\sigma_1, \ldots, \sigma_{n-1}\rangle$. Then, by Corollary 4.4.11,

$$\langle\eta_1, \ldots, \eta_{n-2}\rangle = \Gamma^+(\mathcal{P}'_{(\Phi')}\Diamond\mathcal{Q}'_{(\Psi')}) \cong \Gamma^+(\mathcal{P}'_{(\Phi')}) = \langle\sigma_1, \ldots, \sigma_{n-2}\rangle.$$

Therefore the quotient given by $\eta_j \mapsto \sigma_j$ is one-to-one on $\Gamma^+_{n-1} = \langle\eta_1, \ldots, \eta_{n-2}\rangle$. By Lemma 3.1.63, Γ^+ satisfies the intersection condition (3.10), and $\mathcal{P}_{(\Phi)}\Diamond\mathcal{Q}_{(\Psi)}$ is also a polytope. ■

Theorem 4.4.23 and its dual version imply the following result.

Corollary 4.4.24 *Let $\mathcal{P}_{(\Phi)}$ and $\mathcal{Q}_{(\Psi)}$ be orientable rotary rooted n-polytopes with isomorphic facets or vertex-figures. Then $\mathcal{P}_{(\Phi)}\Diamond\mathcal{Q}_{(\Psi)}$ is also a polytope.*

Describing the rotation group of the mix of two polytopes is often time consuming. The following theorem shows a case when the mix of two orientable rotary polytopes is guaranteed to be polytopal, and its rotation group easily describable.

Theorem 4.4.25 *Let $\mathcal{P}$ and $\mathcal{Q}$ be orientable rotary polytopes with types $\{p_1,\dots,p_{n-1}\}$ and $\{q_1,\dots,q_{n-1}\}$, respectively. If p_i and q_i are coprime for every $i \in \{1,\dots,n-1\}$ then $\mathcal{P}\Diamond\mathcal{Q}$ is an orientable rotary polytope of type $\{p_1q_1,\dots,p_{n-1}q_{n-1}\}$, and $\Gamma^+(\mathcal{P}\Diamond\mathcal{Q}) \cong \Gamma^+(\mathcal{P}) \times \Gamma^+(\mathcal{Q})$, regardless of the choice of base flags.*

Proof Let $\Gamma^+(\mathcal{P}) = \langle\sigma_1,\dots,\sigma_{n-1}\rangle$ and $\Gamma^+(\mathcal{Q}) = \langle\sigma'_1,\dots,\sigma'_{n-1}\rangle$. Since p_i and q_i are coprime for every i, it follows that $\langle(\sigma_i,\sigma'_i)\rangle = \langle(\sigma_i,\mathrm{id})\rangle \times \langle(\mathrm{id},\sigma'_i)\rangle$. In other words, the order of the generator (σ_i,σ'_i) of $\Gamma^+(\mathcal{P})\Diamond\Gamma^+(\mathcal{Q})$ is precisely p_iq_i. This implies that

$$\begin{aligned}\langle\sigma_1,\dots,\sigma_{n-1}\rangle \times \langle\sigma'_1,\dots,\sigma'_{n-1}\rangle &= \langle(\sigma_i,\mathrm{id}),(\mathrm{id},\sigma'_i)\colon i \in \{1,\dots,n-1\}\rangle\\ &= \langle(\sigma_i,\sigma'_i)\colon i \in \{1,\dots,n-1\}\rangle\\ &= \langle\sigma_1,\dots,\sigma_{n-1}\rangle\Diamond\langle\sigma'_1,\dots,\sigma'_{n-1}\rangle\\ &\subseteq \langle\sigma_1,\dots,\sigma_{n-1}\rangle \times \langle\sigma'_1,\dots,\sigma'_{n-1}\rangle,\end{aligned}$$

from where $\Gamma^+(\mathcal{P})\Diamond\Gamma^+(\mathcal{Q}) = \Gamma^+(\mathcal{P}) \times \Gamma^+(\mathcal{Q})$. This proves the second part. As a consequence of the previous analysis we have that for $\eta_i = (\sigma_i,\sigma'_i)$ and for every $I \subseteq \{1,\dots,n-1\}$,

$$\langle\eta_i : i \in I\rangle \cong \langle\sigma_i : i \in I\rangle \times \langle\sigma'_i : i \in I\rangle. \tag{4.3}$$

Next we prove the intersection condition by induction over n, with the help of Proposition 3.1.60 (a). The base case holds since $\langle\eta_1\rangle$ satisfies the intersection condition.

Now assume that the intersection condition holds if there are at most $k-2$ generators η_i. In particular, it holds in $\langle\eta_1,\dots,\eta_{k-2}\rangle$. We only need to show that for all $j \in \{2,\dots,k-1\}$, the group $\langle\eta_1,\dots,\eta_{k-2}\rangle\cap\langle\eta_j,\dots,\eta_{k-1}\rangle$ equals $\langle\eta_j,\dots,\eta_{k-2}\rangle$. By (4.3), $\langle\eta_1,\dots,\eta_{k-2}\rangle \cap \langle\eta_j,\dots,\eta_{k-1}\rangle$ equals

$$\langle\sigma_1,\dots,\sigma_{k-2}\rangle \times \langle\sigma'_1,\dots,\sigma'_{k-2}\rangle \cap \langle\sigma_j,\dots,\sigma_{k-1}\rangle \times \langle\sigma'_j,\dots,\sigma'_{k-1}\rangle.$$

The intersection condition in $\Gamma^+(\mathcal{P})$ and $\Gamma^+(\mathcal{Q})$ implies that this in turn is $\langle\sigma_j,\dots,\sigma_{k-2}\rangle \times \langle\sigma'_j,\dots,\sigma'_{n-2}\rangle$, which by (4.3) equals $\langle\eta_j,\dots,\eta_{k-2}\rangle$. ■

Weaker conditions can be asked for $\mathcal{P}$ and $\mathcal{Q}$ in the previous theorem to guarantee that $\mathcal{P}\Diamond\mathcal{Q}$ is a polytope, but when requiring them we lose control on the automorphism group (see [64, Theorem 3.7]).

Theorem 4.4.26 *Let $\mathcal{P}$ and $\mathcal{Q}$ be orientable rotary n-polytopes ($n \geq 3$) with types $\{p_1,\dots,p_{n-1}\}$ and $\{q_1,\dots,q_{n-1}\}$, respectively. If p_i and q_i are coprime* *for every $i \in \{2, \ldots, n-2\}$ then $\mathcal{P} \diamondsuit \mathcal{Q}$ is a polytope, regardless of the choice of base flags.*

Proof We proceed by induction over n. The result for $n = 3$ follows from Theorem 4.4.22.

Assume that the theorem is true for polytopes with rank at most $n-1$. Let $\Gamma^+(\mathcal{P}) = \langle \sigma_1, \ldots, \sigma_{n-1} \rangle$, $\Gamma^+(\mathcal{Q}) = \langle \sigma_1', \ldots, \sigma_{n-1}' \rangle$ and let $\eta_i = (\sigma_i, \sigma_i')$ for $i \in \{1, \ldots, n-1\}$. Then $\Gamma^+(\mathcal{P}) \diamondsuit \Gamma^+(\mathcal{Q}) = \langle \eta_1, \ldots, \eta_{n-1} \rangle$. The inductive hypothesis holds for the groups $\langle \sigma_1, \ldots, \sigma_{n-2} \rangle \diamondsuit \langle \sigma_1', \ldots, \sigma_{n-2}' \rangle$ and $\langle \sigma_2, \ldots, \sigma_{n-1} \rangle \diamondsuit \langle \sigma_2', \ldots, \sigma_{n-1}' \rangle$. Hence the intersection condition holds for $\langle \eta_1, \ldots, \eta_{n-2} \rangle$ and for $\langle \eta_2, \ldots, \eta_{n-1} \rangle$. By Proposition 3.1.60 (b), it only remains to be shown that

$$\langle \eta_1, \ldots, \eta_{n-2} \rangle \cap \langle \eta_2, \ldots, \eta_{n-1} \rangle \subseteq \langle \eta_2, \ldots, \eta_{n-2} \rangle.$$

Now, from (4.3) we know that

$$\begin{aligned}
&\langle \eta_2, \ldots, \eta_{n-2} \rangle \\
&= \langle \sigma_2, \ldots, \sigma_{n-2} \rangle \times \langle \sigma_2', \ldots, \sigma_{n-2}' \rangle \\
&= (\langle \sigma_1, \ldots, \sigma_{n-2} \rangle \cap \langle \sigma_2, \ldots, \sigma_{n-1} \rangle) \times (\langle \sigma_1', \ldots, \sigma_{n-2}' \rangle \cap \langle \sigma_2', \ldots, \sigma_{n-1}' \rangle) \\
&= (\langle \sigma_1, \ldots, \sigma_{n-2} \rangle \times \langle \sigma_1', \ldots, \sigma_{n-2}' \rangle) \cap (\langle \sigma_2, \ldots, \sigma_{n-1} \rangle \times \sigma_2', \ldots, \sigma_{n-1}' \rangle \\
&\supseteq \langle (\sigma_1, \sigma_1'), \ldots, (\sigma_{n-2}, \sigma_{n-2}') \rangle \cap \langle (\sigma_2, \sigma_2'), \ldots, (\sigma_{n-1}, \sigma_{n-1}') \rangle \\
&= \langle \eta_1, \ldots, \eta_{n-2} \rangle \cap \langle \eta_2, \ldots, \eta_{n-1} \rangle,
\end{aligned}$$

and the intersection condition holds. ■

Constructions of families of chiral 3-, 4- and 5-polytopes with explicit automorphism groups can be found in [78]. The main technique in that work is to mix an orientably regular n-polytope and a totally chiral n-polytope with isomorphic facets (or vertex-figures), and use the ideas behind Lemma 4.4.15 to force chirality when $n \in \{3, 4\}$. For the construction of 5-polytopes they use Theorem 4.4.25; Theorems 4.4.25 and 4.4.26 are used in [64] for the same purpose.

Self-dual chiral polytopes are constructed in [62] by mixing a chiral polytope $\mathcal{P}$ with any of the chirality forms of $\mathcal{P}^\delta$. When rooting $\mathcal{P}^\delta$ on the same flag as $\mathcal{P}$, the outcome of the mix is properly self-dual. Otherwise, if the base flags of $\mathcal{P}$ and of $\mathcal{P}^\delta$ are in different flag orbits then the mix is either orientably regular or improperly self-dual. In each case, some care is needed to determine whether such mix is polytopal and whether it is regular. One instance when polytopality is guaranteed and the automorphism group is easy to compute is when $\mathcal{P}$ is an n-polytope with n odd, and has type $\{p_1, \ldots, p_{n-1}\}$ with p_i and p_{n-i} relatively prime for all i (see Theorem 4.4.25). Explicit examples of the construction of a self-dual chiral polyhedron and of a self-dual chiral 4-polytope are presented in [62, Section 6].

A similar treatment can be applied to obtain *self-Petrial* rotary polytopes, that is, rotary polytopes that are isomorphic to their Petrials. Results about rotary polyhedra that are both self-dual and self-Petrial can be found in [63].

As mentioned in the previous section, the connection group of the chiral 5-polytope with type $\{3,4,4,3\}$ described in Exercise 2.1.21 is not a string C-group. Here we give a non-constructive proof of this fact (see [191, Example 7.6]).

Example 4.4.27 Let $\mathcal{P}$ be the chiral 5-polytope with type $\{3,4,4,3\}$ described in Exercise 2.1.21 and let $\mathcal{Q}$ be the 4-polytope with type $\{3,4,4\}$ isomorphic to any facet of $\mathcal{P}$. Then the vertex-figures of $\mathcal{Q}$ are isomorphic to $\{4,4\}_{(2,1)}$. We recall that $\Gamma(\mathcal{P}) \cong S_6$ and can be represented by assigning permutations $\sigma_1, \sigma_2, \sigma_3, \sigma_4$ as in (4.1). We use this representation also for the groups $\Gamma(\mathcal{Q}) \cong \langle \sigma_1, \sigma_2, \sigma_3 \rangle$ and $\Gamma(\{4,4\}_{(2,1)}) \cong \langle \sigma_2, \sigma_3 \rangle$.

For convenience we introduce the following notation for permutation groups Γ and Λ:

$$\Gamma \perp \Lambda := \{(\gamma, \lambda) \in \Gamma \times \Lambda : \gamma\lambda \text{ is an even permutation}\}.$$

That is, $(\gamma, \lambda) \in \Gamma \perp \Lambda$ if and only if γ and λ are both odd permutations or both even permutations. Since for any rooted chiral polytope $\mathcal{R}_{(\Phi)}$ the standard generators of $\Gamma(\mathcal{R}_{(\Phi^0)})$ are conjugates of those of $\Gamma(\mathcal{R}_{(\Phi)})$ or of their inverses, they have the same cyclic structure as permutations. Therefore $\Gamma(\mathcal{R}_{(\Phi)}) \Diamond \Gamma(\mathcal{R}_{(\Phi^0)}) \subseteq \Gamma(\mathcal{R}_{(\Phi)}) \perp \Gamma(\mathcal{R}_{(\Phi^0)})$.

We first recall that

$$|\Gamma([\{4,4\}_{(2,1)}]_{(\Phi)}) \Diamond \Gamma([\{4,4\}_{(2,1)}]_{(\Phi^0)})| = |\Gamma^+(\{4,4\}_{(5,0)})| = 100$$

for any choice of flag Φ. On the other hand,

$$|\Gamma([\{4,4\}_{(2,1)}]_{(\Phi)}) \perp \Gamma([\{4,4\}_{(2,1)}]_{(\Phi^0)})| = \frac{1}{2} \cdot 20^2 = 200.$$

Therefore there exists

$$\mu \in \Gamma([\{4,4\}_{(2,1)}]_{(\Phi)}) \perp \Gamma([\{4,4\}_{(2,1)}]_{(\Phi^0)}) \setminus \Gamma([\{4,4\}_{(2,1)}]_{(\Phi)}) \Diamond \Gamma([\{4,4\}_{(2,1)}]_{(\Phi^0)}).$$

Now, $\Gamma(\mathcal{Q}) \cong S_5$ and therefore, by Proposition 3.3.14, $\chi(\mathcal{Q}) \in \{A_5, S_5\}$. Lemma 4.4.20 then implies that for some flag Ψ,

$$A_5 \times A_5 \leq \Gamma(\mathcal{Q}_{(\Psi)}) \Diamond \Gamma(\mathcal{Q}_{(\Psi^0)}) \leq \Gamma(\mathcal{Q}_{(\Psi)}) \perp \Gamma(\mathcal{Q}_{(\Psi^0)}) = S_5 \perp S_5.$$

However, $\Gamma(\mathcal{Q}_{(\Psi)})\Diamond\Gamma(\mathcal{Q}_{(\Psi^0)})$ contains elements (γ,λ) with γ odd. Since $A_5 \times A_5$ is an index-2 subgroup of $S_5 \perp S_5$, we have that $\Gamma(\mathcal{Q}_{(\Psi)})\Diamond\Gamma(\mathcal{Q}_{(\Psi^0)}) = \Gamma(\mathcal{Q}_{(\Psi)}) \perp \Gamma(\mathcal{Q}_{(\Psi^0)})$.

The group $\Gamma([\{4,4\}_{(2,1)}]_{(\Phi)}) \perp \Gamma([\{4,4\}_{(2,1)}]_{(\Phi^0)})$ is contained in $\Gamma(\mathcal{Q}_{(\Psi)}) \perp \Gamma(\mathcal{Q}_{(\Psi^0)})$, and hence $\mu \in \Gamma(\mathcal{Q}_{(\Psi)}) \perp \Gamma(\mathcal{Q}_{(\Psi^0)})$. In other words, if $\Gamma(\mathcal{Q}_{(\Psi)})\Diamond\Gamma(\mathcal{Q}_{(\Psi^0)}) = \langle \xi_1,\xi_2,\xi_3,\xi_4\rangle$, then $\mu \in \langle \xi_1,\xi_2,\xi_3\rangle$. A similar argument with the vertex-figure instead of $\mathcal{Q}$ shows that $\mu \in \langle \xi_2,\xi_3,\xi_4\rangle$. However, $\mu \notin \langle \xi_2,\xi_3\rangle \cong \Gamma([\{4,4\}_{(2,1)}]_{(\Phi)})\Diamond\Gamma([\{4,4\}_{(2,1)}]_{(\Phi^0)})$, contradicting the intersection condition.

Some chiral polytopes, like the chiral 5-polytope with type $\{3,4,4,3\}$ described in Exercise 2.1.21, do not have smallest regular covers given by their connection groups. On the other hand, every chiral n-polytope is covered by the universal n-polytope of type $\{\infty,\dots,\infty\}$. We next show that every finite chiral polytope has a finite regular cover. We follow the idea in [197], although the requirement of chirality makes it much simpler.

Theorem 4.4.28 *Every finite chiral n-polytope is a quotient of a finite regular n-polytope.*

Proof From Proposition 4.3.9 we know that if $\mathcal{Q}$ is a chiral polytope with regular facets or vertex-figures then $\mathrm{Conn}(\mathcal{Q})$ is the automorphism group of a regular polytope that covers $\mathcal{Q}$. It only remains to consider the case where the facet and vertex-figures of $\mathcal{Q}$ are chiral.

Assume that $\mathcal{Q}$ is a chiral n-polytope with chiral facets isomorphic to $\mathcal{P}$. Proposition 2.3.8 implies that the facets of $\mathcal{P}$ are regular, and therefore $\mathcal{P}$ has a (finite) smallest regular cover $\tilde{\mathcal{P}}$. Let $\mathcal{R}$ be the (regular) trivial extension of $\tilde{\mathcal{P}}$, described in Exercise 2.2.4. Then the facets of $\mathcal{R}$ cover the facets of $\mathcal{Q}$ and, by Theorem 4.4.23, $\mathcal{Q}\Diamond\mathcal{R}$ is a finite orientable rotary polytope that covers $\mathcal{Q}$.

If $\mathcal{Q}\Diamond\mathcal{R}$ is regular then it is the desired finite cover of $\mathcal{Q}$. Otherwise, $\mathcal{Q}\Diamond\mathcal{R}$ is chiral but has regular facets (isomorphic to $\tilde{\mathcal{P}}$). Then, by Proposition 4.3.9, $\mathcal{Q}\Diamond\mathcal{R}$ has a smallest regular cover which is also finite and covers $\mathcal{Q}$. ■

Example 4.4.4 is a special case of double covers of polytopes. In [18, 153] other natural double covers of maps and hypermaps are studied. We extend these ideas to abstract polytopes in our terminology as follows.

Let $I \subset \{0,\dots,n-1\}$ and let $\mathcal{D}_I$ be the premaniplex consisting of two flags with the property that each of them is i-adjacent to itself for every $i \notin I$, and is adjacent to the other flag otherwise. Swapping the two flags of $\mathcal{D}_I$ determines an automorphism, implying that $\mathcal{D}_I$ is regular. This allows us to write mixes of the unrooted premaniplex $\mathcal{D}_I$ with other maniplexes.

Remark 4.4.29 If $\mathcal{P}_{(\Phi)}$ is a rooted n-polytope and $\mathcal{P}_{(\Phi)}\Diamond\mathcal{D}_I$ is not isomorphic to $\mathcal{P}_{(\Phi)}$ then it is a double cover of $\mathcal{P}_{(\Phi)}$. The projection to the first coordinate serves as a covering.

The *I-double* of a 3-maniplex $\mathcal{P}$ was defined in [153] to be $\mathcal{P}_{(\Phi)}\Diamond\mathcal{D}_I$ whenever it is not isomorphic to $\mathcal{P}_{(\Phi)}$. There it was also denoted by $I \succ \mathcal{P}$. Here we briefly explore this concept for n-maniplexes.

In the case $I = \varnothing$, in the structure $\mathcal{D}_I$ as defined above each flag is j-adjacent to itself for all j, and therefore it is disconnected. We may instead consider only one flag, and regard $\mathcal{P}\Diamond\mathcal{D}_\varnothing$ to be isomorphic to $\mathcal{P}$. This case is not of interest to us, and we shall assume that I is non-empty.

We can visualise $\mathcal{P}_{(\Phi)}\Diamond\mathcal{D}_I$ by considering two copies of the flag set of $\mathcal{P}$, where i-adjacent flags of $\mathcal{P}$ in the same copy are i-adjacent in $\mathcal{P}\Diamond\mathcal{D}_I$ if $i \notin \mathcal{P}$, and a flag Φ in one copy is i-adjacent to the flag Φ^i in the other copy for $i \in I$. This doubles the length of some cycles on the flag graph of $\mathcal{P}$. In particular, if $\mathcal{P}_{(\Phi)}$ is a rooted polyhedron with type $\{p,q\}$ with p odd, then in $\mathcal{P}_{(\Phi)}\Diamond\mathcal{D}_{\{1,2\}}$ the flags Φ and $\Phi^{(1,2)^q}$ still belong to the same copy, but the flags Φ and $\Phi^{(0,1)^p}$ belong to different copies. Therefore $\mathcal{P}_{(\Phi)}\Diamond\mathcal{D}_{\{1,2\}}$ has type $\{2p,q\}$. More generally, if $\mathcal{P}_{(\Phi)}$ is a polytope with type $\{p_1,\ldots,p_{n-1}\}$ with p_j odd for some j then $\mathcal{P}_{(\Phi)}\Diamond\mathcal{D}_{\{0,\ldots,n-1\}\setminus\{j-1\}}$ and $\mathcal{P}_{(\Phi)}\Diamond\mathcal{D}_{\{0,\ldots,n-1\}\setminus\{j\}}$ are still equivelar and the jth entry in their types is $2p_j$.

An equivalence for 3-maniplexes $\mathcal{P}$ satisfying

$$\mathcal{P}_{(\Phi)}\Diamond\mathcal{D}_I \cong \mathcal{P} \tag{4.4}$$

in terms of colourings of flags can be found in [153]. There it is proved that if (4.4) holds for sets I and I' then it also holds for their symmetric difference. In this way we can assign a group of colourings to each polyhedron with the symmetric difference as operation. This fact motivates us to follow this notation where the j-adjacencies for $j \in I$ indicate changing the second coordinate in $\mathcal{P}_{(\Phi)}\Diamond\mathcal{D}_I$, in contrast to the definition of two-orbit polytopes in class 2_I, where the j-adjacencies for $j \in I$ indicate preserving the flag orbit.

We now explore the automorphism group of $\mathcal{P}_{(\Phi)}\Diamond\mathcal{D}_I$. We recall that I-doubles of polytopes are not always polytopes (see Exercise 4.4.7).

Proposition 4.4.30 *Let $\mathcal{P}_{(\Phi)}$ be a rooted polytope, let $I \subset \{0,\ldots,n-1\}$, and assume that $\mathcal{P}_{(\Phi)}\Diamond\mathcal{D}_I$ is not isomorphic to $\mathcal{P}$. Then $\mathcal{P}_{(\Phi)}\Diamond\mathcal{D}_I$ contains a group isomorphic to $\Gamma(\mathcal{P})\times\mathbb{Z}_2$. Furthermore, if $\mathcal{P}$ is chiral then $\mathcal{P}\Diamond\mathcal{D}_I$ is rotary, and if $|\chi(\mathcal{P})| > 2$ then $\mathcal{P}\Diamond\mathcal{D}_I$ is chiral with $\Gamma(\mathcal{P}\Diamond\mathcal{D}_I) \cong \Gamma(\mathcal{P}) \times \mathbb{Z}_2$.*

Proof Let Υ_1 and Υ_2 be the two flags of $\mathcal{D}_I$. Given $\gamma \in \Gamma(\mathcal{P})$, the bijection $\bar{\gamma}$ on the flags of $\mathcal{P}_{(\Phi)}\Diamond\mathcal{D}_I$ defined by $(\Phi,\Upsilon_i)\bar{\gamma} = (\Phi\gamma,\Upsilon_i)$ for every flag Φ

of $\mathcal{P}$ and $i \in \{1, 2\}$, induces an automorphism of $\mathcal{P}_{(\Phi)} \Diamond \mathcal{D}_I$. Also the bijection α mapping (Φ, Υ_i) to (Φ, Υ_{3-i}) is an automorphism of $\mathcal{P}_{(\Phi)} \Diamond \mathcal{D}_I$.

The automorphism α commutes with all automorphisms $\bar{\gamma}$, and $\{\bar{\gamma} : \gamma \in \Gamma(\mathcal{P})\}$ constitutes a group isomorphic to $\Gamma(\mathcal{P})$. This concludes the first part.

Now assume that $\mathcal{P}$ is chiral. Since $\Gamma(\mathcal{P} \Diamond \mathcal{D}_I)$ contains the subgroup isomorphic to $\Gamma(\mathcal{P}) \times \mathbb{Z}_2$ described above, it follows that $\mathcal{P} \Diamond \mathcal{D}_I$ has at most two orbits on flags. Moreover, each of the automorphisms α and $\bar{\gamma}$ with $\gamma \in \Gamma(\mathcal{P})$ fix the flag orbit under $\Gamma(\mathcal{P})$ of the first component. Furthermore, the flags (Φ, Υ_i) and (Ψ, Υ_j) are not adjacent whenever Φ and Ψ are in the same flag orbit under $\Gamma(\mathcal{P})$. Hence, $\mathcal{P} \Diamond \mathcal{D}_I$ is rotary.

Finally, if $|\chi(\mathcal{P})| > 2$ then the cover from the smallest regular maniplex that covers $\mathcal{P}$ to $\mathcal{P}$ has degree at least 3 (see Proposition 4.3.11). Since $\mathcal{P} \Diamond \mathcal{D}_I$ is a rotary double cover, it must be chiral. ■

Remark 4.4.31 The second part of Proposition 4.4.30 holds for every chiral polytope $\mathcal{P}$ with $|\chi(\mathcal{P})| > 2$. We shall see in Proposition 5.2.26 that chiral polytopes with $|\chi(\mathcal{P})| > 2$ do exist.

Problem 31 Are there a set $I \subseteq \{0, \ldots, n-1\}$ and a chiral n-polytope $\mathcal{P}$ such that $\Gamma(\mathcal{P} \Diamond \mathcal{D}_I)$ is regular?

In view of Proposition 4.4.30, I-doubles allow us to construct chiral polytopes as double covers of other chiral polytopes. For example, Sherk's construction in [253] consists of taking the mix $\{3, 6\}_{(a,b)} \Diamond \mathcal{D}_{\{0\}}$. The resulting maps are chiral polyhedra with type $\{6, 6\}$. We know from [49] that $\{3, 6\}_{(a,b)}$ has $a^2 + b^2 + ab$ vertices, $3(a^2 + b^2 + ab)$ edges and $2(a^2 + b^2 + ab)$ triangular faces. Then $\{3, 6\}_{(a,b)} \Diamond \mathcal{D}_{\{0\}}$ has $2(a^2 + b^2 + ab)$ vertices, $6(a^2 + b^2 + ab)$ edges and $2(a^2 + b^2 + ab)$ hexagonal faces. Its Euler characteristic is then $-2(a^2 + b^2 + ab)$. Moreover, if $ab(a - b) \neq 0$ then $\{3, 6\}_{(a,b)} \Diamond \mathcal{D}_{\{0\}}$ is chiral. These maps constitute the first known infinite family of chiral maps on surfaces of genera greater than 1.

Exercise 4.4.32 Let $\mathcal{P}$ be a polyhedron. Show that $\mathcal{P} \Diamond \mathcal{D}_{\{0\}}$ is isomorphic to $\mathcal{P}$ if and only if the 1-skeleton is bipartite.

Exercise 4.4.33 Determine which toroidal polyhedron is $\{4, 4\}_{(2,1)} \Diamond \mathcal{D}_{\{0\}}$.

In the same way that $\mathcal{P}_{(\Phi)} \Diamond \mathcal{Q}_{(\Psi)}$ is the smallest rooted structure covering $\mathcal{P}_{(\Phi)}$ and $\mathcal{Q}_{(\Psi)}$, we may also define the largest rooted premaniplex covered by both $\mathcal{P}_{(\Phi)}$ and $\mathcal{Q}_{(\Psi)}$.

Definition 4.4.34 The *comix* $\mathcal{P}_{(\Phi)}\Box\mathcal{Q}_{(\Psi)}$ of two rooted polytopes $\mathcal{P}_{(\Phi)}$ and $\mathcal{Q}_{(\Phi)}$ is the largest rooted premaniplex covered simultaneously by $\mathcal{P}_{(\Phi)}$ and $\mathcal{Q}_{(\Psi)}$, in the sense that if $\mathcal{M}_{(\Upsilon)}$ is a rooted premaniplex covered by $\mathcal{P}_{(\Phi)}$ and by $\mathcal{Q}_{(\Phi)}$ then it is also covered by $\mathcal{P}_{(\Phi)}\Box\mathcal{Q}_{(\Psi)}$.

It is natural to ask ourselves whethere the comix of two rooted maniplexes always exists. To convince ourselves that it does exist it suffices to give quotient presentations of the rooted maniplexes.

Remark 4.4.35 Let $\mathcal{P}_{(\Phi)} = \mathcal{U}/N$ and $\mathcal{Q}_{(\Psi)} = \mathcal{U}/M$ be quotient presentations. Then $\mathcal{P}_{(\Phi)}\Box\mathcal{Q}_{(\Psi)}$ has $\mathcal{U}/K$ as quotient presentation, where K is the intersection of all subgroups of $\Gamma(\mathcal{U})$ that contain $M \cup N$ (see Proposition 3.2.51).

Corollary 4.4.36 *The comix of maniplexes is commutative and associative.*

As was the case for the mix, the comix of two regular maniplexes does not depend on the base flags chosen.

Proposition 4.4.37 *Let $\mathcal{P}_{(\Phi)}$ and $\mathcal{Q}_{(\Psi)}$ be two rooted maniplexes and assume that $\mathcal{P}$ is regular. Then $\mathcal{P}_{(\Phi)}\Box\mathcal{Q}_{(\Psi)} \cong \mathcal{P}_{(\Upsilon)}\Box\mathcal{Q}_{(\Psi)}$ for every flag Υ of $\mathcal{P}$.*

Proof Let $\mathcal{P}_{(\Phi)} = \mathcal{U}/N$ and $\mathcal{Q}_{(\Psi)} = \mathcal{U}/M$ be quotient presentations. From Corollary 3.2.49 we know that $\mathcal{P}_{(\Upsilon)}$ also has quotient presentation $\mathcal{U}/N$ and the proposition follows. ■

Example 4.4.38 If $\mathcal{P}$ is a square and $\mathcal{Q}$ is a hexagon then $\mathcal{P}\Box\mathcal{Q}$ is the polygon $\{2\}$ with two edges. This follows from the fact that $\{2\}$ is a quotient of both $\mathcal{P}$ and $\mathcal{Q}$ of indices 2 and 3, respectively, and therefore there is no other regular structure covering $\{2\}$ properly which is a quotient of $\mathcal{P}$ and of $\mathcal{Q}$ simultaneously.

In general, the comix of two polytopes is not a polytope. In particular, it may happen that the group K in Definition 4.4.34 equals $\Gamma(\mathcal{U})$ and then $\mathcal{P}_{(\Phi)}\Box\mathcal{Q}_{(\Psi)}$ consists of only one flag i-adjacent to itself for every i.

If $\mathcal{P}$ and $\mathcal{Q}$ are orientable rotary then their comix has a nicer description.

Proposition 4.4.39 *Let $\mathcal{P}_{(\Phi)} = \mathcal{U}/N$ and $\mathcal{Q}_{(\Psi)} = \mathcal{U}/M$ be quotient presentations of orientable rotary polytopes. Then $\mathcal{P}_{(\Phi)}\Box\mathcal{Q}_{(\Psi)} = \mathcal{U}/NM$, and it is an orientable rotary premaniplex.*

Proof Since $N, M \triangleleft [\infty, \dots, \infty]^+$, the product NM is a normal subgroup of $[\infty, \dots, \infty]^+$. Both N and M are subgroups of NM, and satisfy that any group containing M and N must contain NM. Now the proposition follows from Proposition 3.2.51. ■

Next we present an alternative way to compute the rotation group of the comix of rotary maniplexes.

Proposition 4.4.40 *Assume that $\mathcal{P}_{(\Phi)}$ and $\mathcal{Q}_{(\Psi)}$ are orientable rotary and that we have descriptions $\Gamma^+(\mathcal{P}_{(\Phi)}) = \langle \sigma_1, \dots, \sigma_{n-1} \mid R \rangle$ and $\Gamma^+(\mathcal{Q}_{(\Psi)}) = \langle \sigma'_1, \dots, \sigma'_{n-1} \mid R' \rangle$ in terms of generators and relations. Then $\Gamma^+(\mathcal{P}_{(\Phi)} \Box \mathcal{Q}_{(\Psi)}) = \langle \sigma_1, \dots, \sigma_{n-1} \mid R \cup R' \rangle$, where the relations R' are applied to the standard rotations σ_i instead of σ'_i.*

Proof Let $\mathcal{P}_{(\Phi)} = \mathcal{U}/N$ and $\mathcal{Q}_{(\Psi)} = \mathcal{U}/M$ be quotient presentations, so that $\Gamma(\mathcal{P}_{(\Phi)} \Box \mathcal{Q}_{(\Psi)})$ has quotient presentation $\mathcal{U}/NM$ (see Proposition 4.4.39).

As we saw in the proof of Proposition 4.3.11, $\mathcal{P}_{(\Phi)}$ and $\mathcal{Q}_{(\Psi)}$ are the quotients by automorphisms $\mathcal{U}/N$ and $\mathcal{U}/M$. It follows that N and M are the normal closures of R and R', respectively. (We are in the understanding that R can be written either as $\{f_i(\sigma_1, \dots, \sigma_{n-1}) = \text{id}\}_{i \in I}$ or merely as $\{f_i(\sigma_1, \dots, \sigma_{n-1})\}_{i \in I}$, as needed; at this point we choose the latter.) Then NM is the normal closure of $R \cup R'$. ■

Proposition 4.4.40 suggests a reasonable way to determine the comix of two orientable rotary polytopes by computational methods, if they are finite and small enough. This can also be used to compute the size of the mix of two rooted orientable rotary polytopes, as shown in the next proposition (see [64, Proposition 3.3]).

Proposition 4.4.41 *Let $\mathcal{P}_{(\Phi)}$ and $\mathcal{Q}_{(\Psi)}$ be finite orientable rotary polytopes. Then*

$$|\Gamma^+(\mathcal{P}_{(\Phi)} \diamondsuit \mathcal{Q}_{(\Psi)})| \cdot |\Gamma^+(\mathcal{P}_{(\Phi)} \Box \mathcal{Q}_{(\Psi)})| = |\Gamma^+(\mathcal{P})| \cdot |\Gamma^+(\mathcal{Q})|. \tag{4.5}$$

Proof Let $\mathcal{P}_{(\Phi)} = \mathcal{U}/N$ and $\mathcal{Q} = \mathcal{U}/M$ be quotient presentations. Then N and M are normal in $\Gamma^+(\mathcal{U})$, and so is $N \cap M$. We recall that $\Gamma^+(\mathcal{P}_{(\Phi)}) \cong \Gamma^+(\mathcal{U})/N$ and $\Gamma^+(\mathcal{P}_{(\Phi)} \diamondsuit \mathcal{Q}_{(\Psi)}) \cong \Gamma^+(\mathcal{U})/(N \cap M)$. Let π be the natural epimorphism from $\Gamma^+(\mathcal{P}_{(\Phi)} \diamondsuit \mathcal{Q}_{(\Psi)})$ to $\Gamma^+(\mathcal{P}_{(\Phi)})$.

By the isomorphism theorems of groups, we know that

$$\begin{aligned}
|\Gamma^+(\mathcal{P}_{(\Phi)} \diamondsuit \mathcal{Q}_{(\Psi)})| &= |\Gamma^+(\mathcal{P}_{(\Phi)})| \cdot |\ker(\pi)| = |\Gamma^+(\mathcal{P}_{(\Phi)})| \cdot |N/(N \cap M)| \\
&= |\Gamma^+(\mathcal{P}_{(\Phi)})| \cdot |MN/M| \\
&= |\Gamma^+(\mathcal{P}_{(\Phi)})| \cdot |\Gamma^+(\mathcal{U})/M| / |\Gamma^+(\mathcal{U})/MN| \\
&= |\Gamma^+(\mathcal{P}_{(\Phi)})| \cdot |\Gamma^+(\mathcal{Q}_{(\Psi)})| / |\Gamma^+(\mathcal{P}_{(\Phi)} \Box \mathcal{Q}_{(\Psi)}))|.
\end{aligned}$$

■

We observe that the right-hand side of (4.5) is not rooted since the size of the rotation group does not depend on any choice of flag. However, the sizes of $\Gamma^+(\mathcal{P}_{(\Phi)} \Diamond \mathcal{Q}_{(\Psi)})|$ and $|\Gamma^+(\mathcal{P} \Box \mathcal{Q})$ do depend on choices of flags.

Example 4.4.42 Let $\mathcal{P}_{(\Phi)}$ be the polyhedron $\{4,4\}_{(2,1)}$ with any choice of base flag Φ. We first choose $\mathcal{Q}_{(\Psi)} = \mathcal{P}_{(\Phi)}$ to verify that

$$|\Gamma^+(\mathcal{P}_{(\Phi)} \Diamond \mathcal{P}_{(\Phi)})| \cdot |\Gamma^+(\mathcal{P}_{(\Phi)} \Box \mathcal{P}_{(\Phi)})| = |\Gamma^+(\mathcal{P})| \cdot |\Gamma^+(\mathcal{Q})| = 400,$$

since the mix and the comix are precisely $\Gamma^+(\mathcal{P}_{(\Phi)})$ with 20 elements.

On the other hand, if we choose $\mathcal{Q}_{(\Psi)}$ to be $\mathcal{P}_{(\Phi^0)}$ then $\mathcal{P}_{(\Phi)} \Diamond \mathcal{Q}_{(\Psi)}$ is the smallest regular cover of $\{4,4\}_{(2,1)}$, namely $\{4,4\}_{(5,0)}$ (see Example 3.3.13), while $\mathcal{P}_{(\Phi)} \Box \mathcal{Q}_{(\Psi)}$ can be easily seen from Definition 4.4.34 to be the non-polyhedral maniplex $\{4,4\}_{(1,0)}$. Nevertheless, (4.5) is still satisfied:

$$\begin{aligned} |\Gamma^+(\mathcal{P}_{(\Phi)} \Diamond \mathcal{Q}_{(\Psi)})| \cdot |\Gamma^+(\mathcal{P}_{(\Phi)} \Box \mathcal{Q}_{(\Psi)})| &= |\Gamma^+(\{4,4\}_{(5,0)})| \cdot |\Gamma^+(\{4,4\}_{(1,0)})| \\ &= 100 \cdot 4 = 400. \end{aligned}$$

Exercise 4.4.43 Show that the comix of any two distinct Platonic solids is isomorphic to the structure $\mathcal{D}$ of rank 3 with two flags defined in Example 4.4.4.

Exercise 4.4.44 Show that the comix of the octahedron $\{3,4\}$ and of $\{3,6\}_{(a,b)}$ is the unique polyhedron with type $\{3,2\}$. Use this to determine the size of $\{3,4\} \Diamond \{3,6\}_{(a,b)}$.

5

Families of Chiral Polytopes

The aim of this chapter is to explore chiral polytopes, grouped by common properties. It is beyond the scope of this book to provide all the details of the background and proofs of the results presented here, and instead we cite the appropriate sources.

We start by considering in Section 5.1 chiral polyhedra, which are those chiral polytopes of the lowest rank possible. These are included in the more general study of chiral maps, of which more literature is available. Then, in Section 5.2 we analyse chiral polytopes in low ranks with few flags, and explain the small likelihood of finding small chiral polytopes in higher ranks. After that, in Section 5.3 we discuss chiral polytopes obtained from geometric constructions. In Section 5.4.2 we deal with chiral amalgamations of polytopes, that is, chiral polytopes obtained by prescribing the kind of facets and of vertex-figures. Finally, in Section 5.5 we discuss the constructions of chiral polytopes of unbounded ranks published so far. In the process we also discuss chiral extensions of polytope, that is, chiral polytopes with prescribed facets.

5.1 Rank 3

The first chiral polytopes that were studied have rank 3, and hence they can be visualised as maps on surfaces (see Proposition 2.1.37). There is a vast literature on rotary maps (see for example [31, 50, 91, 142, 149, 252, 260, 267, 268]). In this section we include some of the recent directions of research and relevant open problems.

5.1.1 Background

The classification theorem of orientable, closed (compact and without boundary) topological surfaces states that every connected, closed, orientable surface is homeomorphic to a sphere with a finite number g of handles attached (see [90] and the references within). The number g is called the *genus* of the surface. The surface with genus 0 is the sphere and the one with genus 1 is the torus.

Given a map $\mathcal{M}$ on a surface $\mathcal{S}$ we denote by n_v, n_e and n_f the numbers of vertices, edges and faces of $\mathcal{M}$. In each compact closed surface the number $\chi(\mathcal{S}) = n_v - n_e + n_f$ is an invariant called the *Euler characteristic*. In other words, the number $n_v - n_e + n_f$ is the same for all maps on $\mathcal{S}$ (see [125]).

The genus g and the Euler characteristic χ of an orientable closed surface satisfy the relation $\chi = 2 - 2g$. For example, the Euler characteristic of the sphere is 2. As a consequence, every convex polyhedron satisfies that the sum of the numbers of vertices and 2-faces minus the number of edges is 2. The Euler characteristic of the torus is 0.

We recall that finite chiral maps lie on connected, closed, orientable surfaces (see Proposition 2.1.37). This allows us to speak of the genus and the Euler characteristic of a given finite chiral map. Furthermore, chiral maps are equivelar, which allows us to express the numbers of vertices and faces in terms of the number of edges as follows.

Lemma 5.1.1 *Let $\mathcal{M}$ be a finite chiral map with type $\{p,q\}$, and let n_v, n_e, n_f denote the numbers of vertices, edges and faces of $\mathcal{M}$. Then $qn_v = 2n_e = pn_f$.*

The first equality in the statement of Lemma 5.1.1 is one of the most famous results of graph theory. The second one follows from a standard procedure of counting the pairs of incident edge and face in two distinct ways.

5.1.2 Schläfli Types

One of the first questions one can ask about chiral maps on surfaces is: 'what are their types?'. We shall discuss the answer to this question in this section.

Both entries of the type of an equivelar polytope are greater than 1, since the 1-gon does not satisfy the diamond condition. Furthermore, from Lemma 2.3.16 we know that both entries of the type of a chiral polyhedron are greater than 2. Therefore, if $\{p,q\}$ is the type of a chiral polyhedron then $p,q \in \{3,4,\ldots,\} \cup \{\infty\}$. In what follows we shall discard a few more types.

The tetrahedron, the octahedron and the icosahedron are the universal polyhedra with types $\{3,3\}$, $\{3,4\}$ and $\{3,5\}$, respectively (see Exercises 2.1.62

and 2.1.63 and [48, Theorems 8 and 9]). It follows that every equivelar polyhedron with type $\{3,3\}$, $\{3,4\}$ or $\{3,5\}$, or their dual types, must be a quotient of a Platonic solid. Now, $\Gamma^+(\{3,3\}) \cong A_4$, $\Gamma^+(\{3,4\}) \cong S_4$ and $\Gamma^+(\{3,5\}) \cong A_5$. Basic results on symmetric and alternating groups are enough to guarantee that these polyhedra have no chiral quotients. The types $\{2,q\}$, $\{p,2\}$, $\{3,3\}$, $\{3,4\}$, $\{4,3\}$, $\{3,5\}$ and $\{5,3\}$ we have discarded so far as types of chiral polyhedra are precisely those where $1/p + 1/q - 1/2 > 0$, which establishes the next result.

Proposition 5.1.2 *Let $\mathcal{P}$ be a chiral polyhedron with type $\{p,q\}$. Then*

$$1/p + 1/q - 1/2 \leq 0.$$

If $\mathcal{P}$ is an equivelar map with type $\{p,q\}$ then the nature of the number $1/p + 1/q - 1/2$ (positive, 0 or negative) is deeply related to the genus of $\mathcal{P}$. As before, let n_v, n_e and n_f denote respectively the numbers of vertices, edges and faces of $\mathcal{P}$. Then, by using Lemma 5.1.1 we can see that

$$\chi(\mathcal{P}) = n_v - n_e + n_f = \frac{2n_e}{q} - n_e + \frac{2n_e}{p} = 2n_e(1/p + 1/q - 1/2). \quad (5.1)$$

Maps on the sphere have positive Euler characteristic, implying that in those cases $1/p + 1/q - 1/2 > 0$.

Corollary 5.1.3 *There are no polyhedral chiral maps on the sphere.*

The extra condition of the map being a polyhedron in Corollary 5.1.3 is not necessary.

Exercise 5.1.4 Show that there are no chiral maps on the sphere.

If $\mathcal{P}$ is a finite equivelar map on the torus then it has Euler characteristic 0, and hence $1/p + 1/q - 1/2 = 0$. This forces the type of $\mathcal{P}$ to be either $\{3,6\}$, $\{4,4\}$ or $\{6,3\}$. Maps with those types are quotients of the universal polyhedra with the given types; these universal polyhedra are the regular tessellation of the Euclidean plane with equilateral triangles, squares and regular hexagons, respectively (see Section 2.4).

Finally, if $\mathcal{P}$ is a map on a surface of genus at least 2 then its Euler characteristic is negative. In this situation $1/p + 1/q - 1/2 < 0$, and the candidate types $\{p,q\}$ of these maps are all those pairs where $p,q \in \{3,4,\dots\} \cup \{\infty\}$, which are not the types of rotary maps that lie on the sphere or on the torus. For all such values of p and q there is a regular tessellation of the hyperbolic plane with type $\{p,q\}$ (see [51]). This tessellation is actually the universal polyhedron of that type (see for example [48, Theorem 8] and [52, Section 2]), and hence it covers $\mathcal{P}$. The list of regular hyperbolic tessellations includes the degenerate tessellations with infinitely many faces at a vertex, in which the vertices are ideal; it also includes the dual tessellations where the faces are infinite (the vertices of each face belong to some horocircle), and the highly degenerate tessellation with type $\{\infty, \infty\}$. None of these last situations can occur if the map is finite, and so ∞ occurs in the type of chiral maps only if they lie on non-compact surfaces.

The previous discussion motivates the convention of saying that $\{p,q\}$ is a *spherical type*, a *toroidal type* or a *hyperbolic type* if $1/p + 1/q - 1/2 > 0$, $1/p + 1/q - 1/2 = 0$ or $1/p + 1/q - 1/2 < 0$, respectively. It also links the study of orientable rotary maps with that of Riemann surfaces through the uniformisation theorem, since the objects of both categories correspond to quotients of the sphere, Euclidean plane or hyperbolic plane by subgroups of isometries (see [99, Chapter 2] and [149, Sections 4 and 5]).

Rotary maps on the torus were discussed in detail in Section 2.4. We recall that for each type $\{p,q\} \in \{\{3,6\},\{4,4\},\{6,3\}\}$ the chiral polyhedra with type $\{p,q\}$ are parametrised by the pairs (a,b) with $a,b \in \mathbb{Z}$ such that $a > b > 0$. In terms exclusively of the types we have the following result.

Theorem 5.1.5 *There are infinitely many chiral polyhedra with each of the types* $\{3,6\}$, $\{4,4\}$ *and* $\{6,3\}$.

The sets of rotary maps on surfaces of genus greater than 1 are radically different from the rotary maps on the torus, as will be explained next (see Lemmas 5.1.6, 5.1.7 and Proposition 5.1.8). If $p,q < \infty$ and $\mathcal{P}$ is a map with type $\{p,q\}$ on a surface with Euler characteristic χ then (5.1) tells us that $\mathcal{P}$ has precisely $\chi/2\omega$ edges, where $\omega = 1/p + 1/q - 1/2$. It follows from Lemma 5.1.1 that $\mathcal{P}$ has $\chi/q\omega$ vertices and $\chi/p\omega$ faces. But there can only be finitely many maps with $\chi/q\omega$ vertices, $\chi/2\omega$ edges and $\chi/p\omega$ faces of type $\{p,q\}$, up to isomorphism. We can conclude the following lemma.

Lemma 5.1.6 *Each orientable closed surface with genus greater than* 1 *carries finitely many maps of each type.*

Now, the number of flags of $\mathcal{P}$ is four times the number of edges, and can be rewritten as

$$\frac{4\chi}{2(1/p + 1/q - 1/2)} = \frac{-2\chi}{1/2 - 1/p - 1/q}.$$

In the expression on the right, the numerator and denominator are positive. By increasing p or q the denominator increases, and so the presumed number of flags decreases. This implies that the maximum number of flags must be attained with some hyperbolic type $\{p,q\}$ for which there are no $p' < p$ or $q' < q$ such that there exist maps with types $\{p',q\}$ or $\{p,q'\}$ on the same surface. It follows that the extreme case must be attained when the surface carries maps with type $\{3,7\}$, $\{4,5\}$ or the dual of either of these types. The number of flags of a map with type $\{4,5\}$ or $\{5,4\}$ is -40χ, whereas that of a map with type $\{3,7\}$ or $\{7,3\}$ is -84χ. In any case, the bound stated next holds for all closed surfaces with genus greater than 1.

Lemma 5.1.7 *Let $\mathcal{P}$ be an equivelar map on a compact surface with Euler characteristic $\chi > 1$. Then $\mathcal{P}$ has at most -84χ flags.*

As a consequence of Lemma 5.1.7, if $\mathcal{P}$ is a map with hyperbolic type $\{p,q\}$ on the orientable surface with genus χ then $p,q \leq -42\chi$, since each face must have at least $2p$ distinct flags, and each vertex must belong to at least $2q$ distinct flags. (We note that this bound on p and q can be improved substantially if we require $\mathcal{P}$ to be a polyhedron.) It follows that each orientable surface with genus greater than 1 carries equivelar maps with finitely many types. Combining this fact with Lemma 5.1.6 we obtain the following result.

Proposition 5.1.8 *Each connected closed orientable surface with genus at least 2 carries finitely many equivelar maps.*

Maps with types $\{3,7\}$ and $\{7,3\}$ have attracted much attention due to the fact that they have the most possible flags among all equivelar maps on the same surface. Rotary maps with these types are intimately related to Hurwitz's automorphism theorem of compact Riemann surfaces (see [150, Subsection 5.1.2]).

Theorem 5.1.5 and Proposition 5.1.8 indicate a striking difference between the torus and the closed orientable surfaces of higher genus. The torus carries infinitely many rotary maps, even if they are easily describable with only four integer parameters (the type together with the vector in the subindex); on the other hand, each closed orientable surface carries finitely many rotary maps. This naturally raises the following questions. Are there any rotary maps on closed surfaces of a given hyperbolic type? In case there are, are there finitely many or infinitely many rotary maps of that type?

If we leave aside the possibility of entries ∞ in the type, the answers to the previous questions were given in [259, Theorem 6.3], where it is proven that for every type $\{p,q\}$ with $p,q < \infty$ and $1/p + 1/q - 1/2 < 0$ there are infinitely many regular maps with type $\{p,q\}$ on closed surfaces. The proof relies on results about Coxeter groups and finitely generated matrix groups; they can be used to show the result for polyhedra and not only for maps. Furthermore, instead of using the Coxeter groups $[p,q]$ it is possible to follow their ideas with their even subgroups $[p,q]^+$, to ensure the existence of infinitely many orientable rotary maps of each hyperbolic type, in the expectation that many of them will be chiral.

Theorem 5.1.9 *For every $p,q \in \{3,4,\dots\}$ with $1/p + 1/q - 1/2 < 0$ there exist infinitely many orientable rotary polyhedra with type $\{p,q\}$ on closed surfaces.*

Exercise 5.1.10 Prove Theorem 5.1.9 using the ideas in [259, Theorem 6.3].

The techniques used in [259] do not suggest how to determine which of the rotary maps with a given type are regular and which are chiral. In order to determine whether there are any chiral polytopes of a given hyperbolic type and, in case there are, whether there are infinitely many, other techniques are necessary.

The first hyperbolic type for which the existence of infinitely many chiral maps with that type was proven is $\{6,6\}$ (see [253]). Later, the existence of infinitely many chiral maps on closed surfaces with type $\{3,k\}$ for each integer $k \geq 7$ was established through Schreier diagrams of actions of the alternating groups (see [24]).

The existence of infinitely many chiral maps on closed surfaces with each hyperbolic type $\{p,q\}$ with $p,q < \infty$ was proved independently in [42, 147]. Both use an idea of Macbeath to construct infinitely many finite index subgroups with special properties of a given group. The proof in [147] relies also on spaces of differentials on Riemann surfaces and representation theory, while the proof in [42] has its basis in Schreier diagrams of alternating and symmetric groups. The statements of the respective theorems in these publications are more general, but they imply the following.

Theorem 5.1.11 *Let $p,q \in \mathbb{Z}$ with $p,q \geq 3$ and $1/p + 1/q - 1/2 < 0$. Then there are infinitely many chiral maps with type $\{p,q\}$ on connected closed surfaces.*

Much less attention has been given to chiral maps on non-compact surfaces, including those whose type has an entry ∞.

The first infinite chiral polyhedra (also the first chiral maps on non-compact surfaces) to be studied seem to be the chiral polyhedra in Euclidean space $\mathbb{E}^3$ with finite faces (see [245]). They include infinitely many non-isomorphic polyhedra of each of the types $\{4,6\}$, $\{6,4\}$ and $\{6,6\}$ (see Section 6.2.3).

They were proven to be chiral as abstract polyhedra in [228]. Years later, one infinite chiral polyhedron with each of the types $\{8,4\}$, $\{\infty,6\}$ and $\{\infty,\infty\}$ was constructed in [230, Section 4].

In view of Theorem 5.1.11, it is to be expected that there are infinitely many chiral infinite polyhedra with each hyperbolic type. The techniques used in [42, 147, 259] seem to be insufficient to arrive to this conclusion.

Problem 32 For each $p,q \in \{3,\ldots,\infty\}$ with $1/p+1/q-1/2 < 0$ determine whether there exist infinitely many chiral infinite polyhedra with type $\{p,q\}$.

5.1.3 Surfaces Carrying Chiral Polyhedra

When reading Proposition 5.1.8 we wonder whether there exists a number g_0 such that every orientable closed surface with genus greater than g_0 carries chiral maps, even if for some of these surfaces there are few of them. We shall see that this is not the case, even if for each hyperbolic type $\{p,q\}$ with $p,q < \infty$ there exist infinitely many chiral maps of type $\{p,q\}$, and they all lie on orientable closed surfaces of genus at least 2 (see Theorem 5.1.11). We already mentioned in Section 2.3.1 that there are no chiral maps on the orientable closed surface with genus g for $g \in \{2,3,4,5,6\}$ (see [92]). A closer look to the available lists of chiral maps by genera tells us that there are chiral maps on surfaces with genera $7,8,10,11,12,14$ and 15, but not on surfaces with genera 9 or 13 (see [36, 41]).

If we prescribe a type, the situation becames more drastic. Suppose that we are looking for orientable equivelar maps $\mathcal{P}$ with type $\{8,5\}$. Then we know from (5.1) that if $\mathcal{P}$ lies on the surface with Euler characteristic χ then

$$\chi = 2n_e(1/8 + 1/5 - 1/2) = -7n_e/20,$$

where n_e is the number of edges of $\mathcal{P}$. This tells us right away that n_e is divisible by 20, and hence $\chi \leq -7$. It follows that our map $\mathcal{M}$ lives on a surface with genus $g \geq 5$. Moreover, 7 must divide 20χ, forcing χ to be a multiple of 14 (here we use that χ must be even since $\mathcal{M}$ is orientable). Hence, the genera of the surfaces that may carry our map $\mathcal{M}$ are $\{8,15,22,29,\ldots\}$. This argument makes it clear that there must occur infinitely many gaps in the spectrum of equivelar maps with type $\{8,5\}$.

Now assume that we want all equivelar maps $\mathcal{M}$ with type $\{8,5\}$ on the orientable surface with genus 15 (so that $\chi = -28$). Then $\mathcal{M}$ has 80 edges, 32 vertices and 20 octagonal faces. We may try to take 20 disjoint octagons, label the 160 edges and try to identify them by pairs preserving orientability, in such a way that every vertex ends up being in precisely five octagons. Since at this point we are not asking them to be polyhedra, we may allow two or more vertices of the same octagon to get identified. Carrying out this procedure is not an easy task, and to fulfil it we may end up identifying two edges of the same octagon, or performing any other identification that prevents the map from being rotary. Let us say that we get N such maps, up to automorphism. There may be many candidate symmetry type graphs for these N maps, out of which rotary maps correspond only to two of them. Should we expect them to be predominant? If so, why? In fact, the lists [33, 34] indicate that there are no rotary maps on this surface at all. Furthermore, the only orientable surfaces with genus at most 101 carrying regular maps with type $\{8,5\}$ have genera 22, 29, 57 and 64, giving six such maps in total (there are two maps on the surface with genus 29 and two on that of genus 64). There are no chiral maps with type $\{8,5\}$ on orientable surfaces of genus at most 101.

Exercise 5.1.12 Use (5.1) to produce one list of candidate genera of orientable surfaces that carry equivelar maps with type $\{6,6\}$, and one list for maps with type $\{15,16\}$.

We know that the maps devised by Sherk in [253] correspond to polyhedra with type $\{6,6\}$, and for each $a > b > 0$ there is a chiral polyhedron on the surface with genus $a^2 + b^2 + ab + 1$. This establishes the first known infinite family of integers that are the genera of surfaces carrying chiral maps.

The list of chiral maps in [41] includes the following genera g at most 101 such that there is no chiral map on the orientable closed surface with genus g:

$$2, 3, 4, 5, 6, 9, 13, 23, 24, 30, 36, 47, 48, 54, 60, 66, 84, 95.$$

From this sequence it is not obvious whether the list is finite or infinite.

The existence of the lists of rotary maps in [41] made it possible to examine the behaviour of the maps on certain specific families of surfaces. The first surfaces to be considered were those whose Euler characteristic is the negative of a prime number (see [17]) and the negative of a square prime number (see [39]); except for the orientable ones with Euler characteristic -2 and -4, the surfaces in question are non-orientable. The natural continuation was the study of the surfaces whose Euler characteristic is -2 times a prime number. This was performed for orientable surfaces in [43]. The classifications of the rotary maps on those surfaces are achieved through the analysis of properties that their rotation groups must satisfy; they rely on some previously known results on group theory, and also on the classification of rotary maps on surfaces of Euler characteristic at least -28. We highlight the following two consequences of the results from [43] (see specifically theorems 3.1 and 8.4 in that paper).

Theorem 5.1.13 *Let $\mathcal{P}$ be a chiral map on a surface with Euler characteristic χ. Then $\chi/2$ is not coprime with $|\Gamma^+(\mathcal{P})|$.*

This result may prove useful in several ways. For example, if we want to determine all chiral maps on a surface whose Euler characteristic is twice a negative odd number (not necessarily prime), then the candidate automorphism groups do not include 2-groups. Given the high percentage of 2-groups in the list of groups of bounded order, this fact may reduce the search considerably if the procedure includes a search through a list of groups.

Theorem 5.1.14 *Let $\mathcal{P}$ be a chiral polyhedron on a surface with Euler characteristic $-2p$ for some prime $p > 13$. Then one of the following holds.*

(a) $p \equiv 1$ (modulo 8) and $\mathcal{P}$ has type $\{8, 8\}$.
(b) $p \equiv 1$ (modulo 10) and $\mathcal{P}$ has type $\{5, 10\}$ or $\{10, 5\}$.
(c) $p \equiv 1$ (modulo 6) and $\mathcal{P}$ has type $\{6, 6\}$.

Moreover, up to isomorphism and duality there is only one polyhedron for each p in case (a), two polyhedra for each p in case (b) and one polyhedron for each p in case (c).

Remark 5.1.15 The property of a map of being a polyhedron was called XY-disjoint in [43].

Remark 5.1.16 It is well known that all primes $p \equiv 1$ (modulo 6) can be expressed as $a^2 + ab + b^2$ for some distinct integers a and b (see for example [141]). Furthermore, the maps found by Sherk in [253] have Euler characteristic $-2(a^2 + ab + b^2)$. Therefore, the maps in case (c) in Theorem 5.1.14 are some of the maps found by Sherk.

Besides being a sound description of all rotary maps in an infinite family of closed orientable surfaces, Theorem 5.1.14 provides the first infinite family of surfaces that carry no chiral maps, namely those whose Euler characteristic is $-2p$ with p a prime that is not congruent to 1 modulo 6, 8 or 12. For instance, from Dirichlet's prime number theorem (see [141, Chapter 16]), there are infinitely many primes congruent to -1 modulo 120. Surfaces whose Euler characteristic is -2 times those prime numbers do not carry any chiral maps.

The main question about determining all compact surfaces carrying chiral maps remains (see Problem 4).

We move on to chiral maps on non-compact surfaces. The first important observation here is that the classification of connected, orientable non-compact surfaces without boundary depends on more parameters. Besides the genus, there is also a set of what are commonly known as 'ends', which correspond intuitively to the distinct ways to go to infinity (the plane has only one, the cylinder has two, etc.). Furthermore, each of the ends may or may not carry infinite genus. This is explained in [236], where the parameters involving the ends are collected in the triple denoted by the 'ideal boundary'. Nowadays the terminology used in [235] is more popular (see [79] for an application of these terms to graph theory).

One of the most studied non-compact surfaces is called the *Loch Ness monster*. This is the surface obtained by attaching countably many handles to the plane, and so it is the only surface with only one end and infinite genus.

Despite the apparent extra complication of studying non-compact surfaces instead of compact ones, the classification of non-compact surfaces that carry chiral maps is much simpler than that of compact surfaces. The corresponding results were given in [4, Theorems 1.1, 1.2, 1.3] and can be rewritten as follows.

Theorem 5.1.17 *If $\mathcal{M}$ is an orientable rotary map on a connected non-compact surface $\mathcal{S}$ without boundary then $\mathcal{S}$ is homeomorphic to either the plane or the Loch Ness monster. Furthermore, if $\mathcal{P}$ is chiral then $\mathcal{S}$ is necessarily homeomorphic to the Loch Ness monster.*

Given the wide variety of non-compact surfaces it is natural to ask the kind of maps that lie on each of them. For instance, the cylinder does admit two-orbit maps. Figure 5.1 shows two examples of this fact (top and bottom are to be identified orientably); the map in Figure 5.1(a) is in class 2_{02} while the map in Figure 5.1(b) is in class 2_1.

Other surfaces might not have maps with finitely many flag orbits.

Problem 33 Determine the set $\mathcal{X}$ of connected non-compact surfaces that carry maps with finitely many flag orbits. For each such surface $\mathcal{S} \in \mathcal{X}$ determine the smallest k such that $\mathcal{S}$ carries k-orbit maps.

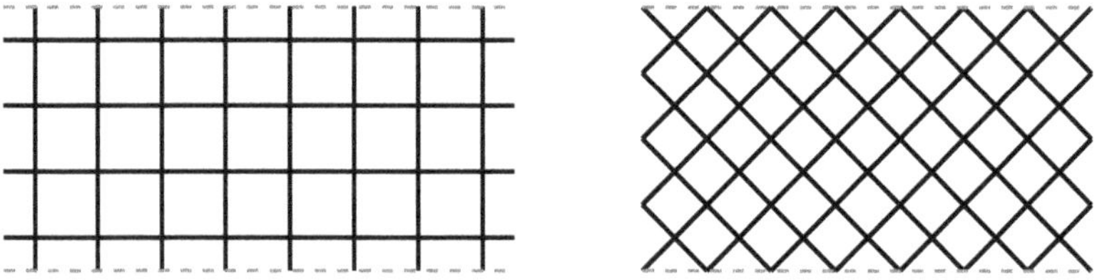

(a) (b)

Figure 5.1 Two two-orbit maps on the cylinder: (a) a map with edges alligned to the axis of the cylinder, (b) a map with no edges parallel to the axis of the cylinder.

5.1.4 Prevalence of Regular or Chiral Polyhedra

In some ways regular maps tend to be easier to find than chiral maps. We next list some examples of this tendency.

- If we prescribe a type $\{p,q\}$, there is a universal map with type $\{p,q\}$ and it is regular.
- If we search for rotary maps by imposing only the lengths of the Petrie paths and/or the lengths of the holes to the universal map of type $\{p,q\}$, the resulting maps are regular (see Exercise 5.1.18).
- The combinatorial construction $2^{\mathcal{K}}$ defined in [75], when applied to the polygon $\{q\}$ yields a regular polyhedron (map) $2^{\{q\}}$.
- There are 59 regular maps on orientable surfaces with genera $2 \leq g \leq 6$ (many of them are not polyhedral), but there are no chiral maps on those surfaces. There are 138 regular maps on orientable surfaces with genera $2 \leq g \leq 10$, and there are only 6 chiral maps on those surfaces.

Exercise 5.1.18 Let $\mathcal{P}$ be a map obtained from the universal map with type $\{p,q\}$ by imposing as relations only the lengths of the Petrie paths and the lengths of the k-holes for $k \in I$ with $I \subseteq \{2, \ldots, \lfloor q/2 \rfloor\}$. Show that $\mathcal{P}$ is regular.

Based on that evidence it is tempting to conjecture that chiral maps are rare, and among rotary maps those that are regular are prevalent. This supposition became less clear when all rotary maps on more surfaces were listed.

In the lists [33, 34] we can see that there are 3378 orientably regular maps versus 594 chiral ones in surfaces of genera less than 102 and greater than 1. The percentage of regular maps among all rotary maps with hyperbolic type on surfaces with genus up to 7 is over 95.8%; and it became about 85% when the genera considered were up to 101. This suggests that the number of chiral maps is slowly catching up with that of regular maps when the bound on the genus is increased.

The notion of prevalency of one or another type of rotary maps needs to be defined properly in order to try to determine the validity of a statement in that direction. This is particularly relevant since there are infinitely many orientably regular and infinitely many chiral maps of each hyperbolic type (see [42, 147, 259]). When we increase the set of maps under consideration by increasing the genus the following is a natural question.

Problem 34 Let N_m and R_m denote the numbers of rotary maps and of regular maps on orientable surfaces with genera $2 \leq g \leq m$, respectively.

Determine whether the sequence $\{R_m/N_m\}_{m\in\mathbb{N}}$ converges, and in that case, determine the limit when m goes to ∞.

There is another natural way to investigate the frequency of regular maps among all rotary ones. This one also includes those on the sphere and on the torus.

Problem 35 Let N_n and R_n denote the number of rotary maps and of regular maps on orientable surfaces with at most n flags. Determine whether the sequence $\{R_n/N_n\}_{n\in\mathbb{N}}$ converges, and in that case, determine the limit when n goes to ∞. Is the limit the same as in Problem 34 (in case it exists)?

It seems more challenging to answer the following more specialised question.

Problem 36 Let $N_{p,q,n}$ and $R_{p,q,n}$ denote the number of rotary maps and of regular maps on orientable surfaces with type $\{p,q\}$ and at most n flags. Determine whether the sequence $\{R_{p,q,n}/N_{p,q,n}\}_{n\in\mathbb{N}}$ converges, and in that case, determine the limit when n goes to ∞. Is the limit the same for all hyperbolic types $\{p,q\}$? Do we get the same answers if we bound by genus instead of by number of flags?

5.2 Tight Chiral Polytopes

When studying families of mathematical objects it is natural to look for the smallest examples, in order to develop intuition of the properties of the family. With this idea, small chiral maps were studied, for example in [93, 271].

This section addresses chiral polytopes with the fewest flags among those with a given Schläfli type. We also discuss lower bounds for the number of flags of chiral polytopes of high ranks.

5.2.1 Flatness and Tightness

When looking for n-polytopes with few flags it is natural to search first for restrictive properties on the facets and vertices, or even on faces of other ranks. One of the first such restrictions was introduced in [187, Section 3] and is the following.

Definition 5.2.1 A polytope is (k,m)-*flat* if every k-face is incident to every m-face. Those n-polytopes that are $(0,n-1)$-flat (so that every vertex is incident to every facet) are merely called *flat*.

Flatness is a feature that cannot be found in the theory of convex polytopes, but does not violate any of the axioms of abstract polytopes. In fact, we have already encountered several flat polytopes.

Example 5.2.2 The following polytopes are flat.

- The toroidal polyhedron $\{6,3\}_{(1,1)}$.
- The hemi-cube (see Exercise 2.2.12).
- The toroidal n-polytope $\{4,3^{n-2},4\}_{(2,0^{n-1})}$ (see Section 2.4.3).
- The trivial extension of a polytope (see Exercise 2.1.61).

We can also restate Lemma 2.1.65 by saying that if an equivelar n-polytope $\mathcal{P}$ has type $\{p_1,\ldots,p_{n-1}\}$ and $p_i = 2$ for some i then $\mathcal{P}$ is (k,m)-flat for every $k < i$ and $m \geq i$.

For some values of k and m the property of being (k,m)-flat implies being also (k',m')-flat for other values k', m'.

Remark 5.2.3 If $0 \leq k_1 \leq k_2 < m_2 \leq m_1 \leq n-1$ and $\mathcal{P}$ is a (k_2,m_2)-flat polytope then $\mathcal{P}$ is also (k_1,m_1)-flat. In particular, all (k_1,m_1)-flat polytopes are flat.

Flatness of some faces can also be used to determine that a polytope is globally flat.

Lemma 5.2.4 *Let $n \geq 3$ and let $0 \leq k < m \leq n-2$. If all facets of an n-polytope $\mathcal{P}$ are (k,m)-flat then $\mathcal{P}$ itself is (k,m)-flat.*

Proof We proceed by contradiction. Assume that there are a k-face F_k and an m-face F_m such that F_k is not incident to F_m. Since $\mathcal{P}$ is connected, there is a sequence of proper faces $\mathbf{S}_0 := (F_k = H_0, H_1, \ldots, H_\ell = F_m)$ such that H_i is incident to H_{i+1} for every i. For each i such that H_i and H_{i+1} are not facets we insert a facet containing H_i and H_{i+1} to construct a sequence $\mathbf{S}_1 := (F_k = H'_0, H'_1, \ldots, H'_{\ell'} = F_m)$ where H'_i is a facet whenever i is odd. If all facets H'_{2j+1} have the same k-faces then the facet $H_{\ell'-1}$ is incident to F_k; and since $H_{\ell'-1}$ is (k,m)-flat, F_k and F_m must be incident, contradicting our hypothesis. Therefore there exist two facets H'_{2j-1} and H'_{2j+1} that are consecutive facets in $\mathbf{S}_1$ and have distinct sets of k-faces.

If the rank of H'_{2j} is $m' < m$ we use that $\mathcal{P}$ is strongly connected to construct a sequence $\mathbf{S}_2$ of proper faces of $\mathcal{P}/H'_{2j}$ from H'_{2j-1} to H'_{2j+1} where any two consecutive faces of $\mathbf{S}_2$ are incident and all faces in $\mathbf{S}_2$ have rank at least $m'+1$. We can repeat this process for each face of $\mathbf{S}_2$ with rank less than m. We can do this repeatedly on the resulting sequences to obtain at the end a sequence $\mathbf{S}_3$ from H'_{2j-1} to H'_{2j+1} where any two consecutive faces are incident and all faces have rank at least m. From Remark 5.2.3 we know that all elements of $\mathbf{S}_3$ are (k,m)-flat. This, together with the fact that the pairs of consecutive faces in the sequence are incident, implies that every two consecutive faces of $\mathbf{S}_3$ will have the same set of k-faces. It follows that all faces of $\mathbf{S}_3$ have the same k-faces, including H'_{2j-1} and H'_{2j+1}, which is a contradiction. ■

By applying Lemma 5.2.4 we obtain the following result.

Corollary 5.2.5 *Let $n \geq 3$ and let $0 \leq k < m < \ell \leq n-1$. If all ℓ-faces of an n-polytope $\mathcal{P}$ are (k,m)-flat then $\mathcal{P}$ itself is (k,m)-flat.*

If $\mathcal{P}$ is a flat polyhedron then the condition of every vertex belonging to every facet forces $\mathcal{P}$ to be equivelar. This is not true for higher ranks.

Exercise 5.2.6 Find a flat 4-polytope that is not equivelar.

The automorphism groups of flat rotary polytopes are characterised as follows.

Proposition 5.2.7 *A rotary n-polytope $\mathcal{P}$ of rank at least 3 with $\Gamma^+(\mathcal{P}) = \langle \sigma_1, \dots, \sigma_{n-1}\rangle$ is flat if and only if*

$$\Gamma^+(\mathcal{P}) = \langle \sigma_2, \dots, \sigma_{n-1}\rangle \cdot \langle \sigma_1, \dots, \sigma_{n-2}\rangle = \langle \sigma_1, \dots, \sigma_{n-2}\rangle \cdot \langle \sigma_2, \dots, \sigma_{n-1}\rangle.$$

Proof Assume first that $\mathcal{P}$ is flat, let Φ be the base flag and let $\gamma \in \Gamma^+(\mathcal{P})$. Let F_{n-1} be the facet contained in Φ, and let G_0 be the vertex contained in $\Phi\gamma$. Since $\mathcal{P}$ is flat we know that $G_0 \leq F_{n-1}$. Let Υ be a flag containing both G_0 and F_{n-1}, in the same flag orbit under $\Gamma^+(\mathcal{P})$ as Φ (this is possible since 1-adjacency does not change vertex or facet). Now we use the strong flag-connectivity to conclude that $\Upsilon = w\Phi$ for some $w \in \mathrm{Conn}^+(F_{n-1}/\varnothing)$, and $\Phi\gamma = x\Upsilon$ for some $x \in \mathrm{Conn}^+(\mathcal{P}/G_0)$. We use Proposition 3.2.19 to write $w = s_{i_k} \cdots s_{i_1}$ with $i_1, \dots, i_k \in \{1, \dots, n-2\}$, and $x = s_{i_{k+\ell}} \cdots s_{i_{k+1}}$ with $i_{k+1}, \dots, i_{k+\ell} \in \{2, \dots, n-1\}$. Then from Remark 3.2.22 we know that

$$\Phi\gamma = x\Upsilon = xw\Phi = s_{i_{k+\ell}} \cdots s_{i_1}\Phi = \Phi\sigma_{i_{k+\ell}} \cdots \sigma_{i_1}.$$

But $\sigma_{i_{k+\ell}} \cdots \sigma_{i_{k+1}} \in \langle \sigma_2, \dots, \sigma_{n-1}\rangle$ and $\sigma_k \cdots \sigma_1 \in \langle \sigma_1, \dots, \sigma_{n-2}\rangle$, proving the first equality. The second equality can be proved in a similar way, by choosing a flag that contains the facet of Φ and the vertex of $\Phi\gamma$ instead of Υ.

Now assume that $\Gamma^+(\mathcal{P}) = \langle \sigma_2, \dots, \sigma_{n-1}\rangle \cdot \langle \sigma_1, \dots, \sigma_{n-2}\rangle$. Then, in the coset representation of $\mathcal{P}$ the vertices are of the form $\Gamma_0\gamma$ with $\Gamma_0 := \langle \sigma_2, \dots, \sigma_{n-1}\rangle$ and $\gamma \in \Gamma_{n-1} := \langle \sigma_1, \dots, \sigma_{n-2}\rangle$. Since the base facet F_{n-1}

corresponds to the coset Γ_{n-1}, we can conclude that F_{n-1} is incident to all vertices. The facet-transitive action of $\Gamma^+(\mathcal{P})$ shows that $\mathcal{P}$ is flat. ■

Flat highly symmetric polytopes are small in the sense that their numbers of flags can be computed in the following ways.

Remark 5.2.8 The number of flags of a flat vertex-transitive polytope $\mathcal{P}$ is the number of vertices of a facet times the number of flags of a vertex-figure of $\mathcal{P}$. Similarly, the number of flags of a flat facet-transitive polytope is the number of facets of a vertex-figure times the number of flags of a facet of $\mathcal{P}$.

The computations in Remark 5.2.8 become lower bounds on the number of flags when dealing with rotary polytopes that are not necessarily flat.

According to Remark 5.2.8, a flat polyhedron with type $\{p,q\}$ must have precisely $2pq$ flags (the facets are p-gons and there must be q of them); this holds even if it is neither vertex-transitive nor facet-transitive. On the other hand, when considering flat rotary polytopes with rank at least 4 the number of flags of a facet or of a vertex-figure may be very large with respect to the entries of the type, leaving the impression that some flat polytopes may not be so small after all.

In [32] it was shown that if $\mathcal{P}$ is a regular polytope with type $\{p_1,\dots,p_{n-1}\}$ then it has at least $2p_1\cdots p_{n-1}$ flags. This was used to find the regular n-polytopes with fewest flags among those that have no entry equal to 2 in their types. Regular polytopes attaining that lower bound were called *tight*. The main result of that paper states, among other things, that for $n \geq 4$ the regular n-polytopes with fewest flags and no entry 2 in their types are tight. It also says that there are three polyhedra with no entry 2 in their types with fewer flags than any other such regular polyhedron; two of these three polyhedra are tight (see [32, Theorem 1.1]).

Every flat regular polyhedron is tight, but this does not extend to higher ranks. For example, the toroidal 4-polytope $\{4,3,4\}_{(2,0,0)}$ is flat and it has 384 flags, whereas $2\cdot 4\cdot 3\cdot 4 = 96$.

The notion of tightness was generalised in [65] to equivelar polytopes as follows.

Definition 5.2.9 An equivelar polytope $\mathcal{P}$ with type $\{p_1,\dots,p_{n-1}\}$ is said to be *tight* whenever it has precisely $2p_1\cdots p_{n-1}$ flags.

With the next two results we justify Definition 5.2.9 by showing that $2p_1\cdots p_{n-1}$ is a lower bound for the number of flags of an equivelar polytope with type $\{p_1,\dots,p_{n-1}\}$.

Lemma 5.2.10 *If $\mathcal{P}$ is an equivelar polytope with type $\{p_1, \ldots, p_{n-1}\}$ then it has at least p_1 vertices and at least p_{n-1} facets.*

Proof Each 2-face of $\mathcal{P}$ is a p_1-gon, and the strong flag-connectivity forces it to have p_1 distinct vertices of $\mathcal{P}$. A dual argument shows that $\mathcal{P}$ has at least p_{n-1} facets. ■

Proposition 5.2.11 *If $\mathcal{P}$ is an equivelar polytope with type $\{p_1, \ldots, p_{n-1}\}$ then it has at least $2p_1 \cdots p_{n-1}$ flags.*

Proof We proceed by induction over the rank n of $\mathcal{P}$. The base case is when $n = 2$, in which case $\mathcal{P}$ is a p_1-gon and it has precisely $2p_1$ flags.

Assume that $n \geq 3$ and that the result is true for $(n-1)$-polytopes. From Lemma 5.2.10 we know that $\mathcal{P}$ has at least p_{n-1} facets, and the induction hypothesis tells us that each facet has at least $2p_1 \cdots p_{n-2}$ flags. It follows from Remark 5.2.8 that $\mathcal{P}$ has at least $2p_1 \cdots p_{n-1}$ flags. ■

Proposition 5.2.11 makes it clear that tightness is a property that ensures the lowest possible number of flags for each given type, whereas flatness is less restrictive and allows higher numbers of flags in general.

Remark 5.2.12 The notion of flatness applies naturally to maniplexes that are not polytopes. For instance, the non-polyhedral toroidal maps $\{4,4\}_{(1,0)}$ and $\{4,4\}_{(1,1)}$ satisfy that all vertices are in all 2-faces.

By contrast, the bound in Proposition 5.2.11 does not hold for general maniplexes. For example, the maps $\{4,4\}_{(1,0)}$ and $\{4,4\}_{(1,1)}$ have 8 and 16 flags, respectively, whereas the bound for polyhedra with type $\{4,4\}$ is $2\cdot4\cdot4 = 32$ (we note that Lemma 5.2.10 does not hold for $\{4,4\}_{(1,0)}$ nor for $\{4,4\}_{(1,1)}$). As a consequence, if we want to study the equivelar maniplexes with fewest flags we should look for a concept different to the one in Definition 5.2.9. With this in mind we understand tightness as defined here as a property reserved exclusively to polytopes.

Much of the general theory of tight polytopes was settled in [65]. In what follows we include some of the results in that paper.

Proposition 5.2.13 *An equivelar polytope is tight if and only if all its sectons of rank 3 are flat.*

Proof We proceed by induction on the rank n.

The base case is when $n = 3$, and we already know that if $\mathcal{P}$ is a flat polyhedron with type $\{p, q\}$ then it has $2pq$ flags, and so it is tight. Conversely, if $\mathcal{P}$ is a tight polyhedron with type $\{p, q\}$ then it has $2pq$ flags and each facet is a p-gon (with $2p$ flags). It follows that $\mathcal{P}$ must have exactly q facets, and since every vertex must belong to q distinct facets, $\mathcal{P}$ must be flat.

Now assume that $n > 3$, that $\mathcal{P}$ is an equivelar n-polytope with type $\{p_1, \ldots, p_{n-1}\}$, and that the proposition is true for rank $n-1$.

If $\mathcal{P}$ is tight then it has $2p_1 \cdots p_{n-1}$ flags, and by Lemma 5.2.10 it has at least p_{n-1} facets. Since each facet is equivelar with type $\{p_1, \ldots, p_{n-2}\}$, it has at least $2p_1 \cdots p_{n-2}$ flags. It follows that $\mathcal{P}$ must have precisely p_{n-1} facets, each with precisely $2p_1 \cdots p_{n-2}$ flags. From here we conclude that the facets of $\mathcal{P}$ are tight. A dual argument shows that the vertex-figures of $\mathcal{P}$ are tight. The inductive hypothesis can now be used to conclude that all sections of rank 3 are flat, since each of them belongs to a facet or to a vertex-figure.

It remains for us to prove that if all sections of rank 3 are flat then $\mathcal{P}$ is tight. The inductive hypothesis applied to the facets tells us that if all sections of rank 3 of $\mathcal{P}$ are flat then the facets of $\mathcal{P}$ are tight. Now, the base of the induction tells us that every 3-face is $(0,2)$-flat. It follows from Corollary 5.2.5 that $\mathcal{P}$ and all its facets are $(0,2)$-flat. Remark 5.2.3 tells us that $\mathcal{P}$ is flat, and in particular it has precisely p_1 vertices (the same as any of its 3-faces). A dual argument shows that $\mathcal{P}$ has precisely p_{n-1} facets. We know that the number of flags of $\mathcal{P}$ is the number of flags of a facet times the number of facets (this is a weaker version of Remark 5.2.8). The inductive hypothesis tells us that each facet has $2p_1 \cdots p_{n-2}$ flags. Therefore $\mathcal{P}$ has precisely $2p_1 \cdots p_{n-1}$ flags and it is tight. ■

Proposition 5.2.13 gives an equivalence of tightness in terms of a property of all rank 3 sections. This is useful when trying to construct tight polytopes of higher ranks. The next proposition shows a restriction to avoid when looking for tight polytopes.

Proposition 5.2.14 *If an equivelar polytope is tight then it does not have two consecutive odd entries in its type.*

Proof Assume to the contrary that $\mathcal{P}$ is a tight polytope with type $\{p_1, \ldots, p_{n-1}\}$ and that p_i and p_{i+1} are odd. Then it follows from Proposition 5.2.13 that the section F_{i+2}/F_{i-2} between an $(i+2)$-face F_{i+2} and an incident $(i-2)$-face F_{i-2} is a flat polyhedron with type $\{p_i, p_{i+1}\}$. Since every edge of F_{i+2}/F_{i-2} belongs to precisely four flags, the number of flags of F_{i+2}/F_{i-2} must be divisible by 4, but being F_{i+2}/F_{i-2} a flat polyhedron with type $\{p_i, p_{i+1}\}$ it has $2p_i p_{i+1}$ flags, a contradiction. ■

The converse of Proposition 5.2.14 is also true for polyhedra, that is, if $p, q \in \mathbb{Z}$ with $p, q \geq 2$ and at least one of them is even then there is a tight polyhedron with type $\{p, q\}$. This was shown with a constructive method in [65, Theorem 5.1], and leads to the following theorem.

Theorem 5.2.15 *There is a tight polyhedron with type $\{p,q\}$ if and only if p,q are integers greater or equal to 2 and at least one of them is even.*

Tight regular polytopes were studied in [35, 65, 71]. In those works it was established that there are some p and q such that there are tight polyhedra with type $\{p,q\}$ but no tight regular polyhedra with type $\{p,q\}$. For example, there are tight polyhedra with type $\{5,22\}$ but none of them are regular. The main results regarding the types of tight regular polyhedra are the following (see [35, Theorem 1] and [71, Theorem 5.10]).

Theorem 5.2.16 *There is a tight orientably regular polyhedron with type $\{p,q\}$ if and only if either p is odd and q is an even divisor of $2p$, q is odd and p is an even divisor of $2q$, or p and q are even.*

Theorem 5.2.17 *There is a tight non-orientably regular polyhedron with type $\{p,q\}$ if and only if p and q satisfy one of the following:*

- *$p = 4$ and q is a multiple of 3;*
- *p is an odd multiple of 4 and q is an odd multiple of 6;*
- *$q = 4$ and p is a multiple of 3;*
- *q is an odd multiple of 4 and p is an odd multiple of 6.*

In fact, regular tight polyhedra are well understood. They were classified in [71]. Less is known about tight regular polytopes of ranks greater than 3. The types of the orientably regular polytopes were determined in [35, Theorem 1]. It turns out that the only limitation is that every section of rank 3 must have a type satisfying the restrictions in Theorem 5.2.16.

Theorem 5.2.18 *There is a tight orientably regular n-polytope with type $\{p_1,\ldots,p_{n-1}\}$ if and only if p_{i-1} and p_{i+1} (when defined) are even divisors of $2p_i$ whenever p_i is odd.*

For ranks greater than 3 it is still unknown how many tight orientably regular polytopes there are for a given type. The types of non-orientably regular polytopes of ranks 4 and higher are still unknown.

Problem 37 Determine the types of tight non-orientably regular polytopes of ranks 4 and higher.

A comparison of Theorem 5.2.15 with Theorems 5.2.16 and 5.2.17 indicates that there are types of tight polyhedra that are not types of any regular polyhedron. It is still unknown if the same holds in higher ranks.

Problem 38 Determine the types of tight n-polytopes for $n \geq 4$.

5.2.2 Tight Chiral Polytopes

As mentioned above, the regular n-polytopes with fewest flags and no entry 2 in their types were proven to be tight for every $n \geq 4$. The success on listing such regular polytopes suggested we should try our luck with tight chiral polytopes in order to find small chiral polytopes of higher ranks. While doing so, one may express the automorphism groups of tight chiral polytopes as follows.

Proposition 5.2.19 *Let $\mathcal{P}$ be a rotary orientable n-polytope with standard rotations $\sigma_1, \dots, \sigma_{n-1}$. Then $\mathcal{P}$ is tight if and only if*

$$\Gamma^+(\mathcal{P}) = \langle\sigma_1\rangle \cdot \langle\sigma_2\rangle \cdots \langle\sigma_{n-1}\rangle. \tag{5.2}$$

Proof Assume first that $\mathcal{P}$ has type $\{p_1, \dots, p_{n-1}\}$ and that (5.2) holds. Then the number of flags of $\mathcal{P}$ is

$$2|\Gamma^+(\mathcal{P})| = 2|\langle\sigma_1\rangle \cdots \langle\sigma_{n-1}\rangle| \leq 2p_1 \cdots p_{n-1}.$$

Since $2p_1 \cdots p_{n-1}$ is a lower bound on the number of flags, $\mathcal{P}$ has the required number of flags to be tight.

To show the forward implication we proceed by induction over n; the base case is $n = 2$, where the result is trivial.

Assume that $n \geq 3$ and that the proposition holds for ranks at most $n - 1$. Then the automorphism groups of the base facet and vertex-figure of $\mathcal{P}$ are $\langle\sigma_1\rangle \cdots \langle\sigma_{n-2}\rangle$ and $\langle\sigma_2\rangle \cdots \langle\sigma_{n-1}\rangle$, respectively. From Proposition 5.2.7 we know that $\Gamma^+(\mathcal{P}) = \langle\sigma_1\rangle \cdots \langle\sigma_{n-2}\rangle \cdot \langle\sigma_2\rangle \cdots \langle\sigma_{n-1}\rangle$. The result follows from the inductive hypothesis for rank $n - 2$, since $\langle\sigma_2, \dots, \sigma_{n-2}\rangle = \langle\sigma_2\rangle \cdots \langle\sigma_{n-2}\rangle$. ■

In view of Proposition 5.2.19 it is possible to find a presentation for the rotation groups of tight rotary orientable polytopes in terms of the abstract rotations σ_i with relations of the form $\sigma_k^{a_1}\sigma_j^{a_2} = \sigma_j^{b_1}\sigma_{j+1}^{b_2} \cdots \sigma_k^{b_{k-j+1}}$ for $j < k$.

When restricting to tight rotary polyhedra there is a natural order given by coverings as explained next.

Lemma 5.2.20 *Let $\mathcal{P}$ be a tight orientable rotary polyhedron with type $\{p,q\}$ for some $2 < p \leq q$. Then:*

(a) $\langle\sigma_2\rangle$ *is not normal in* $\Gamma^+(\mathcal{P})$*;*
(b) there exists $k < q$ *such that* $\langle\sigma_2^k\rangle \triangleleft \Gamma^+(\mathcal{P})$*;*
(c) $\Gamma^+(\mathcal{P})/\langle\sigma_2^k\rangle$ *is the rotation group of a tight orientable rotary polyhedron with type* $\{p,k\}$.

Proof Assume to the contrary that $\langle\sigma_2\rangle \lhd \Gamma^+(\mathcal{P})$. We know that $\langle\sigma_2\rangle$ is the stabiliser in $\Gamma^+(\mathcal{P})$ of the base vertex F_0. The stabiliser of the vertex $F_0\gamma$ is $\gamma^{-1}\langle\sigma_2\rangle\gamma = \langle\sigma_2\rangle$. It follows that σ_2 fixes every vertex and therefore F_0 has only one neighbour $F_0\sigma_1$. Since $\mathcal{P}$ is vertex-transitive, $F_0\sigma_1$ must also have only one neighbour and it must be F_0. This shows that $\mathcal{P}$ has only those two vertices, contradicting that $p > 2$.

The second item follows from studying right cosets of $H := \langle\sigma_1\rangle$. We know from Proposition 5.2.19 that the q cosets of H are $\{H\sigma_2^j\}_{j\in\{0,\ldots,q-1\}}$. Furthermore, every element in H acts trivially on H on the right, while it permutes the remaining $q-1$ cosets. The facts that $\langle\sigma_2\rangle$ has q elements and that $H\sigma_1\sigma_2^j \neq H$ if $j \neq 0$ imply that $H\sigma_1\sigma_2^{j_1} = H\sigma_1\sigma_2^{j_2}$ for some $j_1 \neq j_2$. This tells us that $H\sigma_1\sigma_2^{j_1-j_2} = H\sigma_1$ or, equivalently, that $\sigma_1\sigma_2^{j_1-j_2}\sigma_1^{-1} \in H$. We can choose k to be $j_1 - j_2$, showing the second item.

Next we show that $\Gamma^+(\mathcal{P})/\langle\sigma_2^k\rangle$ is the rotation group of a rotary polyhedron. Let $\sigma_1' = \sigma_1 \cdot \langle\sigma_2^k\rangle$ and $\sigma_2' = \sigma_2 \cdot \langle\sigma_2^k\rangle$ be the generators of $\Gamma^+(\mathcal{P})/\langle\sigma_2^k\rangle$. It is clear that the order of $\sigma_1'\sigma_2'$ is at most 2, and it cannot be 1 since otherwise

$$\sigma_1 \cdot \langle\sigma_2^k\rangle = \sigma_1' = (\sigma_2')^{-1} = \sigma_2^{-1} \cdot \langle\sigma_2^k\rangle,$$

forcing $\sigma_1 \in \langle\sigma_2\rangle$, and in so doing contradicting the intersection condition of $\Gamma^+(\mathcal{P})$. Therefore (3.7) holds. To verify the intersection condition (3.16) we let $\gamma \cdot \langle\sigma_2^k\rangle \in \langle\sigma_1'\rangle \cap \langle\sigma_2'\rangle$, for some $\gamma \in \Gamma^+(\mathcal{P})$. Proposition 5.2.19 implies that $\gamma = \sigma_1^a\sigma_2^{ks} = \sigma_2^b\sigma_2^{kt}$ for some a, b, s and t, but then $\sigma_1^a \in \langle\sigma_2\rangle$, and since $\Gamma^+(\mathcal{P})$ satisfies the intersection condition, σ_1^a and $\gamma \cdot \langle\sigma_2^k\rangle$ are trivial. Proposition 3.1.58 tells us that $\Gamma^+(\mathcal{P})/\langle\sigma_2^k\rangle$ is the rotation group of an orientable rotary polyhedron $\mathcal{Q}$.

It only remains to show that $\mathcal{Q}$ is tight. If $\gamma' = \gamma \cdot \langle\sigma_2^k\rangle \in \Gamma^+(\mathcal{Q})$ for some $\gamma \in \Gamma^+(\mathcal{P})$ then we write $\gamma = \sigma_1^a\sigma_2^b$ for some a and b, forcing $\gamma' = (\sigma_1')^a(\sigma_2')^b$. Tightness of $\mathcal{Q}$ follows from Proposition 5.2.19. ■

When showing tightness in the proof of Lemma 5.2.20 the fact that $\langle\sigma_2^k\rangle$ is a subgroup of $\langle\sigma_2\rangle$ plays no role. The following statement is general and follows from Proposition 5.2.19.

Remark 5.2.21 Let $\mathcal{P}$ and $\mathcal{Q}$ be rotary polytopes such that $\mathcal{P}$ covers $\mathcal{Q}$. If $\mathcal{P}$ is tight then $\mathcal{Q}$ is also tight.

From Lemma 5.2.20 we know that if $\mathcal{P}$ is a tight rotary polyhedron with type $\{p,q\}$ for some $2 < p \leq q$ then we can always find a quotient with type $\{p,q'\}$ for some proper divisor q' of q. If $p \leq q'$ then we may repeat this step until we obtain a quotient where the second entry of the type is less than the first one, so that we may assume that $q' < p$. If $q' > 2$ we may apply the dual version of Lemma 5.2.20 to find a tight rotary polyhedron with type $\{p', q'\}$ for some proper divisor p' of p. In this way we may keep taking quotients to reduce an entry of the type that is greater than or equal to the other until one of the entries equals 2. We observe that the polyhedra with types $\{2, q\}$ and $\{p, 2\}$ are tight, and we already know that they are regular. This has the following consequence.

Corollary 5.2.22 *Every tight chiral polyhedron covers a tight orientably regular polyhedron.*

We know that some regular polyhedra cover chiral polyhedra; in particular, every chiral polyhedron $\mathcal{P}$ has a smallest regular cover, and it is true that if $\mathcal{P}$ is finite then its smallest regular cover is also finite (see Proposition 4.3.5 and Corollary 4.3.10). It turns out that tight orientably regular polyhedra do not cover chiral polyhedra.

Proposition 5.2.23 *Let $\mathcal{P}$ be a tight orientable rotary polyhedron and assume that it covers a chiral polyhedron $\mathcal{Q}$. Then $\mathcal{P}$ is chiral.*

Proof We know from Remark 5.2.21 that $\mathcal{Q}$ is tight. Assume that it has type $\{p, q\}$ and let $\mathcal{R}$ be the smallest regular cover of $\mathcal{Q}$ as in Definition 4.3.4. Then $\mathcal{R}$ is also a polyhedron with type $\{p, q\}$ but it must have more flags than $\mathcal{Q}$; in particular, $\mathcal{R}$ is not tight. If $\mathcal{P}$ were regular then it would cover $\mathcal{R}$, contradicting Remark 5.2.21. ■

As a consequence of Proposition 5.2.23 there exists a family $\mathcal{A}$ of tight chiral polyhedra that are minimal in the sense that they do not cover other tight chiral polyhedra, and such that the set of tight chiral polyhedra is precisely the set of tight rotary polyhedra that cover elements of $\mathcal{A}$. The polyhedra in $\mathcal{A}$ were the key for the main results of [67] and are defined next.

Definition 5.2.24 A chiral polyhedron of type $\{p, q\}$ is *atomic* if it is tight and it does not cover a tight chiral polyhedron with type $\{p', q\}$ or $\{p, q'\}$ for some proper divisor p' of p or some proper divisor q' of q.

The classification of atomic chiral polyhedra was presented in [67]. The types of these polyhedra are given next (see [67, Theorems 4.11, 4.14 and 4.15]).

Theorem 5.2.25 *Up to duality, the type of an atomic chiral polyhedron is one of the following:*

(a) $\{2m, m^\alpha\}$ *for some odd prime m and some integer $\alpha \geq 2$;*
(b) $\{8, 2^\beta\}$ *for some integer $\beta \geq 5$;*
(c) $\{2^\alpha, 2^{\alpha+1}\}$ *for some integer $\alpha \geq 4$.*

The chirality group of an atomic chiral polyhedron $\mathcal{P}$ is easy to compute. If $\mathcal{P}$ has type $\{2m, m^\alpha\}$ then from Lemma 5.2.20 we know that $\Gamma(\mathcal{P})$ has a normal subgroup $\langle \sigma_2^k \rangle$ for some $1 < k < m^\alpha$. Now, every non-trivial subgroup of $\langle \sigma_2 \rangle$ contains $\langle \sigma_2^{m^{\alpha-1}} \rangle$ and hence $\langle \sigma_2^{m^{\alpha-1}} \rangle \triangleleft \Gamma(\mathcal{P})$. We use that $\mathcal{P}$ is atomic to conclude that $\mathcal{P}/\langle \sigma_2^{m^{\alpha-1}} \rangle$ is regular. Since m is prime, this must be the largest regular quotient of $\mathcal{P}$, showing that $\mathcal{P}$ has cyclic chirality group of order m, and the chirality index of $\mathcal{P}$ is m. A similar argument can be used if $\mathcal{P}$ has type $\{8, 2^\beta\}$, $\{2^\beta, 8\}$, $\{2^\alpha, 2^{\alpha+1}\}$ or $\{2^{\alpha+1}, 2^\alpha\}$.

Proposition 5.2.26 *The chirality group of an atomic chiral polyhedron $\mathcal{P}$ is cyclic of order m if $\mathcal{P}$ or its dual has type $\{2m, m^\alpha\}$, and it is cyclic of order 2 if the type of $\mathcal{P}$ or of its dual is $\{8, 2^\beta\}$ or $\{2^\alpha, 2^{\alpha+1}\}$.*

Chiral atomic polyhedra were used to determine the types of all tight chiral polyhedra. They are summarised next (see [67, Theorem 5.2]).

Theorem 5.2.27 *There are tight chiral polyhedra with types $\{p, q\}$ and $\{q, p\}$ if and only if p and q satisfy one of the following:*

- *q is an odd multiple of m^2 for some odd prime m and p is an even multiple of m that divides $2q$;*
- *$p = 2am$, $q = 2bm^2$ for some odd prime m and some integers $a, b \geq 1$;*
- *$p = 8a$, $q = 32b$ for any integers $a, b \geq 1$.*

The techniques used to describe all tight regular polyhedra in [71] do not apply to chiral polyhedra. To this day, we do not have an easy way to determine how many tight chiral polyhedra of a given type are there, and what their automorphism groups are.

Problem 39 Describe the automorphism groups of all tight chiral polyhedra. Find a procedure to determine how many tight chiral polyhedra are there with a given type.

Atomic chiral n-polytopes can be defined in a way analogous to that for rank 3, so that every tight chiral n-polytope covers an atomic one. The types of all atomic chiral 4-polytopes were found in [73], and are summarised next (see [73, Theorem 16]).

Theorem 5.2.28 *Up to duality, the type of an atomic chiral 4-polytope is one of the following:*

(a) $\{m, 2m, m^\alpha\}$ *for some odd prime m and some integer $\alpha \geq 2$;*
(b) $\{4, 8, 2^\beta\}$ *for some integer $\beta \geq 5$;*
(c) $\{4, 2^{\beta-1}, 2^\beta\}$ *for some integer $\beta \geq 5$;*

(d) $\{2m, m^\alpha, 2m\}$ *for some odd prime* m *and some integer* $\alpha \geq 2$;
(e) $\{8, 2^\beta, 8\}$ *for some integer* $\beta \geq 5$;
(f) $\{2^{\beta-1}, 2^\beta, 2^{\beta-1}\}$ *for some integer* $\beta \geq 5$;
(g) $\{8, 2^\beta, 2^{\beta-1}\}$ *for some integer* $\beta \geq 5$.

The polytopes with types in items (a), (b) and (c) have regular facets and chiral vertex-figures. Those whose types are in the remaining four items have chiral facets and chiral vertex-figures.

The proof of Theorem 5.2.28 included showing that the facets and vertex-figures must be rotary tight polyhedra with some extra properties. At least one of them must be an atomic chiral polyhedron. The other one is either an atomic chiral polyhedron or a tight regular polyhedron satisfying that it has no multiple edges, or that its dual has no multiple edges. These conditions on facets and vertex-figures are a necessary condition, but they are not sufficient. For instance, there is no atomic chiral 4-polytope with type $\{8, 2^\beta, 2^{\beta+1}\}$.

The set of types of all tight chiral 4-polytopes has not been determined so far. We only know that if $\mathcal{P}$ is a tight chiral 4-polytope with type $\{p,q,r\}$ then one of the types $\{p',q',r'\}$ listed in Theorem 5.2.28, or its dual type, is such that p' divides p, q' divides q and r' divides r.

Problem 40 Determine the types of all tight chiral 4-polytopes.

The following problem is even more challenging.

Problem 41 Determine how many tight chiral 4-polytopes are there for each type; also determine their automorphism groups.

The previous results are as far as we can reach regarding the rank of tight chiral polytopes. The non-existence of tight chiral 5-polytopes was proved in [73, Theorem 1]. It was proved by showing that some conditions on the facets and vertex-figures for $\mathcal{P}$ to be an atomic chiral 5-polytope are impossible to satisfy.

Theorem 5.2.29 *There are no tight chiral 5-polytopes.*

5.2.3 Lower Bounds on the Size of Chiral n-Polytopes

Before Theorem 5.2.29 was proved it was already known that there are no tight chiral n-polytopes for $n \geq 6$. In fact, better lower bounds on the number of flags of chiral n-polytopes were given in [66]. In what follows we discuss the results in that work. We start with three sets of conditions that force regularity.

Proposition 5.2.30 *If $\mathcal{P}$ is a finite orientable rotary polytope with regular flat facets and regular vertex-figures then $\mathcal{P}$ is regular.*

Proof Let n be the rank of $\mathcal{P}$, let $\mathcal{Q}$ and $\mathcal{R}$ be the base facet and vertex-figure of $\mathcal{P}$, and let $\overline{\mathcal{P}}$ be any orientable rotary polytope with facets isomorphic to $\mathcal{Q}$ and vertex-figures isomorphic to $\mathcal{R}$. Our hypothesis tells us that $\mathcal{Q}$ is flat (that is, $(0, n-2)$-flat). Then Lemma 5.2.4 implies that $\overline{\mathcal{P}}$ itself is $(0, n-2)$-flat, and from Remark 5.2.8 we know that $\overline{\mathcal{P}}$ is flat. This information is enough to count the number of flags of $\overline{\mathcal{P}}$ as the number of vertices times the number of flags containing the base vertex (see Remark 5.2.8); that is, the number of vertices of $\mathcal{Q}$ times the number of flags of $\mathcal{R}$. Therefore, all orientable rotary n-polytopes with facets isomorphic to $\mathcal{Q}$ and vertex-figures isomorphic to $\mathcal{R}$ have the same number of flags.

The smallest regular cover $\widetilde{\mathcal{P}}$ of $\mathcal{P}$ is a regular n-polytope (see Section 4.3). Furthermore, its facets are isomorphic to $\mathcal{Q}$ and its vertex-figures are isomorphic to $\mathcal{R}$ since the action of the subgroups $\langle r_0, \ldots, r_{n-2} \rangle$ and $\langle r_1, \ldots, r_{n-2} \rangle$ of $\mathrm{Conn}(\mathcal{P})$ is regular on the flags of $\mathcal{Q}$ and of $\mathcal{R}$, respectively (see Proposition 3.2.15). This implies that $\widetilde{\mathcal{P}}$ has the same number of flags than $\mathcal{P}$. Then the covering from $\widetilde{\mathcal{P}}$ to $\mathcal{P}$ has degree 1 and $\mathcal{P}$ must be regular. ■

Proposition 5.2.31 *If $\mathcal{P}$ is a finite $(1, n-3)$-flat orientable rotary n-polytope with $n \geq 5$ then it is regular.*

Proof Let F_i denote the i-face of some given flag for $i \in \{-1, \ldots, n\}$. We know from Propositions 2.2.8, 2.3.6 and 2.3.8 that the sections F_{n-3}/F_{-1}, F_{n-2}/F_0, F_{n-1}/F_1 and F_n/F_2 are regular, since they are the facet of the facet or the vertex-figure of the vertex-figure of a section of rank $n-1$. Furthermore the section F_{n-2}/F_0 is flat, since it is $(0, n-4)$-flat. These two facts together with Proposition 5.2.30 and its dual version imply that the sections F_{n-2}/F_{-1} and F_{n-1}/F_0 are regular, and since the former is flat (it is $(1, n-3)$-flat), the section F_{n-1}/F_{-1} is also regular and flat. From Proposition 2.3.8 we also know that the section F_n/F_1 is regular, and combined with the fact that F_{n-1}/F_0 is regular and flat we know that the vertex-figures must be regular. Hence, $\mathcal{P}$ has flat, regular facets and regular vertex-figures; using Proposition 5.2.30 once more we conclude that $\mathcal{P}$ is regular. ■

Corollary 5.2.32 *If $n \geq 6$ then every finite orientable rotary tight n-polytope is regular.*

Proof Let $\mathcal{P}$ be a finite orientable rotary tight n-polytope. Proposition 5.2.13 implies that every section F_4/F_0 between a 4-face F_4 and a vertex F_0 is flat. From Remark 5.2.3 and Lemma 5.2.4 we can conclude that $\mathcal{P}$ is $(1,3)$-flat, and Proposition 5.2.31 forces $\mathcal{P}$ to be regular. ■

The following information can be verified in [33].

- The number of flags of the chiral polyhedron with fewest flags is 40.
- The number of flags of a chiral 4-polytope with chiral facets or chiral vertex-figures and with the fewest flags is 240.
- The number of flags of a chiral 4-polytope with regular facets and regular vertex-figures is 384.
- The number of flags of a chiral 5-polytope with fewest flags is 1440, and such a polytope has chiral facets and chiral vertex-figures.

In this book we have already encountered some of the small chiral polytopes mentioned in the list. There is only one chiral polyhedron with 40 flags, and it is $\{4,4\}_{(2,1)}$ (see Proposition 3.1.84). One of the two chiral 4-polytopes with 240 flags is the one mentioned in Exercise 2.1.20; it is flat, but its sections of rank 3 are not flat. The other chiral 4-polytope with 240 flags has facets and vertex-figures isomorphic to $\{4,4\}_{(2,1)}$ and it is not flat. We shall study this polytope in Section 5.4.2.

The only chiral 5-polytope with 1440 flags is the one mentioned in Exercise 2.1.21; its sections of rank 3 are not flat, but all sections of rank 4 are flat.

The chiral 6-polytopes (up to duality) with fewest flags are a dual pair found in [45]. One of them has type $\{3,3,4,6,3\}$ and 18 432 flags. Its sections of rank 3 are tetrahedra, octahedra, a regular polyhedron with type $\{4,6\}$ and 192 flags, and the toroidal regular polyhedron $\{6,3\}_{(2,0)}$; none of them are flat. Its 4-faces are the hemi-cross polytope (obtained from the cross polytope by identifying antipodal faces); they are regular and flat. The medial sections (vertex-figures of the facets) are flat chiral 4-polytopes with type $\{3,4,6\}$ and 1152 flags. The coedges are flat regular 4-polytopes with type $\{4,6,3\}$ and 768 flags. The facets and vertex-figures are flat chiral 5-polytopes with types $\{3,3,4,6\}$ and $\{3,4,6,3\}$, respectively, both with 4608 flags. We remark that Proposition 5.2.30 is not violated for the 5-faces even though their facets are regular and flat, because their vertex-figures are chiral.

Indeed, the chiral 6-polytopes with fewest flags are not $(1,3)$-flat, as was announced in Proposition 5.2.31. However, they are $(0,3)$- and $(1,4)$-flat and their facets (of rank 5) are chiral. This suggests that in order to construct a small chiral n-polytope we may look for small chiral $(n-1)$-polytopes $\mathcal{Q}$ that are $(0,n-3)$- and $(1,n-2)$-flat with regular facets and chiral vertex-figures. For $\mathcal{Q}$ to have a chance of being the facet of a chiral n-polytope we need the vertex-figures of $\mathcal{Q}$ to have regular facets and regular vertex-figures, so none of the sections with rank $(n-3)$ of the vertex-figure of $\mathcal{Q}$ can be flat.

Based on the available data on regular n-polytopes with few flags (see [34, 116]), a lower bound on the number of flags of regular n-polytopes that are not flat was established in [66, Section 4]. This was used in combination with Proposition 5.2.30 to determine lower bounds on the numbers of flags of chiral n-polytopes, depending on the symmetry type of facets and vertex-figures (see [66, Theorems 5.2, 5.3 and 5.4]).

Proposition 5.2.33 *Every chiral* 5*-polytope with regular facets or regular vertex-figures has at least* 4004 *flags. Every chiral* 6*-polytope with regular facets and chiral vertex-figures (or vice versa) has more than* 18 432 *flags. Every chiral* 7*-polytope with chiral facets and chiral vertex-figures has more than* 55 296 *flags.*

Theorem 5.2.34 *Let M be the number of flags of a chiral n-polytope $\mathcal{P}$.*

(a) If $\mathcal{P}$ has regular facets and regular vertex-figures, and $n \geq 6$, then

$$M \geq (16/3)n \cdot n!\,.$$

(b) If $\mathcal{P}$ or its dual has regular facets and chiral vertex-figures, and $n \geq 7$, then

$$M \geq 16(n-1)(n-1)!\,.$$

(c) If $\mathcal{P}$ has chiral facets and chiral vertex-figures, and $n \geq 8$, then

$$M \geq 48(n-2)(n-2)!\,.$$

Regardless of the symmetry type of the facets and vertex-figures, if $n \geq 8$ then $M \geq 48(n-2)(n-2)!$.

According to [32], for $n \geq 9$ the smallest regular n-polytope (in the sense of smallest number of flags) has precisely $2 \cdot 4^{n-1}$ flags. The lower bound in Theorem 5.2.34 grows much faster as a function of n. Moreover, when evaluating the last bound in Theorem 5.2.34 on $n = 6$ we obtain the number $48 \cdot 4 \cdot 4! = 4608$. A comparison with the number 18 432 of flags of the smallest chiral 6-polytope leaves the uncertainty of how far the bound is from being sharp. We still do not have a good idea of how fast the number of flags of the smallest chiral n-polytope grows with the rank n (see Problem 6).

The results in [45, 66] suggest that, for $n \geq 5$, the smallest chiral n-polytopes have chiral facets and chiral vertex-figures. In that case, one of their $(n-1)$- or $(n-2)$-sections must be a chiral polytope with regular facets and regular vertex-figures. This justifies splitting the study of the smallest chiral polytopes depending on the symmetry type of their facets and vertex-figures (see Problem 7 and problems 4 and 5 of [66]).

5.3 Chiral Polytopes from Geometric Constructions

Given the geometric nature of the motivation of the study of abstract polytopes it is to be expected that we should look for examples of these objects in geometric spaces. From this viewpoint it is not surprising that many of the first regular abstract polytopes to be described were defined as structures in underlying geometric spaces. This is the case of the convex regular polytopes and the regular tessellations of the Euclidean spaces.

5.3.1 From Euclidean Geometry

The first difficulty when searching for chiral polytopes from structures in a geometric space is given by the following result, whose proof can be seen in [163, 168].

Theorem 5.3.1 *There are neither chiral convex polytopes nor chiral tessellations of Euclidean spaces.*

Chirality seems to be particularly incompatible with convexity when we take into account that two-orbit polyhedra in other classes did appeared in the early days of the study of symmetries of convex bodies. This is the case of the cuboctahedron and the icosidodecahedron in class 2_{01}, and their duals the rhombic dodecahedron and the rhombic triacontahedron in class 2_{12} (see [55, Chapter II]). Among the highly symmetric tessellations of Euclidean 3-space by convex polyhedra there are two of them with two orbits of flags: the tessellation with octahedra and tetrahedra where every triangle belongs to one polyhedron of each kind (this is tiling #1 in [106], see also Appendix A.3); and its dual, the tessellation with rhombic dodecahedra. They belong to classes 2_{012} and 2_{123}, respectively. More information about two-orbit convex polytopes and two-orbit Euclidean tessellations can be found in [163, 164].

We do obtain some chiral polyhedra as quotients of regular tessellations of the Euclidean plane to the torus (see Section 2.4.2). However, this only applies for rank 3, since there are no chiral toroidal n-polytopes for $n \geq 4$ (see Corollary 2.4.26). Once more, regular polytopes are present under hypotheses in which chiral polytopes are impossible. By contrast, there are two-orbit tessellations of the n-torus in class $2_{\{1,\dots,n-1\}}$ for every even n (see [30]).

The n-torus is an example of a compact Euclidean *space form*, that is, a compact quotient of $\mathbb{E}^n$ by a discrete group of isometries that acts freely on $\mathbb{E}^n$. It is a consequence of Bieberbach's theorem in [6] that for each $n \geq 2$ there are finitely many n-dimensional compact Euclidean space forms up to some natural kind of equivalence called *affine equivalence*. The compact Euclidean space forms for $n = 2$ are the 2-torus and the Klein bottle. In three dimensions there are 10 compact Euclidean space forms, out of which 6 are orientable and 4 non-orientable (see [46, 110, 211, 274]).

It was proven in [188, Section 6B] that every tessellation of a compact Euclidean space form of dimension n is isomorphic to a quotient of a tessellation of $\mathbb{E}^n$ by a discrete group of isometries that acts freely on $\mathbb{E}^n$. This was used to show that there are no rotary tessellations of compact Euclidean space forms other than the torus (see [122, Theorems 3.2 and 4.3], [188, Section 6G]). The next result follows directly from the discussions in these two sources.

Theorem 5.3.2 *The only chiral tessellations of compact Euclidean space forms are the chiral maps described in Section 2.4.2.*

To get a flavour of the range of symmetry types of equivelar tessellations of other Euclidean space forms, one can look at the classifications of equivelar maps on the Klein bottle (see [272]) and of equivelar tessellations of non-toroidal orientable three-dimensional Euclidean space forms (see [131, 132]). One feature to point out here is that in each of these space forms there are equivelar tessellations with arbitrarily large numers of flag orbits. By contrast, the number of orbits of an equivelar tessellation of the n-torus is bounded by a function of n (see [136, Lemma 6(c)]).

5.3.2 Twisted Honeycombs

A natural idea of where to look next for chiral polytopes is as quotients of regular tessellations of spherical and hyperbolic spaces. The analysis in [188, Section 6B] also shows that such quotients correspond to tessellations of spherical or hyperbolic space forms (obtained from the spherical or hyperbolic n-spaces by discrete isometry groups that act freely). The study of such spaces is beyond the scope of this book; we shall mention them only in general terms without any rigorous treatment. Readers interested in space forms are referred to [188, Chapter 6], [234, 274]. Tessellations of spherical n-dimensional space with convex tiles are in one-to-one correspondence with convex regular n-polytopes (see [55], [188, Section 6B]). Details on tessellations of hyperbolic spaces can be found in [51, 54, 95, 97].

The only isometry groups that act freely on the two-dimensional sphere are the group generated by the antipodal involution and the trivial group (see [103]). Since the quotient of the sphere by the antipodal involution is the projective plane, and it is a non-orientable surface, we can conclude from Corollary 5.1.3 that there are no chiral polyhedra realised as tessellations of spherical space forms.

The two dodecahedral three-dimensional spaces constructed in [262] can be understood as chiral 4-maniplexes whose order complexes are a spherical space form in one case, and a hyperbolic space form in the other. As maniplexes, each of them has only one facet, and hence they are not polytopal. Other chiral 4-maniplexes arising from tessellations of spherical space forms are listed in the first seven rows of [54, Table 3] (the fifth one corresponds to the dodecahedral space); none of them are polytopal. It is still open whether a chiral polytopal tessellation of a spherical space form exists (see [220, Problem 13]).

Exercise 5.3.3 Show that none of the maniplexes in the first seven rows of table 3 in [54] is polytopal.

Problem 42 Enumerate all chiral maniplexes whose order complex is a spherical space form. Determine whether any of them is polytopal.

When looking for chiral polytopes in hyperbolic space forms we face the fact that, unlike spherical and Euclidean spaces, there are regular tessellations of the hyperbolic d-dimensional space with compact tiles and compact vertex-figures only for $d \leq 4$ (see [51]). As a consequence, we may find rotary n-polytopes as tessellations of compact hyperbolic space forms only for $n \in \{3,4,5\}$. By contrast, there are regular tessellations of spherical and Euclidean spaces in every dimension, from which we can derive regular n-polytopes as tessellations of the $(n+1)$-dimensional projective space and the $(n+1)$-torus (compact spherical and Euclidean space forms, respectively).

Chiral maps on compact surfaces with hyperbolic types have compact hyperbolic space forms as their order complexes, and many of these maps are actually polyhedra. We discussed some of their features in Section 5.1.

In some way, the first chiral 4-maniplex in a hyperbolic three-dimensional space form to be studied is the hyperbolic dodecahedral space described in [262], in the understanding that the initial dodecahedral fundamental region has the additional structure of edges, pentagons, and a unique facet.

The first systematic search for chiral tessellations of hyperbolic space forms of dimension 3 was carried out in [54]. The idea there is based on the fact that if $\mathcal{P}$ is an orientable rotary 4-polytope then there are two orbits of Petrie paths under $\Gamma^+(\mathcal{P})$ (see Proposition 3.2.69 and its proof). According to Lemma 3.2.59, if Φ is the base flag of $\mathcal{P}$, there is a Petrie path π_Φ whose kth vertex and kth edge are the vertex and edge of $(r_3r_2r_1r_0)^k\Phi = \Phi(\sigma_3^{-1}\sigma_1^{-1})^k$. Hence, $\sigma_1\sigma_3$ acts like a one-step rotation for π_Φ. We shall refer to the Petrie paths in the orbit of π_Φ as *right Petrie paths*. (Note that the choice of which paths are the right ones is as arbitrary as the choice of base flag.) The remaining

Petrie paths will be called *left Petrie paths* and are those in the orbit of $\pi_{r_0\Phi}$ under $\Gamma^+(\mathcal{P})$. The vertices and edges of $\pi_{r_0\Phi}$ are the vertices and edges of $(r_3r_2r_1r_0)^k r_0\Phi$. Since

$$(r_3r_2r_1r_0)^k r_0\Phi = r_0(r_0r_3r_2r_1)^k\Phi = r_0(r_3r_2r_0r_1)^k\Phi = r_0\Phi(\sigma_3^{-1}\sigma_1)^k,$$

we conclude that the lengths of the right and left Petrie paths are the orders of $\sigma_1\sigma_3$ and of $\sigma_1^{-1}\sigma_3$, respectively.

Starting with a tessellation $\{p,q,r\}$ of the spherical, Euclidean or hyperbolic 3-space $\mathcal{X}$ (combinatorially, this is the universal 4-polytope with that type) it is possible to impose the extra relation given by the length t of the right Petrie paths. The formal expression for the resulting structure is $\{p,q,r\}/Cl((\sigma_1\sigma_3)^t)$, where $Cl(\gamma)$ denotes the normal closure of γ in $\Gamma^+(\{p,q,r\})$; the structure $\{p,q,r\}/Cl((\sigma_1\sigma_3)^t)$ was denoted $\{p,q,r\}_t$ in [54]. This induces a quotient η from $\mathcal{X}$ to some space $\mathcal{X}'$. Sometimes there is a collapse in the vertex-figures or facets of $\{p,q,r\}_t$ in the sense that two or more flags of a vertex-figure or of a facet are identified in the process. In this situation the quotient η is not smooth (and the resulting topological space is not a space form) since there will be no neighbourhood $\mathcal{B}$ of a vertex or of the centre of a facet such that $\eta_{|\mathcal{B}}$ is a homeomorphism. The idea in [54] is to consider only smooth quotients η so that it induces a tessellation with type $\{p,q,r\}$ of a space form $\mathcal{X}'$ with facets and vertex-figures isomorphic to the convex polyhedra $\{p,q\}$ and $\{q,r\}$, respectively.

We choose the image of the base flag of $\{p,q,r\}$ as the base flag of $\{p,q,r\}_t$; this choice forces the sizes of the right Petrie paths of $\{p,q,r\}_t$ to be a divisor of t. We know from Corollary 2.2.7 that if the length t' of the left Petrie paths of $\{p,q,r\}_t$ is different from the length of the right Petrie paths then $\{p,q,r\}_t$ is chiral. Sometimes these chiral quotients are polytopal and sometimes they are not.

In [54] such structures $\{p,q,r\}_t$ were called *twisted honeycombs* in the chiral case, although it was clarified in [56] that this term was intended for all tessellations $\{p,q,r\}_t$ of space forms, regardless of their reflexibility or chirality. The word 'twisted' was chosen after the isometry used to generate the quotient, which is commonly known as a 'twist' in spherical, Euclidean or hyperbolic space. Here shall use the term 'twisted honeycomb' for every quotient $\{p,q,r\}_t$ of the tiling $\{p,q,r\}$ that induces a tessellation of a space form, without requiring regularity or chirality.

In what follows we shall describe geometrically one of the simplest twisted honeycombs, namely $\{3,5,3\}_4$.

The angle between the two triangles at an edge of a Euclidean icosahedron is greater than $2\pi/3$, implying that we cannot fit three icosahedra around an edge. In hyperbolic three-dimensional space $\mathbb{H}^3$, the angle formed by the two triangles at an edge of a regular icosahedron is strictly less than the corresponding angle of a Euclidean icosahedron. In particular, there are icosahedra such that the angle formed by the two triangles at one of their edges is precisely $2\pi/3$ (see [51], [54, Section 10]). Hence, three such icosahedra fit perfectly around an edge and in fact, they tile $\mathbb{H}^3$. The type of the tiling is $\{3,5,3\}$, and therefore the vertex-figures are dodecahedra. Since this tiling is isomorphic to the universal polytope of that type, we shall denote it simply by $\{3,5,3\}$.

The symmetry group of $\{3,5,3\}$ is the Coxeter group $[3,5,3]$, and we shall denote the ith generator by ρ_i. The element $\rho_0\rho_1\rho_2\rho_3$ that acts like a one-step rotation along a Petrie path has infinite order, and so the Petrie paths of $\{3,5,3\}$ are infinite.

We choose the length $t = 4$ for the right Petrie paths of the desired quotient of $\{3,5,3\}$; that is, we look for $\{3,5,3\}_4$, whose even subgroup has presentation

$$\langle \sigma_1, \sigma_2, \sigma_3 \colon \sigma_1^3 = \sigma_2^5 = \sigma_3^3 = (\sigma_1\sigma_2)^2 = (\sigma_2\sigma_3)^2 = (\sigma_1\sigma_2\sigma_3)^2 = (\sigma_1\sigma_3)^4 = \mathrm{id}\rangle.$$

We emphasise that no assumption has been made on the order of $\sigma_1^{-1}\sigma_3$. The relation $(\sigma_1\sigma_3)^4 = \mathrm{id}$ identifies the base flag Φ_0 with the flag

$$\Phi_0(\sigma_1\sigma_3)^4 = (r_0r_1r_2r_3)^4\Phi_0 = (r_0r_1r_2r_0r_1r_2r_3r_2)^2\Phi_0. \tag{5.3}$$

(Here we used that the last entry of the type is 3, and so $r_3r_2r_3 = r_2r_3r_2$.)

We label the vertices of an icosahedron $\mathcal{I}_1$ of $\{3,5,3\}$ as in Figure 5.2, and let $\mathcal{I}_2$ and $\mathcal{I}_2'$ be the icosahedra that share with $\mathcal{I}_1$ the triangles cdk and aef, respectively. We choose as base flag Φ_0 the flag containing vertex d, the edge cd, the icosahedron $\mathcal{I}_2$ and the triangle in $\mathcal{I}_2$ containing cd

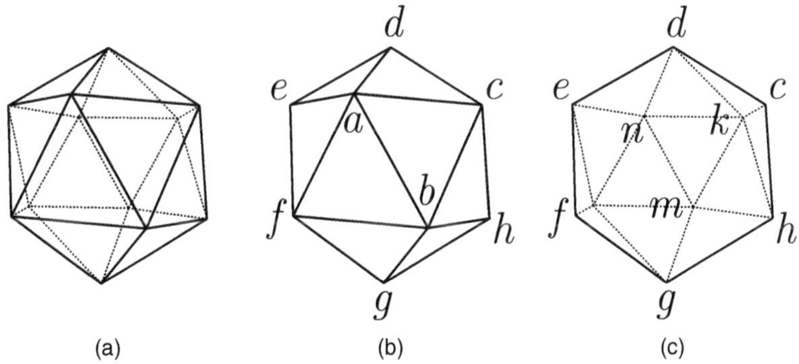

Figure 5.2 (a) The icosahedron $\mathcal{I}_1$ with labels on its vertices: (b) front and (c) back.

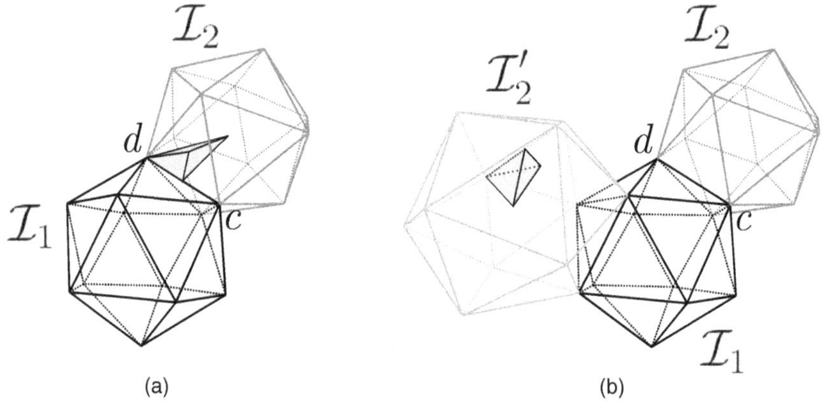

Figure 5.3 (a) The flag Φ_0 and (b) its image under $(\sigma_1\sigma_3)^4$.

other than cdk; this flag is indicated as a tetrahedron in Figure 5.3(a). Then we use (5.3) to determine that $\Phi_0(\sigma_1\sigma_3)^4$ is the flag of $\mathcal{I}_2'$ shown in Figure 5.3(b).

We shall denote by $\widetilde{x}$ the image of each vertex, edge, triangle or icosahedron x of $\{3,5,3\}$ under the quotient by $Cl((\sigma_1\sigma_3)^4)$. Our previous analysis shows that $\widetilde{\mathcal{I}_2} = \widetilde{\mathcal{I}_2'}$, and that $\widetilde{\mathcal{I}_1}$ and $\widetilde{\mathcal{I}_2}$ share at least the triangles $\widetilde{cdk}$ and $\widetilde{aef}$. The vertex of the triangle in Φ_0 other than c and d is mapped to vertex $\widetilde{e}$ of $\{3,5,3\}_4$, since the vertex of the triangle in $\Phi_0(\sigma_1\sigma_3)^4$ that is not in the edge $(cd)\sigma_1\sigma_3$ is e. Furthermore, since vertex e is the preimage in $\mathcal{I}_2'$ of $\widetilde{e}$, we see from Figures 5.2 and 5.3(b) that the vertex-figure at $\widetilde{e}$ in $\widetilde{\mathcal{I}_2}$ has vertices $\widetilde{d}$, $\widetilde{c}$, $\widetilde{a}$ and $\widetilde{f}$ (and some fifth vertex) in that cyclic order.

Let γ be the 3-fold rotation in $[3,5,3]^+$ with its axis through the centres of $\mathcal{I}_1$ and $\mathcal{I}_2$. This rotation preserves the triangle cdk. We use Corollary 3.2.9 on the flags $\Phi_0\gamma$ and $\Phi_0\gamma^2$ to find two more icosahedra that get identified to $\mathcal{I}_2$ in $\{3,5,3\}_4$. These identifications allow us to conclude that $\widetilde{\mathcal{I}_2}$ shares also the triangles $\widetilde{bgh}$ and $\widetilde{jmn}$ with $\widetilde{\mathcal{I}_1}$. The previous analysis on the vertex-figure of $\widetilde{e}$ can be used on other vertices to show that, after these identifications, the labels of the vertices of $\widetilde{\mathcal{I}_2}$ become those in Figure 5.4 (where the tildes of the labels of the vertices are omitted). As a consequence, we have that $\widetilde{\mathcal{I}_1}$ and $\widetilde{\mathcal{I}_2}$ have the same vertex set. (We are not claiming that $\widetilde{\mathcal{I}_1}$ has 12 distinct vertices; we shall see later that the 12 vertices of $\mathcal{I}_1$ get identified in pairs.) Since these are the two facets of $\{3,5,3\}_4$ at the triangle $\widetilde{cdk}$, and $\Gamma(\{3,5,3\}_4)$ acts transitively on the 2-faces, we can conclude that every pair of facets of $\{3,5,3\}_4$ that share a triangle have the same vertex set. Strong connectivity of $\{3,5,3\}$ forces $\{3,5,3\}_4$ to be flat as defined in Definition 5.2.1.

A close inspection of Figure 5.4 shows that the labels of the icosahedron $\widetilde{\mathcal{I}}_2$ can be obtained from those of $\widetilde{\mathcal{I}}_1$ in the following way. We consider the four triangles mentioned previously that $\widetilde{\mathcal{I}}_1$ shares with $\widetilde{\mathcal{I}}_2$; these are disjoint triangles and hence every vertex of $\widetilde{\mathcal{I}}_1$ belongs to one of them. Given two of these triangles Δ_1 and Δ_2 there exist a pair of vertices $\widetilde{u_1}, \widetilde{v_1}$ of Δ_1 and a pair of vertices $\widetilde{u_2}, \widetilde{v_2}$ of Δ_2 such that $\widetilde{u_1}\widetilde{v_1}\widetilde{u_2}$ and $\widetilde{v_1}\widetilde{u_2}\widetilde{v_2}$ are triangles of $\widetilde{\mathcal{I}}_1$. In $\widetilde{\mathcal{I}}_2$, for each such choice of Δ_1 and Δ_2 we keep the triangles Δ_1 and Δ_2 along with the edges $\widetilde{u_1}\widetilde{u_2}$ and $\widetilde{v_1}\widetilde{v_2}$, remove the edge $\widetilde{v_1}\widetilde{u_2}$, and replace it by $\widetilde{u_1}\widetilde{v_2}$, so to form triangles $\widetilde{u_1}\widetilde{v_1}\widetilde{v_2}$ and $\widetilde{u_1}\widetilde{u_2}\widetilde{v_2}$.

On the left-hand side of Figure 5.5 we show the four triangles shared by $\widetilde{\mathcal{I}}_1$ and $\widetilde{\mathcal{I}}_2$. The centres of these triangles are the vertices of a tetrahedron $\mathcal{T}$. The orbit of $\mathcal{T}$ under $\Gamma^+(\widetilde{\mathcal{I}}_1)$ (under the current assumption that it is an icosahedron) consists of a famous compound of five tetrahedra (see [55, Plate III(6)]). The remaining four icosahedra in Figure 5.5 show the 4-tuples of

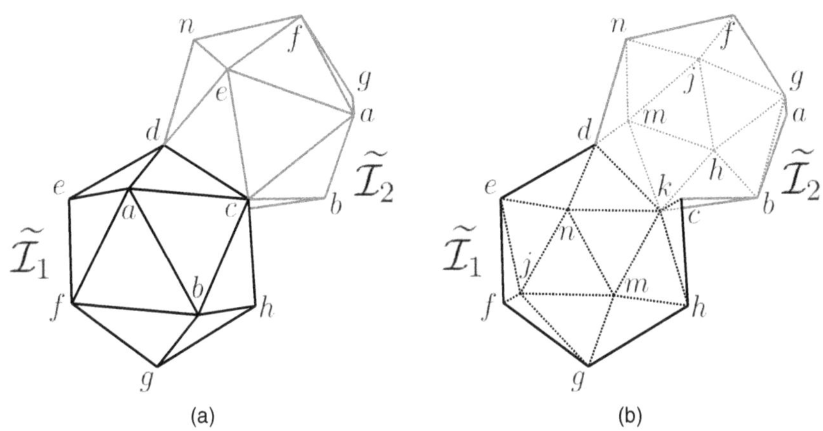

Figure 5.4 Labels of the vertices of $\widetilde{\mathcal{I}}_2$ without the tildes: (a) front and (b) back.

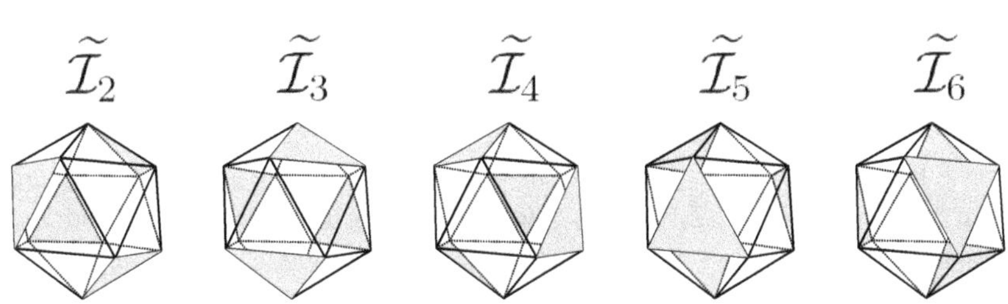

Figure 5.5 Triangles in the intersection of $\widetilde{\mathcal{I}}_1$ with $\widetilde{\mathcal{I}}_j$ for $j \in \{2, 3, 4, 5, 6\}$.

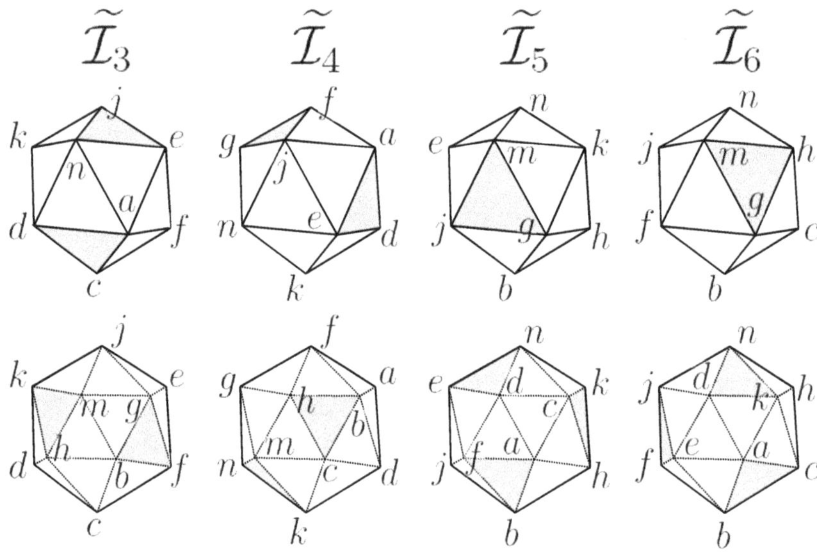

Figure 5.6 Labels without tildes of the vertices of $\widetilde{\mathcal{I}}_j$ for $j \in \{3,4,5,6\}$ (front on the upper row and back on the lower row). The triangles in their intersections with $\widetilde{\mathcal{I}}_1$ are shown in grey.

triangles whose centres are the vertices of the remaining four tetrahedra of the compound.

Now we use Corollary 3.2.9 on the flags $\Phi_0\alpha$ for every $\alpha \in \Gamma^+(\mathcal{I}_1)$ to search for more icosahedra of $\{3,5,3\}$ that share a triangle with $\mathcal{I}_1$ and get identified in the same facet of $\{3,5,3\}_4$. The result is that the 20 icosahedra of $\{3,5,3\}$ that share triangles with $\mathcal{I}_1$ are mapped to only 5 facets of $\{3,5,3\}_4$, each of which shares 4 triangles with $\widetilde{\mathcal{I}}_1$. The centres of these four triangles are the vertices of one of the images of $\mathcal{T}$ under $\Gamma^+(\widetilde{\mathcal{I}}_1)$. We label the remaining four facets of $\{3,5,3\}_4$ that share triangles with $\widetilde{\mathcal{I}}_1$ by $\widetilde{\mathcal{I}}_3$, $\widetilde{\mathcal{I}}_4$, $\widetilde{\mathcal{I}}_5$ and $\widetilde{\mathcal{I}}_6$, as shown in Figure 5.5. The labels of their vertices are shown in Figure 5.6 (again, without the tildes); they can be obtained from procedures analogous to those used to obtain the labels of the vertices of $\widetilde{\mathcal{I}}_2$.

At this point we can observe that antipodal vertices of $\mathcal{I}_1$ are identified when taking the quotient to $\{3,5,3\}_4$. For example, we can see that $\widetilde{b} = \widetilde{n}$ with the following argument. The edge ad belongs to three icosahedra of $\{3,5,3\}$ that we can name $\mathcal{I}_1$, $\mathcal{I}_3$ and $\mathcal{I}_4$ (the last two are preimages of $\widetilde{\mathcal{I}}_3$ and $\widetilde{\mathcal{I}}_4$, respectively). We know that $\mathcal{I}_1$ and $\mathcal{I}_3$ share the triangle acd while $\mathcal{I}_1$, $\mathcal{I}_4$ share the triangle ade. It follows that $\mathcal{I}_3$ and $\mathcal{I}_4$ must share the other triangle that has the edge ad. In the quotient to $\{3,5,3\}_4$ that triangle of $\mathcal{I}_3$ is mapped to $\widetilde{a}\widetilde{b}\widetilde{d}$, whereas the triangle of $\mathcal{I}_4$ is mapped to $\widetilde{a}\widetilde{d}\widetilde{n}$. We can conclude that $\widetilde{b} = \widetilde{n}$. The identification of the remaining pairs of antipodal vertices of $\mathcal{I}_1$ can be verified in a similar fashion.

As a consequence, we know that $\{3,5,3\}_4$ has at most six vertices, and that every pair of vertices of $\{3,5,3\}_4$ are adjacent. For example, $\widetilde{a}$ is adjacent to $\widetilde{h} = \widetilde{e}$. We are also ready to show that $\{3,5,3\}_4$ has at most six facets. This is a consequence of the fact that every edge belongs to three facets. For example, to see that $\widetilde{\mathcal{I}_2}$ shares triangles with $\widetilde{\mathcal{I}_j}$ for $j \in \{3,4,5,6\}$ we study the three facets at the edges $\widetilde{cd}$, $\widetilde{bh}$, $\widetilde{dk}$ and $\widetilde{ck}$. These four edges belong to $\mathcal{I}_1$ and to $\mathcal{I}_2$; the third facet of $\{3,5,3\}$ at these edges is a preimage of $\widetilde{\mathcal{I}_3}$, $\widetilde{\mathcal{I}_4}$, $\widetilde{\mathcal{I}_5}$ and $\widetilde{\mathcal{I}_6}$, respectively. This forces $\widetilde{\mathcal{I}_2}$ to share a triangle with each of the remaining five facets defined so far. Similar arguments show that each facet $\widetilde{\mathcal{I}_j}$ shares a triangle (and hence, four triangles) with each facet $\widetilde{\mathcal{I}_k}$ for every $j \neq k$, $j,k \in \{1,2,3,4,5,6\}$. It follows that every 2-face of the facets $\widetilde{\mathcal{I}_k}$ for $k \in \{1,2,3,4,5,6\}$ is already in two facets; strong connectivity of $\{3,5,3\}$ can be used to show that $\{3,5,3\}_4$ cannot have any other facet.

In order to show that there are no further identifications of vertices or facets of $\{3,5,3\}_4$ we start by exhibiting the resulting poset so far, denoted by $\mathcal{P}$. The poset is shown through its incidences. Overall $\mathcal{P}$ has 6 vertices, 60 edges, 60 triangular 2-faces and 6 facets.

We shall denote the vertices of $\mathcal{P}$ by $\widetilde{a}$, $\widetilde{b}$, $\widetilde{c}$, $\widetilde{d}$, $\widetilde{e}$ and $\widetilde{f}$, and the facets by $\widetilde{\mathcal{I}_1}, \widetilde{\mathcal{I}_2}, \widetilde{\mathcal{I}_3}, \widetilde{\mathcal{I}_4}, \widetilde{\mathcal{I}_5}$ and $\widetilde{\mathcal{I}_6}$, as earlier. For convenience we recall the equivalence of the images of the vertices of $\mathcal{I}_1$:

$$\widetilde{a} = \widetilde{m}, \quad \widetilde{b} = \widetilde{n}, \quad \widetilde{c} = \widetilde{j}, \quad \widetilde{d} = \widetilde{g}, \quad \widetilde{e} = \widetilde{h}, \quad \widetilde{f} = \widetilde{k}.$$

Each edge and each triangle is completely determined by the sets of vertices and facets incident to it. We shall denote each edge and triangle by five characters indicating the vertices incident to it (letters in $\{a,b,c,d,e,f\}$) and the facets incident to it (numbers in $\{1,2,3,4,5,6\}$). For example, the edge $\widetilde{bg}$ and the triangle $\widetilde{bgh}$ in $\widetilde{\mathcal{I}_1}$ (see Figure 5.4) shall be called $bd123$ and $bde12$, respectively. This notation directly indicates the incidences involving vertices and facets (recall that every vertex is incident to every facet). Table 5.1 shows the three triangles incident to each edge of $\widetilde{\mathcal{I}_1}$; in the column 'name' we include the previous name we used for that edge. The incidences between the remaining edges and triangles are given in Table 5.2.

It is routine to verify that every facet of the poset $\mathcal{P}$ is isomorphic to the result of identifying every pair of antipodal vertices of an icosahedron, but keeping the 30 edges and 20 triangles (so that the facets still have 120 flags); and also that every vertex-figure of $\mathcal{P}$ is isomorphic to a dodecahedron with opposite pentagons identified as 2-faces, each of which consists of two

Table 5.1. *Incidences between edges of* $\widetilde{\mathcal{I}}_1$ *and triangles of* $\{3,5,3\}_4$.

Edge	Name	Triangles	Edge	Name	Triangles
$ab156$	$\widetilde{ab}$	$abc16, abf15, abe56$	$ab124$	$\widetilde{mn}$	$abc12, abf14, abd24$
$ac136$	$\widetilde{ac}$	$abc16, acd13, acf36$	$ac125$	$\widetilde{jm}$	$abc12, acd15, ace25$
$ad134$	$\widetilde{ad}$	$acd13, ade14, abd34$	$ad156$	$\widetilde{gm}$	$acd15, ade16, adf56$
$ae124$	$\widetilde{ae}$	$ade14, aef12, ace24$	$ae136$	$\widetilde{hm}$	$ade16, aef13, abe36$
$af125$	$\widetilde{af}$	$abf15, aef12, adf25$	$af134$	$\widetilde{km}$	$abf14, aef13, acf34$
$bc146$	$\widetilde{bc}$	$abc16, bce14, bcd46$	$bc123$	$\widetilde{jn}$	$abc12, bce13, bcf23$
$bd123$	$\widetilde{bg}$	$bde12, bdf13, abd23$	$bd156$	$\widetilde{dn}$	$bde15, bdf16, bcd56$
$be124$	$\widetilde{bh}$	$bce14, bde12, bef24$	$be135$	$\widetilde{en}$	$bce13, bde15, abe35$
$bf135$	$\widetilde{bf}$	$abf15, bdf13, bcf35$	$bf146$	$\widetilde{kn}$	$abf14, bdf16, bef46$
$cd123$	$\widetilde{cd}$	$acd13, cdf12, cde23$	$cd145$	$\widetilde{gj}$	$acd15, cdf14, bcd45$
$ce145$	$\widetilde{ch}$	$bce14, cef15, ace45$	$ce136$	$\widetilde{ej}$	$bce13, cef16, cde36$
$cf125$	$\widetilde{ck}$	$cdf12, cef15, bcf25$	$cf146$	$\widetilde{fj}$	$cdf14, cef16, acf46$
$de145$	$\widetilde{de}$	$ade14, bde15, def45$	$de126$	$\widetilde{gh}$	$ade16, bde12, cde26$
$df126$	$\widetilde{dk}$	$bdf16, cdf12, adf26$	$df134$	$\widetilde{fg}$	$bdf13, cdf14, def34$
$ef126$	$\widetilde{ef}$	$aef12, cef16, bef26$	$ef135$	$\widetilde{hk}$	$aef13, cef15, def35$

Table 5.2. *Incidences between the remaining edges and triangles of* $\{3,5,3\}_4$.

Edge	Triangles	Edge	Triangles
$ab236$	$abc26, abd23, abe36$	$ab345$	$abd34, abe35, abf45$
$ac246$	$abc26, ace24, acf46$	$ac345$	$acd35, acf34, ace45$
$ad235$	$abd23, adf25, acd35$	$ad246$	$abd24, adf26, ade46$
$ae235$	$ace25, aef23, abe35$	$ae456$	$ace45, ade46, abe56$
$af236$	$adf26, aef23, acf36$	$af456$	$abf45, acf46, adf56$
$bc256$	$abc26, bcf25, bcd56$	$bc345$	$bce34, bcf35, bcd45$
$bd245$	$abd24, bde25, bcd45$	$bd346$	$abd34, bdf36, bcd46$
$be256$	$bde25, bef26, abe56$	$be346$	$abe36, bce34, bef46$
$bf245$	$bcf25, bef24, abf45$	$bf236$	$bcf23, bef26, bdf36$
$cd246$	$cde26, cdf24, bcd46$	$cd356$	$acd35, cde36, bcd56$
$ce234$	$ace24, cde23, bce34$	$ce256$	$ace25, cde26, cef56$
$cf234$	$bcf23, cdf24, acf34$	$cf356$	$acf36, bcf35, cef56$
$de235$	$bde25, cde23, def35$	$de346$	$cde36, def34, ade46$
$df245$	$adf25, cdf24, def45$	$df356$	$bdf36, def35, adf56$
$ef234$	$aef23, bef24, def34$	$ef456$	$bef46, def45, cef56$

disconnected cycles, but keeping the 30 edges and 20 vertices (so that also the vertex-figures have 120 flags each). The nature of the 2-faces of the vertex-figures directly implies that $\mathcal{P}$ does not satisfy the strong connectivity and therefore it is not a polytope. On the other hand, the quotient from $\{3,5,3\}$ to $\mathcal{P}$ is smooth in the sense that it does not identify flags that share hyperbolic points. This means that $\mathcal{P}$ is a tessellation of a hyperbolic space form by 'collapsed icosahedra' (as they were called in [60, Section 2], see also [263]) with dodecahedral vertex-figures. Therefore, $\mathcal{P}$ is a twisted honeycomb.

Being a tessellation of a space form, every flag of $\mathcal{P}$ has a unique i-adjacent flag for $i \in \{0,1,2,3\}$. It follows that $\mathcal{P}$ is a 4-maniplex. Moreover, it has $720 = 6 \cdot 120$ flags (tetrahedra in the barycentric subdivision of $\mathcal{P}$ as a tessellation of a three-dimensional space form), which correspond precisely to the 720 maximal chains of its poset. Therefore $\mathcal{P}$ is a faithful maniplex as defined in Section 2.1.3.

By choosing as the base flag $\{\widetilde{d}, cd123, cde23, \widetilde{\mathcal{I}_2}\}$ (the image of our original base flag Φ_0) we can verify that $\mathcal{P}$ is rotary since the permutations of vertices and facets

$$
\begin{aligned}
\sigma_1 &= (\widetilde{a}, \widetilde{f}, \widetilde{b})(\widetilde{c}, \widetilde{d}, \widetilde{e})(\widetilde{\mathcal{I}_1}, \widetilde{\mathcal{I}_5}, \widetilde{\mathcal{I}_4}),\\
\sigma_2 &= (\widetilde{a}, \widetilde{b}, \widetilde{e}, \widetilde{c}, \widetilde{f})(\widetilde{\mathcal{I}_1}, \widetilde{\mathcal{I}_6}, \widetilde{\mathcal{I}_4}, \widetilde{\mathcal{I}_5}, \widetilde{\mathcal{I}_3}),\\
\sigma_3 &= (\widetilde{a}, \widetilde{f}, \widetilde{e})(\widetilde{\mathcal{I}_1}, \widetilde{\mathcal{I}_2}, \widetilde{\mathcal{I}_3})(\widetilde{\mathcal{I}_4}, \widetilde{\mathcal{I}_6}, \widetilde{\mathcal{I}_5})
\end{aligned}
$$

induce automorphisms of $\mathcal{P}$. Furthermore, these automorphisms serve as the standard rotations. Here each σ_i is understood to act on each character of every edge and of every facet; for instance, $ab156\sigma_2 = be346$.

We can finally show that $\mathcal{P}$ is indeed $\{3,5,3\}_4$. To do this we first observe that the lengths of the right and left Petrie paths of $\mathcal{P}$ are 4 and 5, since these are the orders of $\sigma_1\sigma_3 = (\widetilde{a}, \widetilde{e}, \widetilde{c}, \widetilde{d})(\widetilde{b}, \widetilde{f})(\widetilde{\mathcal{I}_1}, \widetilde{\mathcal{I}_4}, \widetilde{\mathcal{I}_2}, \widetilde{\mathcal{I}_3})(\widetilde{\mathcal{I}_5}, \widetilde{\mathcal{I}_6})$ and $\sigma_1^{-1}\sigma_3 = (\widetilde{a}, \widetilde{b}, \widetilde{e}, \widetilde{d}, \widetilde{c})(\widetilde{\mathcal{I}_1}, \widetilde{\mathcal{I}_6}, \widetilde{\mathcal{I}_5}, \widetilde{\mathcal{I}_2}, \widetilde{\mathcal{I}_3})$, respectively. This conclusion could also be achieved by exhibiting directly a right and a left Petrie path.

Exercise 5.3.4 Find a right and a left Petrie path of the maniplex $\mathcal{P}$ defined above, by using Definition 2.1.66.

The facts that $\mathcal{P}$ is a rotary maniplex with type $\{3,5,3\}$ and that its right Petrie paths have size 4 imply that $\mathcal{P}$ is a quotient of $\{3,5,3\}_4$. Since $\mathcal{P}$ has 720 flags and we showed above that $\{3,5,3\}_4$ has at most six facets, each with at most 120 flags, we can conclude that $\mathcal{P} = \{3,5,3\}_4$. We summarise the previous discussion in the next result.

Proposition 5.3.5 *The structure* $\{3,5,3\}_4$ *constructed from the hyperbolic tessellation* $\{3,5,3\}$ *by imposing as the only relation that the right Petrie paths have length* 4 *is a twisted honeycomb, and also a non-polytopal faithful maniplex with* 720 *flags. It has* 6 *vertices,* 60 *edges,* 60 *triangular 2-faces and* 6 *facets. The underlying graph is the complete graph* K_6 *with four edges* *between each pair of vertices. The facets have* 6 *vertices,* 30 *edges and* 20 *triangular* 2*-faces; they are obtained from icosahedra by identifying opposite vertices without identifying edges or triangles. Similarly, the vertex-figures have* 20 *vertices,* 30 *edges and* 6 *(disconnected)* 2*-faces; they are obtained from dodecahedra by identifying opposite pentagons without identifying their vertices or edges. The left Petrie paths of* $\{3,5,3\}_4$ *have length* 5 *and hence* $\{3,5,3\}_4$ *is chiral.*

The quotients $\{3,5,3\}_2$ and $\{3,5,3\}_3$ do show the kind of collapse that prevents them from being tessellations of space forms (and so they are not twisted honeycombs).

Exercise 5.3.6 Determine the number of vertices, edges, 2-faces, facets and flags of $\{3,5,3\}_t$ for $t \in \{2,3\}$. Are they faithful maniplexes?

The procedure shown above is impractical to study $\{3,5,3\}_t$ for higher values of t. In fact, much of the work on the study of twisted honeycombs has been done in a group-theoretical manner, by analysing coset enumeration tables (this is an algorithm used to determine the index of a subgroup in a group, in case the index is finite). At first this was done by hand, and eventually this was performed with the use of computers.

The values of $t \geq 4$ for which $\{3,5,3\}_t$ is known to be finite are 4, 5, 6, 7, 8 and 9. In what follows we summarise the findings in [54, 60, 264] regarding these quotients of $\{3,5,3\}$.

We already described $\{3,5,3\}_4$, and it turns out that $\{3,5,3\}_5$ is its enantiomorph (this is not automatic; in principle $\{3,5,3\}_4$ could have been a proper rooted quotient of the enantiomorph of $\{3,5,3\}_5$ in case the latter had left Petrie paths of size greater than 4). The next value $t = 6$ yields a reflexible twisted honeycomb $\{3,5,3\}_6$ with 1320 flags. The twisted honeycombs $\{3,5,3\}_7$ and $\{3,5,3\}_9$ are both chiral and they have 24 360 and 6840 flags, respectively. It is unexpected that a larger value $t = 9$ yields a smaller twisted honeycomb with respect to the smaller value $t = 7$. Part of this may be explained by the lengths of the left Petrie paths of these twisted honeycombs; these lengths are 29 for $t = 7$ and 10 for $t = 9$. When imposing the relation $(\sigma_1\sigma_3)^8 = \mathrm{id}$ to those of $[3,5,3]$ the remaining relations force $(\sigma_1\sigma_3)^4 = \mathrm{id}$, and so $\{3,5,3\}_8 = \{3,5,3\}_4$. It has not been established if $\{3,5,3\}_{10}$ is finite; in any case it covers (the enantiomorph of) $\{3,5,3\}_9$ since the left Petrie paths of the latter have size 10.

Problem 43 For every $t \geq 10$ determine whether

(a) $\{3,5,3\}_t$ is a twisted honeycomb or there is a collapse in facets and/or vertex-figures;
(b) $\{3,5,3\}_t$ is finite or infinite;
(c) $\{3,5,3\}_t$ is regular or chiral.

We move on to the search for twisted honeycombs with other types, for which the main reference is [54] (see [51] for detailed descriptions of the tessellations with those types). Hyperbolic three-dimensional space also admits a tessellation by cubes where every edge is in five cubes. The angle between the two squares at an edge of these cubes is $2\pi/5$. The type of this tessellation is $\{4,3,5\}$. Its dual tessellation has type $\{5,3,4\}$; its facets are dodecahedra with dihedral angle $2\pi/4$, four of them meeting at every edge. The list of regular tessellations of hyperbolic three-dimensional space with compact facets and finite vertex-figures is completed by the self-dual tessellation with dodecahedra with dihedral angle $2\pi/5$, so that there are five of them around every edge; its type is $\{5,3,5\}$. Each of these tessellations is isomorphic to the universal polytope with its type, and so we shall refer to them simply by $\{4,3,5\}$, $\{5,3,4\}$ and $\{5,3,5\}$. Their symmetry groups are the Coxeter groups $[4,3,5]$, $[5,3,4]$ and $[5,3,5]$, respectively.

We may take the quotient $\{p,q,r\}_t$ of each hyperbolic tessellation $\{p,q,r\}$ by the normal closure of the isometry $(\sigma_1\sigma_3)^t$, as we did for the tessellation $\{3,5,3\}$ (here σ_i stands for the ith standard rotation). This forces the size of the right Petrie paths of $\{p,q,r\}_t$ to be a divisor of t. If the quotient from $\{p,q,r\}$ to $\{p,q,r\}_t$ is one-to-one on the sets of flags of the base facet and of the base vertex-figure, then the resulting structure is a tessellation of a hyperbolic space form and, hence, a twisted honeycomb. The fact that $\{5,3,4\}$ is the geometric dual of $\{4,3,5\}$ implies that we may understand the flags of these two tessellations as the same hyperbolic tetrahedra. It follows that $\{4,3,5\}_t$ induces a tessellation of a hyperbolic space form if and only if $\{5,3,4\}_t$ induces a tessellation of the same hyperbolic space form. Once more, if the length of the right Petrie paths of $\{p,q,r\}_t$ differs from that of the left Petrie paths then $\{p,q,r\}_t$ is chiral.

The known values of t for which $\{4,3,5\}_t$ (and hence also $\{5,3,4\}_t$) is a twisted honeycomb are 6 and 8. Both values yield chiral twisted honeycombs, and the length of their left Petrie paths is 18 and 30, respectively. For $t = 7$ the quotient is not smooth, while for $t > 9$ it is unknown whether the quotient yields a finite maniplex.

Exercise 5.3.7 Determine the number of vertices, edges, 2-faces, facets and flags of $\{4,3,5\}_t$ for $t \in \{2,3,4,5\}$. Are they faithful maniplexes?

Problem 44 For every $t \geq 9$ determine whether

(a) $\{4,3,5\}_t$ is a twisted honeycomb or there is a collapse in facets and/or vertex-figures;
(b) $\{4,3,5\}_t$ is finite or infinite;
(c) $\{4,3,5\}_t$ is regular or chiral.

Besides the twisted honeycomb $\{5,3,5\}_2 \cong \{5,3,5\}_3 \cong \{5,3,5\}_4$ given by the one-facet tessellation of the dodecahedral space, the only value of t for which $\{5,3,5\}_t$ is known to be a finite twisted honeycomb is $t = 5$. The result is a chiral twisted honeycomb whose left Petrie paths have length 9.

Problem 45 For every $t \geq 6$ determine whether

(a) $\{5,3,5\}_t$ is a twisted honeycomb or there is a collapse in facets and/or vertex-figures;
(b) $\{5,3,5\}_t$ is finite or infinite;
(c) $\{5,3,5\}_t$ is regular or chiral.

Other tessellations of three-dimensional hyperbolic space forms have been found by group theoretical methods, although little has been said about their geometric aspects. This includes infinitely many chiral tessellations whose symmetry groups are related with the projective linear groups $PSL_2(q)$ (see [100, 121, 148, 208, 209, 215]). To date, there seems not to be a systematic search for chiral tessellations of four-dimensional hyperbolic space forms. The types of these tessellations can only be $\{5,3,3,3\}$, $\{5,3,3,4\}$, $\{5,3,3,5\}$ and the duals of the first two.

Problem 4 asks for all two-dimensional compact hyperbolic space forms carrying chiral tessellations. As explained in [188, Section 6J], the equivalent question in dimensions 3 and 4 seems to be a much harder problem.

Problem 46 Determine all hyperbolic three-dimensional and four-dimensional space forms that admit chiral tessellations.

We mentioned in Theorem 5.1.11 that there are infinitely many chiral maps with type $\{p,q\}$ for each hyperbolic type $\{p,q\}$. It was shown in [188, Theorem 6J7] that if $\{p_1,\ldots,p_n\}$ is the type of a regular tessellation of the n-dimensional hyperbolic space (so here we only consider $n \in \{3,4\}$) then there are infinitely many regular tessellations of compact hyperbolic space forms with type $\{p_1,\ldots,p_n\}$. Those techniques can be modified to achieve the same conclusion for rotary tessellations, but other methods would be necessary to determine whether infinitely many of those tessellations are chiral.

The findings in [148] imply that there are infinitely many chiral tessellations of hyperbolic space forms with type $\{5,3,5\}$. It remains to be determined whether infinitely many of them are polytopal. It has not been determined yet whether there are infinitely many chiral tessellations of compact hyperbolic space forms with the other candidate hyperbolic types of ranks 4 and 5 (see [220, Problems 14 and 23]).

Problem 47 For each of the types $\{3,5,3\}$, $\{4,3,5\}$, $\{5,3,5\}$, $\{3,3,3,5\}$, $\{4,3,3,5\}$ and $\{5,3,3,5\}$ determine whether there are infinitely many chiral polytopes with that type.

To conclude this section we make a comparison between the polytopes (and maniplexes) obtained through geometric methods from Euclidean spaces, with respect to those obtained from hyperbolic spaces.

When we are shown polytopes through diagrams like those in Figure 2.30, or even like those in Figure 2.32, we get an immediate understanding of most of the incidences of the poset. The only incidences that are not shown directly are those involving faces in the boundary of the diagrams. In those figures the number of identifications to consider in the boundary is small, and this makes it relatively easy to develop intuition on the global geometry of the polytope. For instance, the analysis of a small region of the figure should be enough to make a reasonable guess of the length of the right Petrie paths of the polytope.

By contrast, the diagram in Figure 3.17(d) involves many identifications. This makes it harder to visualise the incidences, and even the number of vertices of the map is not immediately obvious from the diagram. This makes it much harder to develop intuition on the global geometry of the map. These difficulties become even more obvious when attempting an analogous diagram of a tessellation of a higher-dimensional space form (compare with Figures 5.3, 5.4 and 5.6). This explains why most of the work on rotary tessellations of compact hyperbolic space forms has been done through group-theoretical methods, rather than geometric ones.

This suggests that other techniques are needed to construct and visualise chiral polytopes from hyperbolic geometry through geometric methods.

5.3.3 Roli's Cube

In [10], another geometric idea was used to construct a chiral 4-polytope. On this occasion it is through the real projective three-dimensional space $\mathbb{P}^3 := \mathbb{P}^3(\mathbb{R})$.

We may visualise $\mathbb{P}^3$ in a way analogous to that in which we visualise $\mathbb{P}^2$. We start with the 3-sphere $\mathbb{S}^3$, that is, with the set of points of the Euclidean 4-space $\mathbb{E}^4$ at distance 1 from the origin. The projection of $\mathbb{S}^3$ from the point $(0,0,0,1)$ to the hyperplane $w = 0$ (here w denotes the last coordinate) is called the *stereographic projection*. It maps the points (x, y, z, w) of $\mathbb{S}^3$ with $w = 0$, $w < 0$ and $0 < w < 1$ to the unit sphere, the interior of the unit ball, and the exterior of the unit ball, respectively. The point $(0, 0, 0, 1)$ is associated to a *point at infinity*. To obtain $\mathbb{P}^3$ from $\mathbb{S}^3$ we only need to identify all antipodal pairs of points. We may choose as representatives the points (x, y, z, w) of $\mathbb{S}^3$ with $w < 0$, together with half of the points with $w = 0$. The stereographic projection of these points is the interior of the unit sphere of $\mathbb{E}^3$ with half of the points of its boundary. It becomes apparent that $\mathbb{P}^3$ can be modelled by a three-dimensional ball where antipodal points of the boundary are identified.

The image of the 4-cube $\{4, 3, 3\}$ under the stereographic projection is well-known. The vertices $(x, y, z, -1)$ are mapped to the vertices of a cube inside the unit ball, whereas the vertices $(x, y, z, 1)$ are mapped to the vertices of a cube outside the unit ball. When joining adjacent vertices by straight-line segments of $\mathbb{E}^3$, we get the familiar object shown in Figure 5.7.

The edges of $\{4, 3, 3\}$ admit a proper edge-colouring given by their direction vectors. With the appropriate coordinate system, the edges with direction vectors $(0, 1, 0, 0)$ and $(0, 0, 1, 0)$ are the eight horizontal and the eight vertical edges in Figure 5.7, respectively. The ones with direction vector $(1, 0, 0, 0)$ are illustrated as the four edges of the outer cube and the four edges of the inner cube that go front-to-back. Finally, the edges with direction vector $(0, 0, 0, 1)$ join vertices with last coordinate 1 to vertices with last coordinate -1, and therefore they are the eight edges that cross the unit sphere.

Recall that the hemi-4-cube $\{4, 3, 3\}_4$, as defined in Example 4.1.43, is obtained from the cube by identifying every pair of antipodal i-faces for $i \in \{0, 1, 2, 3\}$. It has 8 vertices, 16 edges, 12 square 2-faces and 4 cubical

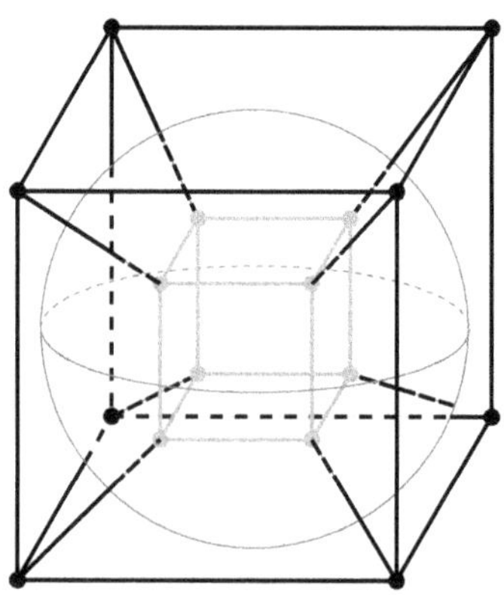

Figure 5.7 Stereographic projection of the 4-cube with edges represented by straight-line segments.

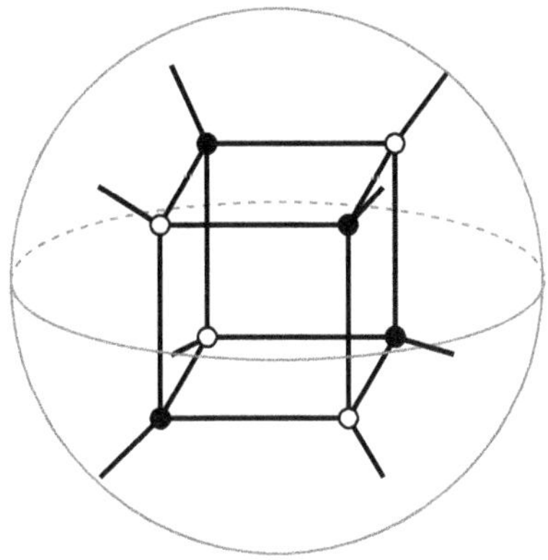

Figure 5.8 Hemi-4-cube in $\mathbb{P}^3$.

facets. It can be visualised as a tessellation of $\mathbb{P}^3$ whose 1-skeleton is that in Figure 5.8 where antipodal pairs of semi-edges are meant to be glued into edges. Six of the squares are represented by the squares of the inner cube. Each pair of antipodal edges of the inner cube are opposite edges of a square whose centre is in the outer sphere of the model. This completes the 12 square 2-faces. The four cubes can be identified as the inner one, the one at the left and right, the one at the front and back and the one at the top and bottom.

The projective space $\mathbb{P}^3$ is an orientable manifold, and accordingly, $\{4,3,3\}_4$ is an orientable 4-polytope.

Exercise 5.3.8 Show that the hemi-4-cube $\{4,3,3\}_4$ is an orientable 4-polytope.

The underlying graph of the hemi-4-cube is the complete bipartite graph $K_{4,4}$. The bipartition of the vertices corresponds to the white and black vertices in Figure 5.8.

The proper edge-colouring of the edges of $\{4,3,3\}$ given by their direction vectors is invariant under the inversion with respect to the centre of $\{4,3,3\}$. Hence, it induces a proper edge colouring of $\{4,3,3\}_4$. The four chromatic classes of this colouring are shown in Figure 5.9.

The combinatorics of $\{4,3,3\}_4$ are completely determined by the colouring in Figure 5.9 in the following sense. Each of the six possible choices of two colours induces two squares of $\{4,3,3\}_4$, and any of the four possible choices of three colours induces a cube of $\{4,3,3\}_4$. It follows that, for $i \in \{0,1,2,3\}$, every i-face of $\{4,3,3\}_4$ is induced by a connected component of

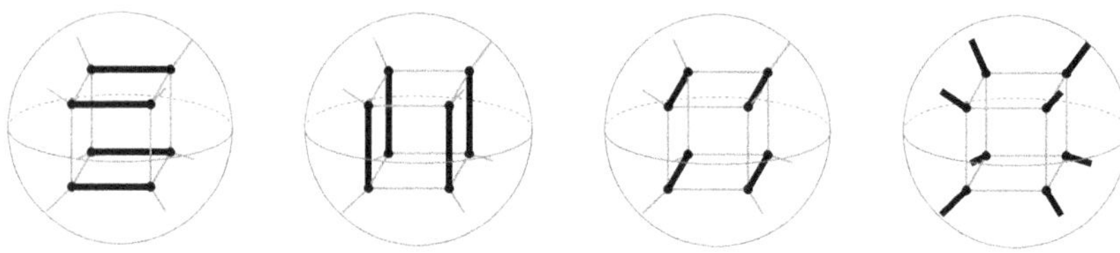

Figure 5.9 Proper colouring of the edges of $\{4,3,3\}_4$.

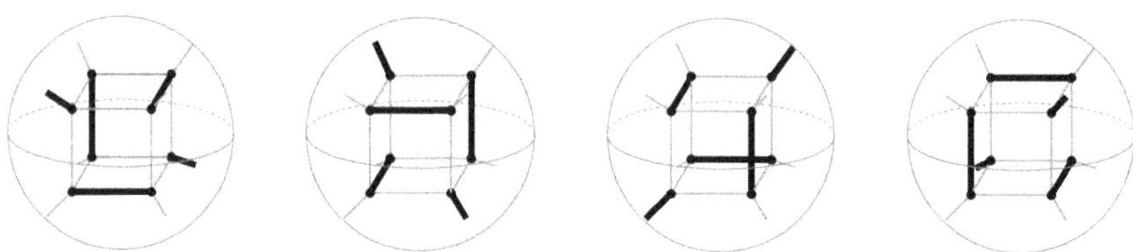

Figure 5.10 A second proper colouring of the edges of $\{4,3,3\}_4$.

the coloured 1-skeleton when only a certain choice of i colours is considered. Conversely, every such connected component corresponds to an i-face of $\{4,3,3\}_4$. Polytopes of rank n constructed from proper n-colourings of n-regular graphs whose poset is determined by the colouring in the way just explained were named *colourful polytopes* in [2].

The edge set of $\{4,3,3\}_4$ also admits the proper edge-colouring in Figure 5.10. Direct inspection shows that the edges of each pair of colours induce two disjoint squares. This is a sufficient condition to guarantee that the colouring is isomorphic to that in Figure 5.9, since it is the key property to construct the 1-skeleton of the cube from that of the square, and the 1-skeleton of the 4-cube from that of the cube.

On the other hand, the colourings in Figures 5.9 and 5.10 are very different geometrically. The cycles induced by two colours in Figure 5.9 correspond to squares, whereas those induced by two colours in Figure 5.10 correspond to Petrie paths of $\{4,3,3\}_4$. This tells us that $\{4,3,3\}_4$ can be visualised in its own 1-skeleton in an alternative way, where the 2-faces are represented by 12 of its Petrie paths, and the 3-faces have as 2-faces the 6 such Petrie paths in the complement of the black edges in Figure 5.10(a), (b), (c) and (d).

Now we return our attention to $\mathbb{E}^4$. We can colour the edges of $\{4,3,3\}$ by assigning every edge the colour of its image in $\{4,3,3\}_4$. When obtaining this colouring from that in Figure 5.9 we recover the colouring with the direction vectors of the edges; the colourful polytope associated to this colouring of

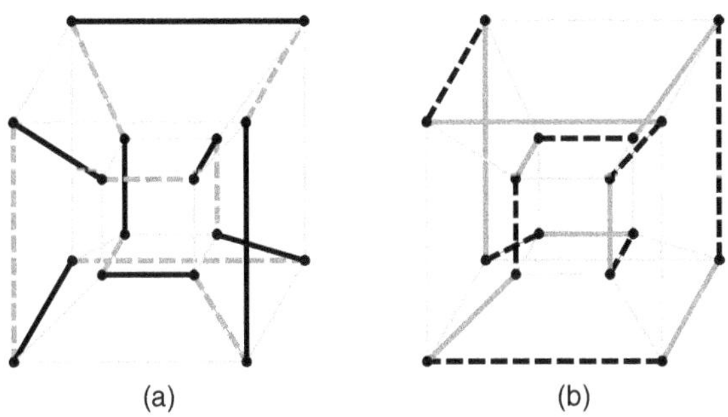

Figure 5.11 A second proper colouring of the edges of $\{4,3,3\}$; two colours are shown in (a) and the remaining two are shown in (b).

$\{4,3,3\}$ is again $\{4,3,3\}$. The object of study of this section arises when we assign the edges of $\{4,3,3\}$ the colours of their images in Figure 5.10 (see Figure 5.11).

We denote by $\mathcal{P}$ the poset whose 1-skeleton is that of $\{4,3,3\}$, and its 2- and 3-faces are respectively the components of the subgraphs induced by two and three colours in Figure 5.11. By adding a minimal face F_{-1} and a maximal face F_4 we obtain a 4-polytope. For some time this polytope $\mathcal{P}$ was informally referred to as *Roli's cube* in honour of the sixtieth birthday of Javier Bracho, one of the authors of [10] who is also called by the nickname 'Roli'. The name of this polytope was formally established in [202].

Exercise 5.3.9 Show that the structure just defined satisfies Definition 2.1.7.

Roli's cube is no longer isomorphic to $\{4,3,3\}$, and so it is not really a cube. Its 1-skeleton does coincide with that of $\{4,3,3\}$, but the 2-faces are some (octagonal) Petrie paths of $\{4,3,3\}$; four of them can be seen in Figure 5.11 as the union of two matchings of distinct colours. Every pair of colours yields two 2-faces, and hence Roli's cube has twelve 2-faces. The 3-faces are given by the edges of three colours, and therefore their vertices are 3-valent. It follows that the 3-faces have type $\{8,3\}$, in contrast to the type $\{4,3\}$ of the facets of $\{4,3,3\}$. Every edge e belongs to the three facets corresponding to the three choices of triples of colours that include the colour of e, implying that Roli's cube has type $\{8,3,3\}$. The subgraph induced by the edges of three colours is connected. It follows that Roli's cube has only four 3-faces; they are shown in Figure 5.12.

Two distinct edges or 2-faces of Roli's cube have distinct vertex sets. Furthermore, two distinct facets have distinct edge sets. This allows us to show

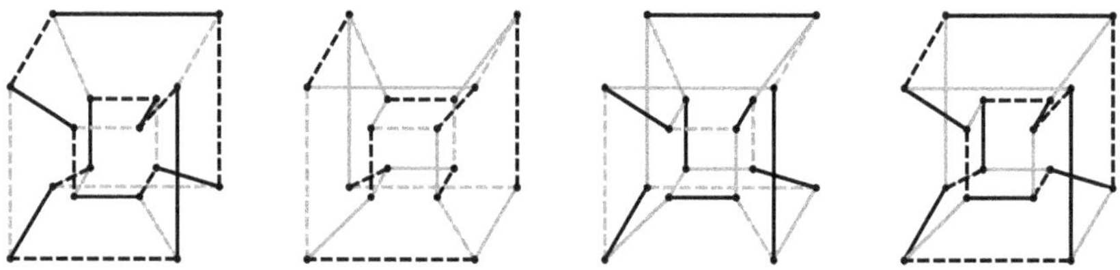

Figure 5.12 The four facets of Roli's cube.

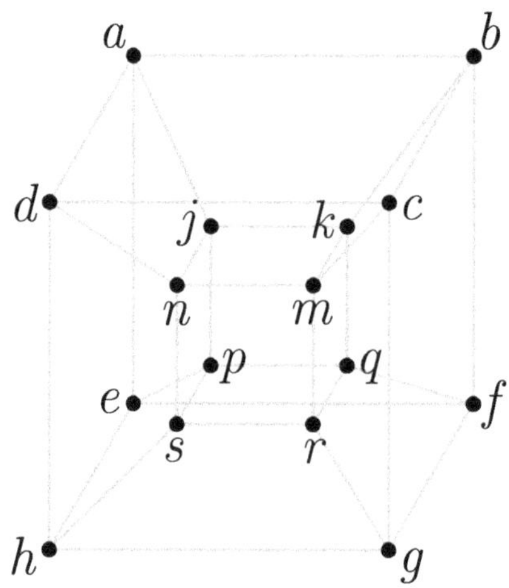

Figure 5.13 Labels of the vertices of Roli's cube.

that Roli's cube is rotary by exhibiting the standard rotations as permutations of the vertices. In order to do this we label the vertices as in Figure 5.13. We choose as the base flag the one containing vertex a, the edge ab, the polygon containing ab in Figure 5.11(a) and the facet in Figure 5.12(a) (see the labels of the vertices in Figure 5.13). Then,

$$\sigma_1 = (a,j,p,s,r,g,c,b)(d,k,e,n,q,h,m,f),$$
$$\sigma_2 = (b,d,j)(c,n,k)(f,h,p)(g,s,q),$$
$$\sigma_3 = (c,k,f)(d,j,e)(g,m,q)(h,n,p).$$

Now we show that Roli's cube is chiral. Assume to the contrary that there exists an automorphism ρ_2. Such an automorphism must preserve the base vertex a and the base edge ab, and so it fixes a and b. The base face (a,b,c,g,r,s,p,j) must be interchanged by ρ_2 with the 2-face

(a, b, f, q, r, s, n, d); this forces $c\rho_2 = f$ and $g\rho_2 = q$. But f and g are adjacent in Roli's cube, whereas $f\rho_2 = c$ and $g\rho_2 = q$ are not. This yields the desired contradiction and we can conclude that Roli's cube is chiral.

Theorem 5.3.10 *Roli's cube is a chiral* 4*-polytope with chirality index* 2*.*

Proof We already showed that Roli's cube is a chiral 4-polytope. Moreover, by construction it has a well-defined quotient by the central inversion. This quotient is isomorphic to the hemi-4-cube, which is regular. It follows that the chirality index is 2. ∎

Here we have described Roli's cube with some emphasis on its combinatorics. Other geometric aspects are stressed in [10]. An alternative construction of Roli's cube is given in [202], where it is related to other famous geometric and combinatorial structures.

The construction of Roli's cube presented here uses many particular features of the 4-cube and of $\{4,3,3\}_4$ that include the following.

- The 1-skeleton of $\{4,3,3\}_4$ is a complete bipartite graph.
- There are at least two isomorphic but non-isometric proper 4-colourings of the edges of $\{4,3,3\}_4$.
- $\{4,3,3\}_4$ is a quotient of $\{4,3,3\}$ of degree 2.

For $n \geq 5$, convex n-polytopes and their quotients by antipodal identification do not satisty analogous conditions, making it impossible to generalise Roli's cube directly to higher ranks. Nevertheless, some of the ideas may admit adaptations to build other chiral polytopes with similar techniques.

Problem 48 Are there other regular colourful n-polytopes whose 1-skeleton admits isomorphic n-colourings of their edges that yield colourful n-polytopes with two or more flag orbits?

Problem 49 Are there chiral n-polytopes for $n \geq 5$ arising from colourful polytopes whose underlying 1-skeleta are well-studied families of graphs?

5.4 Chiral Amalgamations

The two main mechanisms to construct $(n+1)$-polytopes from n-polytopes are amalgamations and extensions. This section is devoted to the former; the latter will be one of the main topics of Section 5.5.

5.4.1 General Amalgamations

It is natural to study each convex regular n-polytope by considering a collection of isometric regular $(n-1)$-polytopes (the facets) and trying to put them together so that the neighbourhoods of any two vertices look alike. This can be restated as requiring an n-polytope all of whose facets are of a certain isomorphism type $\mathcal{P}_1$ and all of whose vertex-figures are also of a certain isomorphism type $\mathcal{P}_2$. Convexity imposes serious restrictions on the possible polytopes $\mathcal{P}_1$ and $\mathcal{P}_2$, and the result is that there are only five and six convex regular polytopes of dimensions 3 and 4, respectively, and only three in each dimension $n \geq 5$.

Moving on to different ambient spaces yields different results. For instance, in the hyperbolic plane it is possible to fit q equilateral triangles at every vertex for every $q \geq 7$ (here $\mathcal{P}_1 = \{3\}$ and $\mathcal{P}_2 = \{q\}$), and to extend this configuration to a regular tessellation. By contrast, if we want to construct a convex polyhedron with triangular faces then at most five triangles must meet at every vertex.

This idea was taken to the context of regular abstract polytopes in [242]. The next definition is a natural first attempt to take that notion to general abstract polytopes.

Definition 5.4.1 Let $\mathcal{P}_1$ and $\mathcal{P}_2$ be two n-polytopes. We say that an $(n+1)$-polytope $\mathcal{Q}$ is an *amalgamation* of $\mathcal{P}_1$ and $\mathcal{P}_2$ if all the facets of $\mathcal{Q}$ are isomorphic to $\mathcal{P}_1$ and all the vertex-figures are isomorphic to $\mathcal{P}_2$.

We emphasise that in Definition 5.4.1 we required $\mathcal{Q}$ to be a polytope. In this work we are only considering polytopal amalgamations of polytopes.

One important fact to keep in mind is that some choices of $\mathcal{P}_1$ and $\mathcal{P}_2$ admit no amalgamation. One simple criterion for this follows from the fact that if $\mathcal{Q}$ is an amalgamation of $\mathcal{P}_1$ and $\mathcal{P}_2$ then the sections between a vertex and a facet of $\mathcal{Q}$ are all isomorphic to vertex-figures of $\mathcal{P}_1$, and also to facets of $\mathcal{P}_2$.

Remark 5.4.2 Let $\mathcal{P}_1$ and $\mathcal{P}_2$ be two n-polytopes such that the set of vertex-figures of $\mathcal{P}_1$ differs from the set of facets of $\mathcal{P}_2$. Then there is no amalgamation of $\mathcal{P}_1$ and $\mathcal{P}_2$.

There are some pairs $(\mathcal{P}_1, \mathcal{P}_2)$ of n-polytopes that do not admit an amalgamation, but with the property that the set of vertex-figures of $\mathcal{P}_1$ equals the set of the facets of $\mathcal{P}_2$.

Example 5.4.3 Let $\mathcal{P}_1$ be the hemi-cube and $\mathcal{P}_2$ be the tetrahedron. We claim that there is no amalgamation of $\mathcal{P}_1$ and $\mathcal{P}_2$.

Assume to the contrary that $\mathcal{Q}$ is an amalgamation of $\mathcal{P}_1$ and $\mathcal{P}_2$. We know from Example 5.2.2 and Remark 5.2.3 that $\mathcal{Q}$ is flat, and hence it only has four vertices. Recall that the underlying graph of $\mathcal{P}_1$ is the complete graph K_4. Assume that the vertex set of $\mathcal{Q}$ is $\{a,b,c,d\}$; then vertex a is adjacent in the base facet to the remaining three vertices. Since the vertex-figure at a must be a tetrahedron, a must be incident to precisely four edges. The only way to achieve this while keeping the polytopality of $\mathcal{Q}$ is through a double edge between vertices a and b (say), preserving single edges to vertices c and d. On the other hand, every triple of vertices of $\mathcal{P}_2$ belongs to a facet of $\mathcal{P}_2$. When reinterpreting $\mathcal{P}_2$ as the vertex-figure at a, this implies that the two edges of $\mathcal{Q}$ between a and b belong to some facet of $\mathcal{Q}$ containing a. But the hemi-cube does not have double edges, a contradiction.

If there exist amalgamations of $\mathcal{P}_1$ and $\mathcal{P}_2$ there may be more than one. In some cases there are finitely many amalgamations and in some cases there are infinitely many.

Exercise 5.4.4 Show that the 4-cube and the hemi-4-cube are the only amalgamations of the 3-cube and the tetrahedron.

Example 5.4.5 The toroidal polytopes $\{4,3,4\}_{(a,0,0)}$ with $a \in \{2,3,\dots\}$, defined in Section 2.4.3, are amalgamations of the cube and the octahedron.

Our previous remarks and examples motivate the following two main questions for any two n-polytopes $\mathcal{P}_1$ and $\mathcal{P}_2$ such that the set of vertex-figures of $\mathcal{P}_1$ equals the set of facets of $\mathcal{P}_1$ (see [220, Problem 31], [250, Problems 8 and 9]).

1. Are there amalgamations of $\mathcal{P}_1$ and $\mathcal{P}_2$?
2. In case of an affirmative answer to question 1, are there infinitely many amalgamations of $\mathcal{P}_1$ and $\mathcal{P}_2$?

The previous two questions are very difficult for general polytopes $\mathcal{P}_1$ and $\mathcal{P}_2$. Most of the attention has been devoted to the case when $\mathcal{P}_1$ and $\mathcal{P}_2$ are regular, and even with this constraint there seems not to be an easy way to answer these questions. Under the assumption of regularity of $\mathcal{P}_1$ and $\mathcal{P}_2$, the set of regular amalgamations of $\mathcal{P}_1$ and $\mathcal{P}_2$ was denoted by $\langle \mathcal{P}_1, \mathcal{P}_2 \rangle$ in [188, Section 4A].

The question of existence of a regular amalgamation of $\mathcal{P}_1$ and $\mathcal{P}_2$ has been solved from the combinatorics of $\mathcal{P}_1$ and $\mathcal{P}_2$ in very few instances. It may involve clever constructions like the following (see [57]).

Exercise 5.4.6 Denote the elements of the field $GF(11)$ with 11 elements by $\{0, \dots, 9, A\}$ and consider the subsets $\mathcal{X}_0 := \{0, 2, 6, 7, 8, A\}$ and $\mathcal{X}_i := \mathcal{X}_0 + i$ for $i \in GF(11)$.

1. Show that if $i \neq j$ then $|\mathcal{X}_i \cap \mathcal{X}_j| = 3$.
2. Let $\mathcal{P}_i$ have as 1-skeleton the complete graph in the elements of $\mathcal{X}_i$, and as 2-faces the 10 triangles that arise from item 1. Show that $\mathcal{P}_i$ is isomorphic to the hemi-icosahedron for every $i \in GF(11)$. (The *hemi-icosahedron* is a non-orientable polyhedron obtained from the icosahedron identifying antipodal i-faces for every $i \in \{0, 1, 2\}$.)
3. Show that the union of the $\mathcal{X}_i$ constitutes a 4-polytope where the vertex-figures are all isomorphic to hemi-dodecahedra. (The *hemi-dodecahedron* is a non-orientable polyhedron obtained from the dodecahedron identifying antipodal i-faces for every $i \in \{0, 1, 2\}$.)

The 4-polytope mentioned in Exercise 5.4.6 is known as the *11-cell*. It is in fact a regular 4-polytope, first mentioned in [105], with its automorphism group isomorphic to the projective linear group $PSL_2(11)$. Its existence shows that $\langle \{3,5\}/2, \{5,3\}/2 \rangle \neq \varnothing$, where $\{3,5\}/2$ and $\{5,3\}/2$ denote the hemi-icosahedron and the hemi-dodecahedron, respectively. Later it was shown in [56] that $\langle \{5,3\}/2, \{3,5\}/2 \rangle \neq \varnothing$ by proving the existence of an abstract polytope, called the 57-*cell*, with 57 facets isomorphic to the hemi-dodecahedron, and 57 vertices, whose vertex-figures are isomorphic to the hemi-icosahedron.

Most of the study of amalgamations of regular polytopes has been done through their automorphism groups. If $\mathcal{Q}$ is a regular amalgamation of $\mathcal{P}_1$ and $\mathcal{P}_2$ with $\Gamma(\mathcal{Q}) = \langle \rho_0, \dots, \rho_n \rangle$ then it follows from the fourth item of Proposition 2.2.10 that $\Gamma(\mathcal{P}_1) \cong \langle \rho_0, \dots, \rho_{n-1} \rangle$ and $\Gamma(\mathcal{P}_2) \cong \langle \rho_1, \dots, \rho_n \rangle$. Furthermore, if $\Gamma(\mathcal{P}_1)$ and $\Gamma(\mathcal{P}_2)$ admit presentations $\langle \rho_0, \dots, \rho_{n-1} : R_1 \rangle$ and $\langle \rho_1, \dots, \rho_n : R_2 \rangle$ in terms of generators and relations, then all relations given by R_1 and R_2 are also satisfied by the generators of $\Gamma(\mathcal{Q})$. It follows that $\Gamma(\mathcal{Q})$ is a quotient of the group

$$\langle \rho_0, \ldots, \rho_n \colon (\rho_0\rho_n)^2 = \mathrm{id}, R_1 \cup R_2 \rangle. \tag{5.4}$$

This group plays the following key role in the study of amalgamations of regular polytopes (see [188, Theorem 4A2 and Proposition 4A8], [242, Theorem 1]).

Theorem 5.4.7 *Let $\mathcal{P}_1$ and $\mathcal{P}_2$ be regular n-polytopes such that the vertex-figures of $\mathcal{P}_1$ are isomorphic to the facets of $\mathcal{P}_2$. Assume that $\Gamma(\mathcal{P}_1)$ and $\Gamma(\mathcal{P}_2)$ admit presentations $\langle \rho_0, \ldots, \rho_{n-1} \colon R_1 \rangle$ and $\langle \rho_1, \ldots, \rho_n \colon R_2 \rangle$, respectively, in terms of generators and relations. If there exists an amalgamation of $\mathcal{P}_1$ and $\mathcal{P}_2$ then the group (5.4) is a string C-group, and it is the automorphism group of a regular amalgamation of $\mathcal{P}_1$ and $\mathcal{P}_2$ that covers every other regular amalgamation of $\mathcal{P}_1$ and $\mathcal{P}_2$.*

The polytope with automorphism group (5.4) in the statement of Theorem 5.4.7 is called the *universal amalgamation* of $\mathcal{P}_1$ and $\mathcal{P}_2$. The idea of this object was first explored in [105] with the name of 'naturally generated polystromata'. We next list some of the known universal amalgamations of regular polytopes.

Example 5.4.8 The universal amalgamation of the universal n-polytopes with types $\{p_1, \ldots, p_{n-1}\}$ and $\{p_2, \ldots, p_n\}$ is the universal $(n+1)$-polytope with type $\{p_1, \ldots, p_n\}$ ([188, Proposition 4A9]).

Example 5.4.9 The 11-cell is the universal amalgamation of the hemi-icosahedron and the hemi-dodecahedron (see [29, 105]).

Example 5.4.10 The 57-cell is the universal amalgamation of the hemi-dodecahedron and the hemi-icosahedron (see [29, 56]).

As mentioned above, it is not easy to determine whether two regular n-polytopes $\mathcal{P}_1$ and $\mathcal{P}_2$ such that the vertex-figures of $\mathcal{P}_1$ are isomorphic to the facets of $\mathcal{P}_2$ have an amalgamation, and hence, a universal amalgamation. We may restate this in terms of the group (5.4) as follows. For an amalgamation of $\mathcal{P}_1$ and $\mathcal{P}_2$ to exist, it is necessary that the group (5.4) is a string C-group such that its subgroups generated by the first and last n generators are isomorphic to $\Gamma(\mathcal{P}_1)$ and $\Gamma(\mathcal{P}_2)$, respectively. In general, it is not an easy task to verify if these properties are satisfied. One of the main problems is determining whether the group (5.4) is finite or infinite. This adds a third question to the list we started previously, in the case that $\mathcal{P}_1$ and $\mathcal{P}_2$ admit amalgamations.

3. Is the universal amalgamation of $\mathcal{P}_1$ and $\mathcal{P}_2$ finite or infinite?

Much of the study of amalgamations of regular polytopes was motivated by a problem posed in [105], which in our terminology can be restated as asking for a classification of all polytopes whose facets and vertex-figures are spherical or toroidal, but not both spherical. Such polytopes are called *locally toroidal polytopes*, and they can be alternatively defined as amalgamations of toroidal polytopes with spherical or toroidal polytopes. Besides that class of polytopes, one may also consider those polytopes whose facets and vertex-figures are spherical or projective, but not both spherical. Such polytopes are called *locally projective polytopes*, and they can be understood as amalgamations of projective polytopes with spherical or projective polytopes.

Much is known about locally projective and locally toroidal regular polytopes. Readers interested in locally toroidal regular polytopes are referred to [188, Chapters 8 and 10–12], [183–186, 192, 193, 195, 199, 216, 244, 248, 265]. Those interested in locally projective polytopes are referred to [114, 115, 117, 119–121, 170], [188, Chapter 14].

5.4.2 Chiral Amalgamations

The unrooted definition of amalgamation above is good enough for some polytopes that are not regular. In what follows we study the unrooted amalgamations of $\{4,4\}_{(2,1)}$ with itself. We start by establishing the number of vertices of any such amalgamation.

Lemma 5.4.11 *Every amalgamation of* $\{4,4\}_{(2,1)}$ *with itself has exactly six vertices and its underlying graph is the complete graph* K_6.

Proof Let n_v denote the number of vertices of an amalgamation $\mathcal{P}$ of $\{4,4\}_{(2,1)}$ with itself. We first show that $n_v \geq 6$. Each facet of $\mathcal{P}$ has five vertices, implying that $n_v \geq 5$. Furthermore, every vertex-figure has five vertices, implying that each vertex F_0 of $\mathcal{P}$ is incident to five edges. Assume to the contrary that $\mathcal{P}$ has only five vertices. Since the underlying graph of the facets is the complete graph K_5, F_0 is adjacent to every other vertex. To complete the degree, there must be a double edge between F_0 and some other vertex F_0'. We use again that in $\{4,4\}_{(2,1)}$ (now seen as the vertex-figure) all pairs of vertices are adjacent to conclude that all pairs of edges at F_0 are consecutive in a square of $\mathcal{P}$; in particular, there must be a square that uses the two edges between F_0 and F_0', contradicting the polytopality of $\mathcal{P}$.

The previous paragraph shows that every vertex of $\mathcal{P}$ has five distinct neighbours. Furthermore, if F_0' and F_0'' are two neighbours of F_0 then there is a square of $\mathcal{P}$ containing the edges from F_0 to F_0' and to F_0''. In particular, there is a square, and hence a facet, containing F_0' and F_0''. Since the underlying graph of the facets is complete, F_0' and F_0'' must be adjacent. It follows that F_0 and its five neighbours induce a complete graph K_6. Since the vertex-figures have five vertices each, the underlying graph of $\mathcal{P}$ must be 5-regular and the lemma follows. ■

We stress the fact that no assumption was made on the symmetry group of the amalgamation in Lemma 5.4.11. The following result follows from the fact that $\{4,4\}_{(2,1)}$ has 40 flags.

Corollary 5.4.12 *Every amalgamation of* $\{4,4\}_{(2,1)}$ *with itself has* 240 *flags,* 15 *edges,* 15 *squares and* 6 *facets.*

Assume that $\mathcal{Q}$ is an amalgamation of $\{4,4\}_{(2,1)}$ with itself. We label the vertices of the base facet F_3 as in Figure 5.14, and the remaining vertex by 5. We shall study the four squares and the four facets that contain the edge F_1 between 0 and 1. Each of the four neighbours of 0 other than 1 must be adjacent to 0 in one of these four squares (see the proof of Lemma 5.4.11). The two squares in F_3 that contain F_1 are illustrated in Figure 5.15(a). One feature of these two squares is that precisely one vertex repeats in opposite corners of the outer rectangle. This needs to be the case since a facet has only five vertices while the union of two squares at an edge has at most six vertices, and any two squares of a facet use the five vertices of the facet.

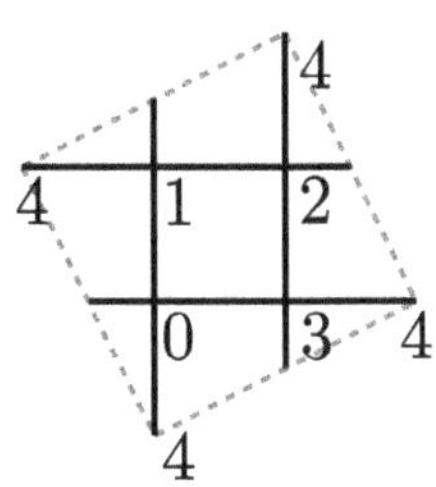

Figure 5.14 Labels of the vertices of $\{4,4\}_{(2,1)}$.

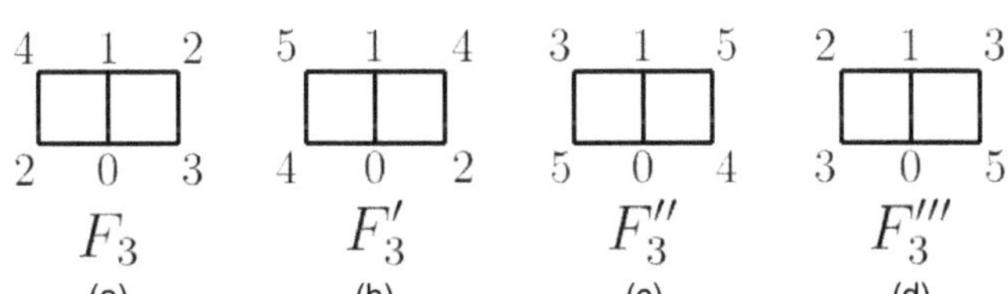

Figure 5.15 (a) The two squares in F_3 that contain F_1; (b), (c) and (d) illustrate the two squares that contain F_1 in the remaining three facets at that edge.

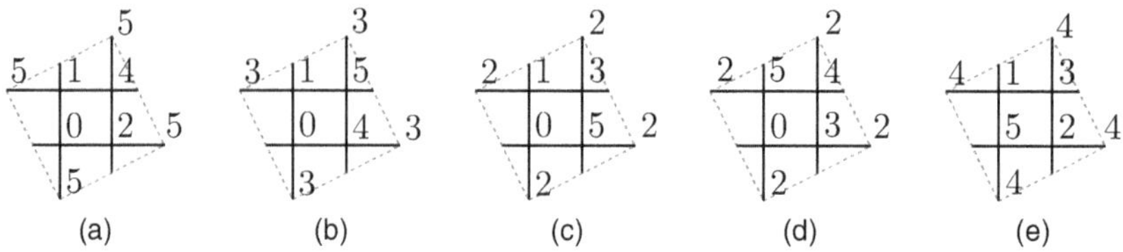

Figure 5.16 The remaining five facets of $\mathcal{Q}$; (a), (b), (c), (d) and (e) miss vertex 3, 2, 4, 1 and 0, respectively.

Let F_3' be the facet that shares with F_3 the square F_2 with vertices 0, 1, 4 and 2, and let F_2' be the other square of F_3' containing F_1. Then the neighbour of 0 in F_2' other than 1 must be either 4 or 5, since the remaining vertices already appear as neighbours of 0 in F_3. Similarly, the neighbour of 1 in F_2' other than 0 must be either 3 or 5. Since there must be precisely five vertices in the union of F_2 and F_2', and 0 and 1 must appear only once, we can conclude that 4 must appear twice, and so 0 is adjacent to 4 in F_2'. Finally, all facets at 0 other than F_3 contain vertex 5, so the vertices of F_2' are 0, 1, 5, 4 in cyclic order. In a similar fashion we can see that the two squares containing F_1 in the four facets containing F_1 are those in Figure 5.15. We label F_3, F_3', F_3'' and F_3''' the four facets around F_1 in cyclic order.

The two squares in each of F_3', F_3'' and F_3''' in Figure 5.15(b), (c) and (d) can be completed in a unique way to a copy of $\{4,4\}_{(2,1)}$. The labels of the vertices of these three facets are shown in the three tori of Figure 5.16(a), (b) and (c). Proceeding in a similar fashion with other edges we obtain the five tori in Figure 5.16, each missing one of the vertices.

It is routine to verify that the structure of rank 4 given by the six facets in Figures 5.14 and 5.16 and their faces is a 4-polytope. We next list some natural consequences of the analysis above. The first one follows from the fact that no choice or extra assumption was made in the construction of this 4-polytope.

Proposition 5.4.13 *There is only one amalgamation of* $\{4,4\}_{(2,1)}$ *with itself, namely that given by the six facets in Figures 5.14 and 5.16. It has* 240 *flags,* 6 *vertices,* 15 *edges,* 15 *squares and* 6 *facets. This* 4*-polytope, and the one mentioned in Exercise 2.1.20 are the two chiral* 4*-polytopes with fewest flags.*

Remark 5.4.14 Let $\mathcal{Q}$ denote the unique amalgamation of $\{4,4\}_{(2,1)}$ with itself. The vertex-figure at 0 of $\mathcal{Q}$ has as vertices the five edges at 0. Given that the 1-skeleton of $\mathcal{Q}$ is K_6, it is safe to label the vertices of the vertex-figure 1, 2, 3, 4 and 5. Direct inspection of Figures 5.14 and 5.16 shows that the five squares can be arranged to the toroidal polyhedron in Figure 5.17(b). We choose as base flag Φ_0 of $\mathcal{Q}$ the one with vertex 0, edge 01, 2-face 0123

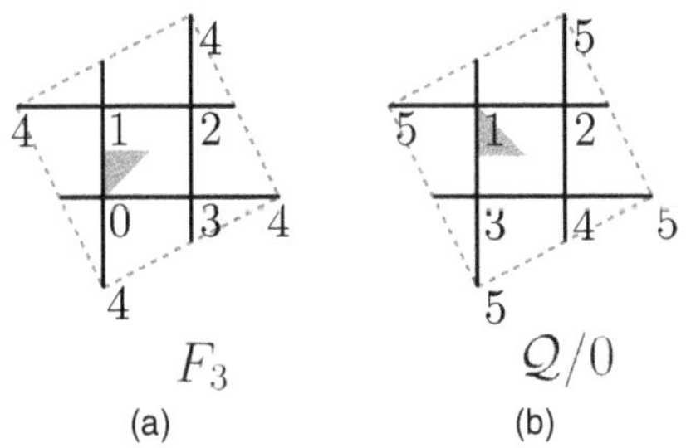

Figure 5.17 (a) The facet F_3 rooted at Φ_0'; (b) the vertex-figure $\mathcal{Q}/0$ rooted at Φ_0''.

and facet F_3 given in Figure 5.14. Comparing the restrictions Φ_0' and Φ_0'' to F_3 and the vertex-figure $\mathcal{Q}/0$ at 0, we see that the rooted polyhedra $(F_3)_{(\Phi_0')}$ and $(\mathcal{Q}/0)_{(\Phi_0'')}$ are enantiomorphs (see Figure 5.17).

Remark 5.4.15 The permutations $(0,3,2,1)$, $(1,2,4,3)$ and $(2,3,5,4)$ on the vertex set of the unique amalgamation $\mathcal{Q}$ of $\{4,4\}_{(2,1)}$ with itself induce automorphisms that act like the standard rotations. It follows that $\mathcal{Q}$ is rotary and, since its facets are chiral, $\mathcal{Q}$ itself is chiral. In particular, it is orientable.

Remark 5.4.16 From Remark 5.4.15 we know that the flags of the unique amalgamation $\mathcal{Q}$ of $\{4,4\}_{(2,1)}$ with itself admit a bipartition where adjacent flags belong to distinct parts. Let $\mathcal{E}$ be the part containing the base flag Φ_0, and recall that the facet in Φ_0 is F_3. If $\Psi \in \mathcal{E}$ and G_3 is the facet in Ψ then $(F_3)_{(\Phi_0)}$ and $(G_3)_{(\Psi)}$ are isomorphic as rooted polytopes. Similarly, any two vertex-figures of $\mathcal{Q}$ are isomorphic when rooted at the restrictions of two flags in $\mathcal{E}$.

In the spirit of Remark 5.4.16 we define the rooted amalgamation of two rotary polytopes as follows. This formalises the idea in [247, Section 6].

Definition 5.4.17 Let $(\mathcal{P}_1)_{(\Phi)}$ and $(\mathcal{P}_2)_{(\Psi)}$ be two rooted orientable rotary polytopes. We say that $\mathcal{Q}_{(\Omega)}$ is a rooted amalgamation of $(\mathcal{P}_1)_{(\Phi)}$ and $(\mathcal{P}_2)_{(\Psi)}$ if it satisfies the following.

- $\mathcal{Q}$ is orientable, and so the part $\mathcal{E}$ of the bipartition of flags containing the base flag Ω of $\mathcal{Q}$ is well defined.
- For every facet F of $\mathcal{Q}$ and every flag Υ in $\mathcal{E}$ containing F, the rooted polytope $F_{(\Upsilon')}$ is isomorphic to $(\mathcal{P}_1)_{(\Phi)}$, where Υ' is the restriction of Υ to F.
- For every vertex-figure G of $\mathcal{Q}$ and every flag Υ in $\mathcal{E}$ containing G, the rooted polytope $G_{(\Upsilon'')}$ is isomorphic to $(\mathcal{P}_2)_{(\Psi)}$, where Υ'' is the restriction of Υ to G.

Following [247], we denote the class of chiral rooted amalgamations of $(\mathcal{P}_1)_{(\Phi)}$ and $(\mathcal{P}_2)_{(\Psi)}$ by $\langle(\mathcal{P}_1)_{(\Phi)}, (\mathcal{P}_2)_{(\Psi)}\rangle^{ch}$.

Example 5.4.18 With the notation above, the unique amalgamation of $\{4,4\}_{(2,1)}$ with itself is the only member of $\langle(\{4,4\}_{(2,1)})_{(\Phi_0)}, (\{4,4\}_{(2,1)})_{(\Phi_0^0)}\rangle^{ch}$.

On the other hand, $\langle(\{4,4\}_{(2,1)})_{(\Phi_0)}, (\{4,4\}_{(2,1)})_{(\Phi_0)}\rangle^{ch} = \emptyset$.

All chiral unrooted amalgamations of $\mathcal{P}_1$ and $\mathcal{P}_2$ arise from a choice of the rooting flags of chiral rooted amalgamations. The details are left as an exercise. With this in mind, in what follows we shall concentrate only on rooted amalgamations of rotary polytopes.

Exercise 5.4.19 Let $\mathcal{P}_1$ and $\mathcal{P}_2$ be two orientable rotary n-polytopes. Show that a chiral $(n+1)$-polytope $\mathcal{Q}$ is an (in principle unrooted) amalgamation of $\mathcal{P}_1$ and $\mathcal{P}_2$ if and only if there exists a flag Υ of $\mathcal{Q}$ such that

$$\mathcal{Q}_{(\Upsilon)} \in \langle(\mathcal{P}_1)_{(\Phi)}, (\mathcal{P}_2)_{(\Psi)}\rangle^{ch}$$

for some $\Phi \in \mathcal{F}l(\mathcal{P}_1)$, $\Psi \in \mathcal{F}l(\mathcal{P}_2)$.

Proposition 5.4.20 *Given any two rooted rotary polytopes $(\mathcal{P}_1)_{(\Phi)}$ and $(\mathcal{P}_2)_{(\Psi)}$, the set $\langle(\mathcal{P}_1)_{(\Phi^1)}, (\mathcal{P}_2)_{(\Psi^1)}\rangle^{ch}$ consists of the enantiomorphs of the elements in $\langle(\mathcal{P}_1)_{(\Phi)}, (\mathcal{P}_2)_{(\Psi)}\rangle^{ch}$.*

Proof Let $\mathcal{Q} \in \langle(\mathcal{P}_1)_{(\Phi)}, (\mathcal{P}_2)_{(\Psi)}\rangle^{ch}$ and let $\mathcal{E}$ be as in Definition 5.4.1. A flag Υ is in $\mathcal{E}$ if and only if $\Upsilon^1 \in \mathcal{F}l(\mathcal{Q}) \setminus \mathcal{E}$. It follows that the restrictions of Υ to its facet and vertex-figure are in the orbits of Φ and Ψ, respectively, if and only if the restrictions of Υ^1 to its facet and vertex-figure are in the orbits of Φ^1 and Ψ^1, respectively. The remark follows from switching the roles of $\mathcal{E}$ and $\mathcal{F}l(\mathcal{Q}) \setminus \mathcal{E}$. ■

Remark 5.4.21 When considering chiral toroidal polyhedra we may distinguish $\{p,q\}_{(a,b)}$ (for some $a,b > 0$) from its enantiomorph by calling the latter $\{p,q\}_{(b,a)}$. This is the convention in [29], where the only amalgamation of $\{4,4\}_{(2,1)}$ with itself is understood as an amalgamation of $\{4,4\}_{(1,2)}$ with $\{4,4\}_{(2,1)}$. Since we do not have analogous notation for chiral polytopes that are not toroidal we prefer to follow the notation of rooted polytopes.

If instead of considering the polyhedron $\{4,4\}_{(2,1)}$ we look at $\{4,4\}_{(3,1)}$, the situation changes drastically. On one hand, it is no longer easy to establish combinatorial properties of an arbitrary amalgamation of $\{4,4\}_{(3,1)}$ with itself. On the other hand, both $\langle(\{4,4\}_{(3,1)})_{(\Phi_0)}, (\{4,4\}_{(3,1)})_{(\Phi_0)}\rangle^{ch}$ and $\langle(\{4,4\}_{(3,1)})_{(\Phi_0)}, (\{4,4\}_{(3,1)})_{(\Phi_0^0)}\rangle^{ch}$ are non-empty, although both of them are finite. Moreover, each of these classes of chiral polytopes has a universal element in the sense that every other member of a class is a quotient of the universal one. The numbers of facets of the universal elements in $\langle(\{4,4\}_{(3,1)})_{(\Phi_0)}, (\{4,4\}_{(3,1)})_{(\Phi_0)}\rangle^{ch}$ and $\langle(\{4,4\}_{(3,1)})_{(\Phi_0)}, (\{4,4\}_{(3,1)})_{(\Phi_0^0)}\rangle^{ch}$ are 50 and 24, respectively (see [29, Table 1]). This suggests that these two universal objects have very different properties. Furthermore, these are universal amalgamations only in the rooted version. It is not clear whether there are non-rooted amalgamations having the two enantiomorphs as facets.

Problem 50 Determine whether there exists an orientable 4-polytope $\mathcal{Q}$ all of whose facets are isomorphic to $\{4,4\}_{(3,1)}$ such that there are two flags Φ and Ψ with the following properties.

- There is a path with even length in the flag graph of $\mathcal{Q}$ between Φ and Ψ.
- The facet containing Φ, when rooted at Φ, is the enantiomorph of the facet containing Ψ rooted at Ψ.

If Problem 50 has an affirmative answer, such a polytope cannot be rotary. The two items of Problem 50 will be satisfied automatically if the 4-polytope $\mathcal{Q}$ is non-orientable.

Problem 51 Are there non-orientable non-rooted amalgamations of two chiral polytopes?

When considering rooted amalgamations of $(\mathcal{P}_1)_{(\Phi)}$ and $(\mathcal{P}_2)_{(\Psi)}$ the flags Φ and Ψ are, in effect, representatives of their orbits, in the sense that distinct choices of flags within the same orbits yield equivalent results.

Lemma 5.4.22 *Let $(\mathcal{P}_1)_{(\Phi)}$ and $(\mathcal{P}_2)_{(\Psi)}$ be two rooted rotary n-polytopes and let $\alpha \in \Gamma(\mathcal{P}_1)$, $\beta \in \Gamma(\mathcal{P}_2)$. Then*

$$\langle(\mathcal{P}_1)_{(\Phi)}, (\mathcal{P}_2)_{(\Psi)}\rangle^{ch} = \langle(\mathcal{P}_1)_{(\Phi\alpha)}, (\mathcal{P}_2)_{(\Psi\beta)}\rangle^{ch}.$$

Proof This follows directly from Definition 5.4.1, since $(\mathcal{P}_1)_{(\Phi)} \cong (\mathcal{P}_1)_{(\Phi\alpha)}$ and $(\mathcal{P}_2)_{(\Psi)} \cong (\mathcal{P}_2)_{(\Psi\beta)}$. ■

With the help of Lemma 5.4.22 we can show that if $\mathcal{P}_1$ or $\mathcal{P}_2$ is regular then the choices of flags play no substantial role in the set of rooted amalgamations.

Proposition 5.4.23 *Let $(\mathcal{P}_1)_{(\Phi)}$ and $(\mathcal{P}_2)_{(\Psi)}$ be two rooted rotary n-polytopes with at least one of them regular. Then $\langle(\mathcal{P}_1)_{(\Phi)}, (\mathcal{P}_2)_{(\Psi)}\rangle^{ch}$ contains the same* *set of unrooted polytopes as* $\langle (\mathcal{P}_1)_{(\Phi')}, (\mathcal{P}_2)_{(\Psi')} \rangle^{ch}$ *for every* $\Phi, \Phi' \in \mathcal{F}l(\mathcal{P}_1)$ *and* $\Psi, \Psi' \in \mathcal{F}l(\mathcal{P}_2)$.

Proof Assume first that $\mathcal{P}_1$ is regular. The statement follows directly from Lemma 5.4.22 if Ψ and Ψ' are in the same flag orbit. In case that Ψ and Ψ' are in distinct orbits then Ψ^1 is in the same orbit as Ψ' and the result follows from Proposition 5.4.20 and Lemma 5.4.22.

The proof when $\mathcal{P}_2$ is regular follows in a dual fashion. ■

We may understand Proposition 5.4.23 as saying that there is essentially one local arrangement of facets and vertex-figures of the chiral amalgamations of two rotary polytopes, if at least one of them is regular. We already know from the amalgamations of the polyhedra $\{4,4\}_{(2,1)}$ and $\{4,4\}_{(3,1)}$ with themselves that Proposition 5.4.23 is no longer true if $\mathcal{P}_1$ and $\mathcal{P}_2$ are chiral. It follows from Proposition 5.4.20 and Lemma 5.4.22 that there are two essentially different local arrangement of facets and vertex-figures of the chiral amalgamations of two chiral polytopes.

The main advantage of studying amalgamations $\mathcal{Q}$ of rotary polytopes $\mathcal{P}_1$ and $\mathcal{P}_2$ from the rooted viewpoint is the relationship between $\Gamma^+(\mathcal{Q})$ with $\Gamma^+(\mathcal{P}_1)$ and $\Gamma^+(\mathcal{P}_2)$, stated in Remark 5.4.24. Its proof is straightforward; it is based on the facts that for $n \geq 3$, σ_{n-1} fixes the base vertex and σ_1 fixes the base facet.

Remark 5.4.24 Let $n \geq 3$, let $(\mathcal{P}_1)_{(\Phi)}$ and $(\mathcal{P}_2)_{(\Psi)}$ be two orientable rotary rooted n-polytopes, and let $\mathcal{Q}_{(\Upsilon)} \in \langle (\mathcal{P}_1)_{(\Phi)}, (\mathcal{P}_2)_{(\Psi)} \rangle^{ch}$. If $\Gamma^+(\mathcal{Q}_{(\Upsilon)}) = \langle \sigma_1, \ldots, \sigma_n \rangle$ then $\Gamma^+((\mathcal{P}_1)_{(\Phi)}) \cong \langle \sigma_1, \ldots, \sigma_{n-1} \rangle$ and $\Gamma^+((\mathcal{P}_2)_{(\Psi)}) \cong \langle \sigma_2, \ldots, \sigma_n \rangle$ (using the notation in (4.2)).

With the help of Remark 5.4.24 we can show easily in an alternative way that $\langle (\{4,4\}_{(2,1)})_{(\Phi)}, (\{4,4\}_{(2,1)})_{(\Phi^0)} \rangle^{ch} \neq \varnothing$ and $\langle (\{4,4\}_{(2,1)})_{(\Phi)}, (\{4,4\}_{(2,1)})_{(\Phi)} \rangle^{ch} = \varnothing$.

The rotation group $\langle \sigma_1, \sigma_2, \sigma_3 \rangle$ of any polytope in $\langle (\{4,4\}_{(2,1)})_{(\Phi)}, (\{4,4\}_{(2,1)})_{(\Phi^0)} \rangle^{ch}$ must satisfy

$$\sigma_1^4 = \sigma_2^4 = \sigma_3^4 = (\sigma_1\sigma_2)^2 = (\sigma_2\sigma_3)^2 = (\sigma_1\sigma_2\sigma_3)^2 = \text{id}, \tag{5.5}$$

as well as the relations

$$\sigma_1^{-1}\sigma_2(\sigma_1\sigma_2^{-1})^2 = \text{id} \quad \text{and} \quad (\sigma_2^{-1}\sigma_3)^2\sigma_2\sigma_3^{-1} = \text{id}, \tag{5.6}$$

given by the symmetry type of the facets and by the symmetry type of the vertex-figures, respectively. Standard procedures of coset enumeration, or a computer software, will show that the group Γ with three generators subject only to (5.5) and (5.6) has order 120, that $\langle \sigma_1, \sigma_2 \rangle \cong \Gamma(\{4,4\}_{(2,1)}) \cong \langle \sigma_2, \sigma_3 \rangle$, and that the intersection condition in (3.17) and (3.18) is satisfied. It follows that the 4-polytope $\mathcal{Q}$ with $\Gamma(\mathcal{Q}) = \Gamma$ and standard rotations σ_1, σ_2, σ_3 is an element of $\langle(\{4,4\}_{(2,1)})_{(\Phi)}, (\{4,4\}_{(2,1)})_{(\Phi^0)}\rangle^{ch}$, and that every other amalgamation in $\langle(\{4,4\}_{(2,1)})_{(\Phi)}, (\{4,4\}_{(2,1)})_{(\Phi^0)}\rangle^{ch}$ is a quotient of $\mathcal{Q}$.

On the other hand, if we replace $(\sigma_2^{-1}\sigma_3)^2\sigma_2\sigma_3^{-1}$ in (5.6) by $\sigma_2^{-1}\sigma_3(\sigma_2\sigma_3^{-1})^2$ then in the resulting group there is only one coset of $\langle\sigma_1,\sigma_2\rangle$. This tells us that every element in $\langle(\{4,4\}_{(2,1)})_{(\Phi)}, (\{4,4\}_{(2,1)})_{(\Phi)}\rangle^{ch}$ has at most one facet, and therefore it cannot be a polytope.

Exercise 5.4.25 Let Γ be the group with generators σ_1, σ_2, σ_3 subject only to the relations (5.5), $\sigma_1^{-1}\sigma_2(\sigma_1\sigma_2^{-1})^2 = \mathrm{id}$ and $\sigma_2^{-1}\sigma_3(\sigma_2\sigma_3^{-1})^2 = \mathrm{id}$. Show that $\langle\sigma_1,\sigma_2\rangle$ has only one coset in Γ. What is the order of Γ?

Given two orientable rooted rotary n-polytopes $(\mathcal{P}_1)_{(\Phi)}$ and $(\mathcal{P}_2)_{(\Psi)}$ with at least one rooted rotary amalgamation there exists a universal amalgamation of $\mathcal{P}_1$ and $\mathcal{P}_2$ in the sense that it covers every other rooted amalgamation of $(\mathcal{P}_1)_{(\Phi)}$ and $(\mathcal{P}_2)_{(\Psi)}$. In [247] it was called the *universal polytope with facet type $\mathcal{P}_1$ and vertex-figure type $\mathcal{P}_2$*.

Theorem 5.4.26 *Let $n \geq 2$ and let $(\mathcal{P}_1)_{(\Phi)}$ and $(\mathcal{P}_2)_{(\Psi)}$ be two rooted orientable rotary n-polytopes such that they have at least one rooted rotary amalgamation. Then there exists an $(n+1)$-polytope, denoted by $\{(\mathcal{P}_1)_{(\Phi)}, (\mathcal{P}_2)_{(\Psi)}\}$ with the property that every other rooted rotary amalgamation of $(\mathcal{P}_1)_{(\Phi)}$ and $(\mathcal{P}_2)_{(\Psi)}$ is a quotient of $\{(\mathcal{P}_1)_{(\Phi)}, (\mathcal{P}_2)_{(\Psi)}\}$. Furthermore, $\{(\mathcal{P}_1)_{(\Phi)}, (\mathcal{P}_2)_{(\Psi)}\}$ is regular if and only if both $(\mathcal{P}_1)_{(\Phi)}$ and $(\mathcal{P}_2)_{(\Psi)}$ are regular.*

Proof Since there are amalgamations of $\mathcal{P}_1$ and $\mathcal{P}_2$, the vertex-figures of $\mathcal{P}_1$ must be isomorphic to the facets of $\mathcal{P}_2$, which is an $(n-1)$-polytope $\mathcal{R}$, say. If $\Gamma^+(\mathcal{R}) = \langle\hat{\sigma}_2, \ldots, \hat{\sigma}_{n-1}\rangle$ we may abuse notation and use the same set of generators for the rotation groups of the vertex-figures of $\mathcal{P}_1$ and of the facets of $\mathcal{P}_2$.

Let $\langle\hat{\sigma}_1, \ldots, \hat{\sigma}_{n-1} \colon R_1\rangle$ and $\langle\hat{\sigma}_2, \ldots, \hat{\sigma}_n \colon R_2\rangle$ be presentations of $\Gamma^+(\mathcal{P}_1)$ and $\Gamma^+(\mathcal{P}_2)$ in terms of generators and relations. We denote by Γ the group

$$\langle\sigma_1, \ldots, \sigma_n \colon R_1 \cup R_2, (\sigma_1 \cdots \sigma_n)^2 = \mathrm{id}\rangle, \tag{5.7}$$

where R_1 and R_2 are understood in terms of the σ_i instead of the $\hat{\sigma}_i$.

We want to show that Γ is a string even group. Since $\Gamma^+(\mathcal{P}_1)$ and $\Gamma^+(\mathcal{P}_2)$ satisfy (3.7), we have that if $j - i < n - 1$ then $(\sigma_{i+1} \cdots \sigma_j)^2 = \mathrm{id}$ is included in $R_1 \cup R_2$. The only missing relation $(\sigma_1 \cdots \sigma_n)^2$ is given explicitly in (5.7). Therefore, Γ satisfies (3.7).

Now let $\mathcal{Q}$ be a rotary rooted amalgamation of $\mathcal{P}_1$ and $\mathcal{P}_2$ with $\Gamma^+(\mathcal{Q}) = \langle \sigma_1', \ldots, \sigma_n' \rangle$. The definition of Γ implies that the mapping $\sigma_i \mapsto \sigma_i'$ defines an epimorphism η to $\Gamma^+(\mathcal{Q})$. This, in turn, induces an epimorphism η_1 from $\Gamma_n := \langle \sigma_1, \ldots, \sigma_{n-1} \rangle$ to $\langle \sigma_1', \ldots, \sigma_{n-1}' \rangle \cong \Gamma^+(\mathcal{P}_1)$. Since $\langle \sigma_1, \ldots, \sigma_{n-1} \rangle$ satisfies the defining relations of $\Gamma^+(\mathcal{P}_1)$, η_1 is an isomorphism. Similarly, there is an isomorphism η_2 from $\Gamma_0 := \langle \sigma_2, \ldots, \sigma_{n-1} \rangle$ to $\Gamma^+(\mathcal{P}_2)$. It follows that Γ_0 and Γ_n satisfy the intersection condition.

According to Proposition 3.1.60 (b), to demonstrate that Γ satisfies the intersection condition it only remains to show that $\Gamma_n \cap \Gamma_0 = \langle \sigma_2, \ldots, \sigma_{n-1} \rangle$. Let $\gamma \in \Gamma_n \cap \Gamma_0$ and then we use that $\Gamma^+(\mathcal{Q})$ is a string even group to conclude that

$$\eta(\gamma) \in \langle \sigma_1', \ldots, \sigma_{n-1}' \rangle \cap \langle \sigma_2', \ldots, \sigma_n' \rangle = \langle \sigma_2', \ldots, \sigma_{n-1}' \rangle.$$

Since $\eta_1 = \eta|_{\langle \sigma_1, \ldots, \sigma_{n-1} \rangle}$ is an isomorphism, $\gamma \in \langle \sigma_2, \ldots, \sigma_{n-1} \rangle$. The other inclusion is trivial. Therefore, Γ is a string even group and there exists an $(n+1)$-polytope $\{(\mathcal{P}_1)_{(\Phi)}, (\mathcal{P}_2)_{(\Psi)}\}$ with $\Gamma^+(\{(\mathcal{P}_1)_{(\Phi)}, (\mathcal{P}_2)_{(\Psi)}\}) = \Gamma$.

The isomorphisms η_1 and η_2 guarantee that $\{(\mathcal{P}_1)_{(\Phi)}, (\mathcal{P}_2)_{(\Psi)}\}$ is a rooted rotary amalgamation of $\mathcal{P}_1$ and $\mathcal{P}_2$. Since the even subgroup of every rooted rotary amalgamation of $\mathcal{P}_1$ and $\mathcal{P}_2$ must satisfy the defining relations of Γ, $\{(\mathcal{P}_1)_{(\Phi)}, (\mathcal{P}_2)_{(\Psi)}\}$ covers every rotary rooted amalgamation of $\mathcal{P}_1$ and $\mathcal{P}_2$.

If either $\mathcal{P}_1$ or $\mathcal{P}_2$ is chiral then so is $\{(\mathcal{P}_1)_{(\Phi)}, (\mathcal{P}_2)_{(\Psi)}\}$, since its facets of vertex-figures are chiral (see the first item of Proposition 2.2.8). On the other hand, if $\mathcal{P}_1$ and $\mathcal{P}_2$ are regular we claim that all defining relations of Γ are satisifed when replacing σ_1, σ_2 and σ_3 by σ_1^{-1}, σ_2^{-1} and $\sigma_2^2\sigma_3$, respectively. Relations in R_1 and R_2 hold because of the regularity of $\mathcal{P}_1$ and $\mathcal{P}_2$ (see Theorem 3.1.53), while direct computation is enough to verify that $\sigma_1^{-1}\sigma_2^{-1}(\sigma_2^2\sigma_3)\sigma_4 \cdots \sigma_n$ maps the base flag Φ to $\Phi^{1,0,n,1}$, and so $(\sigma_1^{-1}\sigma_2^{-1}(\sigma_2^2\sigma_3)\sigma_4 \cdots \sigma_n)^2 = \mathrm{id}$. In this situation, regularity of $\{(\mathcal{P}_1)_{(\Phi)}, (\mathcal{P}_2)_{(\Psi)}\}$ follows from Theorem 3.1.53. ■

In fact, the universal polytope with facet type $\mathcal{P}_1$ and vertex-figure type $\mathcal{P}_2$ covers all rooted amalgamations of $\mathcal{P}_1$ and $\mathcal{P}_2$, not only the rotary ones.

Proposition 5.4.27 *Let $(\mathcal{P}_1)_{(\Phi)}$ and $(\mathcal{P}_2)_{(\Psi)}$ be two rooted orientable rotary n-polytopes and let $\mathcal{Q}_{(\Upsilon)}$ be a rooted amalgamation of $(\mathcal{P}_1)_{(\Phi)}$ and $(\mathcal{P}_2)_{(\Psi)}$. Then $\{(\mathcal{P}_1)_{(\Phi)}, (\mathcal{P}_2)_{(\Psi)}\}$ covers $\mathcal{Q}_{(\Upsilon)}$.*

Proof The strategy is to show that there is an epimorphism from $\mathrm{Conn}(\{(\mathcal{P}_1)_{(\Phi)}, (\mathcal{P}_2)_{(\Psi)}\})$ to $\mathrm{Conn}(\mathcal{Q})$, and use Proposition 3.2.11.

Let $r_0, \ldots, r_n$ and $\hat{r}_0, \ldots, \hat{r}_n$ denote the generators of $\mathrm{Conn}(\{(\mathcal{P}_1)_{(\Phi)}, (\mathcal{P}_2)_{(\Psi)}\})$ and $\mathrm{Conn}(\mathcal{Q})$, respectively. In order to show that the mapping

$r_i \mapsto \hat{r}_i$ induces an epimorphism it suffices to show that if $r_{j_1} \cdots r_{j_k} = \text{id}$ then $\hat{r}_{j_1} \cdots \hat{r}_{j_k} = \text{id}$. Since $\{(\mathcal{P}_1)_{(\Phi)}, (\mathcal{P}_2)_{(\Psi)}\}$ is orientable, an expression $r_{j_1} \cdots r_{j_k} = \text{id}$ implies that k is even and the expression may be written as $s_{i_1} \cdots s_{i_\ell} = \text{id}$, with s_i defined as in Proposition 3.2.19.

Let Φ be the base flag of $\{(\mathcal{P}_1)_{(\Phi)}, (\mathcal{P}_2)_{(\Psi)}\}$. From Proposition 3.2.28 we know that $s_{i_1} \cdots s_{i_\ell} = \text{id}$ if and only if $s_{i_1} \cdots s_{i_\ell}\Phi = \Phi$ and $s_{i_1} \cdots s_{i_\ell}\Phi^1 = \Phi^1$. Furthermore, Remark 3.2.22 implies that $s_{i_1} \cdots s_{i_\ell}\Phi = \Phi$ holds if and only if $\Phi\sigma_{i_1} \cdots \sigma_{i_\ell} = \Phi$, where σ_i is the ith standard rotation of $\{(\mathcal{P}_1)_{(\Phi)}, (\mathcal{P}_2)_{(\Psi)}\}$. On the other hand, we may use Proposition 3.1.65 to conclude that $s_{i_1} \cdots s_{i_\ell}\Phi^1 = \Phi^1$ holds if and only if $\Phi\sigma'_{i_1} \cdots \sigma'_{i_\ell} = \Phi$, where

$$\sigma'_i = \begin{cases} \sigma_i^{-1} & \sigma_1^{-1}\text{if } i \in \{1,2\}, \\ \sigma_2^2\sigma_3 & \text{if } i = 3, \\ \sigma_i & \text{if } i > 3. \end{cases} \tag{5.8}$$

Let $\rho_0, \ldots, \rho_n$ be the generators of the Coxeter group $[\infty, \ldots, \infty]$. Let $w_1, \ldots, w_{m-1}$ be words on the set $\{\rho_0, \ldots, \rho_n\}$ such that the relations $R_1 \cup R_2$ in (5.7) transform into the list of relations $\{w_i = \text{id}\}_{i \in \{1,\ldots,m-1\}}$ when replacing σ_i by $\rho_{i-1}\rho_i$. Let $w_m = (\rho_0\rho_n)^2$, and let $\mathcal{N}$ be the normal closure of $\{w_i\}_{i \in \{1,\ldots,m\}}$ in $[\infty, \ldots, \infty]^+$. From the presentation of $\Gamma^+(\{(\mathcal{P}_1)_{(\Phi)}, (\mathcal{P}_2)_{(\Psi)}\})$ in (5.7) we know that $\{(\mathcal{P}_1)_{(\Phi)}, (\mathcal{P}_2)_{(\Psi)}\}$ is the quotient of the universal polytope with type $\{\infty, \ldots, \infty\}$ by $\mathcal{N}$. Then $\Phi\sigma_{i_1} \cdots \sigma_{i_\ell} = \Phi$ and $\Phi\sigma'_{i_1} \cdots \sigma'_{i_\ell} = \Phi$ if and only if $\sigma_{i_1} \cdots \sigma_{i_\ell} \in \mathcal{N} \cap \rho_1\mathcal{N}\rho_1$. By again using Remark 3.2.22 we can translate this as $s_{i_1} \cdots s_{i_\ell} \in \mathcal{N} \cap \rho_1\mathcal{N}\rho_1$, when interpreting the connection r_i as the flag action of ρ_i.

If $i < m$ then $w_i \in \langle \rho_0, \ldots, \rho_{n-1} \rangle$ if w_i is equivalent to a relation in R_1, and $w_i \in \langle \rho_1, \ldots, \rho_n \rangle$ if w_i is equivalent to a relation in R_2. In other words, w_i is a connection of the facet or of the vertex-figure, and it acts trivially on each flag of the flag orbit containing the base flag. Since $\{(\mathcal{P}_1)_{(\Phi)}, (\mathcal{P}_2)_{(\Psi)}\}$ and $\mathcal{Q}$ have isomorphic facets and isomorphic vertex-figures, and the flag action of w_i fixes every flag of $\{(\mathcal{P}_1)_{(\Phi)}, (\mathcal{P}_2)_{(\Psi)}\}$ in the orbit of Φ, it also fixes every flag of $\mathcal{Q}$ in the same part of the bipartition as Υ. Since all facets of $\{(\mathcal{P}_1)_{(\Phi)}, (\mathcal{P}_2)_{(\Psi)}\}$ and of $\mathcal{Q}$ are isomorphic and both are rooted amalgamations, it is also true that every conjugate tw_it^{-1} with $t \in [\infty, \ldots, \infty]^+$ acts trivially on all flags in the same part of the bipartition as the base flag of $\{(\mathcal{P}_1)_{(\Phi)}, (\mathcal{P}_2)_{(\Psi)}\}$ and of $\mathcal{Q}$.

Similarly, if w'_i is obtained from w_i by replacing each σ_i by σ'_i as in (5.8), and $t \in [\infty, \ldots, \infty]^+$, then tw'_it^{-1} acts trivially on every flag of $\{(\mathcal{P}_1)_{(\Phi)}, (\mathcal{P}_2)_{(\Psi)}\}$ and of $\mathcal{Q}$ that is not in the same part of the base flag.

The group $\mathcal{N}$ is the subgroup of $[\infty, \ldots, \infty]$ generated by all conjugates of the elements w_i with $i < m$, since w_m is equivalent to a relation in $[\infty, \ldots, \infty]$. By rewriting every element $u \in \mathcal{N} \cap \rho_1\mathcal{N}\rho_1$ as a product of conjugates of elements in $\mathcal{N}$ or a product of conjugates of elements in $\rho_1\mathcal{N}\rho_1$, as needed, we see that u acts trivially on every flag of $\mathcal{Q}$. This shows the desired epimorphism η from $\mathrm{Conn}(\{(\mathcal{P}_1)_{(\Phi)}, (\mathcal{P}_2)_{(\Psi)}\})$ to $\mathrm{Conn}(\mathcal{Q})$.

In order to use Proposition 3.2.11 we need to show that the image under η of the stabiliser of Φ is contained in the stabiliser of Υ. But the stabiliser of Φ is precisely $\mathcal{N}$, and this subgroup stabilises all flags of $\mathcal{Q}$ in the same part of the bipartition of Υ. ■

As in the case of regular polytopes, we may ask ourselves whether there exist rooted chiral amalgamations of two given rotary polytopes $\mathcal{P}_1$ and $\mathcal{P}_2$. In case of an affirmative answer, we may also ask whether the universal polytope with facet type $\mathcal{P}_1$ and vertex-figure type $\mathcal{P}_2$ is finite.

It is natural to start the study of amalgamations of rotary polytopes with the locally toroidal ones. In this direction, much has been done in the regular case, where the determination of the finite universal polytope with facet type $\mathcal{P}_1$ and vertex-figure type $\mathcal{P}_2$ is almost complete (see [188, Chapters 10–12] and [216]). As mentioned in [188, Lemma 10A1], regular locally toroidal polytopes can have rank 4, 5 or 6.

By contrast, for $\mathcal{P}_1$ and $\mathcal{P}_2$ rotary, spherical or toroidal with at least one chiral and at least one toroidal, only a few finite universal polytopes with facet type $\mathcal{P}_1$ and vertex-figure type $\mathcal{P}_2$ are known to be finite. The known chiral locally toroidal polytopes are summarised in [29, Table 1] (see also [207, 208, 210, 248]).

If $\mathcal{Q}$ is a chiral universal locally toroidal polytope then, up to duality, its facets are chiral. By definition, the facets must be either spherical or toroidal. But there are no spherical chiral polytopes, hence we can conclude that the facets are chiral toroidal polytopes. In view of Theorem 2.4.27, the facets are in fact polyhedra, and $\mathcal{Q}$ is a 4-polytope.

Assume first that $\mathcal{Q}$ has spherical vertex-figures. If the facets of $\mathcal{Q}$ have type $\{6,3\}$ then $\mathcal{Q}$ has type $\{6,3,3\}$, $\{6,3,4\}$ or $\{6,3,5\}$. If the facets have type $\{4,4\}$ then $\mathcal{Q}$ has type $\{4,4,3\}$. Since there is no spherical polyhedron with a 6 in the first entry of its type, there are no locally toroidal polyhedra with facets with type $\{3,6\}$ and spherical vertex-figures. If $\mathcal{Q}$ has toroidal facets and toroidal (regular or chiral) vertex-figures then the type of the facet determines that of the vertex-figure. The only possibilities for the type of $\mathcal{Q}$ are $\{6,3,6\}$, $\{4,4,4\}$ and $\{3,6,3\}$.

Proposition 5.4.28 *All chiral universal locally toroidal polytopes have rank* 4. *Their types are* $\{6,3,3\}$, $\{6,3,4\}$, $\{6,3,5\}$, $\{4,4,3\}$, $\{6,3,6\}$, $\{4,4,4\}$ *and* $\{3,6,3\}$.

The following two problems are highly related to [220, Problems 15 and 16] and to [250, Problem 19].

Problem 52 For each rooted toroidal chiral polyhedron $(\mathcal{P}_1)_{(\Phi)}$ determine the rooted rotary, spherical or toroidal polyhedra $(\mathcal{P}_2)_{(\Psi)}$ such that $\langle(\mathcal{P}_1)_{(\Phi)}, (\mathcal{P}_2)_{(\Psi)}\rangle^{ch} \neq \varnothing$.

Problem 53 For each $(\mathcal{P}_1)_{(\Phi)}$ and $(\mathcal{P}_2)_{(\Psi)}$ giving a positive answer to Problem 52 determine if $\{(\mathcal{P}_1)_{(\Phi)}, (\mathcal{P}_2)_{(\Psi)}\}$ is finite.

Even in the case when $\langle(\mathcal{P}_1)_{(\Phi)}, (\mathcal{P}_2)_{(\Psi)}\rangle^{ch} \neq \varnothing$ and $\{(\mathcal{P}_1)_{(\Phi)}, (\mathcal{P}_2)_{(\Psi)}\}$ is infinite, it is interesting to study the family $\langle(\mathcal{P}_1)_{(\Phi)}, (\mathcal{P}_2)_{(\Psi)}\rangle^{ch}$. The next question tries to determine if there is a similarity with the infinite families of rotary toroidal polyhedra with a given type, where the entire family is described completely by two parameters additional to the type.

Problem 54 Let $\mathcal{P}_1$ and $\mathcal{P}_2$ be as in Problem 52 such that $\{(\mathcal{P}_1)_{(\Phi)}, (\mathcal{P}_2)_{(\Psi)}\}$ is infinite. Is there a natural finite set of parameters $a_1, \ldots, a_k$ such that every element in $\{(\mathcal{P}_1)_{(\Phi)}, (\mathcal{P}_2)_{(\Psi)}\}$ can be expressed as $\{(\mathcal{P}_1)_{(\Phi)}, (\mathcal{P}_2)_{(\Psi)}\}_{(a_1,\ldots,a_k)}$?

As mentioned in Theorem 5.4.26, if $\mathcal{P}_1$ and $\mathcal{P}_2$ are regular and $\{(\mathcal{P}_1)_{(\Phi)}, (\mathcal{P}_2)_{(\Psi)}\}$ exists then it is also regular. In this situation it is still interesting to study $\langle(\mathcal{P}_1)_{(\Phi)}, (\mathcal{P}_2)_{(\Psi)}\rangle^{ch}$.

Problem 55 Let $\mathcal{P}_1$, $\mathcal{P}_2$ be regular n-polytopes such that $\langle(\mathcal{P}_1)_{(\Phi)}, (\mathcal{P}_2)_{(\Psi)}\rangle^{ch} \neq \varnothing$. Does the partial order in $\langle(\mathcal{P}_1)_{(\Phi)}, (\mathcal{P}_2)_{(\Psi)}\rangle^{ch} \neq \varnothing$ given by $\mathcal{Q}_1 \leq \mathcal{Q}_2$ if and only if $\mathcal{Q}_2$ covers $\mathcal{Q}_1$ have finitely many or infinitely many maximum elements? In case they are finitely many, how many are there? In particular, are there regular polytopes $\mathcal{P}_1$ and $\mathcal{P}_2$ such that $\langle(\mathcal{P}_1)_{(\Phi)}, (\mathcal{P}_2)_{(\Psi)}\rangle^{ch} \neq \varnothing$ and it has a greatest element with the partial order just defined?

5.5 Chiral Polytopes of High Ranks

Since the formalisation of the idea of chiral polytopes in [247] there was interest in having examples of such objects in all ranks in which they exist. This proved to be a difficult task, to the point that for almost 20 years it was unknown if chiral polytopes existed in ranks higher than 6, and if finite chiral polytopes existed in ranks higher than 5. This was finally proven in [218].

In this section we address chiral extensions of polytopes; these consist of constructions of chiral n-polytopes by preassigning some chosen $(n-1)$-polytopes as facets. This has been the main idea to construct chiral n-polytopes for $n \geq 7$, and finite chiral n-polytopes for $n \geq 6$.

5.5.1 Extensions of Polytopes

For $n \geq 5$ it is impractical to construct rotary n-polytopes by building their posets from scratch. It is also difficult to work with their even subgroups as quotients of $[\infty, \ldots, \infty]^+$ by a normal subgroup $\mathcal{N}$, since the index of $\mathcal{N}$ must be large in order to ensure polytopality, particularly when we require that no entry of the type equals 2. As mentioned in Section 5.3, geometric methods to construct n-polytopes did not work well when chirality of the outcome was requested. A logical strategy to construct an n-polytope $\mathcal{Q}$ with specific properties is to start with some chosen $(n-1)$-polytopes as the candidates of the facet type of $\mathcal{Q}$. As we shall see, this provides a reasonable starting point when we require $\mathcal{Q}$ to be rotary.

Definition 5.5.1 An *extension* of an n-polytope $\mathcal{P}$ is an $(n+1)$-polytope whose facets are all isomorphic to $\mathcal{P}$.

Example 5.5.2 For $n \geq 3$ the $(n+1)$-simplex and the cross polytope of rank $n+1$ are extensions of the n-simplex, while the $(n+1)$-cube is an extension of the n-cube.

Example 5.5.3 The trivial extension of $\mathcal{P}$ is an extension of $\mathcal{P}$ (see Exercise 2.1.61).

When we want to construct an $(n+1)$-polytope $\mathcal{Q}$ satisfying certain properties (for instance, chirality), and we choose the isomorphism type $\mathcal{P}$ of the candidate facets of $\mathcal{Q}$, we get an initial structure from which it becomes easier to construct $\mathcal{Q}$, under the assumption that polytopes with those properties do exist. This initial structure may have different natures, as we shall explain next.

First, we may try to merely consider a given number k of copies of $\mathcal{P}$ and glue them by pairs at their facets, in so doing trying to obtain an $(n+1)$-polytope with the desired properties. On its own, this does not seem like a good procedure; but if more information is available (like the number k of facets, the kind of vertex-figures, etc.) this approach may become plausible. This was the idea in the analysis in Section 5.4.2 to determine the unique amalgamation of $\{4,4\}_{(2,1)}$ with itself.

We may also try to start with the poset given by $\mathcal{P}$, and complete it to a poset of rank $(n+1)$, corresponding to $\mathcal{Q}$. This was the approach used to construct the structure $2^{\mathcal{P}}$ in [75]. This construction was formulated for incidence complexes, which are more general than abstract polytopes. When restricting it to polytopes it takes as the input an n-polytope $\mathcal{P}$ and produces an $(n+1)$-polytope called $2^{\mathcal{P}}$, where all vertex-figures are isomorphic to $\mathcal{P}$. The 2-faces of $2^{\mathcal{P}}$ are squares. The polytope $2^{\mathcal{P}}$ is exhibited as a partially ordered set with 2^k vertices, where k is the number of vertices of $\mathcal{P}$. The faces of rank 1 and higher are expressed as specific sets of vertices, and the incidence of faces is given by inclusion. This construction is not an extension in the sense of Definition 5.5.1. A dual construction produces the $(n+1)$-polytope $(2^{\mathcal{P}^\delta})^\delta$, which does have facets isomorphic to $\mathcal{P}$.

The construction in [75] required that the polytope $\mathcal{P}$ satisfies the property that every face F is completely determined by the elements less than F. This is true for most of the polytopes that we can think of, but flat polytopes, for instance, do not satisfy that property. This requirement was dropped in [82, Section 6], where the construction of $(2^{\mathcal{P}^\delta})^\delta$ was performed in terms of maniplexes (edge-coloured graphs). The input now is any maniplex $\mathcal{P}$, although the 2-faces may fail to be squares if a ridge of $\mathcal{P}$ belongs to only one facet. The construction requires 2^m copies of the flag graph of $\mathcal{P}$, where m is the number of facets of $\mathcal{P}$, and then the edges coloured n are added. As was illustrated by this version of the construction, the set of flag graphs of the initial polytope $\mathcal{P}$ constitutes a third possible initial structure to build extensions of $\mathcal{P}$.

When considering rotary polytopes, as we shall do here, it is also natural to search for string even groups $\langle \sigma_1, \ldots, \sigma_n \rangle$ that extend the set of standard rotations in $\Gamma^+(\mathcal{P}) = \langle \sigma_1, \ldots, \sigma_{n-1} \rangle$ in one generator. In this setting, the initial structure to construct $\mathcal{Q}$ is $\Gamma^+(\mathcal{P})$. This is the approach we shall follow here for rotary extensions of polytopes.

Several works have addressed regular extensions of regular polytopes (see [188, Sections 4D, 8C], [217, 219, 239–243]). In what follows we list some of the results in these publications. We start with a universal extension in the spirit of the universal amalgamation in Theorem 5.4.26 (see [188, Theorem 4D4], [239, Theorem 2]).

Theorem 5.5.4 *Let $\mathcal{P}$ be a regular polytope with type $\{p_1, \ldots, p_{n-1}\}$. Then there exists a regular extension $\mathcal{Q}$ of $\mathcal{P}$ with type $\{p_1, \ldots, p_{n-1}, \infty\}$ that covers every regular extension of $\mathcal{P}$.*

If $\mathcal{Q}$ is a regular extension of a regular n-polytope $\mathcal{P}$ then the first $n-1$ entries of the type of $\mathcal{Q}$ are those in the type of $\mathcal{P}$. The last entry in the extension in Theorem 5.5.4 is ∞, but there are extensions for which the last entry of the type are finite (see [217, Theorem 5.1], [241, Theorem 2]).

Proposition 5.5.5 *For every even integer $p_n \geq 4$ and every regular polytope $\mathcal{P}$ with type $\{p_1, \ldots, p_{n-1}\}$, there exists a regular extension of $\mathcal{P}$ with type $\{p_1, \ldots, p_n\}$.*

It is not possible to relax the requirement of p_n being even in Proposition 5.5.5, as shown in [115, Theorem 3.6].

Proposition 5.5.6 *For $n \geq 3$ there is no extension (regular or non-regular) of the hemi-cube of rank n with type $\{4, 3^{n-2}, p\}$ for any odd integer p.*

For a given polytope $\mathcal{P}$ it seems a difficult question to determine if for every odd $m \geq 3$ there is a regular extension $\mathcal{Q}_m$ of $\mathcal{P}$ with last entry m in its type.

Problem 56 Determine for which regular polytopes $\mathcal{P}$ it is true that for every odd integer m there is a regular extension of $\mathcal{P}$ whose last entry of its type is m.

5.5.2 Chiral Extensions of Chiral Polytopes

In this and the next section we address constructions of chiral $(n+1)$-polytopes $\mathcal{Q}$ as extensions of rotary n-polytopes $\mathcal{P}$. We start with the case when $\mathcal{P}$ is itself chiral, and leave for the next section the case when $\mathcal{P}$ is regular.

Given a rotary n-polytope $\mathcal{P}$ and $\mathcal{X}$ a generating set of $\Gamma^+(\mathcal{P})$, in order to find a rotary extension of $\mathcal{P}$ it suffices to find a group Γ containing $\Gamma^+(\mathcal{P})$, generated by $\mathcal{X} \cup \{\gamma\}$ for some $\gamma \in \Gamma \setminus \Gamma^+(\mathcal{P})$, such that it satisfies some relations equivalent to (3.7), as well as some intersection condition equivalent to (3.10). There are several natural choices of generating set $\mathcal{X}$ and of γ.

The standard rotations of $\mathcal{P}$ constitute an obvious candidate for $\mathcal{X}$, together with some element γ playing the role of the standard rotation σ_n. In that case the extra relations needed for Γ are $(\sigma_i \cdots \sigma_n)^2 = \text{id}$ for $i \in \{1, \ldots, n-1\}$. This has not been the most popular choice. In general, it has been preferred to choose γ as an involution, and the natural candidates are the automorphisms $\tau_{i,n}$, defined in (3.12).

For example, in [37] and [230, Section 5] permutation groups were used to extend $\Gamma^+(\mathcal{P})$, expressed as a permutation group, to a larger permutation group Γ. The generating set chosen for $\Gamma^+(\mathcal{P})$ in [230] was still $\mathcal{X} = \{\sigma_1, \sigma_2\}$ if $n = 3$, but the extra generator was chosen to be $\gamma = \tau_{1,3}$. In this situation, the verification of (3.7) required only to check that $(\sigma_1\tau_{1,3})^2 = \text{id}$. For $n = 4$ the generating set was $\mathcal{X} = \{\tau_{0,2}, \sigma_2, \sigma_3\}$ and the extra generator was $\gamma = \tau_{2,4}$. In order to verify (3.7) it suffices to check that $(\tau_{0,2}\tau_{2,4})^2 = (\sigma_2\tau_{2,4})^2 = \text{id}$. By contrast, in [37] the rank of $\mathcal{P}$ was $n = 3$, the generating set was $\mathcal{X} = \{\sigma_1 = \tau_{0,1}, \tau_{0,2}\}$, and the extra generator was $\tau_{0,3}$. Here (3.7) is satisfied as long as $(\sigma_1^{-1}\tau_{0,3})^2 = \text{id}$. These two ideas have natural generalisations to higher ranks.

Exercise 5.5.7 For each generating set $\mathcal{X} \cup \{\gamma\}$ listed below determine a set of relations equivalent to (3.7).

- $\mathcal{X} = \{\sigma_{n-1}\} \cup \{\sigma_{n-k}, \tau_{n-k-2,n-k} : k \geq 2 \text{ even}\}$, $\gamma = \tau_{n-2,n}$.
- $\mathcal{X} = \{\tau_{0,k} : k \in \{1, \ldots, n-1\}\}$, $\gamma = \tau_{0,n}$.

In order to prove that Γ satisfies the intersection condition it has been customary to express the standard rotations in terms of the elements in $\mathcal{X} \cup \{\gamma\}$ and use Proposition 3.1.60.

If we want the extension of $\mathcal{P}$ to be rotary then it follows from Proposition 2.3.8 that the facets of $\mathcal{P}$ must be regular.

Proposition 5.5.8 *If $\mathcal{P}$ is a chiral polytope with chiral facets then it has no rotary extensions.*

Proposition 5.5.8 makes it impossible to obtain chiral n-polytopes through succesive extensions of chiral $(n-k)$-polytopes for $k \geq 2$. If $\mathcal{P}$ is chiral and has an extension $\mathcal{Q}$ then $\mathcal{P}$ must have regular facets; in this situation $\mathcal{Q}$ has chiral facets isomorphic to $\mathcal{P}$, and therefore $\mathcal{Q}$ has no chiral extensions.

The first result regarding chiral extensions of chiral polytopes is the following (see [249, Theorems 1 and 2]).

Theorem 5.5.9 *Let $\mathcal{P}$ be a chiral polytope with regular facets. Then there is an extension $\mathcal{Q}$ of $\mathcal{P}$ that covers all chiral extensions of $\mathcal{P}$. If the type of $\mathcal{P}$ is $\{p_1, \ldots, p_{n-1}\}$ then the type of $\mathcal{Q}$ is $\{p_1, \ldots, p_{n-1}, \infty\}$.*

The proof of Theorem 5.5.9 can be done by showing that the group

$$\Gamma = \langle \sigma_1, \ldots, \sigma_n : R, (\sigma_i \cdots \sigma_n)^2 = \mathrm{id}, i \in \{1, \ldots, n-1\} \rangle$$

is a string even group, where R is a set of relations such that $\langle \sigma_1, \ldots, \sigma_{n-1} : R \rangle$ is a presentation for $\Gamma(\mathcal{P})$. It is clear that if $\mathcal{Q}$ is a chiral extension of $\mathcal{P}$ then $\Gamma(\mathcal{Q})$ is a quotient of Γ, and it can be shown that $\sigma_n \neq \sigma_j$ for $j < n$ through the epimorphism from Γ to $\mathbb{Z}_2$ given by $\sigma_i \mapsto 0$ if $i < n$, $\sigma_n \mapsto 1$. In order to show that the subgroup $\langle \sigma_1, \ldots, \sigma_{n-1} \rangle$ of Γ is isomorphic to $\Gamma(\mathcal{P})$, that the intersection condition holds in Γ, and that the last entry of the type is ∞, in [249] a different presentation of Γ was given so that it is a free product of $\Gamma(\mathcal{P})$ and the full automorphism group of a facet $\mathcal{F}$ of $\mathcal{P}$, with amalgamation of $\Gamma^+(\mathcal{F})$. We shall not reproduce the details here.

Later it was shown that every finite chiral polytope with regular facets has a finite chiral extension (see [70, Theorem 1]).

Theorem 5.5.10 *Every finite chiral n-polytope with regular facets has a finite chiral extension.*

Theorem 5.5.10 was proved through Schreier diagrams of the automorphism groups of chiral polytopes. We shall give the main ideas of this and other results on extensions in Section 5.5.3.

In comparison to the known constructions of regular extensions of regular polytopes, in the construction in [70] we have no control on the last entry of the types of the extensions. Two improvements in that direction were performed in [205], but in both cases there are some extra requirements on the starting chiral polytope $\mathcal{P}$.

We say that a polytope $\mathcal{P}$ is *dually bipartite* if the 1-skeleton of $\mathcal{P}^{\delta}$ is bipartite, or equivalently, if it is possible to colour the facets of $\mathcal{P}$ black and white with the property that two facets that share a ridge have distinct colours. For example, every map of type $\{3, 6\}$ on the torus is dually bipartite, and so are the maps $\{4, 4\}_{(a,b)}$ with $a \equiv b$ (modulo 2). The octahedron is the only dually bipartite Platonic solid.

The bipartition of the facets of a dually bipartite polytope was used as an extra tool in [219] to construct regular extensions of dually bipartite polytopes. In [205, Theorem 3.1] it was used to construct infinitely many chiral extensions of dually bipartite chiral polytopes with regular facets.

Theorem 5.5.11 *Let $\mathcal{P}$ be a dually bipartite chiral polytope with regular facets and type $\{p_1, \ldots, p_{n-1}\}$. Then, for every even integer $s \geq 1$ there exists a chiral extension of $\mathcal{P}$ with type $\{p_1, \ldots, p_n\}$ for some positive integer p_n such that $2s$ divides p_n.*

The proof of Theorem 5.5.11 also relies on Schreier diagrams of the automorphism groups of chiral polytopes. Theorem 5.5.11 has the following corollary.

Corollary 5.5.12 *Given a chiral dually bipartite polytope $\mathcal{P}$ with regular facets and type $\{p_1, \ldots, p_{n-1}\}$ there are infinitely many integers p_n such that there exists a chiral extension of $\mathcal{P}$ with type $\{p_1, \ldots, p_n\}$.*

The next result was proven by using a specific extension of regular maniplexes together with the mix of rotary maniplexes defined in Section 4.4 (see [205, Theorem 4.3]). This construction does not produce a chiral extension of a given chiral polytope $\mathcal{P}$ on its own, but requires an extension as part of its hypotheses.

Theorem 5.5.13 *Let $\mathcal{P}$ be a chiral n-polytope such that it has as a quotient a regular n-maniplex with at least two facets. Assume also that $\mathcal{P}$ has a chiral* *extension of type* $\{p_1, \ldots, p_n\}$. *Then for every integer* $s \geq 1$ *there is a chiral extension of* $\mathcal{P}$ *with type* $\{p_1, \ldots, p_{n-1}, lcm\{p_n, 2s\}\}$.

The constructions in [70] and [205, Sections 3 and 4] do not imply that there exists an integer p_n with $3 \leq p_n < \infty$ such that every chiral polytope with regular facets has a chiral extension where p_n is the last entry of the type (compare with Proposition 5.5.5).

Problem 57 Is it true that there exists an integer p_n with $3 \leq p_n < \infty$ such that every chiral polytope with regular facets has a chiral extension where p_n is the last entry of the type?

Problem 58 Is it true that there exists an integer p_n with $3 \leq p_n < \infty$ such that every finite chiral polytope with regular facets has a finite chiral extension where p_n is the last entry of the type?

Problem 59 Are there chiral polytopes $\mathcal{P}$ such that, for every $p_n \geq 3$, $\mathcal{P}$ has a chiral extension whose last entry of its type is p_n? If the answer is affirmative, find necessary and sufficient conditions for such polytopes $\mathcal{P}$.

5.5.3 Chiral Extensions of Regular Polytopes

In order to construct chiral polytopes of high ranks it is desirable to be able to find chiral extensions of regular polytopes, since we do have many examples of regular polytopes of each rank.

As mentioned in Section 5.5.1, every regular polytope has many regular extensions. This result does not have an exact analogue for chiral extensions of regular polytopes. Indeed, Proposition 5.2.31 tells us that finite $(1, n-2)$-flat rotary $(n+1)$-polytopes are regular. It follows that if $\mathcal{P}$ is a finite $(1, n-2)$-flat regular n-polytope then every rotary extension of $\mathcal{P}$ is regular. Examples of such polytopes $\mathcal{P}$ can be found in [35]. They can also be constructed from flat regular $(n-2)$-polytopes by extending them once, taking the dual of the outcome, and extending this $(n+1)$-polytope by using any of the available constructions for finite regular extensions of regular polytopes.

In what follows we illustrate the main idea of building a chiral extension (when possible) through Schreier diagrams (compare with the discussion about the generating set $\mathcal{X}$ of $\Gamma^+(\mathcal{P})$ and extra generator γ in Section 5.5.2).

Let $\mathcal{P} = \{3,6\}_{(3,0)}$ and let $\mathcal{Y}$ be the orbit of grey flags in Figure 5.18. Recall that $\langle s_1, s_2 \rangle = \text{Conn}^+(\{3,6\}_{(3,0)})$ acts regularly on $\mathcal{Y}$. We define the involutory permutation ω on $\mathcal{Y}$ as the one swapping every pair of flags in Figure 5.18 that are connected by a vertical dark grey segment, while fixing the remaining flags. Note that one third of the flags of $\mathcal{Y}$ are fixed by ω; half of those flags

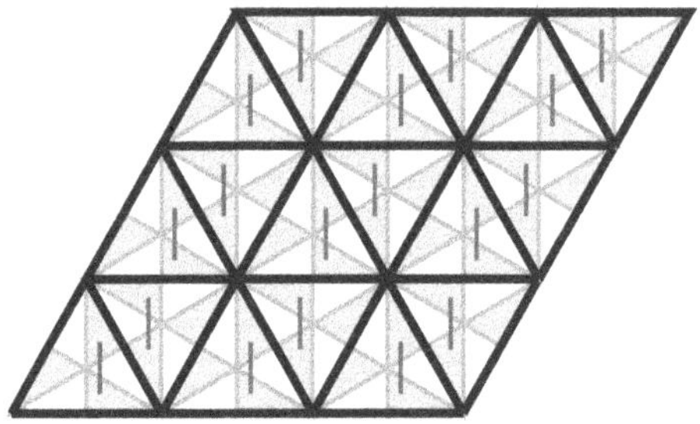

Figure 5.18 A permutation on the grey flags of $\{3,6\}_{(3,0)}$.

are translates of the flag containing the lower left corner, while the other half are translates of the flag containing the upper right corner.

A first idea is to consider the permutation group $\Gamma_\omega := \langle s_1, s_2, \omega\rangle$ on $\mathcal{Y}$ as a candidate of the automorphism group of a rotary extension $\mathcal{Q}_\omega$ of $\mathcal{P}$. In so doing, we interpret s_1 and s_2 as the first two standard rotations σ_1 and σ_2 of $\mathcal{Q}_\omega$, while ω plays the role of $\tau_{1,3}$. Then (3.7) holds since $(s_1s_2)^2 = \omega^2 = \mathrm{id}$, while $(s_1\omega)^2 = \mathrm{id}$ can be verified directly (recall that we apply connections on the left). The intersection condition also holds and it is left as an exercise.

Exercise 5.5.14 Let $\mathcal{Y}$ and Γ_ω be as previously.

1. Describe $\sigma_3 = s_2^{-1}\omega$ as a permutation of $\mathcal{Y}$; use that $\langle s_1, s_2\rangle$ acts regularly on $\mathcal{Y}$ to show that $\langle\sigma_1,\sigma_2\rangle \cap \langle\sigma_3\rangle = \{\mathrm{id}\}$.
2. Determine the orbits of $\langle\sigma_2,\sigma_3\rangle$; show that $\langle\sigma_1,\sigma_2\rangle \cap \langle\sigma_2,\sigma_3\rangle = \{\sigma_2\}$ by observing that if $\gamma \in \langle\sigma_1,\sigma_2\rangle \setminus \langle\sigma_2\rangle$ then γ does not preserve the orbits of $\langle\sigma_2,\sigma_3\rangle$.
3. Conclude that Γ_ω is a string even group with respect to the generating set $\{\sigma_1,\sigma_2,\sigma_3\}$.

The 4-polytope $\mathcal{Q}_\omega$ constructed earlier is regular. Perhaps the permutation ω is invariant under too much symmetry, and in some way this provides combinatorial reflections to the resulting combinatorial structure.

Exercise 5.5.15 Find an involutory permutation ξ in the set $\mathcal{Y}$ defined previously such that $(\sigma_1\xi)^2 = (\sigma_2\xi)^2 = (\omega\xi)^2 = \mathrm{id}$. Use Theorem 3.1.53 to conclude that the 4-polytope $\mathcal{Q}_\omega$ defined earlier is regular.

Remark 5.5.16 The group automorphism η in Theorem 3.1.53 took the form of a permutation on $\mathcal{Y}$ in Exercise 5.5.15. In general this needs not be the case.

So far we only looked at one of many possible permutations to play the role of $\tau_{1,3}$ of an extension of $\mathcal{P}$. That choice was easy to describe, but it seems like 'being easy to describe' makes it more likely to yield a regular outcome. We shall make only one more choice of such permutation to construct a chiral 4-polytope, but it involves a permutation group of a higher degree.

Exercise 5.5.17 Determine if there is a permutation $\hat{\omega}$ of $\mathcal{Y}$ such that $\langle s_1, s_2, \hat{\omega}\rangle$ is the automorphism group of a chiral polytope with standard rotations s_1, s_2 and $s_2^{-1}\hat{\omega}$.

We still let $\mathcal{P}$ be the regular toroidal polyhedron $\{3,6\}_{(3,0)}$, and now we consider the set $\mathcal{Z} = \mathcal{Y} \cup \mathcal{Y}'$, where $\mathcal{Y} = \{y_1, \ldots, y_{54}\}$ and $\mathcal{Y}' = \{y'_1, \ldots, y'_{54}\}$ are two disjoint copies of the grey flags in Figure 5.18. We keep the same notation σ_1 and σ_2 for s_1 and s_2, acting on each copy (so that $y_i\sigma_k := s_k y_i = y_j$ if and only if $y'_i\sigma_k = y'_j$, for $k \in \{1,2\}$). From Corollary 4.4.11 we know that $\langle \sigma_1, \sigma_2\rangle \cong \Gamma^+(\mathcal{P})$. We define an involutory permutation ν of $\mathcal{Z}$ given by the product of the transpositions represented by the grey lines in Figure 5.19 together with the transpositions of flags with the same label. The permutation ν can be understood as acting in the same way as ω in each copy of $\mathcal{Y}$, with the exception of two triangles T_1 and T_2 in each copy (the ones with flags labelled a, b, c, d, e and f in Figure 5.19). On each flag y_i of these triangles, $\nu y_i = (\omega y_i)'$ and $\nu y'_i = \omega y_i$. We define $\Gamma_\nu := \langle s_1, s_2, \nu\rangle$, a group acting transitively on $\mathcal{Z}$.

The permutations ν of $\mathcal{Z}$ and ω of $\mathcal{Y}$ act similarly in the sense that if $\eta\colon \mathcal{Z} \to \mathcal{Y}$ is the natural projection ($y_i \mapsto y_i$ and $y'_i \mapsto y_i$) then $\eta(\nu x) = \omega(\eta(x))$. Furthermore, abusing notation we also have that $\eta(x\sigma_i) = (\eta(x))\sigma_i$

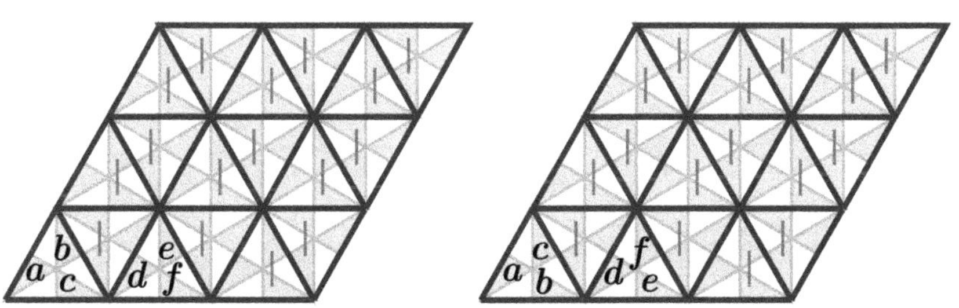

Figure 5.19 A permutation on the grey flags of two copies of $\{3,6\}_{(3,0)}$.

for $i \in \{1,2\}$. It follows that the permutations of $\mathcal{Z}$ given by the elements in Γ_ν are the same as those in Γ_ω, except for the addition or removal of primes to some elements of $\mathcal{Z}$. As a consequence, the sets $\{y_i, y_i'\}$ are blocks of imprimitivity of Γ_ν. With this in mind it is possible to prove that Γ_ν is a string even group in a similar way to the one we suggested for Γ_ω.

Exercise 5.5.18 Use the ideas in Exercise 5.5.14 to conclude that the group Γ_ν defined above is a string even group with respect to s_1, s_2 and $s_2^{-1}\nu$.

From Exercise 5.5.18 we know that there is a rotary 4-polytope $\mathcal{Q}_\nu$ with $\Gamma^+(\mathcal{Q}_\nu) = \Gamma_\nu$. Next we show that $\mathcal{Q}_\nu$ is chiral.

Let $\mu_\omega := s_2^2 s_1 \omega \in \Gamma_\omega$. Direct inspection shows that the local action of μ_ω on the flags of a map with type $\{3,6\}$ is that illustrated in Figure 5.20. Note that μ_ω acts like a one-step 'horizontal' translation in one third of the flags, while in the remaining two thirds of the flags it is $(\mu_\omega)^2$ that acts like a one-step 'horizontal' translation. It follows that the orbits of $\langle \mu_\omega \rangle$ in $\mathcal{Y}$ have sizes 3 and 6.

The action of μ_ω can be used to understand the action of $\mu_\nu := s_2^2 s_1 \nu$ on $\mathcal{Z}$. If $\mu_\omega y_i = y_j$ then $\mu_\nu y_i = y_j$ and $\mu_\nu y_i' = y_j'$, unless y_i is in one of the two triangles T_1 and T_2 where ν involves the change of copy of $\mathcal{Y}$, in which case $\mu_\nu y_i = y_j'$ and $\mu_\nu y_i' = y_j$. In order to compute the sizes of the orbits of $\langle \mu_\nu \rangle$ it suffices to consider the following observation. If an even number of flags in $\langle \mu_\omega \rangle y_i$ are in the two triangles T_1 and T_2 then $|\langle \mu_\nu \rangle y_i| = |\langle \mu_\omega \rangle y_i|$; on the other hand, if an odd number of flags in $\langle \mu_\omega \rangle y_i$ are in T_1 and T_2 then $|\langle \mu_\nu \rangle y_i| = 2|\langle \mu_\omega \rangle y_i|$. Direct inspection shows that every orbit of $\langle \mu_\omega \rangle$ contains an even number of flags in T_1 and T_2. It follows that the orbits of $\langle \mu_\nu \rangle$ also have sizes 3 and 6.

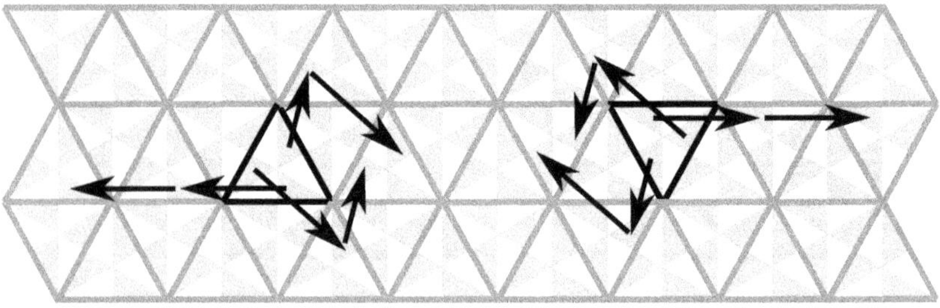

Figure 5.20 The action of μ_ω on some flags.

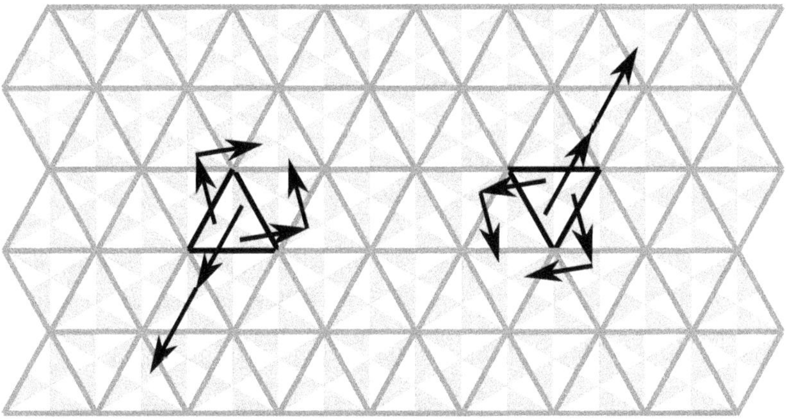

Figure 5.21 The action of $\mu_\nu\eta$ on some flags.

If $\mathcal{Q}_\nu$ were regular then there would exist a group automorphism $\eta\colon \Gamma_\nu \to \Gamma_\nu$ mapping σ_1, σ_2 and ν to $\sigma_1^{-1}, \sigma_2^{-1}$ and ν, respectively (see Theorem 3.1.53 with $i = 1$). Then $\eta(\mu_\nu) = s_2^{-1}s_1^{-1}\nu$. The local action of $s_2^{-1}s_1^{-1}\omega$ and of $\eta(\mu_\nu)$ on a map with type $\{3,6\}$ is shown in Figure 5.21. The orbits of $\langle s_2^{-1}s_1^{-1}\omega\rangle$ still have sizes 3 and 6. In order to compute the sizes of the orbits of $\langle\eta(\mu_\nu)\rangle$ the same considerations as for the orbits of $\langle\mu_\nu\rangle$ apply. Now it becomes apparent that some orbits of $\langle s_2^{-1}s_1^{-1}\omega\rangle$ with size 6 contain precisely one flag in a triangle T_1 or T_2, and therefore the corresponding orbits of $\langle\eta(\mu_\nu)\rangle$ have size 12. This forces $(\mu_\nu)^6 = \mathrm{id}$ and $(\eta(\mu_\nu))^6 \neq \mathrm{id}$, contradicting the regularity of $\mathcal{Q}_\nu$.

The argument to show that $\mathcal{Q}_\nu$ is chiral has been one of the most popular. For instance, this is the idea used in [54]. If $\mathcal{P}$ is a regular twisted honeycomb with $\Gamma^+(\mathcal{P}) = \langle\sigma_1,\sigma_2,\sigma_3\rangle$, then the group automorphism η in Theorem 3.1.53 for $i = 3$ maps σ_1, σ_2 and σ_3 to $\sigma_1, \sigma_2\sigma_3^2$ and σ_3^{-1}, respectively. Therefore $\sigma_1\sigma_3$ and $\sigma_1\sigma_3^{-1}$ must have the same order. When the orders of these two elements differ then the twisted honeycomb must be chiral. In the discussion above for the polytope $\mathcal{Q}_\nu$ we assumed the existence of the group automorphism η in Theorem 3.1.53 for $i = 1$, and found that it would map an element of order 6 to one of order 12.

The choices of ω and ν make it easy to understand the groups $\Gamma^+(\mathcal{Q}_\omega)$ and $\Gamma^+(\mathcal{Q}_\nu)$. We have already explained that the sets $\{y_i, y_i'\}$ constitute blocks of imprimitivity for $\Gamma^+(\mathcal{Q}_\nu)$. For both groups we can also consider the flags in triangles pointing up and those in triangles pointing down as a pair of blocks of imprimitivity. Finally, $\mathcal{Y}$ (resp. $\mathcal{Z}$) admits a partition into six blocks of imprimitivity, each consisting of an orbit under translation (resp. translation and adding/removing a prime). These partitions allow us to understand $\Gamma^+(\mathcal{Q}_\omega)$ and $\Gamma^+(\mathcal{Q}_\nu)$ as subgroups of certain wreath products, rather than subgroups of a (much larger) symmetric group.

Exercise 5.5.19 Describe the groups $\Gamma^+(\mathcal{Q}_\omega)$ and $\Gamma^+(\mathcal{Q}_\nu)$ as subgroups of wreath products. What are the orders of $\Gamma^+(\mathcal{Q}_\omega)$ and $\Gamma^+(\mathcal{Q}_\nu)$?

When trying the aforementioned technique to find chiral extensions of regular n-polytopes $\mathcal{P}$ for $n \geq 4$ we find the following comparison with the case $n = 3$.

1. Expressing $\Gamma^+(\mathcal{P})$ as an intransitive permutation group on a set $\mathcal{Y}$ and finding another permutation ω so that $\langle \Gamma^+(\mathcal{P}), \omega \rangle$ acts transitively on $\mathcal{Y}$ and satisfies (3.7) (for some suitably defined σ_n in terms of $\Gamma^+(\mathcal{P})$ and ω) can be achieved in different ways with relative ease (see Exercise 5.5.7).
2. Showing that some of these constructions satisfy the intersection property is not hard if $\Gamma^+(\mathcal{P})$ acts freely on $\mathcal{Y}$ (as was the case in the constructions of $\mathcal{Q}_\omega$ and $\mathcal{Q}_\nu$ above); in this way the intersection property can be shown by analysing certain orbits under the relevant subgroups of $\langle \Gamma^+(\mathcal{P}), \omega \rangle$ (see [218, Theorem 3.7]). For actions that are not free the difficulty increases.
3. The main difficulty to construct chiral extensions of $\mathcal{P}$ is to force the extensions to be chiral. We already know that not all regular polytopes have chiral extensions (see Proposition 5.2.31), and so $\mathcal{P}$ must satisfy some property additional to regularity. The main technique used so far to extend $\mathcal{P}$ chirally has been to construct $\mathcal{Y}$ and ω so that there is an element $\mu = \sigma_{i_1} \cdots \sigma_{i_k}$ in $\langle \Gamma^+(\mathcal{P}), \omega \rangle$ whose order is greater than the order of $\eta(\sigma_{i_1}) \cdots \eta(\sigma_{i_k})$, for some function η on $\{\sigma_1, \ldots, \sigma_n\}$ defined as those in Theorem 3.1.53. When the rank increases, the number of factors σ_i in such an element μ must necessarily increase as well (see Exercise 5.5.20), and it is unlikely that there is a product $\sigma_{i_1} \cdots \sigma_{i_k}$ that works for all regular n-polytopes that admit chiral extensions. So far, this step has been performed in a case-by-case manner. We still do not know a useful characterisation of those regular n-polytopes $\mathcal{P}$ for which such a construction yields a chiral extension.
4. Since we do not have a recipe for ω, we do not have control over the size and structure of the candidate automorphism group $\langle \Gamma(\mathcal{P}), \omega \rangle$ of the extension.

Exercise 5.5.20 Let $\mathcal{Q}$ be a chiral $(n+1)$-polytope with regular facets and with standard rotations $\sigma_1, \ldots, \sigma_n$. For $I \subseteq \{0, \ldots, n\}$ let Γ_I be as in Lemma 3.1.39. Assume that for some $\sigma_{i_1}, \ldots, \sigma_{i_k}$ and some function η on $\{\sigma_1, \ldots, \sigma_n\}$ defined as those in Theorem 3.1.53, the orders of $\sigma_{i_1} \cdots \sigma_{i_k}$ and $\eta(\sigma_{i_1}) \cdots \eta(\sigma_{i_k})$ are distinct, and let $I \subseteq \{0, \ldots, n\}$ be such that $\omega \in \Gamma_I$ but $\omega \notin \Gamma_J$ for every set J containing I properly. Show that I is either $\{0\}$ or $\emptyset$.

In what follows we explain the general idea of how the third item in our list was handled to construct chiral polytopes of arbitrarily large rank.

Let $\mathcal{Y}$ consist of $2m$ disjoint copies $C_1, \ldots, C_{2m}$ of the set of white flags of some regular n-polytope $\mathcal{P}$, and define $\sigma_1, \ldots, \sigma_{n-1}$ to act on them like $s_1, \ldots, s_{n-1}$, respectively, on each copy. The goal is to add a suitable involutory permutation ν intending to play the role of $\tau_{n-2,n}$. Each transposition in the cyclic structure of ν is required to permute either two elements of the same copy C_i, or two elements of consecutive copies C_i and C_{i+1} (modulo $2m$). The way in which ν connects the C_i is roughly sketched in Figure 5.22.

The key idea is to define the transpositions of ν so that there exists $\mu = \sigma_{i_1} \cdots \sigma_{i_k} \in \langle \sigma_1, \ldots, \sigma_{n-1}, \nu \rangle$ with the property that every orbit of $\langle \mu \rangle$ is contained in two consecutive copies C_i and C_{i+1}, with the exception of some orbits that contain precisely one element in each C_j for j even, and some orbits that contain precisely one element in each C_j for j odd. This should be done while forcing every orbit of $\eta(\sigma_{i_1}) \cdots \eta(\sigma_{i_k})$ to be contained in two consecutive copies C_i and C_{i+1}, for some function η like those given in Theorem 3.1.53. Under these conditions the order of μ is a multiple of m, while the order of

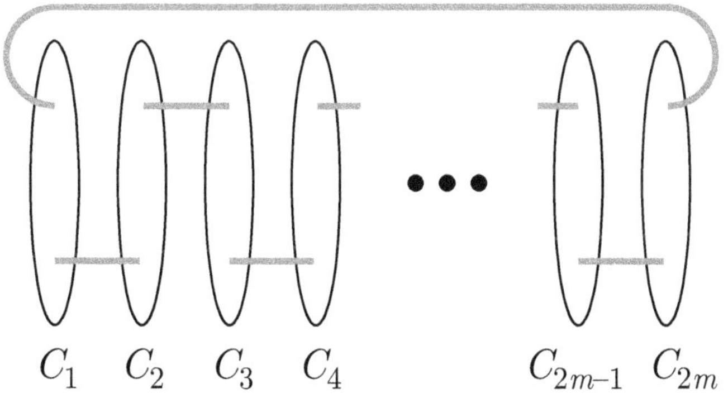

Figure 5.22 Sketch of the action of the involutory permutation ν.

$\eta(\sigma_{i_1}) \cdots \eta(\sigma_{i_k})$ has as an upper bound the factorial of twice the size of C_1. It follows that if the number $2m$ of copies is larger than that factorial, the resulting structure is forced to be chiral. An obvious drawback of this idea is that the degree of the permutation group is huge, and so it is expected that the resulting chiral $(n+1)$-polytope has a very large automorphism group.

In order for such permutations ν and μ to exist, the polytope $\mathcal{P}$ was required in [218] to be the smallest regular cover of a so-called *scattered* chiral polytope $\mathcal{R}$; this means that the automorphism group of $\mathcal{R}$ satisfies certain technical group-theoretical properties. These properties are not satisfied if $\mathcal{R}$ has few facets with respect to their rank, and so one first intuition is that scattered chiral polytopes have many facets. The result of the construction on the smallest regular cover of any scattered chiral n-polytope is a scattered chiral $(n+1)$-polytope with regular facets, and so the construction could be applied again. This was the main idea in the first appearance of the next theorem (see [218, Theorem 6.5]).

Theorem 5.5.21 *For each $n \geq 3$ there exist finite chiral n-polytopes.*

The way to construct chiral polytopes suggested in [218] is highly impractical given the size of the automorphism groups of the resulting chiral $(n+1)$-polytopes. From a given scattered chiral n-polytope $\mathcal{R}$ we must construct the regular n-polytope $\mathcal{P}$ as the smallest regular cover of $\mathcal{R}$, and from there we must perform the construction suggested above. Each of these two steps is expected to increase the number of flags substantially. When using computer software to follow this procedure starting from a scattered chiral polyhedron, it was possible to study one of the resulting chiral 4-polytopes, but the chiral 5-polytopes arising from the latter were already out of reach.

A major improvement was given in [204] where similar techniques were used to show that there exist chiral extensions of many toroidal regular polytopes (see [204, Theorem 3.3.1]).

Theorem 5.5.22 *For each odd $n \geq 3$, all but finitely many regular toroidal n-polytopes $\{4, 3^{n-3}, 4\}_{(a,0^{n-2})}$ have chiral extensions.*

As a consequence of Theorem 5.5.22 we have an infinite family of well-known regular polytopes of high ranks that admit chiral extensions. Still, there is a large gap to fill in order to characterise those regular polytopes that have chiral extensions.

Problem 60 Provide necessary and sufficient conditions for a regular polytope $\mathcal{P}$ to have chiral extensions.

The construction in [204] also has the following advantage over that in [218]. The number of flags of the chiral $(n+1)$-polytope resulting from the construction is necessarily very large, but it is bounded above by a function on the number N of flags of the starting regular polytope. The starting regular n-polytope in [204] is well understood, and in particular we know its number N of flags, whereas with the procedure suggested in [218] the number N is out of reach even for $n = 5$.

Still, further generalisations to the construction in [204] may be explored. The requirement on the parity of n in Theorem 5.5.22 is due to the fact that the central inversion in $\mathbb{E}^n$ is a product of an even number of hyperplane reflections if and only if n is even, and this expression of the central inversion was required in the arguments. Perhaps some modifications in the chosen Schreier diagrams yield extensions of regular toroidal n-polytopes for all even $n \geq 4$.

The flags of the toroidal polytopes $\{4, 3^{n-3}, 4\}_{(a,0^{n-2})}$ admit a useful labelling that was used in the proofs of the results in [204]. It is sensible to look for other families of regular polytopes that admit similar labellings of flags, and try to reach the conclusion that all but finitely many polytopes in those families admit chiral extensions.

In what follows we outline candidate alternative ways to search for chiral extensions of regular polytopes.

When using Schreier diagrams starting with a permutation presentation for the even subgroup $\Gamma^+(\mathcal{P})$ of a regular n-polytope $\mathcal{P}$, we may try to choose the new permutation $\tau_{j,n}$ so that $\langle \Gamma^+(\mathcal{P}), \tau_{j,n} \rangle$ is some prescribed group Δ. The structure of Δ could be used to show chirality and the intersection condition of the outcome. In practice, $\tau_{j,n}$ would be chosen so that the generators $\sigma_1, \ldots, \sigma_{n-1}, \sigma_n := \sigma_{n-1}\tau_{j,n}$ satisfy (3.7), that $\langle \Gamma^+(\mathcal{P}), \tau_{j,n} \rangle \cong \Delta$, and that certain other group-theoretical conditions are satisfied so that the intersection condition holds and the resulting $(n+1)$-polytope is chiral. This was the idea followed in [37] to construct chiral 4-polytopes with tetrahedral facets and symmetric or alternating automorphism groups.

Rather than using permutation groups as a framework to extend the even subgroup of a regular n-polytope $\mathcal{P}$ we may instead try considering $\Gamma^+(P)$ as a subgroup of the linear group $GL_m(K)$ (resp. the projective linear group $PGL_m(K)$) for some $m \geq 2$ and some field K. In order to find a rotary extension of $\mathcal{P}$ we may look for an involution $\tau_{j,n}$ in $GL_{m+1}(K)$ (resp. in $PGL_{m+1}(K)$) such that $\langle \Gamma^+(\mathcal{P}), \tau_{j,n} \rangle$ is a string even group with respect to $\sigma_1, \ldots, \sigma_{n-1}, \sigma_{n-1}^{-1}\tau_{j,n}$.

Another approach to constructing chiral polytopes of arbitrarily large rank is by using presentations in terms of generators and relations. Let $\mathcal{P}_i$ be a family of regular polytopes where the set of ranks is unbounded. Given a presentation

$$\Gamma^+(\mathcal{P}_i) = \langle \sigma_1, \ldots, \sigma_{n-1} : R_i \rangle \tag{5.9}$$

of the even subgroup of the n-polytope $\mathcal{P}_i$, we may devise a new set of relations R_i' containing R_i on the generators $\sigma_1, \ldots, \sigma_n$ of the even subgroup of the universal extension of $\mathcal{P}_i$ (see Theorem 5.5.4), so that $\langle \sigma_1, \ldots, \sigma_n : R_i' \rangle$ satisfies the intersection condition and it is the automorphism group of a chiral polytope. It is important to find a simple recipe to express these sets of relations R_i', and not to try to derive each of them separately. It is also desirable that the subgroup of (5.9) generated by the first $n-1$ generators is isomorphic to $\Gamma^+(\mathcal{P}_i)$, so that the resulting $(n+1)$-polytope is indeed an extension of $\mathcal{P}_i$, and not of a proper quotient.

The last approach to finding chiral polytopes of arbitrarily large rank that we suggest here is by analysing well-known families of groups. A full classification of regular polytopes whose automorphism group is the symmetric groups S_m was achieved only for certain ranks depending on m ([89]); following with this direction for other families of groups seems like a difficult task. Rather than attempting to find all chiral polytopes whose automorphism group is each group in a given family, we may try to find the ranks of all chiral polytopes whose automorphism groups are in that family. This was done in the regular setting for alternating groups ([26, 86–88]), linear groups ([23]), projective linear groups ([21, 158, 159]), symplectic and orthogonal groups ([20, 22]), and other simple groups ([154, 155, 160]). Comparatively, very little has been done in that direction for chiral polytopes ([130, 255]). The following problem is of particular interest due to its connection with the convex regular polytopes.

Problem 61 For each $n \geq 3$ determine the following.

1. The smallest m so that the symmetric group S_m is the automorphism group of a chiral n-polytope.
2. The smallest m so that the alternating group A_m is the automorphism group of a chiral n-polytope.
3. The smallest m so that the full automorphism group of the m-cube is the automorphism group of a chiral n-polytope.
4. The smallest m so that the rotation group of the m-cube is the automorphism group of a chiral n-polytope.

We expect that in upcoming years new ideas will lead to various constructions of chiral polytopes of high ranks, shedding light on many aspects of these polytopes of which little is known today.

We conclude this chapter with a list of open questions related to chiral extensions of regular polytopes and, in general, to chiral polytopes of high ranks. We start with problems 24 and 30 of [220]. The former is about an aspect of the poset of chiral extensions with the order given by the coverings.

Problem 62 Determine whether there exists a regular polytope $\mathcal{P}$ that has a chiral extension which covers every chiral extension of $\mathcal{P}$.

Problem 63 Determine whether every convex regular polytope admits a chiral extension.

There are also interesting questions about the types of the chiral extensions of regular polytopes (see [220, Problems 25 and 26]).

Problem 64 Is it true that if $\mathcal{P}$ is a regular polytope that has chiral extensions then it has a chiral extension that has ∞ as the last entry of its type?

Problem 65 Is there a regular polytope that admits a chiral extension that has q as the last entry of its type for each $q \geq 3$?

Problem 66 Is it true that for every $n \geq 3$ there is a chiral n-polytope with type $\{\infty, \ldots, \infty\}$?

6

Skeletal Polytopes

The geometric objects described in this chapter are inspired by the skeletal polyhedra introduced in [104], and adapted to higher-rank polytopes in [83]. (Skeletal polygons were already mentioned in [61, Section 1], but only planar polygons were used as 2-faces of figures of higher ranks.) This point of view comprises not only convex polytopes, but also star-shaped objects, and more general structures where the j-faces may not have a natural j-manifold associated to them. Convexity, a geometric property of Euclidean sets, is an essential defining component for convex polytopes. The definition of skeletal polytopes below emphasises the combinatorial properties and only then inserts them into a geometric space.

This chapter is organised as follows. In Section 6.1 we survey the main concepts and results related to skeletal polytopes in Euclidean spaces. Then, in Section 6.2, we illustrate the notion of skeletal polytope with the enumeration of the skeletal chiral polyhedra in $\mathbb{E}^3$. We conclude in Section 6.3 with a summary of the work that has been done so far on skeletal chiral polytopes, with emphasis on those called polytopes of full rank.

6.1 Regular and Chiral Skeletal Polytopes

A *skeletal polygon* in the Euclidean d-dimensional space $\mathbb{E}^d$ is an immersion of a connected 2-regular graph in $\mathbb{E}^d$, where the images of edges are allowed to intersect in interior points. A finite polygon consists of a set of distinct points $\{v_1, \dots, v_n\}$ together with distinct line segments $v_1v_2, v_2v_3, \dots, v_nv_1$; an infinite polygon (also called *apeirogon*) is composed of a sequence of distinct points $\{\dots, v_{-2}, v_{-1}, v_0, v_1, \dots\}$ together with the line segments v_iv_{i+1}. The points and line segments are called *vertices* and *edges*, respectively. The vertex set of any infinite polygon is required to be a discrete set. The following remark follows from the fact that the line segments corresponding to two distinct edges have distinct endpoints.

Remark 6.1.1 Every skeletal polygon has at least three edges.

The vertices of a skeletal polygon need not lie on a plane or on any other particular geometric structure. The presence of line segments is useful for visualisation, although for practical purposes each edge may be replaced by its set of two endpoints (see [169]).

The *vertex-figure* of a skeletal polygon at a vertex v is defined as the line segment between the two neighbours of v.

We define skeletal polytopes in $\mathbb{E}^d$ recursively as follows.

Definition 6.1.2 A *skeletal 2-polytope* in $\mathbb{E}^d$ is a skeletal polygon in $\mathbb{E}^d$; its 0- and 1-faces are its vertices and edges, respectively.

For $n \geq 3$ a *skeletal n-polytope* $\mathcal{P}$ in $\mathbb{E}^d$ is a collection of skeletal $(n-1)$-polytopes in $\mathbb{E}^d$ (also called $(n-1)$-faces or *facets*) satisfying the following:

1. The union of the vertex sets of the facets is discrete.
2. The graph determined by the union of the vertices and edges of all facets is connected.
3. The vertex-figure at every vertex is a skeletal $(n-1)$-polytope in $\mathbb{E}^d$. (The *vertex-figure* at a vertex v has as vertices the neighbours of v, and has as edges the vertex-figures at v of the polygons containing v. For $2 \leq j \leq n-2$, the j-faces of the vertex-figure are the vertex-figures at v of the $(j+1)$-faces of $\mathcal{P}$ containing v.)

Example 6.1.3 Convex polytopes and face-to-face tessellations of $\mathbb{E}^d$ naturally yield skeletal polytopes.

Example 6.1.4 The tessellation $\{4,3,4\}$ by cubes of $\mathbb{E}^3$ is a skeletal infinite 4-polytope. When assuming that the vertex set of $\{4,3,4\}$ is $\mathbb{Z}^3$, the vertices of the vertex-figure $\{4,3,4\}/o$ at the origin o are $(\pm 1,0,0)$, $(0,\pm 1,0)$ and $(0,0,\pm 1)$. Two of these six points are adjacent in $\{4,3,4\}/o$ if and only if there is a square of $\{4,3,4\}$ that contains them. Hence, each vertex $\overline{x}$ of $\{4,3,4\}/o$ is adjacent to every other vertex of $\{4,3,4\}/o$ except to $-\overline{x}$. Finally, the 2-faces of $\{4,3,4\}/o$ are the vertex-figures at o of each of the eight cubes that contain it; that is, the triangles formed by all triples of vertices containing no pair $\{\overline{x},-\overline{x}\}$. We can conclude that $\{4,3,4\}/o$ is the standard octahedron.

Example 6.1.5 The small stellated dodecahedron is a skeletal polyhedron in $\mathbb{E}^3$ where some pairs of distinct edges meet in interior points. In Figure 2.1(b) the planar region delimited by each 2-face is coloured with a shade of grey, but with the above definition the 2-faces are merely the cycles consisting only of the five vertices and five edges that form the pentagrams.

We are interested in the geometric properties of skeletal polytope on their own, regardless of the placement of the coordinate axes. With this in mind we shall regard similar skeletal polytopes as equal. (A *similarity* is a composition of a homotety and an isometry.)

Remark 6.1.6 The condition of discreteness in [104] is in general not equivalent to condition 1 in Definition 6.1.2. However, when working with highly symmetric polytopes they do become equivalent (see Exercise 6.1.19).

Remark 6.1.7 Condition 3 in Definition 6.1.2 is stronger than condition (ii) for polyhedra in [104, Section 3], and also than flag-connectivity, local flag-connectivity and local flatness in [83, Section 1]. Item 3 in Definition 6.1.2, under the assumption of item 2, is equivalent to strong flag-connectivity in [83, 167]. When studying regular polyhedra in $\mathbb{E}^3$ the definitions in all these works are equivalent. Here we decided to require condition 3 in order to resemble the idea that every interval of a polytope is a polytope on its own (see Remark 2.1.12).

Remark 6.1.8 Definition 6.1.2 can be adapted to polytopes in many different spaces, in particular in the d-dimensional projective space, and hyperbolic space. When considering a space where there are multiple geodesic segments between pairs of points (like projective spaces), it is convenient to insist that the edges are line segments. Furthermore, the vertices of the vertex-figure at a vertex v must be taken as interior points of the edges containing v. This ensures that the skeletal incidence agrees with the combinatorial incidence (see for example [3, 11]). In this book we shall only work on Euclidean spaces; for skeletal polytopes in other spaces see [3, 9, 11, 12, 51, 96, 266].

Whenever it is not necessary to specify the dimension of the ambient Euclidean space we shall refer to a skeletal polytope in $\mathbb{E}^d$ simply by 'a skeletal polytope'.

When we manufacture a convex polyhedron out of cardboard we glue together polygons in such a way that there are precisely two of them at every edge. The fact that in a polyhedron every edge belongs to two polygons is a consequence of the polygonal nature of the vertex-figures. If $\mathcal{P}$ is a skeletal polyhedron and e is the edge connecting the vertex v_0 with v_1, then v_1 represents the vertex associated to e in the vertex-figure at v_0. Since the vertex-figure at v_0 is a polygon, the vertex v_1 of the vertex-figure at v_0 belongs to precisely two edges. This means that when looking at all polygons of $\mathcal{P}$ containing v_0 there are precisely two of them that contain v_1 as a neighbour of v_0. Since all polygons of the vertex-figure at v_0 that contain v_1 as a neighbour of v_0 also contain e and all polygons of $\mathcal{P}$ containing e are polygons of the vertex-figure at v_0 that contain v_1, the edge e belongs to precisely two 2-faces.

The previous argument can be extended recursively to show that every $(n-2)$-face of an n-polytope belongs to precisely two $(n-1)$-faces. This property is a special case of the diamond condition. The proof of the fact that skeletal polytopes in $\mathbb{E}^d$ satisfy the diamond condition is part of Exercise 6.1.9.

Every skeletal polytope $\mathcal{P}$ induces an abstract polytope by considering its faces together with the empty set and $\mathcal{P}$ itself, with the obvious partial order. The proof of this statement is left as an exercise. When there is no possibility of confusion we shall abuse notation and denote by $\mathcal{P}$ both the skeletal polytope and its underlying abstract polytope.

Exercise 6.1.9 We say that an $(i-1)$-face F and an i-face G of a skeletal polytope are incident whenever F is a facet of G, and this situation is expressed as $F \leq_1 G$. We extend this relation to its transitive closure by establishing that two arbitrary faces F and G satisfy that $F \leq G$ if and only if there exist faces $F_1, \ldots, F_k$ such that $F = F_1$, $F_k = G$, and $F_{i-1} \leq_1 F_i$ for $2 \leq i \leq k$. Show that if $\mathcal{P}$ is a skeletal polytope then the set containing the empty set, $\mathcal{P}$ itself and all the faces of $\mathcal{P}$, constitutes an abstract polytope ordered by $\leq$.

Not every abstract polytope can be represented as a skeletal polytope. Multiple edges are an obstruction (see Remark 6.1.1). In particular, this imposes a lower bound on the number of edges at any given 2-face.

Remark 6.1.10 Every 2-face of a skeletal polytope has at least three edges.

The concepts of *flags* and *i-adjacent flags* extend naturally from the abstract setting to skeletal polytopes. The vertex-figure at v of a skeletal polytope $\mathcal{P}$, as in Definition 6.1.2, has a close proximity to the vertex-figure at v of its underlying abstract polytope, since in both cases there is a one-to-one correspondence between the i-faces of $\mathcal{P}$ containing v and the $(i-1)$-faces of the vertex-figure at v.

Some of the operations defined in Section 4.1 admit a geometric meaning. The geometric Petrial of a skeletal n-polytope $\mathcal{P}$ ($n \geq 3$) has the same $(n-2)$-skeleton as $\mathcal{P}$, but the $(n-1)$-faces are defined recursively as the collections of $(n-2)$-faces determined by the combinatorial Petrie operation.

Example 6.1.11 Recall from Example 4.1.1 that the hemi-cube is the Petrial of the tetrahedron. It follows that we can devise a skeletal polyhedron whose underlying abstract polyhedron is the hemi-cube by considering the 1-skeleton of the standard tetrahedron (say, with vertices $(1,1,1)$, $(1,-1,-1)$, $(-1,1,-1)$ and $(-1,-1,1)$), with the understanding that the 2-faces are the three non-planar quadrilaterals corresponding to the Petrie polygons of the tetrahedron.

The k-facetting operation Φ_k also translates naturally to the geometric setting since the 1-skeleton of $\mathcal{P}^{\Phi_k}$ is contained in the 1-skeleton of $\mathcal{P}$. To perform this operation it suffices to specify the k-holes that serve as 2-faces, whenever the set of all k-holes yields a structure that is not flag-connected. As an example, the great dodecahedron is the result of the 2-facetting operation on the icosahedron.

Given a base vertex, the halving operation can also be applied geometrically to polyhedra with quadrangular 2-faces, since the vertices of $\mathcal{P}^{\eta}$ are a subset of those of $\mathcal{P}$. Here it is possible that the halving operation applied to a skeletal polyhedron $\mathcal{P}$ does not yield a skeletal polyhedron even if the halving operation applied to the underlying abstract polyhedron is an abstract polyhedron. This situation occurs when $\mathcal{P}$ is a polyhedron with quadrangular 2-faces such that there is a pair of vertices that are opposite in two or more distinct squares; in that situation the resulting structure would require a multiple edge, and hence it will not be a skeletal polyhedron.

The dual operation requires more care. There is no natural point of the ambient space to assign as the dual of a given infinite facet. Henceforth we shall assume that skeletal polytopes with infinite facets have no geometric dual. Discreteness is also a problem, since the combinatorial dual of a skeletal polytope may have vertices with infinite degree.

Some skeletal polytopes do admit a geometric dual. Polar reciprocation of convex polytopes, as defined in [107], is an example of geometric duality. When saying that a polytope $\mathcal{P}$ is the geometric dual of a polytope $\mathcal{Q}$ we shall specify how this duality is defined.

All convex polytopes are orientable as abstract polytopes, but there are skeletal polytopes that are not orientable as abstract polytopes. The simplest example is $\{4,3\}_3$, the Petrial of the tetrahedron, which is combinatorially isomorphic to the hemi-cube (see Figure 4.1 and Example 6.1.11).

6.1.1 Symmetries of Skeletal Polytopes and Regularity

A *symmetry* of a skeletal polytope $\mathcal{P}$ is an isometry of the ambient space that preserves $\mathcal{P}$, in the sense that it maps every i-face of $\mathcal{P}$ to some i-face of $\mathcal{P}$, for all i. We shall denote the symmetry group by $G(\mathcal{P})$. If $\mathcal{P}$ is a skeletal polytope such that $G(\mathcal{P})$ is finite then there must exist a point fixed by every element in $G(\mathcal{P})$. In such cases we shall assume that the origin o is one fixed point.

Some symmetries of abstract polytopes fix an i-face F_i for some i. If a skeletal polytope is invariant under such a symmetry, then it must fix some subspace that contains F_i. As we shall see, the fixed-point sets of isometries play an important role in the study of highly symmetric skeletal polytopes. We shall refer to the set of fixed points of an isometry as its *mirror*.

Given a polytope $\mathcal{P}$ in $\mathbb{E}^d$, the *point group* $G_0(\mathcal{P})$ of $\mathcal{P}$ (also called *special group*) consists of all isometries S' fixing the origin o that satisfy that some symmetry $S \in G(\mathcal{P})$ can be expressed as $S = S'T$ for some translation T. This is precisely the group consisting of the linear parts of all symmetries of $\mathcal{P}$.

Example 6.1.12 Let $\mathcal{P}$ be the skeletal apeirogon in $\mathbb{E}^2$ with vertex set $\{(z,0) : z \in \mathbb{Z}\}$ and edges between $(z,0)$ and $(z+1,0)$ for every $z \in \mathbb{Z}$. Then $\mathcal{P}$ has four kinds of symmetries, as explained next. It has translations, one for each vector $(z,0)$ (see Figure 6.1(a)). It also has half-turns about the vertices and about the midpoints of edges (see Figure 6.1(b)). The remaining symmetries are orientation reversing. They include reflections about vertical lines through vertices or midpoints of edges (see Figure 6.1(c)), and also glide reflections along the X-axis with translation component $(z,0)$ for every $z \in \mathbb{Z}$ (see Figure 6.1(d)). The verification of the completeness of $G(\mathcal{P})$ is left as an exercise.

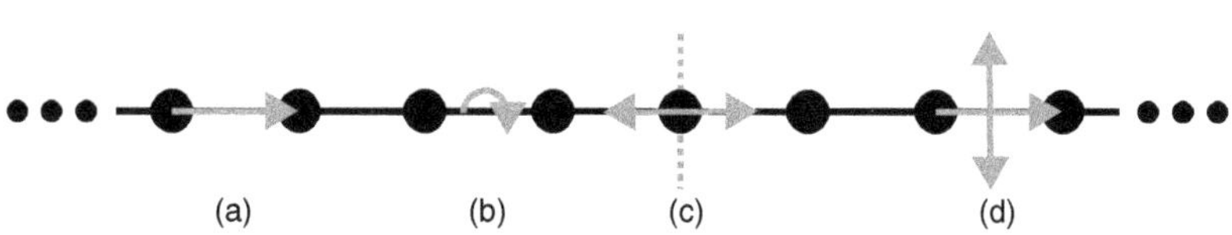

Figure 6.1 The symmetries of the linear apeirogon: (a) a translation, (b) a half-turn, (c) a reflection and (d) a glide reflection.

Standard procedures show that every half-turn can be expressed as a composition of the half-turn H_0 about $(0,0)$ with a translation, that every reflection in a vertical line is a composition of the reflection R_Y with respect to the Y-axis with a translation, and that every glide reflection along the X-axis is a composition of the reflection R_X with respect to the X-axis with a translation. Therefore, $G_0(\mathcal{P}) = \{\mathrm{id}, H_0, R_X, R_Y\} \cong \mathbb{Z}_2 \times \mathbb{Z}_2$.

Exercise 6.1.13 Show that the list of symmetries of the linear apeirogon $\mathcal{P}$ in $\mathbb{E}^2$ given in Example 6.1.12 is complete.

Exercise 6.1.14 Show that the point group of the tessellation $\{4,4\}$ by squares of $\mathbb{E}^2$ is the symmetry group of a square centred at o. For each $S' \in G_0(\{4,4\})$ describe the set of all symmetries S of $\{4,4\}$ such that $S = S'T$ for some translation T.

By preserving the set of faces of $\mathcal{P}$, every symmetry of a skeletal polytope $\mathcal{P}$ preserves the incidence relation given by inclusion, and therefore induces an automorphism of the underlying abstract polytope. Nevertheless, the automorphism group and the symmetry group of a skeletal polytope may have some essential differences. It may happen that $G(\mathcal{P})$ induces more flag orbits on $\mathcal{P}$ than $\Gamma(\mathcal{P})$; one of the easiest instances of this fact is when $\mathcal{P}$ is a triangle with unequal side lengths. Another difference between the symmetry and automorphism groups of a skeletal polytope $\mathcal{P}$ is that $G(\mathcal{P})$ may not act freely on the flags. For example, if $\mathcal{P}$ is a convex planar square in $\mathbb{E}^3$ then the reflection with respect to the plane containing $\mathcal{P}$ is a non-trivial symmetry of $\mathcal{P}$ fixing all flags.

Proposition 2.1.80 (item 1) tells us that all flag stabilisers of an abstract polytope are trivial, and in particular they are equal. The following proposition generalises this last statement for skeletal polytopes.

Proposition 6.1.15 *Let $\mathcal{P}$ be a skeletal polytope. Then all flag stabilisers in $G(\mathcal{P})$ are equal.*

Proof Skeletal polytopes are flag-connected (see Exercise 6.1.9). Then, given any two flags Φ and Ψ, there exist $i_1, \ldots, i_m$ such that $\Psi = \Phi^{i_1,i_2,\ldots,i_m}$. If $T \in G(\mathcal{P})$ is such that $\Phi T = \Phi$ then

$$\Psi T = \Phi^{i_1,i_2,\ldots,i_m} T = (\Phi T)^{i_1,i_2,\ldots,i_m} = \Phi^{i_1,i_2,\ldots,i_m} = \Psi,$$

since every symmetry acts like an automorphism on the set of flags. Therefore a symmetry fixes a flag if and only if it fixes all flags. ∎

If the flag stabilisers in $G(\mathcal{P})$ are trivial then $G(\mathcal{P})$ is isomorphic to the group of all automorphisms that are represented by symmetries of $\mathcal{P}$. In this situation every symmetry is completely determined by its image on any flag (see Proposition 2.1.80 (item 2)). Here the requirement of trivial flag stabilisers is essential. If $\mathcal{P}$ is a planar square in $\mathbb{E}^3$ then it has 16 symmetries, but only 8 automorphisms, making it impossible for $G(\mathcal{P})$ to be isomorphic to a subgroup of $\Gamma(\mathcal{P})$.

In case there exists $T \in G(\mathcal{P})$ that acts trivially on a flag then Proposition 6.1.15 tells us that T fixes every flag, and in particular, it fixes every vertex. It follows that the affine span $\text{Aff}(\mathcal{P})$ of $\mathcal{P}$ is contained in the mirror of T. This justifies the following lemma.

Lemma 6.1.16 *Let $\mathcal{P}$ be a skeletal polytope in $\mathbb{E}^d$. The flag stabilisers of $\mathcal{P}$ are trivial if and only if $dim(\text{Aff}(\mathcal{P})) = d$.*

Some relevant results require skeletal polytopes to have trivial flag stabilisers. The next result is an example of this fact, since it fails for the polyhedron in Example 6.1.12 when seen as a polyhedron in $\mathbb{E}^3$.

Lemma 6.1.17 *Let $\mathcal{P}$ be a skeletal polyhedron with trivial flag stabilisers. Then the vertex stabilisers in $G(\mathcal{P})$ are finite.*

Proof Let v be a vertex of $\mathcal{P}$ and let w be a neighbour of v. Then $wG(\mathcal{P})$ is contained in a sphere centred at v, and the first item of Definition 6.1.2 forces this set to be finite. On the other hand, the edge e between w and v belongs to precisely four flags, and the fact that $\mathcal{P}$ has trivial flag stabilisers implies that the pair (v, e) is stabilised by at most two elements in $G(\mathcal{P})$. It follows that the stabiliser of v in $G(\mathcal{P})$ has at most $2|wG(\mathcal{P})|$ elements, and therefore it is finite. ■

Recall that a group G of isometries is *discrete* if the orbit of every point under G is a discrete set.

Exercise 6.1.18 Show that if $\mathcal{P}$ is a skeletal polytope with trivial flag stabilisers then $G(\mathcal{P})$ is a discrete group of isometries. Is the conclusion still true when we remove the requirement of trivial flag stabilisers?

The notion of *even subgroup* also extends to the geometric setting. If $\mathcal{P}$ is combinatorially non-orientable then its even subgroup $G^+(\mathcal{P})$ is defined simply as $G(\mathcal{P})$, whereas if $\mathcal{P}$ is orientable then $G^+(\mathcal{P})$ is defined as the set of symmetries that preserve the bipartition of flags. When dealing with skeletal polytopes we shall not refer to the even subgroup by 'rotation subgroup' to avoid confusion with the subgroup of orientation-preserving symmetries.

Five of the six regular polyhedra described in Section 6.1.5 are examples in which the even subgroup differs from the subgroup of orientation-preserving symmetries.

In what follows we shall work exclusively with skeletal polytopes with a high degree of symmetry. The next exercise is an example of the consequences of requiring a skeletal polytope to have finitely many flag orbits.

Exercise 6.1.19 Restate Definition 6.1.2 for polyhedra (3-polytopes) and denote by $\mathcal{A}$ its axioms 1, 2 and 3. Denote by $\mathcal{A}'$ the set of axioms 2 and 3 together with the axiom $1'$. Every compact set meets only finitely many polygons.

(a) Show that there are polyhedra that do not satisfy condition $\mathcal{A}'$.
(b) Show that if we require the symmetry group to induce finitely many flag orbits then conditions $\mathcal{A}$ and $\mathcal{A}'$ become equivalent.

We say that $\mathcal{P}$ is *geometrically regular* (resp. *combinatorially regular*) whenever $G(\mathcal{P})$ (resp. $\Gamma(\mathcal{P})$) acts transitively on the flags.

If the flag stabilisers in the symmetry group of a geometrically regular polytope $\mathcal{P}$ are trivial then every symmetry induces a unique automorphism. It follows that $G(\mathcal{P}) \cong \Gamma(\mathcal{P})$, and $G(\mathcal{P})$ is generated by n involutions R_i, where $\Phi R_i = \Phi^i$ for some base flag Φ. In other words, the isometries R_i are geometric realisations of the automorphisms ρ_i; this implies that they satisfy

$$(R_i R_j)^2 = \text{id} \tag{6.1}$$

whenever $|i - j| \geq 2$. If the flag stabilisers are non-trivial then there are multiple isometries acting like ρ_i.

Remark 6.1.20 The involutory isometries of $\mathbb{E}^3$ are plane reflections, half-turns around a line and rotatory reflections by an angle of π (see Appendix A.1). These last isometries are also called *central inversions*. Hence each generator R_i of the symmetry group of a geometrically regular polytope in $\mathbb{E}^3$ must be of one of these three kinds of isometries. We still shall refer to them as *abstract reflections* or *combinatorial reflections*.

It is well known that if T is an involutory isometry then it is completely determined by its mirror. In this situation we shall abuse notation and denote the mirror of T also by T. For example, $x \in T$ indicates that the point x belongs to the mirror of T, and $T_1 \cap T_2$ stands for the intersection of the mirrors of T_1 and T_2.

The Platonic solids are examples of geometrically regular polyhedra in $\mathbb{E}^3$. Their symmetry groups act freely on their flags and all generators R_i are plane reflections. A detailed explanation of the symmetry groups of all convex regular polytopes can be found in [103]. The Petrials of the Platonic solids are also geometrically regular polyhedra; their generators R_0 are half-turns, but their generators R_1 and R_2 are still plane reflections.

The even subgroup of a geometrically regular polytope $\mathcal{P}$ with trivial flag stabilisers is generated by the symmetries $S_i := R_{i-1}R_i$ for $i \in \{1, \ldots, n-1\}$, realising the action of the standard rotations σ_i; that is, for some base flag Φ and every $i \in \{1, \ldots, n-1\}$ they satisfy $\Phi S_i = \Phi^{i,i-1}$. These isometries need not be rotations (although they are still referred to as *abstract rotations* or *combinatorial rotations*), and they can even be orientation reversing isometries. If $\mathcal{P}$ is a Platonic solid then S_1 and S_2 are rotations (each being the product of two plane reflections). On the other hand, if $\mathcal{P}$ is the Petrial of a Platonic solid then S_2 is still a rotation, but S_1 is a rotatory reflection.

Given that every symmetry of $\mathcal{P}$ induces an automorphism, if $\mathcal{P}$ is geometrically regular then there are automorphisms mapping any given flag into any other flag, and hence $\mathcal{P}$ is also combinatorially regular. The converse is not true in general; polygons with unequal side lengths are examples of combinatorially regular polytopes that are not geometrically regular.

6.1.2 Wythoff's Construction of Regular Polytopes

Every geometrically regular polytope $\mathcal{P}$ with trivial flag stabilisers can be recovered from the generators R_i of its symmetry group together with a base vertex v_0 (also called *initial vertex*) by the so-called *Wythoff's construction*, described next.

The faces of the base flag are obtained recursively as follows. The base vertex is $F_0 := v_0$ and the base edge joins it to $v_0 R_0$. For $i \geq 2$ the base j-face F_j is the set of $(j-1)$-faces $F_{j-1}\langle R_0, \ldots, R_{j-1}\rangle$. Once we have determined the j-face in the base flag, the set of all j-faces of $\mathcal{P}$ consists of the images of F_j under all elements of $G(\mathcal{P})$. This is the natural generalisation to skeletal polytopes of the original notion of Wythoff's construction explained in [58, Section 2.4]. The verification of the fact that $\mathcal{P}$ coincides with the structure obtained from Wythoff's construction is left as an exercise (Exercise 6.1.32).

Example 6.1.21 Let $\mathcal{P}$ be the square in Figure 6.2. We choose the base flag to contain the vertex v_0 as well as the horizontal edge containing it. In this setting, R_0 and R_1 are the reflections about the vertical and diagonal lines, respectively, in Figure 6.2(a). Then Wythoff's construction tells us that the base vertex and edge are the ones labelled F_0 and F_1 in Figure 6.2(b). The remaining three vertices and three edges are respectively the orbits of F_0 and F_1 under the dihedral group $G(\mathcal{P}) = \langle R_0, R_1\rangle$.

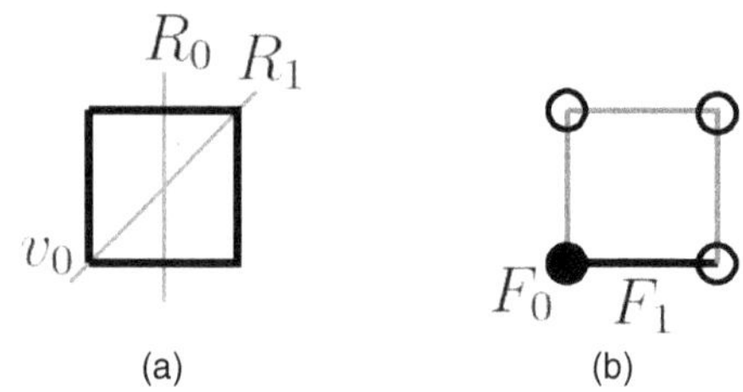

Figure 6.2 (a) The base vertex and generators of the symmetry group of a square; (b) the outcome of Wythoff's construction.

Remark 6.1.22 If $\mathcal{P}$ has non-trivial flag stabilisers it can still be recovered by Wythoff's construction by choosing arbitrary symmetries $\hat{R}_0, \dots, \hat{R}_{n-1}$ acting like $R_0, \dots, R_{n-1}$. In the process, the j-face F_j is still chosen as $F_{j-1}\langle \hat{R}_0, \dots, \hat{R}_{j-1}\rangle$ for $j \geq 2$. Here it may be the case that $G(\mathcal{P})$ is larger than $\langle \hat{R}_0, \dots, \hat{R}_{n-1}\rangle$.

By definition, the vertex v_0 in the base flag is fixed by R_j for all $j \geq 1$, but it is not fixed by R_0. Indeed, if v_0 were also fixed by R_0 then F_1 would have only one vertex, contradicting that it is the base edge. Hence, when applying Wythoff's construction, the initial vertex always belongs to $\cap_{j\geq 1} R_j$, but not to R_0. This allows us to try to reconstruct the polytope $\mathcal{P}$ from $G(\mathcal{P})$ without knowing the base vertex v_0 beforehand, by using as the base vertex any chosen point in the *Wythoff space* $(\cap_{j\geq 1} R_j) \setminus R_0$. If the Wythoff space W has only one point then this point must be v_0 and the result of Wythoff's construction is again $\mathcal{P}$. If $G(\mathcal{P})$ is finite and W is a one-dimensional subspace (minus the origin), then every choice of v_0 yields a polytope similar to $\mathcal{P}$. However, if W is k-dimensional for some $k \geq 2$, or W is one-dimensional and $G(\mathcal{P})$ is infinite, then some choices of base vertex in W may yield polytopes that are essentially different from $\mathcal{P}$.

The previous idea can be used to find all skeletal polytopes whose symmetry group is isomorphic to any given group Γ.

Example 6.1.23 In order to determine all finite geometrically regular polygons in $\mathbb{E}^2$ we may apply Wythoff's construction to all groups of isometries isomorphic to the dihedral group D_k with $2k$ elements ($k \geq 3$). All such groups are generated by two reflections R_0 and R_1 whose mirrors intersect at an angle of $2s\pi/k$ for some $1 \leq s < k/2$ relatively prime to k. If $s = 1$ then the polygon is the standard convex regular k-gon; otherwise, it is a star-shaped polygon that winds s times around its centre. Note that if one generator is a half-turn then the dihedral group must have at most four elements, contradicting that $k \geq 3$.

Example 6.1.24 The symmetry group of a geometrically regular apeirogon $\mathcal{P}$ in $\mathbb{E}^2$ is the infinite dihedral group, generated by two involutions whose product has infinite order.

If R_0 and R_1 are half-turns then the base vertex is the fixed point of R_1 and $\mathcal{P}$ lies on the line spanned by the centres of rotation of R_0 and R_1 (this is the apeirogon in Figure 6.1). We reach the same outcome if R_0 is a plane reflection and R_1 is a half-turn, or if R_0 and R_1 are reflections about parallel lines; in such situations $\mathcal{P}$ lies on the line perpendicular to R_0 through the chosen base vertex in R_1.

Two reflections about non-parallel lines ℓ_1 and ℓ_2 generete an infinite dihedral group if and only if the angle between their mirrors is not a rational multiple of π, but this forces the orbit of every point different from $\ell_1 \cap \ell_2$ to be a non-discrete set. Hence, if R_0 and R_1 are chosen to be the abstract reflections of an apeirogon then their mirrors must be parallel. In that situation the polygon is a linear apeirogon.

The only case left to examine is when R_0 is a half-turn and R_1 is a plane reflection. If the base vertex is chosen to be in the perpendicular to R_1 containing R_0 then we again obtain a linear apeirogon. For every other choice of base vertex $\mathcal{P}$ is a zigzag like that in Figure 6.3.

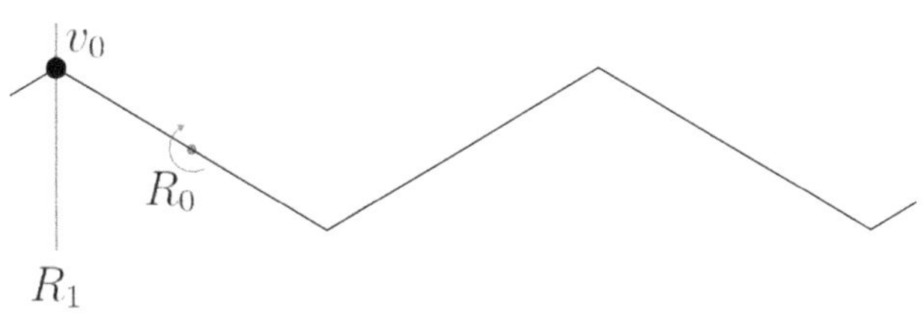

Figure 6.3 The symmetry group of the regular zigzag.

In what follows we shall enumerate all geometrically regular polygons in $\mathbb{E}^3$.

If $\mathcal{P}$ is a geometrically regular polygon in $\mathbb{E}^3$ and lies on a plane Π then every $T \in G(\mathcal{P})$ restricted to Π is an isometry of Π that preserves $\mathcal{P}$, and hence $\mathcal{P}$ can be understood as a geometrically regular polygon in $\mathbb{E}^2$. From Examples 6.1.23 and 6.1.24 we know that $\mathcal{P}$ is either a finite (convex or star-shaped) polygon, a linear apeirogon, or a zigzag.

We shall derive the geometrically regular polygons that are genuinely 3-dimensional through the generator S_1 of their even subgroup, by using the next observation.

Remark 6.1.25 Let T be an isometry of $\mathbb{E}^n$ and let $v \in \mathbb{E}^n$ such that its orbit under $\langle T \rangle$ is discrete and has at least three points. Then the sets $\{vT^k : k \in \mathbb{Z}\}$ and $\{\{vT^k, vT^{k+1}\} : k \in \mathbb{Z}\}$ are the vertices and edges of a polygon.

In order to find genuinely three-dimensional regular polygons, the symmetry S_1 must be an isometry having at least one orbit whose affine span is the entire $\mathbb{E}^3$. The only possibilities are rotatory reflections and twists.

Polygons $\mathcal{P}$ satisfying that $G^+(\mathcal{P})$ is generated by a rotatory reflection S_1 of finite order have their vertices arranged in two parallel planes, both at the same distance of the invariant plane Δ of S_1, but located on opposite sides of Δ. The vertices of any such $\mathcal{P}$ project to Δ into a finite polygon $\mathcal{P}_\Delta$. The convex hull $Conv(\mathcal{P})$ of $\mathcal{P}$ may be a prism or an antiprism. If $Conv(\mathcal{P})$ is a prism over an m-gon then m is odd and S_1 has order $2m$; in that situation $\mathcal{P}$ is said to be a *prismatic* skew polygon, and it is a double cover of $\mathcal{P}_\Delta$. Otherwise, $\mathcal{P}$ is said to be an *antiprismatic skew polygon*, and it is combinatorially isomorphic to $\mathcal{P}_\Delta$. Figure 6.4 shows a prismatic skew hexagon and an antiprismatic skew octagon.

The generator S_1 of the even group of a skeletal polygon cannot be a rotatory reflection of infinite order since otherwise its orbits would be non-discrete (except for the orbit that consists exclusively of its unique fixed point and the orbits of the points in the invariant line, each of which consists of only two points.).

All twists have infinite order, and hence, if $\mathcal{P}$ is a non-planar polygon whose even group is generated by a twist, then $\mathcal{P}$ itself is infinite. Its vertices are at a constant distance to the axis of the twist. Such polygons are called *helical polygons*, or simply *helices*. An example of a portion of a helical polygon is shown in Figure 6.5.

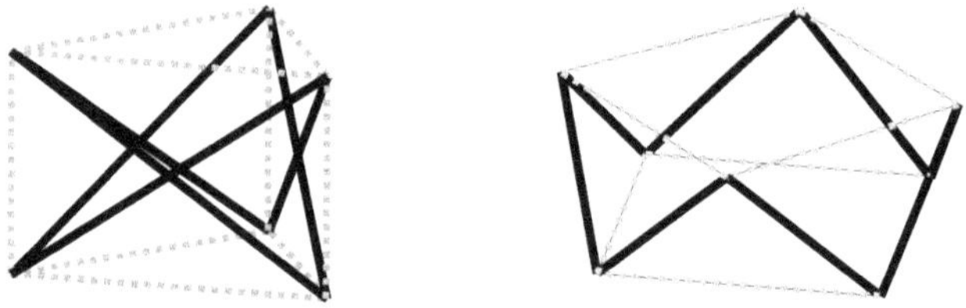

(a) (b)

Figure 6.4 A prismatic skew hexagon (a) and an antiprismatic skew octagon (b).

Figure 6.5 Helical polygon.

Some helical polygons can be constructed from planar finite polygons as follows. Let $\mathcal{P}$ be a planar k-gon on the XY-plane with vertices $v_1, \ldots, v_k$ in cyclic order. We may take infinitely many copies of $\mathcal{P}$, one at each height $z \in \mathbb{Z}$, all projecting orthogonally onto $\mathcal{P}$. We label the vertex in the polygon at height z projecting to v_i as $v_{i,z}$. The polygon whose vertices are $v_{z,z}$, where the first subindex is taken modulo k, such that the edges join $v_{z,z}$ with $v_{z+1,z+1}$ for every z, is called a *helix over a k-gon*. Figure 6.5 shows a helix over a triangle; a helix over a square can be seen in Figure 6.7(c).

From the previous discussion we can conclude the following result.

Proposition 6.1.26 *There are exactly five kinds of regular polygons in $\mathbb{E}^3$: finite planar, finite skew, linear, zigzags and helices. The last three kinds correspond to infinite polygons.*

The grouping of polygons in Proposition 6.1.26 is according to whether they are finite or infinite, and to the nature of their affine spans. In [104] there are seven groups instead of five. There the planar polygons are further divided into convex and star polygons, and the skew finite polygons are divided into prismatic and antiprismatic.

Helices are the only polygons, among those listed in Proposition 6.1.26, that are 'handed' in the sense that a helix cannot be superposed into its image under a plane reflection by a continuous rigid movement. To distinguish a helix from its mirror image we shall look at its behaviour in a neighbourhood of a plane perpendicular to the axis. Informally, we imagine two solid, three-dimensional clocks located each on its side of some plane Π perpendicular to the axis A of the helix H, in such a way that the clocks are facing away from Π, and their centres coincide with $A \cap \Pi$. An observer standing on A away from Π looks at the movement that an object makes to traverse along H from $H \cap \Pi$ towards the observer. If the object can be moved in such a way that the observer can always see the clock hands behind the object, we say that H is a *right helix*. If the clock hands need to make an anticlockwise movement for that to be possible, we say that H is a *left helix*. It is clear that to determine if a helix is right or left it is enough to take any arbitrary plane Π perpendicular to A; the reader should convince himself/herself that two observers located on different sides of Π would reach the same conclusion on whether H is left or right.

If a polygon $\mathcal{Q}$ is a 2-face of some regular polytope $\mathcal{P}$ in $\mathbb{E}^3$ then $\mathcal{Q}$ itself must be regular. Furthermore, there must exist $R_0, R_1 \in G(\mathcal{P})$ that act like the standard reflections on $\mathcal{Q}$. In what follows we shall study the possibilities for such R_0 and R_1.

Let $\mathcal{P}$ be a finite planar polygon or a zigzag on a plane $\Pi \subseteq \mathbb{E}^3$. Each generator R of the symmetry group of $\mathcal{P}$ is the restriction to Π of precisely

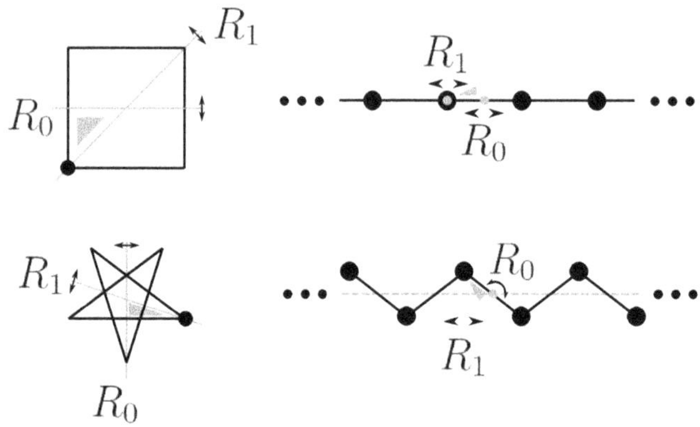

Figure 6.6 Generators of the symmetry group of planar polygons.

two isometries of $\mathbb{E}^3$. When R is the reflection about the line ℓ then it is the restriction of the half-turn about ℓ, and also of the reflection with respect to the plane perpendicular to Π containing ℓ. When R is the half-turn about the point x, it is the restriction of the half-turn about the line perpendicular to Π through x, and also of the central inversion about x. If, instead, $\mathcal{P}$ is a planar apeirogon in the line ℓ then each R_i is the restriction of infinitely many isometries: the central inversion about its fixed point u (vertex or midpoint of edge), the half-turn with respect to any line perpendicular to ℓ through u and the plane reflection about the plane perpendicular to ℓ through u. Figure 6.6 illustrates the generators R_0 and R_1 of $G(\mathcal{P})$ restricted to Π, for each planar polygon $\mathcal{P}$; in each case, the base flag is indicated by a grey triangle.

By contrast, if $\mathcal{P}$ is a non-planar polygon in $\mathbb{E}^3$ then the generators of $G(\mathcal{P})$ are completely determined by $\mathcal{P}$ itself. If $\mathcal{P}$ is skew finite with base flag $\{F_0, F_1\}$ then R_0 must fix the centre o of $\mathcal{P}$ and the midpoint of F_1 while swapping its endpoints. This could be done by a half-turn about the line through o and the midpoint of F_1, or by the plane reflection about the bisector of F_1, but only the former preserves $\mathcal{P}$. The generator R_1 must fix o and F_0 while swapping the two edges at F_0, and this can only be achieved by the reflection with respect to the bisector of the segment between the two neighbours of F_0. The generators R_0 and R_1 when $G(\mathcal{P})$ is a prismatic or an antiprismatic skew polygon are shown in Figure 6.7(a) and (b); the base flag is indicated by the bold dot and the bold edge. When $\mathcal{P}$ is a helix then it is not preserved under any orientation reversing isometry, since isometries of this kind map right helices into left helices and vice versa. It follows that R_0 and R_1 are half-turns; their axes are perpendicular to the axis of $\mathcal{P}$, the

Table 6.1. *Possible kinds of generators of the symmetry groups of regular polygons in* $\mathbb{E}^3$.

Polygon	Generator	Central inversion	Half-turn	Reflection
Finite planar	R_0		○	○
	R_1		○	○
Finite skew	R_0		○	
	R_1			○
Linear	R_0	○	○	○
	R_1	○	○	○
Zigzag	R_0	○	○	
	R_1		○	○
Helix	R_0		○	
	R_1		○	

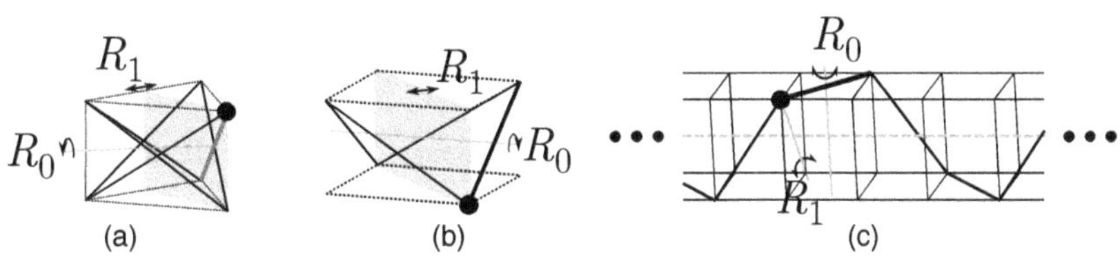

Figure 6.7 Generators of the symmetry group of (a) a prismatic skew polygon, (b) an antiprismatic skew polygon and (c) a helix.

former goes through the midpoint of the base edge whereas the latter through the base vertex. The generators R_0 and R_1 when $G(\mathcal{P})$ is a helix are shown in Figure 6.7(c).

The previous discussion is summarised in Table 6.1.

It is interesting that the mirror of R_0 does not determine completely the nature of the polygon; however, if the mirror of R_1 is a point then the polygon is linear.

Next we analyse the generator S_1 of the even subgroup of a regular skeletal polygon $\mathcal{P}$. We already mentioned that S_1 is a rotatory reflection or a twist, if $\mathcal{P}$ is finite skew or a helix, respectively. In such cases, the fact that $\mathcal{P}$ is genuinely three-dimensional leaves only one choice for the kind of symmetry of S_1.

Now assume that $\mathcal{P}$ is a planar finite polygon in $\mathbb{E}^3$. The symmetry S_1 of $G(\mathcal{P})$ acts on the plane Π containing $\mathcal{P}$ as a rotation around the centre o of $\mathcal{P}$, mapping every vertex to one of its neighbours. As an isometry of $\mathbb{E}^3$, S_1 may be either a rotation about the axis ℓ through o perpendicular to Π, or a rotatory reflection with respect to ℓ and Π, depending on whether S_1 preserves the hemi-spaces determined by Π or not. If R_0 and R_1 are both plane reflections or both half-turns then S_1 is a rotation; S_1 is a rotatory reflection whenever the mirrors of R_0 and R_1 have distinct dimensions.

If $\mathcal{P}$ is a zigzag on a plane Π then S_1 acts like a glide reflection on Π whose axis ℓ is precisely that of $\mathcal{P}$, and its translation component $\bar{v}$ is half the vector between two vertices of $\mathcal{P}$ that have a common neighbour. This action is induced by two isometries of $\mathbb{E}^3$. These two isometries are the glide reflection about the plane through ℓ perpendicular to Π and with translation vector $\bar{v}$, and the twist with axis ℓ, translation vector $\bar{v}$ and rotation component of π. If one generator R_i reverses orientation and the other one preserves it then S_1 is a glide reflection, otherwise it is a twist.

The next result relates the abstract rotation S_1 with the candidate abstract reflections of planar polygons.

Lemma 6.1.27 *Let $\mathcal{P}$ be a planar finite or zigzag polygon in $\mathbb{E}^3$ with rotation group $\langle S_1 \rangle$. Let ℓ be the line through the base vertex v_0 and the midpoint of the line segment between its two neighbours, and let Σ be the plane through ℓ perpendicular to the plane Π containing $\mathcal{P}$ (see Figure 6.8). Let R_ℓ and R_Σ be the half-turn about ℓ and the reflection with respect to Σ, respectively. Then $R_\ell S_1 R_\ell = S_1^{-1}$ and $R_\Sigma S_1 R_\Sigma = S_1^{-1}$.*

Proof Both isometries R_ℓ and R_Σ preserve Π, and act on it like the reflection with respect to ℓ (see Figure 6.9(a) and (b) for the finite polygon and the zigzag, respectively). This is precisely the standard reflection R_1 of $\mathcal{P}$ when restricted to $\mathbb{E}^2$. From there we can conclude that $(R_\ell S_1 R_\ell)_{|\Pi} = (S_1^{-1})_{|\Pi}$ and $(R_\Sigma S_1 R_\Sigma)_{|\Pi} = (S_1^{-1})_{|\Pi}$. Equalities hold in $\mathbb{E}^3$ if and only if the isometries in both sides of the equations preserve the two hemi-spaces determined by Π, or if both swap the hemi-spaces. This last condition is always satisfied regardless of the geometric nature of S_1, and the lemma follows. ■

Exercise 6.1.28 Describe all possibilities for S_1 for a linear polygon.

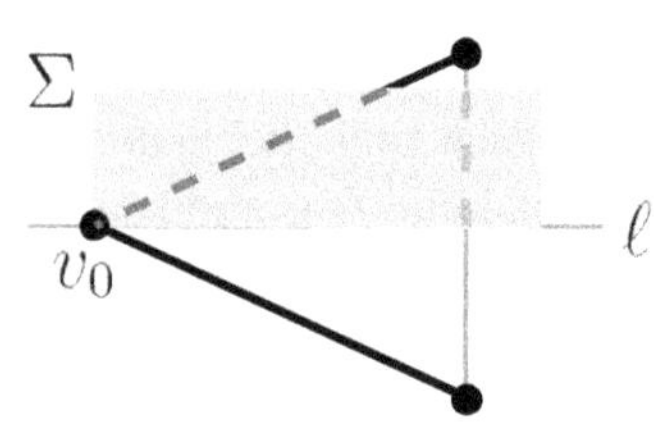

Figure 6.8 The line ℓ and the plane Σ.

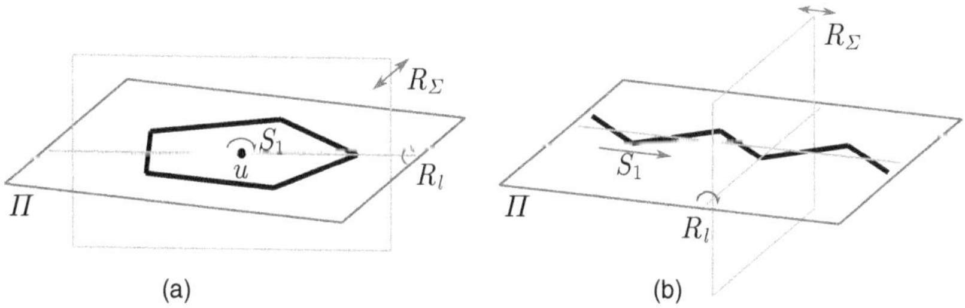

Figure 6.9 Symmetries of planar polygons.

6.1.3 Chiral Skeletal Polytopes

A skeletal n-polytope $\mathcal{P}$ is *geometrically chiral* (resp. *combinatorially chiral*) if $G(\mathcal{P})$ (resp. $\Gamma(\mathcal{P})$) induces two orbits on flags with adjacent flags in distinct orbits. There are no examples of chiral skeletal polytopes in classical theory, and in general they seem hard to build. We shall see some examples of chiral skeletal polyhedra in Section 6.2. Roli's cube is an example of a chiral skeletal 4-polytope. We keep the term *rotary* to denote skeletal polytopes that are either regular or chiral.

A geometrically chiral polytope is either combinatorially chiral or combinatorially regular. Both situations are exemplified in Section 6.2. For the same reasons as in the abstract setting, the facets and vertex-figures of a geometrically chiral n-polytope are either geometrically regular or geometrically chiral, but the $(n-2)$-faces and the vertex-figures of the vertex-figures are always geometrically regular (see Proposition 2.3.8).

If $\mathcal{P}$ is geometrically chiral and $G(\mathcal{P})$ has trivial flag stabilisers then $G(\mathcal{P})$ is generated by isometries $S_i, i \in \{1, \ldots, n-1\}$, where $\Phi S_i = \Phi^{i,i-1}$ for some base flag Φ. The isometries S_i are realisations of the combinatorial rotations σ_i, and hence they satisfy

$$(S_i \cdots S_j)^2 = \text{id} \tag{6.2}$$

for all $1 \leq i < j \leq n-1$.

Proposition 6.1.29 *Let $n \geq 3$, let $\mathcal{P}$ be a geometrically rotary n-polytope that is orientable as an abstract polytope and let $G^+(\mathcal{P}) = \langle S_1, \ldots, S_{n-1} \rangle$. If $\mathcal{P}$ has trivial flag stabilisers then the stabiliser in $G^+(\mathcal{P})$ of the base vertex is $\langle S_2, \ldots, S_{n-1} \rangle$.*

Proof By definition, every element in $\langle S_2, \ldots, S_{n-1} \rangle$ is in the stabiliser of the base vertex.

Now assume that $T \in G^+(\mathcal{P})$ stabilises the base vertex. Since the flag stabilisers of $\mathcal{P}$ are trivial, orientability of $\mathcal{P}$ implies that there exists a unique $\gamma \in \Gamma^+(\mathcal{P})$ (the even subgroup of the combinatorial automorphism group of $\mathcal{P}$) such that T and γ act in the same way on the faces of $\mathcal{P}$. By Proposition 2.3.18, $\gamma = \sigma_{i_1} \cdots \sigma_{i_m}$ for some $i_1, \ldots, i_m \in \{2, \ldots, n-1\}$, and we can conclude that $T = S_{i_1} \cdots S_{i_m}$. ■

Corollary 6.1.30 *Let $\mathcal{P}$ be a geometrically rotary polyhedron that is orientable as an abstract polytope, and let $G^+(\mathcal{P}) = \langle S_1, S_2 \rangle$. If $\mathcal{P}$ has trivial flag stabilisers then $\langle S_1 \rangle \cap \langle S_2 \rangle = \langle \mathrm{id} \rangle$.*

Proof We know from Proposition 6.1.29 that the stabiliser of the base vertex v_0 is $\langle S_2 \rangle$. On the other hand, S_1 acts on the base 2-face f_0 by cyclically permuting its vertices. It follows that $v_0 S_1^k = v_0$ if and only if S_1^k acts like id on f_0. In that situation S_1^k fixes the chain (v_0, f_0); orientability of $\mathcal{P}$ forces S_1^k to fix the base flag and therefore $S_1^k = \mathrm{id}$. ■

Remark 6.1.31 The requirement in Proposition 6.1.29 of $\mathcal{P}$ being orientable is necessary. If $\mathcal{P}$ is a geometrically regular polytope that is non-orientable as an abstract polytope then the stabiliser of the base vertex contains a symmetry R_1 mapping the base flag Φ to Φ^1. Such a symmetry is in $\langle S_1, \ldots, S_{n-1} \rangle$, but it may not belong to $\langle S_2, \ldots, S_{n-1} \rangle$. The hemi-cube is an example of a geometrically rotary skeletal polyhedron that is combinatorially non-orientable, for which the conclusion of Proposition 6.1.29 fails.

There is a variant of Wythoff's construction that can be used to recover a rotary n-polytope $\mathcal{P}$ from the group $G = \langle S_1, \ldots, S_{n-1} \rangle$ and an initial vertex v_0 (here $n \geq 2$). In this setting, v_0 belongs to the mirror of S_j for all $j \geq 2$, but not to the mirror of S_1.

As in the case for regular polytopes and generators R_i, we obtain the faces of the base flag recursively. The base vertex is $F_0 := v_0$, the endpoints of the base edge are F_0 and $F_0 S_1 S_2 = F_0 S_1^{-1}$ and for $i \geq 2$ the base j-face F_j is the set of $(j-1)$-faces $F_{j-1}\langle S_1, \ldots, S_{j-1} \rangle$. Once we have determined the j-face in the base flag, the set of all j-faces of $\mathcal{P}$ consists of the images of F_j under all elements of G.

Exercise 6.1.32 Let $n \geq 2$ and let $\mathcal{P}$ be a skeletal n-polytope with base vertex v_0.

(a) Show that if $\mathcal{P}$ is geometrically regular then it coincides with the structure obtained by Wythoff's construction from the group $\langle R_0, \ldots, R_{n-1} \rangle$ with base vertex v_0.

(b) Show that if $\mathcal{P}$ is geometrically rotary and its underlying abstract polytope is orientable then it coincides with the structure obtained by Wythoff's construction from the group $\langle S_1, \ldots, S_{n-1} \rangle$ with base vertex v_0.

Remark 6.1.33 If we denote by G_j the stabiliser in $G = G^+(\mathcal{P})$ of the base j-face of a skeletal rotary polytope $\mathcal{P}$ then the j-faces are in correspondence with the right cosets of G_j in G. This is the geometric analogue of the correspondence of the j-faces of an abstract orientable rotary polytope with the right cosets of the stabiliser of the base j-face under the rotation subgroup of the automorphism group.

Furthermore, if $\mathcal{P}$ has rank $n \geq 3$ and trivial flag stabilisers then the incidence of faces is recovered as follows. Two faces are incident if and only if the corresponding right cosets intersect non-trivially. The group G acts by right multiplication preserving the incidence just defined. This is a way of recovering the underlying abstract polytope of $\mathcal{P}$ from G.

One way of looking for geometrically rotary n-polytopes in $\mathbb{E}^d$ is through Wythoff's construction. The first step is to search for all sets of isometries $R_0, \ldots, R_{n-1}$ satisfying (6.1), or $S_1, \ldots, S_{n-1}$ satisfying (6.2). These will be our candidates for standard generators of the symmetry groups or their even subgroups of our polytopes. Then we must choose the initial vertex v_0 among all points that are not in the mirror of the first generator but are in the mirrors of all other generators. If the set of points meeting these two requirements is empty then the set of isometries must be disregarded as candidates; otherwise, the resulting structure from any such initial vertex v_0 already satisfies one of the axioms to be a polyhedron, as shown next.

Proposition 6.1.34 *Let* $\{S_1, \ldots, S_{n-1}\}$ *be a set of isometries satisfying* (6.2) *and let* $G = \langle S_1, \ldots, S_{n-1} \rangle$. *Assume that there exists a point* v_0 *invariant under* S_i *for all* $i \geq 2$, *but not invariant under* S_1. *Let* e_0 *be the line segment between* v_0 *and* $v_0 S_1 S_2$. *Then the graph with vertex set* $v_0 G$ *and edge set* $e_0 G$ *is connected.*

Proof Every vertex of the graph can be expressed as $v_0 T$ for some $T \in G$. We shall prove that there is a path from v_0 to $v_0 T$ for every $T \in G$.

We proceed by induction on the number m_T of factors of a shortest expression of T as a product of elements in $\{S_j, S_j^{-1} : j \in \{1, \ldots, n-1\}\}$. If $m_T = 0$ then $T = \mathrm{id}$ and $v_0 T = v_0$. Assume that $m_T \geq 1$ and that there is a path from v_0 to $v_0 T$ for all T with $m_T \leq k-1$. Let $T_1 = S_i^{\varepsilon} T_2 \in G$ with $i \in \{1, \ldots, n-1\}$, $\varepsilon \in \{1, -1\}$, $T_2 \in G$ with $m_{T_2} = k-1$.

If $i \neq 1$ then $v_0 S_i^{\varepsilon} = v_0$ and by induction hypothesis there is a path from v_0 to $v_0 T_2 = v_0 T_1$. If $i = 1$ and $\varepsilon = 1$ then we express T_1 as $S_1 S_2 S_2^{-1} T_2$. We know from the case $i \neq 1$ that there is a path from v_0 to $v_0 S_2^{-1} T_2$. Since e_0 joins v_0 and $v_0 S_1 S_2$ then there is an edge between $v_0 S_2^{-1} T_2$ and $v_0 S_1 S_2 S_2^{-1} T_2 = v_0 T_1$. Concatenating this edge with the path from v_0 to $v_0 S_2^{-1} T_2$, we obtain the desired path.

Finally, if $i = 1$ and $\varepsilon = -1$ then we express T_1 as $S_2 S_2^{-1} S_1^{-1} T_2 = S_2 S_1 S_2 T_2$. It follows that $v_0 T_1 = v_0 S_2 S_1 S_2 T_2 = v_0 S_1 S_2 T_2$. By the same argument as in the case $i = 1$ and $\varepsilon = 1$, the desired path can be constructed from a path from v_0 to $v_0 T_2$, and the edge from $v_0 T_2$ to $v_0 S_1 S_2 T_2 = v_0 T_1$. ■

Proposition 6.1.34 shows that the graph obtained by applying Wythoff's construction to a set of generators $S_1, \ldots, S_{n-1}$ and a suitable initial vertex is connected. An analogous proof shows that this is also true when the generators are $R_0, \ldots, R_{n-1}$. However, the structure resulting from Wythoff's construction may not be discrete; some of the polyhedra $P(a,b)$ described in [245, Section 5] (see also Section 6.2.3) are examples of this phenomenon. Also it is possible that the facets or the vertex-figures are not skeletal polytopes; some of the polyhedra $P_3(a,b)$ in [246, Section 6] (see also Section 6.2.2) are examples of this. Hence, after applying Wythoff's construction we still need to verify discreteness and polytopality of the facets and vertex-figures. When the resulting structure is a polytope we can determine whether it is regular or chiral by determining whether there exists an isometry mapping a flag to any of its adjacent flags.

Proposition 6.1.29 can be used as a criterion to determine that some structures arising from Wythoff's construction from a group G are not skeletal orientable rotary polytopes. This criterion becomes a necessary and sufficient condition, whenever the group G is known to be the subgroup of combinatorial rotations of the symmetry group of a regular polyhedron in $\mathbb{E}^3$ with trivial flag stabilisers.

Theorem 6.1.35 *Let $\mathcal{P}$ be a geometrically regular polyhedron in $\mathbb{E}^3$ with trivial flag stabilisers, and let $G^+(\mathcal{P}) = \langle S_1, S_2 \rangle$. Let $v_0 \in \mathbb{E}^3$ such that its stabiliser in $G^+(\mathcal{P})$ is $\langle S_2 \rangle$.*

(a) The structure $\mathcal{P}'$ constructed from Wythoff's construction from the group G and initial vertex v_0 is a skeletal rotary polyhedron, with $G^+(\mathcal{P}) = G^+(\mathcal{P}')$.

(b) If $\mathcal{P}$ is combinatorially orientable then $\mathcal{P}'$ is combinatorially isomorphic to $\mathcal{P}$.

(c) If $\mathcal{P}$ is combinatorially non-orientable then $\mathcal{P}'$ is combinatorially isomorphic to the orientable double cover of $\mathcal{P}$ as defined in Example 4.4.4.

Proof We know from Proposition 6.1.34 that the 1-skeleton of $\mathcal{P}'$ is connected. Since $G^+(\mathcal{P})$ is the rotation group of a rotary polytope, it is a discrete group of isometries and hence the vertex set of $\mathcal{P}'$ is discrete (see Exercise 6.1.18). To show that $\mathcal{P}'$ is a polyhedron it only remains to prove that the 2-faces and vertex-figures of $\mathcal{P}'$ are polygons.

The vertex and edge sets of the base 2-face f_0 of $\mathcal{P}'$ are respectively the orbit of v_0 and the orbit of the line segment between v_0 and $v_0 S_1 S_2 = v_0 S_1^{-1}$ under the action of $\langle S_1 \rangle$. Since the base 2-face of $\mathcal{P}$ is a polygon, S_1 must have order at least 3 (see Remark 6.1.1). On the other hand, Corollary 6.1.30 tells us that $\langle S_1 \rangle \cap \langle S_2 \rangle = \varnothing$, implying that the orbit of v_0 under $\langle S_1 \rangle$ has at least three points. The fact that f_0 is a polygon follows now from Remark 6.1.25.

We know from Lemma 6.1.17 that S_2 has finite order. It follows that it is either a rotation or a rotatory reflection. If S_2 is a rotatory reflection then it has only one fixed point; in this situation $\mathcal{P}' = \mathcal{P}$, which is a polyhedron. The remaining case is when S_2 is a rotation, and by Remark 6.1.1 it must have order at least 3. The vertex-figure at v_0 of $\mathcal{P}'$ has as vertex set $v_0 S_1^{-1} \langle S_2 \rangle$, and two of these vertices are joined by an edge if and only if they are $v_0 S_1^{-1} S_2^k$ and $v_0 S_1^{-1} S_2^{k+1}$ for some k. We claim that $v_0 S_1^{-1}$ is not invariant under S_2, since otherwise

$$v_0 S_1^{-1} = v_0 S_1^{-1} (S_2 S_1)^2 = v_0 S_1^{-1} S_1 (S_2 S_1) = v_0 S_2 S_1 = v_0 S_1,$$

contradicting that f_0 is a polygon. It follows that the orbit of $v_0 S_1^{-1}$ under S_2 has at least three points, and from Remark 6.1.25 we can conclude that the vertex-figure at v_0 is also a polygon.

From construction, S_1 and S_2 act like the standard rotations of $\mathcal{P}'$, implying that it is a rotary polyhedron. This concludes part (a).

Part (b) follows from Remark 6.1.33. Indeed, the fact that $\mathcal{P}$ has trivial flag stabilisers implies that $\Gamma^+(\mathcal{P}) \cong G^+(\mathcal{P})$, and the stabilisers of the base i-faces of $\mathcal{P}$ and $\mathcal{P}'$ under $G^+(\mathcal{P})$ are the same.

If $\mathcal{P}$ is non-orientable then the stabiliser in $G^+(\mathcal{P})$ of the base vertex of $\mathcal{P}$ contains $\langle S_2 \rangle$ as a proper subgroup, and hence $\mathcal{P}'$ is different from $\mathcal{P}$.

In order to show that $\mathcal{P}'$ is isomorphic to the orientable double cover of $\mathcal{P}$ we rely on the universal property given in Proposition 4.4.10. We only need to show that $\mathcal{P}'$ is a double cover of $\mathcal{P}$ and that it is orientable.

Let Φ be the base flag of $\mathcal{P}'$. As in part (b), the i-faces of $\mathcal{P}$ (resp. of $\mathcal{P}'$) are represented by the cosets of the stabiliser of the base i-face F_i (resp. F_i') under $G(\mathcal{P})$. For each $i \in \{0, 1, 2\}$ the stabiliser $\mathrm{Stab}_{G(\mathcal{P})}(F_i)$ is precisely the set $\mathrm{Stab}_{G(\mathcal{P})}(F_i') \cup [R_0 \mathrm{Stab}_{G(\mathcal{P})}(F_i')]$, where R_0 is the first standard reflection of $\mathcal{P}$. This allows us to define a mapping

$$\begin{aligned} \phi : \mathcal{P}' &\to \mathcal{P} \\ F_i' T &\mapsto F_i T, \end{aligned}$$

corresponding to the mapping of cosets

$$[\mathrm{Stab}_{G(\mathcal{P})}(F_i')]T \mapsto [\mathrm{Stab}_{G(\mathcal{P})}(F_i') \cup [R_0\mathrm{Stab}_{G(\mathcal{P})}(F_i')]]T.$$

It follows from Remark 6.1.33 that the mapping ϕ preserves incidence and hence $\mathcal{P}'$ is indeed a double cover of $\mathcal{P}$.

To verify orientability it is enough to note that Wythoff's construction naturally provides a two colouring of the flags. One chromatic class $\mathcal{C}$ consists of the set of images of the base flag under $G(\mathcal{P})$, while the other chromatic class is the complement of $\mathcal{C}$. This concludes the proof of (c). ∎

Remark 6.1.36 Under the notation in Theorem 6.1.35, it is possible that there exist points in the mirror of S_2 whose stabiliser under G is larger than $\langle S_2 \rangle$. In such cases it is expected that $\mathcal{P}'$ is not a polyhedron.

Proposition 6.1.26 lists all geometrically regular polygons in $\mathbb{E}^3$. It is easy to see that all possible pairs of involutions and suitable initial vertices are considered in the analysis preceding that result. Wythoff's construction applied to the group $\langle S_1 \rangle$ for every possible isometry S_1 and every possible initial vertex with orbit of size at least 3 yields the same family of polygons. The next result follows.

Corollary 6.1.37 *There are no chiral polygons in* $\mathbb{E}^3$.

Exercise 6.1.38 Show that there are no geometrically chiral polygons in $\mathbb{E}^n$ for every $n \in \mathbb{N}$.

In general the generators S_i of the symmetry group of a chiral polytope $\mathcal{P}$ may be orientation reversing isometries. Nevertheless, $\mathcal{P}$ itself must be orientable as an abstract polytope.

Proposition 6.1.39 *Every skeletal chiral polytope is combinatorially orientable.*

Proof The symmetry group of a geometrically chiral polytope induces a bicolouring on the flags such that adjacent flags are assigned different colours. There is no such colouring on non-orientable abstract polytopes. ∎

6.1.4 Skeletal Polytopes as Realisations of Abstract Polytopes

So far we have defined skeletal polytopes by constructing them directly in some pre-established space. An alternative approach to skeletal polytopes is through realisations. They pre-establish a polytope and determine the potential ambient spaces for it, together with the different ways in which this polytope can appear as a skeletal polytope in each of those spaces.

To realise a polytope $\mathcal{P}$ means to give a geometric interpretation to all its faces. Naturally each vertex (element of rank 0) of $\mathcal{P}$ is associated with a point (set of dimension 0) of some space $\mathcal{S}$. Then the edge between vertices v_1 and v_2 must correspond to some subset of $\mathcal{S}$ highly related with the points u_1 and u_2 associated to v_1 and v_2. The simplest way to do this is to think of the edge as the set $\{u_1, u_2\}$, although we may choose to take a line segment between u_1 and u_2 instead. This process can be continued recursively by associating to every j-face F the set of images of the $(j-1)$-faces incident to F or some subset of $\mathcal{S}$ highly related with these images.

Formally, a *realisation* of an n-polytope $\mathcal{P}$ in a space $\mathcal{S}$ is a function β_0 of the vertex set $\mathcal{P}_0$ to $\mathcal{S}$. This function makes it possible to recursively construct functions β_j from the set $\mathcal{P}_j$ of j-faces of $\mathcal{P}$ to the power set of $\mathcal{P}_{j-1}\beta_{j-1}$ given by

$$F \mapsto \{G\beta_{j-1} \colon G \in \mathcal{P}_{j-1}, G \leq F\}$$

for $1 \leq j \leq n-1$.

Under this definition, an edge of $\mathcal{P}$ is represented by a pair of points (the images of its endpoints under β_0). When convenient, it is possible to reinterpret it as the line segment between these two points.

This definition is still very general. The combinatorics of $\mathcal{P}$ may not be reflected in the geometry of the realisation. For example, β_0 can be defined as a constant function mapping every vertex to the same point of $\mathcal{S}$, in which case every pair of faces of distinct ranks are incident in the realisation. A realisation is said to be *faithful* whenever every β_i is injective. In that case there is a bijection between the set of j-faces of $\mathcal{P}$ with the set of their images.

Example 6.1.40 Let $\mathcal{X}_0 = \{a, b, c, d\}$ denote the vertex set of the tetrahedron and let $\beta_0, \delta_0 \colon \mathcal{X}_0 \mapsto \mathbb{E}^2$ be given by $a\beta_0 = a\delta_0 = (0,0)$, $b\beta_0 = b\delta_0 = (1,0)$, $c\beta_0 = c\delta_0 = d\delta_0 = (0,1)$, $d\beta_0 = (1,1)$ (see Figure 6.10). The realisation β_0 is faithful since every two distinct faces of the tetrahedron are mapped to two distinct sets of points. On the other hand, δ_0 is not faithful due to the fact that the triangles $\{a, b, c\}$ and $\{a, b, d\}$ have the same image.

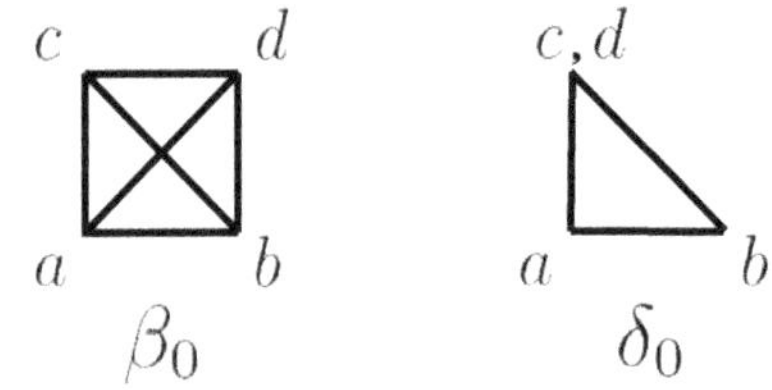

Figure 6.10 Two realisations of the tetrahedron.

Realisations may not relate the automorphisms of $\mathcal{P}$ with isometries of $\mathcal{S}$. A realisation is said to be *symmetric* if for every $\gamma \in \Gamma(\mathcal{P})$ there exists an isometry T_γ such that $\gamma\beta_j = \beta_j T_\gamma$. None of the realisations in Example 6.1.40 is symmetric; indeed, the permutation (a,b) induces an automorphism of the tetrahedron, but the permutations $(a\beta_0, b\beta_0)$ and $(a\delta_0, b\delta_0)$ cannot be extended to isometries of $\mathbb{E}^2$ that fix $c\beta_0$ and $d\beta_0$, and $c\delta_0$ and $d\delta_0$, respectively.

So far most of the works on realisations have dealt with symmetric realisations of regular polytopes in Euclidean spaces. Realisations that are not faithful are included in these studies since they can serve as building blocks of other realisations.

A wealth of results about realisations of regular polytopes can be found in [169, 171, 175, 178, 179, 182] and [188, Chapter 5].

Here we do not intend to present a complete theory of realisations of chiral polytopes. This is an interesting topic on its own and appears as problem 36 in [220] and as problem 21 in [250]; we restate it here for sake of completeness.

Problem 67 Develop the details of the realisation theory for chiral polytopes.

6.1.5 Six Geometrically Regular Polyhedra in $\mathbb{E}^3$

The remainder of this section will be devoted to describing six geometrically regular polyhedra that play an important role in Section 6.2. We adopt the names given in [167] to these polyhedra.

The complete list of geometrically regular polyhedra in $\mathbb{E}^3$, finished in [84], was achieved by analysing all possible discrete groups of isometries of $\mathbb{E}^3$ with three generators that make it possible to obtain geometrically regular polyhedra through Wythoff's construction. The proof of completeness of the list was improved in [167], where all regular n-polytopes in $\mathbb{E}^3$ with $n \geq 4$ are also included. In fact, there are no regular n-polytopes in $\mathbb{E}^3$ for $n \geq 5$ (see Proposition 6.2.2 and [167, Theorem 2.6 and Corollary 2.7]).

The polyhedra $\{4,6|4\}$, $\{6,4|4\}$ and $\{6,6|3\}$ are known as the Petrie–Coxeter polyhedra. The first two were discovered by Petrie in 1926, and soon after Coxeter found the third one. They are the only infinite non-planar regular polyhedra with finite planar 2-faces. Their vertex-figures are skew. They appear as the polyhedra in Class 4 in [104, Section 3] (see also [61]).

The polyhedra $\{\infty,3\}^{(a)}$ and $\{\infty,3\}^{(b)}$ were first described by Grünbaum in [104] with the names $\{\infty^{2\pi/3,2\pi/3},3\}$ and $\{\infty^{2\pi/3,\pi/2},3\}$, as part of the 47 regular polyhedra known at that time. The polyhedron $\{\infty,4\}_{\cdot,*3}$ was the only skeletal regular polyhedron in $\mathbb{E}^3$ missing in Grünbaum's list. It was found by Dress in [84] with the name $\{\infty^{\pi/2,2\pi/3},4\}$. They are the only regular polyhedra in $\mathbb{E}^3$ with helical 2-faces and planar vertex-figures.

We shall describe the polyhedron $\{\infty,3\}^{(a)}$ in full detail, and rely on previous work for the description of the remaining five polyhedra.

As in Example 2.4.24, we denote by $\mathbf{\Lambda}_{(1,0,0)}$, $\mathbf{\Lambda}_{(1,1,0)}$ and $\mathbf{\Lambda}_{(1,1,1)}$ the cubic lattice, face-centred cubic lattice, and body-centred cubic lattice, respectively. We also denote by $\Lambda_{(s^k,0^{3-k})}$ the lattice group with translation vectors in $\mathbf{\Lambda}_{(s^k,0^{3-k})}$.

The Polyhedron $\{4,6|4\}$

The polyhedron $\{4,6|4\}$ was described in Section 4.1 as the Petrie–Coxeter polyhedron of the cubic tessellation $\{4,3,4\}$ of $\mathbb{E}^3$. A portion of this infinite polyhedron is shown in Figure 6.11. To complement that description we give coordinates to its vertices and exhibit explicit generators for $G(\{4,6|4\})$.

We start with the tessellation by cubes with edge length 2 whose centres are in $\frac{1}{2}(1,1,1)+2\Lambda_{(1,0,0)}$, and follow the recipe in Section 4.1 to construct its Petrie–Coxeter polyhedron. The set $\mathcal{S}$ consists of the cubes with centres in $\frac{1}{2}(1,1,1)+2\Lambda_{(1,0,0)}$ and edge length 1. The union of the vertex sets of these cubes, and hence the vertex set of $\{4,6|4\}$, is $\mathbf{\Lambda}_{(1,0,0)}$. Every 2-face of $\{4,6|4\}$ is a square that contains two edges of cubes in $\mathcal{S}$, and two edges that belong to no cube in $\mathcal{S}$. These squares have vertices

$$(a,b,c),(a+1,b,c),(a+1,b+1,c),(a,b+1,c);\quad a+b\equiv 1 \text{ (modulo 2)},$$
$$(a,b,c),(a+1,b,c),(a+1,b,c+1),(a,b,c+1);\quad a+c\equiv 1 \text{ (modulo 2)},$$
$$(a,b,c),(a,b+1,c),(a,b+1,c+1),(a,b,c+1);\quad b+c\equiv 1 \text{ (modulo 2)}.$$

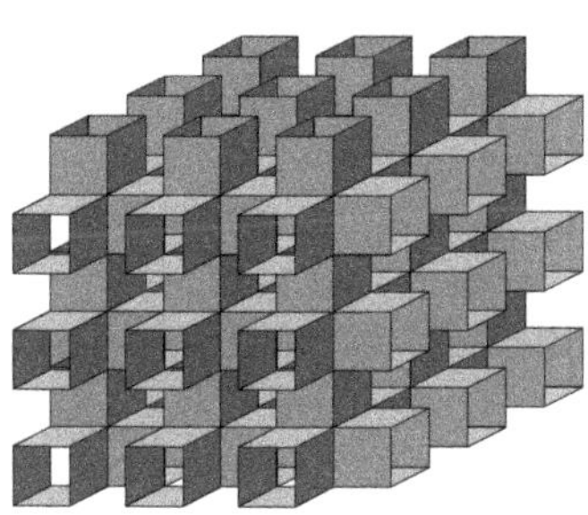

Figure 6.11 The polyhedron $\{4,6|4\}$.

It transpires that the edge set coincides with the edge set of the tessellation by cubes with vertices in $\mathbf{\Lambda}_{(1,0,0)}$, and therefore $\{4,6|4\}$ has the same 1-skeleton as the aforementioned tessellation.

By considering o as the base vertex, the edge between o and $(0,0,1)$ as the base edge, and the 2-face with set of vertices $\{o,(0,0,1),(0,-1,1),(0,-1,0)\}$ as the base 2-face, the standard generators of $G(\{4,6|4\})$ are

$$R_0: \quad (x,y,z) \mapsto (x,y,1-z), \tag{6.3}$$

$$R_1: \quad (x,y,z) \mapsto (-x,-z,-y), \tag{6.4}$$

$$R_2: \quad (x,y,z) \mapsto (y,x,z). \tag{6.5}$$

Both, R_0 and R_2 are plane reflections. The mirror of R_0 is the translate of the XY-plane by $(0,0,1/2)$ whereas that of R_2 is the plane with equation $x = y$. By contrast, R_1 is the half-turn whose axis contains o and has direction vector $(0,1,-1)$.

The polyhedron $\{4,6|4\}$ is invariant under the translations

$$T_1 := R_1R_0R_1R_2R_1R_2R_0R_1R_2R_1R_0, \quad T_2 := R_0T_1R_0, \quad T_3 := R_1R_0T_1R_0R_1$$

by vectors $(1,1,1)$, $(1,1,-1)$ and $(-1,1,-1)$, respectively. Hence, all translations in $\Lambda_{(1,1,1)}$ are symmetries of $\{4,6|4\}$. On the other hand, the set $\mathcal{S}$ is not invariant under the translations by $(1,0,0)$, $(0,1,0)$ or $(1,1,0)$ (representatives of the three non-trivial translates of $\mathbf{\Lambda}_{(1,1,1)}$ in $\mathbf{\Lambda}_{(1,0,0)}$), implying that the group of translations of $\{4,6|4\}$ consists precisely of all translations in $\Lambda_{(1,1,1)}$.

The Polyhedron $\{6,4|4\}$

The polyhedron $\{6,4|4\}$ is the geometric dual of $\{4,6|4\}$ obtained in the following way. The vertices of $\{6,4|4\}$ are the centres of the squares of $\{4,6|4\}$, and two of them are joined by an edge if the corresponding squares share an edge. Each 2-face of $\{6,4|4\}$ is a planar hexagon whose vertices are the centres of the six squares of $\{4,6|4\}$ around some vertex. A portion of this polyhedron is shown in Figure 6.12.

An alternative way of understanding $\{6,4|4\}$ is through the tessellation $\mathcal{T}_1$ by truncated octahedra of $\mathbb{E}^3$ (see Appendix A.3). The vertex and edge sets of $\{6,4|4\}$ coincide with those of $\mathcal{T}_1$, while the 2-faces of $\{6,4|4\}$ are all hexagons of $\mathcal{T}_1$, disregarding the squares. In this way, every edge belongs to precisely two hexagons, and every vertex to four hexagons.

The coordinates of the vertices of $\{6,4|4\}$ when seen as the dual of $\{4,6|4\}$ do not have integer coordinates. For instance, the base vertex is $\frac{1}{2}(0,-1,1)$. However, keeping this geometric placement of $\{6,4|4\}$ simplifies the description of the standard generators R_0, R_1 and R_2 of $G(\{6,4|4\})$ as

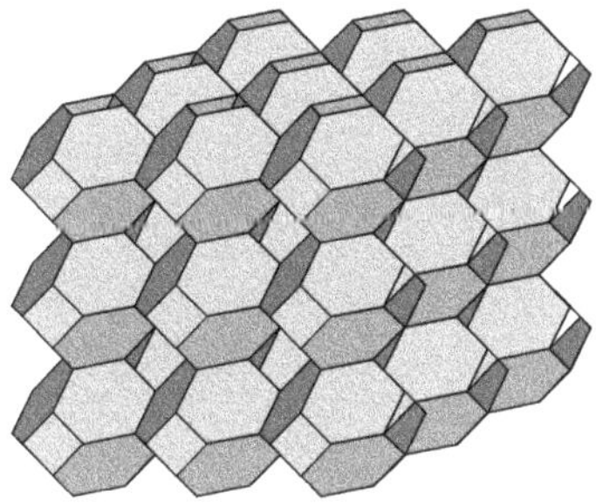

Figure 6.12 Section of the polyhedron $\{6,4|4\}$.

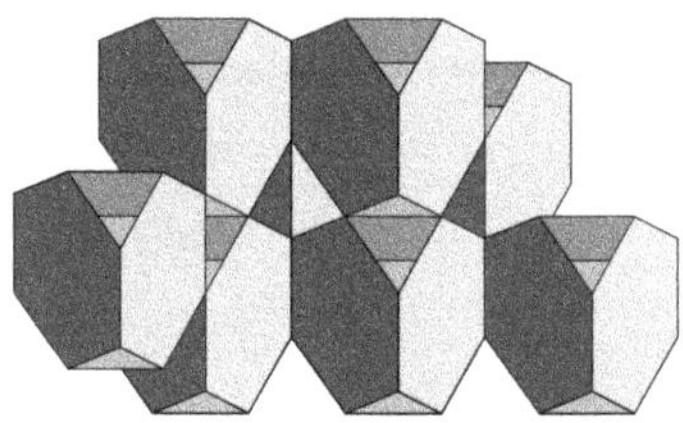

Figure 6.13 Section of the polyhedron $\{6,6|3\}$.

those given in (6.5), (6.4) and (6.3), respectively; that is, they are the standard generators of $G(\{4,6|4\})$ with the order reversed.

Instead, we may take a homotety by a factor of 2 and translate the vertex set by $(0,1,-1)$, so that o is the base vertex. Then the standard generators of $G(\{4,6|4\})$ are

$$R_0\colon\quad (x,y,z)\mapsto(y-1,x+1,z),$$
$$R_1\colon\quad (x,y,z)\mapsto(-x,-z,-y),$$
$$R_2\colon\quad (x,y,z)\mapsto(x,y,-z).$$

Once again, R_0 and R_2 are plane reflections and R_1 is a rotation. Their mirrors are, respectively, the planes with equations $y-x=1$ and $z=0$, and the line through o with direction vector $(0,1,-1)$.

The group of translations consists of all translations by vectors in $2\mathbf{\Lambda}_{(1,1,1)}$.

The Polyhedron $\{6,6|3\}$

We can describe the polyhedron $\{6,6|3\}$ through the tessellation $\mathcal{T}_2$ by tetrahedra and truncated tetrahedra described in Appendix A.3. The vertex and edge sets of $\{6,6|3\}$ coincide with those of $\mathcal{T}_2$. The 2-faces of $\{6,6|3\}$ are all hexagons of $\mathcal{T}_2$, disregarding the triangles (see Figure 6.13).

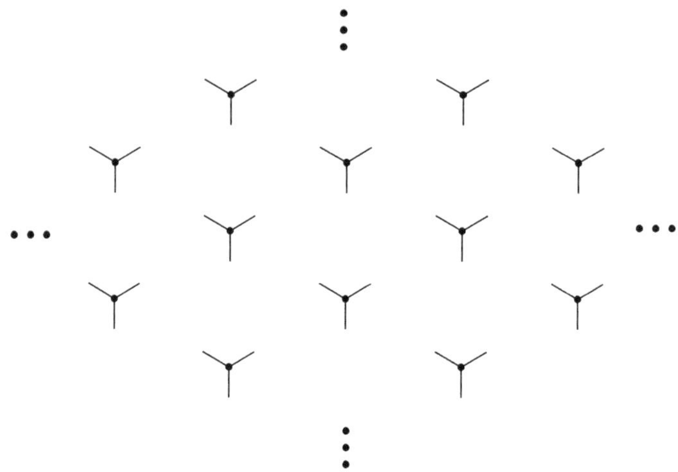

Figure 6.14 Map of the halls in the same floor of a building.

We choose the initial vertex to be o, the base edge to join o and $(1,0,1)$, and the base hexagon to have vertices o, $(1,0,1)$, $(2,-1,1)$, $(2,-2,0)$, $(1,-2,-1)$ and $(0,-1,-1)$. Then the standard generators of $G(\{6,6|3\})$ are

$$\begin{aligned} R_0&: \quad (x,y,z) \mapsto (1-z,y,1-x),\\ R_1&: \quad (x,y,z) \mapsto (-y,-x,-z),\\ R_2&: \quad (x,y,z) \mapsto (z,y,x). \end{aligned}$$

The mirrors of the plane reflections R_0 and R_2 are the planes with equations $x+z=1$ and $x-z=0$, respectively, whereas the mirror of the rotation R_1 is the line through o with direction vector $(1,-1,0)$.

The group of translations consists of all translations by vectors in $2\mathbf{\Lambda}_{(1,1,0)}$.

The Polyhedron $\{\infty,3\}^{(a)}$

Our first (intuitive) description of the polyhedron $\{\infty,3\}^{(a)}$ is as the set of halls and stairways of a certain building. It has infinitely many floors, each of which is infinitely long and wide. The halls of every floor are isometric to each other, and a map of any given floor looks like that in Figure 6.14. (Do not panic! There are ways to access every hall from every other hall.)

There are infinitely many stairways, making it possible to access from one floor to any other. Actually, at the end of every hall there is a stairway. Each stairway has one flight of stairs to go from the nth floor to the $(n+1)$th floor

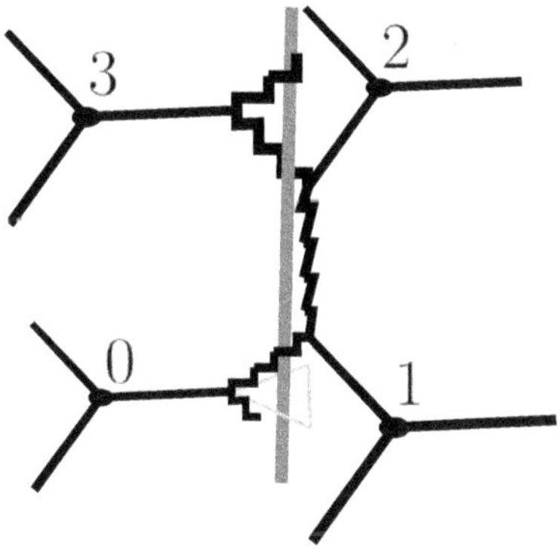

Figure 6.15 Three flights of a stairway with the halls at floors from 0 to 3.

for each $n \in \mathbb{Z}$. The steps in each flight are arranged in a straight line; that is, the stairways make no turns between two consecutive floors. But if we want to go up further to the $(n+2)$th floor, we need to make a turn of 60° to the left at the $(n+1)$th floor in order to take the next flight of steps. In this way, a person going from the nth to the $(n+1)$th floor and a person going from the $(n+3)$th to the $(n+4)$th floor along the same stairway could do so in such a way that the former behaves like a copy of the latter, only three floors below. In other words, while going up or down several floors along the same stairway, a person describes a motion along a helix over a triangle (See Figure 6.15, where the numbers indicate the floor of the halls, the grey line is the axis of the stairway, and the light grey triangle is the projection of the stairway to floor 0).

When arriving to the nth floor along a stairway there is only one hall to follow. If we arrive from the $(n-1)$th floor, we need to make a mild turn of 150° to the right to take the hall (just 30° away from continuing straight); if we arrive from the $(n+1)$th floor, we take the hall by turning 150° to the left. The hall leads to a common area with a circular shape. From the common area there are two new halls; one to the right at an angle of 120° and one at the same angle to the left. Each of these two halls leads to a stairway. All halls have the same length and, surprisingly enough, they have the same length as every flight of stairs.

It is convenient to think that the compasses of the habitants of this building have only three cardinal points at an angle of $2\pi/3$ from each other, instead of the conventional four points (see Figure 6.16). Such an unusual compass together with the local information given so far suffice to be able to navigate the building. Knowing the current location and the location of the destination

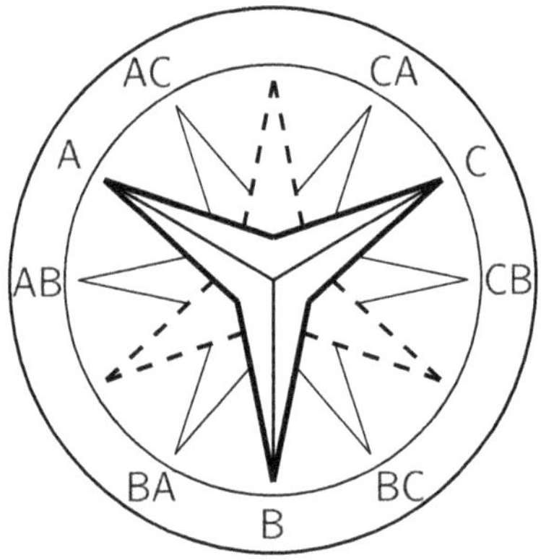

Figure 6.16 A compass with only three cardinal points.

(perhaps by some coordinate system) makes it possible, and not extremely complicated, to devise a path from one of these spots to the other. However, the logic to traverse in such a building is not the same as the one to traverse the buildings we know. One first thing to note is that it may be necessary to go up and down several times to go from one point to another of the same floor (see Exercise 6.1.41).

Standing at the centre of any arbitrary common area, the compass will say that the three halls are perfectly oriented to the three cardinal points, which we shall call A, B and C. The horizontal components of the flights of stairs are aligned precisely with the six directions AB, AC, BA, BC, CA and BC that form angles of $\pi/6$ and $\pi/2$ with the coordinate directions. These directions are illustrated in Figure 6.16. After getting used to the compass we quickly notice that AB, BA, and AC are respectively the same as CB, CA and BC, but in the opposite direction.

Moving long distances in the directions AB, BC and CA is easy and relatively fast, as long as going up or down stairs does not represent a major problem. Starting at a common area, in order to move along the direction AB it is enough to perform repeatedly the following three steps (see Figure 6.17, where the numbers indicate the floor of the halls, and the letters indicate the horizontal component of the direction).

- Take the hall on the direction A to the stair at its end.
- Take the flight of stairs whose horizontal component has direction AB.
- Take the hall to the common area (its direction will be the opposite of C).

While doing this movement, in the second step we will always go down. After being close enough to our destination it is then possible to adjust the height by

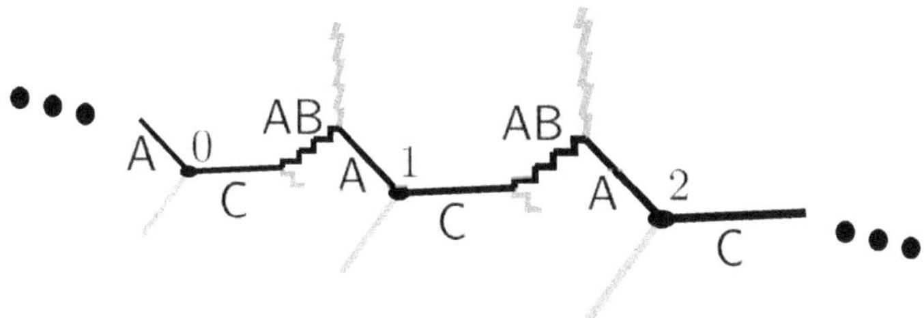

Figure 6.17 Moving long distances in a direction with horizontal component AB.

going up or down the necessary floors along a stairway. To go in the opposite direction it is necessary to start with the hall with direction C instead of A, and then the hall in the third step has the opposite direction to A, instead of the direction of C; the flights of stairs are all upwards. An analogous procedure can be followed to move long distances with a horizontal component in the directions of AC and BC.

After some time of wandering around in the building it is possible to deduce its global structure. We can think that in the axes of the helices described by each stairway there are very firm columns that hold the entire building together. At each floor, the central columns of the stairways are located on what would be the vertex set of a tessellation by triangles of that floor. If we colour properly the vertices of that tessellation with three colours (see Figure A.1(b)), we will notice that the directions of the halls at a given floor exiting from two different stairways are equal only when the stairways correspond to vertices in the same colour class. The common areas are all located on the centres of triangles facing down (say) of that tiling, but on each floor they occur only in one third of the triangles, namely in those where the corresponding colours of the vertices associated to the stairways are aligned in a certain way with respect to the three cardinal points.

It is a little harder to discover that there are four systems of parallel paths that one can use to build an easy recipe for moving from one point to another. One consists of the stairways, while the other three are the movements described above that allow people to move fast in the three directions AB, BC and CA with some vertical component. These are all left helices. Moreover, if the steps of the building are at a certain proportion (each flight of stairs would form approximately an angle of 54.74° with the horizontal, which is too high for us to go up and down comfortably), these helices are all congruent. Someone working on that building would not have any trouble understanding the polyhedron $\{\infty,3\}^{(a)}$ since it would be just a representation of the stairs and halls of the building. Nevertheless, that person would find unnatural that there are symmetries permuting the vertical direction with the directions of the other helices described in the movements in the directions AB, BC and CA; as we shall see, such symmetries do exist!

Someone standing at a common area may realise that there is no access to the point in the same horizontal location but only one floor below or one floor above. Perhaps the building was designed to host three isometric systems of stairs and halls that expand in all directions without being there any access from one system to the other two.

Exercise 6.1.41 Is it possible to move from one common area of the nth floor to any other common area of the same floor by moving exclusively along the nth and $(n+1)$th floor?

Now we proceed with the formal description of the polyhedron $\{\infty,3\}^{(a)}$. With initial vertex at the origin o, the generators of $G(\{\infty,3\}^{(a)})$ are

$$R_0\colon\quad (x,y,z)\mapsto(1-y,1-x,-z), \tag{6.6}$$

$$R_1\colon\quad (x,y,z)\mapsto(z,-y,x), \tag{6.7}$$

$$R_2\colon\quad (x,y,z)\mapsto(y,x,-z). \tag{6.8}$$

They are all half-turns and their mirrors are illustrated in Figure 6.18(a). It follows that $G(\{\infty,3\}^{(a)})$ has no orientation-reversing isometry. The mirrors of R_1 and R_2 intersect at an angle of $\pi/3$ and therefore S_2 is a 3-fold rotation around the line $\{\mu(1,-1,-1)\colon \mu\in\mathbb{R}\}$ perpendicular to both mirrors. The direction vectors of the mirrors of R_0 and R_1 also form an angle of $\pi/3$, although the mirrors do not intersect. Therefore S_1 is a 3-turn twist and S_1^3 is a translation. The axes of S_1 and S_2 are shown in Figure 6.18(b). The axis

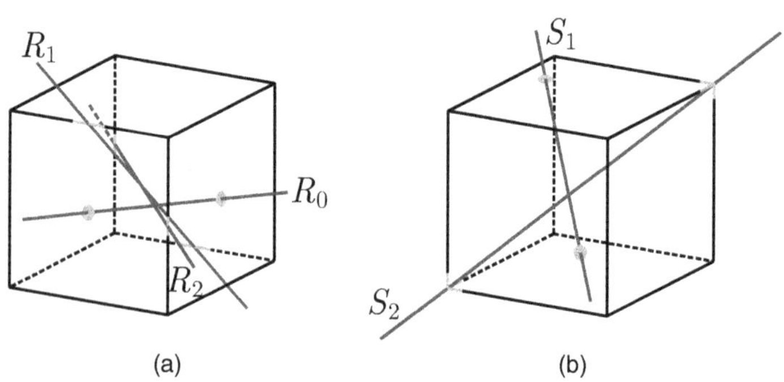

Figure 6.18 Mirrors of the generators (a) of $G(\{\infty,3\}^{(a)})$ and (b) of the even subgroup of $G(\{\infty,3\}^{(a)})$, with respect to the cube with vertices $(\pm1,\pm1,\pm1)$.

of S_1 is the line $(0, -\frac{1}{3}, \frac{2}{3}) + \{\mu(-1,-1,1): \mu \in \mathbb{R}\}$, and it intersects the front and top squares of the cube in that figure; the axis of S_2 contains one of the main diagonals of that cube.

The base 2-face H_1 is the helix with vertices $o\langle R_1 R_0\rangle$, that is,

$$\dots,(-2,-2,2),(-1,-1,2),(0,-1,1),(0,0,0),(1,1,0),(2,1,-1),(2,2,-2),\dots.$$

Regularity of $\{\infty,3\}^{(a)}$ forces all 2-faces to be helices isometric to H_1. The three neighbours of o are $(1,1,0)$, $(0,-1,1)$ and $(0,-1,1)R_2 = (-1,0,-1)$; all belonging to the plane $\Pi^{(1,-1,-1)}$ given by the equation $x-y-z=0$. The angle formed by the edges from o to any two of its neighbours is $2\pi/3$.

The symmetry S_1^3 is the translation by the vector $(-2,-2,2)$. Then $R_2S_1^3R_2$ and $R_1R_2S_1^3R_2R_1$ are the translations by the vectors $(-2,-2,-2)$ and $(-2,2,-2)$, respectively. Hence $G(\{\infty,3\}^{(a)})$ contains the group of all translations by vectors in $2\mathbf{\Lambda}_{(1,1,1)}$.

The plane $\Pi^{(1,-1,-1)}$ intersects $2\mathbf{\Lambda}_{(1,1,1)}$ in the set

$$\mathbf{\Lambda}^{(a)} := \{k(4,4,0)+m(4,0,4): k,m \in \mathbb{Z}\}. \tag{6.9}$$

Therefore all translates of the three edges incident with o by vectors in $\mathbf{\Lambda}^{(a)}$ are edges of $\{\infty,3\}^{(a)}$ that lie in $\Pi^{(1,-1,-1)}$. Any other translate of these three edges by vectors in $2\mathbf{\Lambda}_{(1,1,1)}$ lies on a plane parallel to $\Pi^{(1,-1,-1)}$ at a distance $2\sqrt{3}m/3$ for some $m \in \mathbb{Z}$; that is, on the plane $\Pi^{(1,-1,-1)} + \frac{m}{3}(2,-2,-2)$, which we shall denote by $\Pi_m^{(1,-1,-1)}$. The planes $\Pi_m^{(1,-1,-1)}$ play the role of the floors of the building in the previous visualisation of $\{\infty,3\}^{(a)}$. The edges between vertices in (6.9) correspond to those illustrated in Figure 6.14. The halls of the remaining floors are the translates of those in $\Pi^{(1,-1,-1)}$ by vectors in $2\mathbf{\Lambda}_{(1,1,1)}$.

The thin lines in Figure 6.19(a) represent the orthogonal projection of the 1-skeleton of the cubic tessellation $\{4,3,4\}$ to $\Pi^{(1,-1,-1)}$, with the bold dot representing the origin. The projection is taken so that we are looking at $\Pi^{(1,-1,-1)}$ from a point $k(2,-2,-2)$ for a large (positive) value of k. The edges of $\{\infty,3\}^{(a)}$ at o and their translates by vectors in $\mathbf{\Lambda}^{(a)}$ are the black solid line segments. For every $m \in \mathbb{Z}$, each translate in $\Pi_{3m}^{(1,-1,-1)}$ of the three edges at o project orthogonally to $\Pi^{(1,-1,-1)}$ into the three black edges at the bold dot. The translates in $\Pi_{3m+1}^{(1,-1,-1)}$ and in $\Pi_{3m+2}^{(1,-1,-1)}$ of the edges at o project orthogonally to $\Pi^{(1,-1,1)}$ into solid grey edges and into dashed grey edges, respectively.

The helix $H_4 := H_1R_2R_1R_0$ is the helix with vertices

$$\dots,(-1,3,2),(0,3,1),(1,2,1),(1,1,0),(2,1,-1),(3,0,-1),(3,-1,-2),\dots,$$

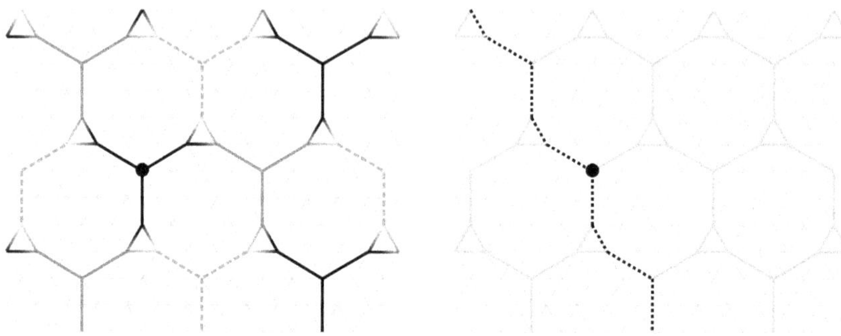

(a) (b)

Figure 6.19 The projection of $\{\infty, 3\}^{(a)}$ to $\Pi^{(1,-1,-1)}$. (a) Edges of the same colour are projections of edges in the same plane $\Pi_m^{(1,-1,-1)}$; (b) the projection of the helix H_1.

and axis in the direction of $(1,-1,-1)$, that is, perpendicular to $\Pi^{(1,-1,-1)}$. It has precisely one vertex in $\Pi_m^{(1,-1,-1)}$ for each $m \in \mathbb{Z}$ and each of its edges has one endpoint in $\Pi_m^{(1,-1,-1)}$ and one in $\Pi_{m+1}^{(1,-1,-1)}$ for some m. This helix projects to $\Pi^{(1,-1,1)}$ into a triangle. The translates of H_4 by vectors in $2\mathbf{\Lambda}_{(1,1,1)}$ are represented by the triangles with fading edges in Figure 6.19(a). The choice of placement of the observer guarantees that H_4 and its translates are left helices.

The base helix H_1 projects to $\Pi^{(1,-1,-1)}$ onto the dotted path D_1 in Figure 6.19(b), with $(0,-1,1)$ projecting into the point below the bold dot. Since the 3-fold rotation S_2 commutes with the projection to $\Pi^{(1,-1,-1)}$, the remaining two helices at o project to the images D_2 and D_3 of D_1 under rotations around the bold dot by an angle of $2\pi/3$ and $4\pi/3$, respectively. The translates of these three helices by vectors in $2\mathbf{\Lambda}_{(1,1,1)}$ project into images of D_1, D_2 and D_3 by translations by vectors in $\Pi^{(1,-1,-1)}$ that map triangles into triangles.

There is only one helix of $\{\infty, 3\}^{(a)}$ projecting to each triangle with fading edges in Figure 6.19(a). On the other hand, if some helix H projects to some translate D of D_0, D_1 or D_2 then also all translates of H by integer multiples of $(2,-2,-2)$ project to D.

For convenience we denote by $H2\Lambda_{(1,1,1)}$ the set of translates of the helix H by vectors in $2\mathbf{\Lambda}_{(1,1,1)}$, by H_2 and H_3 the helices $H_1 S_2$ and $H_1 S_2^2$, respectively, and by $\mathcal{H}$ the set

$$(H_1 2\Lambda_{(1,1,1)}) \cup (H_2 2\Lambda_{(1,1,1)}) \cup (H_3 2\Lambda_{(1,1,1)}) \cup (H_4 2\Lambda_{(1,1,1)}).$$

It is now possible to verify that every edge of H_1 belongs to at least two helices. The edge between $(0,0,0)$ and $(1,1,0)$ belongs also to H_2, the edge between $(0,-1,1)$ and $(0,0,0)$ belongs also to H_3, and the edge between $(1,1,0)$ and $(2,1,-1)$ belongs also to H_4. The remaining edges of H_1 are translates by a multiple of $(2,-2,-2)$ of these three edges and, hence, each of them belongs to a helix in $H_2 2\Lambda_{(1,1,1)}$, $H_3 2\Lambda_{(1,1,1)}$ or $H_4 2\Lambda_{(1,1,1)}$. By translating by suitable vectors in $2\mathbf{\Lambda}_{(1,1,1)}$ we can conclude that every edge in a helix in $H_1 2\Lambda_{(1,1,1)}$ belongs also to a helix in $H_2 2\Lambda_{(1,1,1)}$, in $H_3 \Lambda_{(1,1,1)}$ or in $H_4 \Lambda_{(1,1,1)}$.

The edges in H_2, H_3 and H_4 that are not in a helix of $H_1 \Lambda_{(1,1,1)}$ also belong to at least two helices in $\mathcal{H}$. To verify this it is enough to note that every edge in any of these three helices is either a translate by a vector in $2\mathbf{\Lambda}_{(1,1,1)}$ of one of the edges of H_1, of the edge between $(0,0,0)$ and $(-1,0,-1)$, of the edge between $(1,1,0)$ and $(1,2,1)$, or of the edge between $(0,-1,1)$ and $(1,-2,1)$. The edge between $(0,0,0)$ and $(-1,0,-1)$ belongs to H_2 and to H_3, the edge between $(1,1,0)$ and $(1,2,1)$ belongs to H_2 and to H_4, whereas the edge between $(0,-1,1)$ and $(1,-2,1)$ is in H_3 and a translate by $(-2,-2,2)$ of an edge of H_4. From here it follows that every edge of a helix in $H_2 \Lambda_{(1,1,1)}$, $H_3 \Lambda_{(1,1,1)}$ and $H_4 \Lambda_{(1,1,1)}$ belongs to at least some other helix in $\mathcal{H}$.

It is also true that every edge of a helix in $\mathcal{H}$ belongs to no more than two such helices. One way of verifying this is to note that the direction vectors of all edges of helices in $\mathcal{H}$ are precisely those of the edges of a regular tetrahedron, namely $(1,1,0)$, $(1,-1,0)$, $(1,0,1)$, $(1,0,-1)$, $(0,1,1)$ and $(0,-1,1)$, but each vector occurs in exactly two of the helices H_1, H_2, H_3 and H_4. In fact, the directions of the edges of a helix are precisely those of the three edges incident to one of the vertices of the tetrahedron; hence, every edge of the tetrahedron represents two families of helices, namely the families represented by its two vertices.

Since the 2-faces of $\{\infty,3\}^{(a)}$ are helices over triangles, every vertex of the helices H_1, H_2 and H_3 is a translate by a vector in $2\mathbf{\Lambda}_{(1,1,1)}$ of o or of one of its three neighbours. This statement also holds for the helix H_4, since $(1,2,1)$ and $(2,1,-1)$ are translates of $(-1,0,-1)$ by $(2,2,2)$ and of $(0,-1,1)$ by $(2,2,-2)$, respectively. On the other hand, H_1 only has translates of $(0,0,0)$, $(0,-1,1)$ and $(1,1,0)$; H_2 only has translates of $(0,0,0)$, $(1,1,0)$ and $(-1,0,-1)$; H_3 only has translates of $(0,0,0)$, $(0,-1,1)$ and $(-1,0,-1)$; and H_4 only has translates of $(1,1,0)$, $(-1,0,-1)$ and $(0,-1,1)$.

Each of the neighbours of o belongs to precisely three helices in $\mathcal{H}$ and hence is incident to precisely three edges. Since all other vertices of helices in $\mathcal{H}$ are translates of o or of its neighbours, every such vertex is incident to precisely three edges of helices in $\mathcal{H}$.

The set $\mathcal{H}$ is a set of polygons with a discrete vertex set and a connected underlying graph, where every edge belongs to two polygons and the vertex-figures are all triangles. We can now conclude that this constitutes a skeletal polyhedron. This polyhedron is precisely $\{\infty,3\}^{(a)}$. The formal verification that this is the polyhedron constructed by Wythoff's construction from the generators R_0, R_1 and R_2 is left as an exercise.

Exercise 6.1.42 Show that the polyhedron described above coincides with the one obtained by Wythoff's construction with initial vertex at o and generators given in (6.6), (6.7) and (6.8).

As mentioned earlier, the polyhedron $\{\infty,3\}^{(a)}$ has only orientation-preserving symmetries. Therefore it only contains left helices. Its image under a plane reflection would give the geometric enantiomorph of $\{\infty,3\}^{(a)}$.

The symmetry group of $\{\infty,3\}^{(a)}$ was described in [167, Theorem 7.2] as the quotient of the Coxeter group $[\infty,3] = \langle R_0, R_1, R_2\rangle$ by the single relation

$$(R_0R_1)^3(R_0R_1R_2)^4 = (R_0R_1R_2)^4(R_0R_1)^3,$$

which together with Lemma 3.1.32 implies the following result.

Proposition 6.1.43 *The polyhedron $\{\infty,3\}^{(a)}$ is combinatorially orientable.*

Every symmetry in the stabiliser of o in $G(\{\infty,3\}^{(a)})$ must preserve the plane containing o and its neighbours, and permute the three edges at o. This forces the stabiliser to be the dihedral group $\langle R_1, R_2\rangle$ since these are the only six orientation-preserving isometries fulfilling the previous requirements.

Vertices of $\{\infty,3\}^{(a)}$ in distinct orbits under the group of translations by vectors in $2\mathbf{\Lambda}_{(1,1,1)}$ have their neighbours in distinct directions. This can be used to see that they are not translates of each other, and that the translation subgroup of $G(\{\infty,3\}^{(a)})$ contains only the translations by vectors in $2\mathbf{\Lambda}_{(1,1,1)}$. Since the vertex stabilisers have 6 elements and there are 4 orbits of vertices under the translation group, the latter has index 24 in $G(\{\infty,3\}^{(a)})$.

Exercise 6.1.44 Determine all values of n for which there are n-cycles in the 1-skeleton of $\{\infty,3\}^{(a)}$.

This concludes our description of $\{\infty,3\}^{(a)}$. We encourage the reader to corroborate that the stairs and halls of the building described at the beginning of our discussion matches the skeletal polyhedron just described.

The Polyhedron $\{\infty, 3\}^{(b)}$

The polyhedron $\{\infty, 3\}^{(b)}$ is the Petrial of $\{\infty, 3\}^{(a)}$ and therefore both have the same 1-skeleton and the same symmetry group. With initial vertex at the origin o, the group $G(\{\infty, 3\}^{(b)})$ is generated by

$$R_0: \quad (x, y, z) \mapsto (1 - x, 1 - y, z), \tag{6.10}$$

$$R_1: \quad (x, y, z) \mapsto (z, -y, x), \tag{6.11}$$

$$R_2: \quad (x, y, z) \mapsto (y, x, -z). \tag{6.12}$$

Figure 6.20 shows the mirrors of the generators R_i of $G(\{\infty, 3\}^{(b)})$ as well as the axes of the combinatorial rotations S_i. The generators R_0 and R_1 fix setwise the line l through $(1/2, 0, 1/2)$ with direction vector $(0, 1, 0)$ (marked as S_1 in Figure 6.20(b)), and their mirrors do not intersect. This implies that the base 2-face is a helix with the line l as its axis. The vertices of the base helix H_1 are

$$\ldots, (1, -3, 0), (1, -2, 1), (0, -1, 1), (0, 0, 0), (1, 1, 0), (1, 2, 1), (0, 3, 1), \ldots.$$

We recall that the translations of $\{\infty, 3\}^{(a)}$ (and hence also of $\{\infty, 3\}^{(b)}$) are by vectors in $2\mathbf{\Lambda}_{(1,1,1)}$. The intersection of $2\mathbf{\Lambda}_{(1,1,1)}$ with the XZ-plane (perpendicular to the axis of the base helix) is

$$\mathbf{\Lambda}^{(b)} := \{k(4, 0, 0) + m(0, 0, 4) : k, m \in \mathbb{Z}\}.$$

The directions of the coordinate axes are permuted by R_0, R_1 and R_2. Since the axis of the base helix has direction vector $(0, 1, 0)$, then the axes of all helical 2-faces of $\{\infty, 3\}^{(b)}$ are parallel to coordinate axes. The set $\mathcal{H}_1$ of helices with axes with direction vector $(0, 1, 0)$ consists of H_1, its traslate by $(2, 2, 2)$, and all the translates of these two helices by vectors in $\mathbf{\Lambda}^{(b)}$.

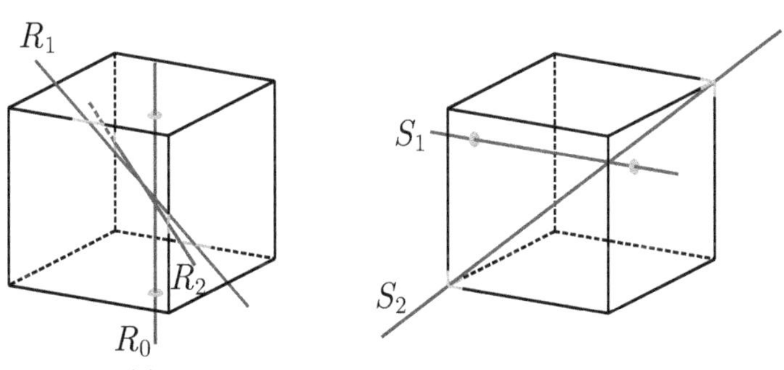

Figure 6.20 Mirrors of the generators (a) of $G(\{\infty, 3\}^{(b)})$ and (b) of the even subgroup of $G(\{\infty, 3\}^{(b)})$.

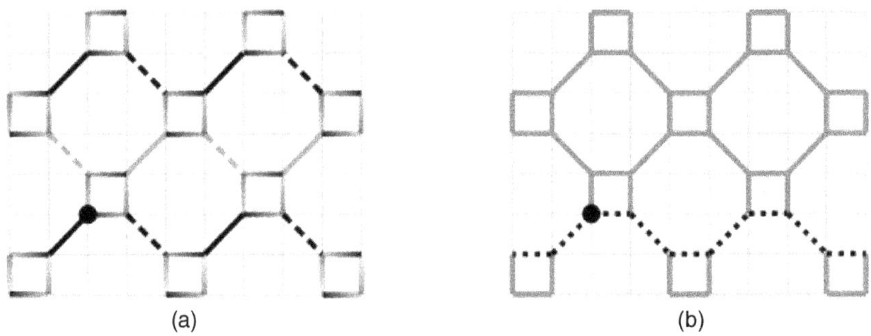

Figure 6.21 The projection of $\{\infty, 3\}^{(b)}$ to the XZ-plane.

Their orthogonal projections to the plane XZ consists of all squares with fading edges in Figure 6.21(a).

Every edge that is not in a helix of $\mathcal{H}_1$ is incident to vertices in two distinct such helices. Furthermore, the three edges at any vertex are the image under some element in $G(\{\infty, 3\}^{(b)})$ of the three edges at o, and therefore they are coplanar, at an angle of $2\pi/3$ from each other. When searching for the placement of the third edge at a vertex of a helix in $\mathcal{H}_1$ it transpires that every edge that is not in a helix of $\mathcal{H}_1$ is contained in a plane parallel to the XZ-plane.

The polyhedron $\{\infty, 3\}^{(b)}$ can be better understood with the help of layers of planes parallel to the XZ-plane. The neighbour of o that is not in H_1 is $(-1, 0, -1)$. The edge between o and $(-1, 0, -1)$ joins helix H_1 with its translate by $(-2, -2, -2)$. The translates of the edge between o and $(-1, 0, -1)$ by vectors in $\mathbf{\Lambda}^{(b)}$ are all edges of $\{\infty, 3\}^{(b)}$ in the XZ-plane; they are represented with solid black edges in Figure 6.21(a). The images of these edges under S_1^{-1}, S_1^{-2} and S_1^{-3} belong to the planes $y = 1$, $y = 2$ and $y = 3$, respectively, and project to the XZ-plane into the dashed black edges, the solid grey edges and the dashed grey edges, respectively. Since the translation by $(0, 4, 0)$ is a symmetry of $\{\infty, 3\}^{(b)}$, every edge of this polyhedron contained in a plane $y = b$ projects orthogonally to the XZ-plane into a solid or dashed, black or grey edge. More precisely, it projects to a solid black edge if $b \equiv 0$, to a dashed black edge if $b \equiv 1$, to a solid grey edge if $b \equiv 2$ and to a dashed grey edge if $b \equiv 3$, modulo 4.

The 1-skeleton of $\{\infty, 3\}^{(b)}$ can then be interpreted as a collection of towers (or stairs) going up in the direction of the Y-axis, each with the shape of a right helix over a square. Each tower has bridges to the four closest towers. Every bridge is horizontal (parallel to the XZ-plane) and it joins two towers at some floor where the distance between the towers is minimal. Two consecutive bridges between two given towers are separated by four floors. The towers are arranged in such a way that at every floor there is a bridge to one and only one neighbouring tower. In any tower, when going up from a floor with a southwestern bridge we meet first a floor with a northwestern bridge, then a floor with a northeastern bridge, after that we arrive to a floor with a southeastern bridge, and only then to another floor with a southwestern bridge.

Exercise 6.1.45 Is it possible to move from the nth floor of one tower to the nth floor of any other tower by walking only along the nth floor and the $(n+1)$th floor? Is it possible if it is allowed to walk only along the nth, the $(n+1)$th and the $(n-1)$th floors?

The dotted line in Figure 6.21(b) represents the projection of a helix H_2 with axis parallel to the X-axis. Every translate of H_2 projects to a translate of the dotted path. The helices with axis in the direction of $(0,0,1)$ project to images of the dotted path by rotations of $\pi/2$. Every bridge belongs to one helix with axis in the direction $(1,0,0)$ and one in the direction $(0,0,1)$.

Exercise 6.1.46 How many floors does the helix H_2 use? When traversing along H_2, which edges that are not bridges will be traversed with an upward direction and which ones with a downward direction?

Exercise 6.1.47 Is it possible to move from o to $(12,12,0)$ by using only two helices of $\{\infty,3\}^{(a)}$? Is it possible to move between these two points by using only two helices of $\{\infty,3\}^{(b)}$? In case of affirmative answers, which of these two ways uses fewer edges?

The symmetry group of $\{\infty,3\}^{(b)}$ was described in [167, Theorem 7.2] as the quotient of the Coxeter group $[\infty,3] = \langle R_0, R_1, R_2\rangle$ by the single relation

$$(R_0R_1)^4(R_0R_1R_2)^3 = (R_0R_1R_2)^3(R_0R_1)^4.$$

The next proposition follows from Lemma 3.1.32.

Proposition 6.1.48 *The underlying abstract polytope of* $\{\infty,3\}^{(b)}$ *is orientable.*

Before moving on to the description of our last polyhedron of the section, we expand on the isometries in $G(\{\infty,3\}^{(a)}) = G(\{\infty,3\}^{(b)})$. The translation subgroup has index 24 and hence the point group $G_0(\{\infty,3\}^{(a)})$ has precisely 24 elements. By noting that all generators preserve setwise the three coordinate directions we can conclude that $G_0(\{\infty,3\}^{(a)})$ consists of the 24 orientation

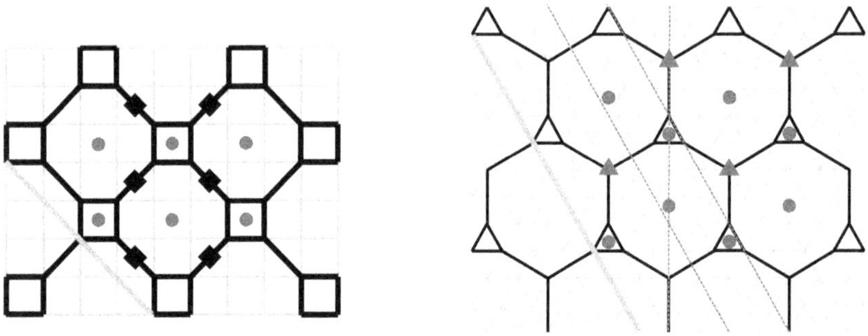

Figure 6.22 Projections of mirrors of rotations in $G(\{\infty,3\}^{(a)})$ (a) to the XZ-plane and (b) to the plane $\Pi^{(1,-1,-1)}$.

preserving symmetries of the octahedron. This allows us to express every element in $G(\{\infty,3\}^{(a)})$ as RT, where R is a rotation of the octahedron $\mathcal{O}$ with vertex set $\{(\pm 1,0,0),(0,\pm 1,0),(0,0,\pm 1)\}$ and T is a translation.

There are nine rotations in $G(\mathcal{O})$ whose axes are parallel to the coordinate axes. Three of these are half-turns, and the remaining six are 4-fold rotations. If $S \in G(\{\infty,3\}^{(a)})$ and $S = RT$ for some translation T and some 4-fold rotation R with axis l parallel to the Y-axis (say), then l must project to the XZ-plane into a point like the bold dark circles in Figure 6.22(a). Since 4-fold rotations about axes parallel to the Y-axis do not do not map bridges to bridges (see also Figure 6.21), such symmetries S must be 4-turn twists. The same must be true for the cases when the axis of R is parallel to the X- or Z-axis. The square power of any of these 4-turn twists is a half-turn twist. On the other hand, the bold dark squares in Figure 6.22(a) are projections of axes of genuine half-turns in $S \in G(\{\infty,3\}^{(a)})$ with axes parallel to the Y-axis.

The group $G(\mathcal{O})$ contains eight rotations of order 3, two of them fixing each pair of opposite triangles of $\mathcal{O}$. If $S \in G(\{\infty,3\}^{(a)})$ and $S = RT$ for some translation T and some 3-fold rotation R with axis l with direction vector $(1,-1,-1)$, then l must project to the plane $\Pi^{(1,-1,-1)}$ into a grey bold point or a grey triangle in Figure 6.22(b). If the projection of l is a bold point then l is a 3-turn twist since the rotations on such lines do not preserve the edge set (see also Figure 6.19). On the other hand, S_2 projects to a grey triangle, and in general there is a 3-fold rotation in $G(\{\infty,3\}^{(a)})$ whose axis projects to any given grey triangle in Figure 6.22(b).

The axes of 3-fold rotations in $G(\{\infty,3\}^{(a)})$ that do not have direction vector $(1,-1,-1,)$ must have direction vectors in $\{(1,1,1),(-1,1,-1),(-1,-1,1)\}$. Each of these axes contains infinitely many vertices of $\{\infty,3\}^{(a)}$

and projects to $\Pi^{(1,-1,-1)}$ to the image of the thick grey line in Figure 6.22(b) under a translation or 3-fold rotation. One can see now that distinct axes of 3-fold rotations of $\{\infty,3\}^{(a)}$ do not intersect. In fact, the set of axes of 3-fold rotations of $\{\infty,3\}^{(a)}$ is similar to the arrangement of non-intersecting lines $\mathcal{L}$ that covers $\mathbf{\Lambda}_{(1,1,1)}$, described in Appendix A.4. The details are left as an exercise.

Exercise 6.1.49 Show that the set consisting of all axes of 3-fold rotations of $\{\infty,3\}^{(a)}$ is similar to the set $\mathcal{L}$ described in Appendix A.4.

The remaining six non-trivial rotations in $G(\mathcal{O})$ are half-turns with respect to axes parallel to the edges of $\mathcal{O}$. All half-turns of $\{\infty,3\}^{(a)}$ whose axes are parallel to an edge of $\mathcal{O}$ are conjugate to some abstract reflection R_i. The details are left as an exercise.

Exercise 6.1.50 Show that every half-turn in $G(\{\infty,3\}^{(a)})$ whose axis is not parallel to a coordinate axis is a conjugate of one of the abstract reflections R_0 or R_2.

It follows from Exercise 6.1.50 that the axis of each rotation $S \in G(\{\infty,3\}^{(a)})$ whose axis is parallel to an edge of $\mathcal{O}$ either contains an edge of $\{\infty,3\}^{(a)}$, or is perpendicular to an edge of $\{\infty,3\}^{(a)}$. In the first case, the axis of S must project to $\Pi^{(1,-1,-1)}$ into a line containing an edge, like the vertical dotted line in Figure 6.22(b). In the second case, the axis of S must project to $\Pi^{(1,-1,-1)}$ into a line perpendicular to an edge (even if the edge of $\{\infty,3\}^{(a)}$ perpendicular to the axis of S does not belong to a plane parallel to $\Pi^{(1,-1,-1)}$). The two non-vertical dotted lines in Figure 6.22(b) are examples of the projections of the axes of this kind of rotations.

The axes of rotations in $G(\{\infty,3\}^{(a)})$ that intersect the axis of S_2 are of particular interest in Section 6.2.

Lemma 6.1.51 *Let $T \neq S_2$ be a rotation in $G(\{\infty,3\}^{(a)})$ such that its mirror intersects the miror of the rotation S_2. Then T is a half-turn with axis parallel to an edge of $G(\{\infty,3\}^{(a)})$ whose direction vector is perpendicular to the mirror of S_2.*

Proof Let l_2 denote the mirror of S_2.

As noted above, $G(\{\infty,3\}^{(a)})$ contains no 4-fold rotations. The thick grey line in Figure 6.22(a) shows the projection of l_2 to the XZ-plane. From that figure it becomes apparent that no half-turn has its axis perpendicular to the

XZ-plane and simultaneously intersecting l_2. Since the projections to the XY-plane and to the YZ-plane of $\{\infty, 3\}^{(a)}$ and of l_2 are isometric to that shown in Figure 6.22(a), l_2 does not intersect the axis of any half-turn if its direction is parallel to a coordinate axis.

The projection of l_2 to the plane $\Pi^{(1,-1,-1)}$ is a point represented by a triangle in Figure 6.22(b). If $S \neq S_2$ is a 3-fold rotation in $G(\{\infty, 3\}^{(a)})$ then its mirror is either parallel to l_2, or projects to $\Pi^{(1,-1,-1)}$ into an image of the thick grey line in Figure 6.22(b) by some symmetry of the projection. Since the thick grey line does not contain points represented by triangles, the axis of no 3-fold rotation in $G(\{\infty, 3\}^{(a)})$ intersects the axis of S_2 (see also Exercise 6.1.49).

It only remains to discard half-turns in $G(\{\infty, 3\}^{(a)})$ whose axes are parallel to the edges of the octahedron $\mathcal{O}$ but not perpendicular to l_2. This can be done by noting that the projections to $\Pi^{(1,-1,-1)}$ of these lines do not contain points represented by triangles in Figure 6.22(b). ■

The Polyhedron $\{\infty, 4\}_{\cdot,*3}$

The vertex set of $\{\infty, 4\}_{\cdot,*3}$ is isometric to $\mathbf{\Lambda}_{(1,0,0)} \setminus \mathbf{\Lambda}_{(1,1,1)}$. We choose to translate that set by $(1,0,0)$ so that o can serve as the base vertex. Two vertices of $\{\infty, 4\}_{\cdot,*3}$ are adjacent if and only if they differ by a vector in the canonical basis. Every vertex v has valency four and the neighbours of v all belong to a plane that contains v. The 2-faces are all the right helices over triangles contained in the 1-skeleton.

With initial vertex at the origin o, we take the following generators for $G(\{\infty, 4\}_{\cdot,*3})$:

$$R_0: \quad (x, y, z) \mapsto (y, x, -1 - z), \tag{6.13}$$

$$R_1: \quad (x, y, z) \mapsto (-x, z, y), \tag{6.14}$$

$$R_2: \quad (x, y, z) \mapsto (-x, -y, z). \tag{6.15}$$

The mirrors of the generators R_i and the axes of the combinatorial rotations S_i are illustrated in Figure 6.23. The base helix H_1 has vertices

$$\ldots, (1,-1,1), (1,-1,0), (0,-1,0), (0,0,0), (0,0,-1), (-1,0,-1), (-1,1,-1), \ldots$$

and axis $\{(0, -\frac{1}{3}, -\frac{1}{3}) + \mu(1,-1,1) : \mu \in \mathbb{R}\}$. The remaining three helices at o are the images of H_1 under powers of a 4-fold rotation around the X-axis.

The translation group consists of all translations in $\Lambda_{(1,1,1)}$, and hence there are only three orbits of vertices under the translation group. The set $\{o, (0,1,0), (0,0,1)\}$ is a system of representatives of these orbits. The four

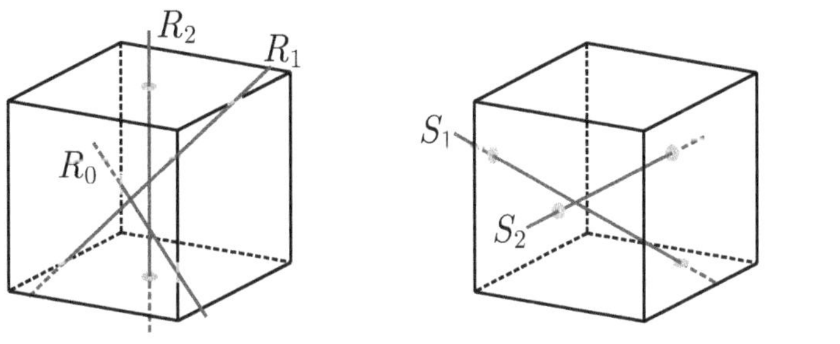

Figure 6.23 Mirrors of the generators (a) of $G(\{\infty,4\}_{\cdot,*3}$ and (b) of the even subgroup of $G(\{\infty,4\}_{\cdot,*3}$.

Figure 6.24 The projection to the plane $\Pi^{(1,-1,1)}$ (a) of the helices with axis in the direction of $(1,-1,1)$ and (b) of a helix whose axis is not in that direction.

edges at a vertex in the orbit of o (resp. of $(0,1,0)$ and of $(0,0,1)$) are parallel to the YZ-plane (resp. XY-plane and XZ-plane).

Figure 6.24(a) shows in black the projection of $\{\infty,4\}_{\cdot,*3}$ to the plane $\Pi^{(1,-1,1)}$ with equation $x-y+z=0$, perpendicular to the axis of H_1; the grey edges are projections of edges of the cubic tiling that are not edges of $\{\infty,4\}_{\cdot,*3}$. The black triangles with thick edges are the projections of all translates of H_1 by elements of $\Lambda_{(1,1,1)}$. The remaining helices project to paths like the dotted one in Figure 6.24(b) and its images under 3-fold rotations.

Remark 6.1.52 The images of the helices of $\{\infty,4\}_{\cdot,*3}$ under the reflection about the XY-plane also have their vertices and edges in the 1-skeleton of $\{\infty,4\}_{\cdot,*3}$; those with axes in the direction of $(1,-1,1)$ project to $\Pi^{(1,-1,1)}$ precisely into the thin black triangles in Figure 6.24(a). However, they are not 2-faces of this polyhedron since $G(\{\infty,4\}_{\cdot,*3})$ has no orientation-reversing symmetries. The 1-skeleton of $\{\infty,4\}_{\cdot,*3}$ together with the images of all its helices under that reflection constitute the geometric enantiomorph of $\{\infty,4\}_{\cdot,*3}$.

The polyhedron $\{\infty,4\}_{\cdot,*3}$ has no edges parallel or perpendicular to $\Pi^{(1,-1,1)}$. The direction vectors of all its edges make the same angle with $(1,-1,1)$. If we intend to model $\{\infty,4\}_{\cdot,*3}$ as a building or set of buildings as we did for $\{\infty,3\}^{(a)}$ and $\{\infty,3\}^{(b)}$, where $\Pi^{(1,-1,1)}$ represents a horizontal plane, then every movement along an edge represents a change of height with respect to this plane. When traversing a helix H whose axis is not vertical, a traveller would find the following pattern that repeats every three edges; the traveller would go upwards along two consecutive edges and downward along the third one. Traversing H in the opposite direction, the traveller would go down along two consecutive edges and go up along the third edge.

If we model a building as $\{\infty,4\}_{\cdot,*3}$ where the horizontal plane is spanned by the neighbours of some vertex, then at every floor the halls behave like a grid. Every hall is infinitely long in both directions; half of the halls have north–south direction and the other half have east–west direction. There are elevators located in the midpoint of every section of every hall between the intersections with two other perpendicular halls (see Figure 6.25, where the squares indicate the location of the elevators). The elevators move vertically and can be used to access any floor from any other floor.

If we enter an elevator from a north–south hall at an even-numbered floor then we will exit to any other even-numbered floor to another north–south hall. However, if we exit to an odd-numbered floor we will find ourselves on an east–west hall. In other words, two floors with the same parity are translates of each other by a vertical vector, but two floors with distinct parities are not.

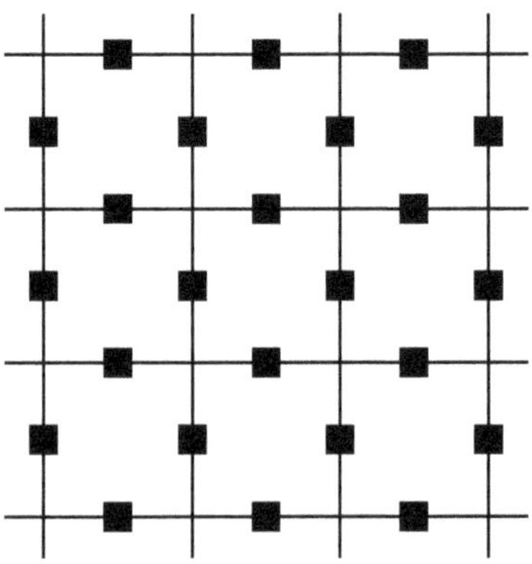

Figure 6.25 The location of elevators in the grid of halls of a floor.

If we think of a floor as a tessellation by squares where the edges represent the halls, then every even-numbered floor is a vertical translate of the geometric dual of the tessellation in any odd-numbered floor, and vice versa. Here the dual of the tessellation is constructed by taking as vertices the centres of the squares with two of them joined by an edge when the corresponding squares share an edge.

Locating the 2-faces of $\{\infty,4\}_{\cdot,*3}$ in this building is not straightforward. Their axes are neither vertical nor horizontal. Coming out from an elevator from the $(n+1)$th floor to a north–south hall on the nth floor we follow one (right) helix by continuing one section to the north, then one section to the west, then taking the elevator to the $(n-1)$th floor, then to the north, then to the west, etc. The remaining helices can be obtained by replacing north by west and west by south, by replacing north by south and west by east, and by replacing north by east and west by north. After coming out from an elevator from the $(n+1)$th floor to a north–south hall of the nth floor and walking one section to the north, we can turn to the east instead of the west. We can then go down one floor, go one section to the north and one to the east and so on. The path described is a left helix and is isometric to the 2-faces of $\{\infty,4\}_{\cdot,*3}$, but it is not a 2-face of this polyhedron.

Exercise 6.1.53 Draw the projection of the four 2-faces at o to the XZ-plane.

The symmetry group of $\{\infty,4\}_{\cdot,*3}$ is described in [167, Section 7] as the quotient of the Coxeter group $[\infty,4]=\langle R_0,R_1,R_2\rangle$ by the single extra relation

$$(R_0R_1R_0R_1R_2)^3=\mathrm{id}.$$

This shows an essential difference between $\{\infty,4\}_{\cdot,*3}$ and the two polyhedra with type $\{\infty,3\}$ previously described (see Propositions 6.1.43 and 6.1.48).

Proposition 6.1.54 *The polyhedron* $\{\infty,4\}_{\cdot,*3}$ *is combinatorially non-orientable.*

The fact that $\{\infty,4\}_{\cdot,*3}$ is combinatorially non-orientable is in no contradiction with the fact that all symmetries of $\{\infty,4\}_{\cdot,*3}$ are orientation-preserving involutions. To better understand this phenomenom it is helpful to verify directly that if Φ is the base flag then

$$\Phi^2=\Phi^{0,1,0,1,2,0,1,0,1,2,0,1,0,1}=\Phi^{1,0,1,0,2,1,0,1,0,2,1,0,1,0}. \tag{6.16}$$

The symmetry group of $\{\infty,4\}_{\cdot,*3}$ consists of all orientation-preserving symmetries of the cubic tessellation $\{4,3,4\}$ that leave invariant the set

$(1,0,0) + \Lambda_{(1,1,1)}$. Hence $G(\{\infty,4\}_{\cdot,*3})$ can be described simply as the set of orientation-preserving symmetries of $(1,0,0)+\Lambda_{(1,1,1)}$, and contains rotations of orders 4 and 3, like S_2 and $R_2 S_1^2$. It also contains half-turns whose axes are not parallel to a coordinate axis, like R_1.

6.2 Chiral Polyhedra in $\mathbb{E}^3$

In this section we describe all chiral polyhedra in $\mathbb{E}^3$. These polyhedra were found in [245, 246] in 2004 and 2005. Here we provide an alternative way to get to the same classification.

6.2.1 Geometric Aspects of Chiral Polytopes in $\mathbb{E}^3$

Before describing the chiral polyhedra in $\mathbb{E}^3$ we provide some preliminary results on skeletal chiral polyhedra. The main sources for this section are [12, 172].

Because of the relevance of Proposition 6.2.2 we state the next results for general rank.

Lemma 6.2.1 *The vertex-figures of a geometrically rotary n-polytope are finite rotary $(n-1)$-polytopes.*

Proof Since the vertex-figures of every geometrically rotary n-polytope $\mathcal{P}$ are rotary, the stabiliser of a vertex v under $G(\mathcal{P})$ acts transitively on the neighbours of v. This forces the neighbours of v to lie on a sphere $\mathcal{S}$ centred at v whose radius is the edge length of $\mathcal{P}$. Since the vertices of the vertex-figure at v are a discrete subset of $\mathcal{S}$, and $\mathcal{S}$ is a compact subset of $\mathbb{E}^3$, the vertex-figure must have finitely many vertices. A straightforward induction shows that the vertex-figures have finitely many i-faces for every $i \in \{1,\dots,n-2\}$, and therefore they are finite. ∎

Proposition 6.2.2 *Let $\mathcal{P}$ be a geometrically rotary n-polytope in $\mathbb{E}^d$. Then*

- *$d \geq n$, if $\mathcal{P}$ is finite,*
- *$d \geq n-1$, if $\mathcal{P}$ is infinite.*

Proof The proof is by induction over n. When $n \in \{0,1\}$ the result is immediate. In what follows we assume that $n \geq 2$.

By Lemma 6.2.1 every vertex-figure of $\mathcal{P}$ is isomorphic to a finite rotary $(n-1)$-polytope in $\mathbb{E}^d$. Let V be the affine hull of the set of neighbours of v. By the inductive hypothesis, $\dim(V) \geq n-1$.

If $\mathcal{P}$ is infinite then $d \geq \dim(V) \geq n-1$ and the proposition holds. If $\mathcal{P}$ is finite then $G(\mathcal{P})$ is finite and hence it fixes some point o. It follows that every vertex of $\mathcal{P}$ lies on some sphere $\mathcal{S}$ of dimension $d-1$ centred at o. Furthermore, the vertex-figure at v is on the sphere $\mathcal{S}'$ centred at v with radius given by the edge length of $\mathcal{P}$. This implies that the vertex-figure at v is in $\mathcal{S} \cap \mathcal{S}'$. Since the centre of $\mathcal{S}'$ lies on $\mathcal{S}$, $\mathcal{S} \cap \mathcal{S}'$ is a sphere of dimension $\dim(\mathcal{S}')-1 = d-2$. Now, V is spanned by some points in $\mathcal{S} \cap \mathcal{S}'$ and hence $\dim(V) \leq \dim(\mathcal{S} \cap \mathcal{S}')+1 = \dim(\mathcal{S}')$. From here it follows that $d = \dim(\mathcal{S}') + 1 \geq \dim(V) + 1 \geq n$. ■

Proposition 6.2.2 establishes an upper bound on the rank of rotary polytopes in $\mathbb{E}^d$. Rotary polytopes achieving this bound are called polytopes *of full rank.* The d-simplex and the tessellation by d-cubes in $\mathbb{E}^d$ are examples of regular polytopes of full rank in $\mathbb{E}^d$. The entire list of regular polytopes of full rank can be found in [172, Section 14].

From now on we concentrate exclusively on the three-dimensional Euclidean space, where all chiral polyhedra are known. We start the description of chiral polyhedra by excluding linear polygons as 2-faces.

Proposition 6.2.3 *There are no rotary polyhedra in $\mathbb{E}^3$ with linear 2-faces.*

Proof Assume to the contrary that F is a linear 2-face of a rotary polyhedron $\mathcal{P}$. Let Φ be a flag of $\mathcal{P}$ containing F. Since $G(\mathcal{P})$ acts transitively on the 2-faces, Φ^2 is a linear polygon isometric to F, and it must contain the base edge. Hence the 2-face of Φ^2 coincides with the base 2-face, contradicting the diamond condition. ■

Remark 6.2.4 Proposition 6.2.3 can be stated for facet transitive polyhedra, and for any dimension $d \geq 3$ of the ambient space.

We next establish a geometric analogue of Theorem 3.1.53 in rank 3.

Lemma 6.2.5 *Let $\mathcal{P}$ be a chiral polyhedron in $\mathbb{E}^3$ and let $G(\mathcal{P}) = \langle S_1, S_2 \rangle$. Then there is no involutory isometry R fixing the base vertex such that $RS_1R = S_1^{-1}$ and $RS_2R = S_2^{-1}$.*

Proof We proceed by contradiction. Let $\hat{v}$ be the base vertex. Suppose that R is an involutory isometry satisfying that $\hat{v}R = \hat{v}$, $RS_1R = S_1^{-1}$ and $RS_2R = S_2^{-1}$.

Let v be a vertex in the base 2-face. Then $v = \hat{v}S_1^k$ for some k, and $vR = \hat{v}S_1^kR = \hat{v}RS_1^{-k} = \hat{v}S_1^{-k}$. It follows that R preserves the base 2-face and it acts on it like the combinatorial reflection that fixes $\hat{v}$. In particular, it interchanges the base edge with the other edge of the base 2-face containing $\hat{v}$.

We next prove that R is a symmetry of $\mathcal{P}$ by showing that it preserves the vertex, edge and 2-face sets. Given any vertex, edge or 2-face F of $\mathcal{P}$ there exists $T \in G(\mathcal{P})$ such that $F = \hat{F}T$, where $\hat{F}$ is the base vertex, edge of 2-face. Then $FR = \hat{F}TR = \hat{F}RT'$, where T' is the symmetry of $\mathcal{P}$ obtained by replacing every factor S_j by S_j^{-1} in any expression of T as a word on S_1 and S_2. Clearly $T' \in G(\mathcal{P})$ and, as explained above, $\hat{F}R$ is a vertex, edge or 2-face of $\mathcal{P}$. This implies that FR is a vertex, edge or 2-face of $\mathcal{P}$.

We have proven that R is a symmetry of $\mathcal{P}$ mapping the base flag to its 1-adjacent flag, contradicting the chirality of $\mathcal{P}$. ■

The 2-faces and vertex-figures of rotary polyhedra in $\mathbb{E}^3$ are regular polygons among those listed in Proposition 6.1.26 (see also Exercise 6.1.38). It follows that the 2-faces of regular polyhedra in $\mathbb{E}^3$ can be planar, skew finite, zigzags or helical (see also [84, 104]). From Lemma 6.2.1, the vertex-figures can only be planar or skew finite.

Among all possible combinations of kinds of 2-faces and kinds of vertex-figures of a regular polyhedron, the only one that does not occur in the classification of regular polyhedra is skew finite 2-faces with skew vertex-figures. Furthermore, none of the three regular polyhedra with helical 2-faces and planar vertex-figures have a vertex that does not belong to the plane of its vertex-figure. In the following results we prove that the two arrangements of kinds of 2-faces and vertex-figures just mentioned are precisely the ones that chiral polyhedra in $\mathbb{E}^3$ can have.

Proposition 6.2.6 *There are no chiral polyhedra in $\mathbb{E}^3$ with planar 2-faces.*

Proof Assume to the contrary that there is a chiral polyhedron $\mathcal{P}$ with planar 2-faces. From Proposition 6.1.26, Corollary 6.1.37 and Proposition 6.2.3 we know that they must be either finite planar or zigzags. Let v_0 be the base vertex of $\mathcal{P}$ and let Π, l, Σ, R_l and R_Σ be as in Lemma 6.1.27 (see Figure 6.9). Then $R_l S_1 R_l = R_\Sigma S_1 R_\Sigma = S_1^{-1}$, and chirality of $\mathcal{P}$ forces $R_l S_2 R_l \neq S_2$ and $R_\Sigma S_2 R_\Sigma \neq S_2$ (see Lemma 6.2.5).

By Lemma 6.2.1, the vertex-figure $\mathcal{P}/v_0$ at v_0 is finite; it may be either planar or skew.

If $\mathcal{P}/v_0$ is planar then we consider the plane Σ, which is the bisector of an edge of $\mathcal{P}/v_0$. Moreover, Σ is perpendicular to the plane Π' containing $\mathcal{P}/v_0$, and contains the line l' through the centre of $\mathcal{P}/v_0$ perpendicular to Π' (see Figure 6.26). This implies that R_Σ commutes with the reflection about Π', and inverts by conjugation every rotation about l'. Recall that S_2 is either a rotation about l' or a rotatory reflection about l' and Π'. In either case $R_\Sigma S_2 R_\Sigma = S_2^{-1}$, contradicting Lemma 6.2.5.

Now assume that $\mathcal{P}/v_0$ is skew. In this situation l is the line through the centre v_0 of $\mathcal{P}/v_0$, and the midpoint of an edge of $\mathcal{P}/v_0$. Then R_l is one of

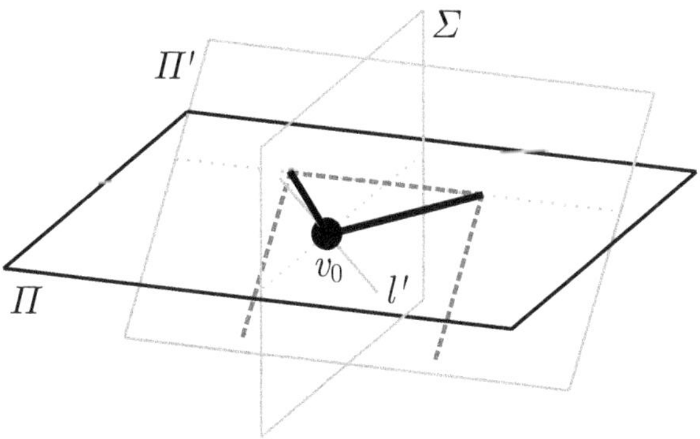

Figure 6.26 Planar 2-face in solid lines and planar vertex-figure in dashed lines.

the generators of the symmetry group of $\mathcal{P}/v_0$ given in Table 6.1, and hence $R_l S_2 R_l = S_2^{-1}$, contradicting again Lemma 6.2.5. ■

Corollary 6.2.7 *There are no planar chiral polyhedra.*

We know from Corollary 6.1.37 that the 2-faces of each chiral polyhedron are regular, and it follows from Proposition 6.2.6 that they are either skew finite or helical. The generators of the symmetry groups of skew and helical polygons are completely determined. This imposes restrictions on the existence of regular or chiral polyhedra with skew or helical 2-faces and a certain types of vertex-figures.

Proposition 6.2.8 *Let $\mathcal{P}$ be a skeletal rotary polyhedron and assume that it has finite skew 2-faces. Then $\mathcal{P}$ is regular if and only if its vertex-figures are planar.*

Proof Assume first that $\mathcal{P}$ is regular. Recall from Lemma 6.2.1 that its vertex-figures are finite, either planar or skew. The first and second standard reflections that generate the symmetry group of a skew polygon are a half-turn and a plane reflection, respectively. Since the 2-faces of $\mathcal{P}$ are skew, the generator R_1 of $G(\mathcal{P})$ must be a plane reflection. But then the vertex-figure cannot be skew, since the first generator of its symmetry group is not a half-turn. This makes it impossible to have regular polyhedra with skew 2-faces and skew vertex-figures. We can conclude that the vertex-figures are planar.

To show the converse, assume to the contrary that $\mathcal{P}$ is a chiral polyhedron in $\mathbb{E}^3$ with skew 2-faces and planar vertex-figures. Let v_0 be the base vertex of $\mathcal{P}$, let Σ be the bisector of the line segment between the two neighbours of v_0 in the base 2-face and let R_Σ be the reflection with respect to Σ (see Figure 6.27).

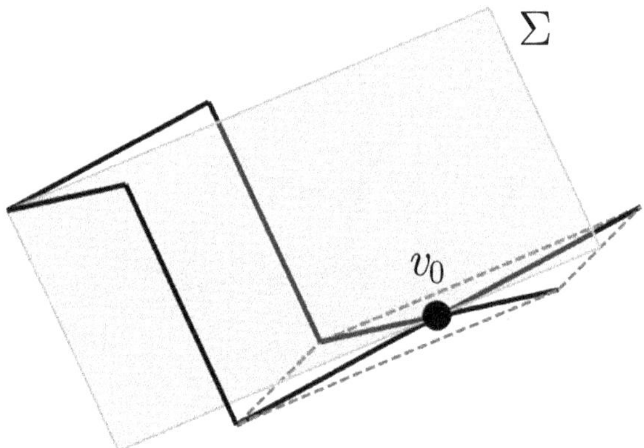

Figure 6.27 Skew 2-face in solid lines and planar vertex-figure in dashed lines.

Since R_Σ is one of the generators given in Table 6.1 of the symmetry group of the base 2-face, S_1 satisfies that $R_\Sigma S_1 R_\Sigma = S_1^{-1}$. Moreover, by Lemma 6.1.27 on the (planar) vertex-figure at v_0 we have that $R_\Sigma S_2 R_\Sigma = S_2^{-1}$, which contradicts Lemma 6.2.5. ■

Proposition 6.2.9 *Let $\mathcal{P}$ be a skeletal rotary polyhedron with helical 2-faces. Then $\mathcal{P}$ is chiral if and only if its vertex-figures are planar and no vertex belongs to the plane determined by its neighbours.*

Proof Assume first that $\mathcal{P}$ is chiral. Let v_0 be the base vertex of $\mathcal{P}$, let l be the line through v_0 and through the midpoint of the line segment between the two neighbours of v_0 in the base 2-face and let R_l be the half-turn about l. Since R_l is one of the generators of the symmetry group of the base 2-face we have that $R_l S_1 R_l = S_1^{-1}$.

The line l contains the midpoint of an edge of the vertex-figure $\mathcal{P}/v_0$ at v_0. We claim that $\mathcal{P}/v_0$ is not skew, since otherwise l contains also the centre of the vertex-figure (namely v_0) and R_l would be a generator listed in Table 6.1 of the symmetry group of $\mathcal{P}$. This would force $R_l S_2 R_l = S_2^{-1}$, contradicting Lemma 6.2.5. Hence, we may assume that $\mathcal{P}/v_0$ is planar.

Let Π' be the plane containing $\mathcal{P}/v_0$, w be the centre of $\mathcal{P}/v_0$, l' be the line through w and through the midpoint of the segment between the two neighbours of v_0 in the base 2-face, and $R_{l'}$ be the half-turn about l' (see Figure 6.28). Since S_2 acts like a rotation on Π', it must be either a rotation about the axis a perpendicular to Π' through w (and through v_0) or a rotatory reflection with respect to a and Π'. The half-turn $R_{l'}$ commutes with the

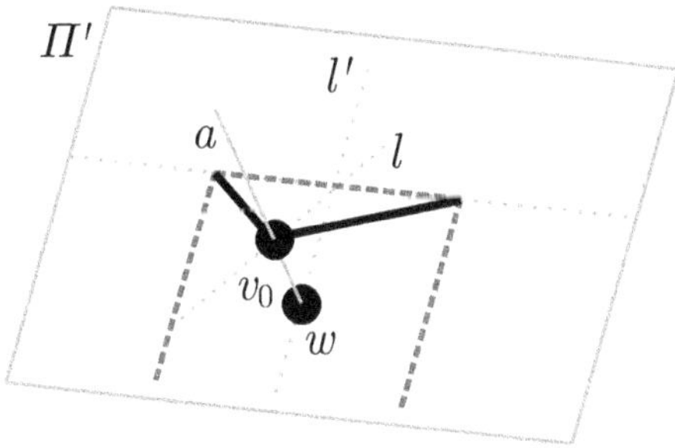

Figure 6.28 Helical 2-face in solid lines and planar vertex-figure in dashed lines.

reflection with respect to Π' since $l \subseteq \Pi'$. Furthermore, l' is perpendicular to a and this implies that $R_{l'} S_2 R_{l'} = S_2^{-1}$. By Lemma 6.2.5 we conclude that $l \neq l'$, and therefore $v_0 \notin \Pi'$. Vertex-transitivity of $\mathcal{P}$ helps us conclude that no vertex of $\mathcal{P}$ belongs to the plane determined by its neighbours.

We next show the contrapositive of the converse. Assume that $\mathcal{P}$ is regular and has planar vertex-figures. As indicated in Table 6.1, the symmetry R_1 is a half-turn.

If R_2 is a plane reflection then the mirror of R_1 cannot be contained in the mirror of R_2, since otherwise $S_2 = R_1 R_2$ would be a plane reflection, contradicting Remark 6.1.10 for the vertex-figure of $\mathcal{P}$. Since the mirror of R_2 intersects the mirror of R_1 in the base vertex v_0, S_2 must be a rotatory reflection and the vertex-figure at v_0 must be contained in the invariant plane of S_2. In this situation v_0 also belongs to that plane, since it is the only fixed point of S_2.

On the other hand, if R_2 is a half-turn then Lemma 6.2.1 implies that the mirrors of R_1 and R_2 intersect in only one point. This point must be the base vertex v_0. Since R_2 fixes pointwise the base edge e_0, the two vertices v_0 and u_0 of e_0 belong to the plane Π' spanned by the mirrors of R_1 and R_2. Furthermore, the vertex set of the vertex-figure $\mathcal{P}/v_0$ at v_0 is the orbit of u_0 under the rotation S_2, whose axis is perpendicular to Π'. It follows that v_0 and the vertex-figure at v_0 belong to Π'. ■

The following theorem follows directly from Propositions 6.2.6, 6.2.8 and 6.2.9.

Theorem 6.2.10 *Every chiral polyhedron in $\mathbb{E}^3$ has skew 2-faces and skew vertex-figures, or helical faces and planar vertex-figures.*

Corollary 6.2.11 *There are no finite chiral polyhedra in $\mathbb{E}^3$.*

Proof Assume to the contrary that $\mathcal{P}$ is a finite chiral polyhedron in $\mathbb{E}^3$. Then every symmetry of $\mathcal{P}$ fixes o, and all vertices belong to some sphere $\mathcal{S}_1$ centred at o. If v is the base vertex of $\mathcal{P}$ then the vertex set of the vertex-figure $\mathcal{P}/v$ is contained in $\mathcal{S}_1 \cap \mathcal{S}_2$, where $\mathcal{S}_2$ is the sphere centred at v with radius equal to the edge length of $\mathcal{P}$. It follows that all vertices of $\mathcal{P}/v$ lie on a plane. Theorem 6.2.10 implies that the facets of $\mathcal{P}$ must be helical, contradicting the hypothesis of finiteness. ∎

In the remainder of this section we shall describe all chiral polyhedra in $\mathbb{E}^3$.

6.2.2 Chiral Polyhedra with Helical 2-Faces

We first study the chiral polyhedra with helical 2-faces in $\mathbb{E}^3$. We start by constructing three families of polyhedra through Wythoff's construction from certain groups $\langle S_1, S_2\rangle$ that are familiar to us. Then we shall show that all chiral polyhedra with helical 2-faces can be constructed in this way.

We recommend the readers interested in understanding the chiral polyhedra with helical 2-faces to follow the details in [246], and compare their results with those stated here.

The Polyhedra $P_1(\lambda)$

In what follows we construct chiral polyhedra $\mathcal{P}$ with $G(\mathcal{P}) = G^+(\{\infty, 3\}^{(a)})$.

In Section 6.1.5 we described $G(\{\infty, 3\}^{(a)})$ as the group generated by R_0, R_1 and R_2 as in (6.6), (6.7) and (6.8), respectively. The abstract rotation $S_1 = R_0R_1$ is the twist that maps a point (x, y, z) to $(-z, x-1, 1-y)$. Its axis is

$$\left\{\tfrac{1}{3}(1,0,1) + \mu(1,1,-1) \colon \mu \in \mathbb{R}\right\}$$

and it has translation component $\frac{1}{3}(-2,-2,2)$. Recall that $S_2 = R_1R_2$ is the 3-fold rotation that maps (x, y, z) to $(-y, z, -x)$. Its axis is the line

$$\{\mu(1,-1,-1) \colon \mu \in \mathbb{R}\}$$

(see Figure 6.18). Let $\hat{G}_1$ be the group $\langle S_1, S_2\rangle$.

Lemma 6.2.12 *The translation subgroup of $G(\{\infty, 3\}^{(a)})$ is contained in $\hat{G}_1$.*

Proof The translation subgroup of $G(\{\infty, 3\}^{(a)})$ is the group of translations by vectors in $2\mathbf{\Lambda}_{(1,1,1)}$, and it is generated by S_1^3, $R_2S_1^3R_2$, and $R_1R_2S_1^3R_2R_1$, all belonging to $\hat{G}_1$ (see Section 6.1.5). ∎

In order to construct a rotary polyhedron $P_1(\lambda)$ from $\hat{G}_1$ by Wythoff's construction we need to determine an initial vertex $v_0 = (\lambda, -\lambda, -\lambda)$ for some $\lambda \in \mathbb{R}$, in the mirror of S_2. The base edge joins $(\lambda, -\lambda, -\lambda)$ and its image $(1-\lambda, 1+\lambda, -\lambda)$ under S_1S_2. The base 2-face is the helix over a triangle whose vertex set is the orbit of v_0 under $\langle S_1\rangle$. We first show that this object is indeed a polyhedron for every parameter λ.

Theorem 6.2.13 *For every $\lambda \in \mathbb{R}$, the structure $P_1(\lambda)$ we have defined is a polyhedron combinatorially isomorphic to $\{\infty, 3\}^{(a)}$. Furthermore, it is regular if and only if $\lambda = 0$.*

Proof Recall that $G(\{\infty, 3\}^{(a)})$ consists only of orientation preserving isometries. Since $\hat{G}_1 \subseteq G(\{\infty, 3\}^{(a)})$, every non-trivial element in $\hat{G}_1$ stabilising a point in $\mathbb{E}^3$ must be a rotation about a line.

For a given point $v_0 = v_0(\lambda)$ in the mirror of S_2, Lemma 6.1.51 tells us that $\mathrm{Stab}_{\hat{G}_1}(v_0) = \langle S_2\rangle$ except, perhaps, if v_0 is invariant under a half-turn $T \in G(\{\infty, 3\}^{(a)})$ whose mirror l_T is parallel to an edge of $\{\infty, 3\}^{(a)}$ and perpendicular to the mirror of S_2. If l_T contains an edge e of $\{\infty, 3\}^{(a)}$ then T is the conjugate of R_1 by a symmetry in $G(\{\infty, 3\}^{(a)})$ mapping e_1 to e. If l_T contains no edge of $\{\infty, 3\}^{(a)}$ then it must be the composition of a translation with the conjugate of R_1 by some symmetry mapping e_1 to an edge parallel to l_T. It follows from Lemma 6.2.12 that T is a product of an odd number of generators R_i of $G(\{\infty, 3\}^{(a)})$ and, by Proposition 6.1.43, $T \notin \hat{G}_1$. Therefore, $\mathrm{Stab}_{\hat{G}_1}(v_0) = \langle S_2\rangle$ for every choice of v_0. Theorem 6.1.35 (b) implies that $P_1(\lambda)$ is a polyhedron combinatorially isomorphic to $\{\infty, 3\}^{(a)}$.

The vertex-figure of $P_1(\lambda)$ at $v_0 = (\lambda, -\lambda, -\lambda)$ is the triangle determined by its neighbours

$$
\begin{aligned}
&(1-\lambda, 1+\lambda, -\lambda),\\
&(1-\lambda, 1+\lambda, -\lambda)S_2 = (-1-\lambda, -\lambda, \lambda-1),\\
&(1-\lambda, 1+\lambda, -\lambda)S_2^2 = (\lambda, \lambda-1, 1+\lambda).
\end{aligned}
$$

It is realised in the plane $x - y - z = -\lambda$, which contains $(\lambda, -\lambda, -\lambda)$ if and only if $\lambda = 0$. Thanks to Proposition 6.2.9 we can conclude that $P(\lambda)$ is regular if and only if $\lambda = 0$. ■

In order to visualise the polyhedra $P_1(\lambda)$ we may start with $P_1(0) = \{\infty, 3\}^{(a)}$, and consider the effect of continuously modifying the value of λ.

For each given vertex w of $\{\infty, 3\}^{(a)}$, the axis of a generator of its stabiliser in $\hat{G}_1$ is the line through w perpendicular to its three edges. Recall from Exercise 6.1.49 that the set of such lines is similar to the set $\mathcal{L}$ described in Appendix A.4.

Every vertex v of $P_1(\lambda)$ can be expressed as $v_0 T_v$ for some $T_v \in \hat{G}_1$, and it belongs to the mirror of the 3-fold rotation $T_v^{-1} S_2 T_v$. Hence its stabiliser in $\hat{G}_1$ is $T_v^{-1}\mathrm{Stab}_{\hat{G}_1}(v_0) T_v$. As noted in the proof of Theorem 6.2.13,

$\mathrm{Stab}_{\hat{G}_1}(v_0) = \langle S_2 \rangle$ and therefore $\mathrm{Stab}_{\hat{G}_1}(v) = \langle T_v^{-1} S_2 T_v \rangle$, which does not depend on λ or v_0. This implies that when variating the value of λ, every vertex of $P_1(\lambda)$ moves continuously along the axis of a generator of its stabiliser. The fact that T_v is an isometry forces all vertices to move at the same speed as v_0 when continuously increasing or decreasing the value of λ. Furthermore, when two distinct vertices describe such movement they never collide, since that would force the size of their stabilisers to be greater than 3. In particular, vertices invariant under the same 3-fold rotation S move along the axis of S in the same direction, when modifying the value of λ.

Let $\eta_{(1,-1,-1)}$ denote the orthogonal projection to the plane $\Pi^{(1,-1,-1)}$. The projections $(P_1(\lambda))\eta_{(1,-1,-1)}$ for $\lambda \in \{0, 1/5, 2/5, 3/5\}$ are shown in Figure 6.29. The thick black lines represent the projection of $P_1(0)$ (compare with Figure 6.19), whereas lines of lighter shades of grey are used to represent the projection of $P_1(\lambda)$ for larger values of λ. Since the group of translations preserving $P_1(\lambda)$ is just the translation subgroup of $\hat{G}_1$, it does not depend on the value of λ. The four orbits of vertices of $P_1(\lambda)$ under the translation subgroup of $\hat{G}_1$ project to the four orbits of vertices of Figure 6.29 under its 2-dimensional translation group. Since the initial vertex v_0 projects to o for all

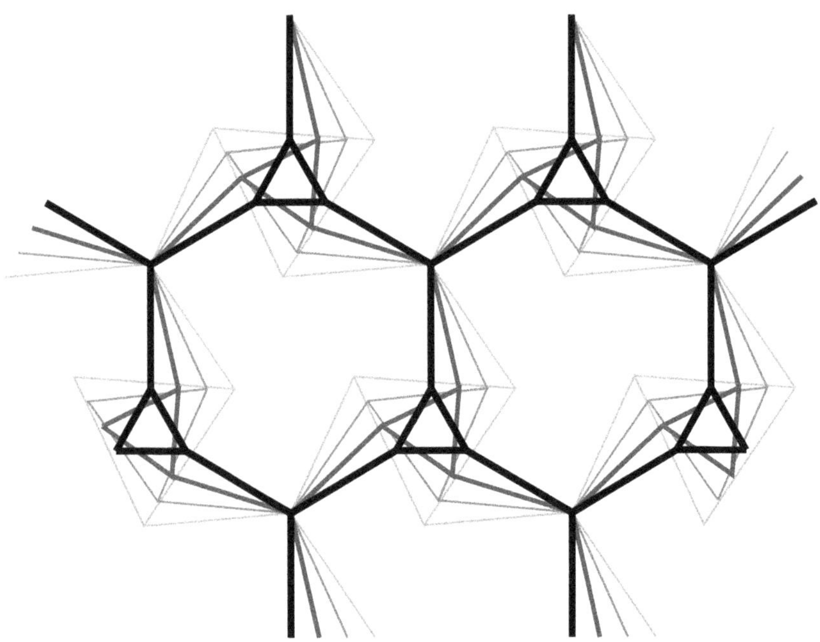

Figure 6.29 The projection to $\Pi^{(1,-1,-1)}$ of $P_1(\lambda)$.

values of λ, the projection of each of its translates by a vector in $2\mathbf{\Lambda}_{(1,1,1)}$ is the same regardless of the value of λ. These are precisely the vertices that project to $\Pi^{(1,-1,-1)}$ into vertices that are not vertices of triangles in Figure 6.29. Even though their projections are invariant, the translates of v_0 do move when increasing or decreasing λ. Moreover, they move at the same speed as any other vertex, but their movement is not shown in the projection to $\Pi^{(1,-1,-1)}$ since it is in the direction of $(1,-1,-1)$.

The base 2-face of $P_1(\lambda)$ is a helix with the same axis l as the base 2-face of $G(\{\infty,3\}^{(a)})$, and in these two helices the distance between pairs of vertices that are three edges apart is the same, because one vertex is the image of the other by a translation in $\hat{G}_1$. However, the distances of the vertices of the base helix to l is different for different values of λ. The line $\{\mu(1,0,1)\colon \mu \in \mathbb{R}\}$ intersects l in $\frac{1}{3}(1,0,1)$ and the mirror of S_2 in o; moreover, it is perpendicular to both lines. Since l is not parallel to the mirror of S_2, the distance from o to l is smallest possible among the distances of points in the mirror of S_2 to l. Therefore the distance from a vertex of the base helix of $P_1(\lambda)$ to l is greater than the distance from a vertex of the base helix of $G(\{\infty,3\}^{(a)})$ to l whenever $\lambda \neq 0$. In fact, increasing the absolute value of λ has the effect of increasing the width of the helical 2-faces of $P_1(\lambda)$ with respect to the translation vector determined by a three-step walk along a 2-face. This can be visualised in Figure 6.29, where triangles with a lighter shade of grey (projections of helices with axis in the direction of $(1,-1,-1)$) have longer edges.

The translates of o are the only vertices of $\{\infty,3\}^{(a)}$ that do not belong to a helix in the direction of $(1,-1,-1)$. In the same way, the translates of v_0 are the only vertices of $P_1(\lambda)$ that do not belong to a helix in the direction of $(1,-1,-1)$. On the other hand, all neighbours of v_0 belong to a helix in the direction of $(1,-1,-1)$. We shall need one such helix for each λ. Let H_4^{λ} be the helix in the direction of $(1,-1,-1)$ containing the neighbour $v_0S_1S_2$ of the base vertex v_0, and let l_4 be the axis of H_4^{λ}. (Note that H_4^0 is the helix H_4 defined in Section 6.1.5.)

So far we know that when increasing the absolute value of λ the helices become wider while keeping the distance between any pair of vertices that are three edges apart. It is also true that the modification of λ also produces a rotation component on the helices around their axes. This can be seen in $\Pi^{(1,-1,-1)}$ as follows. When variating λ, the projection $(v_0S_1S_2)\eta_{(1,-1,-1)}$ of $v_0S_1S_2 = (1-\lambda, 1+\lambda, -\lambda)$ is $(1-2\lambda/3, 1+2\lambda/3, -4\lambda/3)$, and it describes a line through $(oS_1S_2)\eta_{(1,-1,-1)} = oS_1S_2 = (1,1,0)$ perpendicular to the edge between o and $(1,1,0)$. This induces a continuous change in the angle formed by the segment between $o = v_0\eta_{(1,-1,-1)}$ and $(l_4)\eta_{(1,-1,-1)}$ (this segment is invariant under changes of λ) and the segment between $(v_0S_1S_2)\eta_{(1,-1,-1)}$

and $(l_4)\eta_{(1,-1,-1)}$. This is illustrated in Figure 6.29, where the triangles with different shades of grey have non-parallel sides.

The increase or decrease of the value of λ has the same effect on helices of $P_1(\lambda)$ whose axes are not in the direction of $(1,-1,-1)$. When tracing the projection to $\Pi^{(1,-1,-1)}$ of such a helix for different values of λ (starting with the one in Figure 6.19(b) for $\lambda = 0$, say), one can get the impression that the corresponding helix is also getting wider and rotating. We suggest tracing the projected helices in distinct shades of grey in Figure 6.29. The rotation and the increase of width must occur, since $\hat{G}_1$ acts transitively on the set of helices of $P_1(\lambda)$.

The translation subgroup of $G(\{\infty,3\}^{(a)})$ has index 24, implying that it has index 12 in $\hat{G}_1$. Since $\mathrm{Stab}_{\hat{G}_1}(v_0) = \langle S_2\rangle$ and S_2 does not fix the base edge, the stabiliser in $\hat{G}_1$ of the base edge of $P_1(\lambda)$ consists only of $\langle S_1S_2\rangle$. It follows that there are six orbits of edges of $P_1(\lambda)$ under the translation subgroup. The edges listed in Table 6.2 are a system of representatives of these orbits. We also indicate the vector determining every edge. An easy computation shows that the length of the edges of $P_1(\lambda)$ is $\sqrt{2}\sqrt{4\lambda^2+1}$.

Table 6.2 suggests that the polyhedron $P_1(\frac{1}{2})$ is special, since the directions of its edges are those of the coordinate axes. This polyhedron is described in detail in [223, Section 3].

If $|\lambda_1| \neq |\lambda_2|$ then $P_1(\lambda_1)$ and $P_1(\lambda_2)$ are not isometric since they have distinct edge lengths. On the other hand, $P_1(\lambda)$ is isometric to $P_1(-\lambda)$; more specifically, $P_1(\lambda) = P_1(-\lambda)R_1$. Indeed, R_1 maps the base vertex $(\lambda,-\lambda,-\lambda)$ of $P_1(\lambda)$ to the base vertex $(-\lambda,\lambda,\lambda)$ of $P_1(-\lambda)$. Furthermore, it is straightforward to verify that if Φ and Φ' are the base flags of $P_1(\lambda)$ and of $P_1(-\lambda)$, respectively, then $\Phi R_1 = (\Phi')^1$, implying that R_1 maps i-faces of $P_1(\lambda)$ to i-faces of $P_1(-\lambda)$ for all $i \in \{0,1,2\}$. From here we remark that, although $P_1(\lambda)$ and $P_1(-\lambda)$ are isometric and all their 2-faces are left helices,

Table 6.2. *Representatives of the edges of $P_1(\lambda)$ under the translation group.*

Vertices	Vector
$(\lambda,-\lambda,-\lambda),(1-\lambda,1+\lambda,-\lambda)$	$(2\lambda-1,-2\lambda-1,0)$
$(\lambda,-\lambda,-\lambda),(-1-\lambda,-\lambda,\lambda-1)$	$(2\lambda+1,0,1-2\lambda)$
$(\lambda,-\lambda,-\lambda),(\lambda,\lambda-1,1+\lambda)$	$(0,1-2\lambda,-2\lambda-1)$
$(\lambda,\lambda-1,1+\lambda),(-\lambda-1,\lambda-1,2-\lambda)$	$(2\lambda+1,0,2\lambda-1)$
$(1-\lambda,1+\lambda,-\lambda),(1-\lambda,2-\lambda,1+\lambda)$	$(0,2\lambda-1,-2\lambda-1)$
$(-1-\lambda,-\lambda,\lambda-1),(\lambda-2,\lambda+1,\lambda-1)$	$(2\lambda-1,2\lambda+1,0)$

they are geometric enantiomorphs of each other. In other words, if $\lambda \neq 0$ then no isometry mapping $P_1(\lambda)$ to $P_1(-\lambda)$ maps the base flag of one polyhedron to the base flag of the other. We summarise these facts in the next result.

Proposition 6.2.14 *With the notation above, two polyhedra $P_1(\lambda_1)$ and $P_1(\lambda_2)$ are isometric if and only if $\lambda_1 = \pm\lambda_2$. Furthermore, $P_1(-\lambda) = P_1(\lambda)R_1$.*

Instead of considering the family consisting of the polyhedra $P_1(\lambda)$ we may be interested in the polyhedra $\tilde{P}_1(\lambda) := \frac{1}{\sqrt{4\lambda^2+1}} P_1(\lambda)$ obtained from $P_1(\lambda)$ by a homotety with factor $\frac{1}{\sqrt{4\lambda^2+1}}$. The polyhedron $\tilde{P}_1(\lambda)$ cannot be obtained by Wythoff's construction from $\tilde{G}_1$ if $\lambda \neq 0$, since S_1, as described previously, is no longer the symmetry $\tilde{S}_1$ mapping the base flag $\tilde{\Phi}$ of $\tilde{P}_1(\lambda)$ to $\tilde{\Phi}^{1,0}$. More specifically, the axis of $\tilde{S}_1$ moves closer to o when the absolute value of λ increases.

The main interest on the family $\{\tilde{P}_1(\lambda)\}$ is that the edges have length $\sqrt{2}$, regardless of the parameter λ. When increasing the value of λ starting from $\lambda = 0$, the vertices and edges of the polyhedra $\tilde{P}_1(\lambda)$ move continuously while preserving the edge length at all times. The rotation group is also preserved as an abstract group, although geometrically the axes of the rotations and twists continuously move towards o. When varying the value of λ, the vertices of $\{\tilde{P}_1(\lambda)\}$ no longer move at the same speed. Vertices that are farther away from o move faster than those closer to o.

The vectors of the translations preserving $\tilde{P}_1(\lambda)$ are in $\frac{2}{\sqrt{4\lambda^2+1}}\mathbf{\Lambda}_{(1,1,1)}$. This allows us to consider the limit polyhedron $\tilde{P}_1(\infty)$, where the mirrors of all rotations contain o, and the translation group is trivial. Its edges all have length $\sqrt{2}$ and the vectors between their endpoints are $(1,-1,0)$, $(1,0,-1)$, $(0,-1,-1)$, $(1,0,1)$, $(0,1,-1)$ and $(1,1,0)$ (see Table 6.2). They are the vectors between the endpoints of the edges of a tetrahedron with vertices at $\frac{1}{2}(1,1,1)$, $\frac{1}{2}(1,-1,-1)$, $\frac{1}{2}(-1,1,-1)$ and $\frac{1}{2}(-1,-1,1)$. These vertices are precisely the vertices that we obtain as the limit of the base vertex of $\tilde{P}_1(\lambda)$ and its neighbours, when λ tends to ∞. Hence, $\tilde{P}_1(\infty)$ is defined as such tetrahedron.

Note that the vectors between the endpoints of the edges of $P_1(0) = \tilde{P}_1(0)$ are also the vectors of the edges of a tetrahedron. The vectors of the three edges of $P_1(0)$ meeting at a vertex correspond to the vectors of the three edges of the 2-face of the tetrahedron (they are coplanar), and hence any two of them form an angle of $2\pi/3$. By contrast, the vectors of the three edges meeting at a vertex of $\tilde{P}_1(\infty)$ correspond to the vectors of the three edges at a vertex of the tetrahedron, and therefore any two of them form an angle of $\pi/3$. A straightforward computation shows that when increasing the absolute value of λ the angle between two edges at a vertex of $\tilde{P}_1(\lambda)$ decreases. The following result follows.

Proposition 6.2.15 *There are infinitely many skeletal chiral polyhedra $\mathcal{P}$ with $G(\mathcal{P}) = \hat{G}_1$, one for each $\lambda > 0$.*

The first appearance of the polyhedra $P_1(\lambda)$ was in [246, Section 4]. The strategy to approach them was also through Wythoff's construction, but taking o as initial vertex. A 3-fold rotation was fixed to serve as S_2, and all possible isometries that could serve as S_1 were considered. As a result, in that paper the polyhedra in this family are called $P_1(a,b)$ where a and b are any two real parameters, not both 0, and determine the symmetry S_1. Lemma 4.7 of [246] establishes that $P_1(sa,sb)$ is similar to $P_1(a,b)$ for every $s \neq 0$, and hence the family can be described by only one parameter, up to similarities. The regular polyhedra in that family are $P_1(1,1) \cong \{3,3\}$ and $P_1(1,-1) \cong \{\infty,3\}^{(a)}$.

Many ideas and results in this section are merely a reformulation of those in [246, Section 4]. The polyhedron $P_1(\lambda)$ described earlier is isometric to the polyhedron $P_1(2\lambda+1, 2\lambda-1)$ in [246]. The verification of this fact is left as an exercise. The polyhedron $\tilde{P}_1(\lambda)$ is then $P_1\left(\sqrt{\frac{(2\lambda+1)^2}{4\lambda^2+1}}, \sqrt{\frac{(2\lambda-1)^2}{4\lambda^2+1}}\right)$ in the notation of [246]; the notation and results in that article also justify the interpretation of $\tilde{P}_1(\infty)$ as the tetrahedron $P_1(1,1)$.

Exercise 6.2.16 Show that $P_1(\lambda)$ is isometric to the polyhedron $P_1(2\lambda+1, 2\lambda-1)$ described in [246].

The translation between the results in the current section and those in [246] can be done through a change of coordinates. Here we exhibit explicitly only one instance of this. The generating twist $\bar{S}_1$ and rotation $\bar{S}_2$ for $G(P_1(1,-1))$ map (x,y,z) to $(-z-1, -x+1, y)$ and to (y,z,x), respectively. The reader can verify that $S_i = T^{-1}\bar{S}_i T$ for $i \in \{1,2\}$, where T is the isometry mapping (x,y,z) to $(-z,x,y)$, and S_1 and S_2 are the standard generators of $\hat{G}_1$.

The Polyhedra $P_2(\lambda)$

We next construct polyhedra $P_2(\lambda)$ related to $\{\infty,3\}^{(b)}$.

From R_0, R_1 and R_2 in (6.10), (6.11) and (6.12) we derive the abstract rotations $S_1 := R_0R_1$ and $S_2 := R_1R_2$. The symmetry S_1 is a twist that maps (x,y,z) to $(z, y-1, 1-x)$. Its axis is $\{(1/2,0,1/2)+\mu(0,1,0) : \mu \in \mathbb{R}\}$ and has translation vector $(0,-1,0)$. The symmetry S_2 is the same rotation as for $P_1(\lambda)$. The axes of S_1 and S_2 are shown in Figure 6.20.

We next apply Wythoff's construction to the group $\hat{G}_2 := \langle S_1, S_2 \rangle$ of even products of the generators R_i of $G(\{\infty,3\}^{(b)})$. In order to do this we must choose a base vertex $v_0 = (\lambda, -\lambda, -\lambda)$ (for some $\lambda \in \mathbb{R}$) in the mirror of S_2. The resulting structure will be denoted by $P_2(\lambda)$.

Recall that the translation subgroup of $G(\{\infty,3\}^{(b)})$ is the group of translations by vectors in $2\mathbf{\Lambda}_{(1,1,1)}$. It contains the group $\mathcal{T}_{4\mathbf{\Lambda}_{(1,0,0)}}$ of translations by vectors in $4\mathbf{\Lambda}_{(1,0,0)}$ as an index-2 subgroup. The latter is generated by S_1^4, $S_2 S_1^4 S_2^{-1}$ and $S_2^{-1} S_1^4 S_2$, all belonging to $\hat{G}_2$. Therefore the translation subgroup of $G(P_2(\lambda))$ contains at least $\mathcal{T}_{4\mathbf{\Lambda}_{(1,0,0)}}$.

On the other hand, the translation $T_{(2,-2,2)}$ by $(2,-2,2)$ can be expressed as $R_2 S_1^{-2} S_2^{-1} S_1$. Since $\{\infty,3\}^{(b)}$ is orientable, $T_{(2,-2,2)}$ does not belong to $\hat{G}_2$. This forces the translation subgroup of $\hat{G}_2$ to be precisely $\mathcal{T}_{4\mathbf{\Lambda}_{(1,0,0)}}$. In particular, $\hat{G}_1$ is different from $\hat{G}_2$. The fact that these two groups are different has some consequences, as can be seen when comparing the statements of Theorems 6.2.13 and 6.2.17.

Theorem 6.2.17 *The structure $P_2(\lambda)$ defined above is a polyhedron if and only if λ is not an odd integer. In the cases where $P_2(\lambda)$ is a polyhedron, it is combinatorially isomorphic to $\{\infty,3\}^{(b)}$. Furthermore, it is geometrically regular if and only if $\lambda = 0$, and chiral otherwise.*

Proof As a subgroup of $G(\{\infty,3\}^{(b)})$, the group $\hat{G}_2$ consists only of orientation preserving isometries. By Lemma 6.1.51, the stabiliser of a point $v_0 = v_0(\lambda)$ in the mirror of S_2 is $\langle S_2 \rangle$ except, perhaps, when v_0 is fixed by a half-turn with mirror parallel to an edge of $\{\infty,3\}^{(b)}$ that is perpendicular to the mirror of S_2.

The edge e_1 of $\{\infty,3\}^{(b)}$ between o and $(-1,0,-1)$ is fixed pointwise by R_1. Furthermore, the base edge e_0 between o and $(1,1,0)$ is fixed by R_2, and the edge e_2 between o and $(0,-1,1)$ is fixed by $R_2 R_1 R_2$. It follows from Proposition 6.1.48 that none of the three half-turns R_1, R_2 and $R_2 R_1 R_2$ belongs to $\hat{G}_2$.

From the projection of the 1-skeleton of $\{\infty,3\}^{(b)}$ to the plane $\Pi^{(1,-1,-1)}$ it becomes clear that all edges that intersect the mirror of S_2 are parallel to either e_0, e_1 or e_2 (see Figure 6.19). If v_0 is fixed by a half-turn $\hat{R}$ then one of the compositions $R_1\hat{R}$, $R_2\hat{R}$ or $R_2 R_1 R_2 \hat{R}$ is a translation $T_{\bar{u}}$ by a vector $\bar{u}$ in the direction of the mirror of S_2. When writing $\bar{u} = m(2,-2,-2)$ for some $m \in \mathbb{Z}$, the point v_0 must be $\frac{1}{2}\bar{u}$, and hence $\lambda = m$. We remark that $\hat{R}$ belongs to $\hat{G}_2$ if and only if $T_{\bar{u}}$ belongs to $G(\{\infty,3\}^{(b)}) \setminus \hat{G}_2$. In particular, if $\lambda \notin \mathbb{Z}$ then $\hat{R} \notin \hat{G}_2$.

If λ is even then $T_{\bar{u}} \in \mathcal{T}_{4\mathbf{\Lambda}_{(1,0,0)}}$, implying that $\hat{R} \notin \hat{G}_2$. However, if λ is odd then $T_{\bar{u}} \notin \mathcal{T}_{4\mathbf{\Lambda}_{(1,0,0)}}$, and $\hat{R}$ must be in $\hat{G}_2$. This shows that the only situation when $\mathrm{Stab}_{\hat{G}_2}(v_0)$ is not $\langle S_2 \rangle$ is when λ is an odd integer.

We can deduce now from Theorem 6.1.35 (b) that if λ is not an odd integer then $P_2(\lambda)$ is a polyhedron isomorphic to $\{\infty,3\}^{(b)}$. On the other hand, if λ is an odd integer then Proposition 6.1.29 implies that $P_2(\lambda)$ is not a polyhedron.

The base edge of $P_2(\lambda)$ joins v_0 with $v_0 S_1 S_2 = (1+\lambda, 1-\lambda, \lambda)$. The remaining two neighbours of v_0 are

$$(1+\lambda, 1-\lambda, \lambda)S_2 = (\lambda-1, \lambda, -1-\lambda),$$
$$(1+\lambda, 1-\lambda, \lambda)S_2^2 = (-\lambda, -1-\lambda, 1-\lambda).$$

The three neighbours of o belong to the plane given by the equation $x - y - z = \lambda$, and v_0 belongs to that plane only when $\lambda = 0$. Now Proposition 6.2.9 implies the last part of the statement of the theorem. ∎

In what follows we describe the polyhedra $P_2(\lambda)$ through the impact of a continuous variation on the parameter λ. Initially, we may take $P_2(\lambda) = \{\infty,3\}^{(b)}$.

When modifying continuously the value of λ we may think that each vertex v is moving along the axis of a generator of $\mathrm{Stab}_{\hat{G}_2}(v)$, and all vertices are moving at the same speed. These ideas are justified by arguments analogous to those exposed for the family of polyhedra $\{P_1(\lambda)\}_{\lambda\in\mathbb{R}}$.

Any vertex of $\{\infty,3\}^{(b)}$ in the mirror of S_2 is the image of o under a translation $T \in G(\{\infty,3\}^{(b)})$ by a vector $m(2,-2,-2)$ for some $m \in \mathbb{Z}$. The translation T belongs to $\hat{G}_2$ if and only if m is even. This implies that there are two orbits of vertices of $P_2(\lambda)$ in the mirror of S_2 under the translation subgroup. When $\lambda = 0$ one of these orbits contains o and the other contains $(2,-2,-2) = oS_1^{-3}S_2^{-1}S_1^2S_2$. As a consequence, for general λ one orbit contains $v_0 = (\lambda,-\lambda,-\lambda)$ and the other contains

$$v_0 S_1^{-3}S_2^{-1}S_1^2S_2 = (2,-2,-2) - (\lambda,-\lambda,-\lambda). \tag{6.17}$$

One can see from (6.17) that vertices in distinct orbits move in opposite directions. The odd integer values of λ correspond precisely to the instances when two vertices of distinct orbits under the translation group collide, creating a structure with disconnected vertex-figures.

The distance between vertices that are four edges apart in any helix is constant, since they are translates in $\hat{G}_2$. The line $\{\mu(1,0,1)\}$ intersects the axis of S_2 at o and the axis of S_1 at $\left(\frac{1}{2},0,\frac{1}{2}\right)$, while being perpendicular to both. This implies that the distance from o to the axis of S_1 is smallest among the distances of all points in the mirror of S_2 to the axis of S_1. As a consequence, the helices of $P_2(\lambda)$ become wider for larger absolute values of λ. This can be seen in the projection of $P_2(\lambda)$ to the XZ-plane for different values of λ. In Figure 6.30 this projection is shown for $\lambda \in \left\{0,\frac{1}{8},\frac{2}{8},\frac{3}{8}\right\}$.

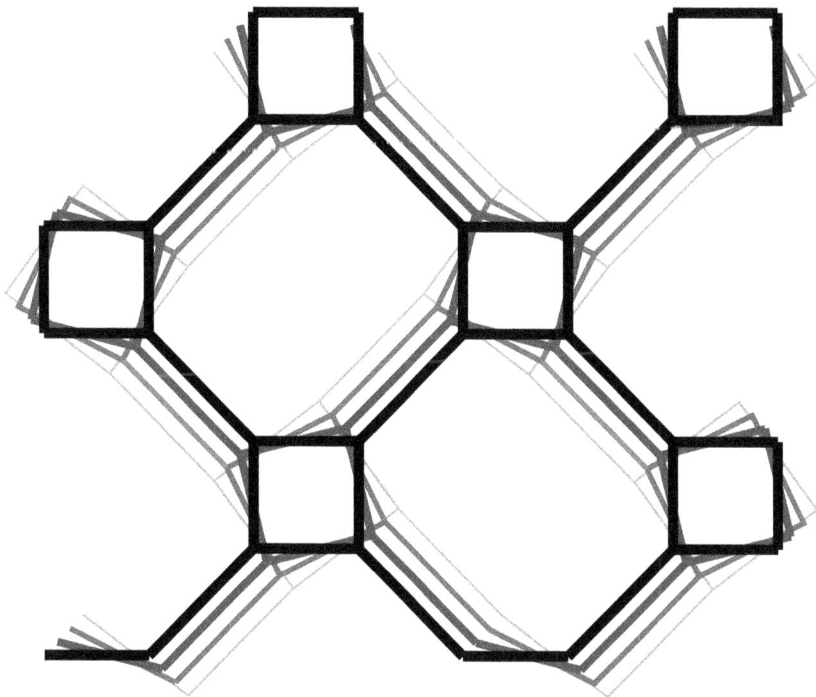

Figure 6.30 The projection to the plane $y = 0$ of $P_2(\lambda)$ for $\lambda \in \left\{0, \frac{1}{8}, \frac{2}{8}, \frac{3}{8}\right\}$.

Since the 1-skeleta of $\{\infty, 3\}^{(a)}$ and $\{\infty, 3\}^{(b)}$ are the same, there are four orbits of vertices and six orbits of edges of $\{\infty, 3\}^{(b)}$ under its translation subgroup. Therefore there are eight orbits of vertices and 12 orbits of edges of $P_2(\lambda)$ under $\mathcal{T}_{4\mathbf{\Lambda}_{(1,0,0)}}$. A system of representatives of the orbits of edges can be found in Table 6.3.

As shown in Table 6.3, the edges of $P_2\left(\frac{1}{2}\right)$ have direction vectors in $\Lambda_{(1,1,1)}$. This suggests interesting properties of this polyhedron; for example, its 1-skeleton is a subset of the well-known diamond net. A more detailed description of this polyhedron was given in [224, Subsection 4.2].

It can be seen in the projection of $\{\infty, 3\}^{(b)}$ to the XZ-plane that the translation group $\mathcal{T}_{4\mathbf{\Lambda}_{(1,0,0)}}$ splits the set of helices with axes parallel to the Y-axis into two orbits (see Figure 6.21). We shall denote the orbit containing the base helix H_1 by $\mathcal{H}_1$, and the other orbit by $\mathcal{H}_2$. The sets of helices with axis parallel to the X-axis and to the Z-axis are also split into two orbits by the group $\mathcal{T}_{4\mathbf{\Lambda}_{(1,0,0)}}$ and will follow an analogous behaviour to the one we shall explain next for those with axis in the direction of the Y-axis.

Table 6.3. *Representatives of the edges of $P_2(\lambda)$ under the translation group.*

Endpoints	Direction vector
$(\lambda, -\lambda, -\lambda), (1+\lambda, 1-\lambda, \lambda)$	$(1, 1, 2\lambda)$
$(-\lambda, -\lambda-1, 1-\lambda), (1-\lambda, -\lambda-2, \lambda+1)$	$(1, -1, 2\lambda)$
$(-\lambda-1, \lambda-1, 2-\lambda), (-\lambda-2, \lambda-2, \lambda+2)$	$(-1, -1, 2\lambda)$
$(\lambda+3, \lambda, -1-\lambda), (2+\lambda, \lambda+1, \lambda-1)$	$(-1, 1, 2\lambda)$
$(1-\lambda, -\lambda-2, \lambda+1), (2-\lambda, \lambda-2, 2+\lambda)$	$(1, 2\lambda, 1)$
$(1+\lambda, 1-\lambda, \lambda), (\lambda+2, \lambda+1, \lambda-1)$	$(1, 2\lambda, -1)$
$(\lambda, -\lambda, -\lambda), (\lambda-1, \lambda, -\lambda-1)$	$(-1, 2\lambda, -1)$
$(-\lambda, -\lambda-1, 1-\lambda), (-\lambda-1, \lambda-1, 2-\lambda)$	$(-1, 2\lambda, 1)$
$(-1-\lambda, \lambda-1, -\lambda-2), (\lambda-1, \lambda, -\lambda-1)$	$(2\lambda, 1, 1)$
$(-\lambda, -\lambda-1, 1-\lambda), (\lambda, -\lambda, -\lambda)$	$(2\lambda, 1, -1)$
$(1-\lambda, -\lambda-2, \lambda+1), (1+\lambda, -\lambda-3, \lambda)$	$(2\lambda, -1, -1)$
$(2-\lambda, \lambda-2, 2+\lambda), (\lambda+2, \lambda-3, \lambda+3)$	$(2\lambda, -1, 1)$

The 12 direction vectors of the edges of $P_2(\lambda)$ can be grouped as follows. The four direction vectors $(2\lambda, 1, -1)$, $(2\lambda, -1, -1)$, $(1, 1, 2\lambda)$ and $(1, -1, 2\lambda)$ correspond to the edges of helices in $\mathcal{H}_1$. The vectors $(2\lambda, 1, 1)$, $(2\lambda, -1, 1)$, $(-1, 1, 2\lambda)$ and $(-1, -1, 2\lambda)$ belong to edges of the helices in $\mathcal{H}_2$. Finally, the vectors $(1, 2\lambda, 1)$, $(1, 2\lambda, -1)$, $(-1, 2\lambda, 1)$ and $(-1, 2\lambda, -1)$ are precisely the direction vectors of bridges joining a helix in $\mathcal{H}_1$ with a helix in $\mathcal{H}_2$.

When increasing the absolute value of λ, the projections to the XZ-plane of the helices in $\mathcal{H}_1$ are squares with increasing edge length. The centre of the (square) projection of each helix in $\mathcal{H}_1$ is invariant under modifications in λ, but the angles of the projections of the edges with respect to the X-axis do change. The movement of a helix in $\mathcal{H}_1$ when increasing λ, starting from 0, can be described as an increase in the Y-coordinate of every vertex together with an increase on the width of the helix, and a continuous rotation around the axis of S_1. A similar movement is described by helices in $\mathcal{H}_2$, but the rotation component is in the other direction. This is easily visible in the projection to the XZ-plane (see Figure 6.30), or from the first and third coordinates of the direction vectors of the edges shown in Table 6.3.

The edge length of $P_2(\lambda)$ is $\sqrt{4\lambda^2+2}$. It follows that if $|\lambda_1| \neq |\lambda_2|$ then $P_2(\lambda_1)$ and $P_2(\lambda_2)$ are not isometric. Moreover, R_1 maps the base vertex and facet of $P_2(\lambda)$ to the base vertex and facet of $P_2(-\lambda)$, while mapping the base edge of $P_2(\lambda)$ to an edge of $P_2(-\lambda)$ that is not in the base flag. This shows that $P_2(\lambda)$ is isometric to $P_2(-\lambda)$, and they represent two geometric enantiomorphs. We summarise this discussion in the following result.

Proposition 6.2.18 *With the notation above, two polyhedra $P_2(\lambda_1)$ and $P_2(\lambda_2)$ are isometric if and only if $\lambda_1 = \pm\lambda_2$. Furthermore, $P_2(-\lambda) = P_2(\lambda)R_1$.*

For every λ we define the normalised polyhedron $\tilde{P}_2(\lambda)$ as $\frac{2}{\sqrt{4\lambda^2+2}}P_2(\lambda)$. The family $\{\tilde{P}_2(\lambda)\}_\lambda$ consists of polyhedra that are all combinatorially isomorphic to $\{\infty,3\}^{(b)}$, all have a group of symmetries isomorphic to $\hat{G}_2$ (only for $\lambda = 0$ the symmetry group is larger), and all have edge length 2.

When λ becomes arbitrarily large, the translation subgroup of $\tilde{P}_2(\lambda)$ contains arbitrarily small translations, and the direction vectors of the edges approach those of the edges of the cube $\mathcal{C}$ with vertices in $(\pm 1, \pm 1, \pm 1)$. The vertices of helices in $\mathcal{H}_1$ approach the points $(1,-1,-1)$, $(1,-1,1)$, $(-1,-1,1)$ and $(-1,-1,-1)$, whereas the vertices of helices in $\mathcal{H}_2$ approach the remaining four vertices of $\mathcal{C}$. This suggests that the limit when λ tends to ∞ of $P_2(\lambda)$ is precisely $\mathcal{C}$, where helices in the same orbit under translations tend to the same square, and two helices with parallel axes that are not translates under the symmetry group tend to opposite squares of $\mathcal{C}$. Hence, we define $P_2(\infty)$ as $\mathcal{C}$.

The angle between two edges at a vertex of $\tilde{P}_2(0)$ is $2\pi/3$, while that between two edges at a vertex of $\tilde{P}_2(\infty)$ is $\pi/2$. The fact that the angle between two edges at a vertex of $\tilde{P}_2(\lambda)$ decreases when increasing the absolute value of λ implies the following.

Proposition 6.2.19 *There are infinitely many skeletal chiral polyhedra $\mathcal{P}$ with $G(\mathcal{P}) = \hat{G}_2$, one for each $\lambda > 0$.*

The polyhedra $P_2(\lambda)$ appeared first in [246, Section 5]. They were constructed by Wythoff's construction with initial vertex o from the group generated by a certain twist S_1 and a certain rotation S_2. In that work the polyhedra $P_2(\lambda)$ are called $P_2(c,d)$, where c,d are two real parameters, not both 0, that determine S_1. For each $s \neq 0$, the polyhedra $P_2(c,d)$ and $P_2(sc,sd)$ are similar, and hence the family can be understood to depend on only one parameter. The regular polyhedra in that family are $P_2(1,0) = \{\infty,3\}^{(b)}$ and $P_2(0,1) = \{4,3\}$. We leave it as an exercise to determine which polyhedra $P_2(\lambda)$ correspond to which polyhedra $P_2(c,d)$.

Exercise 6.2.20 Determine c and d in terms of λ, such that $P_2(c,d)$ is isometric to $P_2(\lambda)$.

Exercise 6.2.21 Determine λ in terms of c,d such that $P_2(\lambda)$ is similar to $P_2(c,d)$.

Exercise 6.2.22 Find an isometry T such that the generators S_1 and S_2 of $P_2(\lambda)$ are conjugates by T of the generators S_1 and S_2 of $P_2(c,d)$ for c and d as in Exercise 6.2.20.

The Polyhedra $P_3(\lambda)$

Our third family of chiral polyhedra with helical 2-faces consists of polyhedra constructed by Wythoff's construction from the group $G(\{\infty,4\}_{\cdot,*3})$ generated by $S_1 = R_0R_1$ and $S_2 = R_1R_2$, with R_0, R_1 and R_2 as in (6.13), (6.14) and (6.15), respectively.

Recall that $\{\infty,4\}_{\cdot,*3}$ is combinatorially non-orientable (see Proposition 6.1.54). This implies that $G(\{\infty,4\}_{\cdot,*3}) = \langle S_1, S_2\rangle$. Nevertheless, we find it convenient to denote this group by $\hat{G}_3$. The orientable double cover of the abstract polyhedron $\{\infty,4\}_{\cdot,*3}$ is $\{\infty,4\}_{\cdot,*6}$, the group obtained from the Coxeter group $[\infty,4] = \langle R_0, R_1, R_2\rangle$ by the single extra relation $(R_0R_1R_0R_1R_2)^6 = \mathrm{id}$. The details of this fact are left as an exercise.

Exercise 6.2.23 Show that the orientable double cover of $\{\infty,4\}_{\cdot,*3}$ is $\{\infty,4\}_{\cdot,*6}$, as defined above. (Hint. Use the definition given in Example 4.4.4 and Proposition 4.4.16 (a).)

The symmetry S_1 is the 3-turn twist that maps (x,y,z) to $(-y,-1-z,x)$. Its axis is the line $\{\frac{1}{3}(0,-1,-1)+\mu(1,-1,1) : \mu\in\mathbb{R}\}$ and has translation vector $\frac{1}{3}(1,-1,1)$. The rotation S_2 maps (x,y,z) to $(x,-z,y)$. It has order 4 and its axis is the X-axis (see Figure 6.23).

In order to construct rotary polyhedra $P_3(\lambda)$ by Wythoff's construction from $\hat{G}_3$ we take an initial vertex $(\lambda,0,0)$ in the axis of S_2. As we shall see, the structure $P_3(\lambda)$ so obtained is not always a polyhedron.

Theorem 6.2.24 *Let $P_3(\lambda)$ be structure obtained from $\hat{G}_3$ by Wythoff's construction with initial vertex $v_0 = (\lambda,0,0)$. Then we have the following.*

(a) $P_3(\lambda)$ is a polyhedron if and only if $\lambda = 0$ or $\lambda\notin\mathbb{Z}$.
(b) $P_3(\lambda)$ is a regular polyhedron only when $\lambda = 0$.
(c) If $\lambda\notin\mathbb{Z}$ then $P_3(\lambda)$ is geometrically chiral and combinatorially isomorphic to the double cover $\{\infty,4\}_{\cdot,*6}$ of $\{\infty,4\}_{\cdot,*3}$.

Proof As noted in Section 6.1.5, the group $G(\{\infty,4\}_{\cdot,*3}) = \hat{G}_3$ consists of all orientation-preserving symmetries of the set $(1,0,0)+\Lambda_{(1,1,1)}$. There are several kinds of rotations in $\hat{G}_3$ whose axes intersect the X-axis:

- 4-fold rotations with axes parallel to the Y- or Z-axis, intersecting the X-axis at $(\lambda, 0, 0)$ for odd integer values of λ;
- 3-fold rotations with axes parallel to $(1,1,1)$, $(1,1,-1)$, $(1,-1,1)$ or $(-1,1,1)$ intersecting the X-axis at $(\lambda, 0, 0)$ for odd integer values of λ;
- half-turns with axes parallel to a coordinate axis or with direction vectors $(0,1,1)$ or $(0,1,-1)$, intersecting the X-axis at $(\lambda, 0, 0)$ for $\lambda \in \mathbb{Z}$.

The previous analysis also shows that $\mathrm{Stab}_{\hat{G}_3}(\lambda, 0, 0) = \langle S_2 \rangle$ for $\lambda \notin \mathbb{Z}$. In such cases Theorem 6.1.35 tells us that the structure $P_3(\lambda)$ is a polyhedron combinatorially isomorphic to the double cover $\{\infty, 4\}_{\cdot,*6}$ of $\{\infty, 4\}_{\cdot,*3}$.

The base edge of $P_3(\lambda)$ joins the base vertex $v_0 = (\lambda, 0, 0)$ with $v_0 S_1 S_2 = (0, -\lambda, -1)$. Hence the vertices of the vertex-figure $P_3(\lambda)/v_0$ are $(0, -\lambda, -1)$ and its images $(0, 1, -\lambda)$, $(0, \lambda, 1)$ and $(0, -1, \lambda)$ under $\langle S_2 \rangle$. They all belong to the plane $x = 0$, which contains v_0 only when $\lambda = 0$. From Proposition 6.2.9 we can conclude that if $\lambda \notin \mathbb{Z}$ then the polyhedron $P_3(\lambda)$ is chiral.

Since $P_3(0) = \{\infty, 4\}_{\cdot,*3}$, it remains only to show that if $\lambda \in \mathbb{Z} \setminus \{0\}$ then $P_3(\lambda)$ is not a polyhedron.

Assume to the contrary that $P_3(\lambda_0)$ is a polyhedron for some $\lambda_0 \in \mathbb{Z} \setminus \{0\}$. As noted above, the fact that $\lambda_0 \neq 0$ implies that $P_3(\lambda_0)$ is chiral. Furthermore, the flag stabilisers of $P_3(\lambda_0)$ are trivial since $P_3(\lambda_0)$ has helical 2-faces and the only isometry fixing a flag of a helix is id. Then Proposition 6.1.29 implies that $\mathrm{Stab}_{\hat{G}_3}(v_0) = \langle S_2 \rangle$. As noted previously, $(\lambda_0, 0, 0)$ is also invariant under the half-turn about the line through $(\lambda_0, 0, 0)$ with direction vector $(0, 1, 0)$; but this half-turn does not belong to $\langle S_2 \rangle$, a contradiction. ■

The axes of the 4-fold rotations in $\hat{G}_3$ are the lines with direction vectors $(1,0,0)$, $(0,1,0)$ and $(0,0,1)$ through points in $(1,0,0) + \Lambda_{(1,1,1)}$. Every point in $(1,0,0) + \Lambda_{(1,1,1)}$ belongs to three such lines, one in each of the three directions. On the other hand, every vertex of $\{\infty, 4\}_{\cdot,*3}$ belongs to the axis of only one 4-fold rotation in $\hat{G}_3$. Indeed, the four edges at a given vertex v of $\{\infty, 4\}_{\cdot,*3}$ have the direction vectors of two coordinate axes, and span lines of $\mathbb{E}^3$ tessellated by edges. Such lines do not contain points in $(1,0,0) + \Lambda_{(1,1,1)}$. The direction vector of the axis of a generator of the stabiliser of v is that of the remaining coordinate axis. The list of axes of 4-fold rotations of $\{\infty, 4\}_{\cdot,*3}$ through its vertices is presented in Table 6.4, where for each orbit under the translation group (first column) we indicate in the second column the direction vector of the axis containing it.

In order to describe further the polyhedra $P_3(\lambda)$ we consider the impact of a continuous modification of the parameter λ, starting from the polyhedron $P_3(0) = \{\infty, 4\}_{\cdot,*3}$.

Table 6.4. *Vertices and axes of 4-fold rotations of* $\{\infty,4\}_{\cdot,*3}$.

Vertices in	Direction vector
$o+\Lambda_{(1,1,1)}$	$(1,0,0)$
$(0,1,0)+\Lambda_{(1,1,1)}$	$(0,0,1)$
$(0,0,1)+\Lambda_{(1,1,1)}$	$(0,1,0)$

When continuously increasing λ, the base vertex increases its X-coordinate while remaining in the X-axis. As was the case in the previous two families of skeletal chiral polyhedra, every vertex moves along the axis of the 4-fold rotations in its stabiliser in $\hat{G}_3$, all at the same speed.

The symmetry R_2 of $\{\infty,4\}_{\cdot,*3}$ preserves the X-axis while interchanging its positive and negative ends. This implies that $oR_2 = o$, but for $\lambda \neq 0$, $v_0R_2 = (\lambda,0,0)R_2 = (-\lambda,0,0) \neq v_0$. Furthermore, $R_2 \in \hat{G}_3$ and can be expressed as $S_1^3S_2S_1^{-1}S_2S_1^2$ (see (6.16)). Therefore, when $\lambda \neq 0$ the vertices $v_0 = (\lambda,0,0)$ and $v_0S_1^3S_2S_1^{-1}S_2S_1^2 = (-\lambda,0,0)$ are distinct, and a continuous variation on λ induces in these two vertices a movement in opposite directions.

By vertex-transitivity of $P_3(\lambda)$, every vertex experiences the same phenomenom as the base vertex v_0 when λ moves continuously starting from 0. This phenomenom can be explained as a duplication of the every vertex of $\{\infty,4\}_{\cdot,*3}$, where the two duplicates of a given vertex v move in opposite directions along the axis of a 4-fold rotation fixing v. The duplicates of v correspond combinatorially to the two vertices of the double cover $\{\infty,4\}_{\cdot,*6}$ of $\{\infty,4\}_{\cdot,*3}$ that project to v under the natural quotient.

From the previous argument we can conclude that $P_3(\lambda)$ is a realisation of the abstract polyhedron $\{\infty,4\}_{\cdot,*6}$, where values of λ that are close to each other induce similar realisations. The realisation given by a parameter λ is faithful if and only if $\lambda \notin \mathbb{Z}$.

Each edge and helix of $\{\infty,4\}_{\cdot,*3}$ also duplicates to the two edges and the two helices determined by their corresponding vertex sets. Nevertheless, the underlying graph remains connected. The axis of the two duplicates of the base helix of $\{\infty,4\}_{\cdot,*3}$ is the axis of S_1. In general, the two duplicates of every helix of $\{\infty,4\}_{\cdot,*3}$ have the same axis.

The line through o with direction vector $(0,1,1)$ is perpendicular to the axes of S_1 and S_2, intersecting the former at $\frac{1}{3}(0,-1,-1)$ and the latter at o. Hence o is the closest point to the axis of S_1 among all points in the axis of S_2. It follows that the helices of $P_3(\lambda)$ become wider when increasing the absolute value of λ. The translations in $\hat{G}_3$ preserving a helix do not depend on λ, implying that the distance between pairs of vertices that are three edges apart in the same helix remains constant.

The projection of the base vertex $(\lambda, 0, 0)$ of $P_3(\lambda)$ to the plane $\Pi^{(1,-1,1)}$ is the point $\frac{\lambda}{3}(2, 1, -1)$, whereas the axis of S_1 projects to $\frac{1}{3}(0, -1, -1)$. The difference between these two points is $\frac{1}{3}(2\lambda, \lambda+1, -\lambda+1)$. This indicates that, when continuously modifying λ, the projection of the base vertex moves along a line that does not contain o, implying that the base helix, in addition to the modification of its width, experiences some rotation around its axis. Since R_2 is a half-turn that maps the base helix or $P_3(\lambda)$ to its duplicate, the rotation factors on these helices when continuously modifying λ are in opposite directions. By 2-face transitivity, the same behaviour can be seen in all helices of $P_3(\lambda)$ when modifying λ.

Figure 6.31 illustrates the effect of a continuous variation of λ on the projections of $P_3(\lambda)$ to the YZ-plane (a) and to the plane $\Pi^{(1,-1,1)}$ (b). Note that in the projection to $\Pi^{(1,-1,1)}$ there are two isometric triangles with the same shade of grey centred at the centre of each black triangle. It is an easy exercise to verify that it is possible to move along the graph of $P_3\left(\frac{1}{4}\right)$ from one duplicate of a helix of $\{\infty, 4\}_{\cdot,*3}$ to the other duplicate.

When λ is an odd integer, six vertices of $\{\infty, 4\}_{\cdot,*6}$ collapse into a point w of $(1, 0, 0) + \Lambda_{(1,1,1)}$. Such six vertices can be split into three pairs, where the two vertices in each pair move along one of the three axis of 4-fold rotations in $\hat{G}$ containing w.

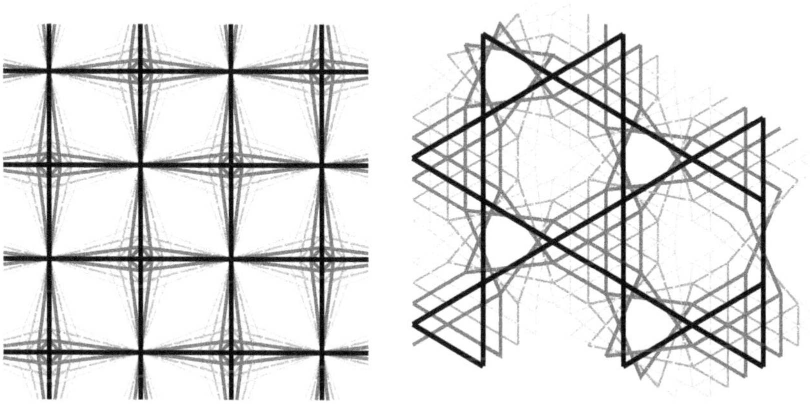

(a) (b)

Figure 6.31 The projection of $P_3(\lambda)$ (a) to the plane $x = 0$ for $\lambda \in \left\{0, \frac{1}{8}, \frac{2}{8}, \frac{3}{8}\right\}$, and (b) to the plane $\Pi^{(1,-1,1)}$ for $\lambda \in \left\{0, \frac{1}{4}, \frac{2}{4}, \frac{3}{4}\right\}$.

A collapse involving fewer vertices occurs when $\lambda \neq 0$ is an even integer. In such cases, two vertices of $\{\infty, 4\}_{\cdot,*6}$ that move in opposite directions along the same axis collapse in one of the vertices of $\{\infty, 4\}_{\cdot,*3}$. The vertex-figure in $P_3(\lambda)$ at this vertex consists of two disconnected squares.

Exercise 6.2.25 Draw the projections to the plane $x = 0$ and to the plane $\Pi^{(1,-1,1)}$ of $P_3(1)$ and of $P_3(2)$. Compare them with the projections to the same two planes of the polyhedron $P_3\left(\frac{3}{2}\right)$.

The translation subgroup of $\hat{G}_3$ is the group of translations by vectors in $\mathbf{\Lambda}_{(1,1,1)}$. This translation group induces three orbits on the vertex set of $\{\infty, 4\}_{\cdot,*3}$, with $\{o, (0,1,0), (0,0,1)\}$ being a set of representatives. Since $\mathrm{Stab}_{\hat{G}_3}(o) = \langle R_1, R_2 \rangle$ is dihedral with eight elements, it follows that the translation subgroup has index 24 in $\hat{G}_3$.

The stabiliser in $\hat{G}_3$ of the base edge of $P_3(0) = \{\infty, 4\}_{\cdot,*3}$ is $\langle R_0, R_2 \rangle$ and has precisely four elements. For $\lambda \neq 0$, the stabiliser of the base edge of $P_3(\lambda)$ is $\langle R_0 R_2 \rangle$ and hence it has only two elements. It follows that there are 12 orbits of edges under the translation subgroup of $\hat{G}_3$. A set of representatives of these edges is given in Table 6.5.

The edge length of $P_3(\lambda)$ is $\sqrt{2\lambda^2 + 1}$, which implies that $P_3(\lambda_1)$ is not isometric to $P_3(\lambda_2)$ when $|\lambda_1| \neq |\lambda_2|$. Furthermore, for $\lambda \neq \mathbb{Z}$ the polyhedra $P_3(\lambda)$ and $P_3(-\lambda)$ are the two geometric enantiomorphs of the same geometrically chiral polyhedron. The details are left as an exercise.

Table 6.5. *Representatives of the edges of $P_3(\lambda)$ under the translation group.*

Endpoints	Direction vector
$(\lambda,0,0)$, $(0,-\lambda,-1)$	$(\lambda,\lambda,1)$
$(\lambda,0,0)$, $(0,1,-\lambda)$	$(\lambda,-1,\lambda)$
$(\lambda,0,0)$, $(0,\lambda,1)$	$(\lambda,-\lambda,-1)$
$(\lambda,0,0)$, $(0,-1,\lambda)$	$(\lambda,1,-\lambda)$
$(0,-\lambda,-1)$, $(-1,0,-1+\lambda)$	$(1,-\lambda,-\lambda)$
$(0,1,-\lambda)$, $(-1,1-\lambda,0)$	$(1,\lambda,-\lambda)$
$(-\lambda,0,0)$, $(0,\lambda,-1)$	$(-\lambda,-\lambda,1)$
$(-\lambda,0,0)$, $(0,-1,-\lambda)$	$(-\lambda,1,\lambda)$
$(-\lambda,0,0)$, $(0,-\lambda,1)$	$(-\lambda,\lambda,-1)$
$(-\lambda,0,0)$, $(0,1,\lambda)$	$(-\lambda,-1,-\lambda)$
$(0,\lambda,-1)$, $(1,0,-1+\lambda)$	$(-1,\lambda,-\lambda)$
$(0,-1,-\lambda)$, $(1,-1+\lambda,0)$	$(-1,-\lambda,-\lambda)$

Exercise 6.2.26 Determine the image under R_1 of the base flag of $P_3(\lambda)$ in terms of the base flag of $P_3(-\lambda)$. Use this to show that $P_3(\lambda)$ and $P_3(-\lambda)$ are the two geometric enantiomorphs of the same geometrically chiral polyhedron.

We now define the normalised poyhedron $\tilde{P}_3(\lambda)$ as $\frac{\sqrt{2}}{\sqrt{2\lambda^2+1}}P_3(\lambda)$. The edge length of $\tilde{P}_3(\lambda)$ is $\sqrt{2}$, regardless of the value of λ.

When increasing the absolute value of λ, the translation subgroup of $\tilde{P}_3(\lambda)$ is generated by translations by smaller vectors, and the direction vectors of the edges of $\tilde{P}_3(\lambda)$ approach those of the edges of the octahedron $\mathcal{O}$ with vertex set $\{(\pm 1,0,0),(0,\pm 1,0),(0,0,\pm 1)\}$. In particular, the angle between two edges at a vertex is $\pi/2$ when $\lambda = 0$, and decreases when increasing the absolute value of λ, tending to $\pi/3$ when λ tends to ∞. This motivates us to define $P_3(\infty)$ as $\mathcal{O}$. When making λ tend to ∞, the edges of the two duplicates of a helix tend to the edges of two opposite triangles of an octahedron. In this way each of the four pairs of opposite triangles of $\mathcal{O}$ correspond to a set of helices whose axes have the same direction vector. The previous discussion implies the next result.

Proposition 6.2.27 *There are infinitely many skeletal chiral polyhedra $\mathcal{P}$ with $G(\mathcal{P}) = \hat{G}_3$, one for each $\lambda > 0$.*

The polyhedra $P_3(\lambda)$ were described first in [246, Section 6]. They were obtained by Wythoff's construction with initial vertex o from the group generated by certain twist S_1 and certain rotation S_2. The polyhedra $P_3(\lambda)$ were denoted $P_3(c,d)$, where c,d are two real parameters, not both 0, that determine S_1. The regular polyhedra in that family are $P_3(1,0) = \{3,4\}$ and $P_3(0,1) = \{\infty,4\}_{\cdot,*3}$.

Exercise 6.2.28 Determine c and d in terms of λ, such that $P_3(c,d)$ is isometric to $P_3(\lambda)$.

Exercise 6.2.29 Determine λ in terms of c,d such that $P_3(\lambda)$ is similar to $P_3(c,d)$.

Exercise 6.2.30 Find an isometry T such that the generators S_1 and S_2 of $P_3(\lambda)$ are conjugates by T of the generators S_1 and S_2 of $P_3(c,d)$, where c and d are as in Exercise 6.2.28.

It was not mentioned in [246] whether the polyhedra $P_1(a,b)$, $P_2(c,d)$ and $P_3(c,d)$ (and hence the polyhedra $P_1(\lambda)$, $P_2(\lambda)$ and $P_3(\lambda)$) were combinatorially regular or chiral, except for the geometrically regular instances in each family. The first proof of the fact that all geometrically chiral polyhedra in a family $P_i(a,b)$ are isomorphic to a regular polyhedron was given in [228]. The continuous movement of the polyhedra in each family when modifying λ is also described there from a slightly different viewpoint. The method used there relies on the theory developed in [246]. This section provides an independent proof of the same facts. We encourage the reader to compare the results in this section with the results in [228] for a better understanding of these families of chiral polyhedra.

The Complete List

To conclude the classification of chiral polyhedra in $\mathbb{E}^3$ with helical 2-faces it remains to show that the polyhedra $P_1(\lambda)$, $P_2(\lambda)$ and $P_3(\lambda)$ are the only such polyhedra. This was done implicitly in [246], and here we present an independent proof of the same fact. To this end we first describe some properties of the symmetries S_1 and S_2 of chiral polyhedra with helical 2-faces. Recall from Proposition 6.2.9 that the generators S_1 and S_2 of the symmetry group of a chiral polyhedron with helical 2-faces are a twist and a rotation, respectively. The first remark is a consequence of Remark 6.1.10 applied to the vertex-figure of a polyhedron.

Remark 6.2.31 The type of a chiral polyhedron with helical 2-faces is $\{\infty, q\}$ for some $q \geq 3$.

Lemma 6.2.32 *Let $\mathcal{P}$ be a chiral polyhedron in $\mathbb{E}^3$ with helical 2-faces and let $G(\mathcal{P}) = \langle S_1, S_2\rangle$. Then the axes of S_1 and S_2 are not parallel.*

Proof Assume to the contrary that the axes of S_1 and S_2 are parallel, and without loss of generality assume that they are parallel to the Z-axis. Then the translation component of S_1 is the vector $(0,0,c)$ for some $c \in \mathbb{R}\backslash\{0\}$, implying that the third coordinate of $(x,y,z)S_1$ is $z+c$, and that of $(x,y,z)S_2$ is z for every point $(x,y,z) \in \mathbb{E}^3$. It follows that the third coordinate of $(x,y,z) = (x,y,z)(S_1S_2)^2$ is $z+2c$, contradicting that $(S_1S_2)^2 = \mathrm{id}$. ■

Lemma 6.2.33 *Let $\mathcal{P}$ be a chiral polyhedron in $\mathbb{E}^3$ with helical 2-faces and let $G(\mathcal{P}) = \langle S_1, S_2\rangle$. Then the axes of the twist S_1 and the rotation S_2 do not intersect.*

Proof Let v_0 be the base vertex of $\mathcal{P}$ and l_i be the axis of S_i for $i \in \{1,2\}$. Suppose to the contrary that l_2 intersects l_1, so that l_2 is precisely the line determined by v_0 and $l_1 \cap l_2$.

Let $u := v_0 S_1^{-1}$ and $w := v_0 S_1$ be the two neighbours of v_0 in the base helix H_0, and let Π be the plane bisecting the segment between u and w. We note that l_1 intersects Π in exactly one point, since otherwise S_1 would have trivial translation component, and the 2-faces of $\mathcal{P}$ would not be helical. Since $wS_2 = v_0 S_1 S_2 = v_0 S_2^{-1} S_1^{-1} = v_0 S_1^{-1} = u$, the axis l_2 must be in Π. The fact that l_2 intersects l_1 forces this intersection to occur in $l_1 \cap \Pi$. We can conclude that l_2 is the only line in Π containing v_0 and $\Pi \cap l_1$.

The symmetry R of H_0 fixing v_0 is a half-turn that interchanges u and v. It follows that the axis l_R of F is also contained in Π. Furthermore, l_R contains v_0 and $\Pi \cap l_1$. We can conclude that $l_2 = l_R$ and therefore that S_2 is a half-turn, contradicting Remark 6.2.31. ∎

Lemma 6.2.34 *Let S_1 be a twist and let S_2 be a rotation of $\mathbb{E}^3$ such that* $(S_1 S_2)^2 = \mathrm{id}$. *Assume that the axes of S_1 and S_2 are not parallel and do not intersect. Let q be the point in the axis of S_2 whose distance to the axis of S_1 is minimal. Then qS_1 belongs to the plane perpendicular to the axis of S_2 through q (see Figure 6.32).*

Proof Without loss of generality we assume that the axis of S_2 is the Z-axis, that the axis of S_1 is on a plane $x = c$ parallel to the YZ-plane, and that $q = o$. This implies that the X-axis is the only line with non-trivial intersection with the axes of S_1 and S_2 that is simultaneously perpendicular to both. Let R be the half-turn about the X-axis. Then $(x, y, z)R = (x, -y, -z)$ for every $(x, y, z) \in \mathbb{E}^3$, and in particular $oR = o$. Furthermore, one can easily verify that $RS_i R = S_i^{-1}$ for $i \in \{1, 2\}$.

We denote by η_z the projection to the Z-axis, that is, $(x, y, z)\eta_z = z$ for every $(x, y, z) \in \mathbb{E}^3$. In particular, η_z satisfies that $S_2 \eta_z = \eta_z$ and that $R\eta_z = -\eta_z$. We want to show that $(oS_1)\eta_z = 0$ or, equivalently, that

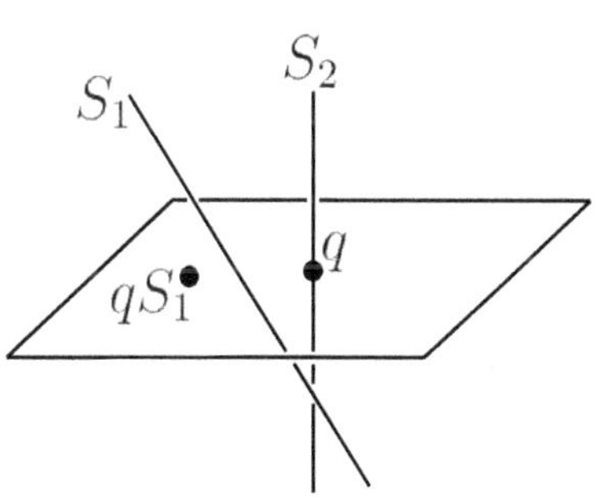

Figure 6.32 The axes of S_1 and S_2 as well as the points q and qS_1.

$(oS_1)\eta_z = -(oS_1)\eta_z$. To do this we rewrite $(S_1S_2)^2 = \text{id}$ as $S_1S_2 = S_2^{-1}S_1^{-1}$ and verify that

$$\begin{aligned}(oS_1)\eta_z &= (oS_1S_2)\eta_z = (oS_2^{-1}S_1^{-1})\eta_z = (oS_1^{-1})\eta_z\\ &= (oRS_1^{-1})\eta_z = (oS_1R)\eta_z = -(oS_1)\eta_z.\end{aligned}$$

■

We are ready for the main theorem of this section.

Theorem 6.2.35 *The polyhedra $P_1(a,b)$, $P_2(a,b)$ and $P_3(a,b)$ are the only chiral polyhedra with helical 2-faces in $\mathbb{E}^3$.*

Proof Let $\mathcal{P}$ be a chiral polyhedron with helical 2-faces in $\mathbb{E}^3$. Let v_0 be the base vertex and let $G(\mathcal{P}) = \langle S_1, S_2\rangle$. By Proposition 6.2.9 we know that S_1 is a twist and S_2 is a rotation.

We know from Lemmas 6.2.32 and 6.2.33 that the axes of S_1 and S_2 do not intersect each other and that they are are not parallel to each other. Let l_1 and l_2 be the axes of S_1 and of S_2, respectively, let l' be the only line intersecting non-trivially l_1 and l_2 that is perpendicular to both, and let R_1 be the half-turn with respect to l'.

We claim that the group $\langle G, R_1\rangle$ is discrete. In fact, since $R_1S_iR_1 = S_i^{-1}$ for $i \in \{1,2\}$, the group $\langle G, R_1\rangle$ is isomorphic either to the semidirect product $G \rtimes \langle R_1\rangle$ of which G is an index-2 subgroup, or to G itself if $R_1 \in G$. It follows that there are at most two points in each orbit of $\langle G, R_1\rangle$ in any fundamental region induced by G, implying the desired discreteness.

For $i \in \{1,2\}$ the group $\langle S_i, R_1\rangle$ is dihedral, and hence $R_0 := S_1R_1$ and $R_2 := R_1S_2$ are involutions. Since S_1, S_2 and R_1 preserve orientation, R_0 and R_2 are half-turns. Furthermore, $\text{id} = (S_1S_2)^2 = (R_0R_2)^2$, implying that the axes of R_0 and R_2 are perpendicular.

The only parallel class $\mathcal{C}$ of lines fixed (setwise) by S_1 is the parallel class $\mathcal{C}_1$ containing the axis of S_1. By Remark 6.2.31 and Lemma 6.2.32, S_2 does not fix the parallel class $\mathcal{C}$. This implies that the group $\langle R_0, R_1, R_2\rangle$ is not affinely reducible, as defined in Appendix A.1.

The possible triples of half-turns R_0, R_1 and R_2 satisfying that $(R_0R_2)^2 = \text{id}$ and that $\langle R_0, R_1, R_2\rangle$ is discrete but not affinely reducible were studied as part of the classification of the regular polyhedra in $\mathbb{E}^3$ (details are provided in [167, Section 6]). They are precisely the generators of the symmetry groups of $\{\infty,3\}^{(a)}$, $\{\infty,3\}^{(b)}$ and $\{\infty,4\}_{\cdot,*3}$.

By Lemma 6.2.34, if q is the point in $l_2 \cap l'$ then qS_1 belongs to the plane through q perpendicular to l_2. Moreover, such a q is the initial vertex to obtain

$\{\infty,3\}^{(a)}$, $\{\infty,3\}^{(b)}$ or $\{\infty,4\}_{\cdot,*3}$ from $\langle S_1, S_2\rangle$ by Wythoff's construction. In particular, the initial vertex v_0 of $\mathcal{P}$ is a point in l_2 distinct from q.

The previous analysis shows that $\mathcal{P}$ can be obtained by Wythoff's construction from the rotation subgroup of $G(\{\infty,3\}^{(a)})$, $G(\{\infty,3\}^{(b)})$ or $G(\{\infty,4\}_{\cdot,*3})$ with initial vertex v_0. This implies that $\mathcal{P}$ is $P_i(\lambda)$ for some $i \in \{1,2,3\}$ and some $\lambda \in \mathbb{R}$. ∎

The method that we used to construct $P_i(\lambda)$ $(i \in \{1,2,3\})$ consists of performing Wythoff's construction from the subgroup $\langle S_1, S_2\rangle$ of the symmetry group $G(\mathcal{P})$ of a regular polyhedron $\mathcal{P}$ by taking as initial vertex all possible vertices in the mirror of S_2. When trying this technique on some regular polyhedra, some vertices of $\mathcal{P}$ are forced to collapse, resulting on a structure that is not a polyhedron (like some instances of $P_2(\lambda)$ and $P_3(\lambda)$). This technique applied to some other polyhedra yields only other regular polyhedra; for example, the regular polyhedra $\mathcal{P}\#\{\}$ can be obtained in this way from the planar regular polyhedra (see [188, Section 7] for definitions and basic theory). Finally, in some cases the mirror of S_2 consists only of one point, making such polyhedra unsuitable for the use of this technique.

To date, there is no publication using this technique to build chiral polyhedra (or chiral higher rank polytopes) in spaces other than $\mathbb{E}^3$.

Problem 68 Search for chiral polyhedra in spaces $\mathcal{S}$ distinct from $\mathbb{E}^3$ by Wythoff's construction on the subgroup $\langle S_1, \dots, S_{n-1}\rangle$ of the symmetry group of regular n-polytopes in $\mathcal{S}$.

6.2.3 Chiral Polyhedra with Skew 2-Faces

The chiral polyhedra with skew 2-faces are less understood than those with helical 2-faces. In this section we provide a way to construct them, along with some of their properties.

We know that chiral polyhedra with skew 2-faces must have skew vertex-figures (see Proposition 6.2.8). It follows that the combinatorial rotations S_1 and S_2 of such a polyhedron are rotatory reflections, and therefore the involution S_1S_2 must be a half-turn.

In Section 6.1.5 we described three regular polyhedra whose combinatorial rotations S_1 and S_2 are rotatory reflections. They are the Petrie–Coxeter polyhedra $\{4,6|4\}$, $\{6,4|4\}$ and $\{6,6|3\}$, and they play an important role in the classification of chiral polyhedra with skew 2-faces. In fact, all chiral polyhedra with skew 2-faces in $\mathbb{E}^3$ can be constructed from $\{4,6|4\}$, $\{6,4|4\}$ and $\{6,6|3\}$ through the geometric operation described next.

A Geometric Operation

Let $\mathcal{P}$ be an infinite regular polyhedron with base vertex o and whose vertices have all integer coordinates. Let $G = G^+(\mathcal{P}) = \langle S_1, S_2 \rangle$ be the even subgroup of $\mathcal{P}$, and assume that S_1 and S_2 are rotatory reflections that preserve $\mathbb{Z}^3$. Let $v = oS_1S_2$, so that the base edge is incident to the vertices o and v.

The combinatorial rotation S_1S_2 is a half-turn whose axis does not contain o. Therefore S_1S_2 can be expressed as $S'_{1,2}T_v$, where $S'_{1,2}$ is the half-turn with axis a' containing o parallel to the axis a of S_1S_2, and T_v is the translation by v. Since T_v and S_1S_2 preserve $\mathbb{Z}^3$, so does $S'_{1,2}$.

As we know, the polyhedron $\mathcal{P}$ can be recovered by Wythoff's construction from G, once we know $S_1S_2 = S'_{1,2}T_v$ and S_2, since S_1 is expressed as $(S_1S_2)S_2^{-1}$. Our goal is to replace the translation T_v by the translation T_w by some vector w, so that we can apply Wythoff's construction to some group $G[w] := \langle S'_{1,2}T_w, S_2 \rangle$ with standard generators $S'_{1,2}T_wS_2^{-1}$ and S_2.

We claim that $S'_{1,2}T_wS_2^{-1}$ is a rotatory reflection. To see this, recall that for every $w \in \mathbb{E}^3$ and every isometry S fixing o the equality $S^{-1}T_wS = T_{wS}$ holds. In our situation, we may rewrite $S'_{1,2}T_wS_2^{-1}$ as

$$S_1S_2T_v^{-1}T_wS_2^{-1} = S_1S_2T_{w-v}S_2^{-1} = S_1S_2S_2^{-1}T_{(w-v)S_2^{-1}} = S_1T_{(w-v)S_2^{-1}}.$$

Being the composition of a rotatory reflection and a translation, the linear part of $S'_{1,2}T_wS_2^{-1}$ is that of S_1. The only isometries of $\mathbb{E}^3$ in the list in Appendix A.1 whose linear parts are those of rotatory reflections are precisely the rotatory reflections.

In order to guarantee that the group $G[w]$ is discrete we shall only consider vectors w with integer coordinates. In that way $oG[w] \subseteq \mathbb{Z}^3$, since $S'_{1,2}$, T_w and S_2 preserve $\mathbb{Z}^3$. We also require $w \neq o$, since otherwise the resulting structure would be finite given that $G[w]$ itself would be finite. We need $S'_{1,2}T_w$ to be an involution, and this forces $w \in a^\perp$, since otherwise $S'_{1,2}T_w$ would be a twist. As we shall see, these three restrictions on w are still insufficient to guarantee that the structure resulting from Wythoff's construction is a polyhedron; we still need to check whether the vertex-figures are polygons.

We denote by $\mathcal{W}(\mathcal{P}, w)$ the outcome of Wythoff's construction applied to $G[w]$ with initial vertex o and standard rotations $S'_{1,2}T_wS_2^{-1}$ and S_2. Even when $\mathcal{W}(\mathcal{P}, w)$ is not a polyhedron we still refer to its substructures as 'vertices', 'edges' and '2-faces'. Since $S'_{1,2}T_wS_2^{-1}$ and S_2 are rotatory reflections, the 2-faces and vertex-figures of $\mathcal{W}(\mathcal{P}, w)$ are planar or skew.

Our goal is to construct chiral polyhedra $\mathcal{W}(\mathcal{P}, w)$ where $\mathcal{P}$ is $\{4,6|4\}$, $\{6,4|4\}$ or $\{6,6|3\}$. To restrict the possible choices for w we use the following proposition.

Proposition 6.2.36 *Let $\mathcal{P}$ be a regular polyhedron with its vertex set in $\mathbb{Z}^3$, and assume that the two standard rotations of $G(\mathcal{P})$ are rotatory reflections. Let a' be as previously and let $w \in (\mathbb{Z}^3 \cap (a')^{\perp}) \setminus \{0\}$. Then, for every $k \neq 0$ such that kw has integer coordinates, the structure $\mathcal{W}(\mathcal{P}, kw)$ is similar to $\mathcal{W}(\mathcal{P}, w)$. In particular, $\mathcal{W}(\mathcal{P}, kw)$ is a polyhedron if and only if $\mathcal{W}(\mathcal{P}, w)$ is a polyhedron.*

Proof Let S_1 and S_2 denote the standard rotations of $\mathcal{P}$. Let $S'_{1,2}$ be as above, let T_w and T_{kw} be the translations by w and kw, respectively, and let $S_w = S'_{1,2}T_w S_2^{-1}$ and $S_{kw} = S'_{1,2}T_{kw}S_2^{-1}$. Then the standard generators of $G[w]$ are S_w and S_2, while those of $G[kw]$ are S_{kw} and S_2.

The proposition follows from the fact that the base vertex and the standard rotations of $\mathcal{W}(\mathcal{P}, kw)$ are those of $\mathcal{W}(\mathcal{P}, w)$ merely rescaled by a factor of k, as explained next.

Let H_k denote the homotety mapping $\bar{x}$ to $k\bar{x}$ for every $\bar{x} \in \mathbb{E}^3$. Then H_k commutes with every linear transformation; in particular,

$$H_k^{-1} S'_{1,2} H_k = S'_{1,2}, \tag{6.18}$$

$$H_k^{-1} S_2 H_k = S_2. \tag{6.19}$$

It follows that

$$S_{kw} = S'_{1,2}T_{kw}S_2^{-1} = (H_k^{-1}S'_{1,2}H_k)(H_k^{-1}T_w H_k)(H_k^{-1}S_2^{-1}H_k) = H_k^{-1}S_w H_k. \tag{6.20}$$

We can conclude that the ith standard rotation of $\mathcal{W}(\mathcal{P}, kw)$ is that of $\mathcal{W}(\mathcal{P}, w)$ conjugated by H_k.

The structures $\mathcal{W}(\mathcal{P}, w)$ and $\mathcal{W}(\mathcal{P}, kw)$ are obtained from Wythoff's construction with initial vertex o from the groups $G[w]$ and $G[kw]$, respectively.

The base edge e_w of $\mathcal{W}(\mathcal{P}, w)$ has o and $w = oS_w S_2$ as its endpoints, while the base edge e_{kw} of $\mathcal{W}(\mathcal{P}, kw)$ has endpoints at o and $kw = oS_{kw}S_2$; hence, $e_{kw} = ke_w$. The base 2-face of $\mathcal{W}(\mathcal{P}, kw)$ is

$$e_{kw}\langle S_{kw}\rangle = e_{kw}H_k^{-1}\langle S_w\rangle H_k = k(e_k\langle S_w\rangle),$$

and therefore the base 2-face of $\mathcal{W}(\mathcal{P}, kw)$ is also that of $\mathcal{W}(\mathcal{P}, w)$ merely rescaled by a factor of k.

Finally, an arbitrary i-face F_{kw} of $\mathcal{W}(\mathcal{P}, kw)$ can be expressed as $\hat{F}_{kw}T$ for some $T \in G[kw]$, where $\hat{F}_{kw}$ is the base i-face of $\mathcal{W}(\mathcal{P}, kw)$. From (6.18), (6.19) and (6.20) we can see that

$$F_{kw} = \hat{F}_{kw}T = \hat{F}_{kw}H_k^{-1}T'H_k = k(\hat{F}_w T'),$$

where $\hat{F}_w$ is the base i-face of $\mathcal{W}(\mathcal{P}, w)$, and T' is the element of $G[k]$ obtained by replacing every factor S_{kw} by S_w in any expression of T in terms of S_{kw}

Table 6.6. *The direction vector of the axis a' and two generators of $\mathbb{Z}^3 \cap (a')^{\perp}$, for each Petrie–Coxeter polyhedron $\mathcal{P}$.*

$\mathcal{P}$	a'	Generators of $\mathbb{Z}^3 \cap (a')^{\perp}$
$\{4,6\|4\}$	$(1,1,0)$	$(1,-1,0)$, $(0,0,1)$
$\{6,4\|4\}$	$(1,1,0)$	$(1,-1,0)$, $(0,0,1)$
$\{6,6\|3\}$	$(0,1,0)$	$(1,0,0)$, $(0,0,1)$

and S_2. It follows that F_{kw} is obtained from an i-face of $\mathcal{W}(\mathcal{P},kw)$ by the homotety H_k, and the proposition holds. ∎

In what follows we study the polyhedra $\mathcal{W}(\mathcal{P},w)$ where $\mathcal{P}$ is either $\{4,6|4\}$, $\{6,4|4\}$ and $\{6,6|3\}$.

The symmetry S_1S_2 of each Petrie–Coxeter polyhedron $\mathcal{P}$ is a half-turn. Its axis has direction vector $(1,1,0)$ if $\mathcal{P}$ is either $\{4,6|4\}$ or $\{6,4|4\}$, whereas it has direction vector $(0,1,0)$ if $\mathcal{P}$ is $\{6,6|3\}$. It follows that a' is the line $\{\mu(1,1,0)\}_{\mu\in\mathbb{R}}$ in the first two cases, and the Y-axis in the last case. Table 6.6 indicates, for each Petrie–Coxeter polyhedron $\mathcal{P}$, the direction vector of a' as well as two vectors that generate $\mathbb{Z}^3 \cap (a')^{\perp}$.

The Regular Instances

Here we show the instances where $\mathcal{P}$ is a Petrie–Coxeter polyhedron and $\mathcal{W}(\mathcal{P},w)$ is a regular polyhedron. As mentioned earlier, the 2-faces and vertex-figures of $\mathcal{W}(\mathcal{P},w)$ are finite, either planar or skew. It follows from Theorem 6.2.10 that $\mathcal{W}(\mathcal{P},w)$ is regular if and only if its 2-faces or its vertex-figures are planar. We consider these two cases in the next lemmas.

Lemma 6.2.37 *Let $\mathcal{P} \in \{\{4,6|4\},\{6,4|4\},\{6,6|3\}\}$ and let $\mathcal{Q} = \mathcal{W}(\mathcal{P},w)$ for some w. Then $\mathcal{Q}$ has planar 2-faces if and only if $\mathcal{Q}$ is similar to $\mathcal{P}$.*

Proof The isometry generating the even symmetry group of the base 2-face of $\mathcal{W}(\mathcal{P},w)$ is $S_w := S'_{1,2}T_wS_2^{-1} = S'_{1,2}S_2^{-1}T_{wS_2^{-1}}$. The symmetries $S'_{1,2}S_2^{-1}$ and S_w are rotatory reflections with the same order and rotational component (their linear components are equal). However, in general they have distinct invariant planes; in any case, the invariant planes of these isometries are parallel.

The base 2-face of $\mathcal{W}(\mathcal{P},w)$ contains the base vertex o, which is in the invariant plane of $S'_{1,2}S_2^{-1}$. The other endpoint w of the base edge of $\mathcal{W}(\mathcal{P},w)$ may or may not be in that invariant plane. In fact, w is in the invariant plane of $S'_{1,2}S_2^{-1}$ if and only if all vertices of the base 2-face of $\mathcal{W}(\mathcal{P},w)$ are in the invariant plane of $S'_{1,2}S_2^{-1}$ (if a polygon is skew finite then its vertices alternate in two distinct parallel planes). It follows that the base 2-face of $\mathcal{W}(\mathcal{P}, w)$ is skew if and only if the invariant plane of $S'_{1,2}S_2^{-1}$ does not contain w. Henceforth we assume that w is in the invariant plane of $S'_{1,2}S_2^{-1}$.

The invariant plane of $S'_{1,2}S_2^{-1}$ is that of S_1, since the planar base 2-face of $\mathcal{P}$ contains the origin. It is the YZ-plane if $\mathcal{P}$ is $\{4,6|4\}$, the plane with equation $x+y+z=0$ if $\mathcal{P}$ is $\{6,4|4\}$ and the plane with equation $x+y-z=0$ if $\mathcal{P}$ is $\{6,6|3\}$.

From the construction of $\mathcal{W}(\mathcal{P}, w)$ we know that $w = (w_1, w_2, w_3)$ is perpendicular to the direction vector of a'. Belonging to the invariant plane of $S'_{1,2}S_2^{-1}$ imposes a second linear restriction. It follows that if $\mathcal{P}$ is $\{4,6|4\}$ then $0 = (w_1, w_2, w_3) \cdot (1,1,0) = w_1 + w_2$ and $w_1 = 0$; Proposition 6.2.36 implies that we may take $w = (0,0,1)$, and $\mathcal{W}(\mathcal{P}, w) = \{4,6|4\}$. If $\mathcal{P}$ is $\{6,4|4\}$ then $0 = (w_1, w_2, w_3) \cdot (1,1,0) = w_1 + w_2$ and $w_1 + w_2 + w_3 = 0$; it follows that we may take $w = (-1,1,0)$, and $\mathcal{W}(\mathcal{P}, w) = \{4,6|4\}$. Finally, if $\mathcal{P}$ is $\{6,6|3\}$ then $0 = (w_1, w_2, w_3) \cdot (0,1,0) = w_2$, and $w_1 + w_2 - w_3 = 0$; now we may choose $w = (1,0,1)$, and $\mathcal{W}(\mathcal{P}, w) = \{6,6|3\}$. ■

There are three skeletal regular polyhedra with skew finite 2-faces and planar vertex-figures, and whose symmetry groups are irreducible (it preserves no parallel class of lines, see Appendix A.1). They are described in [104] as the three polyhedra in Class 6 that do not belong to the infinite families. They were called $\{4,6\}_6$, $\{6,4\}_6$ and $\{6,6\}_4$ in [167], where they were derived from the Petrie–Coxeter polyhedra through the following operations:

$$\begin{aligned} \{6,6\}_4 &= \{4,6|4\}^{\eta}, \\ \{4,6\}_6 &= (\{6,6\}_4)^{\pi}, \\ \{6,4\}_6 &= (\{4,6\}_6)^{\delta}. \end{aligned}$$

Lemma 6.2.38 *Let $\mathcal{P}$ be a Petrie–Coxeter polyhedron and let $\mathcal{Q} = \mathcal{W}(\mathcal{P}, w)$ for some w. Then $\mathcal{Q}$ has planar vertex-figures if and only if $\mathcal{Q}$ is the regular polyhedron $\{4,6\}_6$ if $\mathcal{P} = \{4,6|4\}$, the regular polyhedron $\{6,4\}_6$ if $\mathcal{P} = \{6,4|4\}$ and the regular polyhedron $\{6,6\}_4$ if $\mathcal{P} = \{6,6|3\}$.*

Proof The generator of the even symmetry group of the vertex-figure of $\mathcal{W}(\mathcal{P}, w)$ is S_2. Hence, the vertex-figures will be planar if and only if the base vertex $w = oS'_{1,2}T_w$ of the vertex-figure is in the invariant plane Π_2 of S_2. The plane Π_2 is the plane with equation $x+y+z=0$ if $\mathcal{P}$ is either $\{4,6|4\}$ or $\{6,6|3\}$, and it is the YZ-plane if $\mathcal{P}$ is $\{6,4|4\}$.

The combined restrictions of $w \in (a')^{\perp}$ and w belonging to Π_2, together with Proposition 6.2.36, allow us to assume that $w = (1,-1,0)$ if $\mathcal{P} = \{4,6|4\}$, $w = (0,0,1)$ if $\mathcal{P} = \{6,4|4\}$, and $w = (1,0,-1)$ if $\mathcal{P} = \{6,6|3\}$. Now, the

Table 6.7. *The instances where $\mathcal{W}(\mathcal{P}, w)$ is a regular polyhedron.*

$\mathcal{P}$	w	$\mathcal{W}(\mathcal{P}, w)$
$\{4,6\|4\}$	$(0,0,1)$	$\{4,6\|4\}$
	$(1,-1,0)$	$\{4,6\}_6$
$\{6,4\|4\}$	$(-1,1,0)$	$\{6,4\|4\}$
	$(0,0,1)$	$\{6,4\}_6$
$\{6,6\|3\}$	$(1,0,1)$	$\{6,6\|3\}$
	$(1,0,-1)$	$\{6,6\}_4$

rotatory reflections S_w and S_2 and the point o are, respectively, the standard rotations and the base vertex of $\{4,6\}_6$ if $\mathcal{P} = \{4,6|4\}$, of $\{6,4\}_6$ if $\mathcal{P} = \{6,4|4\}$ and of $\{6,6\}_4$ if $\mathcal{P} = \{6,6|3\}$. The details are left as an exercise ∎

Exercise 6.2.39 Show that $\{4,6\}_6$, $\{6,4\}_6$ and $\{6,6\}_4$ can be obtained from Wythoff's construction with base vertex o and standard rotations S_w and S_2 as in the proof of Lemma 6.2.38.

Lemmas 6.2.37 and 6.2.38, together with Theorem 6.2.10 and Proposition 6.2.36, imply the following theorem.

Theorem 6.2.40 *Let $\mathcal{P}$ be $\{4,6|4\}$, $\{6,4|4\}$ or $\{6,6|3\}$. The structure $\mathcal{W}(\mathcal{P}, w)$ is geometrically regular only for the values of w described in Table 6.7. In particular, for such values of w the polyhedron $\mathcal{W}(\mathcal{P}, w)$ is similar to a Petrie–Coxeter polyhedron, to $\{4,6\}_6$, to $\{6,4\}_6$ or to $\{6,6\}_4$.*

Skeletal rotary polyhedra such that the standard rotations of their symmetry groups are both rotatory reflections were studied in [245]. These polyhedra were divided in three families, denoted by $P(a,b)$, $Q(c,d)$ and $Q(c,d)^*$ (see [245, Sections 5, 6 and 7]). Each family has two regular instances, and they correspond to the polyhedra in the last column of Table 6.7.

Exercise 6.2.41 Show that $\mathcal{W}(\{4,6|4\},(1,-1,0))$, $\mathcal{W}(\{6,4|4\},(0,0,1))$ and $\mathcal{W}(\{6,6|3\},(1,0,-1))$ are similar to the polyhedra $Q(1,0)$, $Q(1,0)^*$ and $P(1,-1)$, described in [245]. Can we replace the word similar by 'congruent', or by 'equal'?

Three Chiral Polyhedra with Skew 2-Faces

Now we provide a geometric description with combinatorial taste of one chiral polyhedron $\mathcal{W}(\mathcal{P}, w)$ for each Petrie–Coxeter polyhedron $\mathcal{P}$.

The Polyhedron $\mathcal{W}(\{4,6|4\},(1,-1,1))$

First, we shall give a rather combinatorial description of $\mathcal{W}(\{4,6|4\},(1,-1,1))$. Then we shall show that it is indeed the polyhedron obtain from Wythoff's construction from the group $G(\{4,6|4\})[(1,-1,1)]$.

There is a pair $\mathcal{L}$, $\mathcal{L}'$ of interesting configurations of pairwise disjoint lines in $\mathbb{E}^3$ with four distinct directions that cover $\mathbf{\Lambda}_{(1,1,1)}$. They are described in Appendix A.4 as an arrangement and its image under a plane reflection. Each of them has infinitely many lines with direction vector v for each v in the set

$$\{(1,1,1),(1,1,-1),(1,-1,1),(-1,1,1)\}. \tag{6.21}$$

In particular, the set $\mathcal{L}'$ of lines, defined after Theorem A.4.1, contains the lines with direction $(1,1,1)$ through all points in $2\mathbf{\Lambda}_{(1,1,1)}$, all lines with direction $(1,1,-1)$ through all points in $(2,0,0)+2\mathbf{\Lambda}_{(1,1,1)}$, all lines with direction $(-1,1,1)$ through all points in $(0,2,0)+2\mathbf{\Lambda}_{(1,1,1)}$ and all lines with direction $(1,-1,1)$ through all points in $(0,0,2)+2\mathbf{\Lambda}_{(1,1,1)}$. The lines in $\mathcal{L}'$ are a subset of the set $\overline{\mathcal{L}'}$ of all lines containing main diagonals of cubes in the cubic tessellation of $\mathbb{E}^3$ with vertex set in $2\mathbf{\Lambda}_{(1,0,0)}$.

Each line in $\overline{\mathcal{L}'}$ can be naturally partitioned into line segments with length $\sqrt{3}$ with endpoints in $\mathbf{\Lambda}_{(1,1,1)}$. We denote by $\mathcal{G}_{\mathbf{\Lambda}_{(1,1,1)}}$ the graph with vertex set $\mathbf{\Lambda}_{(1,1,1)}$ and edge set given by the previously mentioned line segments. We can think of the edges of $\mathcal{G}_{\mathbf{\Lambda}_{(1,1,1)}}$ as the set of diagonals of all cubes with vertices in $2\mathbf{\Lambda}_{(1,0,0)}$, subdivided at their midpoints.

We define $\mathcal{X}$ to be the graph with vertex set $\mathbf{\Lambda}_{(1,1,1)}$ and edges given by all edges of $\mathcal{G}_{\mathbf{\Lambda}_{(1,1,1)}}$ that are not contained in a line of $\mathcal{L}'$. As we shall see in the next results, the graph $\mathcal{X}$ is the 1-skeleton of $\mathcal{W}(\{4,6|4\},(1,-1,1))$.

Lemma 6.2.42 *The graph $\mathcal{X}$ just defined is* 6*-valent and connected.*

Proof Since every point in $\mathbf{\Lambda}_{(1,1,1)}$ belongs to one and only one line in $\mathcal{L}'$, every vertex in $\mathcal{X}$ has degree 6 (two edges for each direction in (6.21), except for that of the line in $\mathcal{L}'$ containing the vertex).

There are nine vertices of $\mathcal{X}$ in each cube C of the tessellation of $\mathbb{E}^3$ with vertex set $2\mathbf{\Lambda}_{(1,0,0)}$, namely the eight vertices and the centre of C. The subgraph $\mathcal{X}[C]$ of $\mathcal{X}$ induced by these nine vertices consists of three connected components; two of them have only one vertex (those in the diagonal of C contained in a line of $\mathcal{L}'$), while the other one contains the remaining seven vertices (the centre of C is adjacent to the other six vertices). In what follows we show that there is a path in $\mathcal{X}$ joining any isolated vertex in $\mathcal{X}[C]$ to a vertex in the connected component with seven vertices. The connectivity (as a 4-polytope) of the tessellation by cubes will imply the lemma.

Let u_0 be an isolated vertex in $\mathcal{X}[C]$, let $\mathcal{S}$ be a square of C containing u_0, and let C' be the cube, other than C, that contains $\mathcal{S}$. Then there is an edge between u_0 and the centre of $\mathcal{C}'$, since the only neighbours of u_0 in $\mathcal{G}_{\mathbf{\Lambda}_{(1,1,1)}}$ that are not neighbours of u_0 in $\mathcal{X}[\mathcal{C}]$ are the centre of $\mathcal{C}$ and the centre of the cube opposite to $\mathcal{C}$ at u_0. Given that $\mathcal{X}$ is 6-valent, there is also an edge from the centre of $\mathcal{C}'$ to some vertex u_1 of $\mathcal{S}$ other than u_0. Using these two edges we join u_0 to u_1, which belongs to the connected component with seven vertices of $\mathcal{X}[\mathcal{C}]$. ∎

By setting $w = (1,-1,1)$ we can see from equations (6.3), (6.4) and (6.5) that the symmetries S_w and S_2 of the structure $\mathcal{W}(\{4,6|4\},(1,-1,1))$ map (x,y,z) to $(1-x, z-1, -y-1)$ and to $(-z,-x,-y)$, respectively. Standard procedures show that both generators of $G(\{4,6|4\})[(1,-1,1)]$ preserve the graph $\mathcal{G}_{\mathbf{\Lambda}_{(1,1,1)}}$; furthermore, S_w and S_2 permute the sets $\mathcal{L}_1$, $\mathcal{L}_2'$, $\mathcal{L}_3'$ and $\mathcal{L}_4'$ defined in Appendix A.4 by the following rule:

$$S_w \mapsto (\mathcal{L}_1\ \mathcal{L}_4'\ \mathcal{L}_3'\ \mathcal{L}_2'),$$
$$S_2 \mapsto (\mathcal{L}_2'\ \mathcal{L}_3'\ \mathcal{L}_4').$$

Hence, $G(\{4,6|4\})[(1,-1,1)]$ preserves $\mathcal{L}'$, and therefore also $\mathcal{X}$. The next result describes further the action of $G(\{4,6|4\})[(1,-1,1)]$ on $\mathcal{X}$.

Lemma 6.2.43 *The group $G(\{4,6|4\})[(1,-1,1)]$ acts vertex-transitively on the graph $\mathcal{X}$.*

Proof The symmetry $S_w S_2^3 S_w^{-1} S_2^3$ is the translation by $(-2,2,-2)$. Its conjugates by S_2^2 and S_2^{-2} are the translations by $(2,-2,-2)$ and $(-2,-2,2)$, respectively. Hence $G(\{4,6|4\})[(1,-1,1)]$ acts transitively on $2\mathbf{\Lambda}_{(1,1,1)}$.

Recall that $\mathbf{\Lambda}_{(1,1,1)}$ contains precisely eight distinct translates of $2\mathbf{\Lambda}_{(1,1,1)}$. It remains to show that the orbit of o under $G(\{4,6|4\})[(1,-1,1)]$ contains an element in each of these translates. A system of representatives of this partition of $\mathbf{\Lambda}_{(1,1,1)}$, together with an element of $G(\{4,6|4\})[(1,-1,1)]$ mapping o to each of them, is given next:

$$\begin{array}{ll} o = (o)\mathrm{id}, & (1,1,1) = o(S_w S_2^3)^2 S_w^{-1} S_2 S_w, \\ (0,0,2) = oS_w^2 S_2, & (1,1,-1) = oS_w S_2^{-1}, \\ (0,2,0) = oS_w^2 S_2^3, & (1,-1,1) = oS_w S_2, \\ (2,0,0) = oS_w^2 S_2^{-1}, & (-1,1,1) = oS_w S_2^3. \end{array}$$

∎

We denote by $\mathcal{Q}_{\{4,6\}}$ the rank-3 structure with 1-skeleton $\mathcal{X}$ and 2-faces given by all 4-cycles of $\mathcal{X}$.

Theorem 6.2.44 *The structure $\mathcal{Q}_{\{4,6\}}$ just defined is a polyhedron.*

Proof The vertex set is $\mathbf{\Lambda}_{(1,1,1)}$, and hence it is discrete. By Lemma 6.2.42, the 1-skeleton is connected. This proves the first two items of Definition 6.1.2.

It remains to show that the vertex-figures are polygons; since the vertices are 6-valent, these polygons must be hexagons. By Lemma 6.2.43 it suffices to confirm this for o. Moreover, since S_2 preserves $\mathcal{X}$, we only need to verify that there are precisely six quadrilaterals in $\mathcal{X}$ containing o, and that S_2 permutes them cyclically.

Let $\mathcal{C}$ be the cube with vertices in $(\pm 1, \pm 1, \pm 1)$. Then the six neighbours of o are vertices of $\mathcal{C}$. Two neighbours of o that are opposite vertices of $\mathcal{C}$ have o as their only neighbour in $\mathcal{G}_{\mathbf{\Lambda}_{(1,1,1)}}$, and hence there is no quadrilateral in $\mathcal{X}$ containing two opposite vertices of $\mathcal{C}$.

We claim that two edges at o that contain two opposite vertices v_1 and v_2 of a square of $\mathcal{C}$ cannot be contained in a quadrilateral in $\mathcal{X}$. Assume on the contrary that such v_1 and v_2 are contained in a quadrilateral in $\mathcal{X}$. The vertices v_1 and v_2 fully determine the square F that contains them, and hence the hypothetical fourth point w of the quadrilateral containing o, v_1 and v_2 can only be the centre of the cube $\mathcal{C}'$ that shares the square F with $\mathcal{C}$. This limits the possibilities of the main diagonals of $\mathcal{C}$ and $\mathcal{C}'$ that are contained in lines of $\mathcal{L}'$; each must contain a vertex v_3 or v_4 of $\mathcal{C}$ distinct from v_1 and v_2. Furthermore, distinct lines in $\mathcal{L}'$ do not intersect, and hence one of this diagonals contains v_3 and the other contains v_4. This forces the diagonals to be parallel. However, parallel lines in $\mathcal{L}'$ are translates of each other by vectors in $2\mathbf{\Lambda}_{(1,1,1)}$, and therefore they cannot contain diagonals of cubes with vertices in $2\mathbf{\Lambda}_{(1,0,0)}$ that share a square.

We know now that every square containing o and the vertex $(1,-1,1)$ must also contain either $(-1,-1,1)$ or $(1,-1,-1)$, since the remaining neighbours of o are opposite to $(1,-1,1)$ in $\mathcal{C}$ or opposite in a square of $\mathcal{C}$. The only points in $\mathbf{\Lambda}_{(1,1,1)}$ that form a quadrilateral with $(1,-1,1)$, o and $(-1,-1,1)$ in $\mathcal{G}_{\mathbf{\Lambda}_{(1,1,1)}}$ are $(0,-2,0)$, $(0,-2,2)$ and $(0,0,2)$. However, the lines through $(0,-2,0)$ and $(-1,-1,1)$ and through $(0,-2,2)$ and $(1,-1,1)$ belong to $\mathcal{L}'$, implying that the only quadrilateral in $\mathcal{X}$ containing $(1,-1,1)$, o, and $(-1,-1,1)$ is completed with $(0,0,2)$. Its images under the powers of S_2 are the remaining five quadrilaterals in $\mathcal{X}$ at o. ■

Now that we have constructed a polyhedron $\mathcal{Q}_{\{4,6\}}$, it is no hard to verify that it coincides with $\mathcal{W}(\{4,6|4\},(1,-1,1))$. The details are left as an exercise.

Exercise 6.2.45 Show that $\mathcal{Q}_{\{4,6\}}$ is precisely the polyhedron $\mathcal{W}(\{4,6|4\},(1,-1,1))$ constructed by Wythoff's construction from the group $G(\{4,6|4\})[(1,-1,1)]$.

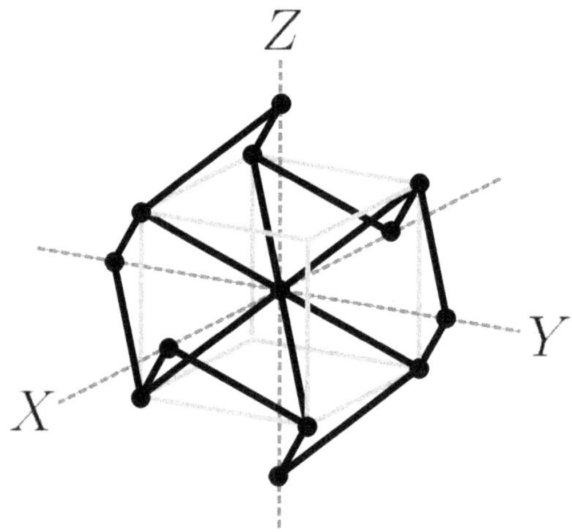

Figure 6.33 The six 2-faces of $\mathcal{W}(\{4,6|4\},(1,-1,1))$ around o.

The base 2-face of $\mathcal{W}(\{4,6|4\},(1,-1,1))$ is the quadrilateral with vertex set given by the orbit

$$\{o,(1,-1,-1),(0,-2,0),(1,-1,1)\}$$

of o under the symmetry S_w. The six 2-faces containing o are the images of the base 2-face under $\langle S_2\rangle$ and are illustrated in Figure 6.33.

Remark 6.2.46 The standard generators S_w and S_2 of $G(\{4,6|4\})[(1,-1,1)]$ are precisely the standard generators of the symmetry group of the polyhedron $Q(1,1)$ described in [245, Section 6], implying that these two are the same polyhedron.

The Polyhedron $\mathcal{W}(\{6,4|4\},(-1,1,2))$

Next we study a dual of the polyhedron $\mathcal{W}(\{4,6|4\},(1,-1,1))$ just described.

We find it convenient to reinterpret the 2-faces of $\mathcal{W}(\{4,6|4\},(1,-1,1))$ in terms of a well-known tessellation of $\mathbb{E}^3$ with vertex set $\mathbf{\Lambda}_{(1,1,1)}$. This tessellation can be understood through the tessellation $\mathcal{T}$ of cubes with vertices in $2\mathbb{Z}^3$ and its dual $\mathcal{T}^\delta$ with vertices in $(1,1,1)+2\mathbb{Z}^3$, as follows. Given an edge e and an incident square s of $\mathcal{T}$ we consider the tetrahedron with vertices at the endpoints of e (two vertices of $\mathcal{T}$) and the centres of the two cubes containing s (two vertices of $\mathcal{T}^\delta$). The orbit of this tetrahedron under the symmetry group of $\mathcal{T}$ yields a tessellation $\hat{\mathcal{T}}$. Every cell of $\hat{\mathcal{T}}$ is a nonregular tetrahedron with two edges of length 2 and four of length $\sqrt{3}$. The symmetry group of each cell has order 8. Each cell can also be described as the region composed of a flag

Φ (understood as a simplex with corners at the barycentres of the elements of Φ) of either $\mathcal{T}$ or $\mathcal{T}^\delta$, together with Φ^0, Φ^3 and $\Phi^{0,3}$. This is the tessellation given by the alcoves of the affine Coxeter group with a square diagram (see [140, Chapter 4]).

Each 2-face of $\mathcal{W}(\{4,6|4\},(1,-1,1))$ consists of the four edges of length $\sqrt{3}$ of a cell of $\hat{\mathcal{T}}$. The two 2-faces at an edge e of $\mathcal{W}(\{4,6|4\},(1,-1,1))$ correspond to two of the six cells of $\hat{\mathcal{T}}$ at e; these two cells do not share a triangle, but each of them shares a triangle with a third cell at e.

At every vertex v of $\mathcal{W}(\{4,6|4\},(1,-1,1))$ there are six 2-faces. For convenience we consider vertex o as the centre of the cube with vertices at $(\pm 1, \pm 1, \pm 1)$. The definition of $\mathcal{W}(\{4,6|4\},(1,-1,1))$ through the arrangement $\mathcal{L}'$ implies that no 2-face at o uses the edge between o and $(1,1,1)$, or the edge between o and $(-1,-1,-1)$. The 12 cells of $\hat{\mathcal{T}}$ at o that do not contain the vertices $\pm(1,1,1)$ describe a cycle where two are consecutive if and only if they share a triangle. The six 2-faces of $\mathcal{W}(\{4,6|4\},(1,-1,1))$ containing o correspond to alternate cells in that cycle. Taking on account that the cube with vertices at $(\pm 1, \pm 1, \pm 1)$ intersects 24 cells of $\hat{\mathcal{T}}$, we may say loosely that the 2-faces of $\mathcal{W}(\{4,6|4\},(1,-1,1))$ correspond to one quarter of the cells of $\hat{\mathcal{T}}$.

For now we shall refer to the dual of $\mathcal{W}(\{4,6|4\},(1,-1,1))$ as $\mathcal{Q}_{\{6,4\}}$. The vertex set of $\mathcal{Q}_{\{6,4\}}$ consists of the barycentres of the convex hulls of all 2-faces of $\mathcal{W}(\{4,6|4\},(1,-1,1))$. It is clear from the interpretation of the 2-faces of $\mathcal{W}(\{4,6|4\},(1,-1,1))$ as cells of the tessellation $\hat{\mathcal{T}}$ that each vertex of $\mathcal{Q}_{\{6,4\}}$ is the barycentre of only one 2-face of $\mathcal{W}(\{4,6|4\},(1,-1,1))$, and hence there is a bijection between the 2-faces of $\mathcal{W}(\{4,6|4\},(1,-1,1))$ and the vertices of $\mathcal{Q}_{\{6,4\}}$.

Alternatively, the vertices of $\mathcal{Q}_{\{6,4\}}$ could be defined as the set of centres of rotatory reflections of order 4 that preserve the arrangement of lines $\mathcal{L}'$.

Exercise 6.2.47 Show that the centres of rotatory reflections of order 4 that preserve the arrangement of lines $\mathcal{L}'$ are precisely the barycentres of the convex hulls of the 2-faces of $\mathcal{W}(\{4,6|4\},(1,-1,1))$.

We define the edges of $\mathcal{Q}_{\{6,4\}}$ as the segments between vertices at distance $\sqrt{3}/\sqrt{2}$ from each other. For instance, the barycentre of the base 2-face of $\mathcal{W}(\{4,6|4\},(1,-1,1))$ is $(1/2, -1, 0)$, and it is at distance $\sqrt{3}/\sqrt{2}$ from the vertices $(1,0,-1/2)$, $(0,-1/2,1)$, $(1,-2,1/2)$ and $(0,-3/2,-1)$, corresponding respectively to the 2-faces with vertices:

$$(0,0,0),(1,-1,-1),(2,0,0),(1,1,-1);$$

$$(0,0,0),(1,-1,1),(0,0,2),(-1,-1,1);$$

$$(1,-1,1),(0,-2,0),(1,-3,1),(2,-2,0);$$
$$(1,-1,-1),(0,-2,0),(-1,-1,-1),(0,-2,-2).$$

Furthermore, those are the only four neighbours of $(1/2,\ -1,0)$.

Exercise 6.2.48 Show that the distance between $(1/2,\ -1,0)$ and any vertex of $\mathcal{Q}_{\{6,4\}}$ other than itself, $(1,0,-1/2)$, $(0,-1/2,1)$, $(1,-2,1/2)$ or $(0,-3/2,-1)$, is greater than $\sqrt{3}/\sqrt{2}$.

Exercise 6.2.49 Show that two vertices of $\mathcal{Q}_{\{6,4\}}$ are adjacent if and only if the convex sets of the corresponding 2-faces of $\mathcal{W}(\{4,6|4\},(1,-1,1))$ share an edge with length $\sqrt{3}/\sqrt{2}$.

The graph determined by the vertices and edges just defined has many hexagons. Among those, we shall consider as 2-faces of $\mathcal{Q}_{\{6,4\}}$ only those that wind once around a line in $\mathcal{L}'$, like for example the one with vertices

$$(1/2,\ -1,0),(0,-1/2,1),(1,0,3/2),(3/2,\ -1,2),(2,-3/2,1),(1,-2,1/2).$$

An example of a hexagon that is not a 2-face is given by that with vertices

$$(1/2,-1,0),(1,0,-1/2),(2,-1/2,-1),(5/2,-1,0),(2,-3/2,1),(1,-2,1/2),$$

since this hexagon is contractible in $\mathbb{E}^3 \setminus \mathcal{L}'$. This choice of 2-faces guarantees that there are precisely two of them at every edge; the details are left as an exercise.

Exercise 6.2.50 Show that with the choice of 2-faces above, there are two 2-faces at every edge of $\mathcal{Q}_{\{6,4\}}$. (Hint. Use that 2-face-transitivity of $\mathcal{W}(\{4,6|4\},(1,-1,1))$ implies vertex-transitivity of the 1-skeleton of $\mathcal{Q}_{\{6,4\}}$.)

Now it is not hard to conclude that $\mathcal{Q}_{\{6,4\}}$ is a skeletal polyhedron. Its vertex set is a subset of the centres of the tiles of $\hat{\mathcal{T}}$, and hence it is discrete. The connectedness of its 1-skeleton follows from Exercise 6.2.49 and from the connectivity of $\mathcal{W}(\{4,6|4\},(1,-1,1))$. Furthermore, its vertex-figures are quadrilaterals, since every vertex of $\mathcal{Q}_{\{6,4\}}$ is 4-valent (again, by Exercise 6.2.49) and every vertex of each vertex-figure belongs to two edges (see Exercise 6.2.50).

In fact, $\mathcal{Q}_{\{6,4\}}$ is a geometric dual of $\mathcal{W}(\{4,6|4\},(1,-1,1))$. The definition of $\mathcal{Q}_{\{6,4\}}$ gives us the bijection between the vertices of $\mathcal{Q}_{\{6,4\}}$ and the 2-faces of $\mathcal{W}(\{4,6|4\},(1,-1,1))$. Exercise 6.2.49 provides the bijection of the edges. Finally, to every vertex v of $\mathcal{W}(\{4,6|4\},(1,-1,1))$ we can associate the hexagon of $\mathcal{Q}_{\{6,4\}}$ with vertices at the centres of the tiles of $\hat{\mathcal{T}}$ that correspond to 2-faces of $\mathcal{W}(\{4,6|4\},(1,-1,1))$ that contain v, since that hexagon winds once around the line of $\mathcal{L}'$ containing v. It is straightforward to verify that these bijections preserve the incidence.

Since $\mathcal{Q}_{\{6,4\}}$ is a chiral polyhedron with type $\{6,4\}$, it may be the case that it is $\mathcal{W}(\{6,4|4\},w)$ for some w. In order to determine if this is the case, we take its image under the translation T by the vector $(-1/2,1,0)$ and then by the homotety H by a factor of 2. The resulting polyhedron has o as the base vertex, and its standard rotations S_w and S_2 are $H^{-1}T^{-1}S_2^{-1}TH$ and $H^{-1}T^{-1}S_w^{-1}TH$, respectively. They map (x,y,z) to $(-y+1,2-z,-1-x)$ and to $(-x,-z,y)$, respectively. It follows that $S'_{1,2}$ maps (x,y,z) to $(y-1,x+1,2-z)$, and therefore $\mathcal{Q}_{\{6,4\}} = \mathcal{W}(\{6,4|4\},w)$ for $w=(-1,1,2)$.

The polyhedron $\mathcal{W}(\{6,4|4\},(-1,1,2))$ is harder to visualise than its dual $\mathcal{W}(\{4,6|4\},(1,-1,1))$ since its edges are larger and they are not aligned with edges or diagonals of cubes of the tessellation $\{4,3,4\}$. The four 2-faces at o of $\mathcal{W}(\{6,4|4\},(-1,1,2))$ are shown in Figure 6.34. The edges of the base 2-face are solid black. A second 2-face has solid grey edges. The remaining two 2-faces have dotted edges; one coloured black and one coloured grey.

As we just saw, the geometric dual of $\mathcal{W}(\{4,6|4\},(1,-1,1))$ is $\mathcal{W}(\{6,4|4\},(-1,1,2))$, and in general we should expect that the geometric dual of $\mathcal{W}(\{4,6|4\},w)$ is $\mathcal{W}(\{6,4|4\},w')$ for some $w' \neq w$.

Exercise 6.2.51 For each w determine w' such that $\mathcal{W}(\{6,4|4\},w')$ is the geometric dual of $\mathcal{W}(\{4,6|4\},w)$.

The Polyhedron $\mathcal{W}(\{6,6|3\},(1,0,0))$

Our third chiral polyhedron is also related to the configuration $\mathcal{L}$ of lines described in Appendix A.4.

Let $\mathcal{T}_2$ be the tessellation of $\mathbb{E}^3$ by cubes with vertices in $2\mathbb{Z}^3$. Then the centre of each cube of $\mathcal{T}_2$ belongs to precisely one line of $\mathcal{L}$.

We define the structure $\mathcal{Q}_{\{6,6\}}$ as the one with the same 1-skeleton as $\mathcal{T}_2$, and with one 2-face for each cube of $\mathcal{T}_2$ obtained as follows. For each cube C of $\mathcal{T}_2$ consider as a 2-face of $\mathcal{Q}_{\{6,4\}}$ the Petrie path of C whose vertices do not belong to the line in $\mathcal{L}$ through the centre of C. This is well defined because every Petrie path of a cube has the vertex set of the cube except for a pair of opposite vertices.

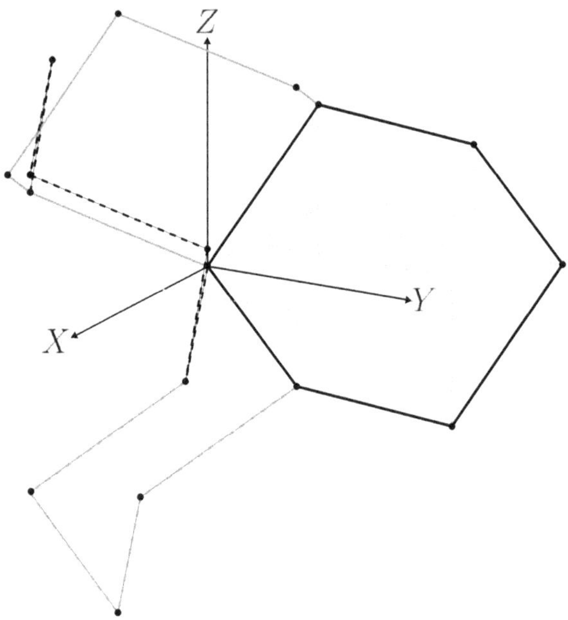

Figure 6.34 The four 2-faces of $\mathcal{W}(\{6,4|4\},(-1,1,2)$ around o.

Theorem 6.2.52 *The structure $\mathcal{Q}_{\{6,6\}}$ just defined is a polyhedron.*

Proof The 1-skeleton is that of $\mathcal{T}_2$, and hence we only need to show that the vertex-figures of $\mathcal{Q}_{\{6,6\}}$ are polygons.

We claim that every edge of $\mathcal{Q}_{\{6,6\}}$ belongs to two 2-faces. Indeed, if u and v are the vertices of an edge e then u and v belong to two lines of $\mathcal{L}$ with distinct slopes. When considering the four cubes C_1, C_2, C_3 and C_4 of $\mathcal{T}_2$ that contain e, and the lines ℓ_i of $\mathcal{L}$ such that ℓ_i contains the centre of C_i, we must regard a Petrie path of C_i containing e as a 2-face of $\mathcal{Q}_{\{6,6\}}$ if and only if $u, v \notin \ell_i$. It follows that Petrie polygons of only two cubes that contain e are 2-faces of $\mathcal{Q}_{\{6,6\}}$. Since $\mathcal{Q}_{\{6,6\}}$ contains as a 2-face precisely one Petrie path of each cube of $\mathcal{T}_2$, we can conclude that e belongs to precisely two 2-faces of $\mathcal{Q}_{\{6,6\}}$.

We also remark that if C_1 and C_2 are the cubes of $\mathcal{T}_2$ whose Petrie paths are the two 2-faces containing an edge e then C_1 and C_2 share a square, since otherwise the centres of the remaining two cubes of $\mathcal{T}_2$ containing e would belong to lines of $\mathcal{L}$ that are parallel, but are not translates by a vector in $\mathbf{\Lambda}_{(1,1,1)}$.

Every vertex v is surrounded by eight cubes of $\mathcal{T}_2$. The line of $\mathcal{L}$ that contains v also contains the centres of two of these cubes. Hence, the six

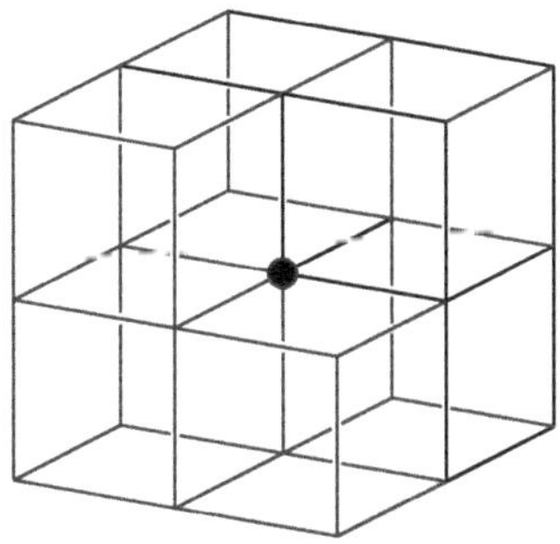

Figure 6.35 Six cubes of $\mathcal{T}_2$ at the black vertex v, after the removal of two opposite cubes at v.

2-faces of $\mathcal{Q}_{\{6,6\}}$ at v are Petrie polygons of the remaining six cubes that contain v. Every edge that contains v belongs to two 2-faces of $\mathcal{Q}_{\{6,6\}}$ that are Petrie paths of two cubes that share a square. These two 2-faces induce an edge of the vertex-figure at v. Examining the six cubes at a vertex after eliminating two opposite cubes we see that the vertex-figure of $\mathcal{Q}_{\{6,6\}}$ at v must be a skew hexagon (see Figure 6.35, where the cubes missing are the ones at the front and back). ■

We next want to prove that $\mathcal{Q}_{\{6,6\}}$ is chiral. In order to do that, we shall show that it is, in fact, $\mathcal{W}(\{6,6|3\},(2,0,0))$. The standard rotations $S_{(2,0,0)}$ and S_2 of $\mathcal{W}(\{6,6|3\},(2,0,0))$ map (x,y,z) to $(-y,z,x-2)$ and $(-z,-x,-y)$, respectively. Then $S_{(2,0,0)}$ and S_2 permute the sets $\mathcal{L}_i$ as follows:

$$\begin{aligned} S_{(2,0,0)} &\mapsto (\mathcal{L}_2,\mathcal{L}_3,\mathcal{L}_4), \\ S_2 &\mapsto (\mathcal{L}_1,\mathcal{L}_4,\mathcal{L}_2). \end{aligned}$$

It follows that $\langle S_{(2,0,0)},S_2\rangle$ preserves $\mathcal{L}$, and hence it preserves $\mathcal{Q}_{\{6,6\}}$. Furthermore, $S_{(2,0,0)}$ and S_2 act like the standard rotations when choosing as the base flag the one with the vertex at o, the edge with endpoints at o and $(2,0,0)$ and the hexagon with vertices o, $(2,0,0)$ $(2,-2,0)$, $(2,-2,-2)$, $(0,-2,-2)$ and $(0,0,-2)$. This proves the following result.

Proposition 6.2.53 *The polyhedron $\mathcal{Q}_{\{6,6\}}$ is chiral and it is congruent to $\mathcal{W}(\{6,6|3\},(2,0,0))$.*

In Figure 6.36 we illustrate the six hexagons of $\mathcal{Q}_{\{6,6\}}$ that contain a vertex. From that figure it is easy to find the lines of $\mathcal{L}$ that contain the centre of each hexagon, and the line of $\mathcal{L}$ that contains the central vertex.

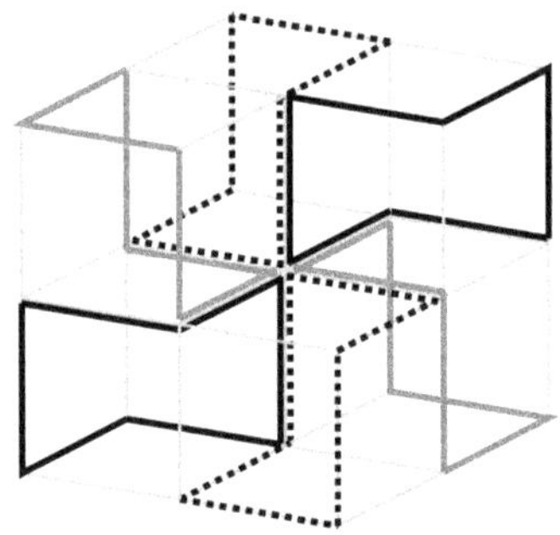

Figure 6.36 The six 2-faces of $\mathcal{W}(\{6,6|3\},(2,0,0))$ around a vertex.

The General Polyhedra with Skew 2-Faces

Now we shall discuss the set of chiral polyhedra with skew 2-faces. We shall not present here the complete proofs of all the results, mainly because of their technical nature. Instead, we shall refer to [228, 245], which are the only references on these polyhedra to date. The preveiling feeling is that there is still much to be discovered about these polyhedra, including new approaches and descriptions.

We know from Remark 6.1.10 and Theorem 6.2.10 that the standard rotations of the symmetry group of every chiral polyhedron with skew 2-faces are both rotatory reflections of orders greater than or equal to 3. We next show that their axes cannot be parallel.

Lemma 6.2.54 *Let $\mathcal{P}$ be a skeletal chiral polyhedron in $\mathbb{E}^3$ with finite skew 2-faces and let $G(\mathcal{P}) = \langle S_1, S_2\rangle$. Then S_1 and S_2 are rotatory reflections with nonparallel axes.*

Proof We know from Theorem 6.2.10 that both generators are rotatory reflections. Let Π_i be the invariant plane of S_i. We only need to show that Π_1 is not parallel to Π_2. The key observation is that two consecutive vertices of a finite skew polygon $\mathcal{Q}$ belong to two distinct planes parallel to the invariant plane of the generator of the symmetry group of $\mathcal{Q}$.

Let u be the base vertex of $\mathcal{P}$, and let v and w be its neighbours in the base 2-face. Then v and w are in the same plane parallel to Π_1, and u is in some other parallel plane. We can conclude that Π_2 cannot be parallel to Π_1, since v and w are adjacent in the finite skew vertex-figure, and hence they must be in distinct planes parallel to Π_2. ■

Remark 6.2.55 Lemma 6.2.54 is a particular case of Lemma 3.7 of [246].

Now we relate the chiral polyhedra with skew 2-faces with the operation $\mathcal{W}$.

Theorem 6.2.56 *Every skeletal chiral polyhedron $\mathcal{P}$ in $\mathbb{E}^3$ with finite skew 2-faces is $\mathcal{W}(\mathcal{Q}, w)$ for some Petrie–Coxeter polyhedron $\mathcal{Q}$ and some $w \in \mathbb{Z}^3$.*

Proof Lemma 6.2.54 implies that the group $G(\mathcal{P})$ is affinely irreducible (see Appendix A.1). This, together with the discreteness of $G(\mathcal{P})$, implies that $G(\mathcal{P})$ is one of the so-called crystallographic groups (see [234, Section 5.4], [245, Lemma 4.1] and [254]).

Every affinely irreducible crystallographic group can be seen as a subgroup of the Coxeter group $[4,3,4]$, the symmetry group of the tessellation $\{4,3,4\}$ by cubes of $\mathbb{E}^3$ (see for example [46, Chapters 22 and 23]). It follows that the rotatory reflections S_1 and S_2 that serve as standard rotations for $G(\mathcal{P})$ can be chosen from those of the group $[4,3,4]$. In particular, each of them has order either 4 or 6.

By choosing the coordinate system appropriately (that is to say, up to conjugacy) we may choose the base vertex at o, and choose S_2 to map (x, y, z) to $(-x, -z, y)$ if it has order 4, or to $(-z, -x, -y)$ if it has order 6. Rather than looking for candidates for S_1, and then verifying whether for each such choice the isometry $S_1 S_2$ is an involution, we may instead choose $S_1 S_2$ as a half-turn whose axis does not contain the fixed point of S_2, so that $S_1 = (S_1 S_2) S_2^{-1}$ is another rotatory reflection in $[4,3,4]$.

The point group $G_0(\mathcal{P})$ of $G(\mathcal{P})$ must be a subgroup of the group $[3,4]$ of symmetries of the octahedron, since $G(\mathcal{P})$ is a subgroup of $[4,3,4]$. We shall consider the nine half-turns in $[3,4]$ for each of the two choices of S_2 mentioned above. Each such choice of half-turn will be a candidate for the half-turn $S'_{1,2}$ with axis through o parallel to that of $S_1 S_2$.

We denote the nine axes of half-turns of $[3,4]$ by the integers from 1 to 9 as in Figure 6.37(b). The half-turns about the coordinate axes are 1, 2 and 3. The remaining axes contain midpoints of opposite edges of the cube in that figure, and the label of each axis is indicated by one of its intersections with the cube.

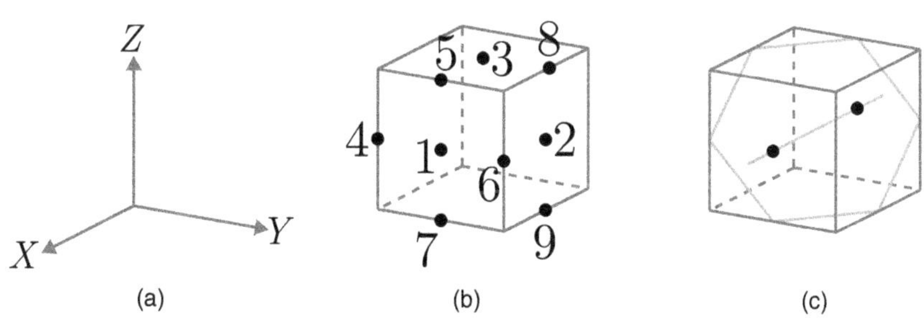

Figure 6.37 (a) The coordinate axes; (b) the nine axes of half-turns of the group $[3,4]$; (c) the axis of the rotatory reflection S_2.

If S_2 is a rotatory reflection of order 4 then its axis is the grey line in Figure 6.37(c). In this situation we may discard axes 1, 2, 3, 8 and 9 as candidates for $S'_{1,2}$, since for each of those choices the axes of S_1 and S_2 would be parallel (see Lemma 6.2.54). The remaining four candidates are in the same orbit under $\langle S_2 \rangle$, and hence they correspond to half-turns that are conjugates by elements in $\langle S_2 \rangle$. Without loss of generality we may choose $S'_{1,2}$ as the half-turn about axis 6, that is, the one with direction vector $(1,1,0)$.

In a similar fashion, if S_2 is a rotatory reflection of order 6 then its invariant plane is illustrated in grey in Figure 6.37(c). Now we can discard axes 4, 7 and 9 as candidates of $S'_{1,2}$ since for such choices the axes of S_1 and S_2 would be parallel. The remaining six axes split into two orbits under the action of $\langle S_2 \rangle$; one contains the coordinate axes, and the other contains the axes labelled 5, 6 and 8. Without loss of generality we may assume that the axis of $S'_{1,2}$ is either 2 (the Y-axis), or 6 (with direction vector $(1,1,0)$).

In total we have three non-equivalent possibilities for the pair consisting of S_2 and $S'_{1,2}$. There is one for which S_2 has order 4, and two for which S_2 has order 6.

We can recover $S_1 S_2$ as a half-turn whose axis is a translate of that of $S_1 S_2$. All such half-turns are compositions of $S'_{1,2}$ with a translation by a vector w perpendicular to its axis. Furthermore, w must be different from 0 (otherwise $\mathcal{P}$ would be finite, contradicting Corollary 6.2.11), and it must have integer coordinates (so that S_1 preserves $\mathbb{Z}^3$).

If S_2 has order 4 then we obtain precisely the base vertex and generators S_w and S_2 of $\mathcal{W}(\{4,6|4\},w)$. Where S_2 has order 6 and $S'_{1,2}$ is the Y-axis we get the generators S_w and S_2 of $\mathcal{W}(\{6,4|4\},w)$. Finally, if S_2 has order 6 and $S'_{1,2}$ is axis 6 we get the generators S_w and S_2 of $\mathcal{W}(\{6,6|3\},w)$. ■

The proof of Theorem 6.2.56 follows the ideas of [245], but they were adapted to relate the polyhedra with the operation $\mathcal{W}$. In that paper the polyhedra $\mathcal{W}(\{4,6|4\},w)$, $\mathcal{W}(\{6,4|4\},w)$ and $\mathcal{W}(\{6,6|3\},w)$ are called $Q(c,d)$, $Q(c,d)^*$ and $P(a,b)$, respectively. In every case either one parameter is 0, or the parameters must be rational multiples of each other. Multiplying both parameters of a polyhedron by the same number different from 0 yields a similar polyhedron, and so the two parameters may be understood to be relatively prime integers.

Exercise 6.2.57 For every $\mathcal{P} \in \{\{4,6|4\},\{6,4|4\},\{6,6|3\}\}$ and every suitable w, determine parameters a, b, c or d such that $\mathcal{W}(\mathcal{P},w)$ is congruent to $P(a,b)$, $Q(c,d)$ or $Q(c,d)^*$.

When constructing $\mathcal{W}(\mathcal{P}, w)$ from a Petrie–Coxeter polyhedron $\mathcal{P}$, not every choice of $w \in \mathbb{Z}^3 \cap (a')^{\perp}$ yields a polyhedron. In some cases the vertex-figures are not connected.

Example 6.2.58 The structure $\mathcal{W}(\{6,4|4\}, w)$ for $w = (1,-1,1)$ is not a polyhedron. To verify this we start with the following two observations. The first one states that the axes of rotatory reflections that stabilise three consecutive vertices u, v and w of a 2-face of $\mathcal{W}(\{6,4|4\},(1,-1,1))$ must be in the three coordinate directions (if S denotes a a rotatory reflection of order 6 in the stabiliser of that 2-face then the axes of the stabilisers of u, v and w are some line ℓ together with ℓS and ℓS^2). The second observation states that if the axis of a 4-fold rotatory reflection that fixes a vertex v has direction $(1,0,0)$ then two of its neighbours are invariant under rotatory reflections with axes having direction $(0,1,0)$ and the other two are invariant under rotatory reflections with axes having direction $(0,0,1)$; furthermore, neighbours of v whose rotatory reflections have parallel axes are not consecutive in the vertex-figure at v (this follows from a similar argument as previously, but now S is a rotatory reflection of order 4 instead of 6). Analogous conclusions apply for vertices invariant under rotatory reflections with axes with direction $(0,1,0)$ or $(0,0,1)$.

The symmetries S_w and S_2 of $\mathcal{W}(\{6,4|4\}, w)$ map (x,y,z) to $(-y-1,-z-1,1-x)$ and $(-x,-z,y)$, respectively. Its base 2-face F_2 has vertices o, $(-1,-1,1)$, $(0,-2,2)$, $(1,-3,1)$, $(2,-2,0)$ and $(1,-1,-1)$ in that order. The combinatorial rotation that preserves the edge between $(2,-2,0)$ and $(1,-1,-1)$ while swapping its endpoints is $S_w^2 S_2 S_w^{-1}$, and maps (x,y,z) to $(3-x, z-1, y+1)$. With this in mind, the 2-face $F_2 S_w^2 S_2 S_w$ has vertices $(3,-1,1)$, $(4,0,0)$, $(3,1,-1)$, $(2,0,-2)$, $(1,-1,-1)$ and $(2,-2,0)$ in that order (see Figure 6.38(a)).

The two-step rotation about o is S_2^2, and it maps (x,y,z) to $(x,-y,-z)$. It follows that the vertices of the 2-face $F_2 S_w^2 S_2 S_w S_2^2$ are $(3,1,-1)$, $(4,0,0)$, $(3,-1,1)$, $(2,0,2)$, $(1,1,1)$ and $(2,2,0)$, in that order, as can be seen in Figure 6.38(a).

Now we use that the axis of the rotatory reflections of order 4 fixing o has direction $(1,0,0)$ while that of the rotatory reflection fixing $(1,-1,-1)$ has direction $(0,1,0)$, in order to identify the direction vector of the axis of the rotatory reflections stabilising every vertex in the partial map in Figure 6.38(a). This is illustrated in Figure 6.38(b). Vertex $(3,1,-1,)$ is simultaneously in 2-faces $F_2 S_w^2 S_2 S_w$ and $F_2 S_w^2 S_2 S_w S_2^2$, and it is indicated by grey circles in Figure 6.38(b). It becomes apparent that it must be invariant under one rotatory reflection with order 4 whose axis is parallel to $(1,0,0)$, and under another one with axis parallel to $(0,0,1)$, contradicting Proposition 6.1.29.

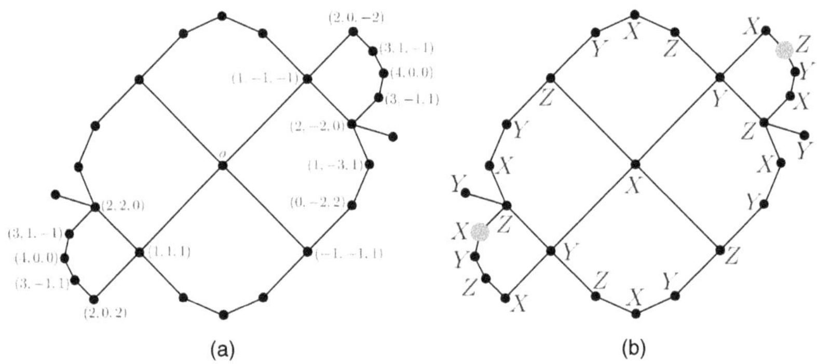

Figure 6.38 Partial representation of the map of $\mathcal{W}(\{6,4|4\},w)$ for $w = (1,-1,1)$. (a) Coordinates of some vertices; (b) direction of the stabilisers of the vertices.

According to [245, Theorem 5.8], $P(a,b)$ is a polyhedron for every integer values of a and b, not both 0. Numerical conditions on c and d are given in [245, Section 6] to indicate when $Q(c,d)$ is a polyhedron. Those for $Q(c,d)^*$ can be analysed through duality. These results were obtained by determining the set of direction vectors of the edges at every vertex of the structure. In some cases there were more edges than the second entry of the corresponding Schläfli type, implying that the structure is not a polyhedron (like the one in Example 6.2.58).

Exercise 6.2.59 For each $\mathcal{P} \in \{\{4,6|4\},\{6,4|4\},\{6,6|3\}\}$ determine the vectors $w \in \mathbb{Z}^3 \cap (a')^{\perp}$ such that $\mathcal{W}(\mathcal{P},w)$ is a polyhedron.

In [245, Theorems 5.18 and 6.13] it was established that there are infinitely many polyhedra in each family $P(a,b)$, $Q(c,d)$ or $Q(c,d)^*$ whose underlying abstract polyhedra are mutually non-isomorphic. Presentations for their automorphism groups can be found in [228, Section 4]. The number of non-isomorphic polyhedra in each family constitutes one of the main differences between the families of skeletal chiral polyhedra with skew 2-faces and those with helical 2-faces. Recall that there are only three abstract polyhedra that are realised as skeletal chiral polyhedra with helical 2-faces (see Theorems 6.2.13, 6.2.17, 6.2.24 and 6.2.35).

Another main difference between chiral polyhedra with skew 2-faces and those with helical 2-faces is given by the following result (see [228, Theorems 4.1 and 4.3] and compare with [245, Theorems 5.17 and 6.12]). Recall that the underlying abstract polyhedron of every skeletal chiral polyhedron with helical 2-faces is regular (see Theorems 6.2.13, 6.2.17, 6.2.24 and 6.2.35).

Theorem 6.2.60 *Every chiral skeletal polyhedron with skew 2-faces is combinatorially chiral.*

The proof of Theorem 6.2.60 in [228] consisted of finding two words on the standard generators S_1 and S_2 that represent the same symmetry, but such that when replacing S_1 by S_1^{-1} and S_2 by S_2^{-1} they represent distinct elements (see Theorem 3.1.53 for $i = j = 1$).

The description of each family of chiral skeletal polyhedra with skew 2-faces in terms of the operation $\mathcal{W}$ allows us to relate them through a geometric version of the embeddings defined in Section 4.2. In the combinatorial setting we said that $\mathcal{Q}$ is embedded symmetrically in $\mathcal{P}$ if there is an injection of the vertex set of $\mathcal{Q}$ into that of $\mathcal{P}$, and $\Gamma(\mathcal{Q})$ is isomorphic to a subgroup of $\Gamma(\mathcal{P})$ where the isomorphism respects the action on the vertex set of $\mathcal{Q}$. When working with skeletal polyhedra we may instead use the following definition.

Definition 6.2.61 A skeletal polytope $\mathcal{Q}$ is *embedded symmetrically* in $\mathcal{P}$ if

- the vertex set of $\mathcal{Q}$ is contained in that of $\mathcal{P}$;
- $G(\mathcal{Q}) \subseteq G(\mathcal{P})$.

Under Definition 6.2.61 it is still true that the tetrahedron with edge length $\sqrt{2}$ can be embedded symmetrically in the cube with edge length 1, but no cube can be embedded symmetrically in a dodecahedron.

The polyhedron $\{4, 6|4\}$ has many half-turns whose axes have direction vector $(1, 1, 0)$. Their intersections with the plane through o with normal vector $(1, 1, 0)$ are illustrated as full black circles in Figure 6.39. In the same figure we indicate with full grey circles the intersections of the axes of half-turns of $\{4, 6\}_6$, and with empty black circles the intersections of the axes of half-turns of $\mathcal{W}(\{4, 6|4\}, (1, -1, 1))$.

Since for every w the base vertex of $\mathcal{W}(\{4, 6|4\}, w)$ is o, and the generator S_2 of $G(\mathcal{W}(\{4, 6|4\}, w))$ is the same for all choices of w, we can conclude that $\mathcal{W}(\{4, 6|4\}, (4, -4, 1))$, $\mathcal{W}(\{4, 6|4\}, (3, -3, 2))$ and $\mathcal{W}(\{4, 6|4\}, (3, -3, -1)$ are embedded in $\{4, 6|4\}$, $\{4, 6\}_6$ and $\mathcal{W}(\{4, 6|4\}, (1, -1, 1))$, respectively.

Exercise 6.2.62 For each Petrie–Coxeter polyhedron $\mathcal{P}$ determine numerical, necessary and sufficient conditions on the entries of v and w so that $\mathcal{W}(\mathcal{P}, w)$ is embedded in $\mathcal{W}(\mathcal{P}, v)$.

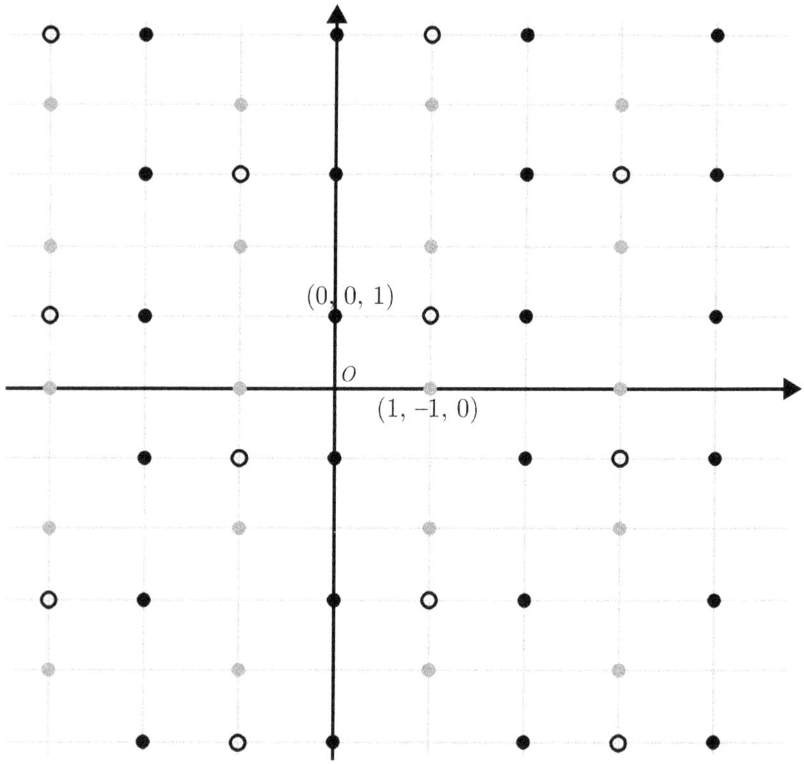

Figure 6.39 Intersection with the plane with normal vector $(1, 1, 0)$ of the half-turns with direction $(1, 1, 0)$.

As mentioned previously, in [245] the three families of chiral polyhedra with skew 2-faces depended on two parameters a and b (for those with type $\{6,6\}$), or c and d (for those with type $\{6,4\}$ or $\{4,6\}$). The condition of the two parameters to be rational multiples of each other, or one of them to be 0, was required since otherwise the resulting vertex set is not discrete. Since multiplication of the two parameters by the same number yields a similar polyhedron, we may think of the first family to consist of the polyhedra $P(1,0)$ and $P(\alpha, 1)$ for $\alpha \in \mathbb{Q}$, up to similarity. The second and third families may be thought to consist of the polyhedra $Q(1,0)$ and $Q(1,0)^*$ together with $Q(\alpha, 1)$ and $Q(\alpha, 1)^*$, respectively, for $\alpha \in \mathbb{Q}$, up to similarity. It still makes sense to ask ourselves the following questions.

Problem 69 Is there anything meaningful to say about polyhedra $P(\alpha, 1)$ and $P(\beta, 1)$ (or $Q(\alpha, 1)$ and $Q(\beta, 1)$, or $Q(\alpha, 1)^*$ and $Q(\beta, 1)^*$) when $|\alpha - \beta| < \epsilon$

for small values of ϵ? If we remove the discreteness condition in the definition of skeletal polyhedra, can we relate combinatorially the structures $P(\alpha, 1)$ and $P(\beta, 1)$ (or $Q(\alpha, 1)$ and $Q(\beta, 1)$, or $Q(\alpha, 1)^*$ and $Q(\beta, 1)^*$) for $\alpha, \beta \in \mathbb{R}$ when $|\alpha - \beta| < \epsilon$ for small values of ϵ?

6.3 Skeletal Chiral Polytopes

In our final section we summarise what is known to this date about skeletal chiral polytopes. In Section 6.3.1 we shall mostly list results and refer to the corresponding sources. Section 6.3.2, the last of this book, comprises what is known to date about chiral polytopes of full rank.

6.3.1 An Update on Chiral Polytopes in $\mathbb{E}^d$

Most of what we know about skeletal rotary polytopes is actually about skeletal regular polytopes. The wide literature about this topic includes [3, 9, 25, 61, 83, 84, 104, 167, 169, 171–178, 181, 182, 198, 200]. Much of this material is exposed in [179].

The first results on skeletal chiral polytopes were those in [245, 246], leading to the classification of skeletal chiral polyhedra in $\mathbb{E}^3$ (see Section 6.2). These works were followed more than a decade later with the determination in [222–224] of all chiral skeletal 4-polytopes in $\mathbb{E}^3$. The main results of [224] can be summarised as follows.

Theorem 6.3.1 *There are precisely three skeletal chiral* 4*-polytopes in* $\mathbb{E}^3$*. These three polytopes are combinatorially chiral.*

The analysis in [222] shows that every rotary 4-polytope in $\mathbb{E}^3$ with geometrically regular facets is itself regular, and so the three chiral 4-polytopes have geometrically chiral facets. Their vertex-figures, being finite rotary polyhedra in $\mathbb{E}^3$, are regular (see Corollary 6.2.11). One of the chiral 4-polytopes has tetrahedral vertex-figures, another one has octahedral vertex-figures and the remaining one has as vertex-figures the Petrial of the cube. In what follows we describe these three 4-polytopes.

The facets of one of the skeletal chiral 4-polytopes in $\mathbb{E}^3$ are isometric to the polyhedron $P_1(\frac{1}{2})$. That is precisely the polyhedron $P_1(\lambda)$ for which the angle between two edges at a vertex is $\pi/2$. The 1-skeletion of the 4-polytope is that of the cubic tessellation $\{4, 3, 4\}$. The vertex-figure at o of the base facet is one of the triangles of the vertex-figure at o of $\{4, 3, 4\}$. In fact, the chiral 4-polytope has only eight facets, which can be obtained as the orbit of the base facet under the group of rotations of the octahedron. The vertex-figures are octahedra, and therefore its type is $\{\infty, 3, 4\}$. In [224] this polyhedron was called $\mathcal{P}_{\{\infty,3,4\}}$.

Every facet of $\mathcal{P}_{\{\infty,3,4\}}$ contains all vertices of $\{4,3,4\}$, implying that $\mathcal{P}_{\{\infty,3,4\}}$ is flat in the sense of Definition 5.2.1. The polyhedron $\mathcal{P}_{\{\infty,3,4\}}$ can be understood as a subdivision of the edges of $\{4,3,4\}$ into eight sets $X_1, \dots, X_8$ (the edge sets of the eight facets) with the following properties.

- X_i is isometric to X_j for every $i, j \in \{1, \dots, 8\}$.
- For every i, each vertex of $\{4,3,4\}$ is contained in precisely three edges of X_i.
- Each edge of $\{4,3,4\}$ belongs to precisely four sets X_i.
- For each i, every edge of X_i belongs to precisely two regular helices over triangles contained in X_i (those helices are the 2-faces of $\mathcal{P}_{\{\infty,3,4\}}$).
- Every regular helix over a triangle contained in X_i belongs to exactly one X_j with $j \in \{1, \dots, 8\} \setminus \{i\}$.

Even though these properties are stated concisely and appear to be very restrictive, it is a difficult riddle to find such eight sets from scratch. The best way to solve this riddle seems to be to get familiarised first with the polyhedron $P_1(\frac{1}{2})$.

The remaining two chiral 4-polytopes pose similar riddles, but in different 1-skeleta.

As in Section 6.2.3, let $\mathcal{G}_{\mathbf{\Lambda}_{(1,1,1)}}$ be the graph with vertex set $\mathbf{\Lambda}_{(1,1,1)}$, where two vertices are adjacent if and only if the distance between them is $\sqrt{3}$. Its vertices are 8-valent and it has girth 4. The neighbours of each vertex of $\mathcal{G}_{\mathbf{\Lambda}_{(1,1,1)}}$ are the vertices of a cube.

Let $\mathcal{G}_{\diamondsuit}$ be the subgraph of $\mathcal{G}_{\mathbf{\Lambda}_{(1,1,1)}}$ induced by the vertex set $2\mathbf{\Lambda}_{(1,1,0)} \cup \big((-1,-1,-1) + 2\mathbf{\Lambda}_{(1,1,0)}\big)$. The graph $\mathcal{G}_{\diamondsuit}$ is the well-known *diamond net*. It is 4-valent and has girth 6. The neighbours of each vertex of $\mathcal{G}_{\diamondsuit}$ are the vertices of a regular tetrahedron.

Let $\mathcal{Q}_{\{\infty,3\}}$ be the image of the polyhedron $P_2(\frac{1}{2})$ under a homotety by a factor of 2. Then the 1-skeleton of $\mathcal{Q}_{\{\infty,3\}}$ is contained in the graph $\mathcal{G}_{\diamondsuit}$. In fact, $\pm 1/2$ are the only values of λ for which the angle between two edges at a vertex of $P_2(\lambda)$ coincides with that between two edges at a vertex of $\mathcal{G}_{\diamondsuit}$. The group generated by the 3-fold rotations about two edges at o of $\mathcal{Q}_{\{\infty,3\}}$ is the rotation subgroup of a tetrahedron, and the orbit of $\mathcal{Q}_{\{\infty,3\}}$ under that group has precisely four elements. These are the four facets of a flat chiral 4-polytope with tetrahedral vertex-figures. It has type $\{\infty,3,3\}$, and it was denoted by $\mathcal{P}_{\{\infty,3,3\}}$ in [224].

The third chiral 4-polytope differs from the previous two in that its facets are combinatorially chiral (and so they have skew 2-faces), and in that its vertex-figures are not isomorphic to a Platonic solid.

In Section 6.2.3 we described the polyhedron $\mathcal{W}(\{4,6|4\},(1,-1,1))$, whose 1-skeleton is contained in the graph $\mathcal{G}_{\mathbf{\Lambda}_{(1,1,1)}}$. The vertex-figure at o of this polyhedron is a skew hexagon, isometric to a Petrie polygon of a cube $\mathcal{C}$ with vertices at $(\pm 1, \pm 1, \pm 1)$. Let G be the subgroup of the symmetry group of $\mathcal{C}$ consisting of all orientation preserving isometries that preserve the two inscribed tetrahedra of $\mathcal{C}$, together with the orientation-reversing isometries that interchange these two tetrahedra. Then the orbit of $\mathcal{W}(\{4,6|4\},(1,-1,1))$ under G consists of four elements (one for each Petrie polygon of $\mathcal{C}$), which are the facets of a flat chiral 4-polytope. Its vertex-figures have type $\{6,3\}$, and it was called $\mathcal{P}_{\{4,6,3\}}$ in [224].

Other attempts to construct chiral 4-polytopes by naively putting together isometric chiral polyhedra fail. For example, when looking for a chiral polyhedron with skew faces and type $\{6,4\}$ whose 1-skeleton is contained in $\mathcal{G}_{\mathbf{\Lambda}_{(1,1,1)}}$, we realise that such a polyhedron could only be $\mathcal{W}(\{6,4|4\},(1,-1,1))$, which is not really a polyhedron (see Example 6.2.58). Hence, we cannot have a 4-polytope with such facets. An attempt to find a 4-polytope whose facets have type $\{6,6\}$ and whose 1-skeleton is that of $\{4,3,4\}$ also fails. In such a situation the facets would have to be $\mathcal{W}(\{6,6|3\},(1,0,0))$. But then the 4-polytope would have as vertex-figures the Petrial of the octahedron (that is the only finite regular polyhedron whose facets are isometric to the vertex-figures of $\mathcal{W}(\{6,6|3\},(1,0,0)))$, which is a non-orientable polyhedron, contradicting Proposition 6.1.39.

In what follows we discuss skeletal chiral polytopes in higher-dimensional Euclidean spaces.

Not every abstract rotary polytope admits a faithful chiral realisation in a Euclidean space. For instance, the flat chiral polyhedron with type $\{6,9\}$ has triple edges (see [67, Theorem 3.6]). However, in our understanding, the three edges between two vertices should be represented by the only line segment between the corresponding pair of Euclidean points. Hence, the combinatorics of the realisation would not match that of the starting polyhedron.

Nevertheless, most chiral polytopes we can think of admit faithful symmetric realisations in Euclidean spaces. The easiest symmetric realisation of a polytope $\mathcal{P}$ to describe is the one in the $(N-1)$-simplex, where N is the number of vertices of $\mathcal{P}$.

Example 6.3.2 The polyhedron $\{4,4\}_{(2,1)}$ has five vertices, and hence it can be realised faithfully and symmetrically in the 4-simplex $\{3,3,3\}$.

Let V be the four-dimensional Euclidean space constructed as the affine hull of the points in $\mathbb{E}^5$

$$e_1 := (1,0,0,0,0), e_2 := (0,1,0,0,0), e_3 := (0,0,1,0,0),$$
$$e_4 := (0,0,0,1,0), e_5 := (0,0,0,0,1).$$

Every permutation σ in $\{1,2,3,4,5\}$ induces an isometry of V determined by $e_i\sigma_i = e_{\sigma(i)}$ for every i. This is in fact the restriction to V of the corresponding permutation of coordinate axes of $\mathbb{E}^5$. We shall abuse notation and represent such isometries as permutations.

We define a polyhedron $\mathcal{P}$ as the structure arising by Wythoff's construction with base vertex e_1 and standard rotations S_1 and S_2 as the permutations $(1,4,3,2)$ and $(2,5,4,3)$, respectively. We claim that $\mathcal{P}$ is a realisation of the polyhedron $\{4,4\}_{(2,1)}$, since S_1 and S_2 are precisely the standard rotation of $\{4,4\}_{(2,1)}$ given in Example 2.3.11. Furthermore, the 1-skeleton of $\mathcal{P}$ coincides with that of $\{3,3,3\}$. The details are left as an exercise.

Exercise 6.3.3 Find all 2-faces of the polyhedron $\mathcal{P}$ in Example 6.3.2. Show that $\mathcal{P}$ is a faithful symmetric realisation of $\{4,4\}_{(2,1)}$ whose 1-skeleton coincides with that of $\{3,3,3\}$.

The embeddings of chiral polytopes in simplices help little in the enumeration of all chiral skeletal polytopes in $\mathbb{E}^n$. On one hand, it only takes into account chiral polytopes with a fixed number of vertices for each dimension, and on the other hand, chiral polytopes that are combinatorially regular will not admit these kinds of realisations (their standard reflections will also be represented by a symmetry of the simplex). Moreover, no infinite chiral polytope will have the 1-skeleton of a simplex. This is already a problem in dimension 3, where all chiral polytopes are infinite (see Proposition 6.2.2, Corollary 6.2.11 and Theorem 6.3.1).

Thanks to the known results about embeddings of abstract toroidal chiral polyhedra in toroidal regular polyhedra (see Section 4.2 and [11, Section 2]), and to the previous works on realisations of toroidal regular polyhedra (see [25, 198, 200]), we can construct skeletal chiral polyhedra in dimensions higher than 3 having toroidal chiral polyhedra as underlying abstract polyhedra. The results in [25, 200] suggest that in $\mathbb{E}^6$ there are infinite families of chiral polyhedra whose underlying abstract polyhedra are $\{3,6\}_{(a,b)}$ and $\{6,3\}_{(a,b)}$ for $a > b > 0$. A family of chiral polyhedra of type $\{4,4\}_{(a,b)}$ in $\mathbb{E}^4$ is included as part of the discussion in [11, Section 4].

Problem 70 Describe the skeletal chiral polyhedra of types $\{3,6\}$ and $\{6,3\}$ in $\mathbb{E}^6$.

Few chiral polytopes in $\mathbb{E}^d$ for $d \geq 4$ have been described so far, apart from the toroidal polyhedra and polytopes whose 1-skeleton is contained in that of a simplex. One of them is Roli's cube, which was described in Section 5.3.3.

Some finite chiral polyhedra in $\mathbb{E}^4$ are described in [12]. The construction in that work is a geometric version of that in Proposition 4.1.42. It was inspired by the construction of $\{4, 6|4\}$ as the Petrie–Coxeter polyhedron of the cubic tessellation, as explained at the beginning of Section 4.1.5. The starting objects in [12] are the regular tessellations of the 3-sphere $\mathbb{S}^3$ with convex and starry cells. Geometric versions of the Petrie–Coxeter, halving and facetting operations (in that order) are applied to those tessellations to obtain continuous families of rotary polyhedra, out of which only finitaly many polyhedra are regular. By considering the 3-sphere as a subset of $\mathbb{E}^4$, the desired chiral polyhedra in $\mathbb{E}^4$ can be obtained. A total of 10 families of chiral polyhedra is given, each family indexed by a continuous parameter.

To date, for $n \geq 5$ only one infinite chiral skeletal n-polytope has been described. It is a 5-polytope in $\mathbb{E}^4$ with type $\{3, 8, 3, 3\}$ (see [226]). Each of its vertex-figures is a Roli's cube, and therefore this polytope is combinatorially chiral.

Roli's cube, the three chiral 4-polytopes in $\mathbb{E}^3$ and the chiral 5-polytope with type $\{3, 8, 3, 3\}$ in $\mathbb{E}^4$ are examples of chiral polytopes of full rank. Polytopes with this property will be addressed in Section 6.3.2.

According to Proposition 6.2.2, if a chiral polytope $\mathcal{P}$ in $\mathbb{E}^d$ is not of full rank then its rank is at most $d - 1$ if $\mathcal{P}$ is finite, and at most d if $\mathcal{P}$ is infinite. Following [176], we say that finite chiral $(d - 1)$-polytopes and infinite chiral d-polytopes in $\mathbb{E}^d$ are chiral polytopes of *nearly full rank.*

All chiral polytopes of nearly full rank in $\mathbb{E}^3$ have been determined. They are the infinite chiral polyhedra mentioned in Section 6.2 (see also Corollary 6.1.37). In dimensions $d \geq 4$ we do not have the complete enumeration of chiral polytopes of nearly full rank (see [72, Problem 21], [220, Problems 38 and 29]).

Problem 71 For each $d \geq 4$ enumerate all chiral polytopes of nearly full rank in $\mathbb{E}^d$.

Problem 71 seems like a difficult task. The natural first step is the enumeration of the finite chiral skeletal polyhedra in $\mathbb{E}^4$, where some partial work can be found in [11, 12].

Problem 72 Enumerate all finite chiral skeletal polyhedra in $\mathbb{E}^4$.

As we shall see in Section 6.3.2, chiral polytopes of full rank have an upper bound in their rank. It is not known whether the same is true for chiral polytopes of nearly full rank.

Problem 73 Does there exist $N \in \mathbb{N}$ such that there is no chiral polytope of nearly full rank in $\mathbb{E}^d$ for every $d \geq N$?

In case Problem 73 has a positive answer, it can be taken even further.

Problem 74 For each $m \in \mathbb{Z}^+$ determine whether there exists $N_m \in \mathbb{N}$ such that there is no chiral $(d-m)$-polytope in $\mathbb{E}^d$ for every $d \geq N$.

6.3.2 Chiral Polytopes of Full Rank

The main goal of this section is to derive the upper bound for the rank of chiral polytopes of full rank. Throughout, we mainly follow [227].

We start with the next natural generalisation of Lemma 6.2.5.

Lemma 6.3.4 *Let $n \geq 3$, let $\mathcal{P}$ be a skeletal rotary n-polytope with base vertex F_0, and let $G^+(\mathcal{P}) = \langle S_1, \ldots, S_{n-1}\rangle$. If there is an involutory isometry R such that*

- $F_0 R = F_0$,
- $S_i R = R S_i^{-1}$ *for* $i \in \{1,2\}$,
- $S_3 R = R S_2^2 S_3$,
- $S_i R = R S_i$ *for* $i \geq 4$,

then $\mathcal{P}$ is regular.

Proof We follow the idea of the proof of Lemma 6.2.5. Let Φ be the base flag of $\mathcal{P}$ and assume that such an isometry R exists.

The vertices of the base edge F_1 are F_0 and $F_0 S_1^{-1}$, and they are mapped by R to F_0 and $F_0 S_1$, which are the vertices of the edge of Φ^1. Furthermore, for $j \geq 2$ the base j-face F_j is defined by the $(j-1)$-faces

$$F_{j-1}\langle S_1, \ldots, S_{j-1}\rangle = F_{j-1}\langle S_1, \ldots, S_{j-1}\rangle R.$$

It follows that R maps Φ to Φ^1.

An arbitrary j-face G_j can be expressed as $F_j T$ for some $T \in G^+(\mathcal{P})$. Hence $G_j R = F_j T R = F_j R T'$, where T' is the isometry obtained by replacing every factor S_1, S_2 and S_3 by S_1^{-1}, S_2^{-1} and $S_2^2 S_3$, respectively, in any expression of T in terms of the standard rotations. Since $F_j R T'$ is a j-face of $\mathcal{P}$, we can conclude that R is a symmetry of $\mathcal{P}$ that maps Φ to Φ^1, forcing the regularity of $\mathcal{P}$. ■

Rotary polytopes of full rank have the following hereditary property.

Lemma 6.3.5 *The vertex-figures of a rotary n-polytope of full rank are $(n-1)$-polytopes of full rank.*

Proof Let $\mathcal{P}$ be a rotary n-polytope of full rank with vertex-figures isomorphic to some polytope $\mathcal{Q}$. From Lemma 6.2.1 we know that $\mathcal{Q}$ is a finite, rotary $(n-1)$-polytope. If $\mathcal{P}$ is infinite then its ambient space is $\mathbb{E}^{n-1}$, implying that $\mathcal{Q}$ is of full rank. If $\mathcal{P}$ is finite and v is a vertex of $\mathcal{P}$ then the vertex-figure at v is contained in the intersection of the $(n-1)$-sphere containing all vertices of $\mathcal{P}$ with the $(n-1)$-sphere centred at v that contains all neighbours of v. Those spheres intersect in an $(n-2)$-sphere, whose affine span is $\mathbb{E}^{n-1}$. We can conclude that $\mathcal{Q}$ is of full rank. ■

The statement of Lemma 6.3.5 fails if we replace 'vertex-figures' by 'facets'. The three chiral 4-polytopes in $\mathbb{E}^3$ are counterexamples, since their facets are infinite polyhedra in $\mathbb{E}^3$.

The next two results show implications of the property of being of full rank on the even subgroup of rotary polytopes.

Lemma 6.3.6 *Let $n \geq 3$, let $\mathcal{P}$ be a finite rotary n-polytope of full rank, and let Σ_i be the mirror of the ith standard rotation of $\mathcal{P}$. Then*

$$\Sigma := \bigcap_{i \in \{2,\ldots,n-1\}} \Sigma_i$$

is one dimensional.

Proof Let S_i be the ith standard rotation, and let v_0 be the base vertex of $\mathcal{P}$. Since o and v_0 are distinct points in Σ, the dimension of Σ must be at least 1. We shall show that it is exactly 1 by induction over n.

Our base case is when $n = 3$. According to [84] and Corollary 6.2.11, the finite rotary polyhedra of full rank are the Platonic solids, the Kepler–Poinsot polyhedra and the Petrials of these nine polyhedra. The second standard rotations of all of them are rotations. In that situation, $\Sigma = \Sigma_2$ is a line.

Assume that the lemma holds for polytopes of rank $n-1$. It follows from Lemma 6.3.5 that the vertex-figure $\mathcal{P}/v_0$ is an $(n-1)$-polytope of full rank. Our inductive hypothesis implies that

$$\bigcap_{i \in \{3,\ldots,n-1\}} \Sigma_i \tag{6.22}$$

is one dimensional in the affine span of $\mathcal{P}/v_0$. Furthermore, it contains the vertex u_0 in the base edge of $\mathcal{P}$, distinct from v_0. Since v_0 is not in the affine span of $\mathcal{P}/v_0$ and it is preserved by S_i for every $i \geq 3$, the set (6.22) must be two dimensional in $\mathbb{E}^n$. Now we use that u_0 is in the set (6.22), but not in Σ, to conclude that the dimension of Σ is 1. ■

Lemma 6.3.7 *Let $n \geq 3$, let $\mathcal{P}$ be a finite rotary n-polytope of full rank with centre at o, and let $G^+(\mathcal{P}) = \langle S_1, \ldots, S_{n-1} \rangle$. Then the only isometries of $\mathbb{E}^n$ that commute with every element of $G^+(\mathcal{P})$ are* id *and the central inversion at o that maps $\bar{x}$ to $-\bar{x}$, for every $\bar{x} \in \mathbb{E}^n$.*

Proof Assume that T is an isometry such that $RT = TR$ for every $R \in G^+(\mathcal{P})$. It follows that if R stabilises a vertex v then $vT = vRT = vTR$, and R also stabilises the point vT. Lemma 6.3.6 implies that for each vertex v of $\mathcal{P}$, either $vT = v$, or $vT = -v$.

Suppose that $T \neq$ id. Since $\mathcal{P}$ is of full rank, there exists a vertex w of $\mathcal{P}$ such that $wT \neq w$, and so $wT = -w$. We claim that $vT = -v$ for every vertex v, and so T is the central inversion. Assume to the contrary that there exists a vertex u such that $uT = u$. By choosing $S \in G^+(\mathcal{P})$ such that $wS = u$ we see that $wST = uT = u$, whereas $wTS = -wS = -u \neq wST$, contradicting that $ST = TS$. ■

When dealing with skeletal rotary polytopes with high ranks, the dimension of the affine span of a face with low rank has a natural upper bound.

Proposition 6.3.8 *Let $n \geq 5$, let $\mathcal{P}$ be a rotary n-polytope in $\mathbb{E}^d$, and let F be a k-face for some $2 \leq k \leq n - 3$. Then the dimension of the affine span of F is at most $d - n + k + 1$.*

Proof Let $G^+(\mathcal{P}) = \langle S_1, \ldots, S_{n-1} \rangle$, and let V be the affine span of F. Since S_i fixes a flag of $F/\varnothing$ for every $i \in \{k+2, \ldots, n-1\}$, it must fix every vertex of F, and we can conclude that $\langle S_{k+2}, \ldots, S_{n-1} \rangle$ fixes V pointwise.

From Definition 6.1.2 we know that the coface $\mathcal{P}/F$ is a skeletal polytope of rank $n-k-1$. The affine span W of $\mathcal{P}/F$ must be a space whose dimension, according to Proposition 6.2.2, must be at least $n-k-1$. Furthermore, no vertex of $\mathcal{P}/F$ is preserved by every element of $\langle S_{k+2}, \ldots, S_{n-1} \rangle$, and so the centre of $\mathcal{P}/F$ is the only point of W fixed by every element of $\langle S_{k+2}, \ldots, S_{n-1} \rangle$. Since $\langle S_{k+2}, \ldots, S_{n-1} \rangle$ fixes V pointwise, $V \cap W$ contains only the centre of $\mathcal{P}/F$. Hence, $d \geq \dim(V) + \dim(W) \geq \dim(V) + n - k - 1$. Therefore, $\dim(V) \leq d - n + k + 1$. ■

The versions of Proposition 6.3.8 that are relevant for our current topic are the following.

Proposition 6.3.9 *Let $n \geq 5$ and let $\mathcal{P}$ be a rotary n-polytope of full rank. Each 2-face F_2 of $\mathcal{P}$ satisfies one of the following.*

- *The affine span of F_2 has dimension 2.*
- *$\mathcal{P}$ is finite, the affine span of F_2 has dimension 3, and the centres of $\mathcal{P}$ and of F_2 coincide.*

Proof We know from Proposition 6.2.3 that the dimension of F_2 is at least 2. The upper bounds are direct consequences of Proposition 6.3.8.

Assume that $\mathcal{P}$ is finite and that the affine span V_2 of F_2 is three dimensional. This forces V_2 to consist of all points in $\mathbb{E}^n$ fixed by every element in $\langle S_4, \ldots, S_{n-1}\rangle$, since the dimension of $\mathcal{P}/F_2$ must have at least dimension $n-3$. It follows that the centre o of $\mathcal{P}$ belongs to V_2. Furthermore, the centre of F_2 is the only point of V_2 invariant under S_1 (the even symmetry group of a skew 2-face is generated by a rotatory reflection). Since $o \in V_2$ and $oS_1 = o$, the centres of $\mathcal{P}$ and of F_2 coincide. ■

Proposition 6.3.10 *Let $n \geq 6$ and let $\mathcal{P}$ be a finite rotary n-polytope of full rank. Then every 3-face F_3 of $\mathcal{P}$ satisfies one of the following.*

- *The affine span of F_3 has dimension 3 and it does not contain the centre o of $\mathcal{P}$.*
- *The affine span of F_3 has dimension 4 and it contains o.*

Proof According to Proposition 6.2.2, the dimension of the affine span V_3 of F_3 is at least 3. The fact that it is at most 4 follows directly from Proposition 6.3.8.

Let $G^+(\mathcal{P}) = \langle S_1, \ldots, S_{n-1}\rangle$. We know from Lemma 6.3.5 and Proposition 6.2.2 that the dimension of the affine span W_3 of $\mathcal{P}/F_3$ is $n-4$, and that the only point of W_3 fixed by $\langle S_5, \ldots, S_{n-1}\rangle$ is the centre of $\mathcal{P}/F_3$.

Assume first that the dimension of V_3 is 4. Since all points in V_3 are fixed by every element in $\langle S_5, \ldots, S_{n-1}\rangle$, it must be true that $\mathbb{E}^n$ is a direct sum of V_3 and W_3 (as a vector space). The fact that o is fixed by every element in $\langle S_5, \ldots, S_{n-1}\rangle$ forces $o \in V_3$.

If V_3 has dimension 3 then F_3 is one of the 18 finite regular polyhedra described in [104]. Suppose that $o \in V_3$, let F_2 and F_4 be respectively a 2-face and a 4-face of $\mathcal{P}$ incident to F_3, and let G_3 be the 3-face of $\mathcal{P}$ that contains F_2 and is contained in F_4, other than F_3. Then $G_3 = F_3T_0$ for some $T_0 \in G^+(\mathcal{P})$, since rotary polytopes are transitive on their i-faces for every i. Since $oT = o$ for every $T \in G^+(\mathcal{P})$, G_3 is contained in the affine span of F_2 and o. The dimension of this affine span is 3 for every choice of finite regular polyhedron in $\mathbb{E}^3$. We can conclude that V_3 contains also G_3. Connectivity of F_4 implies that all its 3-faces are contained in V_3, but then F_4 is a finite rotary 4-polytope in a three-dimensional Euclidean space, contradicting Proposition 6.2.2. ■

In view of Proposition 6.3.9, it is convenient to find properties of rotary polytopes of full rank whose 2-faces are either planar (finite or zigzags, as in Proposition 6.1.26), or skew finite with a three-dimensional affine span (also listed in Proposition 6.1.26) and whose centre coincides with that of the polytope. This condition is already restrictive, as shown by the next proposition.

Proposition 6.3.11 *Let $n \geq 6$ and let $\mathcal{P}$ be a rotary n-polytope of full rank with regular vertex-figures, and whose 2-faces satisfy one of the items of Proposition 6.3.9. Then $\mathcal{P}$ itself is regular.*

Proof Let $G^+(\mathcal{P}) = \langle S_1, \ldots, S_{n-1} \rangle$. Let F_i denote the base i-face of the base flag Φ, and let G_i denote the i-face of Φ^i. Then the vertices of the base edge are F_0 and G_0. Let H_0 be the vertex of G_1 (the edge at F_0 contained in F_2, other than F_1), other than F_0. Then G_0 is the base vertex of the vertex-figure $\mathcal{P}/F_0$, and the base edge of this vertex-figure joins it to H_0.

First assume that $\mathcal{P}$ is infinite, and so the ambient space of $\mathcal{P}$ and of $\mathcal{P}/F_0$ is $\mathbb{E}^{n-1}$. Since $\mathcal{P}/F_0$ is finite and regular, there exists an involutory isometry R of $\mathbb{E}^{n-1}$ that maps the base flag of $\mathcal{P}/F_0$ to its 0-adjacent flag. (When restricting our attention to the vertex-figure, this symmetry is R_0.) The isometry R fixes F_0, since $R \in G(\mathcal{P}/F_0)$. It also satisfies that $RS_1R = S_1^{-1}$ since F_2 is planar, R fixes F_0 while swapping G_0 and H_0 (see Lemma 6.1.27), and F acts like id on the orthogonal complement of the affine span of F_2 at F_0. Furthermore, it satisfies the remaining items of Lemma 6.3.4 since that is the way in which the standard rotations interact with the first standard reflection. Lemma 6.3.4 forces the regularity of $\mathcal{P}$.

Now assume that $\mathcal{P}$ is finite. Then the ambient space of $\mathcal{P}$ is $\mathbb{E}^n$ while that of $\mathcal{P}/F_0$ is an $(n-1)$-dimensional subspace W. Furthermore, the line ℓ of $\mathbb{E}^n$ through the centre of $\mathcal{P}/F_0$ orthogonal to W contains F_0 and the centre of $\mathcal{P}$. If F_2 is planar then its affine span cannot contain o, since otherwise all 2-faces of $\mathcal{P}$ would be congruent regular finite planar polygons whose affine spans contain o (in fact, o would be the centre of every 2-face); but in that case no edge could belong to two distinct 2-faces. We can conclude that the affine span U of $F_2 \cup \{o\}$ is three dimensional.

Let R be the isometry of W mapping the base flag of $\mathcal{P}/F_0$ to its 0-adjacent flag. We define $\hat{R}$ as the only isometry of $\mathbb{E}^n$ acting like R on W, and like id on ℓ. Then $\hat{R}$ is an involutory isometry of $\mathbb{E}^n$. Furthermore, $\hat{R}$ acts like a plane reflection on U, which is spanned by F_0, G_0, H_0 and o, while Lemma 6.3.7 implies that it acts on the affine span W_2 of $\mathcal{P}/2$ either as id or as the central inversion with respect to the centre of F_2. Since W_2 is an orthogonal complement of U, we must have that $S_1\hat{R} = \hat{R}S_1^{-1}$. The remaining items of Lemma 6.3.4 also hold since R and $(S_i)_{|W}$ satisfy the relations of the standard rotations with the standard reflections of $\mathcal{P}/F_0$, and because of the fact that $\hat{R}$ and S_i fix F_0 for $i \geq 2$. Once again, Lemma 6.3.4 forces the regularity of $\mathcal{P}$. ∎

The previous proposition is sufficient to provide an upper bound on the rank of chiral polytopes of full rank.

Proposition 6.3.12 *There are no chiral n-polytopes of full rank for $n \geq 7$.*

Proof Let $\mathcal{P}$ be a rotary n-polytope of full rank for some $n \geq 7$. It follows from Lemma 6.3.5 and Proposition 6.3.9 that the vertex-figures of $\mathcal{P}$ are rotary $(n-1)$-polytopes of full rank whose 2-faces satisfy one of the items of Proposition 6.3.9; furthermore, its vertex-figures are regular. We can use Proposition 6.3.11 to conclude that $\mathcal{P}$ has regular vertex-figures. Also the 2-faces of $\mathcal{P}$ satisfy one of the hypotheses of Proposition 6.3.9; this, together with Proposition 6.3.11, forces $\mathcal{P}$ to be regular. ■

The bound in Proposition 6.3.12 can be improved.

Theorem 6.3.13 *Chiral polytopes of full rank exist only in ranks* 4 *and* 5.

Proof Corollaries 6.1.37 and 6.2.11 and Propositions 6.2.6 and 6.3.12 imply that a chiral polytope of full rank must have rank 4, 5 or 6. Roli's cube and the chiral 5-polytope in $\mathbb{E}^4$ described in [226] are examples of polytopes of full rank of ranks 4 and 5. It remains only to show that there are no chiral 6-polytopes of full rank.

Assume to the contrary that there exists a chiral 6-polytope $\mathcal{P}$ of full rank. Proposition 6.3.9 tells us that either the 2-faces of $\mathcal{P}$ are planar, or they are three dimensional and have their centres at o. Proposition 6.3.11 will show that $\mathcal{P}$ is regular, reaching the desired contradiction, once we can prove that the vertex-figures of $\mathcal{P}$ are regular. Since the rank of the vertex-figures is 5, we cannot use Proposition 6.3.11 directly.

Let F_i be the ith face of the base flag of $\mathcal{P}$, let V_i be the affine span of F_i and let W_i be the affine span of $\mathcal{P}/F_i$. It follows from Lemma 6.3.5 that the dimension of W_i is $5-i$ for $i \in \{0,1,2,3\}$ and, in particular, W_3 is a two-dimensional space.

Assume first that $\mathcal{P}$ is finite. Let U be the affine span of $V_3 \cup \{o\}$. We know from Proposition 6.3.10 that U is a four-dimensional space. The generator S_5 of the even symmetry group of $\mathcal{P}/F_3$ fixes U pointwise. Furthermore, it acts like a rotation around the centre of $\mathcal{P}/F_3$ on W_3. Since the dimension of W_3 is 2, the subspaces U and W_3 are orthogonal complements of each other, and S_5 is completely determined by its action on these two subspaces.

We know that S_1 commutes with S_4 and S_5, and that S_2 commutes with S_5. We can use Lemma 6.3.7 on the polyhedron $\mathcal{P}/F_2$ to conclude that S_1 acts on W_2 either like id or like a central inversion about the centre of W_2. Since $W_3 \subseteq W_2$, the action of S_1 on W_3 is the restriction of the action of S_1 on

W_2, and hence S_1 acts on W_3 either like id or like a central inversion about the centre of $\mathcal{P}/F_3$. (Note here that S_1 preserves W_3, since it preserves its orthogonal complement U.) We cannot use Lemma 6.3.7 on S_2 restricted to W_3 and the polytope $\mathcal{P}/F_3$, since the latter has rank 2. In fact, the isometries that commute with $(S_5)_{|W_3}$ are all the rotations of W_3 with the same fixed point as $(S_5)_{|W_3}$ (including id as a rotation of angle 0, and the central inversion as a rotation of angle π). However,

$$\mathrm{id}_{W_3} = ((S_1S_2)^2)|W_3 = [(S_1)_{|W_3}(S_2)_{|W_3}]^2 = (S_2)^2_{|W_3},$$

which forces S_2 to act on W_3 either as id, or as the central inversion at the centre of $\mathcal{P}/F_3$.

Now we can proceed to show that the base vertex-figure $\mathcal{Q}$ of $\mathcal{P}$ is regular. Let R be the involutory isometry of W_1 that maps the base flag of $\mathcal{P}/F_1$ to its 0-adjacent flag, and let $\hat{R}$ be the isometry of W_0 that acts like R on W_1 and fixes the endpoint G_0 of F_1 that is not F_0 (that is, the base vertex of $\mathcal{P}/F_0$). Then we can prove that $\hat{R}$ satisfies all items in the statement of Lemma 6.3.4 (our ambient space is W_0 and so the subindices of the generators S_i are shifted by 1). The first one is satisfied by definition, while the last two are satisfied since they hold on W_1 and on G_0, and W_0 is the affine span of $W_1 \cup \{G_0\}$. The second item for S_3 holds for that same reason. It only remains to show that the second item holds for S_2. Here we use that S_2 acts on W_3 like id or like a central inversion, and this makes it possible to proceed like in the last paragraph of the proof of Proposition 6.3.11.

In case $\mathcal{P}$ is infinite we proceed as for the finite case, except in two details. One of the differences is that now U must be defined simply as V_3, and hence it is three dimensional. The other difference is that S_1 may not preserve W_2 or W_3, but instead we can use that it acts on the projections of W_2 and of W_3 to the orthogonal complements of V_2 and of V_3, respectively.

Regardless of whether $\mathcal{P}$ is finite or infinite, we conclude from Lemma 6.3.4 that $\mathcal{P}/F_0$ is regular, and then from Proposition 6.3.11 that $\mathcal{P}$ itself is regular. This is the desired contradiction. ■

Infinite chiral 4-polytopes of full rank are already enumerated (see Theorem 6.3.1). To date, Roli's cube is the only finite chiral 4-polytope of full rank known to exist. Similarly, the 5-polytope described in [226] is the only infinite chiral 5-polytope of full rank known to exist. We still do not know whether there exist finite chiral 5-polytopes of full rank (see [72, Problem 20]).

Problem 75 Enumerate all chiral polytopes of full rank.

Appendix A

A Few Treats on Euclidean Geometry

A.1 Euclidean Isometries

Here we briefly review the isometries of Euclidean 3-space $\mathbb{E}^3$ (see for example [103, Section 2.3]).

There are four kinds of isometries of $\mathbb{E}^3$ that have fixed points, according to the dimension $d \in \{0, 1, 2, 3\}$ of their fixed space. There is only one isometry for which $d = 3$, namely id.

A non-trivial isometry with a plane Π of fixed points (here $d = 2$) is completely determined by Π, and it is precisely the reflection about Π.

Isometries whose space of fixed points is a line ℓ (here $d = 1$) are rotations about ℓ. They are determined once the angle of rotation and the direction of rotation are established.

There are isometries that fix only one point v (so that $d = 0$). Each of these isometries T can be described as a composition of a reflection about a plane Π containing v and a rotation about the line ℓ perpendicular to Π through v. If the angle of rotation around ℓ is π then T can be seen as a *central inversion*, which is represented by the mapping $x \mapsto -x$ when establishing v as the origin of $\mathbb{E}^3$. If the angle of rotation is not π then there is a unique plane Π containing v such that T decomposes as $T = R_\Pi R_\ell = R_\ell R_\Pi$, where R_Π is the reflection about Π and R_ℓ is the rotation about the perpendicular ℓ to Π through v. This decomposition justifies the name *rotatory reflection* of these isometries.

The remaining isometries have no fixed point. There are three kinds of such isometries, each related to the isometries T with fixed space of dimension $d \in \{1, 2, 3\}$ in the following way. Every such isometry T can be composed with a translation by a defining vector of its fixed space, yielding an isometry with no fixed points. In all the cases, such a translation commutes with the isometry T.

The first kind of isometry with no fixed points corresponds to $T = \text{id}$. The composition of this isometry with a translation is the translation itself.

A three-dimensional *glide reflection* is the composition of a reflection in a plane Π with a translation by a defining vector of Π. This is the natural three-dimensional analogue of the two-dimensional glide reflection, arising as the composition of a reflection on a line $\ell \subsetneq \mathbb{E}^2$ and the translation by a direction vector of ℓ.

The last kind of isometry of $\mathbb{E}^3$ corresponds to the composition of a rotation R_ℓ about a line ℓ with a translation by a direction vector of ℓ. They are called *twists* or *screw motions*. If the rotation R_ℓ has order k then the screw motions are also called *k-turn twists*, and *half-turn twists* when $k = 2$.

Every isometry with no fixed points has infinite order. In fact, for each such isometry T there is a line of $\mathbb{E}^3$ on which T acts like a translation. Reflections and id have orders 2 and 1, respectively. Rotations and rotatory reflections have finite order whenever the rotational component is a rational multiple of π, and infinite order otherwise.

Rotations, twists and id can be expressed as compositions of an even number of plane reflections, and hence they preserve orientation. Reflections, rotatory reflections and glide reflections can be expressed as compositions of an odd number of plane reflections, and therefore they reverse orientation.

A finite group G of isometries of $\mathbb{E}^d$ is said to be *reducible* if there is a proper linear subspace that is invariant under every element of G. We say that G is *irreducible* if it is not reducible.

For example, the symmetry group of a prism is an example of a reducible group of isometries of $\mathbb{E}^3$, since the line through the centres of the top and bottom of the prism is preserved under all such symmetries. The group generated by a rotation of order at least 3 and the symmetry group of the cube are examples of irreducible groups of isometries of $\mathbb{E}^2$ and of $\mathbb{E}^3$, respectively.

The notion of reducibility has been taken to discrete infinite groups of isometries under various names. Here we follow [167, Section 2] and say that an infinite discrete group G of isometries of $\mathbb{E}^d$ is *affinely reducible* if there is a proper subspace V of $\mathbb{E}^d$ such that Vg is a translate of V for every $g \in G$, and we say that G is *affinely irreducible* otherwise.

For each $d \geq 2$, every group of translations of $\mathbb{E}^d$ is affinely reducible. The symmetry group of the tessellation by d-cubes of $\mathbb{E}^d$ is affinely irreducible.

A.2 Tessellations by Triangles

The 1-skeleton of $\{4,3,4\}$ projects orthogonally to the plane $x + y + z = 0$ into the 1-skeleton of a tessellation by equilateral triangles. Figure A.1(a) represents this projection, with the full dark circle being the projection of

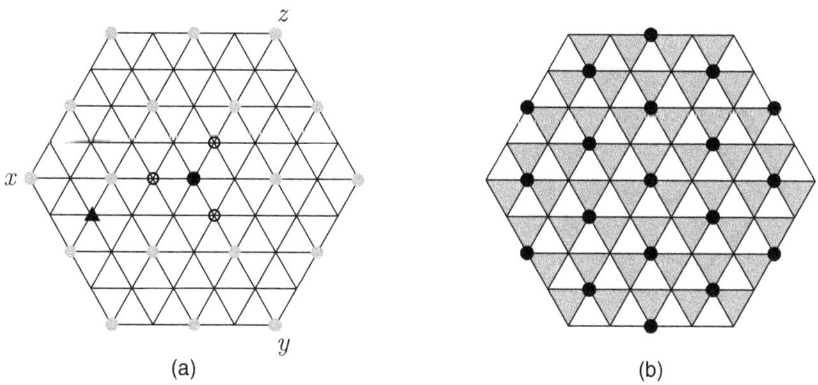

Figure A.1 Projection of the 1-skeleton of $\{4,3,4\}$ to the plane $x + y + z = 0$. (a) The projection of $\mathbf{\Lambda}_{(1,1,1)}$ and the coordinate axes; (b) a 2-colouring of the triangles of the tessellation, and a chromatic class of a proper 3-colouring of the vertices.

the origin. The projections of $(1,0,0)$, $(0,1,0)$ and $(0,0,1)$ are surrounded by hollow dark circles, each in the direction of the axis indicated by x, y, z. Then $\mathbf{\Lambda}_{(1,1,1)}$ projects to all grey circles and the full black circle. To find the projection of (a,b,c) it suffices to walk from the dark circle a steps in the direction of x (to the left), b steps in the direction of y and c steps in the direction of z. The black triangle is then the projection of $(2,0,-1)$, and of all its translates by integer multiples of $(1,1,1)$.

Exercise A.2.1 Show that the orthogonal projection of the 1-skeleton of $\{4,3,4\}$ to the plane $x - y - z = 0$ is also the 1-skeleton of the tessellation by equilateral triangles. Determine and draw a picture of the images of $(1,0,0)$, $(0,1,0)$ and $(0,0,1)$ under this projection.

In Figure A.1(b) we exhibit the natural bicolouring of the tessellation by triangles into triangles pointing up and triangles pointing down. Every edge of the tessellation belongs to one triangle of each colour. We also exhibit as black circles one chromatic class of the unique proper 3-colouring of the 1-skeleton of the tessellation by triangles. The remaining two classes can be obtained by translating the black circles one and two steps to the right, respectively.

Exercise A.2.2 Determine the smallest positive integer m such that the projection of every point in $m\mathbf{\Lambda}_{(1,1,1)}$ to the plane $x + y + z = 0$ is a black circle in Figure A.1(b).

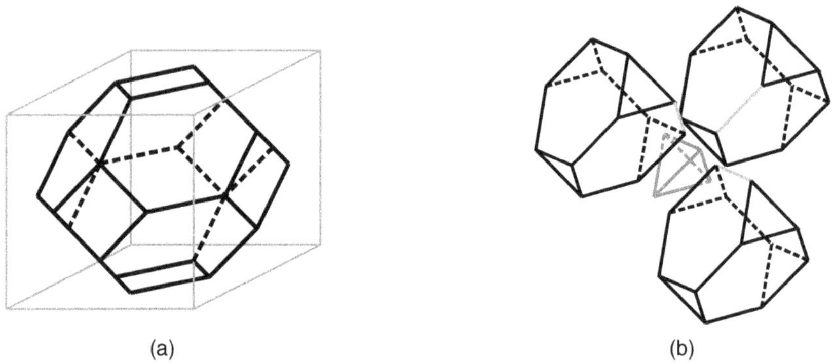

Figure A.2 (a) Truncated octahedron inscribed in a cube and (b) the tessellation $\mathcal{T}_2$.

A.3 Tessellations of $\mathbb{E}^3$

Our aim is to introduce two highly symmetric tessellations of $\mathbb{E}^3$. The tessellation $\mathcal{T}_1$ has truncated octahedra as tiles and is labelled #28 in [106]. The tiles of the tessellation $\mathcal{T}_2$ are tetrahedra and truncated tetrahedra, and can be found in [106] as #6.

We denote by $\{4,3,4\}$ the tessellation of $\mathbb{E}^3$ by cubes with vertex set $\Lambda_{(1,0,0)}$, and by $\{4,3,4\}^\delta$ the geometric dual of $\{4,3,4\}$ constructed by taking the centres of the cubes of $\{4,3,4\}$ as vertices of $\{4,3,4\}^\delta$.

Every cube of $\{4,3,4\}$ has a unique inscribed truncated octahedron in such a way that the squares of the truncated octahedron are contained in the squares of the cube (see Figure A.2(a)). When considering all truncated octahedra inscribed in cubes of $\{4,3,4\}$ together with all truncated octahedra inscribed in cubes of $\{4,3,4\}^\delta$ we obtain the tessellation $\mathcal{T}_1$ of $\mathbb{E}^3$ where all the tiles are truncated octahedra. Indeed, every hexagon belongs to the plane bisecting the segment between a vertex and centre of a cube of $\{4,3,4\}$, while the squares belong to the bisectors of segments between neighbouring vertices or between centres of neighbouring cubes of $\{4,3,4\}$. It is remarkable that every truncated octahedron can be mapped to each other by a translation preserving $\mathcal{T}_1$.

To define the tessellation $\mathcal{T}_2$ we let $\mathcal{X}$ be the set of cubes of the cubic tessellation with centres in

$$\left(\frac{1}{2}(1,1,1)+2\Lambda_{(1,1,0)}\right)\cup\left(\frac{1}{2}(3,3,-1)+2\Lambda_{(1,1,0)}\right).$$

Each cube in $\mathcal{X}$ touches precisely four other cubes in $\mathcal{X}$. The vertices of each cube C that intersect other cubes in $\mathcal{X}$ determine a convex tetrahedron $\mathcal{P}_C$ inscribed in C.

Let $\mathcal{Y}$ be $\cup\{\mathcal{P}_\mathcal{C} : \mathcal{C} \in \mathcal{X}\}$. (Here each tetrahedron $\mathcal{P}_\mathcal{C}$ is considered as a covex polyhedron and hence it includes its interior.) Then the complement of $\mathcal{Y}$ in $\mathbb{E}^3$ can be tessellated by truncated tetahedra. (The truncated tetrahedron is a polyhedron consisting of four convex regular hexagons and four equilateral triangles in such a way that at every vertex there are two hexagons and one triangle.) Indeed, the convex hulls of the sets

$$\begin{aligned} A_1 = \{&(1,1,0),(1,0,1),(2,2,0),(2,0,2),(3,2,1),(3,1,2),\\ &(0,1,1),(0,2,2),(2,3,1),(2,1,3),(1,3,2),(1,2,3)\},\\ A_2 = \{&(1,1,0),(1,0,1),(2,2,0),(2,0,2),(3,2,1),(3,1,2),\\ &(2,-1,1),(2,1,-1),(4,1,1),(3,0,-1),(3,-1,0),(4,0,0)\}, \end{aligned}$$

are truncated tetrahedra contained in the closure of the complement of $\mathcal{Y}$, and so are all their translates by vectors in $2\mathbf{\Lambda}_{(1,0,0)}$. The tessellation $\mathcal{T}_2$ consists of all tetrahedra in $\mathcal{Y}$ together with all truncated tetrahedra tessellating $\mathbb{E}^3 \setminus \mathcal{Y}$. Every triangle belongs to one tetrahedron and to one truncated tetrahedron, and every hexagon to two truncated tetrahedra. It follows that every truncated tetrahedron shares faces with four other truncated tetrahedra and with four tetrahedra, whereas each tetrahedron shares its faces with four truncated tetrahedra. The arrangement of three truncated tetrahedra of $\mathcal{T}_2$ sharing triangles with a tetrahedron is indicated in Figure A.2(b). The grey edges of the truncated tetrahedra correspond to three of the edges of some other tetrahedron of $\mathcal{T}_2$.

A.4 An Interesting Arrangement of Lines in $\mathbb{E}^3$

In what follows we introduce a particular arrangement of lines in $\mathbb{E}^3$ that covers $2\mathbb{Z}^3$.

Given $x, v \in \mathbb{E}^3$ let $\ell_v(x)$ be the line $\{x + \mu v\}_{\mu\in\mathbb{R}}$. Consider the set

$$\mathcal{L}_1 := \{\ell_{(1,1,1)}(x) : x \in 2\mathbf{\Lambda}_{(1,1,1)}\}$$

consisting of parallel lines that cover $2\mathbf{\Lambda}_{(1,1,1)}$. Every line in $\mathcal{L}_1$ can be seen as the union of main diagonals of cubes of the cubic tessellation $\mathcal{T}^{(2)}$ with edge length 2 and vertices at $2\mathbf{\Lambda}_{(1,0,0)}$. Each of these diagonals joins the two vertices in $2\mathbf{\Lambda}_{(1,1,1)}$ of the given cube. Among the cubes of $\mathcal{T}^{(2)}$, only the cubes with centres in $(1,1,1) + 2\Lambda_{(1,1,1)}$ have a main diagonal contained in $\mathcal{L}_1$. On the other hand, every line in $\mathcal{L}_1$ is completely contained in the union of all cubes of $\mathcal{T}^{(2)}$ with centres in $(1,1,1) + 2\Lambda_{(1,1,1)}$.

As noted in Section 2.4, $2\mathbf{\Lambda}_{(1,0,0)}$ can be partitioned into the sets $2\mathbf{\Lambda}_{(1,1,1)}$, $(2,0,0)+2\mathbf{\Lambda}_{(1,1,1)}$, $(0,2,0)+2\mathbf{\Lambda}_{(1,1,1)}$ and $(0,0,2)+2\mathbf{\Lambda}_{(1,1,1)}$. We cover each of the last three sets by the sets of lines $\mathcal{L}_i$ with $i \in \{2,3,4\}$ as follows:

$$\mathcal{L}_2 := \{\ell_{(1,-1,1)}(x) \colon x \in (2,0,0) + 2\Lambda_{(1,1,1)}\},$$
$$\mathcal{L}_3 := \{\ell_{(1,1,-1)}(x) \colon x \in (0,2,0) + 2\Lambda_{(1,1,1)}\},$$
$$\mathcal{L}_4 := \{\ell_{(-1,1,1)}(x) \colon x \in (0,0,2) + 2\Lambda_{(1,1,1)}\}.$$

Furthermore, we let $\mathcal{L} := \mathcal{L}_1 \cup \mathcal{L}_2 \cup \mathcal{L}_3 \cup \mathcal{L}_4$. Alternatively, for $i \in \{1,2,3\}$ we can define $\mathcal{L}_{i+1}$ as $\mathcal{L}_i S_0$, where S_0 is the rotatory reflection mapping (x,y,z) to $(1-y, x+1, -1-z)$. We can deduce from here that $\mathcal{L}_i$ is isometric to $\mathcal{L}_j$ for any $i,j \in \{1,2,3,4\}$.

Observing carefully we can see that $\mathcal{L}_2$ (resp. $\mathcal{L}_3$ and $\mathcal{L}_4$) is contained in the union of all cubes of $\mathcal{T}^{(2)}$ with centres in $(1,1,-1) + 2\Lambda_{(1,1,1)}$ (resp. in $(-1,1,1) + 2\Lambda_{(1,1,1)}$ and in $(1,-1,1) + 2\Lambda_{(1,1,1)}$). These inclusions have two nice implications. On one hand it shows that no two distinct lines in $\mathcal{L}$ intersect. On the other hand, the fact that $(1,1,1) + 2\mathbf{\Lambda}_{(1,0,0)}$ consists of all centres of cubes of $\mathcal{T}^{(2)}$, and that

$$\mathbf{\Lambda}_{(1,1,1)} = 2\mathbf{\Lambda}_{(1,0,0)} \cup \big((1,1,1) + 2\mathbf{\Lambda}_{(1,0,0)}\big)$$

imply that every point in $\mathbf{\Lambda}_{(1,1,1)}$ is covered by a line in $\mathcal{L}$.

We have proven the following theorem.

Theorem A.4.1 *The arrangement of lines $\mathcal{L}$ satisfies that:*

1. *there are infinitely many lines in $\mathcal{L}$ with direction vector $\overline{v}$ for every $\overline{v} \in \{(1,1,1),(1,1,-1),(1,-1,1),(-1,1,1)\}$,*
2. *the lines in $\mathcal{L}$ cover $\mathbf{\Lambda}_{(1,1,1)}$, and hence also $2\mathbf{\Lambda}_{(1,0,0)}$,*
3. *no two lines in $\mathcal{L}$ intersect.*

The set $\mathcal{L}$ is highly symmetric, since it is invariant under the lattice group $2\Lambda_{(1,1,1)}$. The rotatory reflection S_0 defined earlier is also one of its symmetries, and therefore $\mathcal{L}$ is not chiral in the sense that it has orientation reversing symmetries. However, it was constructed in a 'handed' manner. Compare the definition of $\mathcal{L}_2$, $\mathcal{L}_3$ and $\mathcal{L}_4$ with

$$\mathcal{L}_2' := \{\ell_{(1,1,-1)}(x) \colon x \in (2,0,0) + 2\Lambda_{(1,1,1)}\},$$
$$\mathcal{L}_3' := \{\ell_{(-1,1,1)}(x) \colon x \in (0,2,0) + 2\Lambda_{(1,1,1)}\},$$
$$\mathcal{L}_4' := \{\ell_{(1,-1,1)}(x) \colon x \in (0,0,2) + 2\Lambda_{(1,1,1)}\}.$$

The arrangement $\mathcal{L}' := \mathcal{L}_1 \cup \mathcal{L}_2' \cup \mathcal{L}_3' \cup \mathcal{L}_4'$ is isometric to $\mathcal{L}$, since we can obtain one from the other by reflecting on the plane $y = z$. However, $\mathcal{L}$ itself is

Figure A.3 Portion of the hexastix.

not invariant under any plane reflection. This hints that chiral structures, in one sense or another, can be directly related to $\mathcal{L}$ (see Sections 6.1.5 and 6.2.3).

The lines in $\mathcal{L}$ can be substituted by isometric cylinders over circles as wide as possible without overlapping. If instead we use cylinders over hexagons, accomodated in a certain specific way, they cover an even larger portion of space than the cylinders over circles. The arrangement of prisms over hexagons was called a *hexastix* in [46, Chapter 23], and has a relation with crystallography (see [212]). Figure A.3 illustrates a portion of the hexastix with cylinders over circles.

Appendix B

A Few Words about Numbers

When working with toroidal $(n + 1)$-polytopes we can visualise them as quotients of a Eucidean tessellation $\mathcal{T}$ by an isometry group Λ generated by n independent translations by vectors $\bar{u}_1, \ldots, \bar{u}_n$ (see Section 2.4.1). By assigning appropriate coordinates we may ensure that the translation vectors have integer coordinates. Since the polytopes themselves are completely determined by $\mathcal{T}$ and Λ, and the latter is completely determined by $\bar{u}_1, \ldots, \bar{u}_n$, it is convenient to have handy some results on number theory.

We start with two well-known results. The first one is Dirichlet's theorem, whose proof can be found in [128, Section 5.12].

Theorem B.0.1 *If a and b are coprime integers then there are infinitely many primes of the form $ak + b$.*

The following result is an immediate consequence of the fact that the multiplicative group of every finite field is cyclic. An alternative proof can be found in [128, Section 3.2].

Lemma B.0.2 *Let p be an odd prime. Then -1 is a square in $\mathbb{Z}_p$ if and only if $p \equiv 1$ (modulo 4).*

Chiral toroidal polyhedra with type $\{4,4\}$ are quotients of the tessellation $\{4,4\}$ by squares of $\mathbb{E}^2$ by the translations by (a,b) and $(b,-a)$ for some positive a,b satisfying that $ab(a - b) \neq 0$. The fundamental region of the translation group has $a^2 + b^2$ squares. Here we present a few results about prime numbers p that admit an expression as a sum of two squares (we mainly follow [128, Section 6.7]).

Proposition B.0.3 *Let p be a prime number satisfying that $p \equiv 1$ (modulo 4). Then there exist $a,b \in \mathbb{N}$ such that $p = a^2 + b^2$.*

Proof From Lemma B.0.2 we know that there exists $l \in \mathbb{Z}$ such that $l^2 \equiv -1$ (modulo p). For such an l we consider all expressions $x - ly$ where $0 \leq x < \sqrt{p}$ and $0 \leq y < \sqrt{p}$. There are $(\lfloor\sqrt{p}\rfloor+1)^2 > p^2$ such expressions, implying the existence of x_1, x_2, y_1, y_2 such that $x_1 - ly_1 \equiv x_2 - ly_2$ (modulo p) with $0 \leq x_1 < x_2 < \sqrt{p}$ and $0 \leq y_i < \sqrt{p}$. We let $x_0 := x_2 - x_1$ and $y_0 := y_2 - y_1$. From the construction we know that $0 < x_0 < \sqrt{p}$ and $0 < |y_0| < \sqrt{p}$. It follows that $x_0^2 < p$ and $y_0^2 < p$, and consequently

$$0 < x_0^2 + y_0^2 < 2p. \tag{B.1}$$

Furthermore, $x_0 \equiv ly_0$ (modulo p) and therefore $x_0^2 \equiv -y_0^2$ (modulo p), or equivalently, $x_0^2 + y_0^2 \equiv 0$ (modulo p). This, together with (B.1) implies that $x_0^2 + y_0^2 = p$. ■

Lemma B.0.4 *The set $\{a, b\}$ of natural numbers in Proposition B.0.3 is unique. Furthermore, these integers satisfy that $a \equiv lb$ (modulo p) where $l^2 \equiv -1$ (modulo p).*

Proof Without loss of generality assume that $a < b$. We denote by l one square root of -1; then the only $d \in \mathbb{Z}_p$ such that $d^2 = -1$ are l and $-l$.

If $a^2 + b^2 = p$ then $a^2 \equiv -b^2 \equiv (-1)b^2$ (modulo p). Using that p is prime and that $0 < a < b < p$, we have that $a \equiv \pm lb$. If necessary, we rename l and $-l$ to obtain the second part of the statement.

Now we prove uniqueness. Assume that $a^2 + b^2 = p = c^2 + d^2$ with $0 < a < b$, $0 < c$, $0 < d$, $a \equiv lb$ and $c \equiv ld$ (modulo p). Then

$$p^2 = (a^2+b^2)(c^2+d^2) = a^2c^2+b^2d^2+a^2d^2+b^2c^2 = (ac+bd)^2+(ad-bc)^2.$$

But $ac + bd \equiv bd(1 + l^2) \equiv 0$ (modulo p), which forces $ac + bd = p$ and $ad - bc = 0$. This last equality is equivalent to $\frac{a}{c} = \frac{b}{d}$, and thus

$$p = a^2 + b^2 = \left(\frac{a}{c}\right)^2 (c^2 + d^2) = \left(\frac{a}{c}\right)^2 p.$$

This suffices to conclude that $a = c$ and $b = d$. ■

Given a prime p congruent to 1 modulo 4 there exists an algorithm to determine positive integers a and b such that $a^2 + b^2 = p$ with the help of a generator of the multiplicative group of $\mathbb{Z}_p$. The details can be found in [141, Chapter 8].

If a chiral toroidal polyhedron has type $\{3,6\}$ or $\{6,3\}$ then it is a quotient of the tessellation $\{3,6\}$ by equilateral triangles, or of the tessellation $\{6,3\}$ by regular hexagons. The quotient is taken with respect to a group of isometries generated by the translations by (a,b) and $(-b, a+b)$ for some positive integers a, b satisfying that $ab(a-b) \neq 0$ (with respect to some suitable basis).

In these cases a fundamental region of the translation group has $a^2 + ab + b^2$ vertices or hexagons. Chiral 4-polytopes whose facets have type $\{3,6\}$ or $\{6,3\}$ were found in [248] by using some analogues of Proposition B.0.3 and Lemma B.0.4 for integers n that can be expressed as $a^2 + ab + b^2$.

Appendix C

Open Problems

In the chapters of this book we explained many aspects of chiral polytopes. We inserted a sample of open problems as evidence of the existence of many other aspects that remain unknown. We believe that these problems, together with those in [220, 250], will serve to inspire active research on chiral polytopes over the upcoming decades.

For convenience, we include our 75 problems next. They are grouped by the sections where they are stated, so that the reader interested in a specific problem may look for the context in which it was included in this book.

Section 2.2

Problem 1 Determine whether there exist two-orbit n-polytopes in class 2_I for every $n \geq 4$ and for every proper subset I of $\{0, \ldots, n-1\}$.

Problem 2 For every $n \geq 4$ determine whether there exists a two-orbit polytope in class $2_{\{0,\ldots,n-2\}}$ whose facets are all isomorphic.

Problem 3 Determine if there are self-dual n-polytopes in class 2_I for every set I satisfying that $i \in I$ if and only if $n - i - 1 \in I$.

Section 2.3

Problem 4 Determine all positive integers g such that there exists a chiral map on a surface of genus g.

Problem 5 Characterise all chiral maps on surfaces that are not polytopal.

Problem 6 Determine the number of flags of the smallest chiral n-polytope for every n.

Problem 7 Let $f_{rr}(n)$, $f_{cr}(n)$ and $f_{cc}(n)$ denote the number of flags of the smallest chiral n-polytopes with regular facets and regular vertex-figures, chiral facets and regular vertex-figures, and chiral facets and chiral vertex-figures, respectively.

1. Determine whether $f_{rr}(n) \geq f_{cr}(n) \geq f_{cc}(n)$ for every $n \geq 3$.
2. Determine the functions $f_{rr}(n)$, $f_{cr}(n)$ and $f_{cc}(n)$.

Problem 8 Exhibit a properly self-dual and an improperly self-dual chiral n-polytope for every $n \geq 6$.

Problem 9 Determine all values of n for which there exist self-dual chiral n-polytopes that possess no polarity.

Section 3.1

Problem 10 Develop the theory of the coset representation of fully transitive polytopes that are not 2-chain transitive.

Section 3.2

Problem 11 Determine whether there exists a non-regular infinite polytope $\mathcal{P}$ such that $\Gamma(\mathcal{P}) \cong \mathrm{Conn}(\mathcal{P})$.

Problem 12 Determine the pairs $\{a,b\} \subseteq \mathbb{Z} \cup \{\infty\}$ such that there exists a chiral 4-polytope with Petrie paths of sizes a and b.

Problem 13 Determine all $a \in \mathbb{Z} \cup \{\infty\}$ such that there is a chiral 4-polytope whose Petrie paths all have size a.

Problem 14 For each $m \geq 2$ find an example of a chiral $4m$-polytope whose Petrie paths have distinct sizes, and an example of a chiral $4m$-polytope whose Petrie paths are all of the same size. Else, show the corresponding non-existence.

Section 3.3

Problem 15 Determine sufficient conditions for a group G to be the chirality group of a chiral polytope.

Problem 16 Determine necessary conditions for a group G to be the chirality group of a chiral polytope.

Problem 17 Determine which finite simple groups are the automorphism groups of chiral polytopes.

Problem 18 Determine which finite non-simple groups are the chirality groups of totally chiral polytopes.

Problem 19 Determine whether there are infinite chiral polytopes with finite chirality index, and whether there are infinite chiral polytopes with infinite chirality index.

Problem 20 Are there infinite totally chiral polytopes?

Section 4.1

Problem 21 Find operations on regular polytopes that produce chiral n-polytopes from regular polytopes for $n \geq 5$.

Section 4.2

Problem 22 Is there a symmetric embedding of a regular n-polytope in a chiral n-polytope for some n?

Problem 23 Is there a symmetric embedding of a chiral polytope in a non-orientably regular polytope?

Problem 24 Is there a symmetric embedding of a non-orientably regular polytope in a chiral polytope?

Problem 25 Are there chiral embeddings of regular n-polytopes in other regular n-polytopes?

Section 4.3

Problem 26 For each $n \geq 4$ let $f_n(x)$ (resp. $g_n(x)$) determine the number of chiral n-polytopes with chiral facets and chiral vertex-figures that have at most x flags, and whose connection group is a string C-group (resp. is not a string C-group). For each n, is either of

$$\lim_{x\to\infty} \frac{f_n(x)}{g_n(x)}, \qquad \lim_{x\to\infty} \frac{g_n(x)}{f_n(x)}$$

equal to 0? If not, does it converge?

Problem 27 Determine whether there exists a chiral polytope whose connection group is not a string C-group but it still has a unique regular (polytopal) cover.

Problem 28 Determine whether there exists a chiral polytope with infinitely many minimal regular covers.

Problem 29 For $n \geq 3$, is $\mathcal{U}_n$ the only regular polytope of rank n that does not admit a chiral cover?

Problem 30 Is there a regular polytope that has a smallest chiral cover?

Section 4.4

Problem 31 Are there a set $I \subseteq \{0, \ldots, n-1\}$ and a chiral n-polytope $\mathcal{P}$ such that $\Gamma(\mathcal{P} \Diamond \mathcal{D}_I)$ is regular?

Section 5.1

Problem 32 For each $p, q \in \{3, \ldots, \infty\}$ with $1/p + 1/q - 1/2 < 0$, determine whether there exist infinitely many chiral infinite polyhedra with type $\{p, q\}$.

Problem 33 Determine the set $\mathcal{X}$ of connected non-compact surfaces that carry maps with finitely many flag orbits. For each such surface $\mathcal{S} \in \mathcal{X}$, determine the smallest k such that $\mathcal{S}$ carries k-orbit maps.

Problem 34 Let N_m and R_m denote the numbers of rotary maps and of regular maps on orientable surfaces with genera $2 \leq g \leq m$, respectively. Determine whether the sequence $\{R_m/N_m\}_{m \in \mathbb{N}}$ converges, and in that case, determine the limit when m goes to ∞.

Problem 35 Let N_n and R_n denote the number of rotary maps and of regular maps on orientable surfaces with at most n flags. Determine whether the sequence $\{R_n/N_n\}_{n \in \mathbb{N}}$ converges, and in that case, determine the limit when n goes to ∞. Is the limit the same as in Problem 34 (in case it exists)?

Problem 36 Let $N_{p,q,n}$ and $R_{p,q,n}$ denote the number of rotary maps and of regular maps on orientable surfaces with type $\{p, q\}$ and at most n flags. Determine whether the sequence $\{R_{p,q,n}/N_{p,q,n}\}_{n \in \mathbb{N}}$ converges, and in that case, determine the limit when n goes to ∞. Is the limit the same for all hyperbolic types $\{p, q\}$? Do we get the same answers if we bound by genus instead of by number of flags?

Section 5.2

Problem 37 Determine the types of tight non-orientably regular polytopes of ranks 4 and higher.

Problem 38 Determine the types of tight n-polytopes for $n \geq 4$.

Problem 39 Describe the automorphism groups of all tight chiral polyhedra. Find a procedure to determine how many tight chiral polyhedra are there with a given type.

Problem 40 Determine the types of all tight chiral 4-polytopes.

Problem 41 Determine how many tight chiral 4-polytopes are there for each type; also determine their automorphism groups.

Section 5.3

Problem 42 Enumerate all chiral maniplexes whose order complex is a spherical space form. Determine whether any of them is polytopal.

Problem 43 For every $t \geq 10$ determine whether

(a) $\{3,5,3\}_t$ is a twisted honeycomb or there is a collapse in facets and/or vertex-figures;
(b) $\{3,5,3\}_t$ is finite or infinite;
(c) $\{3,5,3\}_t$ is regular or chiral.

Problem 44 For every $t \geq 9$ determine whether

(a) $\{4,3,5\}_t$ is a twisted honeycomb or there is a collapse in facets and/or vertex-figures;
(b) $\{4,3,5\}_t$ is finite or infinite;
(c) $\{4,3,5\}_t$ is regular or chiral.

Problem 45 For every $t \geq 6$ determine whether

(a) $\{5,3,5\}_t$ is a twisted honeycomb or there is a collapse in facets and/or vertex-figures;
(b) $\{5,3,5\}_t$ is finite or infinite;
(c) $\{5,3,5\}_t$ is regular or chiral.

Problem 46 Determine all hyperbolic three-dimensional and four-dimensional space forms that admit chiral tessellations.

Problem 47 For each of the types $\{3,5,3\}$, $\{4,3,5\}$, $\{5,3,5\}$, $\{3,3,3,5\}$, $\{4,3,3,5\}$ and $\{5,3,3,5\}$ determine whether there are infinitely many chiral polytopes with that type.

Problem 48 Are there other regular colourful n-polytopes whose 1-skeleton admits isomorphic n-colourings of their edges that yield colourful n-polytopes with two or more flag orbits?

Problem 49 Are there chiral n-polytopes for $n \geq 5$ arising from colourful polytopes whose underlying 1-skeleta are well-studied families of graphs?

Section 5.4

Problem 50 Determine whether there exists an orientable 4-polytope $\mathcal{Q}$ all whose facets are isomorphic to $\{4,4\}_{(3,1)}$ such that there are two flags Φ and Ψ with the following properties.

- There is a path with even length in the flag graph of $\mathcal{Q}$ between Φ and Ψ.
- The facet containing Φ, when rooted at Φ, is the enantiomorph of the facet containing Ψ rooted at Ψ.

Problem 51 Are there non-orientable non-rooted amalgamations of two chiral polytopes?

Problem 52 For each rooted toroidal chiral polyhedron $(\mathcal{P}_1)_{(\Phi)}$ determine the rooted rotary, spherical or toroidal polyhedra $(\mathcal{P}_2)_{(\Psi)}$ such that $\langle(\mathcal{P}_1)_{(\Phi)}, (\mathcal{P}_2)_{(\Psi)}\rangle^{ch} \neq \varnothing$.

Problem 53 For each $(\mathcal{P}_1)_{(\Phi)}$ and $(\mathcal{P}_2)_{(\Psi)}$ giving a positive answer to Problem 52 determine if $\{(\mathcal{P}_1)_{(\Phi)}, (\mathcal{P}_2)_{(\Psi)}\}$ is finite.

Problem 54 Let $\mathcal{P}_1$ and $\mathcal{P}_2$ be as in Problem 52 such that $\{(\mathcal{P}_1)_{(\Phi)}, (\mathcal{P}_2)_{(\Psi)}\}$ is infinite. Is there a natural finite set of parameters $a_1, \ldots, a_k$ such that every element in $\{(\mathcal{P}_1)_{(\Phi)}, (\mathcal{P}_2)_{(\Psi)}\}$ can be expressed as $\{(\mathcal{P}_1)_{(\Phi)}, (\mathcal{P}_2)_{(\Psi)}\}_{(a_1,\ldots,a_k)}$?

Problem 55 Let $\mathcal{P}_1, \mathcal{P}_2$ be regular n-polytopes such that $\langle(\mathcal{P}_1)_{(\Phi)}, (\mathcal{P}_2)_{(\Psi)}\rangle^{ch} \neq \varnothing$. Does the partial order in $\langle(\mathcal{P}_1)_{(\Phi)}, (\mathcal{P}_2)_{(\Psi)}\rangle^{ch} \neq \varnothing$ given by $\mathcal{Q}_1 \leq \mathcal{Q}_2$ if and only if $\mathcal{Q}_2$ covers $\mathcal{Q}_1$ have finitely many or infinitely many maximum elements? In case they are finitely many, how many are there? In particular, are there regular polytopes $\mathcal{P}_1$ and $\mathcal{P}_2$ such that $\langle(\mathcal{P}_1)_{(\Phi)}, (\mathcal{P}_2)_{(\Psi)}\rangle^{ch} \neq \varnothing$, and it has a greatest element with the partial order just defined?

Section 5.5

Problem 56 Determine for which regular polytopes $\mathcal{P}$ it is true that for every odd integer m there is a regular extension of $\mathcal{P}$ whose last entry of its type is m.

Problem 57 Is it true that there exists an integer p_n with $3 \leq p_n < \infty$ such that every chiral polytope with regular facets has a chiral extension where p_n is the last entry of the type?

Problem 58 Is it true that there exists an integer p_n with $3 \leq p_n < \infty$ such that every finite chiral polytope with regular facets has a finite chiral extension where p_n is the last entry of the type?

Problem 59 Are there chiral polytopes $\mathcal{P}$ such that, for every $p_n \geq 3$, $\mathcal{P}$ has a chiral extension whose last entry of its type is p_n? If the answer is affirmative, find necessary and sufficient conditions for such polytopes $\mathcal{P}$.

Problem 60 Provide necessary and sufficient conditions for a regular polytope $\mathcal{P}$ to have chiral extensions.

Problem 61 For each $n \geq 3$ determine the following.

1. The smallest m so that the symmetric group S_m is the automorphism group of a chiral n-polytope.
2. The smallest m so that the alternating group A_m is the automorphism group of a chiral n-polytope.
3. The smallest m so that the full automorphism group of the m-cube is the automorphism group of a chiral n-polytope.

4. The smallest m so that the rotation group of the m-cube is the automorphism group of a chiral n-polytope.

Problem 62 Determine whether there exists a regular polytope $\mathcal{P}$ that has a chiral extension which covers every chiral extension of $\mathcal{P}$.

Problem 63 Determine whether every convex regular polytope admits a chiral extension.

Problem 64 Is it true that if $\mathcal{P}$ is a regular polytope that has chiral extensions then it has a chiral extension that has ∞ as the last entry of its type?

Problem 65 Is there a regular polytope that admits a chiral extension that has q as the last entry of its type for each $q \geq 3$?

Problem 66 Is it true that for every $n \geq 3$ there is a chiral n-polytope with type $\{\infty, \ldots, \infty\}$?

Section 6.1

Problem 67 Develop the details of the realisation theory for chiral polytopes.

Section 6.2

Problem 68 Search for chiral polyhedra in spaces $\mathcal{S}$ distinct from $\mathbb{E}^3$ by Wythoff's construction on the subgroup $\langle S_1, \ldots, S_{n-1} \rangle$ of the symmetry group of regular n-polytopes in $\mathcal{S}$.

Problem 69 Is there anything meaningful to say about polyhedra $P(\alpha, 1)$ and $P(\beta, 1)$ (or $Q(\alpha, 1)$ and $Q(\beta, 1)$, or $Q(\alpha, 1)^*$ and $Q(\beta, 1)^*$) when $|\alpha - \beta| < \epsilon$ for small values of ϵ? If we remove the discreteness condition in the definition of skeletal polyhedra, can we relate combinatorially the structures $P(\alpha, 1)$ and $P(\beta, 1)$ (or $Q(\alpha, 1)$ and $Q(\beta, 1)$, or $Q(\alpha, 1)^*$ and $Q(\beta, 1)^*$) for $\alpha, \beta \in \mathbb{R}$ when $|\alpha - \beta| < \epsilon$ for small values of ϵ?

Section 6.3

Problem 70 Describe the skeletal chiral polyhedra of types $\{3,6\}$ and $\{6,3\}$ in $\mathbb{E}^6$.

Problem 71 For each $d \geq 4$ enumerate all chiral polytopes of nearly full rank in $\mathbb{E}^d$.

Problem 72 Enumerate all finite chiral skeletal polyhedra in $\mathbb{E}^4$.

Problem 73 Does there exist $N \in \mathbb{N}$ such that there is no chiral polytope of nearly full rank in $\mathbb{E}^d$ for every $d \geq N$?

Problem 74 For each $m \in \mathbb{Z}^+$ determine whether there exists $N_m \in \mathbb{N}$ such that there is no chiral $(d - m)$-polytope in $\mathbb{E}^d$ for every $d \geq N$.

Problem 75 Enumerate all chiral polytopes of full rank.

References

[1] Norbert A'Campo, Lizhen Ji, and Athanase Papadopoulos. Actions of the absolute Galois group. In *Handbook of Teichmüller Theory, Vol. VI*, volume 27 of IRMA Lect. Math. Theor. Phys., pages 397–435. European Mathematical Society, 2016. 165

[2] Gabriela Araujo-Pardo, Isabel Hubard, Deborah Oliveros, and Egon Schulte. Colorful polytopes and graphs. *Israel J. Math.*, 195(2):647–675, 2013. 306

[3] Jorge L. Arocha, Javier Bracho, and Luis Montejano. Regular projective polyhedra with planar faces. I. *Aequationes Math.*, 59(1–2):55–73, 2000. 344, 437

[4] John A. Arredondo, Camilo Ramírez Maluendas, and Ferrán Valdez. On the topology of infinite regular and chiral maps. *Discrete Math.*, 340(6):1180–1186, 2017. 272

[5] Leah Wrenn Berman, Mark Mixer, Barry Monson, Deborah Oliveros, and Gordon Williams. The monodromy group of the n-pyramid. *Discrete Math.*, 320:55–63, 2014. 243

[6] Ludwig Bieberbach. Über die Bewegungsgruppen der Euklidischen Räume (Zweite Abhandlung.) Die Gruppen mit einem endlichen Fundamentalbereich. *Math. Ann.*, 72(3):400–412, 1912. 289

[7] N. L. Biggs and A. T. White. *Permutation Groups and Combinatorial Structures*, volume 33 of London Math. Soc. Lecture Note Ser. Cambridge University Press, 1979. 23

[8] Garrett Birkhoff. *Lattice Theory*, volume 25 of Amer. Math. Soc. Colloq. Publications. American Mathematical Society, 3rd ed. 1979. 8

[9] Javier Bracho. Regular projective polyhedra with planar faces. II. *Aequationes Math.*, 59(1–2):160–176, 2000. 344, 437

[10] Javier Bracho, Isabel Hubard, and Daniel Pellicer. A finite chiral 4-polytope in $\mathbb{R}^4$. *Discrete Comput. Geom.*, 52(4):799–805, 2014. 303, 307, 309

[11] Javier Bracho, Isabel Hubard, and Daniel Pellicer. Realising equivelar toroids of type $\{4,4\}$. *Discrete Comput. Geom.*, 55(4):934–954, 2016. 233, 344, 440, 441

[12] Javier Bracho, Isabel Hubard, and Daniel Pellicer. Chiral polyhedra in 3-dimensional geometries and from a Petrie–Coxeter construction. *Discrete Comput. Geom.*, 66(3):1025–1052, 2021. 344, 388, 441

[13] H. R. Brahana. Regular maps on an anchor ring. *Amer. J. Math.*, 48(4):225–240, 1926. 4, 63, 64, 91, 144

[14] H. R. Brahana. Regular maps and their groups. *Amer. J. Math.*, 49(2):268–284, 1927. 64, 91

[15] Ana Breda, Antonio Breda D'Azevedo, and Roman Nedela. Chirality group and chirality index of Coxeter chiral maps. *Ars Combin.*, 81:147–160, 2006. 194, 201, 203, 242

[16] Antonio Breda D'Azevedo, Gareth Jones, Roman Nedela, and Martin Škoviera. Chirality groups of maps and hypermaps. *J. Algebraic Combin.*, 29(3):337–355, 2009. 165, 193, 194, 203, 205

[17] Antonio Breda d'Azevedo, Roman Nedela, and Jozef Širáň. Classification of regular maps of negative prime Euler characteristic. *Trans. Amer. Math. Soc.*, 357(10):4175–4190, 2005. 270

[18] Antonio J. Breda d'Azevedo and Gareth A. Jones. Double coverings and reflexive abelian hypermaps. *Beiträge Algebra Geom.*, 41(2):371–389, 2000. 257

[19] Ulrich Brehm and Wolfgang Kühnel. Equivelar maps on the torus. *European J. Combin.*, 29(8):1843–1861, 2008. 91

[20] P. A. Brooksbank, J. T. Ferrara, and D. Leemans. Orthogonal groups in characteristic 2 acting on polytopes of high rank. *Discrete Comput. Geom.*, 63(3): 656–669, 2020. 340

[21] P. A. Brooksbank and D. Leemans. Polytopes of large rank for $\mathrm{PSL}(4,\mathbb{F}_q)$. *J. Algebra*, 452:390–400, 2016. 340

[22] Peter A. Brooksbank. On the ranks of string C-group representations for symplectic and orthogonal groups. In *Polytopes and Discrete Geometry*, volume 764 of Contemp. Math., pages 31–41. American Mathematical Society, 2021. 340

[23] Peter A. Brooksbank and Deborah A. Vicinsky. Three-dimensional classical groups acting on polytopes. *Discrete Comput. Geom.*, 44(3):654–659, 2010. 340

[24] Emilio Bujalance, Marston D. E. Conder, and Antonio F. Costa. Pseudo-real Riemann surfaces and chiral regular maps. *Trans. Amer. Math. Soc.*, 362(7):3365–3376, 2010. 268

[25] H. Burgiel and D. Stanton. Realizations of regular abstract polyhedra of types $\{3,6\}$ and $\{6,3\}$. *Discrete Comput. Geom.*, 24(2-3): 241–255, 2000. The Branko Grünbaum birthday issue. 437, 440

[26] Peter J. Cameron, Maria Elisa Fernandes, Dimitri Leemans, and Mark Mixer. Highest rank of a polytope for A_n. *Proc. Lond. Math. Soc. (3)*, 115(1):135–176, 2017. 340

[27] Professor Cayley. Desiderata and suggestions: No. 2. The theory of groups: Graphical representation. *Amer. J. Math.*, 1(2):174–176, 1878. 156

[28] Arjeh M. Cohen. Coxeter groups and three related topics. In *Generators and Relations in Groups and Geometries (Lucca, 1990)*, volume 333 of NATO Adv. Sci. Inst. Ser. C: Math. Phys. Sci., pages 235–278. Kluwer Academic Publishers, 1991. 112

[29] Charles J. Colbourn and Asia Ivić Weiss. A census of regular 3-polystroma arising from honeycombs. *Discrete Math.*, 50(1):29–36, 1984. 313, 318, 319, 324

[30] José Collins and Antonio Montero. Equivelar toroids with few flag-orbits. *Discrete Comput. Geom.*, 65(2):305–330, 2021. 85, 289

[31] Marston Conder. Regular maps with small parameters. *J. Austral. Math. Soc. Ser. A*, 57(1):103–112, 1994. 263

[32] Marston Conder. The smallest regular polytopes of given rank. *Adv. Math.*, 236:92–110, 2013. 66, 277, 288

[33] Marston Conder. Chiral polytopes with up to 4000 flags. www.math.auckland.ac.nz/~conder/ChiralPolytopesWithUpTo4000Flags-ByOrder.txt, 2014. 150, 164, 270, 273, 287

[34] Marston Conder. Regular polytopes with up to 4000 flags. www.math.auckland.ac.nz/~conder/RegularPolytopesWithUpTo4000Flags-ByOrder.txt, 2014. 150, 164, 270, 273, 287

[35] Marston Conder and Gabe Cunningham. Tight orientably-regular polytopes. *Ars Math. Contemp.*, 8(1):68–81, 2015. 280, 331

[36] Marston Conder and Peter Dobcsányi. Determination of all regular maps of small genus. *J. Combin. Theory Ser. B*, 81(2):224–242, 2001. 269

[37] Marston Conder, Isabel Hubard, Eugenia O'Reilly-Regueiro, and Daniel Pellicer. Construction of chiral 4-polytopes with alternating or symmetric automorphism group. *J. Algebraic Combin.*, 42(1):225–244, 2015. 75, 328, 339

[38] Marston Conder, Isabel Hubard, and Tomaž Pisanski. Constructions for chiral polytopes. *J. Lond. Math. Soc. (2)*, 77(1):115–129, 2008. 65, 66, 236, 238

[39] Marston Conder, Primož Potočnik, and Jozef Širáň. Regular maps with almost Sylow-cyclic automorphism groups, and classification of regular maps with Euler characteristic $-p^2$. *J. Algebra*, 324(10):2620–2635, 2010. 270

[40] Marston D. E. Conder. Generators for alternating and symmetric groups. *J. London Math. Soc. (2)*, 22(1):75–86, 1980. 157

[41] Marston D. E. Conder. Regular maps and hypermaps of Euler characteristic -1 to -200. *J. Combin. Theory Ser. B*, 99(2):455–459, 2009. 51, 59, 64, 115, 162, 163, 164, 174, 178, 236, 269, 270

[42] Marston D. E. Conder, Veronika Hucíková, Roman Nedela, and Jozef Širáň. Chiral maps of given hyperbolic type. *Bull. Lond. Math. Soc.*, 48(1):38–52, 2016. 64, 268, 269, 273

[43] Marston D. E. Conder, Jozef Širáň, and Thomas W. Tucker. The genera, reflexibility and simplicity of regular maps. *J. Eur. Math. Soc. (JEMS)*, 12(2):343–364, 2010. 64, 270, 271

[44] Marston D. E. Conder and Wei-Juan Zhang. Abelian covers of chiral polytopes. *J. Algebra*, 478:437–457, 2017. 66

[45] Marston D. E. Conder and Wei-Juan Zhang. The smallest chiral 6-polytopes. *Bull. Lond. Math. Soc.*, 49(4):549–560, 2017. 66, 287, 288

[46] John Conway, Heidi Burgiel, and Chaim Goodman-Strauss. *The Symmetries of Things*. A K Peters, 2008. 290, 431, 455

[47] Robert Cori and Antonio Machì. Maps, hypermaps and their automorphisms: a survey. I, II, III. *Exposition. Math.*, 10(5):403–427, 429–447, 449–467, 1992. 194

[48] H. S. M. Coxeter. Discrete groups generated by reflections. *Ann. of Math. (2)*, 35(3):588–621, 1934. 4, 113, 147, 265, 266

[49] H. S. M. Coxeter. Configurations and maps. *Rep. Math. Colloquium (2)*, 8:18–38, 1949. 4, 51, 63, 64, 144, 259

[50] H. S. M. Coxeter. Self-dual configurations and regular graphs. *Bull. Amer. Math. Soc.*, 56:413–455, 1950. 263

[51] H. S. M. Coxeter. Regular honeycombs in hyperbolic space. In *Proceedings of the International Congress of Mathematicians, 1954, Amsterdam, vol. III*, pages 155–169. Erven P. Noordhoff N. V., 1956. 265, 290, 291, 293, 301, 344

[52] H. S. M. Coxeter. Regular compound tessellations of the hyperbolic plane. *Proc. Roy. Soc. London Ser. A*, 278:147–167, 1964. 266

[53] H. S. M. Coxeter. *Introduction to Geometry*. John Wiley & Sons, Inc., 2nd ed., 1969. 64, 178

[54] H. S. M. Coxeter. *Twisted Honeycombs*. American Mathematical Society, 1970. Conference Board of the Mathematical Sciences Regional Conference Series in Mathematics, No. 4. 65, 193, 290, 291, 292, 293, 300, 301, 335

[55] H. S. M. Coxeter. *Regular Polytopes*. Dover Publications, Inc., 3rd ed., 1973. 11, 13, 51, 80, 81, 99, 112, 113, 213, 219, 228, 229, 230, 289, 290, 295

[56] H. S. M. Coxeter. Ten toroids and fifty-seven hemidodecahedra. *Geom. Dedicata*, 13(1):87–99, 1982. 292, 312, 313

[57] H. S. M. Coxeter. A symmetrical arrangement of eleven hemi-icosahedra. In *Convexity and Graph Theory (Jerusalem, 1981)*, volume 87 of North-Holland Math. Stud., pages 103–114. North-Holland, 1984. 312

[58] H. S. M. Coxeter. *Regular Complex Polytopes*. Cambridge University Press, 2nd ed., 1991. 351

[59] H. S. M. Coxeter and W. O. J. Moser. *Generators and Relations for Discrete Groups*, 2nd ed. Ergebnisse der Mathematik und ihrer Grenzgebiete, Neue Folge, Band 14. Springer-Verlag, 1965. 51, 64, 79, 91

[60] H. S. M. Coxeter and Asia Ivić Weiss. Twisted honeycombs $\{3,5,3\}_t$ and their groups. *Geom. Dedicata*, 17(2):169–179, 1984. 65, 299, 300

[61] Harold Scott MacDonald Coxeter. Regular skew polyhedra in three and four dimensions, and their topological analogues. *Proc. London Math. Soc.*, 43:33–62, 1937. 2, 220, 342, 367, 437

[62] Gabe Cunningham. Constructing self-dual chiral polytopes. *European J. Combin.*, 33(6):1313–1323, 2012. 255, 256

[63] Gabe Cunningham. Internal and external invariance of abstract polytopes. 2012. PhD thesis, Northeastern University. 207, 256

[64] Gabe Cunningham. Mixing chiral polytopes. *J. Algebraic Combin.*, 36(2):263–277, 2012. 243, 254, 255, 261

[65] Gabe Cunningham. Minimal equivelar polytopes. *Ars Math. Contemp.*, 7(2):299–315, 2014. 277, 278, 279, 280

[66] Gabe Cunningham. Non-flat regular polytopes and restrictions on chiral polytopes. *Electron. J. Combin.*, 24(3):Paper No. 3.59, 14, 2017. 66, 70, 285, 288

[67] Gabe Cunningham. Tight chiral polyhedra. *Combinatorica*, 38(1):115–142, 2018. 283, 284, 439

[68] Gabe Cunningham. Chiral polytopes whose smallest regular cover is a polytope. *J. Combin. Theory Ser. A*, 204:Paper No. 105839, 2024. 240, 243

[69] Gabe Cunningham, María Del Río-Francos, Isabel Hubard, and Micael Toledo. Symmetry type graphs of polytopes and maniplexes. *Ann. Comb.*, 19(2):243–268, 2015. 54

[70] Gabe Cunningham and Daniel Pellicer. Chiral extensions of chiral polytopes. *Discrete Math.*, 330:51–60, 2014. 76, 329, 330, 331

[71] Gabe Cunningham and Daniel Pellicer. Classification of tight regular polyhedra. *J. Algebraic Combin.*, 43(3):665–691, 2016. 280, 284

[72] Gabe Cunningham and Daniel Pellicer. Open problems on k-orbit polytopes. *Discrete Math.*, 341(6):1645–1661, 2018. 207, 441, 448

[73] Gabe Cunningham and Daniel Pellicer. Tight chiral polytopes. *J. Algebraic Combin.*, 54(3):837–878, 2021. 284, 285

[74] Gabe Cunningham, Daniel Pellicer, and Gordon Williams. Stratified operations on maniplexes. *Algebr. Comb.*, 5(2):267–287, 2022. 208

[75] L. Danzer. Regular incidence-complexes and dimensionally unbounded sequences of such. I. In *Convexity and Graph Theory (Jerusalem, 1981)*, volume 87 of North-Holland Math. Stud., pages 115–127. North-Holland, 1984. 273, 326, 327

[76] L. Danzer and E. Schulte. Reguläre Inzidenzkomplexe. I. *Geom. Dedicata*, 13(3):295–308, 1982. 3

[77] Michael W. Davis. Regular convex cell complexes. In *Geometry and Topology (Athens, Ga., 1985)*, volume 105 of Lecture Notes in Pure and Appl. Math., pages 53–88. Dekker, 1987. 80

[78] Antonio Breda D'Azevedo, Gareth A. Jones, and Egon Schulte. Constructions of chiral polytopes of small rank. *Canad. J. Math.*, 63(6):1254–1283, 2011. 66, 194, 201, 205, 243, 255

[79] Reinhard Diestel. Ends and tangles. *Abh. Math. Semin. Univ. Hambg.*, 87(2):223–244, 2017. 272

[80] Manfredo P. do Carmo. *Differential Geometry of Curves and Surfaces*. Dover Publications, Inc., 2016. Revised and updated second edition. 80

[81] Stephen Doro and Stephen E. Wilson. Rotary maps of type $\{6, 6\}_4$. *Quart. J. Math. Oxford Ser. (2)*, 31(124):403–414, 1980. 62

[82] Ian Douglas, Isabel Hubard, Daniel Pellicer, and Steve Wilson. The twist operator on maniplexes. In *Discrete Geometry and Symmetry*, volume 234 of Springer Proc. Math. Stat., pages 127–145. Springer, 2018. 66, 327

[83] Andreas W. M. Dress. A combinatorial theory of Grünbaum's new regular polyhedra. I. Grünbaum's new regular polyhedra and their automorphism group. *Aequationes Math.*, 23(2–3):252–265, 1981. 3, 342, 344, 437

[84] Andreas W. M. Dress. A combinatorial theory of Grünbaum's new regular polyhedra. II. Complete enumeration. *Aequationes Math.*, 29(2–3):222–243, 1985. 3, 366, 367, 390, 437, 443

[85] Euclid. *Euclid's Elements*. Green Lion Press, 2002. All thirteen books complete in one volume, the Thomas L. Heath translation, edited by Dana Densmore. 2

[86] Maria Elisa Fernandes and Dimitri Leemans. String C-group representations of alternating groups. *Ars Math. Contemp.*, 17(1):291–310, 2019. 340

[87] Maria Elisa Fernandes, Dimitri Leemans, and Mark Mixer. All alternating groups A_n with $n \geq 12$ have polytopes of rank $\lfloor \frac{n-1}{2} \rfloor$. *SIAM J. Discrete Math.*, 26(2):482–498, 2012.

[88] Maria Elisa Fernandes, Dimitri Leemans, and Mark Mixer. Polytopes of high rank for the alternating groups. *J. Combin. Theory Ser. A*, 119(1):42–56, 2012. 340

[89] Maria Elisa Fernandes, Dimitri Leemans, and Mark Mixer. An extension of the classification of high rank regular polytopes. *Trans. Amer. Math. Soc.*, 370(12):8833–8857, 2018. 340

[90] Jean Gallier and Dianna Xu. *A Guide to the Classification Theorem for Compact Surfaces*, volume 9 of Geometry and Computing. Springer, 2013. 264

[91] Dietmar Garbe. A generalization of the regular maps of type $\{4, 4\}_{b,\,c}$ and $\{3, 6\}_{b,\,c}$. *Canad. Math. Bull.*, 12:293–297, 1969. 263

[92] Dietmar Garbe. Über die regulären Zerlegungen geschlossener orientierbarer Flächen. *J. Reine Angew. Math.*, 237:39–55, 1969. 64, 269

[93] Dietmar Garbe. A remark on nonsymmetric compact Riemann surfaces. *Arch. Math. (Basel)*, 30(4):435–437, 1978. 274

[94] A. Gardiner, R. Nedela, J. Širáň, and M. Škoviera. Characterisation of graphs which underlie regular maps on closed surfaces. *J. London Math. Soc. (2)*, 59(1):100–108, 1999. 51

[95] C. W. L. Garner. Coordinates for vertices of regular honeycombs in hyperbolic space. *Proc. Roy. Soc. London Ser. A*, 293:94–107, 1966. 290

[96] Cyril W. L. Garner. Regular skew polyhedra in hyperbolic three-space. *Canad. J. Math.*, 19:1179–1186, 1967. 344

[97] Cyril Wilbur Luther Garner. *Polyhedra and honetcombs in hyperbolic space*. ProQuest LLC, 1964. PhD thesis, University of Toronto (Canada). 290

[98] Jorge Garza-Vargas and Isabel Hubard. Polytopality of maniplexes. *Discrete Math.*, 341(7):2068–2079, 2018. 29, 31, 32, 35

[99] Ernesto Girondo and Gabino González-Diez. *Introduction to Compact Riemann Surfaces and Dessins d'Enfants*, volume 79 of London Math. Soc. Student Texts. Cambridge University Press, 2012. 266

[100] Monique Gradolato and Bruno Zimmermann. Finite quotients of hyperbolic tetrahedral groups. In *Proceedings of the Eleventh International Conference of Topology (Trieste, 1993)*, volume 25, pages 211–222, 1993. 302

[101] Jack E. Graver and Mark E. Watkins. Locally finite, planar, edge-transitive graphs. *Mem. Amer. Math. Soc.*, 126(601):vi+75, 1997. 54

[102] Alexandre Grothendieck. Esquisse d'un programme. In *Geometric Galois Actions, 1*, volume 242 of London Math. Soc. Lecture Note Ser., pages 5–48. Cambridge University Press, 1997. With an English translation on pp. 243–283. 165

[103] L. C. Grove and C. T. Benson. *Finite Reflection Groups*, volume 99 of Graduate Texts in Mathematics. Springer-Verlag, 2nd ed., 1985. 290, 351, 449

[104] Branko Grünbaum. Regular polyhedra – old and new. *Aequationes Math.*, 16(1–2):1–20, 1977. 2, 27, 220, 342, 344, 355, 367, 390, 419, 437, 445

[105] Branko Grünbaum. Regularity of graphs, complexes and designs. In *Problèmes combinatoires et théorie des graphes (Colloq. Internat. CNRS, Univ. Orsay, Orsay, 1976)*, volume 260 of Colloq. Internat. CNRS, pages 191–197. CNRS, 1978. 312, 313, 314

[106] Branko Grünbaum. Uniform tilings of 3-space. *Geombinatorics*, 4(2):49–56, 1994. 289, 452

[107] Branko Grünbaum. *Convex Polytopes*, volume 221 of Graduate Texts in Mathematics. Springer-Verlag, 2nd ed., 2003. Prepared and with a preface by Volker Kaibel, Victor Klee, and Günter M. Ziegler. 27, 346

[108] Branko Grünbaum and G. C. Shephard. Edge-transitive planar graphs. *J. Graph Theory*, 11(2):141–155, 1987. 54

[109] Branko Grünbaum and Geoffrey Colin Sheppard. *Tilings and Patterns*. Freeman, 1987. 80

[110] W. Hantzsche and H. Wendt. Dreidimensionale euklidische Raumformen. *Math. Ann.*, 110(1):593–611, 1935. 290

[111] Michael Hartley. All polytopes are quotients, and isomorphic polytopes are quotients by conjugate subgroups. *Discrete Comput. Geom. 21*, 2:289–298, 1999. 164, 180, 183

[112] Michael Hartley, Daniel Pellicer, and Gordon Williams. Minimal covers of the prisms and antiprisms. *Discrete Math.*, 20:3046–3058, 2012. 207, 243

[113] Michael I. Hartley. More on quotient polytopes. *Aequationes Math.*, 57(1):108–120, 1999. 164, 183

[114] Michael I. Hartley. Quotients of some finite universal locally projective polytopes. *Discrete Comput. Geom.*, 29(3):435–443, 2003. 314

[115] Michael I. Hartley. Locally projective polytopes of type $\{4,3,\ldots,3,p\}$. *J. Algebra*, 290(2):322–336, 2005. 314, 328

[116] Michael I. Hartley. An atlas of small regular abstract polytopes. *Period. Math. Hungar.*, 53(1–2):149–156, 2006. 51, 59, 115, 162, 163, 171, 173, 287

[117] Michael I. Hartley. An exploration of locally projective polytopes. *Combinatorica*, 28(3):299–314, 2008. 314

[118] Michael I. Hartley, Isabel Hubard, and Dimitri Leemans. Two atlases of abstract chiral polytopes for small groups. *Ars Math. Contemp.*, 5(2):371–382, 2012. 64, 236

[119] Michael I. Hartley and Dimitri Leemans. Quotients of a universal locally projective polytope of type $\{5,3,5\}$. *Math. Z.*, 247(4):663–674, 2004. 314

[120] Michael I. Hartley and Dimitri Leemans. Errata: "Quotients of a universal locally projective polytope of type $\{5,3,5\}$" [Math. Z. **247** (2004), no. 4, 663–674; mr2077414]. *Math. Z.*, 253(2):433–434, 2006.

[121] Michael I. Hartley and Dimitri Leemans. On locally spherical polytopes of type $\{5,3,5\}$. *Discrete Math.*, 309(1):247–254, 2009. 302, 314

[122] Michael I. Hartley, Peter McMullen, and Egon Schulte. *Symmetric Tessellations on Euclidean Space-Forms. Canad. J. Math.*, 51:1230–1239, 1999. Dedicated to H. S. M. Coxeter on the occasion of his 90th birthday. 4, 80, 290

[123] Allen Hatcher. *Algebraic Topology*. Cambridge University Press, 2002. 24

[124] L. Heffter. Ueber metacyklische Gruppen und Nachbarconfigurationen. *Math. Ann.*, 50(2–3):261–268, 1898. 63, 144

[125] Michael Henle. *A Combinatorial Introduction to Topology*. A Series of Books in Mathematical Sciences. W. H. Freeman and Co., 1979. 264

[126] Edmund Hess. Über die regulären polytope höherer art. *Sitz. Ges. Naturwiss Marburg*, pages 31–57, 1885. 2

[127] Howard Hiller. *Geometry of Coxeter Groups*, volume 54 of Research Notes in Mathematics. Pitman (Advanced Publishing Program), 1982. 112

[128] Loo Keng Hua. *Introduction to Number Theory*. Springer-Verlag, 1982. Translated from the Chinese by Peter Shiu. 456

[129] Isabel Hubard. Two-orbit polyhedra from groups. *European J. Combin.*, 31(3):943–960, 2010. 54, 103, 105, 112

[130] Isabel Hubard and Dimitri Leemans. Chiral polytopes and Suzuki simple groups. In *Rigidity and Symmetry*, volume 70 of Fields Institute Communications, pages 155–175. Springer, 2014. 340

[131] Isabel Hubard, Mark Mixer, Daniel Pellicer, and Asia Ivić Weiss. Cubic tessellations of the didicosm. *Adv. Geom.*, 14(2):299–318, 2014. 290

[132] Isabel Hubard, Mark Mixer, Daniel Pellicer, and Asia Ivić Weiss. Cubic tessellations of the helicosms. *Discrete Comput. Geom.*, 54(3):686–704, 2015. 290

[133] Isabel Hubard and Elías Mochán. All polytopes are coset geometries: Characterizing automorphism groups of k-orbit abstract polytopes. *European J. Combin.*, 113:Paper No. 103746, 2023. 29, 106

[134] Isabel Hubard, Elías Mochán, and Antonio Montero. Voltage operations on maniplexes, polytopes and maps. *Combinatorica*, 43(2):385–420, 2023. 208

[135] Isabel Hubard, Alen Orbanić, and Asia Ivić Weiss. Monodromy groups and self-invariance. *Canad. J. Math.*, 61(6):1300–1324, 2009. 209

[136] Isabel Hubard, Alen Orbanić, Daniel Pellicer, and Asia Ivić Weiss. Symmetries of equivelar 4-toroids. *Discrete Comput. Geom.*, 48(4):1110–1136, 2012. 85, 290

[137] Isabel Hubard, Egon Schulte, and Asia Ivić Weiss. Petrie–Coxeter maps revisited. *Beiträge Algebra Geom.*, 47(2):329–343, 2006. 225, 226

[138] Isabel Hubard and Asia Ivić Weiss. Self-duality of chiral polytopes. *J. Combin. Theory Ser. A*, 111(1):128–136, 2005. 79

[139] Isabel A. Hubard. From geometry to groups and back: The study of highly symmetric polytopes. ProQuest LLC, 2007. PhD thesis, York University (Canada). 54

[140] James E. Humphreys. *Reflection Groups and Coxeter Groups*, volume 29 of Camb. Studies Adv. Math. Cambridge University Press, 1990. 112, 113, 118, 147, 425

[141] Kenneth Ireland and Michael Rosen. *A Classical Introduction to Modern Number Theory*, volume 84 of Graduate Texts in Mathematics. Springer-Verlag, 2nd ed., 1990. 271, 457

[142] Lynne D. James and Gareth A. Jones. Regular orientable imbeddings of complete graphs. *J. Combin. Theory Ser. B*, 39(3):353–367, 1985. 263

[143] Stanislav Jendroľ. A noninvolutory self-duality. *Discrete Math.*, 74(3):325–326, 1989. 46

[144] Norman W. Johnson. *Geometries and Transformations*. Cambridge University Press, 2018. 155

[145] G. A. Jones and J. S. Thornton. Operations on maps, and outer automorphisms. *J. Combin. Theory Ser. B*, 35(2):93–103, 1983. 208

[146] Gareth Jones, Roman Nedela, and Martin Škoviera. Complete bipartite graphs with a unique regular embedding. *J. Combin. Theory Ser. B*, 98(2):241–248, 2008. 63

[147] Gareth A. Jones. Chiral covers of hypermaps. *Ars Math. Contemp.*, 8(2):425–431, 2015. 268, 269, 273

[148] Gareth A. Jones and Cormac D. Long. Epimorphic images of the $[5, 3, 5]$ Coxeter group. *Math. Z.*, 275(1–2):167–183, 2013. 302

[149] Gareth A. Jones and David Singerman. Theory of maps on orientable surfaces. *Proc. London Math. Soc. (3)*, 37(2):273–307, 1978. 23, 263, 266

[150] Gareth A. Jones and Jürgen Wolfart. *Dessins d'Enfants on Riemann Surfaces*. Springer Monographs in Mathematics. Springer, 2016. 165, 267

[151] Mitsuyoshi Kato. On combinatorial space forms. *Sci. Papers College Gen. Ed. Univ. Tokyo*, 30(2):107–146, 1980. 80

[152] Johannes Kepler. *The Harmony of the World*, volume 209 of Memoirs Amer. Philos. Soc. American Philosophical Society, 1997. Translated from the Latin and with an introduction and notes by E. J. Aiton, A. M. Duncan, and J. V. Field, with a preface by Duncan and Field. 2, 7

[153] Hiroki Koike, Daniel Pellicer, Miguel Raggi, and Steve Wilson. Flag bicolorings, pseudo-orientations, and double covers of maps. *Electron. J. Combin.*, 24(1):Paper No. 1.3, 23, 2017. 63, 257, 258

[154] Dimitri Leemans. Almost simple groups of Suzuki type acting on polytopes. *Proc. Amer. Math. Soc.*, 134(12):3649–3651, 2006. 340

[155] Dimitri Leemans. String C-group representations of almost simple groups: a survey. In *Polytopes and Discrete Geometry*, volume 764 of Contemp. Math., pages 157–178. American Mathematical Society, 2021. 340

[156] Dimitri Leemans and Martin W. Liebeck. Chiral polyhedra and finite simple groups. *Bull. Lond. Math. Soc.*, 49(4):581–592, 2017. 205

[157] Dimitri Leemans, Jérémie Moerenhout, and Eugenia O'Reilly-Regueiro. Projective linear groups as automorphism groups of chiral polytopes. *J. Geom.*, 108(2):675–702, 2017. 205

[158] Dimitri Leemans and Egon Schulte. Groups of type $L_2(q)$ acting on polytopes. *Adv. Geom.*, 7(4):529–539, 2007. 340

[159] Dimitri Leemans and Egon Schulte. Polytopes with groups of type $\mathrm{PGL}_2(q)$. *Ars Math. Contemp.*, 2(2):163–171, 2009. 340

[160] Dimitri Leemans, Egon Schulte, and Hendrik Van Maldeghem. Groups of Ree type in characteristic 3 acting on polytopes. *Ars Math. Contemp.*, 14(2):209–226, 2018. 340

[161] Dimitri Leemans and Adrien Vandenschrick. On chiral polytopes having a group $\mathrm{PSL}(3,q)$ as automorphism group. *J. Lond. Math. Soc. (2)*, 106(1):85–111, 2022. 205

[162] Dimitri Leemans and Laurence Vauthier. An atlas of abstract regular polytopes for small groups. *Aequationes Math.*, 72(3):313–320, 2006. 163

[163] Nicholas Matteo. Combinatorially two-orbit convex polytopes. *Discrete Comput. Geom.*, 55(3):662–680, 2016. 80, 289

[164] Nicholas Matteo. Two-orbit convex polytopes and tilings. *Discrete Comput. Geom.*, 55(2):296–313, 2016. 289

[165] P. McMullen. Combinatorially regular polytopes. *Mathematika*, 14:142–150, 1967. 80

[166] P. McMullen and E. Schulte. Self-dual regular 4-polytopes and their Petrie–Coxeter-polyhedra. *Results Math.*, 12(3–4):366–375, 1987. 220, 222, 224

[167] P. McMullen and E. Schulte. Regular polytopes in ordinary space. *Discrete Comput. Geom.*, 17(4):449–478, 1997. Dedicated to Jörg M. Wills. 344, 366, 378, 381, 387, 414, 419, 437, 450

[168] Peter McMullen. On the combinatorial structure of convex polytopes. 1968. PhD thesis, University of Birmingham (UK). 4, 63, 67, 80, 289

[169] Peter McMullen. Realizations of regular polytopes. *Aequationes Math.*, 37(1):38–56, 1989. 343, 366, 437

[170] Peter McMullen. Locally projective regular polytopes. *J. Combin. Theory Ser. A*, 65(1):1–10, 1994. 314

[171] Peter McMullen. Realizations of regular apeirotopes. *Aequationes Math.*, 47(2–3):223–239, 1994. 366, 437

[172] Peter McMullen. Regular polytopes of full rank. *Discrete Comput. Geom.*, 32(1):1–35, 2004. 3, 388, 389

[173] Peter McMullen. Four-dimensional regular polyhedra. *Discrete Comput. Geom.*, 38(2):355–387, 2007.

[174] Peter McMullen. Regular apeirotopes of dimension and rank 4. *Discrete Comput. Geom.*, 42(2):224–260, 2009. 3

[175] Peter McMullen. Realizations of regular polytopes, III. *Aequationes Math.*, 82(1–2):35–63, 2011. 366

[176] Peter McMullen. Regular polytopes of nearly full rank. *Discrete Comput. Geom.*, 46(4):660–703, 2011. 3, 210, 441

[177] Peter McMullen. Regular polytopes of nearly full rank: Addendum. *Discrete Comput. Geom.*, 49(3):703–705, 2013. 3

[178] Peter McMullen. Realizations of regular polytopes, IV. *Aequationes Math.*, 87(1–2):1–30, 2014. 366, 437

[179] Peter McMullen. *Geometric Regular Polytopes*, volume 172 of Encyclopedia of Mathematics and its Applications. Cambridge University Press, 2020. vii, 366, 437

[180] Peter McMullen. Quasi-regular polytopes of full rank. *Discrete Comput. Geom.*, 66(2):475–509, 2021. 61

[181] Peter McMullen. Realizations of the 120-cell. In *Polytopes and Discrete Geometry*, volume 764 of Contemp. Math., pages 193–219. American Mathematical Society, 2021. 437

[182] Peter McMullen and Barry Monson. Realizations of regular polytopes. II. *Aequationes Math.*, 65(1–2):102–112, 2003. 366, 437

[183] Peter McMullen and Egon Schulte. Hermitian forms and locally toroidal regular polytopes. *Adv. Math.*, 82(1):88–125, 1990. 314

[184] Peter McMullen and Egon Schulte. Locally toroidal regular polytopes of rank 4. *Comment. Math. Helv.*, 67(1):77–118, 1992.

[185] Peter McMullen and Egon Schulte. Higher toroidal regular polytopes. *Adv. Math.*, 117(1):17–51, 1996. 79, 100

[186] Peter McMullen and Egon Schulte. Twisted groups and locally toroidal regular polytopes. *Trans. Amer. Math. Soc.*, 348(4):1373–1410, 1996. 314

[187] Peter McMullen and Egon Schulte. Flat regular polytopes. *Ann. Comb.*, 1(3):261–278, 1997. 274

[188] Peter McMullen and Egon Schulte. *Abstract Regular Polytopes*. Cambridge University Press, 2002. vii, 2, 3, 6, 20, 47, 95, 100, 111, 115, 120, 121, 208, 213, 225, 227, 247, 250, 290, 302, 312, 313, 314, 324, 327, 366, 415

[189] Mark Mixer, Daniel Pellicer, and Gordon Williams. Minimal covers of the Archimedean tilings, part II. *Electron. J. Combin.*, 20(2):Paper 20, 19, 2013. 243

[190] Elías Mochán. Abstract polytopes from their symmetry type graphs. 2021. PhD thesis, Universidad Nacional Autónoma de México (Mexico). 58

[191] B. Monson, Daniel Pellicer, and Gordon Williams. Mixing and monodromy of abstract polytopes. *Trans. Amer. Math. Soc.*, 366(5):2651–2681, 2014. 20, 117, 118, 164, 237, 240, 251, 256

[192] B. Monson and Egon Schulte. Reflection groups and polytopes over finite fields. II. *Adv. in Appl. Math.*, 38(3):327–356, 2007. 314

[193] B. Monson and Egon Schulte. Reflection groups and polytopes over finite fields. III. *Adv. in Appl. Math.*, 41(1):76–94, 2008. 314

[194] B. Monson and Egon Schulte. Modular reduction in abstract polytopes. *Canad. Math. Bull.*, 52(3):435–450, 2009. 65

[195] B. Monson and Egon Schulte. Locally toroidal polytopes and modular linear groups. *Discrete Math.*, 310(12):1759–1771, 2010. 314

[196] B. Monson and Egon Schulte. Semiregular polytopes and amalgamated C-groups. *Adv. Math.*, 229(5):2767–2791, 2012. 61

[197] B. Monson and Egon Schulte. Finite polytopes have finite regular covers. *J. Algebraic Combin.*, 40(1):75–82, 2014. 257

[198] B. Monson and A. I. Weiss. Realizations of regular toroidal maps of type $\{4,4\}$. *Discrete Comput. Geom.*, 24:453–465, 2000. The Branko Grünbaum birthday issue. 437, 440

[199] B. Monson and A. Ivić Weiss. Eisentein [Eisenstein] integers and related C-groups. *Geom. Dedicata*, 66(1):99–117, 1997. 65, 314

[200] B. Monson and A. Ivić Weiss. Realizations of regular toroidal maps. *Canad. J. Math.*, 51:1240–1257, 1999. Dedicated to H. S. M. Coxeter on the occasion of his 90th birthday. 437, 440

[201] B. Monson and Asia Ivić Weiss. Polytopes related to the Picard group. *Linear Algebra Appl.*, 218:185–204, 1995. 65

[202] Barry Monson. On Roli's cube. *Art Discrete Appl. Math.*, 5(3):Paper No. 3.10, 17, 2022. 307, 309

[203] Barry Monson, Daniel Pellicer, and Gordon Williams. The tomotope. *Ars Math. Contemp.*, 5(2):355–370, 2012. 237

[204] Antonio Montero. Chiral extensions of toroids. 2019. PhD thesis, Universidad Nacional Autónoma de México (Mexico). 338, 339

[205] Antonio Montero. On the Schläfli symbol of chiral extensions of polytopes. *Discrete Math.*, 344(11):Paper No. 112507, 16, 2021. 155, 330, 331

[206] Roman Nedela and Martin Škoviera. Regular maps on surfaces with large planar width. *European J. Combin.*, 22(2):243–261, 2001. 165

[207] B. Nostrand and E. Schulte. Chiral polytopes from hyperbolic honeycombs. *Discrete Comput. Geom.*, 13(1):17–39, 1995. 144, 324

[208] Barbara Nostrand. *Chiral honeycombs*. ProQuest LLC, 1993. PhD thesis, Northeastern University. 302, 324

[209] Barbara Nostrand. Ring extensions and chiral polytopes. In *Proceedings of the Twenty-fifth Southeastern International Conference on Combinatorics, Graph Theory and Computing (Boca Raton, FL, 1994)*, volume 102, pages 147–153, 1994. 65, 302

[210] Barbara Nostrand, Egon Schulte, and Asia Ivić Weiss. Constructions of chiral polytopes. In *Proceedings of the Twenty-fourth Southeastern International Conference on Combinatorics, Graph Theory, and Computing (Boca Raton, FL, 1993)*, volume 97, pages 165–170, 1993. 65, 324

[211] Werner Nowacki. Die euklidischen, dreidimensionalen, geschlossenen und offenen Raumformen. *Comment. Math. Helv.*, 7(1):81–93, 1934. 290

[212] M. O'Keeffe and S. Andersson. Rod packings and crystal chemistry. *Acta Cryst. A*, 33(6):914–923, 1977. 455

[213] Alen Orbanić. F-actions and parallel-product decomposition of reflexible maps. *J. Algebraic Combin.*, 26(4):507–527, 2007. 165

[214] Alen Orbanić, Daniel Pellicer, and Asia Ivić Weiss. Map operations and k-orbit maps. *J. Combin. Theory Ser. A*, 117(4):411–429, 2010. 59, 185

[215] Luisa Paoluzzi. $\mathrm{PSL}(2,q)$ quotients of some hyperbolic tetrahedral and Coxeter groups. *Comm. Algebra*, 26(3):759–778, 1998. 302

[216] Dmitrii V. Pasechnik. Locally toroidal polytopes of rank 6 and sporadic groups. *Adv. Math.*, 312:459–472, 2017. 314, 324

[217] Daniel Pellicer. Extensions of regular polytopes with preassigned Schläfli symbol. *J. Combin. Theory Ser. A*, 116(2):303–313, 2009. 327

[218] Daniel Pellicer. A construction of higher rank chiral polytopes. *Discrete Math.*, 310(6–7):1222–1237, 2010. 66, 76, 243, 325, 336, 338, 339

[219] Daniel Pellicer. Extensions of dually bipartite regular polytopes. *Discrete Math.*, 310(12):1702–1707, 2010. 327, 330

[220] Daniel Pellicer. Developments and open problems on chiral polytopes. *Ars Math. Contemp.*, 5(2):333–354, 2012. 5, 64, 65, 66, 193, 291, 303, 311, 324, 341, 366, 441, 459

[221] Daniel Pellicer. The higher dimensional hemicuboctahedron. In *Symmetries in Graphs, Maps, and Polytopes*, volume 159 of Springer Proc. Math. Stat., pages 263–271. Springer, 2016. 61

[222] Daniel Pellicer. Regular and chiral polyhedra in Euclidean nets. *Symmetry*, 8(11):Art. 115, 14, 2016. 437

[223] Daniel Pellicer. A chiral 4-polytope in $\mathbb{R}^3$. *Ars Math. Contemp.*, 12(2):315–327, 2017. 398

[224] Daniel Pellicer. Chiral 4-polytopes in ordinary space. *Beitr. Algebra Geom.*, 58(4):655–677, 2017. 403, 437, 438, 439

[225] Daniel Pellicer. Cleaved abstract polytopes. *Combinatorica*, 38(3):709–737, 2018. 226

[226] Daniel Pellicer. A chiral 5-polytope of full rank. *Discrete Math.*, 344(6):Paper No. 112370, 10, 2021. 441, 447, 448

[227] Daniel Pellicer. Chiral polytopes of full rank exist only in ranks 4 and 5. *Beitr. Algebra Geom.*, 62(3):651–665, 2021. 442

[228] Daniel Pellicer and Asia Ivić Weiss. Combinatorial structure of Schulte's chiral polyhedra. *Discrete Comput. Geom.*, 44(1):167–194, 2010. 206, 269, 412, 430, 434, 435

[229] Daniel Pellicer, Primož Potočnik, and Micael Toledo. An existence result on two-orbit maniplexes. *J. Combin. Theory Ser. A*, 166:226–253, 2019. 58

[230] Daniel Pellicer and Asia Ivić Weiss. Generalized CPR-graphs and applications. *Contrib. Discrete Math.*, 5(2):76–105, 2010. 76, 157, 193, 205, 206, 269, 328

[231] Daniel Pellicer and Gordon Williams. Minimal covers of the Archimedean tilings, part 1. *Electron. J. Combin.*, 19(3):paper 6, 37, 2012. 243

[232] Louis Poinsot. Mémoire sur les polygones et les polyèdres. *J. École Polytech.*, 10:16–48, 1810. 2

[233] Primož Potočnik. Primož Potočnik's home page. https://users.fmf.uni-lj.si/potocnik/work.htm. Accessed: 2023-07-13. 64

[234] John G. Ratcliffe. *Foundations of Hyperbolic Manifolds*, volume 149 of Graduate Texts in Mathematics. Springer, 3rd ed., 2019. 79, 80, 290, 431

[235] Frank Raymond. The end point compactification of manifolds. *Pacific J. Math.*, 10:947–963, 1960. 272

[236] Ian Richards. On the classification of noncompact surfaces. *Trans. Amer. Math. Soc.*, 106:259–269, 1963. 272

[237] Ludwig Schläfli. *Theorie der vielfachen Kontinuität.* George & Company, 1901. 2

[238] Otto Schreier. Die Untergruppen der freien Gruppen. *Abh. Math. Sem. Univ. Hamburg*, 5(1):161–183, 1927. 157

[239] Egon Schulte. On arranging regular incidence-complexes as faces of higher-dimensional ones. *European J. Combin.*, 4(4):375–384, 1983. 327

[240] Egon Schulte. Extensions of regular complexes. In *Finite Geometries (Winnipeg, Man., 1984)*, volume 103 of Lecture Notes in Pure and Appl. Math., pages 289–305. Dekker, 1985.

[241] Egon Schulte. Regular incidence-polytopes with Euclidean or toroidal faces and vertex-figures. *J. Combin. Theory Ser. A*, 40(2):305–330, 1985. 327

[242] Egon Schulte. Amalgamation of regular incidence-polytopes. *Proc. London Math. Soc. (3)*, 56(2):303–328, 1988. 310, 313

[243] Egon Schulte. On a class of abstract polytopes constructed from binary codes. *Discrete Math.*, 84(3):295–301, 1990. 327

[244] Egon Schulte. Classification of locally toroidal regular polytopes. In *Polytopes: Abstract, Convex and Computational (Scarborough, ON, 1993)*, volume 440 of NATO Adv. Sci. Inst. Ser. C: Math. Phys. Sci., pages 125–154. Kluwer Academic Publishers, 1994. 314

[245] Egon Schulte. Chiral polyhedra in ordinary space. I. *Discrete Comput. Geom.*, 32(1):55–99, 2004. 4, 206, 268, 362, 388, 420, 424, 430, 431, 432, 434, 435, 436, 437

[246] Egon Schulte. Chiral polyhedra in ordinary space. II. *Discrete Comput. Geom.*, 34(2):181–229, 2005. 4, 362, 388, 394, 400, 405, 411, 412, 430, 437

[247] Egon Schulte and Asia Ivić Weiss. Chiral polytopes. In *Applied Geometry and Discrete Mathematics*, volume 4 of DIMACS Ser. Discrete Math. Theoret. Comput. Sci., pages 493–516. American Mathematical Society, 1991. 1, 62, 65, 67, 111, 122, 317, 318, 321, 325

[248] Egon Schulte and Asia Ivić Weiss. Chirality and projective linear groups. *Discrete Math.*, 131(1–3):221–261, 1994. 65, 77, 144, 154, 155, 193, 314, 324, 458

[249] Egon Schulte and Asia Ivić Weiss. Free extensions of chiral polytopes. *Canad. J. Math.*, 47(3):641–654, 1995. 65, 206, 329

[250] Egon Schulte and Asia Ivić Weiss. Problems on polytopes, their groups, and realizations. *Period. Math. Hungar.*, 53(1–2):231–255, 2006. 5, 311, 324, 366, 459

[251] Brigitte Servatius and Herman Servatius. Self-dual maps on the sphere. *Discrete Math.*, 134(1–3):139–150, 1994. 23

[252] F. A. Sherk. The regular maps on a surface of genus three. *Canad. J. Math.*, 11:452–480, 1959. 263

[253] F. A. Sherk. A family of regular maps of type $\{6, 6\}$. *Canad. Math. Bull.*, 5:13–20, 1962. 64, 144, 259, 268, 270, 271

[254] G. A. Soĭfer. Affine crystallographic groups. In *Algebra and Analysis (Irkutsk, 1989)*, volume 163 of Amer. Math. Soc. Transl. Ser. 2, pages 165–170. American Mathematical Society, 1995. 431

[255] Pablo Spiga. On the rank of Suzuki polytopes: an answer to Hubard and Leemans. *Discrete Math.*, 344(5):Paper No. 112331, 5, 2021. 340

[256] J. Tits. Sur les analogues algébriques des groupes semi-simples complexes. In *Colloque d'algèbre supérieure, tenu à Bruxelles du 19 au 22 décembre 1956*, Centre Belge de Recherches Mathématiques, pages 261–289. Établissements Ceuterick, Louvain; Librairie Gauthier-Villars, 1957. 112

[257] William T. Tutte. What is a map? In *New Directions in the Theory of Graphs (Proc. Third Ann Arbor Conf., Univ. Michigan, Ann Arbor, Mich., 1971)*, pages 309–325. Academic Press, 1973. 23

[258] Andrew Vince. Combinatorial maps. *J. Combin. Theory Ser. B*, 34(1):1–21, 1983. 22, 164

[259] Andrew Vince. Regular combinatorial maps. *J. Combin. Theory Ser. B*, 35(3):256–277, 1983. 64, 267, 268, 269, 273

[260] Jozef Širáň. Regular maps on a given surface: a survey. In *Topics in Discrete Mathematics*, volume 26 of Algorithms Combin., pages 591–609. Springer, 2006. 263

[261] Jozef Širáň, Thomas W. Tucker, and Mark E. Watkins. Realizing finite edge-transitive orientable maps. *J. Graph Theory*, 37(1):1–34, 2001. 54

[262] C. Weber and H. Seifert. Die beiden Dodekaederräume. *Math. Z.*, 37(1): 237–253, 1933. 65, 291

[263] Asia Ivić Weiss. On trivalent graphs embedded in twisted honeycombs. In *Combinatorics '81 (Rome, 1981)*, volume 78 of North-Holland Math. Stud., pages 781–787. North-Holland, 1983. 299

[264] Asia Ivić Weiss. Twisted honeycombs $\{3, 5, 3\}_t$. *C. R. Math. Rep. Acad. Sci. Canada*, 5(5):211–215, 1983. 65, 300

[265] Asia Ivić Weiss. An infinite graph of girth 12. *Trans. Amer. Math. Soc.*, 283(2):575–588, 1984. 314

[266] Asia Ivić Weiss and Zoran Lučić. Regular polyhedra in hyperbolic three-space. *Mitt. Math. Sem. Giessen*, (165):237–252, 1984. 344

[267] Stephen E. Wilson. Non-orientable regular maps. *Ars Combin.*, 5:213–218, 1978. 263

[268] Stephen E. Wilson. Riemann surfaces over regular maps. *Canad. J. Math.*, 30(4):763–782, 1978. 263

[269] Stephen E. Wilson. Operators over regular maps. *Pacific J. Math.*, 81(2):559–568, 1979. 23, 209

[270] Stephen E. Wilson. Parallel products in groups and maps. *J. Algebra*, 167(3):539–546, 1994. 243, 244, 246

[271] Steve Wilson. The smallest nontoroidal chiral maps. *J. Graph Theory*, 2(4):315–318, 1978. 274

[272] Steve Wilson. Uniform maps on the Klein bottle. *J. Geom. Graph.*, 10(2):161–171, 2006. 290

[273] Steve Wilson. Maniplexes: Part 1: Maps, polytopes, symmetry and operators. *Symmetry*, 4(2):265–275, 2012. 28, 29, 165

[274] Joseph A. Wolf. *Spaces of Constant Curvature*. AMS Chelsea Publishing, 6th ed., 2011. 290

[275] Wei-Juan Zhang. Some simplifications of the intersection condition of chiral form for polytopes. *J. Algebra Appl.*, 18(11):1950203, 15, 2019. 66

[276] Wei-Juan Zhang. More on chiral polytopes of type $\{4, 4, \ldots, 4\}$ with solvable automorphism groups. *J. Group Theory*, 26(6):1231–1265, 2023. 66

Index

www.ingramcontent.com/pod-product-compliance
Lightning Source LLC
Chambersburg PA
CBHW071143070126
37801CB00030B/71
* 9 7 8 1 1 0 8 4 9 3 2 4 6 *